INTRACELLULAR
PATHOGENS II

Rickettsiales

INTRACELLULAR PATHOGENS II

Rickettsiales

EDITED BY

Guy H. Palmer
Paul G. Allen School for Global Animal Health,
Washington State University, Pullman, WA

Abdu F. Azad
Department of Microbiology and Immunology, School of Medicine,
University of Maryland—Baltimore, Baltimore, MD

***Lead Editor,* Ming Tan**
Departments of Microbiology & Molecular Genetics, and Medicine,
University of California, Irvine, CA

ASM PRESS

Washington, DC

Cover image: Confocal fluorescent microscopic image of GFPuv-expressing *A. phagocytophilum* infecting a rhesus cell (RF/6A) expressing DsRed2, a red fluorescent protein. Courtesy of Ulrike Munderloh (see chapter 14).

Library of Congress Cataloging-in-Publication Data

Intracellular pathogens II : Rickettsiales / edited by Guy H. Palmer, Abdu F. Azad ; lead editor, Ming Tan.
p. ; cm.
Rickettsiales
Includes bibliographical references and index.
ISBN-13: 978-1-55581-677-3 (alk. paper)
ISBN-10: 1-55581-677-0 (alk. paper)
I. Palmer, Guy (Guy Hughes), 1955- II. Azad, Abdu F. III. Tan, Ming, M.D.
IV. Title:
Rickettsiales.
[DNLM: 1. Rickettsiaceae Infections. 2. Anaplasmataceae. 3. Rickettsieae.
WC 600]

616.9'22—dc23

2011043433

eISBN: 978-1-55581-733-6

10 9 8 7 6 5 4 3 2

Printed in the United States of America

Address editorial correspondence to: ASM Press, 1752 N St., N.W., Washington, DC 20036-2904, U.S.A.

Send orders to: ASM Press, P.O. Box 605, Herndon, VA 20172, USA
Phone: 800-546-2416; 703-661-1593
Fax : 703-661-1501
E-mail: books@asmusa.org
Online: http://estore.asm.org

CONTENTS

CONTRIBUTORS

Jonathon P. Audia
Department of Microbiology and Immunology, University of South Alabama
College of Medicine, Mobile, AL 36688

Abdu F. Azad
Department of Microbiology and Immunology,
University of Maryland School of Medicine,
Baltimore, MD 21201

Kelly A. Brayton
Department of Veterinary Microbiology and Pathology, Paul G. Allen School for Global
Animal Health, Washington State University, Pullman, WA 99164-7040

Wendy C. Brown
Program in Vector-borne Diseases,
Department of Veterinary Microbiology and Pathology,
Washington State University, Pullman, WA 99164

Nicole Y. Burkhardt
Department of Entomology, University of Minnesota,
St. Paul, MN 55108

Jason A. Carlyon
Department of Microbiology and Immunology, Virginia Commonwealth University
School of Medicine, Richmond, VA 23298-0678

Shane M. Ceraul
Department of Microbiology and Immunology, University of Maryland School of
Medicine, Baltimore, MD 21201

Adela S. Oliva Chávez
Department of Entomology, University of Minnesota,
St. Paul, MN 55108

J. Stephen Dumler
Division of Medical Microbiology, Department of Pathology,
The Johns Hopkins University School of Medicine,
Baltimore, MD 21205

Roderick F. Felsheim
Department of Entomology, University of Minnesota,
St. Paul, MN 55108

Joseph J. Gillespie
Cyberinfrastructure Division, Virginia Bioinformatics Institute,
Virginia Tech, Blacksburg, VA 24061

Cat M. Haglund
Department of Molecular & Cell Biology, University of California, Berkeley,
Berkeley, CA 94720

Michael J. Herron
Department of Entomology, University of Minnesota,
St. Paul, MN 55108

Nahed Ismail
Department of Pathology, University of Pittsburgh, Pittsburgh, PA 15261

Timothy J. Kurtti
Department of Entomology, University of Minnesota,
St. Paul, MN 55108

Juan J. Martinez
Department of Microbiology and Howard Taylor Ricketts Laboratory,
University of Chicago, Chicago, IL 60637

Jennifer H. McQuiston
Rickettsial Zoonoses Branch, Division of Vectorborne Diseases, Centers for Disease
Control and Prevention, Atlanta, GA 30333

Ulrike G. Munderloh
Department of Entomology, University of Minnesota,
St. Paul, MN 55108

Curtis M. Nelson
Department of Entomology, University of Minnesota,
St. Paul, MN 55108

Susan M. Noh
Animal Disease Research Unit, Agriculture Research Service,
U.S. Department of Agriculture, Pullman, WA 99164

Eric K. Nordberg
Cyberinfrastructure Division, Virginia Bioinformatics Institute, Virginia Tech,
Blacksburg, VA 24061

Christopher D. Paddock
Infectious Diseases Pathology Branch, Division of High Consequence Pathogens and
Pathology, Centers for Disease Control and Prevention Atlanta, GA 30333

Shawna C. O. Reed
Department of Molecular & Cell Biology, University of California, Berkeley,
Berkeley, CA 94720

Elena Rydkina
Department of Microbiology and Immunology, University of Rochester School of
Medicine and Dentistry, Rochester, NY 14642

Sanjeev K. Sahni
Department of Pathology, University of Texas Medical Branch,
Galveston, TX 77555

Patricia J. Simpson-Haidaris
Departments of Medicine (Hematology-Oncology Division), Microbiology and
Immunology, and Pathology and Laboratory Medicine, University of Rochester School
of Medicine and Dentistry, Rochester, NY 14642

Bruno S. W. Sobral
Cyberinfrastructure Division, Virginia Bioinformatics Institute, Virginia Tech,
Blacksburg, VA 24061

Heather L. Stevenson
Department of Pathology, University of Texas Medical Branch,
Galveston, TX 77555

Gustavo Valbuena
Department of Pathology, University of Texas Medical Branch,
Galveston, TX 77555

Matthew D. Welch
Department of Molecular & Cell Biology, University of California, Berkeley,
Berkeley, CA 94720

PREFACE

QUANTUM PROGRESS IN THE SCIENCE OF RICKETTSIAL PATHOGENS

Just over a century ago, Sir Arnold Theiler reported what came to be known as the first rickettsial pathogen, *Anaplasma marginale*. With this report in 1910, Theiler ushered in a dramatic era of discoveries of rickettsiae as pathogens with a series of independent investigations by scientists who became recognized as giants in the field: Cowdry, Nicolle, and Ricketts, among others. Their investigations of the fundamental epidemiology, immunology, and pathogenesis of rickettsial infections have stood both the test of time and the subsequent reanalysis of their work using tools unimaginable in their era. In the process, their original monographs have become classics in the study of infectious diseases.

Following this initial burst of creative investigation in rickettsiology during roughly the first 3 decades of the 20th century, there have been quantum leaps in knowledge. Some have been spurred by the development of powerful new tools, such as the electron microscope in the 1950s, which confirmed that rickettsiae were indeed bacteria. Other gains in knowledge, more tragically, have come from armed conflicts, which led to the discovery of multiple previously unknown pathogens and understanding of the role that host susceptibility plays in disease epidemiology and severity.

The development of molecular biology as both a field and a corresponding set of tools ushered in a new era of quantum gains in rickettsiology. There has been a virtual explosion of genome sequences, from the first sequencing of *Rickettsia prowazekii* in 2005, through 2012, when not only have species in all six genera of the order *Rickettsiales* been sequenced, but multiple strains with unique phenotypes have now been sequenced for several key pathogens. This knowledge has transformed the field: the discovery of extrachromosomal elements among the *Rickettsiaceae* and the use of lateral gene transfer to acquire genetic material is just one example of major shifts in the paradigm of rickettsial biology. Importantly, rickettsial genome data are publicly and freely available, thus allowing scientists worldwide, especially those in low- and middle-income countries that suffer a disproportionate burden of rickettsial disease, to access sequence data and progress in investigation.

The sudden richness of the genome data and the ability of scientists worldwide to identify rickettsiae molecularly also bring challenges to the field. While the seminal taxonomic reorganization of *Anaplasma*, *Ehrlichia*, and *Neorickettsia* in 2001 brought sense to the family *Anaplasmataceae* and promoted comparative investigations that addressed major knowledge gaps, the criteria required for species definition are increasingly problematic, especially for the family *Rickettsiaceae*, challenging the classical descriptions of spotted fever group versus typhus group rickettsioses. Similarly, the widespread availability and ease of molecular identification can lead to the misplaced assumption that a molecular "tag" is sufficient to infer epidemiologic and pathogenic behavior identical to that of a type strain or species. Indeed, recent work on several rickettsial pathogens has highlighted strain differences in phenotype—hence the continuing need for isolation and phenotypic characterization of a diversity of strains.

The ability to now ask heretofore unanswerable questions regarding rickettsial epidemiology, immunity, and pathogenesis has been accompanied by an influx of new investigators into the field of rickettsiology. It is these individuals, as much as any molecular tool, who have given rise to the next quantum leap in the field that is the subject of the current volume. These young scientists have brought new perspectives, from both within and outside the field, to the study of rickettsiae. To the greatest degree possible, we have selected among these young investigators, now firmly established as independent scientists, to author the chapters in this volume. Their achievements are recent, their viewpoints fresh, and their writing unencumbered by prior authorship of similar chapters. Consequently, this book, *Intracellular Pathogens II: Rickettsiales*, is not an updating of any prior edition but rather a new text that links the quantum increase in our knowledge of rickettsiae over the past decade with the future perspectives of the cohort of scientists who will lead the field forward in the decades to come. Equally we wish to acknowledge the superb contributions of the many superb rickettsiologists, junior and senior, who have not authored a chapter in the current text—we have endeavored to present your achievements faithfully and appreciate your notable contributions.

The authors have been asked to focus on the most exciting advances in their respective topic areas rather than attempt to be encyclopedic. Thus, individual chapters often center on only a few of the rickettsial species, based on where the leading edge is being pursued. As a result, however, five of the six genera in the order *Rickettsiales* are covered extensively in multiple chapters. The exception is *Wolbachia*—this is by no means to diminish the importance of these remarkable bacteria, but rather in acknowledgment that their biology has been discussed in several recent texts, and to allow us to focus on those rickettsiae that cause disease in mammalian hosts.

One of the giants of the past 40 years in rickettsiology, Herb Winkler, cleverly paraphrased Jacques Monod by noting that "what is true for *E. coli* may well be true for elephants—but not for rickettsia." Indeed, what is now happening is that rickettsiology is making novel contributions to microbiology and to biology in general by expanding our understanding of the roles of convergent and divergent evolution in the remarkable diversity of life. With respect for both Monod and Winkler: rickettsiae, *E. coli*, and elephants all teach the same lesson of adaptation to their unique niches on the planet.

GUY HUGHES PALMER
ABDU AZAD

CLINICAL DISEASE: CURRENT TREATMENT AND NEW CHALLENGES

J. Stephen Dumler

1

INTRODUCTION

The order *Rickettsiales* includes a diverse group of bacteria with an obligatory intracellular existence and a mandatory transmission cycle that includes arthropods as hosts, reservoirs, and vectors (Dumler and Walker, 2005a). While the entire range of bacteria and their underlying genetic diversity have not been fully characterized, considerable insight into their ability to cause disease has been gleaned from advanced studies of genomes of an increasing number of *Rickettsia*, *Orientia*, *Anaplasma*, *Ehrlichia*, and *Neorickettsia* species (Georgiades et al., 2011; Gillespie et al., 2008). There is substantial controversy over the capacity of species in the *Rickettsiales* to cause disease, and examples of both human and veterinary pathogens exist for the vast majority of the species described in all of the genera in both the families *Rickettsiaceae* and *Anaplasmataceae*. In fact, a number of species are also recognized to be pathogenic for their arthropod hosts or to alter these hosts in ways that promote survival of the bacterium (Saridaki and Bourtzis, 2010). Yet species not associated with disease in any vertebrate or invertebrate system exist, and evidence of infection in humans and animals without accompanying evidence for clinical disease, even with established pathogenic rickettsiae, is clear (Bakken et al., 1998; Marshall et al., 2003). The extent to which disease is caused; the penetrance of the diseased phenotype for each species, strain, or subspecies; and mechanisms that govern the degree of pathogenicity are areas of vigorous investigation.

Despite the extensive recent study regarding disease in humans infected by members of the family *Anaplasmataceae*, infections in animals are the precedent and basis for all of the work. The vast majority of clinical, ecological, and epidemiological information described for humans is likely applicable to animals. Moreover, animals sustain infections and variations on disease not yet described in humans (Rikihisa, 2006). This is likely in part due to the greater likelihood of an animal encountering an infected tick, and perhaps because of tick-pathogen-animal coevolution. One hypothesis also maintains that coevolution contributes to the reduction in apparent acute virulence and promotion of sustained persistent bacteremia in animals that develop minimal clinical signs and disease. This hypothesis in part stems from the observation that transovarial transmission

J. Stephen Dumler, Division of Medical Microbiology, Department of Pathology, Johns Hopkins University School of Medicine, Baltimore, MD 21205.

Intracellular Pathogens II: Rickettsiales
Edited by G. H. Palmer and A. F. Azad © 2012 ASM Press, Washington, DC
10.1128/9781555817336.ch1

in ticks is very inefficient (Long et al., 2003). This would imply greater fitness for those *Anaplasmataceae* that maintain high blood microbe loads, longer intervals of patent infection, and reduced disease costs to the infected mammalian host (Dumler, 2001). A complete discussion of disease processes that occur in animals is beyond the scope of the current chapter, so comments are largely confined to information relevant to human infection and disease, except where animal models inform human disease as well.

Thus, the purpose of this chapter is to review and discuss (i) the human disease manifestations for major pathogenic groups, with some attention to detail on how these differ between genera and species within a single genus; (ii) the evolution of treatment for rickettsial infections; (iii) the existing challenges for treatment and management in human infections; and (iv) corroborated or suspected treatment failures or persistent clinical complaints.

CLASSIFICATION AND PATHOGENS

The order *Rickettsiales* is divided into two families, *Rickettsiaceae* and *Anaplasmataceae* (Dumler and Walker, 2005a). Pathogenic species are recognized in both. In general, pathogenic members of the *Rickettsiaceae* family, including spotted fever and typhus groups of *Rickettsia* (SFGR and TGR, respectively) and *Orientia tsutsugamushi*, target infection predominantly in endothelial cells in mammals, including humans. In contrast, pathogens in *Anaplasmataceae* have a wider variety of in vivo targets, ranging from hematopoietic cells including professional phagocytes like monocytes, macrophages (*Ehrlichia canis*, *Ehrlichia chaffeensis*, *Ehrlichia muris*, *Ehrlichia ruminantium*, *Anaplasma bovis*, *Neorickettsia helminthoeca*, *Neorickettsia sennetsu*, *Neorickettsia risticii*), and neutrophils (*Anaplasma phagocytophilum*, *Ehrlichia ewingii*); to platelets (*Anaplasma platys*) or erythrocytes (*Anaplasma marginale*, *Anaplasma centrale*); to endothelial cells (*E. ruminantium*) and even intestinal epithelial cells (*N. risticii*) (Dumler and Walker, 2005a).

Rickettsiaceae

There are only two recognized genera in the family *Rickettsiaceae*: *Rickettsia* and *Orientia*. The genus *Rickettsia* has classically been divided into the SFGR and TGR based on serological reactions, for which the basis is expression of surface antigens encoded by the outer membrane protein A (*ompA*) and B (*ompB*) genes, the former possessed uniquely by the SFGR species (Dumler and Walker, 2005a). However, recent characterization and full genome sequences illustrate genetic resolution that allows the SFGR to be divided into up to four clades (Gillespie et al., 2008). This division provides for both SFGR and TGR clades, but also one clade that is sometimes called "ancestral, nonvirulent," including *R. bellii* and *R. canadensis*, and another that appears to be "transitional" between SFGR and TGR and includes *R. akari* and *R. felis*. From a phylogenetic standpoint, such genetic methodology defines evolutionary relationships that may or may not have any bearing on pathogenicity or the ability to cause mammalian infection and disease.

To address this issue and the difficulties in investigating and classifying rickettsiae that may not yet be cultivated, one proposal examines the sequences of five genes present in SFGR species as a surrogate for the genetic relatedness established by complete genome sequencing (Fournier et al., 2003a). By coupling phenotypic data like serological responses and clinical features, this approach has been used to define a large number of new species (Fournier and Raoult, 2009). The controversy with this approach stems from the fact that very little is known about the molecular basis for rickettsial virulence and how this relates to pathogenicity in the genus *Rickettsia* (Fournier et al., 2009; Walker and Ismail, 2008). Moreover, this approach is necessarily confounded by phenotypic data that are in part host- and not pathogen-related. The former approach has resulted in an expansion of the list of accepted and candidate SFGR species, accurately reflecting some genetic diversity that may or may not relate to disease-causing ability. In fact, genetic diversity among SFGR, based on sequence analysis of *ompA*, is similar to that of the major 56-kDa surface protein gene of *O. tsutsugamushi*, the cause of scrub typhus, yet only a single species is proposed and recognized for that microbe (Fournier et al., 2008). One result has been an increased level of confusion regarding rickettsial diseases, as busy practitioners attempt to understand the changing bacterial nomenclature and disease names (Walker et al., 2008). The resolution to this controversy will be assisted by studies that employ genomic data and modern molecular methods to delineate the contributions of candidate virulence genes in *Rickettsia* species and isolates.

In contrast to the marked expansion of SFGR species, the TGR group is represented by only two species: *R. prowazekii*—the cause of louse-borne or epidemic typhus; and *R. typhi*—the cause of murine, flea-borne, or endemic typhus. Both are well-recognized pathogens of humans, and in general have a high degree of genetic homogeneity (Walker and Yu, 2005). The genetic bases for virulence and pathogenicity are not well understood.

O. tsutsugamushi, the cause of scrub typhus, is transmitted by larval trombiculid mites and is sufficiently unique at the genetic level that it occupies a distinct clade and genus (Georgiades et al., 2011). While considerable genetic diversity exists, it is still considered a single species with at least several characteristic strains (Fournier et al., 2008). As for other rickettsiae, little is known regarding the molecular bases for virulence and pathogenicity.

Anaplasmataceae

There are currently five genera classified within the *Anaplasmataceae* family, including *Ehrlichia*, *Anaplasma*, "*Candidatus* Neoehrlichia," *Neorickettsia*, and *Wolbachia* (not including *Aegyptianella*, which is likely within the genus *Anaplasma*) (Dumler and Walker, 2005a; Kawahara et al., 2004). Members of this family were first recognized because of veterinary disease, but human disease is now well recognized for *Ehrlichia*, *Anaplasma*, and *Neorickettsia* (Dumler et al., 2007b; Newton et al., 2009); a few cases of infection with "*Candidatus* Neoehrlichia" are also published (Fehr et al., 2010; Pekova et al., 2011; von Loewenich et al., 2010; Welinder-Olsson et al., 2010), while *Wolbachia* may be a significant cofactor for promoting inflammatory disease in humans with some forms of filariasis (Shakya et al., 2008; Turner et al., 2009). For the purposes of this chapter, discussion is limited predominantly to the former two genera.

Complete genome sequences of *Ehrlichia* and *Anaplasma* species demonstrate their close relationships and genomic synteny (Dunning Hotopp et al., 2006). In contrast, as anticipated from sequence analysis of individual genes, *Neorickettsia* species are more divergent (Dunning Hotopp et al., 2006). The genus "*Candidatus* Neoehrlichia" is now validly published, but lack of a genome sequence confounds appropriate evaluation of its taxonomic position (Kawahara et al., 2004). The current organization of these genera and their species was based on comparisons of partial sequences of *rrs* (16S rRNA genes) and the *groESL* heat shock operons (Dumler et al., 2001). All members described in the genera *Ehrlichia*, *Anaplasma*, "*Candidatus* Neoehrlichia," and *Neorickettsia* are capable of causing disease in humans, animals, or both. Significant pathogens of humans include *E. chaffeensis*, *E. ewingii*, and *A. phagocytophilum*, all of which also cause veterinary disease (Little, 2010; Thomas et al., 2009). Recent investigations implicate an *E. muris*-like bacterium as a human pathogen (Pritt et al., 2011), and individual reports suggest human pathogenicity for *Anaplasma bovis* (J. S. Chae, personal communication) and *E. ruminantium* or a similar organism as well (Reeves et al., 2008). While *E. canis* can cause disease in humans, it is best known as a pathogen of dogs, as is *E. ewingii* (Liddell et al., 2003; Little, 2010; Perez et al., 2006). Except for *A. phagocytophilum* and possibly *A. bovis*, all other *Anaplasma* species, including the erythrocytic *A. marginale* and *A. centrale*, as well as *A. ovis* and *A. platys*, have been recognized as veterinary pathogens only (de la Fuente et al., 2006; Kocan et al., 2010). The precise taxonomic position of the avian/amphibian/reptile anaplasmas currently listed in the *Aegyptianella* genus (e.g., *A. pullorum*) is still not entirely settled, but molecular evidence suggests a close relationship to the genus *Anaplasma* (Rikihisa, 2006). Human infection has not been reported. All members of the *Ehrlichia* and *Anaplasma* genera are transmitted by ticks, usually acarid ticks. The "*Candidatus* Neoehrlichia" genus forms a clade intermediate between *Anaplasma* and *Ehrlichia* (Kawahara et al., 2004). The first example of DNA from this group was obtained from Dutch *Ixodes ricinus* ticks (Schotti variant) and among *I. ricinus* ticks in northeastern Italy ("*Ehrlichia walkeri*") (Brouqui et al., 2003; Dumler et al., 2001), but it was first isolated from *Rattus norvegicus* and *Ixodes ovatus* ticks in Japan (Naitou et al., 2006). A second species, "*Candidatus* Neoehrlichia lotoris" (Yabsley et al., 2008), was identified in raccoons in the United States.

The genus *Neorickettsia* is more distantly related to *Ehrlichia* and *Anaplasma* (Dumler et al., 2001). *N. helminthoeca* was the first species described, and disease is apparently limited to dogs and bears that consume fish infested by infected trematodes, the major vector for this genus (Pretzman et al., 1995). *N. sennetsu*, previously classified as *Rickettsia sennetsu*, then *Ehrlichia sennetsu*, is the first and so far only human pathogen in this genus (Newton et al., 2009; Tachibana, 1986). Although only rarely diagnosed, it also is likely to be vectored by consumption of fluke-infested fish (Tachibana, 1986). The major pathogen in this genus, *N. risticii*, is very closely related to *N. sennetsu*, but has never been reported as a cause of human disease (Rikihisa, 2006). The complex life cycle of *N. risticii* serves as a model for understanding transmission in this genus, where infected trematode cercariae are vectors for fish, possibly other aquatic animals like waterfowl, and insects and their larvae that breed in water containing infected cercariae (Pusterla et al., 2003). Adult insects, such as mayflies, can then carry the infectious agents to terrestrial environments, where they are inadvertently consumed by animals and establish infection in intestinal epithelial cells before disseminating in phagocytes like macrophages.

RICKETTSIOSES

Pathogenesis of Vasculotropic Rickettsial Infections

The underlying theme for disease manifestations in the *Rickettsiaceae* is systemic infection of endothelial cells, leading to increased vascular permeability. Although the presence of the bacterium and its microbe-associated molecular pattern structures likely trigger fever, the predominant clinical manifestation, increased vascular permeability, is likely to be the major pathogenic component, especially for those with complicated or severe disease or in those who die (Walker and Ismail, 2008). Where examinations of human or animal tissues have been conducted, both *Rickettsia* and *Orientia* target endothelial cells in a variety of tissues, after dissemination from the dermal inoculum via tick, mite, flea, or louse bites (Moron et al., 2001; Walker and Ismail, 2008). It is clear that these vasculotropic rickettsiae initially interact with other cells at the site of inoculation, since most inoculation sites become inflamed and large numbers of defense cells, including macrophages, dendritic cells, lymphocytes, and neutrophils, are recruited in response to both pathogen- and arthropod-associated products. It is likely that the earliest interactions with rickettsiae occur with resident dendritic cells and macrophages that both initiate early immune responses but perhaps are also the vehicles by which the bacteria disseminate, first to local draining lymph nodes and then to blood and all tissues and organs (Fang et al., 2007; Murphy et al., 1978). The ability of rickettsiae to disseminate, infect endothelial cells, and induce localized microvascular alterations competes with the waxing immune responses that generally lead to infection control.

Although vasculotropic rickettsiae can kill endothelial cells, whether this is the main determinant of increased vascular permeability or whether induced changes in cytoskeletal structure are major contributors is not well investigated (Walker and Ismail, 2008; Woods and Olano, 2008). In vitro investigations suggest that the combined effects of early rickettsial infection with increasing nitric oxide production and a proinflammatory cytokine milieu all contribute to the increased microvascular permeability (Walker et al., 1997; Woods et al., 2005). These observations are consistent with the concept that infection and host response are in part the cause of early vascular compromise in vasculotropic rickettsial infections. In time, rickettsiae alone have the capacity to kill endothelial cells, but the majority of in vivo lesions examined in humans and animal models have a significant degree of inflammation and, infrequently, thrombosis (Walker et al., 2003). Infection in animal models devoid of the capacity to induce protection, e.g., major histocompatibility complex class I knockout mice, illustrates the beneficial role of the immune and inflammatory response, although contributions to early pathogenesis seem likely (Walker et al., 2001).

Dissemination to and infection of endothelium in diverse anatomical compartments often explains many significant complications, severe disease, and death (Walker et al., 2003). The disseminated rash is the result of rickettsial infection of dermal capillaries and venules, initially resulting in vascular dilatation (macular rash), followed by transudation of edema fluid and exudation of inflammatory cells to create an elevated erythematous (maculopapular) rash (Dumler and Walker, 2005b). Not all infections advance to this state before control by increasing immunity or antimicrobial treatment. However, in some cases, particularly with the severest forms of rickettsial infections like Rocky Mountain spotted fever (RMSF), Mediterranean spotted fever (MSF), louse-borne typhus, murine typhus, and scrub typhus, vascular permeability owing to cytoskeletal retraction and loss of intercellular junctions is replaced by frank endothelial cell injury or death and extravasation of vascular contents, including erythrocytes, into dermal tissues to create a nonblanching lesion called a petechia. Petechiae are considered a hallmark of many rickettsial infections and generally indicate an advanced stage of vascular injury or vasculitis, as illustrated in histopathologic studies of these diseases (Kao et al., 1997). The range of histopathologies characterized by examination of skin biopsies from patients with documented vasculotropic rickettsial infections extends from leukocytoclastic vasculitis with significant degrees of associated necrosis and formation of nonocclusive fibrin-platelet thrombi, to lymphocytic capillaritis and venulitis, to simple vascular dilatation with perivascular inflammatory cell infiltrates that represents the earliest change. Epithelial cells are not infected in vivo, but are affected by the vascular compromise. Basal vacuolization and apoptotic keratinocytes are observed in skin biopsies of RMSF patients, likely a response to local inflammatory mediators and/or ischemic changes with altered local microvascular circulation.

These vascular changes also underlie the systemic manifestations of severe vasculotropic rickettsial diseases. Major complications include gastrointestinal symptoms and signs, such as abdominal pain, nausea, vomiting, and diarrhea, likely owing to similar vascular changes that are documented in autopsy studies. Acute respiratory distress syndrome and noncardiogenic pulmonary edema are life-threatening events with these infections, and occur as a result of infection of the massive microvascular network of the lung (Walker, 2007). As vascular permeability increases, loss of intravascular fluids into pulmonary alveoli leads to significant respiratory compromise that requires very careful hemodynamic management to avoid iatrogenic complications. Similarly, increasing vascular permeability with infection of the meninges or cerebral capillaries can lead to degrees of cerebral edema that could cause brainstem herniation and death. Even in the absence of generalized cerebral edema, the compromise in vascular integrity and moderate levels of ischemia lead to ataxia, seizures, coma, and auditory deficits acutely (Sexton and Kirkland,

1998), and long-term sequelae such as paraparesis; hearing loss; peripheral neuropathy; bladder and bowel incontinence; cerebellar, vestibular, and motor dysfunction; language disorders; learning disabilities; and behavioral problems (Archibald and Sexton, 1995).

Diseases Caused by *Rickettsia* and *Orientia*

SPOTTED FEVER GROUP RICKETTSIOSES (INCLUDING SOME TRANSITIONAL GROUP RICKETTSIOSES)

RMSF. RMSF is caused by infection with *Rickettsia rickettsii*, perhaps the most virulent of all *Rickettsiales* for humans. RMSF occurs after an incubation period that varies from 2 to 14 days (median, 7 days). More than half of patients report a history of tick bite, although no tick bite lesion or eschar is generally evident. Risks include living where RMSF is known to occur, outdoor activities (especially in the woods), occurrence in a season when ticks are active, other family members with similar illnesses, and close dog contact (Dalton et al., 1995). Mild illness is probably infrequent and presentation is often nonspecific, with fever, headache, anorexia, and myalgia. Gastrointestinal manifestations include nausea, vomiting, diarrhea, and abdominal pain, which occur early in disease. The triad of fever, headache, and rash is observed in 44% of patients throughout the infection, but only in 3% at the time of first examination (Helmick et al., 1984).

Rash appears generally after day 3 of illness, and up to 20% of patients do not develop a rash or have an atypical rash (Chapman et al., 2006). The rash initially appears as small blanching macules or maculopapules, often on the extremities, including the ankles, wrists, or legs, and then rapidly spreads to the whole body, including the palms and soles. In the absence of treatment, the rash becomes petechial within several more days, sometimes evolving to palpable purpura. Severe disease can result in ecchymosis and necrosis with confluence of rash. Vascular obstruction secondary to vasculitis and thrombosis is uncommon but can lead to gangrene of the fingers, toes, earlobes, scrotum, nose, and even limbs (Hove and Walker, 1995; Kirkland et al., 1993).

Meningismus and changes in mental status often occur with central nervous system (CNS) infection, and patients can also develop ataxia, auditory deficits, seizures, and coma (Sexton and Kirkland, 1998). Approximately one-third of patients will have CSF pleocytosis, and mononuclear cell- or neutrophil-dominated infiltrates are equally likely; 20% have elevated protein. Imaging studies are generally uninformative, although cerebral edema and meningeal enhan cement can be seen with severe disease.

Pulmonary disease manifests as rales, infiltrates, and noncardiogenic pulmonary edema. In children, conjunctival suffusion, periorbital edema, dorsal hand and foot edema, and hepatosplenomegaly also occur (Helmick et al., 1984). With severe disease, myocarditis, acute renal failure, and vascular collapse with shock and multiorgan failure can occur. Fulminant RMSF, defined as death from *R. rickettsii* infection within 5 days of onset, is characterized by coagulopathy and thrombosis leading to renal, hepatic, and respiratory failure (Walker et al., 1983). Risk for fulminant RMSF is very high in those with glucose-6-phosphate dehydrogenase deficiency. There has been little study of RMSF in immunocompromised subjects (Rallis et al., 1993).

The case fatality rate in the United States has declined over the years, but some question whether this is related to changing practices for laboratory confirmation, as the majority of cases reported to the CDC are no longer considered "confirmed" (Openshaw et al., 2010). When first identified, RMSF had high case fatality rates, even as high as 90%. Even in recent years, the case fatality rate for confirmed RMSF has been higher than estimated by CDC surveillance (Paddock et al., 1999, 2002). For example, the case fatality rate for RMSF in Texas was as high as 8% despite advances in medical care and support (Billings et al., 1998). *R. rickettsii* also causes significant disease in Central and South America. In Brazil, where the disease is known as Brazilian spotted fever, case fatality rates of between 30% and 80% are regularly cited (Nogueira Angerami et al., 2009). Whether the bacterium is distinct and more virulent or other factors, such as delay in diagnosis and treatment, are determinants of severity and fatality still needs investigation.

Laboratory abnormalities are generally nonspecific, and include a normal or low white blood cell count with a counterintuitive left shift in the differential leukocyte count; thrombocytopenia; hyponatremia; and elevated serum aminotransferase activities reflecting underlying hepatic injury.

MSF. MSF is caused by infection with *Rickettsia conorii*, a SFGR species with a close genetic relationship to *R. rickettsii* at the genomic level (Gillespie et al., 2008). The disease is also known by the name "boutonneuse fever," owing to the frequent presence of an eschar, or "tache noire," at the tick bite site. Although not validly published, some rickettsiologists divide the *R. conorii* species clade into genetically distinguishable subspecies, including typical *R. conorii* Malish, Moroccan, and Kenyan strains (subspecies *conorii*) and subspecies *israelensis*, *caspia*, and *indica*, reflecting the geographic origin of these isolates (Zhu et al., 2005). Whether the subspecies divisions reflect significant clinical differences is a matter of some dispute.

Classical MSF has many similarities to RMSF, with the key clinical manifestations of fever, headache, myalgia, and a maculopapular rash that appears 3 to 5 days after onset of fever (Raoult et al., 1986). Between 20 and 87% of patients develop an eschar at the site of tick attachment. Older citations identify eschars associated with draining lymphadenopathy, but more recent publications associate this feature and lymphangitis with distinct *Rickettsia* species (Fournier et al., 2005). As with RMSF, severe forms occur where skin lesions coalesce to purpura; there is CNS involvement, respiratory failure, and renal failure; and the case fatality rate is about 2.5%. Laboratory abnormalities are similar to those observed with RMSF, including normal to low leukocyte counts with a left-shifted differential leukocyte count, and thrombocytopenia that can be profound.

A strain of *R. conorii* first identified in Israel (subspecies *israelensis*), for which the disease was named Israeli spotted fever, and for which eschar is variably a less frequently reported sign, is associated with greater morbidity and mortality where it occurs in Israel and in Portugal (Bacellar et al., 1999; Gross and Yagupsky, 1987; Sousa et al., 2008). Case fatality rates vary from 21 to 33% in Portugal, and alcoholism is an important risk factor. Other signs and symptoms of severe infection are not significantly different between classical MSF caused by *R. conorii* Malish strains (*R. conorii* subsp. *conorii*) and MSF caused by Israeli spotted fever (*R. conorii* subsp. *israelensis*) strains (de Sousa et al., 2005). Whether the increased virulence is a function of the bacterium, the host, or other social, economic, or medical practice factors is unknown.

Similarly, an unknown "viral exanthematous" disease identified in the Caspian Sea region was given the name "Astrakhan fever" (Tarasevich et al., 1991a). Subsequent investigations, including isolation and molecular characterization, showed the etiologic agent to be *R. conorii*, now identified as Astrakhan fever strains or subspecies *caspia* (Fournier et al., 2006). Little is published offering careful clinical evaluation of the disease manifested by infection with this *R. conorii* subspecies; however, the disease appears indistinguishable from MSF except for the notable lack of fatalities. Since the original characterization in Astrakhan, infections and/or infected ticks have been identified in Chad in Africa and in Kosovo in Eastern Europe (Fournier et al., 2003b, 2003c).

Very little is available in terms of clinical description and clear identification of human infection by *R. conorii* Indian tick typhus strains (subspecies *indica*), and descriptions of Indian tick typhus often present no specific evidence of infection by this strain/subspecies of *R. conorii* (Parola et al., 2005). Prior reports often assume that SFGR infections in Pakistan and India are caused by this organism (Parola et al., 2001), although the recent description of rickettsiosis caused by a proposed new SFGR species, *R. kellyi*, brings such assumptions into question (Miah et al., 2007; Padbidri et al., 1984; Parola et al., 2001; Prakash et al., 2012; Rolain et al., 2006). Without further species delineation, these infections are reported to be less often associated with eschars and more often with purpuric rashes; paradoxically, fatalities have not been reported.

Other Spotted Fever Group Rickettsioses. There are a number of other described clinical conditions attributable to SFGR or transitional genomic group *Rickettsia* infections (Gillespie et al., 2008). These include the entities African tick bite fever (*R. africae*), maculatum disease (*R. parkeri*), Siberian/North Asian tick typhus/lymphangitis-associated rickettsiosis ("*R. sibirica* subsp. *mongolotimonae*"), Japanese spotted fever (*R. japonica*), Far Eastern tick-borne rickettsiosis (*R. heilongjiangensis*), Flinders Island spotted fever/Thai tick typhus (*R. honei*), tick-borne lymphadenopathy (*R. slovaca*), Queensland tick typhus (*R. australis*), rickettsialpox (*R. akari*), and flea-borne spotted fever/cat flea typhus ("*R. felis*"), among many others, including many with limited description of clinical disease (Parola et al., 2005; Walker, 2007). Many of these bear geographically descriptive names based on initial characterization of the causative rickettsia, although current knowledge of geographic ranges may differ considerably, and as a result, so do the arthropod vectors. Clinical manifestations also vary considerably, although most patients develop nonspecific findings of fever, headache, and rash at some time during the infection; eschars are relatively common features of these infections, although vesicles seem to occur in selected rickettsial infections as well (Table 1).

TYPHUS GROUP RICKETTSIOSES

Louse-Borne Typhus. Louse-borne or epidemic typhus is caused by infection with *R. prowazekii*, one of two members of the TGR genomic group. The genome of *R. prowazekii* appears to be a reduced set of that in SFGR, such as *R. conorii* (Bechah et al., 2008; Walker et al., 2008). The name "typhus" reflects the frequency with which CNS involvement occurs and underscores the high case fatality rate often observed with this infection. Classically, humans were considered the only reservoir for *R. prowazekii*, in part because of the strict obligate relationship of the human body louse *Pediculus humanus corporis* with humans (Li et al., 2010). Epidemics occurred during times when lice were readily transferred from person to person by clothing and bedding. Because *R. prowazekii* can establish persistent or latent infection, recrudescence at some later time would provide an opportunity for transmission back to lice and epidemic spread (Bechah et al., 2008; Walker and Ismail, 2008). The recognition that ectoparasites of some mammals, such as flying squirrels, can also harbor *R. prowazekii* could shed light on the diversity of its reservoir hosts and the sporadic occurrence of louse-borne typhus not associated with clear louse exposure (Reynolds et al., 2003).

Patients with louse-borne typhus present with disease manifestations very much like those of other severe rickettsioses, like RMSF (Table 2). The characteristic rash is less often recognized in recent times, perhaps in part owing to the disease's occurrence in geographic regions where individuals with dark-colored skin reside (Raoult et al., 1998). It is widely stated that typhus rashes occur more often on the trunk than extremities; however, the objective data for this are meager. The other major manifestation that differs from other rickettsioses is CNS involvement, with findings such as meningoencephalitis, delirium, stupor, coma, and other severe complications (Massung et al., 2001; Perine et al., 1992). Cough and respiratory signs are common, and thus pneumonitis was a frequently cited complication in older textbooks, but the few published clinical studies that objectively evaluate this feature do not support this (Bechah et al., 2008; Walker and Ismail, 2008). Louse-borne typhus has also been traditionally thought of as

TABLE 1 Select spotted fever and transitional group rickettsioses, their causative agents, and key clinical features[a]

Disease	Species	% of patients with:				Multiple eschars	Vesicles or pustules	Draining lymphadenopathy
		Fever	Headache	Rash	Eschar			
RMSF	*R. rickettsii*	100	80	90	0	No	No	No
MSF	"*R. conorii* subsp. *conorii*"	100	56	97	72	Very rare	No	Rare
Israeli tick typhus	"*R. conorii* subsp. *israelensis*"	93	24	96	39	No	No	No
Astrakhan fever	"*R. conorii* subsp. *caspia*"	100	92	100	23	No	No	No
African tick bite fever	*R. africae*	89	15	51	99	Yes	Yes	Yes
Maculatum disease	*R. parkeri*	100	83	83	92	Yes	Yes	Yes
Siberian/North Asian tick typhus	*R. sibirica* sensu stricto	100	75	75	50	No	No	Yes
Lymphangitis-associated rickettsiosis	"*R. sibirica* subsp. *mongolotimonae*"	100	55	78	89	Yes	No	Yes
Far Eastern tick-borne rickettsiosis	*R. heilongjiangensis*	100	100	92	92	No	No	Yes
Japanese spotted fever	*R. japonica*	100	80	100	90	No	No	No
Flinders Island spotted fever	*R. honei*	100	71	43	29	No	No	Yes
Queensland tick typhus	*R. australis*	100	90	94	65	No	Yes	Yes
Rickettsialpox	*R. akari*	97	100	100	90	Yes	Yes	Yes
Cat flea typhus	"*R. felis*"	100	100	84	16	Rare	No	Yes
Tick-borne lymphadenopathy (TIBOLA); *Dermacentor*-borne necrosis, erythema, and lympadenopathy (DEBONEL)	*R. slovaca*	38	33	10	100	No	No	Yes

[a]Data from Brouqui et al., 2007; de Sousa et al., 2005; Fournier et al., 2005; Jensenius et al., 2003; Lewin et al., 2003; Paddock et al., 2003, 2008; Pérez-Osorio et al., 2008; and Tarasevich et al., 1991a, 1991b.

TABLE 2 Common clinical features of rickettsioses, scrub typhus, ehrlichiosis, and anaplasmosis in humans[a]

Clinical finding	Frequency of finding (%) in patients infected with:						
	R. rickettsii (RMSF)	*R. prowazekii* (typhus)	*R. typhi* (murine typhus)	*R. conorii* (MSF)	*O. tsutsugamushi* (scrub typhus)	*E. chaffeensis* (HME)	*A. phagocytophilum* (HGA)
Fever	100	100	99	100	100	96	93
Rash	90	32	37	98	49	26	6
Rash on palms and soles	82			79			
Headache	91	100	59	70	100	72	76
Myalgia	72	70	46	58	32	68	77
Nausea or vomiting	60	45	40	36	28	57	38
Conjunctivitis	30	34		36	29		
Pneumonitis	15	8	15	10	28		
Any severe neurologic complication	26	50	6	11	10	20	17

[a]Data from Bernabeu-Wittel et al., 1999; Billings et al., 1998; Dumler et al., 1991; Fergie et al., 2000; Font-Creus et al., 1985; Gikas et al., 2002; Helmick et al., 1984; Hernandez Cabrera et al., 2004; Kaplowitz et al., 1983; Koliou et al., 2007b; Older, 1970; Perine et al., 1992; Raoult et al., 1986, 1998; Sexton and Burgdorfer, 1975; Silpapojakul et al., 1991a, 1993; Taylor et al., 1986, Tselentis et al., 1992, and Wilde et al., 1991.

one of the most severe rickettsioses; case fatality rates as high as 12% have been documented in recent times, underscoring the gravity of infection by *R. prowazekii* still (Raoult et al., 1997, 1998).

Murine Typhus. Murine typhus derives its name from the main reservoir hosts for the vector fleas: rats and small rodents. The infection is caused by *R. typhi*, the other member of the TGR genomic group. Transmission from the infected rodent host occurs when *R. typhi* is acquired by the flea vector, usually the rat flea (*Xenopsylla cheopis*) or cat flea (*Ctenocephalides felis*), where it replicates in the midgut epithelial cells and is shed in flea feces that contaminate flea bite wounds in humans (Azad et al., 1997; Vaughan and Azad, 1990). Owing to the ubiquitous presence of rats, murine typhus most often occurs in warm climates in urban and suburban regions where there is considerable human activity (Azad, 1990; Boostrom et al., 2002; Schriefer et al., 1994; Williams et al., 1992). As for other rickettsioses, the clinical presentation is most often nonspecific and complicated by the lack of a rash in up 50% of patients (Dumler et al., 1991; Gikas et al., 2002; Silpapojakul et al., 1993; Whiteford et al., 2001) (Table 2). When present, the rash is maculopapular most often and petechial only infrequently. While murine typhus is often thought to be a "mild" infection, in fact, hospitalization occurs frequently, up to 17% of patients have severe neurologic complications, and the case fatality rate in the United States is as high as 4% (Bernabeu-Wittel et al., 1998; Dumler et al., 1991). Supportive laboratory findings include a normal leukocyte count with a left shift, mild to moderate thrombocytopenia, and elevations in hepatic transaminases.

SCRUB TYPHUS

Scrub typhus is a distinct entity among rickettsioses mostly because the etiologic agent is assigned to a unique genus and transmitted by larval-stage trombiculid mites in the genus *Leptotrombidium*, not ticks, lice, or fleas. *O. tsutsugamushi* is antigenically and biologically different than *Rickettsia* species, and the vector also dictates an ecological pattern thought to be distinct from rickettsioses like murine typhus (Phongmany et al., 2006). The infection is largely confined to a region bounded by Korea on the east, northern Australia on the south, and Pakistan to the west, an area with more than 2 billion human inhabitants (Kelly et al., 2009). It is estimated that more than 1 million patients acquire scrub typhus yearly (Kelly et al., 2009). Despite such unique features, the clinical disease observed in scrub typhus has many nonspecific features also observed with other rickettsioses, including rapid onset of fever, headache, myalgia, rash, and eschar, with other features like pneumonitis and severe neurologic complications occurring in less than 30 and 10% of patients, respectively (Table 2). Lymphadenopathy is suggested as a differentiating feature for scrub typhus versus murine typhus (Phongmany et al., 2006).

EHRLICHIOSIS AND ANAPLASMOSIS

Pathogenesis of *Anaplasmataceae* Infections

The underlying theme for disease manifestations caused by the *Anaplasmataceae* is infection of cells derived from bone marrow, chiefly phagocytic cells, leading to localized or systemic inflammatory responses, including the production of vasoactive cytokines and vascular permeability (Dumler, 2005; Walker et al., 2008). For some veterinary pathogens, infection of platelets, erythrocytes, and endothelial cells also occurs (Dumler et al., 2001). For microbes that rely on inflammatory cells for survival, mechanisms that promote inflammation are plausible evolutionary adaptations, as long as sufficient inactivation or subversion of host antimicrobial factors is also present, and this seems to be so for most *Anaplasmataceae* studied (Carlyon and Fikrig, 2006; Garcia-Garcia et al., 2009a, 2009b; Scorpio et al., 2006; Zhang et al., 2004). Microbe-associated molecular pattern structures likely trigger inflammatory responses through either Toll-like receptor 2 and/or the inflammasome (Choi et al., 2004; Pedra et al., 2007a, 2007b). Animal models, chiefly knockout mice, show that inflammation, the predominant clinical manifestation, is a major pathogenic component, especially for those with complicated or severe disease or in those who die (Browning et al., 2006; Ismail et al., 2006, 2007; Ismail and Walker, 2005; Scorpio et al., 2004, 2005, 2006; Stevenson et al., 2008; Thirumalapura et al., 2008; von Loewenich et al., 2004). Where examinations of human or animal blood and tissues have been conducted, *E. chaffeensis*, *E. canis*, *E. ewingii*, and *A. phagocytophilum* target professional phagocytes such as monocytes, macrophages, and granulocytes in blood and coursing through tissues. The mechanisms by which these cells become infected after dermal inoculum via tick bite and the infection eventually disseminates are not well studied. The key determinant of severity seems to be related not to pathogen burden but to induction of proinflammatory and immune responses in immunocompetent individuals (Dumler et al., 2007a; Ismail and Walker, 2005; Scorpio et al., 2004). However, there is increased disease severity with fulminant infection among those with HIV infection and other processes that lead to immunocompromise (Marty et al., 1995; Paddock et al., 2001; Stone et al., 2004; Thomas et al., 2007). That induction of immunity is also beneficial is shown in findings that increasing immunological responses lead to eventual resolution of clinical disease and infection (Winslow et al., 2005; Yager et al., 2005).

Once the bacteria enter their host cell, replication begins within a vacuole, and the doubling times for *E. chaffeensis* and *A. phagocytophilum* are likely to vary between 8 and 24 hours (Branger et al., 2004; IJdo and Mueller, 2004; Lin et al., 2007). To sustain infection, host cells must also survive to allow sufficient replication of the bacterium for ongoing transmission to other cells and to increase the circulating pool of infected cells accessible to subsequent tick bites in reservoir hosts. Thus, where investigated in nucleated cells, most *Anaplasmataceae* delay apoptosis (Carlyon and Fikrig, 2006; Choi et al., 2005; Ge and Rikihisa, 2006; Lee and Goodman, 2006; Xiong et al., 2008). For *A. phagocytophilum*, altered expression of adhesion ligands on the host cell's surface precludes excessive tissue accumulation (Choi et al., 2003), but where rheologic conditions favor slow blood flow or pooling, tissue inflammation can be readily observed, such as in the liver sinusoids (Browning et al., 2006; Bunnell et al., 1999). This is the likely explanation for the frequent elevations of alanine and aspartate transaminase activities in serum of infected patients. Presumably, similar conditions can exist in other anatomical sites, explaining some of the disseminated manifestations of *Anaplasmataceae* infections. One enigma is the prevalence of leukopenia and thrombocytopenia, which occur in 60 and 80% of human infections, respectively (Dumler, 2005). There are too few infected cells circulating in human blood to account for the degree of leukopenia, and for *E. chaffeensis*, *A. phagocytophilum*, and *E. ewingii*, platelets are not targets for infection. Some have advanced the hypothesis that infection induces myelosuppressive and hematopoiesis-suppressing cytokines (Klein et al., 2000), while other hypotheses suggest macrophage activation and nonspecific consumption of the hematologic factors in a process like that observed in infection-associated hemophagocytic syndromes (Abbott et al., 1991; Dierberg and Dumler, 2006; Dumler et al., 2007a).

A key clinical observation in humans infected with *E. chaffeensis* or *A. phagocytophilum* is the relative infrequency of rash, with the exception of children with *E. chaffeensis* infection (Dumler et al., 2007b; Walker et al., 2008). Macular or erythematous and rarely petechial rashes are described in human monocytic ehrlichiosis (HME), and the rarity of the latter feature is consistent with the observation that these microbes only rarely target endothelial cells. Thus, it is likely that rash, when observed, is the result of local inflammatory reactions that result in vasodilation (erythema) and modest compromise of vascular permeability (edema), or reflect compromise of hemostasis owing to thrombocytopenia (Nyarko et al., 2006). One exception is the description of rash in human granulocytic anaplasmosis (HGA), where it reflects coinfection by *Borrelia burgdorferi* and the presence of classical erythema migrans lesions (Bakken and Dumler, 2008).

What factors determine the development of severe disease? This is an active area of study, but most data now point to the induction of immune pathology (Browning et al., 2006; Dumler et al., 2007a; Ismail et al., 2006, 2007; Ismail and Walker, 2005; Scorpio et al., 2004, 2005, 2006). Severe complications include acute respiratory distress syndrome, disseminated intravascular coagulation- or septic/toxic shock-like syndrome, meningoencephalitis, and renal failure, among other conditions (Demma et al., 2005; Dumler et al., 2007b). The underlying theme of these observations is likely tied back to the proposed major pathogenic factor for these organisms, induction of inflammation. Presumably, localized proinflammatory recruitment of new host cells often leads to minimal tissue injury and disease, while promoting expansion of the bacterial population through subsequent tick bites. However, variations in host inflammatory and immune responses owing to genetically determined or acquired factors or variations in pathogen strain virulence could promote a systemic proinflammatory response that when poorly controlled leads to severe disease (Dumler et al., 2007a; Scorpio et al., 2005). Supporting this concept is the observation that severity of HGA is linked to factors that result in overproduction and poor resolution of inflammatory effectors, as observed in macrophage activation syndromes (Dierberg and Dumler, 2006; Dumler et al., 2007a). Fortunately, most patients do not develop severe complications, but the net result is a disease that leads to hospitalization in over 36%, a requirement of intensive care unit admission in 5 to 10%, and death in between 0.6 and 1.9% (Bakken and Dumler, 2008; Dumler et al., 2007b). The long-term ramifications of infection, including those causing severe disease, are essentially unstudied. In the only long-term study of outcomes of these diseases in humans, HGA patients had a significantly greater likelihood of experiencing fever, chills, sweats, and fatigue for 1 year or longer after resolution of their initial episode compared with healthy controls (Ramsey et al., 2002). Although persistence of infection is readily observed in animal reservoir hosts, this has never been documented in humans. Thus, the long-term consequence in humans of antecedent infections by *Anaplasmataceae* species is still unknown.

Another common concern about these tick-transmitted infections is the safety of the public health blood supply (Leiby and Gill, 2004). Considerable time and study has been extended to this question, and it is clear that these agents can survive under existing storage conditions for intervals as long as several weeks (McKechnie et al., 2000). Yet no clear example of *E. chaffeensis* and only one or two examples of *A. phagocytophilum* transmission via blood products have been reported (Leiby and Gill, 2004). This is likely because of current practices that lead to leukoreduction or UV irradiation prior to transfusion.

Human Ehrlichioses

HME

HME is caused by infection with *E. chaffeensis*. The disease name is derived from the predilection that *E. chaffeensis* has for blood-borne monocytes and tissue macrophages, the favored host cell (Paddock and Childs, 2003; Sehdev and Dumler, 2003). Transmission is achieved with bites of the infected *Amblyomma americanum* (lone star tick) (Childs and Paddock, 2003). This tick species is now highly abundant throughout most of the south central, southeastern, and Mid-Atlantic regions of the United States and is expanding rapidly to other regions including the Northeast, where it has been found as far north as Maine. Since the disease's recognition in 1986, at least 7,500 cases have been passively reported to the CDC. Infections of all severities have been recognized in all age groups, yet rates of infection are highest in those aged 60 to 69, and men are 1.4-fold more often affected than women (Dahlgren et al., 2011).

Initiation of infection and the pathophysiologic effects at the tick bite site are essentially unstudied. Regardless, monocytes become infected and enter the general circulation to allow dissemination to all organs and tissues. The incubation period, that period between which inoculation occurs and clinically relevant disease first appears, is about 9 days (Fishbein et al., 1987). Onset is abrupt, chiefly presenting with fever, headache, and myalgia (Table 2). A range of other symptoms and signs point to systemic involvement, such as in the gastrointestinal, respiratory, renal, and central nervous systems. During this time, laboratory studies initially can show leukopenia, thrombocytopenia, anemia, and elevations in serum transaminase activities that worsen over days to 1 week; serum urea nitrogen and creatinine levels may deteriorate during this interval as well. When treatment is started promptly, leading to rapid reduction in pathogen blood loads, patients rapidly defervesce and report feeling much better within 1 to 2 days. However, the median interval before which patients report return to normal health is approximately 3 weeks (Fishbein et al., 1994).

Reported complications of HME include acute respiratory distress syndrome, meningoencephalitis, septic/toxic shock-like syndromes, increased severity with preexisting conditions, and rarely opportunistic infections or myocarditis/cardiac failure (Dumler, 2005; Dumler et al., 2007b; Thomas et al., 2009). In addition, individuals with preexisting immunocompromise, chiefly HIV infection (Paddock et al., 2001), organ transplantation with ongoing immunosuppressive therapy (Thomas et al., 2007), immune suppressive therapy for rheumatologic and autoimmune conditions (Stone et al., 2004), or other immunosuppressive diseases, are at risk for fulminant infection characterized by extensive tissue injury and necrosis in the presence of high microbial loads (Sehdev and Dumler, 2003).

E. EWINGII EHRLICHIOSIS

E. ewingii is an *Anaplasmataceae* species first demonstrated to cause disease in dogs, in which unlike the situation for *E. canis*, the pathogen infects neutrophils and granulocytes (Buller et al., 1999). The bacterium is genetically and antigenically closely related to *E. chaffeensis* and *E. canis*, but phenotypically resembles *A. phagocytophilum* because of the granulocyte niche (Dumler et al., 2001). Since the reassignment of *A. phagocytophilum* to the *Anaplasma* genus and the resulting change in disease nomenclature, some have suggested using the disease name "human granulocytic ehrlichiosis" to describe infection with *E. ewingii* in humans; however, to avoid confusion with the prior use of this name for disease caused by *A. phagocytophilum*, this is best avoided. Thus, the currently preferred name is *E. ewingii* (human) ehrlichiosis (Walker et al., 2008). Human disease caused by *E. ewingii* is rarely documented because the organism has not yet been cultivated in vitro, precluding the development of simple serological diagnostic tests. When first identified, *E. ewingii* comprised about 7% of the infections originally thought to be *E. chaffeensis* (Buller et al., 1999). Since then, very few human infections have been reported. The vast majority of these infections occur with immunocompromise, such as in HIV-infected patients, although infection in immunocompetent persons is reported (Paddock et al., 2001; Paris et al., 2008). Fever, myalgia, and malaise are frequent, as are leukopenia, thrombocytopenia, and elevated hepatic transaminase activities. Complications with *E. ewingii* infection are infrequent, and no deaths have been reported (Perez et al., 1996).

HUMAN *E. CANIS* EHRLICHIOSIS

Despite the first publication on human infection by an *Ehrlichia* species claiming *E. canis* as its cause, only to be disproven with the isolation of *E. chaffeensis*, human infection by *E. canis* is now recognized, yet is very rare (Perez et al., 1996, 2006). To date, only 20 cases of human *E. canis* infection have been reported, and typical features of ehrlichiosis are present, including fever, headache, and myalgia, among other findings that occur in the minority of human infections. Leukopenia and thrombocytopenia can also occur. No complications or fatalities have been reported. A single subject was asymptomatically infected for at least 1 year (Perez et al., 1996).

HUMAN INFECTION BY AN *E. MURIS*-LIKE BACTERIUM

In 2009, routine PCR diagnostic testing for *E. chaffeensis*, *E. ewingii*, and *A. phagocytophilum* detected aberrant signals for which resulting sequence analysis supported the occurrence of infection by an *E. muris*-like agent (Pritt et al., 2011). Since then, 14 patients were identified among those submitted for testing for ehrlichiosis because of fever, headache, lymphopenia, and thrombocytopenia; all were from the upper Midwest, and 4 were taking immunosuppressive medications. At least 1 patient had a serological response to *E. chaffeensis*. All recovered after doxycycline treatment.

HGA

A. phagocytophilum, a long-known veterinary pathogen (*Ehrlichia phagocytophila*; *Ehrlichia equi*), was first identified in a human patient in 1990 (Bakken et al., 1994; Chen et al., 1994). Since then, at least 9,371 cases have been passively reported to the CDC. HGA has received much attention, in part owing to its emerging status and rapid ascension as the third-most-common human tick-borne disease in the United States, and likely in the world, but also because of its transmission by *Ixodes* spp. ticks that also transmit *B. burgdorferi*, the causative agent of Lyme borreliosis, or Lyme disease. Clear examples of coinfection with culture recovery of both pathogens are established, and it is estimated that coinfection could occur as often as every 10 infections with either agent (De Martino et al., 2001; Mitchell et al., 1996; Moss and Dumler, 2003; Pusterla et al., 1998; Wormser et al., 1997). *Ixodes* species known to carry and potentially transmit *A. phagocytophilum* to man include *I. scapularis* and *I. pacificus* in the United States, *I. ricinus* and *I. persulcatus* in Europe, and *I. persulcatus* in Asia (Cao et al., 2006; Dumler et al., 2005; Kim et al., 2003; Ohashi et al., 2005). Like infection by *E. chaffeensis*, HGA occurs in patients in all age ranges, but does so more frequently with advancing age; rates in the 60- to 69-year-old group are the highest, and the male/female ratio is 1.4:1 (Dahlgren et al., 2011).

A. phagocytophilum is transmitted to humans via *Ixodes* spp. tick bite. Since neutrophils are by far the most frequently infected cell observed in humans and animals, but are generally considered to have a life span of <24 hours, inconsistent with acquisition, replication, and dissemination into blood from the tick bite wound, some investigators have proposed that endothelium provides a bridge cell (Munderloh et al., 2004). However, direct investigation of this in vivo has not supported the concept (Granquist et al., 2010). Presumably, the local inflammation induced by the tick bite attracts many cells, including neutrophils, which then become infected. Whether these cells return directly to the bloodstream or act as vehicles for travel into the lymphatics is not known. However, *A. phagocytophilum* infection significantly reduces surface adhesion molecule expression, possibly allowing infected cells the opportunity to avoid tissue sequestration (Choi et al., 2003). Regardless, infected neutrophils appear in the bloodstream within about 1 week after the tick bite, and on average, at peak levels of infection, about 1% of all neutrophils contain morulae, or intravacuolar colonies (Bakken and Dumler, 2008; Dumler et al., 2005, 2007b).

The disease is characterized by sudden onset of fever, chills, headache, and myalgia (Table 2), with concomitant presence of thrombocytopenia, leukopenia, anemia, and two- to threefold increases in serum transaminase activities beyond the normal range. Other findings that implicate systemic dissemination and tissue involvement include manifestations referable to the gastrointestinal tract, cardiopulmonary system, renal system, and bone marrow. Serological studies suggest that most individuals who sustain infection probably do not develop significant enough disease to warrant any specific treatment and their infections resolve spontaneously (Aguero-Rosenfeld et al., 2002; Bakken et al., 1998; IJdo et al., 1999). However, a small proportion will seek health care after about 4 to 7 days of fever, after which 36% are hospitalized and 3 to 7% have life-threatening conditions that require admission to an intensive care unit (Bakken et al., 1996b; Demma et al., 2005). Complications requiring such interventions include acute respiratory distress syndrome, disseminated intravascular coagulation- or toxic/septic shock-like syndrome, peripheral neuropathy, and myocarditis, among other findings (Dahlgren et al., 2011). Meningoencephalitis has also been reported to the CDC; however, careful observations that support the occurrence of meningoencephalitis in HGA have not been described (Bakken and Dumler, 2008; Dumler et al., 2005).

The case fatality rate based on reports to the CDC is 0.6%. Most fatalities are linked to the development of opportunistic infections, a recognized complication of *A. phagocytophilum* infection in animals (Dumler et al., 2005, 2007b). Likewise, preexisting illness associated with degrees of immune dysfunction, including malignancy, diabetes mellitus, autoimmune arthritis, and asplenia, increases the risk for hospitalization, development of a life-threatening complication, or death (Bakken and Dumler, 2008).

"*CANDIDATUS* NEOEHRLICHIA MIKURENSIS" INFECTION IN HUMANS

Less-well-classified *Anaplasmataceae* organisms within "*Candidatus* N. mikurensis" have also been recently described as human pathogens in six recent cases (Fehr et al., 2010; Kawahara et al., 2004; Pekova et al., 2011; von Loewenich et al., 2010; Welinder-Olsson et al., 2010); infection was identified within peripheral blood granulocytes in one patient, and no isolates were made from any of these patients. All patients were from Europe, including cases from the Czech Republic, Germany, Sweden, and Switzerland. Of the six patients, four had evidence of preexisting immunocompromise (chronic lymphocytic leukemia; mantle cell lymphoma; chronic inflammatory demyelinating polyneuropathy treated with rituximab, cyclophosphamide, and prednisolone; and orthotopic liver transplantation) and developed severe sepsislike syndromes, some of which were protracted over weeks. All cases were diagnosed by broad-range amplification of eubacterial *rrs* (16S rRNA genes). Doxycycline treatment was associated with clinical resolution in five of these cases; the sixth patient died.

MANAGEMENT, ANTIMICROBIAL EFFICACY, AND TREATMENT FAILURES IN RICKETTSIOSES AND EHRLICHIOSES

Once diagnosed, rickettsioses, ehrlichiosis, and anaplasmosis are primarily managed with specific antimicrobial agent therapy. Depending upon complications that vary with length of time before initiation of therapy, age of patient and underlying medical conditions, and likely other factors such as prior medications, other supportive care approaches could also be required (Walker, 2007; Walker et al., 2008). Medical therapy is largely directed at reducing viability and propagation of the organism. Complications are fortunately not common in most rickettsial infections if therapy is instituted promptly (Chapman et al., 2006). Management of complications should always take into consideration the underlying pathophysiology of the vasculotropic rickettsioses, vasculitis, since careful hemodynamic management will be needed to preclude iatrogenic complications like noncardiogenic pulmonary edema that is life-threatening.

Antimicrobial therapy does not substantially differ for various rickettsioses, but there is very little clinical evidence beyond retrospective studies to support the use of specific regimens for many rickettsioses. Tools to assess antimicrobial resistance in these obligate intracellular bacteria have not been well validated against standard methods for other bacteria; thus, interpretation of laboratory-derived MICs of various drugs remains somewhat speculative (Barker, 1968; Brouqui and Raoult, 1992; McDade, 1969; Rolain et al., 1998). Retrospective reviews of outcomes from clinical studies, and on occasion prospective randomized treatment trials, have been conducted, generally examining very small numbers of patients (Bitsori et al., 2002; Smadel et al., 1948). With the advent of molecular characterization of resistance and the increasing availability of rickettsial genome sequences, in silico predictions of resistance can be made and tested, at least in vitro, to provide some guidance (Biswas et al., 2008, 2009; Rolain and Raoult, 2005a, 2005b). With these approaches and technologies, *Rickettsia* species are generally found to be sensitive to tetracyclines, fluoroquinolones, chloramphenicol, and rifamycins, and to a lesser degree sensitive to macrolides (Rolain et al., 1998). In general, *Anaplasmataceae*, including *Ehrlichia* spp. and *A. phagocytophilum*, have the same profile except for predicted and established fluoroquinolone resistance (Branger et al., 2004; Maurin et al., 2001). In general, *Rickettsiales* are genetically resistant to aminoglycosides via aminoglycoside phosphotransferases, β-lactams and cephalosporins via β-lactamases or metallo-β-lactamase, and combination trimethoprim-sulfamethoxazole probably because the genomes lack *folA*, *folP*, or both (Biswas et al., 2008).

The most widely accepted antimicrobial treatment for rickettsial infections is the use of an antibiotic in the tetracycline class, specifically doxycycline. The major advantage of doxycycline is its twice-daily oral dosing, as opposed to three or four doses daily for tetracycline. It is broadly active in vitro against all *Rickettsia* and *Orientia* species examined, although some moderate increases in MIC against *O. tsutsugamushi* isolates have been reported (Rolain et al., 1998; Watt et al., 1999). A prior concern regarding tetracycline use has been its proclivity toward intercalating into developing teeth and bones, leading to yellowed teeth or weakened bone structure, previously limiting use in children and pregnant women. However, as an increasing repertoire of antibiotics with specific activities has evolved, the use of tetracycline antibiotics has significantly diminished, as has the overall dose to which children are exposed (Abramson and Givner, 1990). The lower lifetime dosages coupled with the lower propensity of doxycycline to accumulate in bone have dramatically decreased the perceived likelihood of adverse effects with its use in children. Simultaneously, chloramphenicol, once a favored drug for rickettsial infections, has become increasingly inaccessible. In addition, it is associated with excess morbidity and mortality in RMSF and also has potentially serious complications (Dalton et al., 1995). Thus, doxycycline is now the first-line drug recommended by all major textbooks for treatment of rickettsial infections, even among patients under the age of 8 (Abramson and Givner, 1990). The concerns regarding the use of doxycycline and tetracyclines in pregnancy have not abated. As a result, chloramphenicol is still often believed to be the first choice for treatment of both children and pregnant women (O'Reilly et al., 2003). Chloramphenicol use in RMSF has now been demonstrated to be associated with excess mortality compared to tetracyclines when controlled for all other factors (Dalton et al., 1995). Thus, given the high case fatality rate associated with RMSF, many authorities advocate the use of doxycycline to preclude maternal and fetal demise from an otherwise treatable infection.

In general, tetracycline antibiotics are considered rickettsiastatic, not rickettsiacidal. This concept has led to several specific recommendations regarding treatment in rickettsial disease. First, treatment should be provided for long enough that immune recognition and clearance of active infection can proceed to eliminate viable although nonreplicating rickettsia surviving within intracellular niches. Second, since proper immune induction is in part dependent on rickettsial antigen mass, early therapy can sometimes simply suppress growth in the absence of waxing immunity, leading to relapse after apparent appropriate treatment doses and intervals. This and very limited animal data suggest that prophylactic antibiotics for rickettsial infections after tick bites are contraindicated. The data to support this contention are meager at best, and need more investigation (Kenyon et al., 1978).

Aside from tetracyclines and chloramphenicol, only very limited data support the use of alternative antimicrobial agents. The data have been better developed for treatment of *R. conorii* MSF, where clinical trials are published regarding the comparative efficacy of fluoroquinolones, macrolides, and other classes of antimicrobial agents (Bella et al., 1990, 1991; Bella-Cueto et al., 1987; Cascio and Colomba, 2002; Cascio et al., 2001, 2002; Dzelalija et al., 2002; Gudiol et al., 1989; Muñoz-Espin et al., 1986; Ruiz Beltrán and Herrero Herrero, 1992b). Continuing disfavor has been documented with the use of chloramphenicol, in part because of the excess mortality in RMSF and longer intervals to recovery in MSF, its inaccessibility as an oral agent in the U.S. market, and its profile of adverse effects that include aplastic anemia and gray baby syndrome (Cascio et al., 2001; Colomba et al., 2006; Shaked et al., 1989). Likewise, antimicrobial combinations that include sulfa drugs such as sulfadiazine, sulfamethoxazole, or sulfisoxazole have been widely associated with greater disease severity and adverse outcomes (Dumler et al., 1991; Paddock et al., 2001; Ruiz Beltrán and Herrero Herrero, 1992a); their use is therefore contraindicated in infections caused by members of both *Rickettsiaceae* and *Anaplasmataceae*.

Spotted Fever Group Rickettsioses

Discussion of the treatment of spotted fever group rickettsioses is well described in reviews and textbooks (Yu et al., 2002). Randomized clinical trials for comparison of antimicrobial efficacy have been conducted only to a limited degree on small patient cohorts for spotted fever group rickettsioses. There have been nine randomized clinical trials conducted among patients diagnosed with MSF (Bella et al., 1990, 1991; Bella-Cueto et al., 1987; Cascio and Colomba, 2002; Cascio et al., 2001, 2002; Dzelalija et al., 2002; Gudiol et al., 1989; Muñoz-Espin et al., 1986; Ruiz Beltrán and Herrero Herrero, 1992b), while none have been conducted for RMSF or other spotted fever group rickettsioses. In general, MSF and likely other spotted fevers aside from RMSF can be effectively treated with alternatives to tetracyclines and doxycycline, such as fluoroquinolones (ciprofloxacin, ofloxacin, pefloxacin); however, such antibiotics are also currently contraindicated for use in pediatric populations and during pregnancy. Moreover, delayed defervescence with ciprofloxacin compared to doxycycline has been reported, and the use of fluoroquinolones has been identified as a risk factor for development of severe complications in MSF (Botelho-Nevers et al., 2011). Likewise, the macrolide antibiotics josamycin, azithromycin, and clarithromycin compare favorably with doxycycline, chloramphenicol, and each other, even when used in pediatric age groups (Cascio and Colomba, 2002; Cascio et al., 2002; Colomba et al., 2006; Meloni and Meloni, 1996). The advantage of the latter compounds is that they are more widely accepted for use in children and during pregnancy. Other alternatives include rifampin, which was shown to be as efficacious as doxycycline for MSF, although defervescence was longer for rifampin (Bella et al., 1991). Rifampin has also been suggested as an alternative for treatment of children and pregnant women, although the data supporting its use are restricted to case reports and in vitro susceptibility testing (Cohen et al., 1999).

Typhus Group Rickettsioses

As with spotted fever group rickettsioses, randomized clinical trials have not been conducted in either louse-borne or murine typhus, and the recommendations for antimicrobial therapy are based on retrospective review of treated patients and on short case reports. In contrast to the situation with spotted fever group rickettsioses, a comprehensive review of the treatment of both louse-borne and murine typhus has not been published. Since 1948, at least 11 patients in 7 published case reports (Ackley and Peter, 1981; Centers for Disease Control, 1984; Massung et al., 2001; Niang et al., 1999; Reynolds et al., 2003; Schumann et al., 1993; Turcinov et al., 2000) and 484 patients in 10 case series (Huys et al., 1973a, 1973b; Krause et al., 1975; Payne et al., 1948; Perine et al., 1974, 1992; Raoult et al., 1997, 1998; Sezi et al., 1972; Smadel et al., 1948) have provided treatment details for louse-borne typhus. Among these, 327 patients received at least a single dose of doxycycline, 22 received tetracycline, and 73 were treated with chloramphenicol. A small minority were treated with macrolides, penicillin, or combination trimethoprim-sulfamethoxazole. In general, those who received a tetracycline or chloramphenicol responded favorably; 4 of 6 with no antimicrobial therapy died (Payne et al., 1948; Raoult et al., 1997, 1998), and treatment with azithromycin failed in 3 of 3 patients (Massung et al., 2001; Turcinov et al., 2000). When compared, tetracycline or doxycycline therapy was generally associated with more rapid defervescence and lack of relapse versus chloramphenicol treatment. A single dose of doxycycline was advocated for treatment of louse-borne typhus; however, 35 of 37 patients treated this way relapsed (Bechah et al., 2008).

For murine typhus, 26 case reports and 23 case series were published that provided treatment results for 1,087 patients, of whom 886 could be evaluated because they received single-antibiotic regimens (Basrai et al., 2007; Bernabeu-Wittel et al., 1998, 1999; Bitsori et al., 2002; Centers for Disease Control and Prevention, 2003; Duffy et al., 1990; Dumler et al., 1991; Esperanza et al., 1992; Fergie et al., 2000; Gikas et al., 2002, 2004; Giroud et al., 1951; Gómez et al., 1992; Graves et al., 1992; Gray et al., 2007; Hassan and Ong, 1995; Hernandez Cabrera et al., 2004; Hudson et al., 1997; Jensenius et al., 1997; Jones et al., 2004; Koliou et al., 2007a, 2007b; Laferl et al., 2002; Letaief et al., 2005; Norazah et al., 1995; Older, 1970; Pascual Velasco and Borobio Enciso, 1991; Rozsypal et al., 2006; Sakaguchi et al., 2004; Silpapojakul et al., 1991a, 1991b, 1993; Silva and Papaiordanou, 2004; Smadel et al., 1948; Strand and Stromberg, 1990; Takagi et al., 2001; Taylor et al., 1986; Toumi et al., 2007; Tsiachris et al., 2008; Vallejo-Maroto et al., 2002; Vander et al., 2003; van der Kleij et al., 1998; van Doorn et al., 2006; Whiteford et al., 2001; Wilde et al., 1991; Woo et al., 1990). Of these, 265 received doxycycline, 330 received tetracycline, 6 received minocycline, 97 received chloramphenicol, 37 were treated with ciprofloxacin, 31 were treated with a cephalosporin or β-lactam, 16 received one of eight different antibiotics, and 104 were not treated with any antimicrobial agent. Among those who were not treated, 2 patients died and 5 had prolonged recovery. There are conflicting data regarding interval to defervescence when comparing doxycycline/tetracycline, chloramphenicol, and ciprofloxacin, but in general those treated with doxycycline had shorter febrile periods. Among those treated with chloramphenicol, there were 3 deaths and 2 relapse infections (Duffy et al., 1990; Silpapojakul et al., 1991b; Taylor et al., 1986), whereas there were no deaths among the ciprofloxacin-treated group and only 1 patient who did not have a clinical response required subsequent doxycycline (van der Kleij et al., 1998). Table 3 shows adverse outcomes that occurred with various antibiotics used for murine typhus.

Overall, the majority of data support the use of doxycycline or a tetracycline antibiotic as the first-line treatment for both louse-borne and murine typhus. However, the safety profile and adverse outcomes associated with chloramphenicol use and the lack of experience with ciprofloxacin for murine typhus do not provide sufficient evidence to recommend one over the other. The most significant problem is the lack of blinded randomized clinical trials to assess the efficacy of alternative antimicrobial agents that may have better therapeutic safety profiles than tetracyclines for children or in pregnancy; have better consistent efficacy and safety than chloramphenicol; and would be effective, inexpensive, and broadly applicable for all rickettsial infections.

TABLE 3 Number of murine typhus patients with various adverse outcomes after treatment with potentially useful antibiotics[a]

Adverse outcome	No. of patients with outcome				
	Doxycycline (n = 265)	Doxycycline/tetracycline[b] (n = 595)	Chloramphenicol (n = 97)	Ciprofloxacin (n = 37)	No antibiotic (n = 104)
Death	4	4	5[c]	0	2
Relapse	1	1	2	0	Not applicable
Prolonged recovery	2	2	0	1	5
No response	0	0	0	1	Not available

[a]See references in the text.
[b]Patients were treated with either doxycycline or a tetracycline.
[c]P = 0.049 versus doxycycline; P < 0.001 versus doxycycline/tetracycline; χ^2 test.

Scrub Typhus

Although tetracycline antibiotics and chloramphenicol have long been advocated for the treatment of scrub typhus, recent recognition of delayed or inadequate disease resolution despite doxycycline therapy has raised the possibility of acquired resistance in some strains of *O. tsutsugamushi*, especially in northern Thailand (Tanskul et al., 1998; Watt et al., 1996). In fact, isolates of *O. tsutsugamushi* from northern Thailand and Korea were found to have mouse infection survival correlates of higher MIC for doxycycline and/or azithromycin (Watt et al., 1999). Only limited comparative MIC calculations to examine the in vitro susceptibility of *O. tsutsugamushi* to various antimicrobials have been published over a range of strains, and these do not include newly described doxycycline-resistant strains (Miyamura et al., 1989). In this study, Miyamura et al. show mean MICs of 0.05 ± 0.03 μg/ml for tetracycline, 0.04 ± 0.03 μg/ml for doxycycline, 0.32 ± 0.10 μg/ml for chloramphenicol, 0.04 ± 0.03 μg/ml for rifampin, and levels above 9 μg/ml for the fluoroquinolones ciprofloxacin and ofloxacin; all β-lactams and cephalosporins had MICs exceeding 1,000 μg/ml. The lack of efficacy for fluoroquinolones has been determined to be the result of natural resistance mutations in the quinolone resistance-determining region in *O. tsutsugamushi gyrA* (Tantibhedhyangkul et al., 2010). Moreover, only limited data allow in vitro comparisons of efficacy for growth suppression by doxycycline versus azithromycin (Brown et al., 1978; Song et al., 1995).

An analysis of data published in Strickman et al. (1995), using the plateau of growth suppression (percentage of infected cells × number of rickettsiae/cell) after 3 days' growth in L929 cells as a surrogate for MIC, showed that *O. tsutsugamushi* Karp strain is inhibited by approximately 0.15 and 0.015 μg/ml doxycycline and azithromycin, respectively. For the doxycycline-resistant strain AFSC-4, the doxycycline plateau starts at 0.25 μg/ml, whereas that for azithromycin starts at 0.025 μg/ml. Using twice-daily oral dosing of 100 mg for doxycycline, its maximum concentration is 2.0 μg/ml and its area under the curve is 13 μg · h/ml, at least 8 times the surrogate MIC for even the doxycycline-resistant AFSC-4 strain; similarly, azithromycin given as a single oral 250-mg daily dose achieves a maximum concentration of 0.24 μg/ml and area under the curve of 2.1 μg · h/ml, almost 100 times greater than its MIC for *O. tsutsugamushi* AFSC-4 strain. In addition, azithromycin is concentrated up to 10 to 100 times in cells, providing readily achievable surrogate MICs for both strains of *O. tsutsugamushi*. Given such inclusive in vitro data, it is necessary to rely on in vivo clinical trials for therapeutic guidance.

Comparisons of antimicrobial efficacy for scrub typhus have been conducted in only five randomized clinical trials and two nonblinded randomized studies (Brown et al., 1978; Kim et al., 2004, 2007; Phimda et al., 2007; Sheehy et al., 1973; Song et al., 1995; Watt et al., 2000). Antibiotic treatments evaluated included comparisons between doxycycline and azithromycin, telithromycin, rifampin, and tetracycline. One study compared tetracycline to chloramphenicol. All drugs were judged to be efficacious in the treatment of scrub typhus. A combined 243 patients among these studies received doxycycline, and 2.5% were judged to have no response during the study time frame; 1.2% had a relapse and 2.5% were delayed in recovery. Tetracycline was given to a total of 104 patients, all of whom responded, but 4.8% had delayed responses and 1.9% relapsed (Kim et al., 2004, 2007; Song et al., 1995; Watt et al., 2000). Chloramphenicol was evaluated in 30 individuals; all responded, but 5 (17%) relapsed (Sheehy et al., 1973). Among 77 patients treated with azithromycin, only a single patient did not respond (1.3%), and no relapses or prolonged intervals to defervescence were observed (Kim et al., 2004; Phimda et al., 2007). Both telithromycin (47 patients) and rifampin (50 patients) were efficacious (Kim et al., 2007; Watt et al., 2000) and had no observed relapses or prolonged recovery intervals. Based on data pooled from these studies, doxycycline and azithromycin were equally likely to be associated with no response (P = 0.541; χ^2 test) or prolonged recovery interval (P = 0.164; χ^2 test) among patients with scrub typhus (Table 4). Other general observations from these studies are that adverse effects including gastrointestinal disturbances and rashes were common with doxycycline, tetracycline, and rifampin (Song et al., 1995; Watt et al., 1999). Discrepant results were obtained with regard to time to defervescence for azithromycin versus doxycycline, whereas the single study examining rifampin showed significantly less fever for that drug versus doxycycline (Kim et al., 2004; Phimda et al., 2007; Watt et al., 2000).

The overall results of in vitro susceptibility testing and limited clinical trials provided good evidence (Infectious Diseases Society of America grade AI) to support the use of doxycycline, tetracycline, or azithromycin as first-line antimicrobials for *O. tsutsugamushi* infection. Alternatives (all grade BI evidence) include rifampin, telithromycin, and chloramphenicol. One obvious advantage of azithromycin and rifampin is their potential use in pregnancy, and this concept is underscored by increasing clinical series and reports of effective azithromycin use in this situation (Mathai et al., 2003; Phupong and Srettakraikul, 2004; Suntharasaj et al., 1997; Tantibhedhyangkul et al., 2010; Watt et al., 1999). These findings underscore the common problems with evidence for supporting specific therapeutic recommendations—the lack of sufficient numbers of treated patients in any clinical trial or study, which undermines the power of the analysis.

TABLE 4 Number of scrub typhus patients with various adverse outcomes after treatment with potentially useful antibiotics[a]

Adverse outcome	No. of patients with outcome					
	Doxycycline (N = 243)	Tetracycline (N = 104)	Chloramphenicol (N = 30)	Rifampin (N = 50)	Azithromycin (N = 77)	Telithromycin (N = 47)
Death	0	0	0	0	0	0
Relapse	3	2	5	0	0	0
Prolonged recovery	6[b]	5	0	0	0	0
No response	6[c]	0	0	0	1	0

[a]Data from Brown et al., 1978; Kim et al., 2004, 2007; Phimda et al., 2007; Sheehy et al., 1973; Song et al., 1995; and Watt et al., 2000.

[b]P = 0.164 versus azithromycin; χ^2 test.

[c]P = 0.541 versus azithromycin; χ^2 test.

Human Ehrlichioses and Anaplasmosis

HME and HGA were first described in 1987 and 1994, respectively (Bakken et al., 1994; Chen et al., 1994; Maeda et al., 1987). At the time of their emergence as significant human infections after tick bites, routine use of doxycycline, tetracycline, or chloramphenicol was advocated based on the principles established in part in veterinary practice and their suspected taxonomic relationship to other *Rickettsia* species pathogenic for humans. Since then, doxycycline has remained the predominant treatment for all forms of ehrlichiosis and anaplasmosis in humans, based upon limited retrospective clinical study, anecdotal case reports of effective use, and in vitro susceptibility determinations (Brouqui and Raoult, 1990, 1992; Horowitz et al., 2001; Klein et al., 1997; Maurin et al., 2003). No clinical trials have been conducted to support antimicrobial treatment of HME or HGA. *E. ewingii* has not been cultivated in vitro and thus no data on in vitro antimicrobial susceptibility are available (Buller et al., 1999). The *E. muris*-like ehrlichia identified in the upper midwestern United States has been cultivated in vitro, but no in vitro susceptibility tests have been conducted (Pritt et al., 2011). The in vitro antimicrobial susceptibility of *N. sennetsu* is known, but very limited data support the ongoing occurrence of this infection in humans (Brouqui and Raoult, 1990; Dumler et al., 2001; Newton et al., 2009; Tachibana, 1986). Humans infected with "*Candidatus* N. mikurensis" seemed to respond when doxycycline but not other antibiotics were used (Fehr et al., 2010; Pekova et al., 2011; von Loewenich et al., 2010; Welinder-Olsson et al., 2010).

HME

The current recommendations for treatment of HME and HGA are similar. *E. chaffeensis* is highly susceptible to doxycycline (MIC, ≤0.5 μg/ml) and rifampin (MIC, 0.03 to 0.125 μg/ml), but β-lactams, chloramphenicol, ciprofloxacin, erythromycin and other macrolides, fluoroquinolones, cotrimoxazole, and gentamicin are inactive (Branger et al., 2004; Brouqui and Raoult, 1992). In most published case series where clear treatment data are available, doxycycline is associated with therapeutic efficacy, although treatment failures are documented (Bakken and Dumler, 2002; Dumler et al., 1993; Fishbein et al., 1994). Although some studies show similar therapeutic efficacy for chloramphenicol, in fact, very few patients have been treated with this antibiotic alone, and among those, the interval to defervescence is longer than among doxycycline-treated persons (Bakken and Dumler, 2002). Among chloramphenicol-treated patients, approximately 13% fail to improve, and the case fatality rate is 6%, double the rate for HME in general. Rifampin use in HME has not been reported despite the promising MICs and its potential application to pediatrics and pregnancy. Fluoroquinolones in general lack activity against *E. chaffeensis*, with MICs generally >1 μg/ml, and their use is not advocated (Branger et al., 2004; Brouqui and Raoult, 1992; Maurin et al., 2001). Thus, the lack of any clinical trials in humans precludes definitive advice beyond that established empirically: for treating HME, doxycycline and tetracycline are the drugs of choice (grade AIII), and chloramphenicol has questionable activity and an empirical track record and cannot be recommended. At least one investigation provides compelling data that support the early use of doxycycline as a significant factor in reducing risk for severe disease with HME (Hamburg et al., 2008).

HGA

Evidence regarding treatment options for HGA bears the same problems as for HME, chiefly the lack of any clinical trials to assess efficacy of existing, apparently beneficial antimicrobial regimens. Thus, the majority of evidence is derived from in vitro susceptibility studies and limited retrospective clinical analyses of cases (Aguero-Rosenfeld et al., 1996; Bakken et al., 1994, 1996b; Branger et al., 2004; Goodman et al., 1996; Horowitz et al., 2001; Klein et al.,

1997; Maurin et al., 2003). In general, in vitro testing has shown that *A. phagocytophilum* is susceptible to doxycycline (MIC, 0.03 μg/ml), rifampin (MIC, 0.03 μg/ml), and to a lesser degree fluoroquinolones (MIC for levofloxacin, 0.06 to 1 μg/ml; ciprofloxacin and ofloxacin, 1 to 2 μg/ml). Aminoglycosides (gentamicin, amikacin), cephalosporins, cotrimoxazole, chloramphenicol, and macrolides (telithromycin, azithromycin, erythromycin) are not active in vitro at achievable MICs in vivo.

No clinical trials to evaluate antimicrobial choice in HGA have been conducted. The consensus derived from case series, case reports, and expert opinion favors the use of doxycycline or tetracycline, although treatment failures have occurred even with this option (Bakken and Dumler, 2002; Bakken et al., 1994, 1996b). Chloramphenicol was reported as effective in several patients, while others who received this drug failed to respond (Bakken and Dumler, 2002; Bakken et al., 1996a; Goodman et al., 1996). As chloramphenicol is not active in vitro against *A. phagocytophilum*, the best evidence suggests that it should be avoided for treating HGA. Only limited clinical study of HGA treated with rifampin has been conducted, and among two children and five pregnant women, all successfully resolved infection (Buitrago et al., 1998; Horowitz et al., 1998; Krause et al., 2003); among the five pregnant patients treated with rifampin, four had uneventful deliveries of healthy children. These data suggest that rifampin may be an acceptable alternative to doxycycline for treating HGA in children or during pregnancy. Fluoroquinolones have in vitro activity, and an examination of the *gyrA* quinolone resistance-determining region suggests that resistance should not be a problem, at least at this resistant locus marker (Maurin et al., 2001). However, at least one report of levofloxacin use in HGA demonstrated suppression of infection by this drug and infection relapse when it was discontinued (Wormser et al., 2006). The patient was eventually successfully treated with doxycycline.

Since evidence from clinical trials does not exist, the best data currently support the use of doxycycline in most situations when HGA is diagnosed (grade AIII). Rifampin could be considered an option in children or during pregnancy, but much more study is needed (grade BIII). Neither fluoroquinolones nor chloramphenicol can currently be recommended based on clinical experience and/or in vitro susceptibility tests.

CONTROVERSIES AND NEW CHALLENGES IN RICKETTSIAL TREATMENT

While a number of severe clinical infectious diseases caused by *Rickettsia* and *Orientia* species have been known and studied for years, the list of organisms within the *Rickettsiales* that infect and cause disease in humans and animals continues to expand (Fournier and Raoult, 2009; Georgiades et al., 2011; Li et al., 2009; Parola et al., 2008). It is anticipated that the logarithmic "discovery" of "new pathogens," clinical syndromes, vectors, and ecologies will continue. Several important controversies and challenges are brought forth with these advances. (i) Will rickettsiologists achieve consensus regarding controversies of taxonomy, phylogeny, and clinical classifications to provide both useful scientific and medical frameworks (Fournier et al., 2003a, 2006; Fournier and Raoult, 2009; Walker, 2007; Walker and Ismail, 2008; Walker et al., 2008; Zhu et al., 2005)? (ii) How will the diversity of new pathogenic "species" and their variable susceptibility or innate resistance to existing antimicrobial agents affect therapies and disease management (Biswas et al., 2008; Rolain and Raoult, 2005a)? (iii) Does cotransmission of pathogenic rickettsiae with other vector-borne prokaryote and eukaryote pathogens result in worse disease or more complicated treatments (Belongia, 2002; De Martino et al., 2001; Grab et al., 2007; Holden et al., 2005; Lane et al., 2001; Levin and Fish, 2000; Moss and Dumler, 2003; Thomas et al., 2001; Thompson et al., 2001; Wormser et al., 1997; Zeidner et al., 2000)?

Taxonomy, Phylogeny, and Clinical Disease

What is the definition of a *Rickettsia* species? Some rickettsiologists, the "splitters," apply a genetic or mathematical system for classifying species based on sequencing of modestly to moderately conserved genes, and support this approach with phenotypic features such as clinical findings (Fournier et al., 2003a, 2006, 2007, 2008; Fournier and Raoult, 2009; Shpynov et al., 2004; Zhu et al., 2005). This is the basis of the Linnaean system of classification that is objective, reproducible, and consistent with taxonomic structures for other eubacteria. However, rickettsiae have evolved in very distinct niches, for which evolutionary pressures, even lateral gene transfer, are not well understood (Fournier et al., 2009; Fournier and Raoult, 2009; Georgiades et al., 2011; Gillespie et al., 2007; Simser et al., 2005). Moreover, these genetically based divisions may not delineate clinically relevant structures, since many are based on very few clinical observations that are further modulated by host factors (Fournier et al., 2003a, 2005; Paddock, 2005; Shpynov et al., 2004). Overall, this allows for the creation of a large repertoire of pathogenic organisms and disease names that may only be marginally different in clinical disease manifestations or treatment approaches.

The alternative approach for classification, "lumping," is based upon grouping of bacteria with similar disease mechanisms that underscore their common features and that are likely to respond to similar therapies and management strategies (Walker, 2007; Walker and Ismail, 2008; Walker et al., 2008). Here rickettsiologists continue to be stymied, since *Rickettsia* species are only grudgingly revealing their secrets of cellular and molecular pathogenicity (Gaywee et al., 2003; Gillespie et al., 2008; Rahman et al., 2003, 2005; Sousa et al., 2008; Walker, 2006, 2007; Walker and Ismail, 2008; Walker and Yu, 2005; Woods et al., 2005). Moreover, it becomes difficult to account for emerging differences in pathogenic potential that could arise from recent evidence of lateral gene transfer (Gillespie et al., 2007, 2008). How rickettsiologists resolve this issue will be key not only to inform scientists about the relevance of their work, but to assist clinicians in making appropriate evidence-based decisions regarding therapies and alternative strategies that address not only the bacterium's pathogenic potential but also the contributions of the host to disease manifestations.

Antimicrobial Resistance

The exceptional ecological and clinical work that has been fostered by the introduction of molecular identification and classification tools has also led to an expansion of attempts to recover interesting and unique rickettsiae for biological examination (Apperson et al., 2008; Eremeeva et al., 2006; Kim et al., 2006; Koutaro et al., 2005; Naitou et al., 2006; Ndip et al., 2007; Shpynov et al., 2006). This is critically important if rickettsiologists are to understand how and why these intracellular organisms survive and evolve, and how they permute their hosts to those ends. Understanding this will not only foster improved comprehension of rickettsial biology, but will identify critical interactors with rickettsial host cells, how these interactions influence host cell function, and how such observations might be translated into biological and even therapeutic tools (Garcia-Garcia et al., 2009b; IJdo et al., 2007; Lin et al., 2007; Martinez et al., 2005; Walker and Ismail, 2008). For example, could one engineer rickettsial effectors that direct induction of proinflammatory or antiapoptotic pathways to address human conditions like myocardial infarction or cancer? Indeed, knowing what the rickettsiae know about endothelial or leukocyte cell biology would push science forward in a remarkable way.

One observation among rickettsial isolates is the ability of many rickettsiae to resist the actions of important antimicrobial agents (Biswas et al., 2008; Fournier et al., 2007; Li et al., 2009; Rolain and Raoult, 2005a). It is difficult to understand how antimicrobial resistance occurs, such as with doxycycline-resistant *O. tsutsugamushi* strains, when it is highly unlikely that such drugs would be encountered in the bacterium's natural life cycle to drive selection and fitness (Rajapakse et al., 2011; Strickman et al., 1995; Watt et al., 1999). The identification of innate resistance owing to naturally occurring genetic "mutations" that confer resistance to currently useful antibiotics underscores the degree of genetic diversity observed in *Rickettsiales* genome projects (Biswas et al., 2008; Fournier et al., 2007; Gillespie et al., 2008; McLeod et al., 2004; Renesto et al., 2005; Rolain and Raoult, 2005a, 2005b; Walker and Yu, 2005). Thus, how will one select proper treatments in the future? The genomes themselves may provide clues. The previously difficult and now simple task of genome sequencing and annotation has provided unique opportunities to devise axenic cultures for several intracellular bacteria; could this work for rickettsiae, and if so, could these features be used to devise novel chemotherapeutic agents specific for rickettsiae?

Coinfection

Rickettsiales, by virtue of their intracellular life in arthropods, encounter or are sometimes cotransmitted with other pathogens (Carpenter et al., 1999; Christova et al., 2001; De Martino et al., 2001; Holden et al., 2005; Hulinska et al., 2002; Lane et al., 2001; Levin and Fish, 2000; Moss and Dumler, 2003; Steere et al., 2003; Thomas et al., 2001; Zeidner et al., 2000). In some instances, coinfection of the arthropod results in control of simultaneous infection by a pathogen, as with *Rickettsia peackockii* control of *R. rickettsii* in *Dermacentor andersoni* ticks (Niebylski et al., 1997). In other situations, cotransmission of two or more pathogens that do not influence the arthropod host could contribute to significantly worsened clinical manifestations (Krause et al., 2002; Thompson et al., 2001). While evidence for this situation with *Rickettsiales* is currently lacking, cotransmission of *B. burgdorferi* and *Babesia microti* via *I. scapularis* ticks leads to worse clinical signs, pathogen loads, and duration of infection than with *B. burgdorferi* alone (Thompson et al., 2001). The mere presence of *A. phagocytophilum* in the same tick has triggered a popular and activist storm of coinfections that have been speculated to contribute to increased severity and treatment refractoriness of disease. While limited in vitro biological investigations provide plausibility for such a circumstance (Belongia, 2002; Grab et al., 2007; Steere et al., 2003; Thomas et al., 2001; Zeidner et al., 2000), the critical investigations in animal and human clinical studies are sorely lacking. These and other future directions are in the crosshairs of young rickettsiologists, who with little support but much enthusiasm plow forward like their hero predecessors, Ricketts, Wolbach, Theiler, and Nicolle, among many giants.

REFERENCES

Abbott, K. C., S. J. Vukelja, C. E. Smith, C. K. McAllister, K. A. Konkol, T. J. O'Rourke, C. J. Holland, and M. Ristic. 1991. Hemophagocytic syndrome: a cause of pancytopenia in human ehrlichiosis. *Am. J. Hematol.* **38:**230–234.

Abramson, J. S., and L. B. Givner. 1990. Should tetracycline be contraindicated for therapy of presumed Rocky Mountain spotted fever in children less than 9 years of age? *Pediatrics* **86:**123–124.

Ackley, A. M., and W. J. Peter. 1981. Indigenous acquisition of epidemic typhus in the eastern United States. *South. Med. J.* **74:**245–247.

Aguero-Rosenfeld, M. E., L. Donnarumma, L. Zentmaier, J. Jacob, M. Frey, R. Noto, C. A. Carbonaro, and G. P. Wormser. 2002. Seroprevalence of antibodies that react with *Anaplasma phagocytophila*, the agent of human granulocytic ehrlichiosis, in different populations in Westchester County, New York. *J. Clin. Microbiol.* **40:**2612–2615.

Aguero-Rosenfeld, M. E., H. W. Horowitz, G. P. Wormser, D. F. McKenna, J. Nowakowski, J. Muñoz, and J. S. Dumler. 1996. Human granulocytic ehrlichiosis: a case series from a medical center in New York State. *Ann. Intern. Med.* **125:**904–908.

Apperson, C. S., B. Engber, W. L. Nicholson, D. G. Mead, J. Engel, M. J. Yabsley, K. Dail, J. Johnson, and D. W. Watson. 2008. Tick-borne diseases in North Carolina: is "*Rickettsia amblyommii*" a possible cause of rickettsiosis reported as Rocky Mountain spotted fever? *Vector Borne Zoonotic Dis.* **8:**597–606.

Archibald, L. K., and D. J. Sexton. 1995. Long-term sequelae of Rocky Mountain spotted fever. *Clin. Infect. Dis.* **20:**1122–1125.

Azad, A. F. 1990. Epidemiology of murine typhus. *Annu. Rev. Entomol.* **35:**553–569.

Azad, A. F., S. Radulovic, J. A. Higgins, B. H. Noden, and J. M. Troyer. 1997. Flea-borne rickettsioses: ecologic considerations. *Emerg. Infect. Dis.* **3:**319–327.

Bacellar, F., L. Beati, A. Franca, J. Pocas, R. Regnery, and A. Filipe. 1999. Israeli spotted fever rickettsia (*Rickettsia conorii* complex) associated with human disease in Portugal. *Emerg. Infect. Dis.* **5:**835–836.

Bakken, J. S., and J. S. Dumler. 2002. *Ehrlichia* and *Anaplasma* species (ehrlichioses), p. 875-882. *In* V. L. Yu, R. Weber, and D. Raoult (ed.), *Antimicrobial Therapy and Vaccines*. Apple Trees Production, LLC, New York, NY.

Bakken, J. S., J. S. Dumler, S. M. Chen, M. R. Eckman, L. L. Van Etta, and D. H. Walker. 1994. Human granulocytic ehrlichiosis in the upper Midwest United States. A new species emerging? *JAMA* **272:**212–218.

Bakken, J. S., and S. Dumler. 2008. Human granulocytic anaplasmosis. *Infect. Dis. Clin. North. Am.* **22:**433–448, viii.

Bakken, J. S., P. Goellner, M. Van Etten, D. Z. Boyle, O. L. Swonger, S. Mattson, J. Krueth, R. L. Tilden, K. Asanovich, J. Walls, and J. S. Dumler. 1998. Seroprevalence of human granulocytic ehrlichiosis among permanent residents of northwestern Wisconsin. *Clin. Infect. Dis.* **27:**1491–1496.

Bakken, J. S., J. Krueth, R. L. Tilden, J. S. Dumler, and B. E. Kristiansen. 1996a. Serological evidence of human granulocytic ehrlichiosis in Norway. *Eur. J. Clin. Microbiol. Infect. Dis.* **15:**829–832.

Bakken, J. S., J. Krueth, C. Wilson-Nordskog, R. L. Tilden, K. Asanovich, and J. S. Dumler. 1996b. Clinical and laboratory characteristics of human granulocytic ehrlichiosis. *JAMA* **275:** 199–205.

Barker, L. F. 1968. Determination of antibiotic susceptibility of Rickettsiae and Chlamydiae in BS-C-1 cell cultures. *Antimicrob. Agents Chemother.* **8:**425–428.

Basrai, D., C. Pox, and W. Schmiegel. 2007. Fever of intermediate duration after return from the Canary Islands. *Internist (Berl.)* **48:**413–419. (In German.)

Bechah, Y., C. Capo, J. L. Mege, and D. Raoult. 2008. Epidemic typhus. *Lancet Infect. Dis.* **8:**417–426.

Bella, F., E. Espejo, S. Uriz, J. A. Serrano, M. D. Alegre, and J. Tort. 1991. Randomized trial of 5-day rifampin versus 1-day doxycycline therapy for Mediterranean spotted fever. *J. Infect. Dis.* **164:**433–434.

Bella, F., B. Font, S. Uriz, T. Muñoz, E. Espejo, J. Traveria, J. A. Serrano, and F. Segura. 1990. Randomized trial of doxycycline versus josamycin for Mediterranean spotted fever. *Antimicrob. Agents Chemother.* **34:**937–938.

Bella-Cueto, F., B. Font-Creus, F. Segura-Porta, E. Espejo-Arenas, P. López-Parés, and T. Muñoz-Espin. 1987. Comparative, randomized trial of one-day doxycycline versus 10-day tetracycline therapy for Mediterranean spotted fever. *J. Infect. Dis.* **155:**1056–1058.

Belongia, E. A. 2002. Epidemiology and impact of coinfections acquired from *Ixodes* ticks. *Vector Borne Zoonotic Dis.* **2:**265–273.

Bernabeu-Wittel, M., J. Pachon, A. Alarcon, L. F. Lopez-Cortes, P. Viciana, M. E. Jimenez-Mejias, J. L. Villanueva, R. Torronteras, and F. J. Caballero-Granado. 1999. Murine typhus as a common cause of fever of intermediate duration: a 17-year study in the south of Spain. *Arch. Intern. Med.* **159:**872–876.

Bernabeu-Wittel, M., J. L. Villanueva-Marcos, A. de Alarcon-Gonzalez, and J. Pachon. 1998. Septic shock and multiorganic failure in murine typhus. *Eur. J. Clin. Microbiol. Infect. Dis.* **17:**131–132.

Billings, A. N., J. A. Rawlings, and D. H. Walker. 1998. Tick-borne diseases in Texas: a 10-year retrospective examination of cases. *Tex. Med.* **94:**66–76.

Biswas, S., D. Raoult, and J. M. Rolain. 2008. A bioinformatic approach to understanding antibiotic resistance in intracellular bacteria through whole genome analysis. *Int. J. Antimicrob. Agents* **32:**207–220.

Biswas, S., D. Raoult, and J. M. Rolain. 2009. Molecular mechanism of gentamicin resistance in *Bartonella henselae*. *Clin. Microbiol. Infect.* **15**(Suppl. 2): 98–99.

Bitsori, M., E. Galanakis, A. Gikas, E. Scoulica, and S. Sbyrakis. 2002. *Rickettsia typhi* infection in childhood. *Acta Paediatr.* **91:**59–61.

Boostrom, A., M. S. Beier, J. A. Macaluso, K. R. Macaluso, D. Sprenger, J. Hayes, S. Radulovic, and A. F. Azad. 2002. Geographic association of *Rickettsia felis*-infected opossums with human murine typhus, Texas. *Emerg. Infect. Dis.* **8:**549–554.

Botelho-Nevers, E., C. Rovery, H. Richet, and D. Raoult. 2011. Analysis of risk factors for malignant Mediterranean spotted fever indicates that fluoroquinolone treatment has a deleterious effect. *J. Antimicrob. Chemother.* **66:** 1821–1830.

Branger, S., J. M. Rolain, and D. Raoult. 2004. Evaluation of antibiotic susceptibilities of *Ehrlichia canis*, *Ehrlichia chaffeensis*, and *Anaplasma phagocytophilum* by real-time PCR. *Antimicrob. Agents Chemother.* **48:**4822–4828.

Brouqui, P., P. Parola, P. E. Fournier, and D. Raoult. 2007. Spotted fever rickettsioses in southern and eastern Europe. *FEMS Immunol. Med. Microbiol.* **49:**2–12.

Brouqui, P., and D. Raoult. 1990. In vitro susceptibility of *Ehrlichia sennetsu* to antibiotics. *Antimicrob. Agents Chemother.* **34:**1593–1596.

Brouqui, P., and D. Raoult. 1992. In vitro antibiotic susceptibility of the newly recognized agent of ehrlichiosis in humans, *Ehrlichia chaffeensis*. *Antimicrob. Agents Chemother.* **36:**2799–2803.

Brouqui, P., Y. O. Sanogo, G. Caruso, F. Merola, and D. Raoult. 2003. *Candidatus Ehrlichia walkerii*: a new *Ehrlichia* detected in *Ixodes ricinus* tick collected from asymptomatic humans in Northern Italy. *Ann. N. Y. Acad. Sci.* **990:**134–140.

Brown, G. W., J. P. Saunders, S. Singh, D. L. Huxsoll, and A. Shirai. 1978. Single dose doxycycline therapy for scrub typhus. *Trans. R. Soc. Trop. Med. Hyg.* **72:**412–416.

Browning, M. D., J. W. Garyu, J. S. Dumler, and D. G. Scorpio. 2006. Role of reactive nitrogen species in development of hepatic injury in a C57BL/6 mouse model of human granulocytic anaplasmosis. *Comp. Med.* **56:**55–62.

Buitrago, M. I., J. W. IJdo, P. Rinaudo, H. Simon, J. Copel, J. Gadbaw, R. Heimer, E. Fikrig, and F. J. Bia. 1998. Human granulocytic ehrlichiosis during pregnancy treated successfully with rifampin. *Clin. Infect. Dis.* **27:**213–215.

Buller, R. S., M. Arens, S. P. Hmiel, C. D. Paddock, J. W. Sumner, Y. Rikhisa, A. Unver, M. Gaudreault-Keener, F. A. Manian, A. M. Liddell, N. Schmulewitz, and G. A. Storch. 1999. *Ehrlichia ewingii*, a newly recognized agent of human ehrlichiosis. *N. Engl. J. Med.* **341:** 148–155.

Bunnell, J. E., E. R. Trigiani, S. R. Srinivas, and J. S. Dumler. 1999. Development and distribution of pathologic lesions are related to immune status and tissue deposition of human granulocytic ehrlichiosis agent-infected cells in a murine model system. *J. Infect. Dis.* **180:**546–550.

Cao, W. C., L. Zhan, J. He, J. E. Foley, S. J. de Vlas, X. M. Wu, H. Yang, J. H. Richardus, and J. D. Habbema. 2006. Natural *Anaplasma phagocytophilum* infection of ticks and rodents from a forest area of Jilin Province, China. *Am. J. Trop. Med. Hyg.* **75:**664–668.

Carlyon, J. A., and E. Fikrig. 2006. Mechanisms of evasion of neutrophil killing by *Anaplasma phagocytophilum*. *Curr. Opin. Hematol.* **13:**28–33.

Carpenter, C. F., T. K. Gandhi, L. K. Kong, G. R. Corey, S. M. Chen, D. H. Walker, J. S. Dumler, E. Breitschwerdt, B. Hegarty, and D. J. Sexton. 1999. The incidence of ehrlichial and rickettsial infection in patients with unexplained fever and recent history of tick bite in central North Carolina. *J. Infect. Dis.* **180:** 900–903.

Cascio, A., and C. Colomba. 2002. Macrolides in the treatment of children with Mediterranean spotted fever. *Infez. Med.* **10:**145–150. (In Italian.)

Cascio, A., C. Colomba, S. Antinori, D. L. Paterson, and L. Titone. 2002. Clarithromycin versus azithromycin in the treatment of Mediterranean spotted fever in children: a randomized controlled trial. *Clin. Infect. Dis.* **34:**154–158.

Cascio, A., C. Colomba, D. Di Rosa, L. Salsa, L. di Martino, and L. Titone. 2001. Efficacy and safety of clarithromycin as treatment for Mediterranean spotted fever in children: a randomized controlled trial. *Clin. Infect. Dis.* **33:**409–411.

Centers for Disease Control. 1984. Epidemic typhus—Georgia. *MMWR Morb. Mortal. Wkly. Rep.* **33:**618–619.

Centers for Disease Control and Prevention. 2003. Murine typhus—Hawaii, 2002. *MMWR Morb. Mortal. Wkly. Rep.* **52:**1224–1226.

Chapman, A. S., J. S. Bakken, S. M. Folk, C. D. Paddock, K. C. Bloch, A. Krusell, D. J. Sexton, S. C. Buckingham, G. S. Marshall, G. A. Storch, G. A. Dasch, J. H. McQuiston, D. L. Swerdlow, S. J. Dumler, W. L. Nicholson, D. H. Walker, M. E. Eremeeva, and C. A. Ohl. 2006. Diagnosis and management of tickborne rickettsial diseases: Rocky Mountain spotted fever, ehrlichioses, and anaplasmosis—United States: a practical guide for physicians and other health-care and public health professionals. *MMWR Recomm. Rep.* **55:**1–27.

Chen, S. M., J. S. Dumler, J. S. Bakken, and D. H. Walker. 1994. Identification of a granulocytotropic *Ehrlichia* species as the etiologic agent of human disease. *J. Clin. Microbiol.* **32:**589–595.

Childs, J. E., and C. D. Paddock. 2003. The ascendancy of *Amblyomma americanum* as a vector of pathogens affecting humans in the United States. *Annu. Rev. Entomol.* **48:**307–337.

Choi, K. S., J. Garyu, J. Park, and J. S. Dumler. 2003. Diminished adhesion of *Anaplasma phagocytophilum*-infected neutrophils to endothelial cells is associated with reduced expression of leukocyte surface selectin. *Infect. Immun.* **71:**4586–4594.

Choi, K. S., J. T. Park, and J. S. Dumler. 2005. *Anaplasma phagocytophilum* delay of neutrophil apoptosis through the p38 mitogen-activated protein kinase signal pathway. *Infect. Immun.* **73:** 8209–8218.

Choi, K. S., D. G. Scorpio, and J. S. Dumler. 2004. *Anaplasma phagocytophilum* ligation to Toll-like receptor (TLR) 2, but not to TLR4, activates macrophages for nuclear factor-κB nuclear translocation. *J. Infect. Dis.* **189:**1921–1925.

Christova, I., L. Schouls, I. van De Pol, J. Park, S. Panayotov, V. Lefterova, T. Kantardjiev, and J. S. Dumler. 2001. High prevalence of granulocytic ehrlichiae and *Borrelia burgdorferi* sensu lato in *Ixodes ricinus* ticks from Bulgaria. *J. Clin. Microbiol.* **39:**4172–4174.

Cohen, J., Y. Lasri, and Z. Landau. 1999. Mediterranean spotted fever in pregnancy. *Scand. J. Infect. Dis.* **31:**202–203.

Colomba, C., L. Saporito, V. F. Polara, R. Rubino, and L. Titone. 2006. Mediterranean spotted fever: clinical and laboratory characteristics of 415 Sicilian children. *BMC Infect. Dis.* **6:**60.

Dahlgren, F. S., E. J. Mandel, J. W. Krebs, R. F. Massung, and J. H. McQuiston. 2011. Increasing incidence of *Ehrlichia chaffeensis* and *Anaplasma phagocytophilum* in the United States, 2000–2007. *Am. J. Trop. Med. Hyg.* **85:**124-131.

Dalton, M. J., M. J. Clarke, R. C. Holman, J. W. Krebs, D. B. Fishbein, J. G. Olson, and J. E. Childs. 1995. National surveillance for Rocky Mountain spotted fever, 1981-1992: epidemiologic summary and evaluation of risk factors for fatal outcome. *Am. J. Trop. Med. Hyg.* **52:**405–413.

de la Fuente, J., A. Torina, V. Naranjo, S. Nicosia, A. Alongi, F. La Mantia, and K. M. Kocan. 2006. Molecular characterization of *Anaplasma platys* strains from dogs in Sicily, Italy. *BMC Vet. Res.* **2:**24.

De Martino, S. J., J. A. Carlyon, and E. Fikrig. 2001. Coinfection with *Borrelia burgdorferi* and the agent of human granulocytic ehrlichiosis. *N. Engl. J. Med.* **345:**150–151.

Demma, L. J., R. C. Holman, J. H. McQuiston, J. W. Krebs, and D. L. Swerdlow. 2005. Epidemiology of human ehrlichiosis and anaplasmosis in the United States, 2001-2002. *Am. J. Trop. Med. Hyg.* **73:**400–409.

de Sousa, R., N. Ismail, S. Doria-Nobrega, P. Costa, T. Abreu, A. Franca, M. Amaro, P. Proenca, P. Brito, J. Pocas, T. Ramos, G. Cristina, G. Pombo, L. Vitorino, J. Torgal, F. Bacellar, and D. Walker. 2005. The presence of eschars, but not greater severity, in Portuguese patients infected with Israeli spotted fever. *Ann. N. Y. Acad. Sci.* **1063:**197–202.

Dierberg, K. L., and J. S. Dumler. 2006. Lymph node hemophagocytosis in rickettsial diseases: a pathogenetic role for CD8 T lymphocytes in human monocytic ehrlichiosis (HME)? *BMC Infect. Dis.* **6:**121.

Duffy, P. E., H. Le Guillouzic, R. F. Gass, and B. L. Innis. 1990. Murine typhus identified as a major cause of febrile illness in a camp for displaced Khmers in Thailand. *Am. J. Trop. Med. Hyg.* **43:**520–526.

Dumler, J. S. 2005. *Anaplasma* and *Ehrlichia* infection. *Ann. N. Y. Acad. Sci.* **1063:**361–373.

Dumler, J. S. 2010. Fitness and freezing: vector biology and human health. *J. Clin. Invest.* **120:**3087–3090.

Dumler, J. S., N. C. Barat, C. E. Barat, and J. S. Bakken. 2007a. Human granulocytic anaplasmosis and macrophage activation. *Clin. Infect. Dis.* **45:**199–204.

Dumler, J. S., A. F. Barbet, C. P. Bekker, G. A. Dasch, G. H. Palmer, S. C. Ray, Y. Rikihisa, and F. R. Rurangirwa. 2001. Reorganization of genera in the families *Rickettsiaceae* and *Anaplasmataceae* in the order *Rickettsiales*: unification of some species of *Ehrlichia* with *Anaplasma*, *Cowdria* with *Ehrlichia* and *Ehrlichia* with *Neorickettsia*, descriptions of six new species combinations and designation of *Ehrlichia equi* and 'HGE agent' as subjective synonyms of *Ehrlichia phagocytophila*. *Int. J. Syst. Evol. Microbiol.* **51:**2145–2165.

Dumler, J. S., K. S. Choi, J. C. Garcia-Garcia, N. S. Barat, D. G. Scorpio, J. W. Garyu, D. J. Grab, and J. S. Bakken. 2005. Human granulocytic anaplasmosis and *Anaplasma phagocytophilum*. *Emerg. Infect. Dis.* **11:**1828–1834.

Dumler, J. S., J. E. Madigan, N. Pusterla, and J. S. Bakken. 2007b. Ehrlichioses in humans: epidemiology, clinical presentation, diagnosis, and treatment. *Clin. Infect. Dis.* **45**(Suppl. 1)**:**S45–S51.

Dumler, J. S., W. L. Sutker, and D. H. Walker. 1993. Persistent infection with *Ehrlichia chaffeensis*. *Clin. Infect. Dis.* **17:**903–905.

Dumler, J. S., J. P. Taylor, and D. H. Walker. 1991. Clinical and laboratory features of murine typhus in south Texas, 1980 through 1987. *JAMA* **266:**1365–1370.

Dumler, J. S., and D. H. Walker. 2005a. Order II. Rickettsiales, p. 96–145. *In* G. M. Garrity, D. J. Brenner, N. R. Krieg, and J. T. Staley (ed.), *Bergey's Manual of Systematic Bacteriology*, 2nd ed., vol. 2. Springer-Verlag, New York, NY.

Dumler, J. S., and D. H. Walker. 2005b. Rocky Mountain spotted fever—changing ecology and persisting virulence. *N. Engl. J. Med.* **353:** 551–553.

Dunning Hotopp, J. C., M. Lin, R. Madupu, J. Crabtree, S. V. Angiuoli, J. Eisen, R. Seshadri, Q. Ren, M. Wu, T. R. Utterback, S. Smith, M. Lewis, H. Khouri, C. Zhang, H. Niu, Q. Lin, N. Ohashi, N. Zhi, W. Nelson, L. M. Brinkac, R. J. Dodson, M. J. Rosovitz, J. Sundaram, S. C. Daugherty, T. Davidsen, A. S. Durkin, M. Gwinn, D. H. Haft, J. D. Selengut, S. A. Sullivan, N. Zafar, L. Zhou, F. Benahmed, H. Forberger, R. Halpin, S. Mulligan, J. Robinson, O. White, Y. Rikihisa, and H. Tettelin. 2006. Comparative genomics of emerging human ehrlichiosis agents. *PLoS Genet.* **2:**e21.

Dzelalija, B., M. Petrovec, T. Avsic-Zupanc, J. Strugar, and T. A. Milic. 2002. Randomized trial of azithromycin in the prophylaxis of Mediterranean spotted fever. *Acta Med. Croatica* **56:**45–47.

Eremeeva, M. E., A. Oliveira, J. B. Robinson, N. Ribakova, N. K. Tokarevich, and G. A. Dasch. 2006. Prevalence of bacterial agents in *Ixodes persulcatus* ticks from the Vologda Province of Russia. *Ann. N. Y. Acad. Sci.* **1078:**291–298.

Esperanza, L., D. A. Holt, J. T. T. Sinnott, M. R. Cancio, E. A. Bradley, and M. Deutsch. 1992. Murine typhus: forgotten but not gone. *South. Med. J.* **85:**754–755.

Fang, R., N. Ismail, L. Soong, V. L. Popov, T. Whitworth, D. H. Bouyer, and D. H. Walker. 2007. Differential interaction of dendritic cells with *Rickettsia conorii*: impact on host susceptibility to murine spotted fever rickettsiosis. *Infect. Immun.* **75:**3112–3123.

Fehr, J. S., G. V. Bloemberg, C. Ritter, M. Hombach, T. F. Luscher, R. Weber, and P. M. Keller. 2010. Septicemia caused by tick-borne bacterial pathogen *Candidatus* Neoehrlichia mikurensis. *Emerg. Infect. Dis.* **16:**1127–1129.

Fergie, J. E., K. Purcell, and D. Wanat. 2000. Murine typhus in South Texas children. *Pediatr. Infect. Dis. J.* **19:**535–538.

Fishbein, D. B., J. E. Dawson, and L. E. Robinson. 1994. Human ehrlichiosis in the United States, 1985 to 1990. *Ann. Intern. Med.* **120:** 736–743.

Fishbein, D. B., L. A. Sawyer, C. J. Holland, E. B. Hayes, W. Okoroanyanwu, D. Williams, K. Sikes, M. Ristic, and J. E. McDade. 1987. Unexplained febrile illnesses after exposure to ticks. Infection with an *Ehrlichia*? *JAMA* **257:**3100–3104.

Font-Creus, B., F. Bella-Cueto, E. Espejo-Arenas, R. Vidal-Sanahuja, T. Muñoz-Espin, M. Nolla-Salas, A. Casagran-Borrell, J. Mercade-Cuesta, and F. Segura-Porta. 1985. Mediterranean spotted fever: a cooperative study of 227 cases. *Rev. Infect. Dis.* **7:**635–642.

Fournier, P. E., M. Drancourt, and D. Raoult. 2007. Bacterial genome sequencing and its use in infectious diseases. *Lancet Infect. Dis.* **7:**711–723.

Fournier, P. E., J. S. Dumler, G. Greub, J. Zhang, Y. Wu, and D. Raoult. 2003a. Gene sequence-based criteria for identification of new *Rickettsia* isolates and description of *Rickettsia heilongjiangensis* sp. nov. *J. Clin. Microbiol.* **41:** 5456–5465.

Fournier, P. E., J. P. Durand, J. M. Rolain, J. L. Camicas, H. Tolou, and D. Raoult. 2003b. Detection of Astrakhan fever rickettsia from ticks in Kosovo. *Ann. N .Y. Acad. Sci.* **990:**158–161.

Fournier, P. E., K. El Karkouri, Q. Leroy, C. Robert, B. Giumelli, P. Renesto, C. Socolovschi, P. Parola, S. Audic, and D. Raoult. 2009. Analysis of the *Rickettsia africae* genome reveals that virulence acquisition in *Rickettsia* species may be explained by genome reduction. *BMC Genomics* **10:**166.

Fournier, P. E., F. Gouriet, P. Brouqui, F. Lucht, and D. Raoult. 2005. Lymphangitis-associated rickettsiosis, a new rickettsiosis caused by *Rickettsia sibirica mongolotimonae*: seven new cases and review of the literature. *Clin. Infect. Dis.* **40:**1435–1444.

Fournier, P. E., and D. Raoult. 2009. Current knowledge on phylogeny and taxonomy of *Rickettsia* spp. *Ann. N. Y. Acad. Sci.* **1166:**1–11.

Fournier, P. E., S. Siritantikorn, J. M. Rolain, Y. Suputtamongkol, S. Hoontrakul, S. Charoenwat, K. Losuwanaluk, P. Parola, and D. Raoult. 2008. Detection of new genotypes of *Orientia tsutsugamushi* infecting humans in Thailand. *Clin. Microbiol. Infect.* **14:**168–173.

Fournier, P. E., B. Xeridat, and D. Raoult. 2003c. Isolation of a rickettsia related to Astrakhan fever rickettsia from a patient in Chad. *Ann. N. Y. Acad. Sci.* **990:**152–157.

Fournier, P. E., Y. Zhu, X. Yu, and D. Raoult. 2006. Proposal to create subspecies of *Rickettsia sibirica* and an emended description of *Rickettsia sibirica*. *Ann. N. Y. Acad. Sci.* **1078:**597–606.

Garcia-Garcia, J. C., N. C. Barat, S. J. Trembley, and J. S. Dumler. 2009a. Epigenetic silencing of host cell defense genes enhances intracellular survival of the rickettsial pathogen *Anaplasma phagocytophilum*. *PLoS Pathog.* **5:**e1000488.

Garcia-Garcia, J. C., K. E. Rennoll-Bankert, S. Pelly, A. M. Milstone, and J. S. Dumler. 2009b. Silencing of host cell *CYBB* gene expression by the nuclear effector AnkA of the intracellular pathogen *Anaplasma phagocytophilum*. *Infect. Immun.* **77:**2385–2391.

Gaywee, J., J. B. Sacci, Jr., S. Radulovic, M. S. Beier, and A. F. Azad. 2003. Subcellular localization of rickettsial invasion protein, InvA. *Am. J. Trop. Med. Hyg.* **68:**92–96.

Ge, Y., and Y. Rikihisa. 2006. *Anaplasma phagocytophilum* delays spontaneous human neutrophil apoptosis by modulation of multiple apoptotic pathways. *Cell. Microbiol.* **8:**1406–1416.

Georgiades, K., V. Merhej, K. El Karkouri, D. Raoult, and P. Pontarotti. 2011. Gene gain and loss events in *Rickettsia* and *Orientia* species. *Biol. Direct* **6:**6.

Gikas, A., S. Doukakis, J. Pediaditis, S. Kastanakis, A. Manios, and Y. Tselentis. 2004. Comparison of the effectiveness of five different antibiotic regimens on infection with *Rickettsia* *typhi*: therapeutic data from 87 cases. *Am. J. Trop. Med. Hyg.* **70:**576–579.

Gikas, A., S. Doukakis, J. Pediaditis, S. Kastanakis, A. Psaroulaki, and Y. Tselentis. 2002. Murine typhus in Greece: epidemiological, clinical, and therapeutic data from 83 cases. *Trans. R. Soc. Trop. Med. Hyg.* **96:**250–253.

Gillespie, J. J., M. S. Beier, M. S. Rahman, N. C. Ammerman, J. M. Shallom, A. Purkayastha, B. S. Sobral, and A. F. Azad. 2007. Plasmids and rickettsial evolution: insight from *Rickettsia felis*. *PLoS One* **2:**e266.

Gillespie, J. J., K. Williams, M. Shukla, E. E. Snyder, E. K. Nordberg, S. M. Ceraul, C. Dharmanolla, D. Rainey, J. Soneja, J. M. Shallom, N. D. Vishnubhat, R. Wattam, A. Purkayastha, M. Czar, O. Crasta, J. C. Setubal, A. F. Azad, and B. S. Sobral. 2008. *Rickettsia* phylogenomics: unwinding the intricacies of obligate intracellular life. *PLoS One* **3:**e2018.

Giroud, P., P. Le Gac, M. Rouby, J. Lagarde, and J. A. Gaillard. 1951. Contribution to the study of rickettsial diseases in Oubangui-Chari following brush fire; 4 new cases in Bangui; clinical cases, serological behavior, isolation of strains; spectacular results of antibiotics. *Bull. Soc. Pathol. Exot. Filiales* **44:**871–879. (In French.)

Gómez, J., M. Molina Boix, V. Baños, and M. Sempere. 1992. Murine typhus. Efficacy of treatment with ciprofloxacin. *Enferm. Infecc. Microbiol. Clin.* **10:**377. (In Spanish.)

Goodman, J. L., C. Nelson, B. Vitale, J. E. Madigan, J. S. Dumler, T. J. Kurtti, and U. G. Munderloh. 1996. Direct cultivation of the causative agent of human granulocytic ehrlichiosis. *N. Engl. J. Med.* **334:**209–215.

Grab, D. J., E. Nyarko, N. C. Barat, O. V. Nikolskaia, and J. S. Dumler. 2007. *Anaplasma phagocytophilum-Borrelia burgdorferi* coinfection enhances chemokine, cytokine, and matrix metalloprotease expression by human brain microvascular endothelial cells. *Clin. Vaccine Immunol.* **14:**1420–1424.

Granquist, E. G., M. Aleksandersen, K. Bergström, S. J. Dumler, W. O. Torsteinbø, and S. Stuen. 2010. A morphological and molecular study of *Anaplasma phagocytophilum* transmission events at the time of *Ixodes ricinus* tick bite. *Acta Vet. Scand.* **52:**43.

Graves, S. R., J. Banks, B. Dwyer, and G. K. King. 1992. A case of murine typhus in Queensland. *Med. J. Aust.* **156:**650–651.

Gray, E., P. Atatoa-Carr, A. Bell, S. Roberts, D. Al Mudallal, and G. D. Mills. 2007. Murine typhus: a newly recognised problem in the Waikato region of New Zealand. *N. Z. Med. J.* **120:**U2661.

Gross, E. M., and P. Yagupsky. 1987. Israeli rickettsial spotted fever in children. A review of 54 cases. *Acta Trop.* **44:**91–96.

Gudiol, F., R. Pallares, J. Carratala, F. Bolao, J. Ariza, G. Rufi, and P. F. Viladrich. 1989. Randomized double-blind evaluation of ciprofloxacin and doxycycline for Mediterranean spotted fever. *Antimicrob. Agents Chemother.* **33:**987–988.

Hamburg, B. J., G. A. Storch, S. T. Micek, and M. H. Kollef. 2008. The importance of early treatment with doxycycline in human ehrlichiosis. *Medicine (Baltimore)* **87:**53–60.

Hassan, I. S., and E. L. Ong. 1995. Fever in the returned traveller. Remember murine typhus! *J. Infect. Dis.* **31:**173–174.

Helmick, C. G., K. W. Bernard, and L. J. D'Angelo. 1984. Rocky Mountain spotted fever: clinical, laboratory, and epidemiological features of 262 cases. *J. Infect. Dis.* **150:**480–488.

Hernandez Cabrera, M., A. Angel-Moreno, E. Santana, M. Bolanos, A. Frances, M. S. Martin-Sanchez, and J. L. Perez-Arellano. 2004. Murine typhus with renal involvement in Canary Islands, Spain. *Emerg. Infect. Dis.* **10:**740–743.

Holden, K., E. Hodzic, S. Feng, K. J. Freet, R. B. Lefebvre, and S. W. Barthold. 2005. Coinfection with *Anaplasma phagocytophilum* alters *Borrelia burgdorferi* population distribution in C3H/HeN mice. *Infect. Immun.* **73:**3440–3444.

Horowitz, H. W., T. C. Hsieh, M. E. Aguero-Rosenfeld, F. Kalantarpour, I. Chowdhury, G. P. Wormser, and J. M. Wu. 2001. Antimicrobial susceptibility of *Ehrlichia phagocytophila*. *Antimicrob. Agents Chemother.* **45:**786–788.

Horowitz, H. W., E. Kilchevsky, S. Haber, M. Aguero-Rosenfeld, R. Kranwinkel, E. K. James, S. J. Wong, F. Chu, D. Liveris, and I. Schwartz. 1998. Perinatal transmission of the agent of human granulocytic ehrlichiosis. *N. Engl. J. Med.* **339:**375–378.

Hove, M. G., and D. H. Walker. 1995. Persistence of rickettsiae in the partially viable gangrenous margins of amputated extremities 5 to 7 weeks after onset of Rocky Mountain spotted fever. *Arch. Pathol. Lab. Med.* **119:**429–431.

Hudson, H. L., A. B. Thach, and P. F. Lopez. 1997. Retinal manifestations of acute murine typhus. *Int. Ophthalmol.* **21:**121–126.

Hulinska, D., J. Votypka, J. Plch, E. Vlcek, M. Valesova, M. Bojar, V. Hulinsky, and K. Smetana. 2002. Molecular and microscopical evidence of *Ehrlichia* spp. and *Borrelia burgdorferi* sensu lato in patients, animals and ticks in the Czech Republic. *New Microbiol.* **25:**437–448.

Huys, J., P. Freyens, J. Kayihigi, and G. Van den Berghe. 1973a. Treatment of epidemic typhus. A comparative study of chloramphenicol, trimethoprim-sulphamethoxazole and doxycycline. *Trans. R. Soc. Trop. Med. Hyg.* **67:**718–721.

Huys, J., J. Kayhigi, P. Freyens, and G. Van den Berghe. 1973b. Single-dose treatment of epidemic typhus with doxycyline. *Chemotherapy* **18:**314–317.

IJdo, J. W., A. C. Carlson, and E. L. Kennedy. 2007. *Anaplasma phagocytophilum* AnkA is tyrosine-phosphorylated at EPIYA motifs and recruits SHP-1 during early infection. *Cell. Microbiol.* **9:**1284–1296.

IJdo, J. W., and A. C. Mueller. 2004. Neutrophil NADPH oxidase is reduced at the *Anaplasma phagocytophilum* phagosome. *Infect. Immun.* **72:**5392–5401.

IJdo, J. W., C. Wu, L. A. Magnarelli, and E. Fikrig. 1999. Serodiagnosis of human granulocytic ehrlichiosis by a recombinant HGE-44-based enzyme-linked immunosorbent assay. *J. Clin. Microbiol.* **37:**3540–3544.

Ismail, N., E. C. Crossley, H. L. Stevenson, and D. H. Walker. 2007. Relative importance of T-cell subsets in monocytotropic ehrlichiosis: a novel effector mechanism involved in *Ehrlichia*-induced immunopathology in murine ehrlichiosis. *Infect. Immun.* **75:**4608–4620.

Ismail, N., H. L. Stevenson, and D. H. Walker. 2006. Role of tumor necrosis factor alpha (TNF-α) and interleukin-10 in the pathogenesis of severe murine monocytotropic ehrlichiosis: increased resistance of TNF receptor p55- and p75-deficient mice to fatal ehrlichial infection. *Infect. Immun.* **74:**1846–1856.

Ismail, N., and D. H. Walker. 2005. Balancing protective immunity and immunopathology: a unifying model of monocytotropic ehrlichiosis. *Ann. N. Y. Acad. Sci.* **1063:**383–394.

Jensenius, M., P. E. Fournier, P. Kelly, B. Myrvang, and D. Raoult. 2003. African tick bite fever. *Lancet Infect. Dis.* **3:**557–564.

Jensenius, M., A. Maeland, and S. Vene. 1997. Endemic typhus imported to Norway. *Tidsskr. Nor. Laegeforen.* **117:**2447–2449. (In Norwegian.)

Jones, S. L., E. Athan, D. O'Brien, S. R. Graves, C. Nguyen, and J. Stenos. 2004. Murine typhus: the first reported case from Victoria. *Med. J. Aust.* **180:**482.

Kao, G. F., C. D. Evancho, O. Ioffe, M. H. Lowitt, and J. S. Dumler. 1997. Cutaneous histopathology of Rocky Mountain spotted fever. *J. Cutan. Pathol.* **24:**604–610.

Kaplowitz, L. G., J. V. Lange, J. J. Fischer, and D. H. Walker. 1983. Correlation of rickettsial titers, circulating endotoxin, and clinical features in Rocky Mountain spotted fever. *Arch. Intern. Med.* **143:**1149–1151.

Kawahara, M., Y. Rikihisa, E. Isogai, M. Takahashi, H. Misumi, C. Suto, S. Shibata, C. Zhang, and M. Tsuji. 2004. Ultrastructure and phylogenetic analysis of '*Candidatus* Neoehrlichia mikurensis' in the family *Anaplasmataceae*, isolated from wild rats and found in *Ixodes ovatus* ticks. *Int. J. Syst. Evol. Microbiol.* **54:**1837–1843.

Kelly, D. J., P. A. Fuerst, W. M. Ching, and A. L. Richards. 2009. Scrub typhus: the geographic distribution of phenotypic and genotypic variants of *Orientia tsutsugamushi*. *Clin. Infect. Dis.* **48**(Suppl. 3)**:**S203–S230.

Kenyon, R. H., R. G. Williams, C. N. Oster, and C. E. Pedersen, Jr. 1978. Prophylactic treatment of Rocky Mountain spotted fever. *J. Clin. Microbiol.* **8:**102–104.

Kim, C. M., M. S. Kim, M. S. Park, J. H. Park, and J. S. Chae. 2003. Identification of *Ehrlichia chaffeensis*, *Anaplasma phagocytophilum*, and *A. bovis* in *Haemaphysalis longicornis* and *Ixodes persulcatus* ticks from Korea. *Vector Borne Zoonotic Dis.* **3:**17–26.

Kim, C. M., Y. H. Yi, D. H. Yu, M. J. Lee, M. R. Cho, A. R. Desai, S. Shringi, T. A. Klein, H. C. Kim, J. W. Song, L. J. Baek, S. T. Chong, M. L. O'Guinn, J. S. Lee, I. Y. Lee, J. H. Park, J. Foley, and J. S. Chae. 2006. Tick-borne rickettsial pathogens in ticks and small mammals in Korea. *Appl. Environ. Microbiol.* **72:**5766–5776.

Kim, D. M., K. D. Yu, J. H. Lee, H. K. Kim, and S. H. Lee. 2007. Controlled trial of a 5-day course of telithromycin versus doxycycline for treatment of mild to moderate scrub typhus. *Antimicrob. Agents Chemother.* **51:**2011–2015.

Kim, Y. S., H. J. Yun, S. K. Shim, S. H. Koo, S. Y. Kim, and S. Kim. 2004. A comparative trial of a single dose of azithromycin versus doxycycline for the treatment of mild scrub typhus. *Clin. Infect. Dis.* **39:**1329–1335.

Kirkland, K. B., P. K. Marcom, D. J. Sexton, J. S. Dumler, and D. H. Walker. 1993. Rocky Mountain spotted fever complicated by gangrene: report of six cases and review. *Clin. Infect. Dis.* **16:**629–634.

Klein, M. B., S. Hu, C. C. Chao, and J. L. Goodman. 2000. The agent of human granulocytic ehrlichiosis induces the production of myelosuppressing chemokines without induction of proinflammatory cytokines. *J. Infect. Dis.* **182:**200–205.

Klein, M. B., C. M. Nelson, and J. L. Goodman. 1997. Antibiotic susceptibility of the newly cultivated agent of human granulocytic ehrlichiosis: promising activity of quinolones and rifamycins. *Antimicrob. Agents Chemother.* **41:**76–79.

Kocan, K. M., J. de la Fuente, E. F. Blouin, J. F. Coetzee, and S. A. Ewing. 2010. The natural history of *Anaplasma marginale*. *Vet. Parasitol.* **167:**95–107.

Koliou, M., C. Christoforou, and E. S. Soteriades. 2007a. Murine typhus in pregnancy: a case report from Cyprus. *Scand. J. Infect. Dis.* **39:**625–628.

Koliou, M., A. Psaroulaki, C. Georgiou, I. Ioannou, Y. Tselentis, and A. Gikas. 2007b. Murine typhus in Cyprus: 21 paediatric cases. *Eur. J. Clin. Microbiol. Infect. Dis.* **26:**491–493.

Koutaro, M., A. S. Santos, J. S. Dumler, and P. Brouqui. 2005. Distribution of '*Ehrlichia walkeri*' in *Ixodes ricinus* (*Acari: Ixodidae*) from the northern part of Italy. *J. Med. Entomol.* **42:**82–85.

Krause, D. W., P. L. Perine, J. E. McDade, and S. Awoke. 1975. Treatment of louse-borne typhus fever with chloramphenicol, tetracycline or doxycycline. *East. Afr. Med. J.* **52:**421–427.

Krause, P. J., C. L. Corrow, and J. S. Bakken. 2003. Successful treatment of human granulocytic ehrlichiosis in children using rifampin. *Pediatrics* **112:** e252–e253.

Krause, P. J., K. McKay, C. A. Thompson, V. K. Sikand, R. Lentz, T. Lepore, L. Closter, D. Christianson, S. R. Telford, D. Persing, J. D. Radolf, and A. Spielman. 2002. Disease-specific diagnosis of coinfecting tickborne zoonoses: babesiosis, human granulocytic ehrlichiosis, and Lyme disease. *Clin. Infect. Dis.* **34:**1184–1191.

Laferl, H., P. E. Fournier, G. Seiberl, H. Pichler, and D. Raoult. 2002. Murine typhus poorly responsive to ciprofloxacin: a case report. *J. Travel Med.* **9:**103–104.

Lane, R. S., J. E. Foley, L. Eisen, E. T. Lennette, and M. A. Peot. 2001. Acarologic risk of exposure to emerging tick-borne bacterial pathogens in a semirural community in northern California. *Vector Borne Zoonotic Dis.* **1:**197–210.

Lee, H. C., and J. L. Goodman. 2006. *Anaplasma phagocytophilum* causes global induction of antiapoptosis in human neutrophils. *Genomics* **88:**496–503.

Leiby, D. A., and J. E. Gill. 2004. Transfusion-transmitted tick-borne infections: a cornucopia of threats. *Transfus. Med. Rev.* **18:**293–306.

Letaief, A. O., N. Kaabia, M. Chakroun, M. Khalifa, N. Bouzouaia, and L. Jemni. 2005. Clinical and laboratory features of murine typhus in central Tunisia: a report of seven cases. *Int. J. Infect. Dis.* **9:**331–334.

Levin, M. L., and D. Fish. 2000. Acquisition of coinfection and simultaneous transmission of *Borrelia burgdorferi* and *Ehrlichia phagocytophila* by *Ixodes scapularis* ticks. *Infect. Immun.* **68:**2183–2186.

Lewin, M. R., D. H. Bouyer, D. H. Walker, and D. M. Musher. 2003. *Rickettsia sibirica* infection in members of scientific expeditions to northern Asia. *Lancet* **362:**1201–1202.

Li, W., G. Ortiz, P. E. Fournier, G. Gimenez, D. L. Reed, B. Pittendrigh, and D. Raoult. 2010. Genotyping of human lice suggests multiple emergencies of body lice from local head louse populations. *PLoS Negl. Trop. Dis.* **4:**e641.

Li, W., D. Raoult, and P. E. Fournier. 2009. Bacterial strain typing in the genomic era. *FEMS Microbiol. Rev.* **33:**892–916.

Liddell, A. M., S. L. Stockham, M. A. Scott, J. W. Sumner, C. D. Paddock, M. Gaudreault-Keener, M. Q. Arens, and G. A. Storch. 2003. Predominance of *Ehrlichia ewingii* in Missouri dogs. *J. Clin. Microbiol.* **41:**4617–4622.

Lin, M., A. den Dulk-Ras, P. J. Hooykaas, and Y. Rikihisa. 2007. *Anaplasma phagocytophilum* AnkA secreted by type IV secretion system is tyrosine phosphorylated by Abl-1 to facilitate infection. *Cell. Microbiol.* **9:**2644–2657.

Little, S. E. 2010. Ehrlichiosis and anaplasmosis in dogs and cats. *Vet. Clin. North Am. Small Anim. Pract.* **40:**1121–1140.

Long, S. W., X. Zhang, J. Zhang, R. P. Ruble, P. Teel, and X. J. Yu. 2003. Evaluation of transovarial transmission and transmissibility of *Ehrlichia chaffeensis* (*Rickettsiales: Anaplasmataceae*) in *Amblyomma americanum* (*Acari: Ixodidae*). *J. Med. Entomol.* **40:**1000–1004.

Maeda, K., N. Markowitz, R. C. Hawley, M. Ristic, D. Cox, and J. E. McDade. 1987. Human infection with *Ehrlichia canis*, a leukocytic rickettsia. *N. Engl. J. Med.* **316:**853–856.

Marshall, G. S., G. G. Stout, R. F. Jacobs, G. E. Schutze, H. Paxton, S. C. Buckingham, J. P. DeVincenzo, M. A. Jackson, V. H. San Joaquin, S. M. Standaert, and C. R. Woods. 2003. Antibodies reactive to *Rickettsia rickettsii* among children living in the southeast and south central regions of the United States. *Arch. Pediatr. Adolesc. Med.* **157:**443–448.

Martinez, J. J., S. Seveau, E. Veiga, S. Matsuyama, and P. Cossart. 2005. Ku70, a component of DNA-dependent protein kinase, is a mammalian receptor for *Rickettsia conorii*. *Cell* **123:**1013–1023.

Marty, A. M., J. S. Dumler, G. Imes, H. P. Brusman, L. L. Smrkovski, and D. M. Frisman. 1995. Ehrlichiosis mimicking thrombotic thrombocytopenic purpura. Case report and pathological correlation. *Hum. Pathol.* **26:**920–925.

Massung, R. F., L. E. Davis, K. Slater, D. B. McKechnie, and M. Puerzer. 2001. Epidemic typhus meningitis in the southwestern United States. *Clin. Infect. Dis.* **32:**979–982.

Mathai, E., J. M. Rolain, L. Verghese, M. Mathai, P. Jasper, G. Verghese, and D. Raoult. 2003. Case reports: scrub typhus during pregnancy in India. *Trans. R. Soc. Trop. Med. Hyg.* **97:**570–572.

Maurin, M., C. Abergel, and D. Raoult. 2001. DNA gyrase-mediated natural resistance to fluoroquinolones in *Ehrlichia* spp. *Antimicrob. Agents Chemother.* **45:**2098–2105.

Maurin, M., J. S. Bakken, and J. S. Dumler. 2003. Antibiotic susceptibilities of *Anaplasma (Ehrlichia) phagocytophilum* strains from various geographic areas in the United States. *Antimicrob. Agents Chemother.* **47:**413–415.

McDade, J. E. 1969. Determination of antibiotic susceptibility of *Rickettsia* by the plaque assay technique. *Appl. Microbiol.* **18:**133–135.

McKechnie, D. B., K. S. Slater, J. E. Childs, R. F. Massung, and C. D. Paddock. 2000. Survival of *Ehrlichia chaffeensis* in refrigerated, ADSOL-treated RBCs. *Transfusion* **40:**1041–1047.

McLeod, M. P., X. Qin, S. E. Karpathy, J. Gioia, S. K. Highlander, G. E. Fox, T. Z. McNeill, H. Jiang, D. Muzny, L. S. Jacob, A. C. Hawes, E. Sodergren, R. Gill, J. Hume, M. Morgan, G. Fan, A. G. Amin, R. A. Gibbs, C. Hong, X. J. Yu, D. H. Walker, and G. M. Weinstock. 2004. Complete genome sequence of *Rickettsia typhi* and comparison with sequences of other rickettsiae. *J. Bacteriol.* **186:**5842–5855.

Meloni, G., and T. Meloni. 1996. Azithromycin vs. doxycycline for Mediterranean spotted fever. *Pediatr. Infect. Dis. J.* **15:**1042–1044.

Miah, M. T., S. Rahman, C. N. Sarker, G. K. Khan, and T. K. Barman. 2007. Study on 40 cases of rickettsia. *Mymensingh Med. J.* **16:**85–88.

Mitchell, P. D., K. D. Reed, and J. M. Hofkes. 1996. Immunoserologic evidence of coinfection with *Borrelia burgdorferi*, *Babesia microti*, and human granulocytic *Ehrlichia* species in residents of Wisconsin and Minnesota. *J. Clin. Microbiol.* **34:**724–727.

Miyamura, S., T. Ohta, and A. Tamura. 1989. Comparison of in vitro susceptibilities of *Rickettsia prowazekii*, *R. rickettsii*, *R. sibirica* and *R. tsutsugamushi* to antimicrobial agents. *Nihon Saikingaku Zasshi* **44:**717–721. (In Japanese.)

Moron, C. G., V. L. Popov, H. M. Feng, D. Wear, and D. H. Walker. 2001. Identification of the target cells of *Orientia tsutsugamushi* in human cases of scrub typhus. *Mod. Pathol.* **14:**752–759.

Moss, W. J., and J. S. Dumler. 2003. Simultaneous infection with *Borrelia burgdorferi* and human granulocytic ehrlichiosis. *Pediatr. Infect. Dis. J.* **22:**91–92.

Munderloh, U. G., M. J. Lynch, M. J. Herron, A. T. Palmer, T. J. Kurtti, R. D. Nelson, and J. L. Goodman. 2004. Infection of endothelial cells with *Anaplasma marginale* and *A. phagocytophilum*. *Vet. Microbiol.* **101:**53–64.

Muñoz-Espin, T., P. López-Parés, E. Espejo-Arenas, B. Font-Creus, I. Martinez-Vila, J. Travería-Casanova, F. Segura-Porta, and F. Bella-Cueto. 1986. Erythromycin versus tetracycline for treatment of Mediterranean spotted fever. *Arch. Dis. Child.* **61:**1027–1029.

Murphy, J. R., C. L. Wisseman, Jr., and P. Fiset. 1978. Mechanisms of immunity in typhus infection: some characteristics of *Rickettsia mooseri* infection of guinea pigs. *Infect. Immun.* **21:**417–424.

Naitou, H., D. Kawaguchi, Y. Nishimura, M. Inayoshi, F. Kawamori, T. Masuzawa, M. Hiroi, H. Kurashige, H. Kawabata, H. Fujita, and N. Ohashi. 2006. Molecular identification of *Ehrlichia* species and '*Candidatus* Neoehrlichia mikurensis' from ticks and wild rodents in Shizuoka and Nagano Prefectures, Japan. *Microbiol. Immunol.* **50:**45–51.

Ndip, L. M., R. N. Ndip, V. E. Ndive, J. A. Awuh, D. H. Walker, and J. W. McBride. 2007. *Ehrlichia* species in *Rhipicephalus sanguineus* ticks in Cameroon. *Vector Borne Zoonotic Dis.* **7:**221–227.

Newton, P. N., J. M. Rolain, B. Rasachak, M. Mayxay, K. Vathanatham, P. Seng, R. Phetsouvanh, T. Thammavong, J. Zahidi, Y. Suputtamongkol, B. Syhavong, and D. Raoult. 2009. Sennetsu neorickettsiosis: a probable fish-borne cause of fever rediscovered in Laos. *Am. J. Trop. Med. Hyg.* **81:**190–194.

Niang, M., P. Brouqui, and D. Raoult. 1999. Epidemic typhus imported from Algeria. *Emerg. Infect. Dis.* **5:**716–718.

Niebylski, M. L., M. E. Schrumpf, W. Burgdorfer, E. R. Fischer, K. L. Gage, and T. G. Schwan. 1997. *Rickettsia peacockii* sp. nov., a new species infecting wood ticks, *Dermacentor andersoni*, in western Montana. *Int. J. Syst. Bacteriol.* **47:**446–452.

Nogueira Angerami, R., E. M. Nunes, E. M. M. Nascimento, A. Ribas Freitas, B. Kemp, A. F. C. Feltrin, M. R. Pacola, G. E. C. Perecin, V. Sinkoc, M. Ribeiro Resende, G. Katz, and L. Jacintho da Silva. 2009. Clusters of Brazilian spotted fever in São Paulo State, southeastern Brazil. A review of official reports and the scientific literature. *Clin. Microbiol. Infect.* **15:**202–204.

Norazah, A., A. Mazlah, Y. M. Cheong, and A. G. Kamel. 1995. Laboratory acquired murine typhus—a case report. *Med. J. Malaysia* **50:**177–179.

Nyarko, E., D. J. Grab, and J. S. Dumler. 2006. *Anaplasma phagocytophilum*-infected neutrophils enhance transmigration of *Borrelia burgdorferi* across the human blood brain barrier in vitro. *Int. J. Parasitol.* **36:**601–605.

Ohashi, N., M. Inayoshi, K. Kitamura, F. Kawamori, D. Kawaguchi, Y. Nishimura, H. Naitou, M. Hiroi, and T. Masuzawa. 2005. *Anaplasma phagocytophilum*-infected ticks, Japan. *Emerg. Infect. Dis.* **11:**1780–1783.

Older, J. J. 1970. The epidemiology of murine typhus in Texas, 1969. *JAMA* **214:**2011–2017.

Openshaw, J. J., D. L. Swerdlow, J. W. Krebs, R. C. Holman, E. Mandel, A. Harvey, D. Haberling, R. F. Massung, and J. H. McQuiston. 2010. Rocky Mountain spotted fever in the United States, 2000-2007: interpreting contemporary increases in incidence. *Am. J. Trop. Med. Hyg.* **83:**174–182.

O'Reilly, M., C. Paddock, B. Elchos, J. Goddard, J. Childs, and M. Currie. 2003. Physician knowledge of the diagnosis and management of Rocky Mountain spotted fever: Mississippi, 2002. *Ann. N. Y. Acad. Sci.* **990:**295–301.

Padbidri, V. S., J. J. Rodrigues, P. S. Shetty, M. V. Joshi, B. L. Rao, and R. N. Shukla. 1984. Tick-borne rickettsioses in Pune district, Maharashtra, India. *Int. J. Zoonoses* **11:**45–52.

Paddock, C. D. 2005. *Rickettsia parkeri* as a paradigm for multiple causes of tick-borne spotted fever in the Western Hemisphere. *Ann. N. Y. Acad. Sci.* **1063:**315–326.

Paddock, C. D., and J. E. Childs. 2003. *Ehrlichia chaffeensis*: a prototypical emerging pathogen. *Clin. Microbiol. Rev.* **16:**37–64.

Paddock, C. D., R. W. Finley, C. S. Wright, H. N. Robinson, B. J. Schrodt, C. C. Lane, O. Ekenna, M. A. Blass, C. L. Tamminga, C. A. Ohl, S. L. McLellan, J. Goddard, R. C. Holman, J. J. Openshaw, J. W. Sumner, S. R. Zaki, and M. E. Eremeeva. 2008. *Rickettsia parkeri* rickettsiosis and its clinical distinction from Rocky Mountain spotted fever. *Clin. Infect. Dis.* **47:** 1188–1196.

Paddock, C. D., S. M. Folk, G. M. Shore, L. J. Machado, M. M. Huycke, L. N. Slater, A. M. Liddell, R. S. Buller, G. A. Storch, T. P. Monson, D. Rimland, J. W. Sumner, J. Singleton, K. C. Bloch, Y. W. Tang, S. M. Standaert, and J. E. Childs. 2001. Infections with *Ehrlichia chaffeensis* and *Ehrlichia ewingii* in persons coinfected with human immunodeficiency virus. *Clin. Infect. Dis.* **33:**1586–1594.

Paddock, C. D., P. W. Greer, T. L. Ferebee, J. Singleton, Jr., D. B. McKechnie, T. A. Treadwell, J. W. Krebs, M. J. Clarke, R. C. Holman, J. G. Olson, J. E. Childs, and S. R. Zaki. 1999. Hidden mortality attributable to Rocky Mountain spotted fever: immunohistochemical detection of fatal, serologically unconfirmed disease. *J. Infect. Dis.* **179:**1469–1476.

Paddock, C. D., R. C. Holman, J. W. Krebs, and J. E. Childs. 2002. Assessing the magnitude of fatal Rocky Mountain spotted fever in the United States: comparison of two national data sources. *Am. J. Trop. Med. Hyg* **67:**349–354.

Paddock, C. D., S. R. Zaki, T. Koss, J. Singleton, J. W. Sumner, J. A. Comer, M. E. Eremeeva, G. A. Dasch, B. Cherry, and J. E. Childs. 2003. Rickettsialpox in New York City: a persistent urban zoonosis. *Ann. N. Y. Acad. Sci.* **990:** 36–44.

Paris, D. H., K. Jenjaroen, S. D. Blacksell, R. Phetsouvanh, V. Wuthiekanun, P. N. Newton, N. P. Day, and G. D. Turner. 2008. Differential patterns of endothelial and leucocyte activation in 'typhus-like' illnesses in Laos and Thailand. *Clin. Exp. Immunol.* **153:**63–67.

Parola, P., S. D. Blacksell, R. Phetsouvanh, S. Phongmany, J. M. Rolain, N. P. Day, P. N. Newton, and D. Raoult. 2008. Genotyping of *Orientia tsutsugamushi* from humans with scrub typhus, Laos. *Emerg. Infect. Dis.* **14:**1483–1485.

Parola, P., F. Fenollar, S. Badiaga, P. Brouqui, and D. Raoult. 2001. First documentation of *Rickettsia conorii* infection (strain Indian tick typhus) in a traveler. *Emerg. Infect. Dis.* **7:**909–910.

Parola, P., C. D. Paddock, and D. Raoult. 2005. Tick-borne rickettsioses around the world: emerging diseases challenging old concepts. *Clin. Microbiol. Rev.* **18:**719–756.

Pascual Velasco, F., and M. V. Borobio Enciso. 1991. Acute suppurative arthritis: an unusual manifestation of murine typhus (endemic). *Med. Clin. (Barc.)* **97:**142–143. (In Spanish.)

Payne, E. H., J. A. Knaudt, and S. Palacious. 1948. Treatment of epidemic typhus with chloromycetin. *Trop. Med. Hyg.* **51:**68–71.

Pedra, J. H., F. S. Sutterwala, B. Sukumaran, Y. Ogura, F. Qian, R. R. Montgomery, R. A. Flavell, and E. Fikrig. 2007a. ASC/PYCARD and caspase-1 regulate the IL-18/IFN-γ axis during *Anaplasma phagocytophilum* infection. *J. Immunol.* **179:**4783–4791.

Pedra, J. H., J. Tao, F. S. Sutterwala, B. Sukumaran, N. Berliner, L. K. Bockenstedt, R. A. Flavell, Z. Yin, and E. Fikrig. 2007b. IL-12/23p40-dependent clearance of *Anaplasma phagocytophilum* in the murine model of human anaplasmosis. *FEMS Immunol. Med. Microbiol.* **50:**401–410.

Pekova, S., J. Vydra, H. Kabickova, S. Frankova, R. Haugvicova, O. Mazal, R. Cmejla, D. W. Hardekopf, T. Jancuskova, and T. Kozak. 2011. *Candidatus Neoehrlichia mikurensis* infection identified in 2 hematooncologic patients: benefit of molecular techniques for rare pathogen detection. *Diagn. Microbiol. Infect. Dis.* **69:**266–270.

Perez, M., M. Bodor, C. Zhang, Q. Xiong, and Y. Rikihisa. 2006. Human infection with *Ehrlichia canis* accompanied by clinical signs in Venezuela. *Ann. N. Y. Acad. Sci.* **1078:**110–117.

Perez, M., Y. Rikihisa, and B. Wen. 1996. *Ehrlichia canis*-like agent isolated from a man in Venezuela: antigenic and genetic characterization. *J. Clin. Microbiol.* **34:**2133–2139.

Pérez-Osorio, C. E., J. E. Zavala-Velázquez, J. J. Arias León, and J. E. Zavala-Castro. 2008. *Rickettsia felis* as emergent global threat for humans. *Emerg. Infect. Dis.* **14:**1019–1023.

Perine, P. L., B. P. Chandler, D. K. Krause, P. McCardle, S. Awoke, E. Habte-Gabr, C. L. Wisseman, Jr., and J. E. McDade. 1992. A clinico-epidemiological study of epidemic typhus in Africa. *Clin. Infect. Dis.* **14:**1149–1158.

Perine, P. L., D. W. Krause, S. Awoke, and J. E. McDade. 1974. Single-dose doxycycline treatment of louse-borne relapsing fever and epidemic typhus. *Lancet* **2:**742–744.

Phimda, K., S. Hoontrakul, C. Suttinont, S. Chareonwat, K. Losuwanaluk, S. Chueasuwanchai, W. Chierakul, D. Suwancharoen, S. Silpasakorn, W. Saisongkorh, S. J. Peacock, N. P. Day, and Y. Suputtamongkol. 2007. Doxycycline versus azithromycin for treatment of leptospirosis and scrub typhus. *Antimicrob. Agents Chemother.* **51:**3259–3263.

Phongmany, S., J. M. Rolain, R. Phetsouvanh, S. D. Blacksell, V. Soukkhaseum, B. Rasachack, K. Phiasakha, S. Soukkhaseum, K. Frichithavong, V. Chu, V. Keolouangkhot, B. Martinez-Aussel, K. Chang, C. Darasavath, O. Rattanavong, S. Sisouphone, M. Mayxay, S. Vidamaly, P. Parola, C. Thammavong, M. Heuangvongsy, B. Syhavong, D. Raoult, N. J. White, and P. N. Newton. 2006. Rickettsial infections and fever, Vientiane, Laos. *Emerg. Infect. Dis.* **12:**256–262.

Phupong, V., and K. Srettakraikul. 2004. Scrub typhus during pregnancy: a case report and review of the literature. *Southeast Asian J. Trop. Med. Public Health* **35:**358–360.

Prakash, J. A. J., T. S. Lal, V. Rosemol, V. P. Verghese, S. A. Pulimood, M. Reller, and J. S. Dumler. 2012. Molecular detection and analysis of spotted fever group *Rickettsia* in patients with fever and rash at a tertiary care centre in Tamil Nadu, India. *Pathog. Glob. Health* **106:**40–45.

Pretzman, C., D. Ralph, D. R. Stothard, P. A. Fuerst, and Y. Rikihisa. 1995. 16S rRNA gene sequence of *Neorickettsia helminthoeca* and its phylogenetic alignment with members of the genus *Ehrlichia. Int. J. Syst. Bacteriol.* **45:**207–211.

Pritt, B. S., J. D. McFadden, E. Stromdahl, D. F. Neitzel, D. K. Hoang Johnson, L. M. Sloan, L. Gongping, U. Munderloh, S. M. Paskewitz, M. Kemperman, C. A. Grady, D. Boxrud, K. M. McElroy, J. McQuiston, D. M. Warshauer, J. P. Davis, C. R. Steward, W. E. Irwin, J. J. Franson, M. J. Binnicker, R. Patel, C. M. Nelson, T. K. Miller, M. A. Feist, X. Lee, S. Tongdean, J.Brezinka, M. Skoglund, S. A. Martin, K. Bogumill, M. E. Bjorgaard, S. Cunningham, J. Cope, and M. M. Eremeeva. 2011. Emergence of a novel *Ehrlichia* sp. agent pathogenic for humans in the Midwestern United States, p. 67, abstr. no. P075. *In Abstracts of the 6th International Meeting on Rickettsiae and Rickettsial Diseases.* Hellenic Society for Infectious Diseases, Heraklion, Crete, Greece.

Pusterla, N., E. M. Johnson, J. S. Chae, and J. E. Madigan. 2003. Digenetic trematodes, *Acanthatrium* sp. and *Lecithodendrium* sp., as vectors of *Neorickettsia risticii*, the agent of Potomac horse fever. *J. Helminthol.* **77:**335–339.

Pusterla, N., R. Weber, C. Wolfensberger, G. Schar, R. Zbinden, W. Fierz, J. E. Madigan, J. S. Dumler, and H. Lutz. 1998. Serological evidence of human granulocytic ehrlichiosis in Switzerland. *Eur. J. Clin. Microbiol. Infect. Dis.* **17:**207–209.

Rahman, M. S., J. A. Simser, K. R. Macaluso, and A. F. Azad. 2003. Molecular and functional analysis of the *lepB* gene, encoding a type I signal peptidase from *Rickettsia rickettsii* and *Rickettsia typhi. J. Bacteriol.* **185:**4578–4584.

Rahman, M. S., J. A. Simser, K. R. Macaluso, and A. F. Azad. 2005. Functional analysis of *secA* homologues from rickettsiae. *Microbiology* **151:**589–596.

Rajapakse, S., C. Rodrigo, and S. D. Fernando. 2011. Drug treatment of scrub typhus. *Trop. Doct.* **41:**1–4.

Rallis, T. M., J. D. Kriesel, J. S. Dumler, L. E. Wagoner, E. D. Wright, and S. L. Spruance. 1993. Rocky Mountain spotted fever following cardiac transplantation. *West. J. Med.* **158:** 625–628.

Ramsey, A. H., E. A. Belongia, C. M. Gale, and J. P. Davis. 2002. Outcomes of treated human granulocytic ehrlichiosis cases. *Emerg. Infect. Dis.* **8:**398–401.

Raoult, D., J. B. Ndihokubwayo, H. Tissot-Dupont, V. Roux, B. Faugere, R. Abegbinni, and R. J. Birtles. 1998. Outbreak of epidemic typhus associated with trench fever in Burundi. *Lancet* **352:**353–358.

Raoult, D., V. Roux, J. B. Ndihokubwayo, G. Bise, D. Baudon, G. Marte, and R. Birtles. 1997. Jail fever (epidemic typhus) outbreak in Burundi. *Emerg. Infect. Dis.* **3:**357–360.

Raoult, D., P. J. Weiller, A. Chagnon, H. Chaudet, H. Gallais, and P. Casanova. 1986. Mediterranean spotted fever: clinical, laboratory and epidemiological features of 199 cases. *Am. J. Trop. Med. Hyg.* **35:**845–850.

Reeves, W. K., A. D. Loftis, W. L. Nicholson, and A. G. Czarkowski. 2008. The first report of human illness associated with the Panola Mountain *Ehrlichia* species: a case report. *J. Med. Case Reports* **2:**139.

Renesto, P., H. Ogata, S. Audic, J. M. Claverie, and D. Raoult. 2005. Some lessons from *Rickettsia* genomics. *FEMS Microbiol. Rev.* **29:**99–117.

Reynolds, M. G., J. S. Krebs, J. A. Comer, J. W. Sumner, T. C. Rushton, C. E. Lopez, W. L. Nicholson, J. A. Rooney, S. E. Lance-Parker, J. H. McQuiston, C. D. Paddock, and J. E. Childs. 2003. Flying squirrel-associated typhus, United States. *Emerg. Infect. Dis.* **9:**1341–1343.

Rikihisa, Y. 2006. New findings on members of the family *Anaplasmataceae* of veterinary importance. *Ann. N. Y. Acad. Sci.* **1078:**438–445.

Rolain, J. M., E. Mathai, H. Lepidi, H. R. Somashekar, L. G. Mathew, J. A. Prakash, and D. Raoult. 2006. "*Candidatus Rickettsia kellyi*," India. *Emerg. Infect. Dis.* **12:**483–485.

Rolain, J. M., M. Maurin, G. Vestris, and D. Raoult. 1998. In vitro susceptibilities of 27 rickettsiae to 13 antimicrobials. *Antimicrob. Agents Chemother.* **42:**1537–1541.

Rolain, J. M., and D. Raoult. 2005a. Genome comparison analysis of molecular mechanisms of resistance to antibiotics in the *Rickettsia* genus. *Ann. N. Y. Acad. Sci.* **1063:**222–230.

Rolain, J. M., and D. Raoult. 2005b. Prediction of resistance to erythromycin in the genus *Rickettsia* by mutations in L22 ribosomal protein. *J. Antimicrob. Chemother.* **56:**396–398.

Rozsypal, H., V. Aster, and V. Skokanová. 2006. Murine typhus—rare cause of fever return from Egypt. *Klin. Mikrobiol. Infekc. Lek.* **12:** 244–246. (In Czech.)

Ruiz Beltrán, R., and J. I. Herrero Herrero. 1992a. Deleterious effect of trimethoprim-sulfamethoxazole in Mediterranean spotted fever. *Antimicrob. Agents Chemother.* **36:**1342–1343.

Ruiz Beltrán, R., and J. I. Herrero Herrero. 1992b. Evaluation of ciprofloxacin and doxycycline in the treatment of Mediterranean spotted fever. *Eur. J. Clin. Microbiol. Infect. Dis.* **11:**427–431.

Sakaguchi, S., I. Sato, H. Muguruma, H. Kawano, Y. Kusuhara, S. Yano, S. Sone, and T. Uchiyama. 2004. Reemerging murine typhus, Japan. *Emerg. Infect. Dis.* **10:**964–965.

Saridaki, A., and K. Bourtzis. 2010. *Wolbachia*: more than just a bug in insects genitals. *Curr. Opin. Microbiol.* **13:**67–72.

Schriefer, M. E., J. B. Sacci, Jr., J. P. Taylor, J. A. Higgins, and A. F. Azad. 1994. Murine typhus: updated roles of multiple urban components and a second typhus-like rickettsia. *J. Med. Entomol.* **31:**681–685.

Schumann, V., E. Fritschka, U. Helmchen, K. Wagner, and T. Philipp. 1993. Interstitial nephritis in typhus]. *Dtsch. Med. Wochenschr.* **118:**893–897. (In German.)

Scorpio, D. G., M. Akkounlu, E. Fikrig, and J. S. Dumler. 2004. CXCR2 blockade influences *Anaplasma phagocytophilum* propagation but not histopathology in the mouse model of human granulocytic anaplasmosis. *Clin. Diagn. Lab. Immunol.* **11:**963–968.

Scorpio, D. G., F. D. von Loewenich, C. Bogdan, and J. S. Dumler. 2005. Innate immune tissue injury and murine HGA: tissue injury in the murine model of granulocytic anaplasmosis relates to host innate immune response and not pathogen load. *Ann. N. Y. Acad. Sci.* **1063:**425–428.

Scorpio, D. G., F. D. von Loewenich, H. Gobel, C. Bogdan, and J. S. Dumler. 2006. Innate immune response to *Anaplasma phagocytophilum* contributes to hepatic injury. *Clin. Vaccine Immunol.* **13:**806–809.

Sehdev, A. E., and J. S. Dumler. 2003. Hepatic pathology in human monocytic ehrlichiosis. *Ehrlichia chaffeensis* infection. *Am. J. Clin. Pathol.* **119:**859–865.

Sexton, D. J., and W. Burgdorfer. 1975. Clinical and epidemiologic features of Rocky Mountain spotted fever in Mississippi, 1933–1973. *South. Med. J.* **68:**1529–1535.

Sexton, D. J., and K. B. Kirkland. 1998. Rickettsial infections and the central nervous system. *Clin. Infect. Dis.* **26:**247–248.

Sezi, C. L., E. Nnochiri, E. Nsanzumuhire, and D. Buttner. 1972. A small outbreak of louse typhus in Masaka District, Uganda. *Trans. R. Soc. Trop. Med. Hyg.* **66:**783–788.

Shaked, Y., Y. Samra, M. K. Maier, and E. Rubinstein. 1989. Relapse of rickettsial Mediterranean spotted fever and murine typhus after treatment with chloramphenicol. *J. Infect. Dis.* **18:**35–37.

Shakya, S., P. Bajpai, S. Sharma, and S. Misra-Bhattacharya. 2008. Prior killing of intracellular bacteria *Wolbachia* reduces inflammatory reactions and improves antifilarial efficacy of diethylcarbamazine in rodent model of *Brugia malayi*. *Parasitol. Res.* **102:**963–972.

Sheehy, T. W., D. Hazlett, and R. E. Turk. 1973. Scrub typhus. A comparison of chloramphenicol and tetracycline in its treatment. *Arch. Intern. Med.* **132:**77–80.

Shpynov, S., P. E. Fournier, N. Rudakov, M. Tankibaev, I. Tarasevich, and D. Raoult. 2004. Detection of a rickettsia closely related to *Rickettsia aeschlimannii*, "*Rickettsia heilongjiangensis*," *Rickettsia* sp. strain RpA4, and *Ehrlichia muris* in ticks collected in Russia and Kazakhstan. *J. Clin. Microbiol.* **42:**2221–2223.

Shpynov, S., P. E. Fournier, N. Rudakov, I. Tarasevich, and D. Raoult. 2006. Detection of members of the genera *Rickettsia*, *Anaplasma*, and *Ehrlichia* in ticks collected in the Asiatic part of Russia. *Ann. N. Y. Acad. Sci.* **1078:**378–383.

Silpapojakul, K., P. Chayakul, S. Krisanapan, and K. Silpapojakul. 1993. Murine typhus in Thailand: clinical features, diagnosis and treatment. *Q. J. Med.* **86:**43–47.

Silpapojakul, K., S. Chupuppakarn, S. Yuthasompob, B. Varachit, D. Chaipak, T. Borkerd, and K. Silpapojakul. 1991a. Scrub and murine typhus in children with obscure fever in the tropics. *Pediatr. Infect. Dis. J.* **10:**200–203.

Silpapojakul, K., C. Ukkachoke, S. Krisanapan, and K. Silpapojakul. 1991b. Rickettsial meningitis and encephalitis. *Arch. Intern. Med.* **151:** 1753–1757.

Silva, L. J., and P. M. Papaiordanou. 2004. Murine (endemic) typhus in Brazil: case report and review. *Rev. Inst. Med. Trop. Sao Paulo* **46:**283–285.

Simser, J. A., M. S. Rahman, S. M. Dreher-Lesnick, and A. F. Azad. 2005. A novel and naturally occurring transposon, ISRpe1 in the *Rickettsia peacockii* genome disrupting the *rickA* gene involved in actin-based motility. *Mol. Microbiol.* **58:**71–79.

Smadel, J. E., A. P. Leon, H. L. Ley, and G. Varela. 1948. Chloromycetin in the treatment of patients with typhus fever. *Proc. Soc. Exp. Biol. Med.* **68:**12–19.

Song, J. H., C. Lee, W. H. Chang, S. W. Choi, J. E. Choi, Y. S. Kim, S. R. Cho, J. Ryu, and C. H. Pai. 1995. Short-course doxycycline treatment versus conventional tetracycline therapy for scrub typhus: a multicenter randomized trial. *Clin. Infect. Dis.* **21:**506–510.

Sousa, R. D., A. Franca, S. Doria Nobrega, A. Belo, M. Amaro, T. Abreu, J. Pocas, P. Proenca, J. Vaz, J. Torgal, F. Bacellar, N. Ismail, and D. H. Walker. 2008. Host- and microbe-related risk factors for and pathophysiology of fatal *Rickettsia conorii* infection in Portuguese patients. *J. Infect. Dis.* **198:**576–585.

Steere, A. C., G. McHugh, C. Suarez, J. Hoitt, N. Damle, and V. K. Sikand. 2003. Prospective study of coinfection in patients with erythema migrans. *Clin. Infect. Dis.* **36:**1078–1081.

Stevenson, H. L., E. C. Crossley, N. Thirumalapura, D. H. Walker, and N. Ismail. 2008. Regulatory roles of CD1d-restricted NKT cells in the induction of toxic shock-like syndrome in an animal model of fatal ehrlichiosis. *Infect. Immun.* **76:**1434–1444.

Stone, J. H., K. Dierberg, G. Aram, and J. S. Dumler. 2004. Human monocytic ehrlichiosis. *JAMA* **292:**2263–2270.

Strand, O., and A. Stromberg. 1990. Ciprofloxacin treatment of murine typhus. *Scand. J. Infect. Dis.* **22:**503–504.

Strickman, D., T. Sheer, K. Salata, J. Hershey, G. Dasch, D. Kelly, and R. Kuschner. 1995. In vitro effectiveness of azithromycin against doxycycline-resistant and -susceptible strains of *Rickettsia tsutsugamushi*, etiologic agent of scrub typhus. *Antimicrob. Agents Chemother.* **39:**2406–2410.

Suntharasaj, T., W. Janjindamai, and S. Krisanapan. 1997. Pregnancy with scrub typhus and vertical transmission: a case report. *J. Obstet. Gynaecol. Res.* **23:**75–78.

Tachibana, N. 1986. Sennetsu fever: the disease, diagnosis, and treatment, p. 205–208. *In* L. Lieve (ed.), *Microbiology*. American Society for Microbiology, Washington, DC.

Takagi, K., H. Iwasaki, S. Kishi, T. Nakamura, N. Takada, and T. Ueda. 2001. Murine typhus infected in Oku-etsu area, Fukui Prefecture. *Kansenshogaku Zasshi* **75:**341–344. (In Japanese.)

Tanskul, P., K. J. Linthicum, P. Watcharapichat, D. Phulsuksombati, S. Mungviriya, S. Ratanatham, N. Suwanabun, J. Sattabongkot, and G. Watt. 1998. A new ecology for scrub typhus associated with a focus of antibiotic resistance in rice farmers in Thailand. *J. Med. Entomol.* **35:**551–555.

Tantibhedhyangkul, W., E. Angelakis, N. Tongyoo, P. N. Newton, C. E. Moore, R. Phetsouvanh, D. Raoult, and J. M. Rolain. 2010. Intrinsic fluoroquinolone resistance in *Orientia tsutsugamushi*. *Int. J. Antimicrob. Agents* **35:**338–341.

Tarasevich, I. V., V. A. Makarova, N. F. Fetisova, A. V. Stepanov, E. D. Miskarova, N. Balayeva, and D. Raoult. 1991a. Astrakhan fever, a spotted-fever rickettsiosis. *Lancet* **337:**172–173.

Tarasevich, I. V., V. A. Makarova, N. F. Fetisova, A. V. Stepanov, E. D. Miskarova, and D. Raoult. 1991b. Studies of a "new" rickettsiosis "Astrakhan" spotted fever. *Eur. J. Epidemiol.* **7:**294–298.

Taylor, J. P., T. G. Betz, and J. A. Rawlings. 1986. Epidemiology of murine typhus in Texas. 1980 through 1984. *JAMA* **255:**2173–2176.

Thirumalapura, N. R., H. L. Stevenson, D. H. Walker, and N. Ismail. 2008. Protective heterologous immunity against fatal ehrlichiosis and lack of protection following homologous challenge. *Infect. Immun.* **76:**1920–1930.

Thomas, L. D., I. Hongo, K. C. Bloch, Y. W. Tang, and S. Dummer. 2007. Human ehrlichiosis in transplant recipients. *Am. J. Transplant.* **7:**1641–1647.

Thomas, R. J., J. S. Dumler, and J. A. Carlyon. 2009. Current management of human granulocytic anaplasmosis, human monocytic ehrlichiosis and *Ehrlichia ewingii* ehrlichiosis. *Expert Rev. Anti Infect. Ther.* **7:**709–722.

Thomas, V., J. Anguita, S. W. Barthold, and E. Fikrig. 2001. Coinfection with *Borrelia burgdorferi* and the agent of human granulocytic ehrlichiosis alters murine immune responses, pathogen burden, and severity of Lyme arthritis. *Infect. Immun.* **69:**3359–3371.

Thompson, C., A. Spielman, and P. J. Krause. 2001. Coinfecting deer-associated zoonoses: Lyme disease, babesiosis, and ehrlichiosis. *Clin. Infect. Dis.* **33:**676–685.

Toumi, A., C. Loussaief, S. Ben Yahia, F. Ben Romdhane, M. Khairallah, M. Chakroun, and N. Bouzouaïa. 2007. Meningitis revealing *Rickettsia typhi* infection. *Rev. Med. Interne* **28:** 131–133. (In French.)

Tselentis, Y., T. L. Babalis, D. Chrysanthis, A. Gikas, G. Chaliotis, and D. Raoult. 1992. Clinicoepidemiological study of murine typhus on the Greek island of Evia. *Eur. J. Epidemiol.* **8:** 268–272.

Tsiachris, D., M. Deutsch, D. Vassilopoulos, R. Zafiropoulou, and A. J. Archimandritis. 2008. Sensorineural hearing loss complicating severe rickettsial diseases: report of two cases. *J. Infect. Dis.* **56:**74–76.

Turcinov, D., I. Kuzman, and B. Herendic. 2000. Failure of azithromycin in treatment of Brill-Zinsser disease. *Antimicrob. Agents Chemother.* **44:**1737–1738.

Turner, J. D., R. S. Langley, K. L. Johnston, K. Gentil, L. Ford, B. Wu, M. Graham, F. Sharpley, B. Slatko, E. Pearlman, and M. J. Taylor. 2009. *Wolbachia* lipoprotein stimulates innate and adaptive immunity through Toll-like receptors 2 and 6 to induce disease manifestations of filariasis. *J. Biol. Chem.* **284:**22364–22378.

Vallejo-Maroto, I., S. Garcia-Morillo, M. B. Wittel, P. Stiefel, M. Miranda, E. Pamies, R. Aparicio, and J. Carneado. 2002. Aseptic meningitis as a delayed neurologic complication of murine typhus. *Clin. Microbiol. Infect.* **8:**826–827.

Vander, T., M. Medvedovsky, S. Valdman, and Y. Herishanu. 2003. Facial paralysis and meningitis caused by *Rickettsia typhi* infection. *Scand. J. Infect. Dis.* **35:**886–887.

van der Kleij, F. G., R. T. Gansevoort, H. G. Kreeftenberg, and W. D. Reitsma. 1998. Imported rickettsioses: think of murine typhus. *J. Intern. Med.* **243:**177–179.

van Doorn, H. R., S. M. Lo-A-Njoe, J. Ottenkamp, and D. Pajkrt. 2006. Widened coronary arteries in a feverish child. *Pediatr. Cardiol.* **27:** 515–518.

Vaughan, J. A., and A. F. Azad. 1990. Acquisition of murine typhus rickettsiae by fleas. *Ann. N. Y. Acad. Sci.* **590:**70–75.

von Loewenich, F. D., W. Geissdorfer, C. Disque, J. Matten, G. Schett, S. G. Sakka, and C. Bogdan. 2010. Detection of "*Candidatus* Neoehrlichia mikurensis" in two patients with severe febrile illnesses: evidence for a European sequence variant. *J. Clin. Microbiol.* **48:**2630–2635.

von Loewenich, F. D., D. G. Scorpio, U. Reischl, J. S. Dumler, and C. Bogdan. 2004. Control of *Anaplasma phagocytophilum*, an obligate intracellular pathogen, in the absence of inducible nitric oxide synthase, phagocyte NADPH oxidase, tumor necrosis factor, Toll-like receptor (TLR)2 and TLR4, or the TLR adaptor molecule MyD88. *Eur. J. Immunol.* **34:**1789–1797.

Walker, D. H. 2006. Targeting rickettsia. *N. Engl. J. Med.* **354:**1418–1420.

Walker, D. H. 2007. Rickettsiae and rickettsial infections: the current state of knowledge. *Clin. Infect. Dis.* **45**(Suppl. 1)**:**S39–S44.

Walker, D. H., H. K. Hawkins, and P. Hudson. 1983. Fulminant Rocky Mountain spotted fever. Its pathologic characteristics associated with glucose-6-phosphate dehydrogenase deficiency. *Arch. Pathol. Lab. Med.* **107:**121–125.

Walker, D. H., and N. Ismail. 2008. Emerging and re-emerging rickettsioses: endothelial cell infection and early disease events. *Nat. Rev. Microbiol.* **6:**375–386.

Walker, D. H., J. P. Olano, and H. M. Feng. 2001. Critical role of cytotoxic T lymphocytes in immune clearance of rickettsial infection. *Infect. Immun.* **69:**1841–1846.

Walker, D. H., C. D. Paddock, and J. S. Dumler. 2008. Emerging and re-emerging tick-transmitted rickettsial and ehrlichial infections. *Med. Clin. North Am.* **92:**1345–1361, x.

Walker, D. H., V. L. Popov, P. A. Crocquet-Valdes, C. J. Welsh, and H. M. Feng. 1997. Cytokine-induced, nitric oxide-dependent, intracellular anti-rickettsial activity of mouse endothelial cells. *Lab. Invest.* **76:**129–138.

Walker, D. H., G. A. Valbuena, and J. P. Olano. 2003. Pathogenic mechanisms of diseases caused by *Rickettsia*. *Ann. N. Y. Acad. Sci.* **990:**1–11.

Walker, D. H., and X. J. Yu. 2005. Progress in rickettsial genome analysis from pioneering of *Rickettsia prowazekii* to the recent *Rickettsia typhi*. *Ann. N. Y. Acad. Sci.* **1063:**13–25.

Watt, G., C. Chouriyagune, R. Ruangweerayud, P. Watcharapichat, D. Phulsuksombati, K. Jongsakul, P. Teja-Isavadharm, D. Bhodhidatta, K. D. Corcoran, G. A. Dasch, and D.

Strickman. 1996. Scrub typhus infections poorly responsive to antibiotics in northern Thailand. *Lancet* **348:**86–89.

Watt, G., P. Kantipong, K. Jongsakul, P. Watcharapichat, and D. Phulsuksombati. 1999. Azithromycin activities against *Orientia tsutsugamushi* strains isolated in cases of scrub typhus in northern Thailand. *Antimicrob. Agents Chemother.* **43:**2817–2818.

Watt, G., P. Kantipong, K. Jongsakul, P. Watcharapichat, D. Phulsuksombati, and D. Strickman. 2000. Doxycycline and rifampicin for mild scrub-typhus infections in northern Thailand: a randomised trial. *Lancet* **356:**1057–1061.

Welinder-Olsson, C., E. Kjellin, K. Vaht, S. Jacobsson, and C. Wenneras. 2010. First case of human "*Candidatus* Neoehrlichia mikurensis" infection in a febrile patient with chronic lymphocytic leukemia. *J. Clin. Microbiol.* **48:**1956–1959.

Whiteford, S. F., J. P. Taylor, and J. S. Dumler. 2001. Clinical, laboratory, and epidemiologic features of murine typhus in 97 Texas children. *Arch. Pediatr. Adolesc. Med.* **155:**396–400.

Wilde, H., J. Pornsilapatip, T. Sokly, and S. Thee. 1991. Murine and scrub typhus at Thai-Kampuchean border displaced persons camps. *Trop. Geogr. Med.* **43:**363–369.

Williams, S. G., J. B. Sacci, Jr., M. E. Schriefer, E. M. Andersen, K. K. Fujioka, F. J. Sorvillo, A. R. Barr, and A. F. Azad. 1992. Typhus and typhus-like rickettsiae associated with opossums and their fleas in Los Angeles County, California. *J. Clin. Microbiol.* **30:**1758–1762.

Winslow, G. M., C. Bitsaktsis, and E. Yager. 2005. Susceptibility and resistance to monocytic ehrlichiosis in the mouse. *Ann. N. Y. Acad. Sci.* **1063:**395–402.

Woo, J. H., J. Y. Cho, Y. S. Kim, D. H. Choi, N. M. Lee, K. W. Choe, and W. H. Chang. 1990. A case of laboratory-acquired murine typhus. *Korean J. Intern. Med.* **5:**118–122.

Woods, M. E., and J. P. Olano. 2008. Host defenses to *Rickettsia rickettsii* infection contribute to increased microvascular permeability in human cerebral endothelial cells. *J. Clin. Immunol.* **28:**174–185.

Woods, M. E., G. Wen, and J. P. Olano. 2005. Nitric oxide as a mediator of increased microvascular permeability during acute rickettsioses. *Ann. N. Y. Acad. Sci.* **1063:**239–245.

Wormser, G. P., A. Filozov, S. R. Telford III, S. Utpat, R. S. Kamer, D. Liveris, G. Wang, L. Zentmaier, I. Schwartz, and M. E. Aguero-Rosenfeld. 2006. Dissociation between inhibition and killing by levofloxacin in human granulocytic anaplasmosis. *Vector Borne Zoonotic Dis.* **6:**388–394.

Wormser, G. P., H. W. Horowitz, J. Nowakowski, D. McKenna, J. S. Dumler, S. Varde, I. Schwartz, C. Carbonaro, and M. Aguero-Rosenfeld. 1997. Positive Lyme disease serology in patients with clinical and laboratory evidence of human granulocytic ehrlichiosis. *Am. J. Clin. Pathol.* **107:**142–147.

Xiong, Q., W. Bao, Y. Ge, and Y. Rikihisa. 2008. *Ehrlichia ewingii* infection delays spontaneous neutrophil apoptosis through stabilization of mitochondria. *J. Infect. Dis.* **197:**1110–1118.

Yabsley, M. J., S. M. Murphy, M. P. Luttrell, B. R. Wilcox, E. W. Howerth, and U. G. Munderloh. 2008. Characterization of '*Candidatus* Neoehrlichia lotoris' (family *Anaplasmataceae*) from raccoons (*Procyon lotor*). *Int. J. Syst. Evol. Microbiol.* **58:**2794–2798.

Yager, E., C. Bitsaktsis, B. Nandi, J. W. McBride, and G. Winslow. 2005. Essential role for humoral immunity during *Ehrlichia* infection in immunocompetent mice. *Infect. Immun.* **73:** 8009–8016.

Yu, V. L., T. C. Merigan, and S. L. Barriere (ed.). 2002. *Antimicrobial Therapy and Vaccines.* Apple Trees Production, LLC, New York, NY.

Zeidner, N. S., M. C. Dolan, R. Massung, J. Piesman, and D. Fish. 2000. Coinfection with *Borrelia burgdorferi* and the agent of human granulocytic ehrlichiosis suppresses IL-2 and IFNγ production and promotes an IL-4 response in C3H/HeJ mice. *Parasite Immunol.* **22:**581–588.

Zhang, J. Z., M. Sinha, B. A. Luxon, and X. J. Yu. 2004. Survival strategy of obligately intracellular *Ehrlichia chaffeensis*: novel modulation of immune response and host cell cycles. *Infect. Immun.* **72:**498–507.

Zhu, Y., P. E. Fournier, M. Eremeeva, and D. Raoult. 2005. Proposal to create subspecies of *Rickettsia conorii* based on multi-locus sequence typing and an emended description of *Rickettsia conorii*. *BMC Microbiol.* **5:**11.

PUBLIC HEALTH: RICKETTSIAL INFECTIONS AND EPIDEMIOLOGY

Jennifer H. McQuiston and Christopher D. Paddock

2

INTRODUCTION

Rickettsial diseases have had a profound impact on human lives for centuries but were poorly understood prior to the beginning of the 20th century. The study of these pathogens is challenging and requires an intimate understanding of arthropod vectors, mammalian hosts, and human nature. Pathogenic *Rickettsiales* species exist on all continents except Antarctica, and in diverse habitats associated with a variety of arthropod species that include fleas, lice, ticks, and mites. The clinical spectrum of rickettsial disease in humans varies widely, ranging from mild illness to rapidly fatal disease.

The pioneering efforts of rickettsiologists during the first half of the 20th century and the revolutionary molecular techniques developed during the last 25 years have contributed enormously to our understanding of the rickettsioses and the expanding spectrum of pathogens that cause these varied infections. Our perception of what constitutes a pathogenic *Rickettsiales* species is rapidly changing with the advent of new and powerful molecular tools to aid diagnosis, and a growing list of rickettsiae originally considered nonpathogenic are now recognized as pathogens of humans. With improved insight into the mode of transmission of these pathogens, including the relative roles of arthropod vectors and mammalian hosts, we *should* have the knowledge and tools to apply to the prevention and control of rickettsial infections at the environmental level. Surprisingly, and with few notable exceptions, the public health impact of most rickettsial diseases has not been diminished since the beginning of the 20th century.

In fact, the reported incidence of tick-borne rickettsioses in the United States is higher than it has ever been. During the past 2 decades, new human pathogens have been discovered in the order *Rickettsiales*, including *Ehrlichia chaffeensis* and *Anaplasma phagocytophilum* (the agents of human ehrlichiosis and anaplasmosis). *Rickettsia parkeri* and *Rickettsia* 364D, previously recognized species that were once considered nonpathogenic, are now known to be human pathogens that likely contribute to the spectrum of disease previously labeled Rocky Mountain spotted fever (RMSF). Furthermore, *Rickettsia rickettsii* (the agent of RMSF) has been recently identified

Jennifer H. McQuiston, Rickettsial Zoonoses Branch, Division of Vectorborne Diseases, Centers for Disease Control and Prevention, Atlanta, GA 30333. *Christopher D. Paddock*, Infectious Diseases Pathology Branch, Division of High Consequence Pathogens and Pathology, Centers for Disease Control and Prevention, Atlanta, GA 30333.

Intracellular Pathogens II: Rickettsiales
Edited by G. H. Palmer and A. F. Azad © 2012 ASM Press, Washington, DC
10.1128/9781555817336.ch2

in new geographic regions and tick vectors, forcing us to question our understanding of this enigmatic disease.

Despite the availability of effective antimicrobials, deaths attributed to rickettsial infections continue to occur in the United States and worldwide. The modern history of rickettsial diseases is too brief to properly inform us of what to expect in a world characterized by transforming land use patterns and climate change. Combating rickettsial diseases and reducing the potential morbidity and mortality associated with these infections will continue to be a public health issue facing epidemiologists during the next century. This chapter focuses on rickettsial infections of public health importance to humans, primarily focusing on rickettsiae in the United States, but also briefly addressing the global importance of rickettsial pathogens.

HISTORICAL PUBLIC HEALTH CONCERNS

An understanding of the epidemiology of rickettsial diseases requires not only a look at contemporary public health issues, but also an understanding of the history of these infections and classic approaches to control and prevention. Hippocrates first used the word "typhus" to describe the signs of delirium and confusion that frequently accompanied high fevers (Zinsser, 1996). The term was historically applied to any malady that produced this clinical state rather than to a specific disease. The recognition of typhus as a clinical state characterized by a fever with headache, malaise, and rash became more focused during the period from 1400 to the 1800s, but the understanding that typhus and other rickettsial diseases were distinct entities did not take place until the turn of the 20th century.

Even before scientists began to understand the etiology of the rickettsioses, these diseases were considered important public health problems. Typhus was known to sweep through communities with frightening speed, often striking vulnerable populations and leaving more than a third of patients dead in its wake. The disease tended to surface among populations collected together or displaced by war, or among crowded conditions that contributed to its communicable nature. During the 1600s, the disease was described as following armies and spreading subsequently throughout Europe (Raoult et al., 2004). Typhus was thought to have been a major contributing influence to soldier deaths during the Napoleonic Wars, and forensic analyses of corpses from mass graves dating to the war of Spanish succession in 1710 to 1712 have confirmed the presence of typhus (Nguyen-Hieu et al., 2010; Raoult et al., 2004, 2006). The wars along the European peninsula during the mid-1800s also resulted in typhus epidemics, and efforts to improve hygiene and sanitation and other public health measures to facilitate control of the disease proved contentious even then; the German physician Rudolph Virchow, who advocated for social reform as a means to improve the health of the general public, lost a government position after publishing a controversial account of a typhus outbreak in Upper Silesia (Virchow, 1985).

The control of epidemic typhus can be considered one of the few public health successes associated with rickettsial infections. Advances in sanitation and hygiene during the first part of the 20th century helped reduce body louse infestations as a health issue and simultaneously reduced the public health importance of epidemic typhus. Today, epidemic typhus is considered rare, and the disease is usually associated with either sporadic cases or the occasional outbreak in prison or refugee settings in developing countries.

In contrast, RMSF has proven far more difficult to control. During the 1870s through the 1900s, a disease known as black measles was reported among otherwise healthy settlers of the Bitterroot Valley of Montana (Harden, 1990; Price, 1948; Rucker, 1912; Wilson and Chowning, 1904). The disease occurred in midspring, killed otherwise healthy young lumbermen and farmers, and then mysteriously disappeared by late June. Many theories abounded, some rooted in superstition: the

American Indians in the area attributed it to evil spirits that visited the valley in the spring, while local settlers blamed it alternately on drinking water from melting snow and sawdust from the lumber camps in the valley. At the time, the science of epidemiology was in its infancy, and the germ theory was new and not yet widely accepted among medical experts, much less among the general populace. Early researchers who arrived in the Bitterroot Valley to investigate the disease, including Howard Ricketts, for whom the genus was eventually named, applied the same techniques to the study of this disease that we would apply today: the careful study of patients and an investigation into environmental factors that may have contributed to illness (Harden, 1990; Price, 1948). Once researchers had observed that all patients had been bitten by ticks, they turned to scientific study and showed through a series of groundbreaking experiments that ticks could transmit the disease. These early investigations formed a benchmark by which future public health studies and outbreak investigations into rickettsial illness would follow.

In the Bitterroot Valley, the theory of tick-borne transmission was not widely accepted when first proposed. Not all tick bites proved fatal or even infectious, and an apparent marked difference in case fatality rates from the east side of the valley (<10%) and the west side of the valley (where rates of 80% were observed) lent an air of suspicion to the investigation (Price, 1948). Worse, the public wanted an easy answer for control—preventing tick bites was widely viewed as next to impossible. As with most public health actions throughout history, the story of control of RMSF in this region was exacerbated by concerns that warnings and advertisements of the disease in such an agriculturally important region would lead to social stigma and loss of income due to falling property values (Price, 1948). These same concerns characteristically accompany public health investigations, even today.

The 1910s to 1940s marked a period of intense scientific inquiry into rickettsial infections caused by *R. rickettsii*, *Rickettsia prowazekii*, and *Rickettsia typhi*. Ricketts developed one of the earliest diagnostic tests in the form of guinea pig inoculation. The Weil-Felix and OX-19 agglutination tests developed during the 1920s and 1930s formed an improved, if imprecise, serologic tool to aid the diagnosis of rickettsial infection (Harden, 1990). The development of the yolk sac cultivation technique in the 1940s paved the way for the development of vaccines against RMSF and typhus, and also allowed the larger-scale preparation of rickettsial antigen to drive the development of complement fixation tests (Harden, 1990). Perhaps the biggest advance in this time period, however, was the discovery of antimicrobial therapeutic agents with efficacy against rickettsiae. The first was *para*-aminobenzoic acid, followed by the antibiotics chloromycetin (chloramphenicol) and aureomycin and terramycin (both tetracycline-class drugs) in the later 1940s (Harden, 1990).

The period between the 1950s and 1970s was arguably a time when U.S. public health officials believed they had effectively controlled infectious diseases like typhus and RMSF, with incidence greatly reduced and fatalities restricted to a fraction of what had been observed in the pre-antibiotic era (Harden, 1990). During the 1960s and 1970s, U.S. health officials and scientists widely claimed that the war against infectious disease had been won (Pier, 2008), a perception that was fueled by groundbreaking discoveries in antibiotics and immunizations, and top public health officials understandably saw a future where only chronic diseases would require U.S. public health dollars. While these prominent health scientists neglected the enormous impact of infectious diseases in a global setting, they also missed the mark on the state of public health in the United States. Over 40 years later, the national reported incidence of RMSF remains *higher* than it was in the 1920s, when RMSF was considered a serious public health issue and antibiotics had not yet been discovered (Openshaw et al., 2010).

By the early 1960s, public health opinion was also shifting against the indiscriminate use of the miraculous antibiotics that had changed the course of the fight against rickettsial diseases only 2 decades earlier. Use of these drugs was not without side effects—chloramphenicol was linked to aplastic anemia, tetracyclines to dental staining when used in children whose permanent teeth were still developing (Rheingold and Spurling, 1952; Witkop and Wolf, 1963). While these concerns were reasonable, the lack of alternative antibiotics with the same efficacy created a clinical dilemma for physicians, especially when faced with a need to treat quickly and empirically to avoid fatal outcome. These concerns remain, even as we enter the 21st century.

CONTEMPORARY PUBLIC HEALTH CONCERNS

As the application of new serologic and molecular techniques has become more common and propagation of rickettsiae by tissue culture more widely applied, new discoveries have fueled our understanding of which species in the *Rickettsiales* represent pathogens. In fact, at least 14 new rickettsial species have been identified as emerging human pathogens since the 1980s (Paddock, 2005; Parola et al., 2005).

Infection caused by an *Ehrlichia* species was first recognized in the Western Hemisphere in 1986 and was initially attributed to *E. canis* (Maeda et al., 1987); the true causative agent of human monocytic ehrlichiosis was eventually isolated from a patient at Fort Chaffee, Arkansas, in 1991 and named *E. chaffeensis* (Anderson et al., 1991; Dawson et al., 1991). Disease caused by an *Anaplasma* species, originally classified as an *Ehrlichia* species, was first identified in a case series from Wisconsin and Minnesota in 1994 (Bakken et al., 1994; Chen et al., 1994; Dumler et al., 1994). The causative agent of human granulocytic anaplasmosis was later renamed *A. phagocytophilum*, combining *Ehrlichia phagocytophila* and *Ehrlichia equi* (Dumler et al., 2001). *Ehrlichia ewingii* was first identified as an agent of human disease from four patients from Missouri with granulocytic infections screening positive for assays specific to the *Ehrlichia* genus but negative for assays specific to *E. chaffeensis* (Buller et al., 1999).

In 2004, *R. parkeri*, an organism that had been recognized since 1939 but which was considered nonpathogenic, was reported for the first time as a cause of eschar-associated human illness in the United States; molecular techniques helped establish it as a distinct etiology in patients who would have otherwise been considered to have RMSF (Paddock, et al., 2004). Similarly, the *Rickettsia* species currently called 364D was considered to be a nonpathogenic agent isolated from *Dermacentor occidentalis* ticks until the reported molecular confirmation of the organism in tissues from several patients with eschar-associated disease in 2010 (Shapiro et al., 2010).

In the United States, RMSF, ehrlichiosis, and anaplasmosis are the rickettsial diseases requiring the notification of U.S. public health officials, although in 2010 the reporting category traditionally called "RMSF" was redesignated "Spotted Fever Rickettsioses" to better reflect the scope of etiologies likely responsible for infection. For each of these notifiable conditions, the number of cases reported annually is on the rise (Fig. 1). While RMSF case reports have been tracked for almost a century, with several defined periods of increase, the reported incidence of RMSF has changed dramatically in the last decade to an all-time high by 2008 (Chapman et al., 2006; Openshaw et al., 2010; CDC, unpublished data). While ehrlichiosis and anaplasmosis are somewhat newly tracked, case reports have also risen steadily since the diseases became notifiable (Dahlgren et al., 2011a; CDC, unpublished).

The precise causes for the rapid rise in the reported incidence of rickettsial diseases have not been determined with certainty, but hypotheses range from increased interest in surveillance, to changes in vector ecology, to the marketing of new diagnostic tests of different sensitivities and specificities (Dahlgren et al., 2011a; Openshaw et al., 2010). There also appear to be some significant changes in our understanding of the epidemiology of some of these infections. For example, in 2003, an outbreak of RMSF was first reported in eastern Arizona, in a hot, dry region considered generally

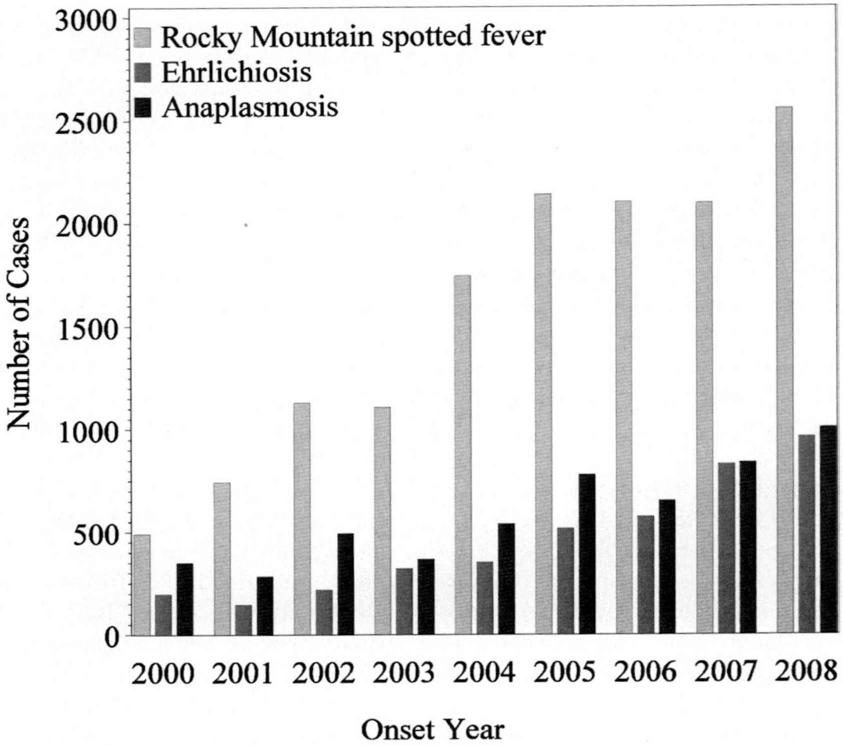

FIGURE 1 Reports of tick-borne rickettsial infections in the United States have increased in recent years. Cases are graphed by onset year, as reported to the Centers for Disease Control and Prevention through the NNDSS. doi:10.1128/9781555817336.ch2f1

unfavorable to support populations of the *Dermacentor* spp. of ticks considered the primary vectors in the United States. A different tick species, *Rhipicephalus sanguineus* (the brown dog tick), was found to be associated with infection (Demma et al., 2005, 2006b; McQuiston et al., 2011). *R. sanguineus*-associated RMSF struck on a larger scale in 2009, infecting over a thousand cases in Mexicali, Mexico (Sanchez et al., 2009).

The phenomenon of disease emergence is not restricted to the tick-borne pathogens. Although not a notifiable condition and reported from only southern Texas and parts of California and Hawaii, murine typhus continues to surprise, with reported case counts on the rise during the last decade. In 2008, an outbreak of murine typhus was confirmed among patients in Austin, Texas, outside the previously recognized enzootic region, affecting over 30 confirmed cases (Adjemian et al., 2010). And while epidemic typhus is largely controlled as a communicable public health threat, sporadic U.S.-acquired cases continue to be reported in association with the sylvatic zoonotic reservoir *Glaucomys volans*, the southern flying squirrel (Reynolds et al., 2003; Chapman et al., 2009).

The 2001 anthrax attack in the United States spurred a decade of scrutiny regarding the possible use of biologic agents as weapons, and *R. rickettsii* and *R. prowazekii* are considered potential agents of bioterrorism. This led to their classification as Select Agents, resulting in restrictions on public access to live organisms (see the National Select Agent Registry at http://www.selectagents.gov/). Beyond the possibility of bioterrorism, the dramatic recent rise in reported incidence of rickettsial infections is also an important point of concern to modern public health officials. Are the reported increases valid, caused by changing land use patterns and increased contact between humans and arthropods? Or do these represent artifacts of changing surveillance techniques and less specific serologic assays? Or, in an age where anyone can look up fever and rash and arrive at a self-diagnosis of rickettsial infection, could these increases be influenced by growing public access to information?

These questions are not easily answered or easily ignored. Whether or not the increase is real, the perception of increased risk fuels public fear and consumes public health dollars. As a result, rickettsial diseases remain as important today as they were at the turn of the century in Montana's Bitterroot Valley.

EPIDEMIOLOGY OF RICKETTSIAL DISEASES IN THE UNITED STATES

Since the early investigations of Ricketts and others into the origins and ecology of RMSF, public health interest in rickettsiae has remained strong in the United States. This chapter, and this section in particular, focuses on the historical and current epidemiology and public health importance of rickettsial pathogens known to occur in the United States (Table 1).

RMSF

RMSF is the most commonly diagnosed spotted fever group rickettsial (SFGR) infection in the United States, although the disease syndrome that is clinically recognized as RMSF is most likely a constellation of infections caused by several related but different bacteria within the genus *Rickettsia*, and specifically within the SFGR subgroup. Because RMSF is treatable and preventable, diligent public health actions can achieve measurable impact. The disease is of special interest to epidemiologists because it is the only rickettsial infection that has almost a century of continuous national surveillance data.

In the United States, three primary tick vectors of *R. rickettsii* have been identified: *Dermacentor variabilis* (the American dog tick) is the primary vector in the eastern half of the United States, while *Dermacentor andersoni* (the Rocky Mountain wood tick) is responsible for most infections in the western United States (Dantas-Torres, 2007). In focal geographic areas such as eastern Arizona, *R. sanguineus* is the primary tick vector (Demma et al., 2005). There has also been a report of RMSF transmission by a *Amblyomma americanum* (lone star tick) in North Carolina (Breitschwerdt et al., 2011). *R. rickettsii* is maintained in tick populations through transovarial transmission (Fig. 2). Small mammals (in the case of *D. variabilis* and *D. andersoni*) and possibly dogs (as has been theorized in the case of *R. sanguineus*) may be involved as transiently rickettsemic hosts by which other ticks may become infected (McQuiston et al., 2011; Piranda et al., 2011; Rudakov et al., 2003).

The seasonality of human RMSF in the United States is closely linked to the seasonal activity of its vectors. Most cases are reported during April to September, corresponding with increased average daily temperatures and increased host-seeking behavior by *D. variabilis* and *D. andersoni*. Thirty-eight percent of human RMSF cases report symptom onset in June and July, while only 4% of cases report onset during December, January, or February (Openshaw et al., 2010). In eastern Arizona, where *R. sanguineus* serves as the primary tick vector responsible for transmission, cases tend to peak later, with most cases occurring in August and September (CDC, unpublished), which may correspond more to seasonal variations in precipitation following regional monsoons than traditional summer models of tick activity.

Surveillance for RMSF dates back to 1873, when the first case was recorded in Montana (Harden, 1990; Price, 1948). During the period from 1873 to 1920, at least 431 cases and 283 deaths (case fatality rate of 66%) were recorded in the western United States, and the disease was initially believed restricted to this region. A national surveillance program was initiated in 1920. Since that time, at least three well-defined peaks in reported incidence of RMSF have been noted (Childs and Paddock, 2002). During the first period of reporting (~1920 to 1950), incidence ranged as high as 4 cases per million persons (Fig. 3). Cases were

TABLE 1 Rickettsiae known to be pathogenic to humans in the United States

Rickettsiales species	Vector	Human disease	Geographic range	Clinical presentation
A. phagocytophilum	Tick-borne: *I. scapularis*	Human granulocytic anaplasmosis	Northeast and upper midwestern United States	Fever, headache, myalgia, moderate systemic signs
E. chaffeensis	Tick-borne: *A. americanum*	Human monocytic ehrlichiosis	Southeast and south-central United States	Fever, headache, myalgia, moderate to severe systemic signs
E. ewingii	Tick-borne: *A. americanum*	*E. ewingii* ehrlichiosis	Southeast and south-central United States	Fever, headache, myalgia, moderate systemic signs
EML agent	Tick-borne: *I. scapularis* (suspected)	EML ehrlichiosis	Wisconsin, Minnesota (upper midwestern United States)	Fever, headache, myalgia, moderate systemic signs
R. akari	Mite-borne: *L. sanguineus*	Rickettsialpox	Urban settings (New York City)	Eschar at site of mite bite, vesicular rash, mild to moderate systemic signs
R. felis	Flea-borne: *C. felis*, *X. cheopis*	*R. felis* rickettsiosis (cat flea rickettsiosis)	Texas, California, Hawaii	Fever, headache, myalgia, maculopapular to petechical rash, mild to moderate systemic signs
R. parkeri	Tick-borne: *A. maculatum*	*R. parkeri* rickettsiosis (maculatum disease)	Coastal and eastern United States	Eschar at site of tick bite, maculopapular rash, mild to moderate systemic signs
R. rickettsii	Tick-borne: *D. variabilis*, *D. andersoni*, *R. sanguineus*	RMSF	Much of the contiguous United States	Fever, maculopapular to petechical rash, headache, moderate to severe systemic signs
R. prowazekii	Louse-borne: *P. humanus corporis* Arthropods (fleas, lice) of the southern flying squirrel (*G. volans*)	Epidemic typhus (louse-borne typhus), Brill-Zinsser disease (recrudescent form) Flying squirrel-associated typhus	Rare in the United States Southern and central United States	Fever, maculopapular to petechical rash, headache, moderate to severe systemic signs
R. typhi	Flea-borne: *X. cheopis*, *C. felis*	Murine typhus	Texas, California, Hawaii	Fever, headache, myalgia, maculopapular to petechical rash, moderate to severe systemic signs
Rickettsia 364D	Tick-borne: *D. occidentalis*	Eschar-associated rickettsial infection	California	Eschar at site of tick bite, mild systemic signs

recognized in the eastern United States as well as in the West. The reported case fatality rate ranged from 20 to 30% prior to the discovery and subsequent availability of antibiotics in the late 1940s (Harden, 1990); from around 1950 onward, the reported case fatality rate rested below 10%. It has been speculated by some researchers that this period of robust reporting was spurred in part by interest from Ralph Parker, one of the leading rickettsial researchers of the time (Childs and Paddock, 2002). Parker's death in 1949 marked the beginning of a period of decline in reported cases of RMSF that extended through around 1970.

An upsurge in reported cases was once again noted in 1970 (Fig. 3). This increase cor-

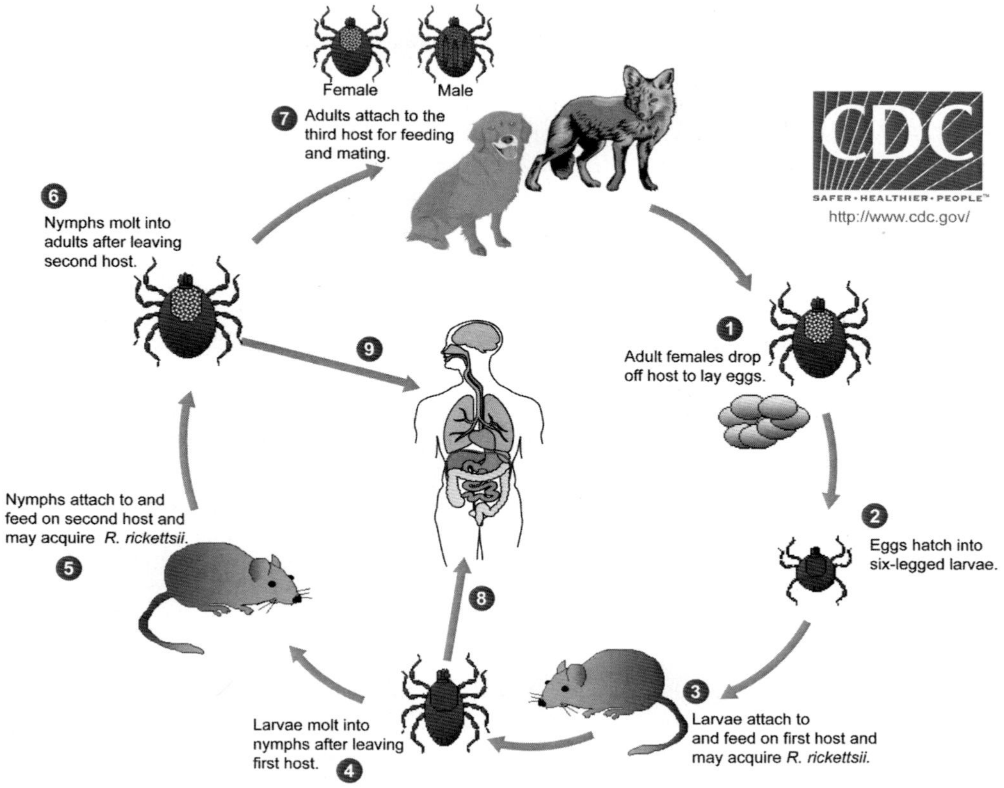

FIGURE 2 Transmission cycle for maintenance of *R. rickettsii* in the United States. doi:10.1128/9781555817336.ch2f2

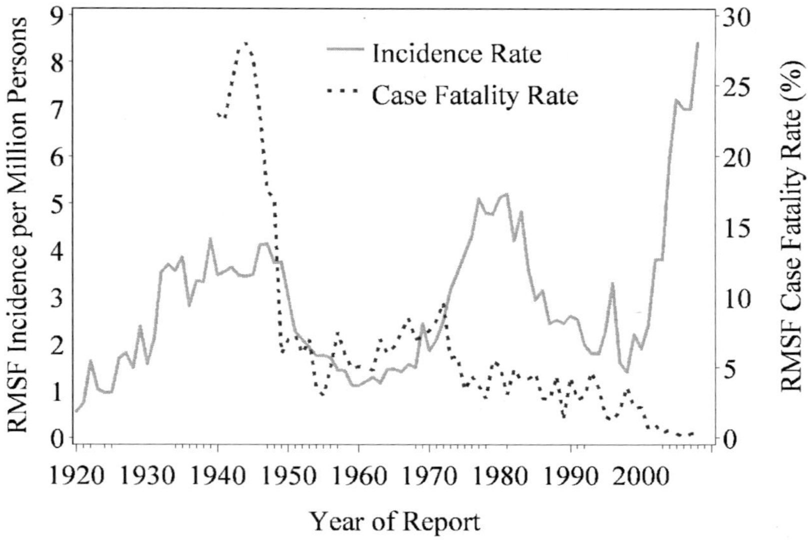

FIGURE 3 Historical trends in reported incidence and case fatality rate of RMSF in the United States, 1920 to 2008. doi:10.1128/9781555817336.ch2f3

responded with the establishment of a national RMSF reporting system to better count and track cases (Helmick and Winkler, 1981) and the gradual adoption of the newer indirect immunofluorescent antibody tests, which were more sensitive than complement fixation (Elisberg and Bozeman, 1969; Philip et al., 1976). During this time, RMSF cases were reported from all over the United States (Hattwick et al., 1973; Helmick and Winkler, 1981; Helmick et al., 1984). The highest incidence was observed in the southeastern and south-central United States. The reported incidence once again decreased through the 1990s, with a low incidence of 1.4 cases per million persons in 1998 (Chapman et al., 2006; Dalton et al., 1995; Treadwell et al., 2000). Case fatality rates averaged around 2 to 6% during much of this period (Dalton et al., 1995; Hattwick et al., 1973; Helmick and Winkler, 1981; Helmick et al., 1984; Treadwell et al., 2000).

Starting in 2000, a new rise in reported cases of RMSF was noted, with an all-time high of >7 cases per million persons reported by 2005. While cases have continued to be reported from across the continental United States (Fig. 4), five states (North Carolina, Oklahoma, Missouri, Arkansas, and Tennessee) accounted for 64% of all cases (Openshaw et al., 2010). This period represented a shift in how cases were classified, including a 2004 reclassification of the case definition to include newly available immunoglobulin M- and enzyme immunoassay-based test results for the diagnosis of RMSF (for the CDC's most recent spotted fever rickettsioses 2010 case definition, see http://www.cdc.gov/osels/ph_surveillance/nndss/casedef/spottedfever_current.htm). Increasing incidence has been attributed to rising reports of RMSF classified as probable cases using a serologic result obtained from a single serum specimen, raising questions about the reporting accuracy using recently implemented case definitions. While this period of increase coincided with remarkable advances in molecular technology, fewer than 0.5% of reported RMSF cases were diagnosed using molecular tests during 2000 to 2007 (Openshaw et al., 2010). Also notable during this time period was a marked decrease in the calculated case fatality rate, averaging only 0.5%. Again, this change is likely influenced by the accuracy of reporting among probable cases, as 3% of cases meeting a confirmed case definition had a reported fatal outcome (Openshaw et al., 2010).

Superficial examination of geographic trends does not properly permit an understanding of the role of RMSF emergence or concentrated focal outbreaks. For example, a cluster of unusually severe RMSF cases characterized by a high proportion of fatalities and hospitalizations with complications has been observed in western Tennessee; possible explanations for this finding include differences in strain pathogenicity by geographic area, differences in health care-seeking behavior by local residents, or even differences in surveillance report capture techniques (Adjemian et al., 2008). Simultaneously, an unusually mild cluster of illness has been observed in North Carolina, with fewer fatalities and less severe outcomes than expected based on national reporting trends (Adjemian et al., 2008). These variations in case severity would be ideally explored through targeted active surveillance projects in these sites.

In another example, prior to 2000, reports of RMSF from the southwestern United States were uncommon, particularly in Arizona (Chapman et al., 2006; Dalton et al., 1995; Treadwell et al., 2000). However, an outbreak of RMSF recently emerged in eastern Arizona, beginning with the 2003 identification of a fatal RMSF case from the region (Demma et al., 2005). Extensive environmental investigations found no evidence of *D. variabilis* or *D. andersoni* ticks but identified large numbers of *R. sanguineus* ticks, including many infected with *R. rickettsii*, and widespread exposure of stray and pet dogs and community inhabitants to SFGR (Demma et al., 2005, 2006b). Since that time, over 200 cases have been reported from Arizona, with a case fatality rate of ~7%, and eastern Arizona has emerged as a region with one of the highest annual incidences of RMSF in the United States (McQuiston et al., 2011; CDC, unpublished).

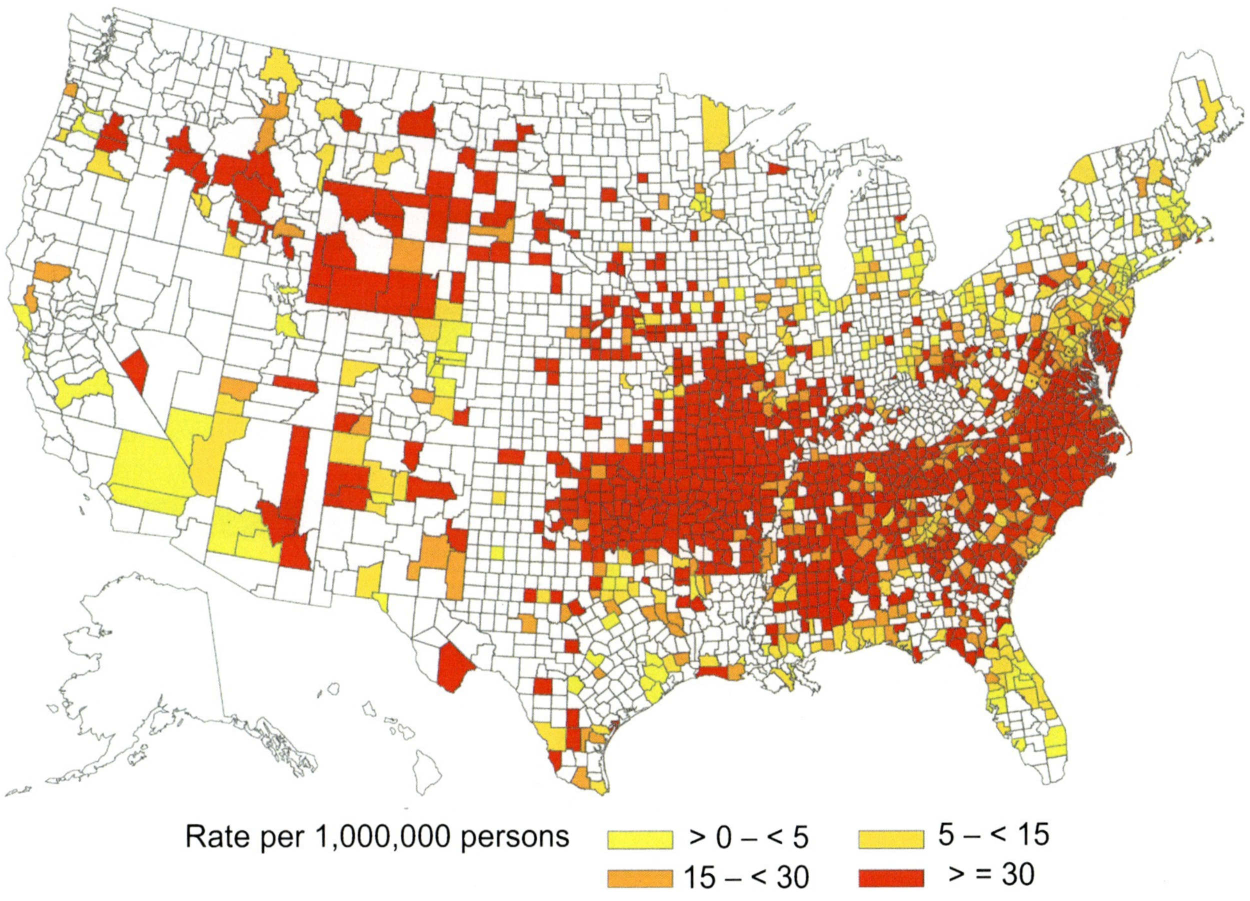

FIGURE 4 Incidence of RMSF by U.S. county, 2000 to 2007. doi:10.1128/9781555817336.ch2f4

When RMSF was first recognized in the Bitterroot Valley, the disease was observed to affect primarily adult males who were exposed during occupational activities that took them into tick-infested habitats (Harden, 1990; Price, 1948). As RMSF became more widely recognized across the United States, a shift was noted toward increased incidence in children, particularly those from rural areas. From 1970 to 1979, over 50% of RMSF cases occurred in children (Hattwick et al., 1973; Helmick and Winkler, 1981; Helmick et al., 1984), and during 1980 to 1993, the highest incidence was observed among children ages 5 to 9 years (Dalton et al., 1995; Treadwell et al., 2000). While the most cases were reported in children, mortality appeared highest among older age groups (Dalton et al., 1995; Treadwell et al., 2000). However, a change in the age-specific epidemiology of reported RMSF has been noted since 2000. The reported incidence has become highest among 50- to 70-year-olds, and the case fatality rate is now highest among children 0 to 9 years of age (Fig. 5 and 6) (Chapman et al., 2006; Openshaw et al., 2010).

The explanation for this inverse relationship in recent reporting periods is not well understood. It may be an artifact of surveillance related to an increase in the reporting of other, less severe spotted fever rickettsioses, such as *R. parkeri* or 364D rickettsiosis (see below), or misdiagnosed cases of other illnesses (i.e., non-RMSF). Studies examining the seroprevalence of antibodies to *R. rickettsii* in healthy people suggest that antibody titers can persist for months or years following acute infection and that the proportion of seropositive persons increases with age (Graf et al., 2008; Hilton et al., 1999). Among contemporary health care providers, there is an increasing tendency to rely on a single serologic test to diagnose RMSF (Openshaw et al., 2010), and the specificity of this technique likely decreases with patient age due to increasing background seroprevalence among older age groups (Hilton et al., 1999; Graf et al., 2008).

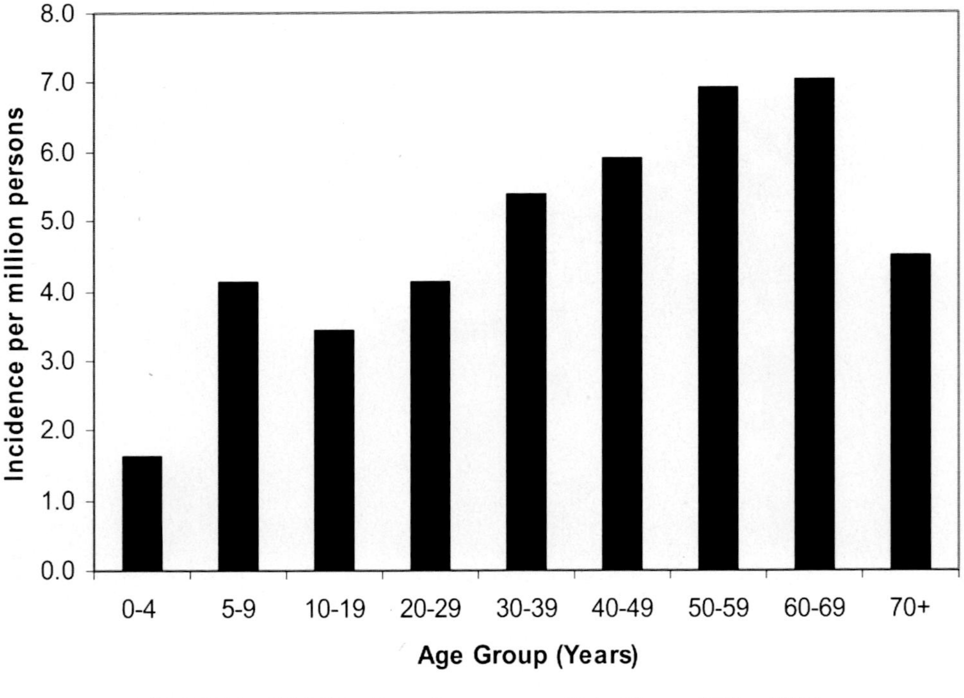

FIGURE 5 RMSF incidence by age group, United States, 2000 to 2007.
doi:10.1128/9781555817336.ch2f5

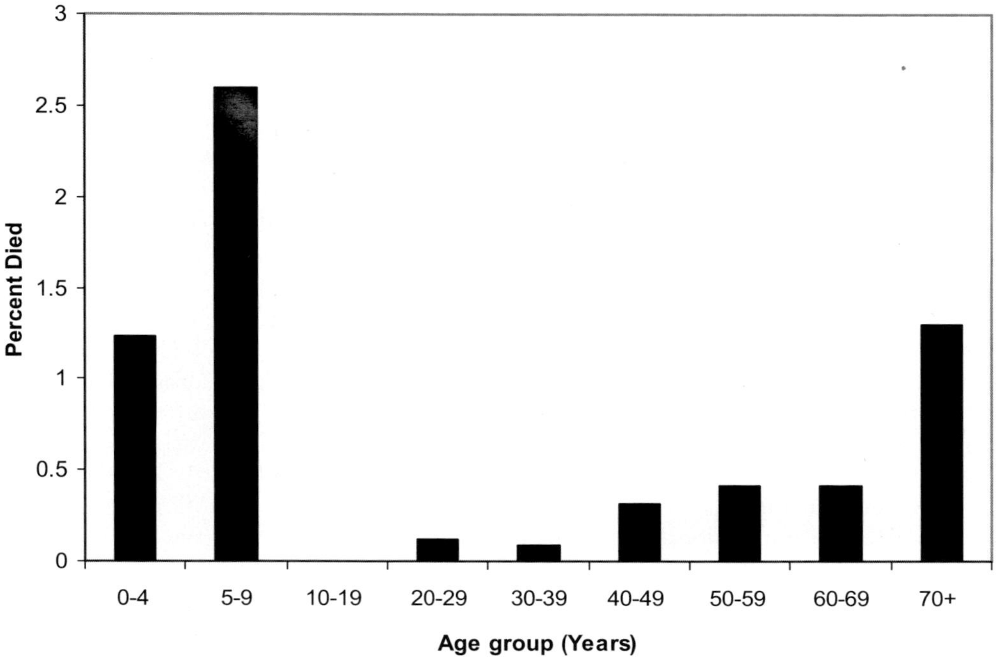

FIGURE 6 RMSF case fatality rate by age group, United States, 2000 to 2007. doi:10.1128/9781555817336.ch2f6

Gender differences in RMSF incidence have been observed throughout the history of reporting. At the beginning of the 20th century, the disease was widely recognized as being more common among males than females, likely due to occupational exposures to ticks (Harden, 1990; Price, 1948). From 1970 to the present, cases in males (56 to 58%) have continued to outnumber cases in females, although perhaps less significantly than was observed in the Bitterroot Valley (Chapman et al., 2006; Dalton et al., 1995; Hattwick et al., 1973; Openshaw et al., 2010; Treadwell et al., 2000). In some studies, the gender differences have appeared more pronounced among 10- to 19-year-olds (Dalton et al., 1995), perhaps reflecting high-risk behaviors for tick exposure between males and females in this age group.

Differences in race and ethnicity have also been reported. Although glucose-6-phosphate dehydrogenase deficiency, a heritable trait that is more common among persons of African or Middle Eastern descent, has been reported as a risk factor for fulminant RMSF (Walker et al., 1983), an increased risk for either infection or fatal outcome among those reporting black race has not been observed (Chapman et al., 2006; Dalton et al., 1995; Hattwick et al., 1973; Openshaw et al., 2010; Treadwell et al., 2000). Nor has Hispanic ethnicity been reported in association with an increased risk for RMSF. In contrast, American Indians have been shown through several models of assessment to have an apparent increased incidence of RMSF compared to other racial groups (Demma et al., 2006a; McQuiston et al., 2000b). In the most recent national surveillance summary, spanning 2000 to 2007, the incidence among American Indians was greater than four times that of other racial groups. This particular group also has a higher reported case fatality rate, at approximately four times that of other races (Openshaw et al., 2010). The reasons for this increased risk for RMSF among American Indians are not well understood. The emergence of RMSF in eastern Arizona among a predominantly American Indian population can account for some recent cases, but the elevated risk appears broadly distributed among American Indian populations in other parts of the country (Openshaw et al., 2010). Explanatory hypotheses include cultural associations of risk, such as increased exposure to outdoor tick habitat through ceremonial practices or subsistence living, differences in health care-seeking behavior, or genetic differences in susceptibility to infection among American Indians.

Because national surveillance for RMSF relies on laboratory testing, case counts may be influenced by access to health care, and by extension to socioeconomic status. Reporting of RMSF cases according to race, and the inferences drawn from national surveillance data, must therefore be conducted with caution. Analyses of reporting trends across U.S. counties suggest that for nonwhite patients, race data are not uniformly reported, and that the actual race-specific risk of infection may actually be higher among blacks and American Indians than current national surveillance data show (Dahlgren et al., 2011b).

R. parkeri Rickettsiosis

In 1937, a team of investigators from the Rocky Mountain Laboratories isolated a unique SFGR species from *Amblyomma maculatum* ticks collected near the Gulf Coast of Texas that was subsequently characterized and formally named *R. parkeri* (Parker et al., 1939). However, disease in humans caused by *R. parkeri* was not described until 2004, when it was first associated with an eschar-associated rickettsiosis in a patient from the Tidewater region of Virgina (Paddock et al., 2004). *R. parkeri* infection can be difficult to distinguish from infection with *R. rickettsii*, as extensive serologic cross-reactivity occurs between this and other SFGR species. In this context, some proportion of cases currently reported as RMSF in the United States are likely infections with *R. parkeri* (Raoult and Paddock, 2005).

While not specifically considered a notifiable disease in the United States, in 2010 the reporting category for RMSF was changed to more broadly reflect disease reporting for all SFGR (for the CDC's case definition of spotted fever group rickettsioses, see http://www.cdc.gov/osels/ph_surveillance/nndss/casedef/spottedfever_current.htm); as a result, *R. parkeri* cases may now be reported to U.S. public health authorities through the category "Other Spotted Fever Rickettsioses." Through 2011, approximately 30 confirmed or suspected cases of *R. parkeri* rickettsiosis have been identified in states along the Gulf Coast and southern Atlantic region, from Texas to Maryland, approximating the U.S. distribution of *A. maculatum* (Cragun et al., 2010; Paddock et al., 2008; CDC, unpublished). In most cases of *R. parkeri* infection, a necrotic inoculation eschar forms several days after the bite of an infected tick and precedes a low to moderate fever by several days. *R. parkeri* rickettsiosis appears to be milder than *R. rickettsii* infections; no severe systemic manifestations or deaths have been described (Cragun et al., 2010; Paddock et al., 2008; Romer et al., 2011), and the illness appears to be self-limiting in most cases. This contrasts with RMSF, a potentially fatal disease for which most patients require admission to a hospital (Paddock et al., 2008).

In the United States, molecular evidence of infection with *R. parkeri* has been identified in 8 to 43% of questing adult ticks collected from locations in Florida, Georgia, Mississippi, North Carolina, and Virginia (Fornadel et al., 2011; Paddock et al., 2010; Sumner et al., 2007; Wright et al., 2011; Varela-Stokes et al., 2011). These percentages are considerably higher than the percentages of human biting ticks infected with *R. rickettsii*, providing support for the notion that at least some of the cases reported as RMSF may in fact represent cases of *R. parkeri* rickettsiosis (Paddock, 2005, 2009; Raoult and Paddock, 2005).

Rickettsia 364D Rickettsiosis

In 2008, the first confirmed infection with *Rickettsia* 364D was identified in an 80-year-old man from northern California. In addition to the index patient, three suspect cases have been described, each with an eschar and mild constitutional complaints that variably include fever, headache, myalgia, and fatigue (Shapiro et al., 2010). As with *R. parkeri*, *Rickettsia* 364D is not specifically considered a notifiable condition in the United States, but may be captured under the broad reporting category "Other Spotted Fever Rickettsioses," which was created in 2010. *Rickettsia* 364D has been detected in approximately 4 to 8% of questing *D. occidentalis* ticks throughout California (Lane et al., 1981; Wikswo et al., 2008). Because *R. rickettsii* is rarely identified in human-biting ticks in this state, it has been suggested that *Rickettsia* 364D may be responsible for many of the illnesses in California classified as RMSF (Lane et al., 1981; Paddock, 2005; Wikswo et al., 2008).

Rickettsialpox

Rickettsialpox, a mite-borne zoonosis caused by *Rickettsia akari*, was discovered in 1946 in the borough of Queens, New York City (Greenberg et al., 1947a, 1947b). In 1949, Soviet investigators in the Ukraine identified a large outbreak of a disease they described as "vesicular rickettsiosis," subsequently recognized as rickettsialpox. The organism cycles among house mice (*Mus musculus*) and house mouse mites (*Liponyssoides sanguineus*). Although now recognized to be worldwide in distribution, rickettsialpox appears to have a tenacious presence in New York City: every case series of rickettsialpox published during the last 50 years describes a patient cohort from the geographic boundaries of New York City (Koss et al., 2003; Paddock et al., 2003; Saini et al., 2004), despite documented occurrence of the pathogen in several other regions of the United States and in many other countries (Radulovic et al., 1996; Choi et al., 2005; Paddock and Eremeeva, 2007; Zavala-Castro et al., 2009a).

Rickettsialpox has been described in patients of all ages, from infants as young as 6 months to adults as old as 92 years. In most patient series, the disease occurs equally among males and females, and cases are documented from all months of the year. Cases of rickettsialpox often cluster in time and space, and simultaneous or consecutive illnesses have been identified among family members or other residents from a single common location. In contrast with many other rickettsioses, most cases of rickettsialpox are described from large metropolitan areas and urban centers, consistent with the important role of peridomestic rodents in the distribution and occurrence of this rickettsia. *R. akari* has been cultured from multiple commensal and wild rodent species, particularly the house mouse; however, transovarial and transstadial transmission of rickettsiae occurs in *L. sanguineus*, implicating the mite as an important reservoir host of *R. akari*. *L. sanguineus* has been collected from house mice and various rodent species in the United States, Eurasia, and Africa.

Despite broad geographic distributions of the vector mite and the house mouse, confirmed reports of rickettsialpox are relatively sparse and sporadic, suggesting that conditions that favor the parasitism of humans by *L. sanguineus* may have the greatest impact on the epidemiological activity of rickettsialpox (Paddock and Eremeeva, 2007).

Rickettsia felis Rickettsiosis

R. felis, formerly known as the "ELB agent," was originally isolated from cat fleas (*Ctenocephalides felis*) collected from a commercial flea colony (Adams et al., 1990). The first human infection with *R. felis* was reported in 1994 (Schriefer et al., 1994a), and since that time the organism has been reported with a broad, worldwide geographic distribution (Parola, 2011; Reif and Macaluso, 2009; Tijsse-Klasen et al., 2011; Williams et al., 2011).

R. felis appears clinically and epidemiologically similar to another flea-borne rickettsiosis, murine typhus (see below), and symptoms may include fever, rash, headache, and myalgia (Brouqui et al., 2007). Although it is generally considered milder than murine typhus, severe manifestations of *R. felis* infection have been reported, including hepatitis and neurologic manifestations (Galvão et al., 2003; Zavala-Castro et al., 2009b).

R. felis is transmitted to humans by way of the cat flea, and in the United States its enzootic cycle may involve cats, dogs, and wildlife such as opossums (*Didelphis virginianus*) (Bouyer et al., 2001; Macaluso et al., 2008; Reif and Macaluso, 2009). In regions considered enzootic for murine typhus, *R. felis* has been detected in the Oriental rat flea (*Xenopsylla cheopis*) at frequencies greater than those observed for *R. typhi* (Eremeeva et al., 2008; Karpathy et al., 2009). Despite this observation, the contribution of cases of *R. felis* infections possibly misclassified as murine typhus is not known (Wiggers et al., 2005). However, *R. felis* and *R. typhi* appear to circulate in similar ecological cycles, and the detection of *R. felis* in fleas, opossums, and cats in areas considered enzootic for *R. typhi* raises the possibility that some cases of "murine typhus" are in fact infections with *R. felis*. Because infections with either agent elicit a serologic response that may react with typhus group rickettsiae in serologic assays, molecular techniques are required to differentiate the causative agent, and as a result, relatively few *R. felis* infections have been diagnosed among U.S. residents.

Epidemic and Flying Squirrel-Associated Typhus

The clinical disease described as typhus in North America is not one entity, but two: epidemic or louse-borne typhus, caused by *R. prowazekii*, and endemic or flea-borne typhus, caused by *R. typhi*. Both organisms are described as belonging to the broader category of typhus group rickettsiae. Although the two diseases have many clinical similarities, infections with *R. prowazekii* are generally considered more severe, with case fatality rates around 30% in the pre-antibiotic era; fatalities associated with murine typhus are rarely reported (Adjemian et al., 2010; Bechah et al., 2008; Dumler et al., 1991; Manea et al., 2001). The relative rarity or regional specificity of both types of typhus in the United States has precluded listing either of these infections as nationally notifiable diseases. Diagnoses and reporting, therefore, lack both provider awareness and a requirement for notification of public health authorities. Nonetheless, public health interest in these diseases remains strong, and reporting requirements exist in some states. Because they have different vectors, reservoirs, and ecological cycles, control and prevention must be undertaken as different approaches.

Epidemic typhus, known also as louse-borne typhus, is a form of exanthematous typhus caused by *R. prowazekii* (Raoult et al., 2004; Bechah et al., 2008). The disease is so named because the disease often causes epidemics when conditions favor human-to-human transmission of the human body louse vector (*Pediculus humanus corporis*) (Bechah et al., 2008). The epidemic form of the disease is rarely reported in the United States, and usually occurs in settings with close crowding and poor sanitary conditions (e.g., prisons, refugee camps, among the homeless), situations where body louse infestations may flourish. *R. prowazekii* enters the human body through contamination of broken skin, conjunctivae, or mucosal membranes with the crushed bodies or feces of infected lice (Fig. 7). Patients who recover from their acute infection may later experience recrudescent infection known as Brill-Zinsser disease, which is speculated to contribute to the interepidemic survival of *R. prowazekii* (Murray et al., 1950; Price et al., 1958; Raoult et al., 2004).

In contemporary U.S. settings, *R. prowazekii* infections in humans are reported sporadically and are most commonly associated with contact with *G. volans* (the southern flying squirrel), which appears to serve as a zoonotic reservoir (Fig. 7) (Bozeman et al., 1975; Chapman et al., 2009; Reynolds et al., 2003). Transmission from flying squirrels is speculated either to involve the aerosolization of feces of infected arthropods found on these animals or to be through contamination of broken skin, conjunctivae, or mucosal membranes with the feces of infected arthropods (Bozeman et al., 1981). Inhalation of the organism is also a suspected route of transmission.

Following the discovery of effective antibiotics and the adaption of more stringent hygiene measures for the control of human body

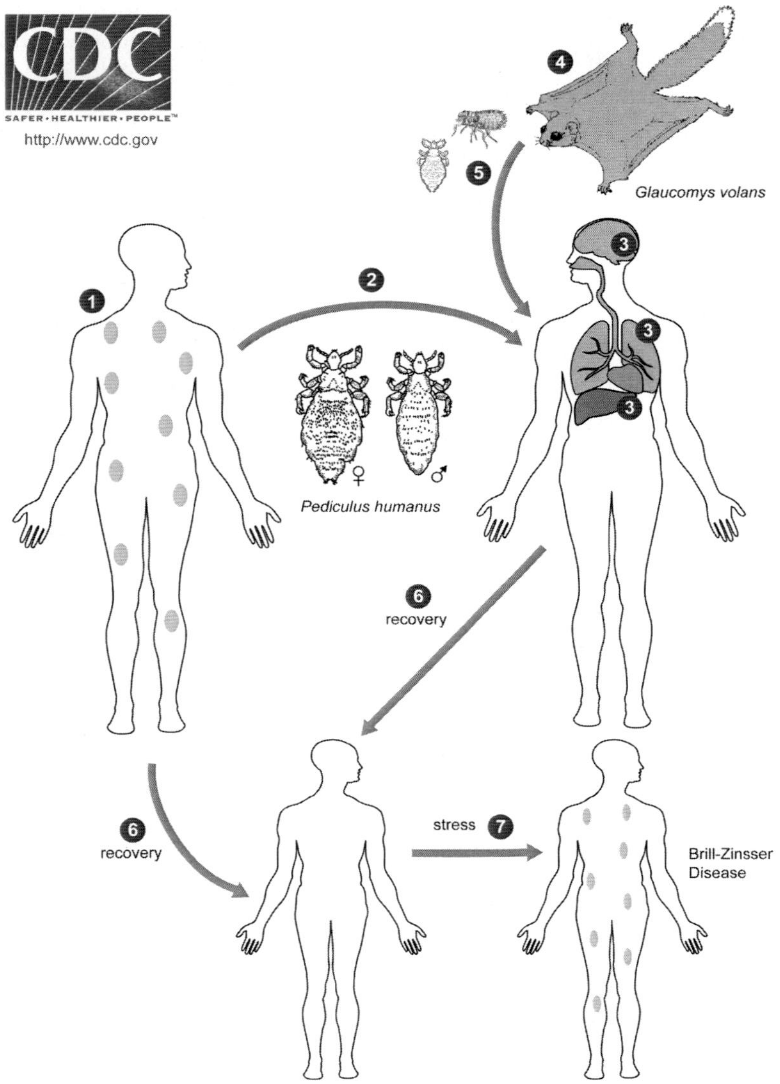

FIGURE 7 Transmission cycle for maintenance of *R. prowazekii* in the United States. doi:10.1128/9781555817336.ch2f7

lice, *R. prowazekii* has ceased to be a significant contemporary public health threat in the United States. While the potential for sporadic infection exists among travelers returning from endemic areas, cases acquired by classic louseborne transmission rarely occur. Confirmed reports of *R. prowazekii* infection acquired in the United States are usually associated with reported recent contact with or residence in homes inhabited by the southern flying squirrel. Since 1976, there have been at least 47 documented cases of flying squirrel-associated *R. prowazekii* infections in the United States, and likely more have occurred that have not been recognized or reported (Bozeman et al., 1975; Chapman et al., 2009; Reynolds et al., 2003; CDC, unpublished). The geographic distribution of reported human cases follows the expected range of the southern flying squirrel, and most human cases have been reported from Georgia (at least 13 cases), Pennsylvania (6 cases), Virginia (5 cases), North Carolina (4 cases), West Virginia (2 cases), and New York (2 cases) (CDC, unpublished).

There are few contemporary data on age-specific risk for *R. prowazekii* infections, particularly among residents of the United States, where traditional louse-borne transmission is rare. The median and mean patient age among U.S. *R. prowazekii* patients who have acquired infection from flying squirrels is 33 and 38 years, respectively; case patients ranged from 11 to 81 years old (CDC, unpublished).

Brill-Zinsser disease is rarely reported, likely because most infections with *R. prowazekii* receive antibiotic treatment, which may mitigate the risk of Brill-Zinsser disease. There have been fewer than 10 Brill-Zinsser cases reported in the contemporary literature, and most patients reported exposure to *R. prowazekii* in World War II concentration camps in Europe (Green et al., 1990; McQuiston et al., 2010; Portnoy et al., 1974; Reilly et al., 1980; Stein et al., 1999). In 2009, a case of Brill-Zinsser disease was diagnosed for the first time in a U.S. patient who had initially been diagnosed with acute *R. prowazekii* infection acquired from flying squirrels a decade prior. Lack of early and appropriate antibiotic therapy may have influenced the later development of Brill-Zinsser disease in this patient (McQuiston et al., 2010).

Brill-Zinsser patients represented in the published literature tend to be male (5/7 cases; 71%) and older, with a median and mean age of 62 and 56 years, respectively (Green et al., 1990; McQuiston et al., 2010; Portnoy et al., 1974; Reilly et al., 1980; Stein et al., 1999). Factors predisposing patients to the development of Brill-Zinsser disease have been speculated to include waning immunity, underlying health conditions, and stress.

Murine Typhus

R. typhi, the cause of murine typhus, is an occasional cause of human febrile illness in the United States, with increased incidence in certain geographic areas, particularly southern Texas, southern California, and Hawaii. It is maintained in an ecological cycle involving commensal rodents (commonly the black rat, *Rattus rattus*) and the Oriental rat flea. Humans are considered accidental hosts (Fig. 8). Recent investigations indicate that a peridomestic cycle involving involving opossums, domestic cats, and cat fleas may also exist, although elucidation of the ecological cycle is complicated by the apparent cocirculation of *R. felis*, in a similar cycle (Boostrom et al., 2002; Civen and Ngo, 2008; Schriefer et al., 1994b; Williams et al., 1992). *R. typhi* DNA has been isolated from opossum spleens, and opossums trapped near sites of human outbreaks often show high seroprevalence of antibodies reactive with typhus group rickettsiae (Adjemian et al., 2010; Boostrom et al., 2002; Williams et al., 1992). Humans become infected when the organism enters the body through contamination of broken skin, conjunctivae, or mucosal membranes with feces from infected fleas.

Because of its association with rodents, murine typhus is found in many regions of the world. It was once widespread in the United States, but aggressive rodent control programs in the 1940s and 1950s, as well as the discovery of highly effective pesticides such as DDT, provided additional tools against the mammalian reservoirs and ectoparasites responsible for its spread (Mohr et al., 1953). In the United States, reports of murine typhus fell dramatically during the second half of the 20th century, from tens of thousands of cases during the period from 1931 to 1946 (White, 1965) to currently less than a few hundred people annually (Fig. 9) (CDC, unpublished).

Endemic typhus continues to occur in a low-level enzootic fashion in the United States in southern Texas, California, and Hawaii, and as a sporadic travel-acquired illness in persons exposed outside the United States. While it is not a reportable condition, 952 cases of murine typhus were reported to the CDC from 1992 to 2008 (Fig. 9 and 10) (CDC, unpublished). Although case reports have increased in recent years, factors for this increase are likely

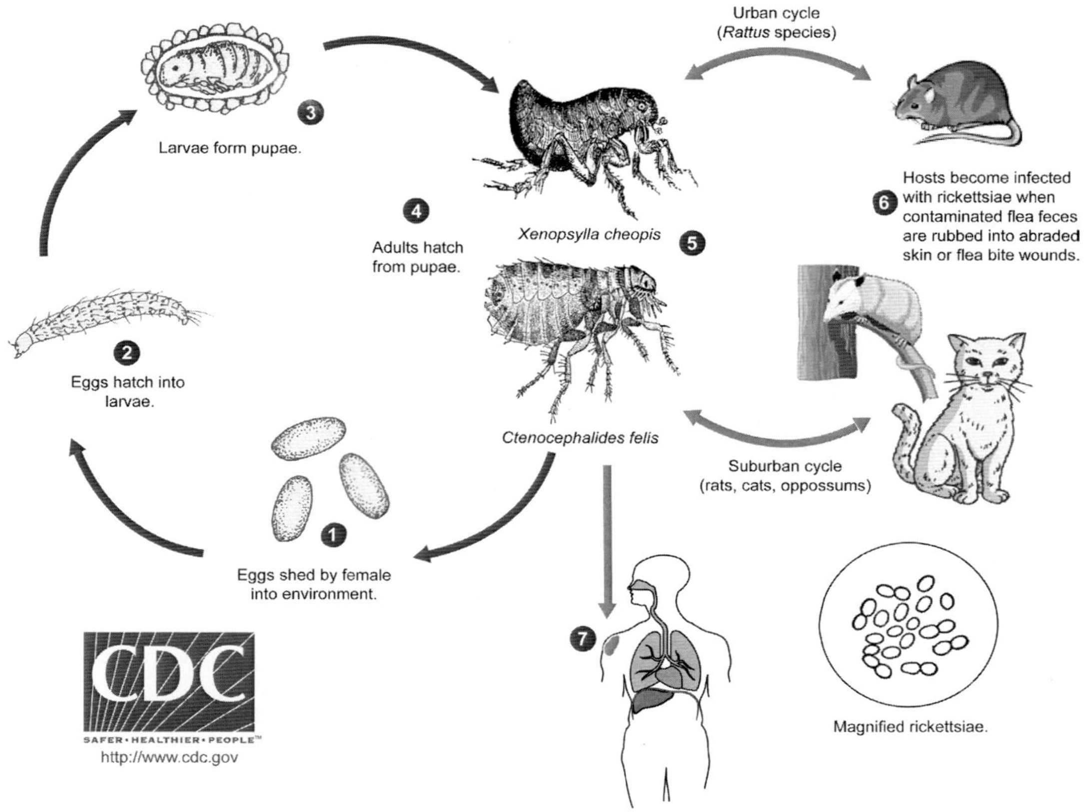

FIGURE 8 Transmission cycle for maintenance of *R. typhi* in the United States.
doi:10.1128/9781555817336.ch2f8

to include increased surveillance efforts; for example, the majority of U.S.-reported cases (n = 868; 91%) come from Texas, where the disease is considered reportable, and case reports from this state increased dramatically following the initiation of enhanced surveillance efforts for murine typhus in 2006.

In Texas, most murine typhus reports come from southern counties near the U.S. border with Mexico (Texas Department of State Health Services, 1998). However, in 2008, an outbreak of murine typhus was recognized in Austin, in an area where cases had not been previously reported (Adjemian et al., 2010). This outbreak resulted in 53 suspected human cases, of which 33 were eventually laboratory confirmed. The environmental investigation found seropositivity among domestic pets (cats and dogs) and opossums near patient homes. This outbreak highlighted the potential for *R. typhi* to spread from established to new areas; since the 2008 outbreak, human *R. typhi* cases have continued to be reported from Austin, suggesting a new enzootic focus.

Among the 952 cases of murine typhus reported to the CDC from 1992 to 2008, 30 cases (3%) were reported from California, where the disease is considered endemic in the Los Angeles area (CDC, unpublished). In addition to Texas and parts of California, enzootic and outbreak transmission of *R. typhi* infection has been reported in Hawaii; in 2002, 47 human cases were reported, against an expected enzootic background of 5 to 6 cases annually (Centers for Disease Control and Prevention, 2003; Manea et al., 2001).

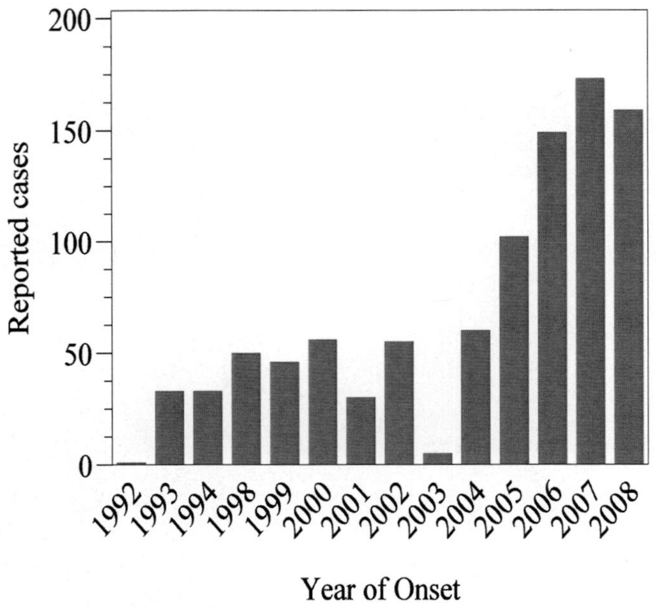

FIGURE 9 Reported cases of *R. typhi*, United States, 1992 to 2008. The majority of the reports (n = 868; 91%) were from Texas (CDC, unpublished).
doi:10.1128/9781555817336.ch2f9

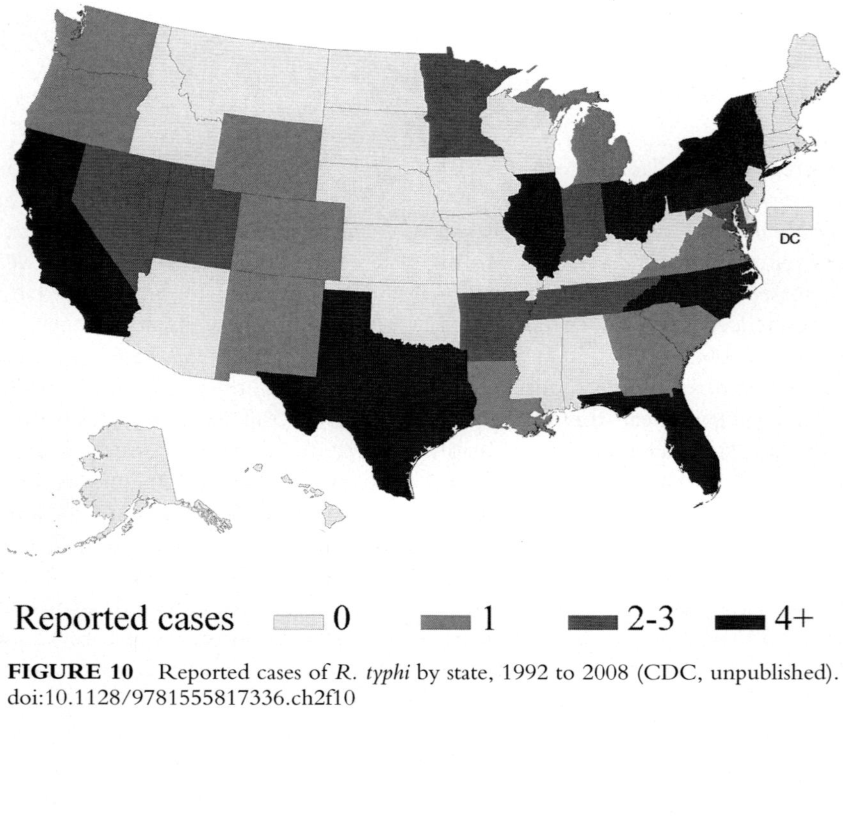

FIGURE 10 Reported cases of *R. typhi* by state, 1992 to 2008 (CDC, unpublished).
doi:10.1128/9781555817336.ch2f10

Among murine typhus case reports received by the CDC from 1992 to 2008, the highest number of cases was reported in children 10 to 19 years old, suggesting that younger age may be a risk factor for *R. typhi* infection (Fig. 11). (CDC, unpublished). Published case studies suggest that children experience an illness similar to that reported in adults (Fergie et al., 2000). During the 2008 Austin, Texas, outbreak of murine typhus, confirmed case-patients averaged 39 years of age (range, 7 to 64 years; 15% <18 years) (Adjemian et al., 2010). The majority of patients were male (56%). While there were no deaths attributed to murine typhus among the Austin cohort of case-patients, severe illness was noted. Seventy percent (23/33) were hospitalized, and 27% (9/33) required intensive care nursing. It is uncertain why the Austin cohort experienced a more severe illness than has traditionally been reported in the literature for murine typhus; reported complications among Austin case-patients included pneumonia, coagulopathy, and renal failure. The emergence of murine typhus in an area not considered enzootic may have contributed to delays in physician recognition and treatment with appropriate and effective antibiotics. Among confirmed case-patients, only 17 of 33 (51%) patients received antibiotics, and only 13 of those (76%) were treated with doxycycline, the treatment of choice. In this outbreak, the mean time from symptom onset to antibiotic treatment was 8.3 days (median, 8; range, 1 to 19) (Adjemian et al., 2010).

Ehrlichiosis Caused by *E. chaffeensis*

E. chaffeensis was first confirmed as a cause of human illness in a patient from Fort Chaffee, Arkansas, in the late 1980s (Anderson et al., 1991; Dawson et al., 1991). The infection caused by this organism is sometimes called human monocytic ehrlichiosis because of the propensity of the organism to replicate in monocytes. Patients with human monocytic ehrlichiosis infection are identifiable through clusters of organisms known as morulae that can be identified within monocytes on peripheral blood smears. *E. chaffeensis* is transmitted by *Amblyomma americanum* (the lone star tick), which is broadly distributed throughout the southeastern and south-central United States. White-tailed deer (*Odocoileus virginianus*) are the primary wildlife reservoir and the preferred host for all feeding stages of *A. americanum*; humans are accidental hosts for *E. chaffeensis* infection (Fig. 12).

From 1986 through 1997, 742 cases of ehrlichiosis were reported across the United

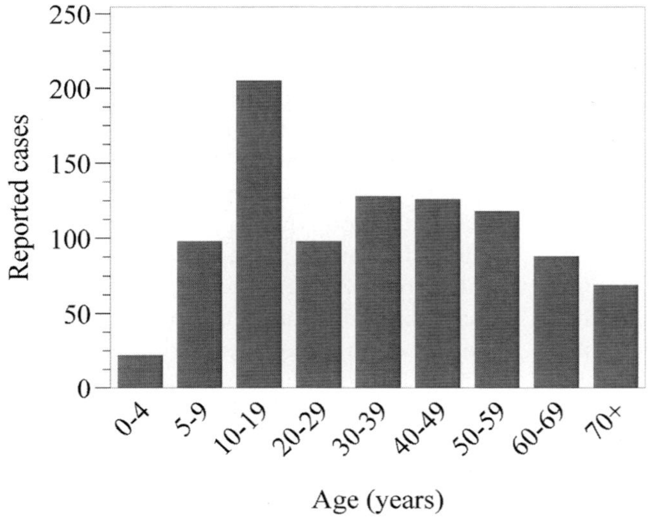

FIGURE 11 Reported cases of *R. typhi* by age group, United States, 1992 to 2008 (CDC, unpublished). doi:10.1128/9781555817336.ch2f11

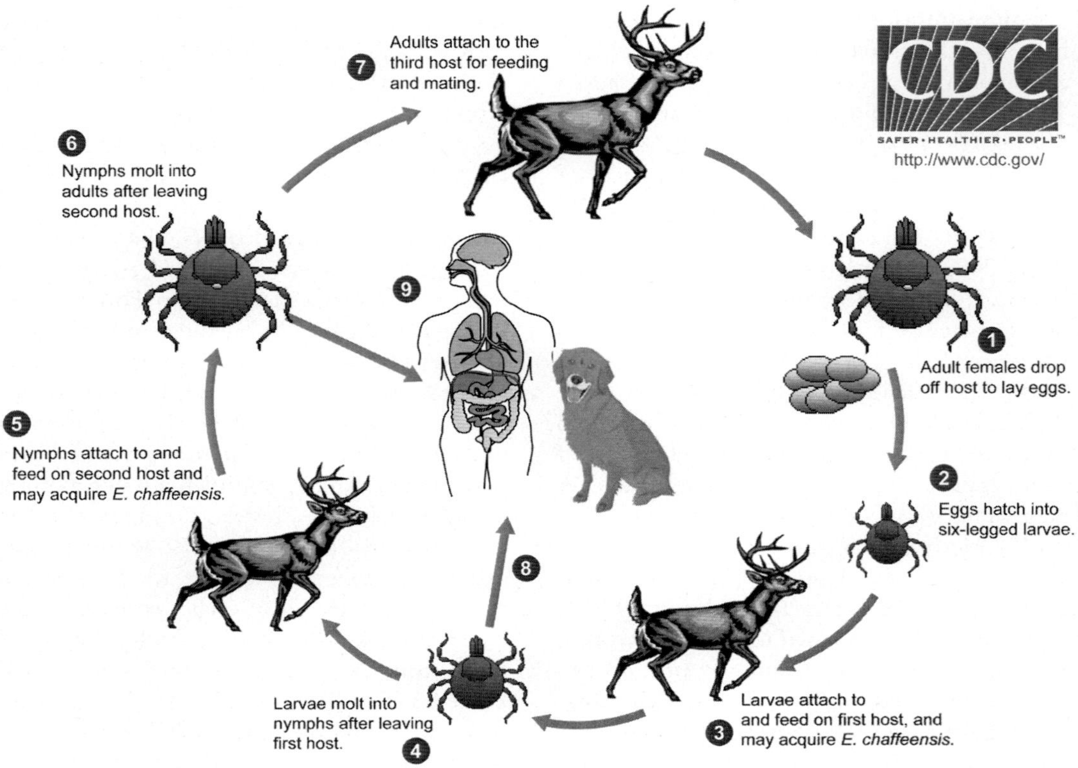

FIGURE 12 Transmission cycle for maintenance of *E. chaffeensis* in the United States. doi:10.1128/9781555817336.ch2f12

States, although data were collected systematically in less than half of state health departments during this time (McQuiston et al., 1999). Ehrlichiosis was added to the U.S. list of notifiable conditions in January 1999 (see the position statement of the Council of State and Territorial Epidemiologists [CSTE] at http://www.cste.org/dnn/AnnualConference/PositionStatementArchive/tabid/398/Default.aspx), effectively establishing a platform for longitudinal surveillance. From 2000 to 2007, 3,126 cases of presumed *E. chaffeensis* and an additional 824 cases classified as "Ehrlichiosis Undetermined/Unspecified/Other Agent" ("Ehrlichiosis UUOA") were reported within the United States. Ehrlichiosis incidence increased from 0.80 cases per million persons in 2000 to 3.0 cases in 2007 (Fig. 13) (Dahlgren et al., 2011a). The observed increase mirrors changes observed in the reporting of other tick-borne diseases like RMSF, but it is difficult to ascertain the significance of these increases when surveillance was so newly established.

During the first decade of reporting, ehrlichiosis cases were reported primarily from the southeastern and south-central United States, mirroring the expected range of the tick vector *A. americanum* (Fig. 14A). Cross-reactivity between *E. chaffeensis* and *A. phagocytophilum* in serologic assays, as well as confusion among health care providers regarding the correct diagnostic test to use in a given geographic area, account for the large number of Ehrlichiosis UUOA cases that were reported during this time period (Fig. 14C) (Centers for Disease Control and Prevention, 2009; Dahlgren et al., 2011a). Most of these Ehrlichiosis UUOA cases were probably *E. chaffeensis* or *A. phagocytophilum* cases that could not be properly classified.

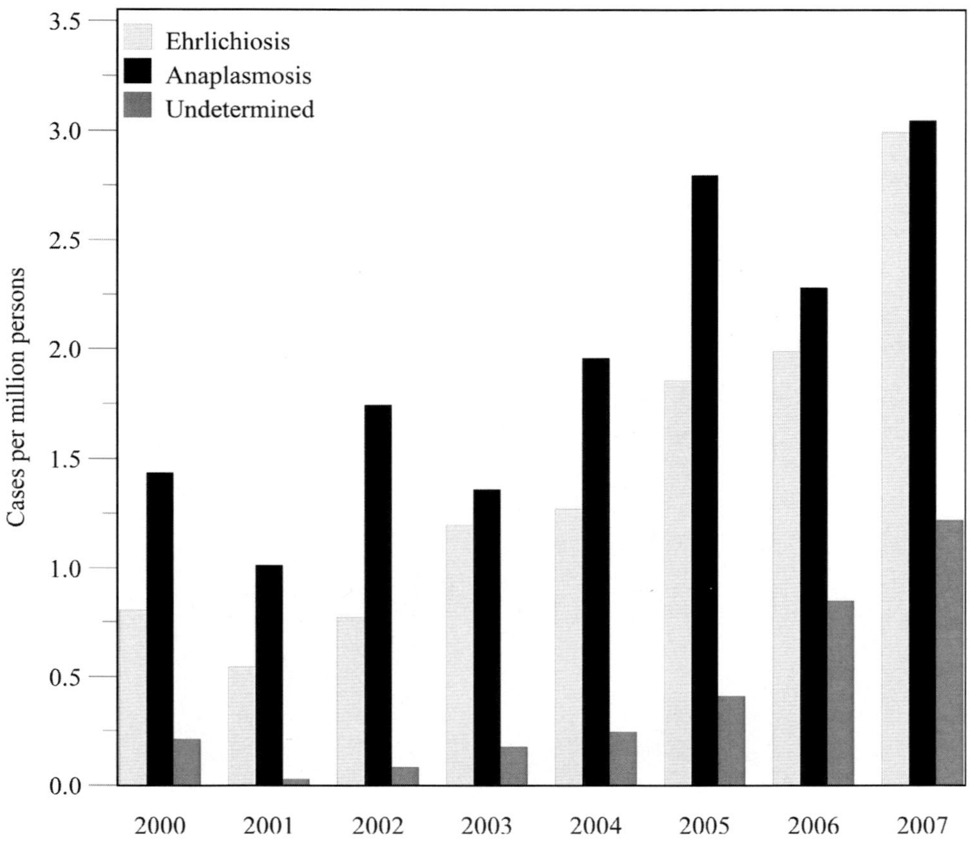

FIGURE 13 Annual incidence of infections with *E. chaffeensis* and *A. phagocytophilum* and Ehrlichiosis UUOA, United States, 2000 to 2007. doi:10.1128/9781555817336.ch2f13

The national reported incidence of ehrlichiosis increased with patient age during 2000 to 2007 (Fig. 15), with the highest incidence in persons over 60 years old. Males appeared to be at higher risk for infection than females, accounting for 59% of ehrlichiosis case reports (Dahlgren et al., 2011a). As with RMSF, the reasons for this difference in gender-associated risk are speculated to relate to occupational and recreational activities that increase exposures to ticks among males. During 2000 to 2007, ehrlichiosis patients were hospitalized at a rate of 49%. Hospitalization rates tended to be highest among children and patients 60 years of age and older (Fig. 16). Over 9% of ehrlichiosis cases reported the development of a life-threatening complication over the course of illness, including acute respiratory distress syndrome, coagulopathies, meningitis, encephalitis, or renal failure. The overall case fatality rate for ehrlichiosis during this time period was 1.9%. Similar to observations from RMSF, case fatality rates for ehrlichiosis followed a pattern suggesting highest risk of death among children ages 5 to 9 years (3.7%), followed by older patients ≥60 years of age (Dahlgren et al., 2011). As has been speculated for RMSF, this higher rate of fatal outcome among children with ehrlichiosis may be related to physician reluctance to prescribe doxycycline to children early during the course of a suspected infection. Similarly to RMSF, doxycycline is the treatment of choice for suspected ehrlichiosis infections, regardless of patient age.

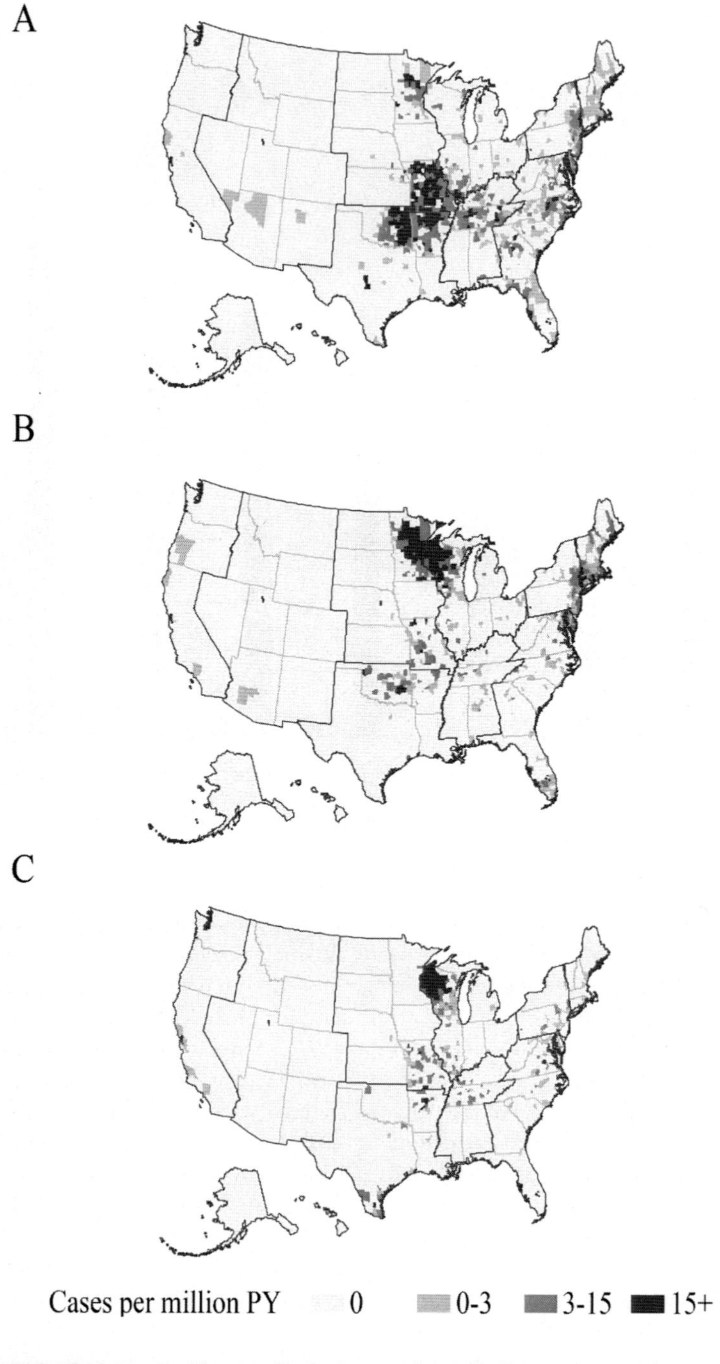

FIGURE 14 Incidence of infections with *E. chaffeensis* (A), *A. phagocytophilum* (B), and Ehrlichiosis UUOA (C) by U.S. county, 2000 to 2007. PY, person-years. doi:10.1128/9781555817336.ch2f14

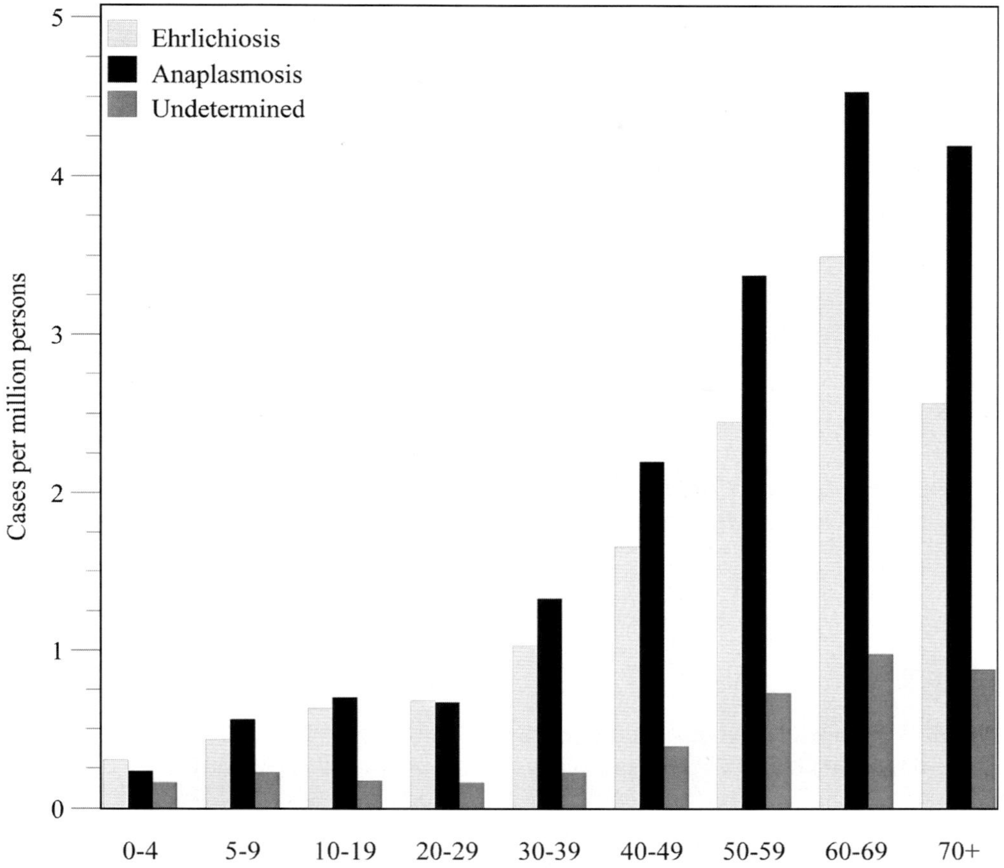

FIGURE 15 Incidence of infections with *E. chaffeensis* and *A. phagocytophilum* and Ehrlichiosis UUOA by age group, United States, 2000 to 2007. doi:10.1128/9781555817336.ch2f15

Immunosuppression appears to influence the course of ehrlichiosis infections. During 2000 to 2007, 12% of ehrlichiosis patients reported concurrent immune-suppressing conditions or medications. The clinical course also appeared more severe for immunosuppressed patients, with almost twice the rate of hospitalization, two to three times the risk of life-threatening complications, and almost four times the rate of death in this cohort compared with immune-intact patients (Dahlgren et al., 2011a).

Human Anaplasmosis

An illness similar to human monocytic ehrlichiosis was identified in patients from the upper Midwest and Northeast in the early 1990s as well, although morulae were identified in granulocytes instead of monocytes, and the causative agent was distinct from *E. chaffeensis*. The disease caused by this newly identified organism was initially called human granulocytic ehrlichiosis (Bakken et al., 1994; Chen et al., 1994; Dumler et al., 1994). A taxonomic review reclassified the pathogen to the genus *Anaplasma* in 2001, and the disease is now known as human granulocytic anaplasmosis (Dumler et al., 2001).

A. phagocytophilum is transmitted by *Ixodes scapularis* (the black-legged tick), which is the same tick vector responsible for transmission of the pathogens that cause Lyme disease and

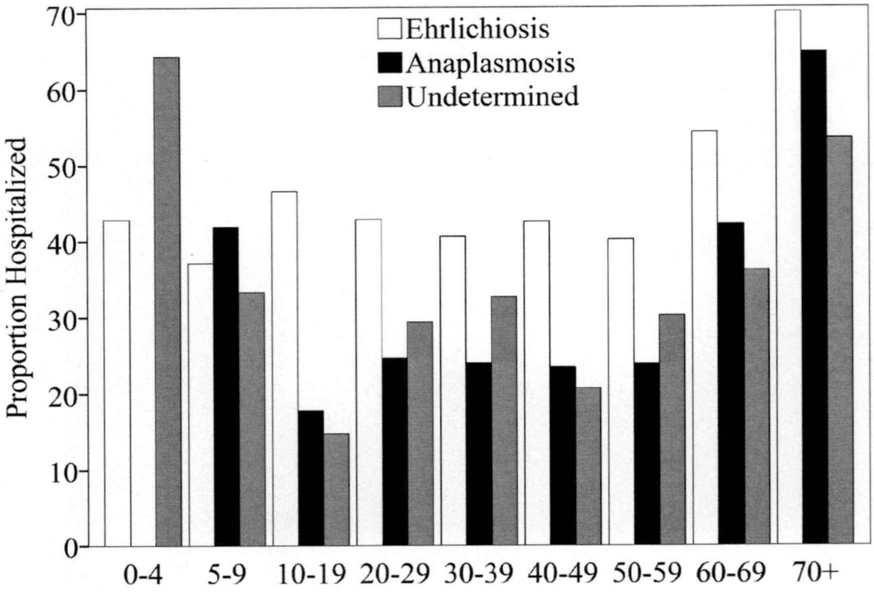

FIGURE 16 Hospitalization rates for infections with *E. chaffeensis* and *A. phagocytophilum* and Ehrlichiosis UUOA infections by age group, United States, 2000 to 2007.
doi:10.1128/9781555817336.ch2f16

babesiosis. This tick is found throughout much of the eastern United States, but reports of human anaplasmosis closely resemble the distribution of Lyme disease and are focused in the northeastern and upper midwestern United States, where immature stages of *I. scapularis* feed on small wild rodents that serve as the primary reservoirs for the pathogen. *A. phagocytophilum* is maintained primarily through transstadial transmission, with new larvae only acquiring the pathogen after feeding on infected rodents (Fig. 17).

From 1986 through 1997, 449 cases of anaplasmosis (at the time called human granulocytic ehrlichiosis) were reported across the United States, although data were not collected uniformly in every state (McQuiston et al., 1999). Like ehrlichiosis, disease caused by *A. phagocytophilum* was added to the national list of notifiable diseases in 1999 (see the position statement of the CSTE at http://www.cste.org/dnn/AnnualConference/PositionStatementArchive/tabid/398/Default.aspx). From 2000 to 2007, reports of anaplasmosis increased from 1.4 to 3.0 cases per million persons (Dahlgren et al., 2011a). Anaplasmosis cases were primarily reported from parts of the country where *I. scapularis*-transmitted diseases like Lyme and babesiosis are commonly reported, with a focus in the Northeast and upper Midwest (Fig. 14B).

The national reported incidence of anaplasmosis increased with patient age during 2000 to 2007 (Fig. 15), with the highest incidence in persons over 60 years old. As with both ehrlichiosis and RMSF, males appeared to be at higher risk for infection than females, accounting for 58% of anaplasmosis case reports (Dahlgren et al., 2011a). Thirty-six percent of anaplasmosis patients were hospitalized during their illness, and 3% of anaplasmosis cases reported the development of life-threatening complications. During 2000 to 2007, 6.5% of anaplasmosis patients reported concurrent immune-suppressing conditions or medications. The overall case fatality rate for anaplasmosis patients was 0.6% during

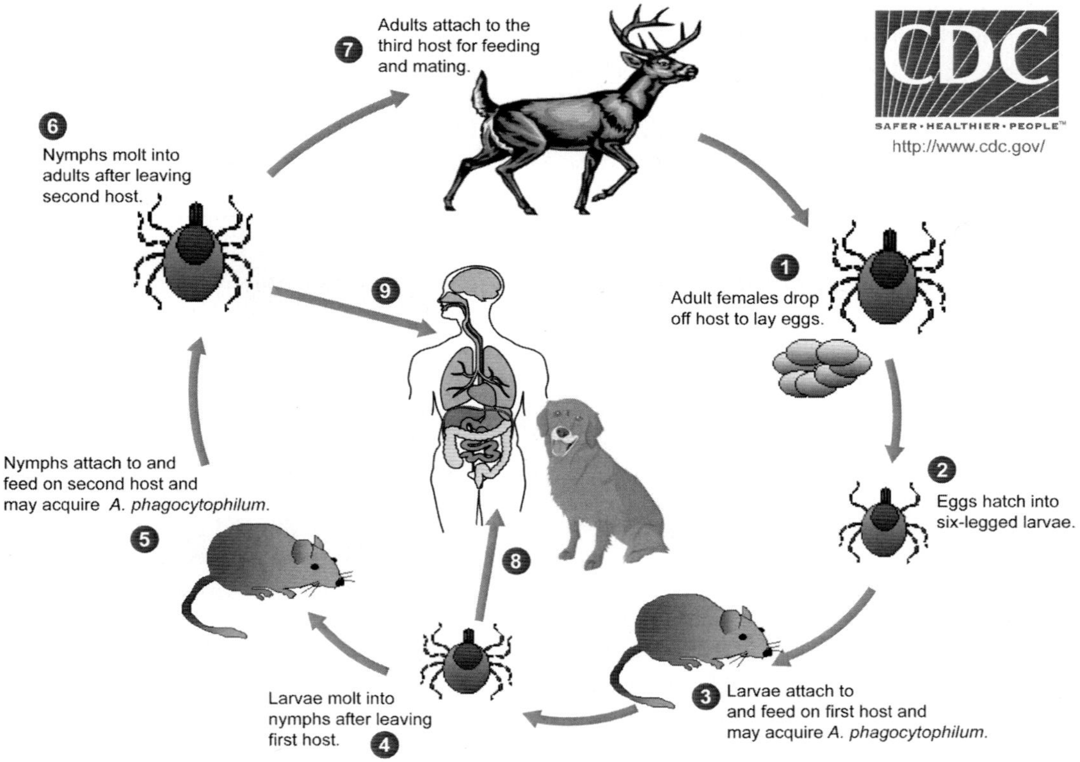

FIGURE 17 Transmission cycle for maintenance of *A. phagocytophilum* in the United States.
doi:10.1128/9781555817336.ch2f17

2000 to 2007, suggesting that the course of illness was less severe overall compared to that of ehrlichiosis (Dahlgren et al., 2011a). As for ehrlichiosis and RMSF, the treatment of choice for anaplasmosis is doxycycline, and delays in treatment may contribute to a more severe or potentially fatal outcome.

Several blood transfusion-associated cases of anaplasmosis have been identified, with the most likely source of transmission linked back to donors who were either asymptomatic or who gave blood shortly before becoming ill (Centers for Disease Control and Prevention, 2008). Because both *A. phagocytophilum* and *E. chaffeensis* are intracellular organisms that are found in white blood cells, transmission risks associated with blood transfusion are present (McQuiston et al., 2000a). Blood transfusion recipients may also be at increased risk for severe outcomes following ehrlichiosis or anaplasmosis due to immunosuppressive or comorbid conditions. There is currently no national blood donor screening program in the United States for ehrlichiosis or anaplasmosis. The use of leukoreduced blood products may reduce the risk of transmission of these agents through blood transfusion, by removing the cells most likely to harbor the organism (Mettille et al., 2000). Similarly, solid organ transplant recipients are at increased risk for ehrlichiosis and anaplasmosis because of the immunosuppression state that accompanies such conditions, but these procedures themselves may in some cases pose a potential risk for transmission of the organism; at least one instance of suspected ehrlichiosis transmission via solid organ transplantation in two recipients from a single donor has been investigated, although donor infection was not able to be confirmed (Dahlgren et al., 2011b).

Assessing a relationship between race or ethnicity and incidence is difficult for both anaplasmosis and ehrlichiosis, for the same reasons that have already been discussed for RMSF. No significant racial associations have been reported for either disease. In the United States, surveillance systems for tick-borne rickettsial diseases collect demographic data on race and ethnicity, but the case definitions used for surveillance require laboratory testing, and therefore are influenced by access to health care, which is not evenly distributed among the different racial and ethnic groups in the United States. Racial associations are likely to be more pronounced than are currently reflected by passive surveillance programs (Dahlgren et al., 2011b).

Other Ehrlichioses

With the application of new molecular tools for diagnosis, we have begun to better understand the role of other species in the family *Anaplasmataceae* as emerging pathogens. For example, although it is not commonly identified as a human pathogen, molecular analyses have permitted the recognition that *E. canis*, a canine pathogen transmitted by the tick vector *R. sanguineus*, can on occasion cause human illness (Perez et al., 1996, 2006).

E. ewingii was first identified as an agent of human disease from four residents of Missouri in the 1990s (Buller et al., 1999). *E. ewingii* shows a predilection for granulocytes and causes a spectrum of illness similar to infection with *E. chaffeensis* and *A. phagocytophilum*. Prior organ transplant and immunosuppression appear to be risk factors for infection. As for *E. chaffeensis*, the tick vector for *E. ewingii* is *A. americanum*, and the expected ecological cycle and area of geographic risk are thought to be similar. To date, the organism has been identified only through molecular assays and has not proven cultivable. Without the ability to culture the organism, serologic diagnosis for this agent remains limited, and current diagnostic techniques are limited to PCR. As a result, few cases of *E. ewingii* have been confirmed in the United States.

In 2009, an unusual *Ehrlichia* species was identified for the first time as a cause of illness among patients in Wisconsin and Minnesota. The finding was unexpected, as the states lie outside the expected geographic range of *A. americanum*, the expected tick vector for the other known *Ehrlichia* species in the United States. The organism shares 98% similarity to *E. muris*, an *Ehrlichia* species that has been reported as a cause of human illness in Russia but which had not been previously identified in the United States (Nefedova et al., 2008; Pritt et al., 2011). Following the identification of human cases, evidence of the *E. muris*-like (EML) agent was identified in *I. scapularis* ticks collected from the environment near case patients, as well as in white-footed mice (*Peromyscus leucopus*) (Pritt et al., 2011). Subsequent testing of archived ticks collected from Wisconsin during the 1990s showed that approximately 1% were infected with the EML agent during that time (Telford et al., 2011).

During 2009 to 2010, the EML agent was identified in 14 patients from Minnesota and Wisconsin. Patients experienced a clinical illness consistent with that of other ehrlichiosis and anaplasmosis patients, including fever, headache, and malaise; 3 (21%) had a history of prior organ transplant, and patients appeared responsive to doxycycline therapy (Pritt et al., 2011). In a subset of tested patients (n = 3), seroreactivity on *E. chaffeensis* assays was observed, as well as on assays for the EML agent; cross-reactivity on *A. phagocytophilum* serologic tests was not observed, although the numbers tested to date are low (Pritt et al., 2011). A recent increase in the number of patients from Minnesota and Wisconsin with reported positive serologic results on *E. chaffeensis* assays raises the interesting possibility that this organism may contribute more substantially to the regional burden of tick-borne rickettsial disease than is currently appreciated (Dahlgren et al., 2011a; CDC, unpublished).

GLOBAL IMPACT OF RICKETTSIAL INFECTIONS

The impact of rickettsial infections in other countries is a topic worthy of its own discussion. While regional geographic specificity does occur based on host and vector distributions, rickettsial pathogens truly have a global reach, and our knowledge regarding the scope of pathogen variation is expanding almost daily. In addition to having an important impact on the health of people worldwide, rickettsial pathogens in other countries pose a potential risk to U.S. travelers (Jensenius et al., 2004, 2009).

Spotted Fever Group Rickettsioses

Tick-borne SFGR species are broadly distributed throughout the world (Table 2). In Europe, *Rickettsia conorii* is the cause of Mediterranean spotted fever, and causes a moderate to severe human illness. *R. conorii* is transmitted by *R. sanguineus* ticks. In addition to fever, headache,

TABLE 2 Tick-borne spotted fever group rickettsioses around the world

Disease	Species	Geographic distribution	Clinical symptoms
Unnamed rickettsiosis	*Rickettsia aeschlimannii*	Africa, Mediterranean region	Fever, eschar, maculopapular rash
African tick bite fever	*R. africae*	Sub-Saharan Africa, West Indies	Fever, often multiple eschars, regional adenopathy
Queensland tick typhus	*R. australis*	Australia, Tasmania	Fever, eschar, regional adenopathy, rash on extremities
Mediterranean spotted fever (boutonneuse fever)	*R. conorii*[a]	Mediterranean region and Africa to Indian subcontinent	Fever, eschar (usually single), regional adenopathy, maculopapular rash on extremities
Far Eastern spotted fever	*Rickettsia heilongjiangensis*	Far East of Russia, northern China, eastern Asia	Fever, eschar, maculopapular rash, regional adenopathy
Flinders Island spotted fever, Thai tick typhus	*R. honei*	Australia, Thailand	Mild spotted fever; eschar and adenopathy are rare
Japanese spotted fever	*R. japonica*	Japan	Fever, eschar(s), regional adenopathy, rash on extremities
Unnamed rickettsiosis	*R. massiliae*	France, Italy, Argentina	Fever, maculopapular rash, necrotic eschar
RMSF, febre maculosa, São Paulo exanthematic typhus, Minas Gerais exanthematic typhus, Brazilian spotted fever	*R. rickettsii*	North, Central, and South America	Fever, headache, abdominal pain, maculopapular rash progressing into papular or petechial rash (generally originating on extremities)
North Asian tick typhus, Siberian tick typhus	*Rickettisa sibirica*	Broadly distributed through northern Asia	Fever, eschar(s), regional adenopathy, maculopapular rash
Lymphangitis-associated rickettsiosis	*R. sibirica* subsp. *mongolotimonae*	Southern France, Portugal, China, Sub-Saharan Africa	Fever, multiple eschars, regional adenopathy and lymphangitis, maculopapular rash
Tick-borne lymphadenopathy; *Dermacentor*-borne necrosis, erythema, and lymphadenopathy	*R. slovaca*, *R. raoultii*	Southern and Eastern Europe, Asia	Necrosis erythema, cervical lymphadenopathy and enlarged lymph nodes, rare maculopapular rash

[a]Includes four different subspecies that are the etiologic agents of boutonneuse fever and Mediterranean tick fever in southern Europe and Africa (*R. conorii* subsp. *conorii*), Indian tick typhus in South Asia (*R. conorii* subsp. *indica*), Israeli tick typhus in southern Europe and the Middle East (*R. conorii* subsp. *israelensis*), and Astrakhan spotted fever in the North Caspian region of Russia (*R. conorii* subsp. *caspia*).

myalgia, and rash, the disease often presents with inoculation eschars at the site of tick attachment (Cascio and Iaria, 2006; Segura-Porta et al., 1989). Other European SFGR species known to be human pathogens include *Rickettsia slovaca*, *Rickettsia sibirica* subsp. *mongolitimonae*, *Rickettsia raoultii*, and *Rickettsia massiliae* (Brouqui et al., 2007). African tick bite fever, caused by *Rickettsia africae*, appears broadly distributed in Africa and is a commonly reported eschar-associated illness reported in travelers returning from southern Africa (Jensenius et al., 2004, 2009). In Australia, *Rickettsia australis* and *Rickettsia honei* are responsible for Queensland tick typhus and Flinders Island spotted fever, respectively, while in Japan, *Rickettsia japonica* is the cause of Japanese spotted fever (Graves and Stenos, 2009; Mahara, 2006). In addition to sporadic case reports in the literature, surveys suggest widespread exposure to SFGR species in many parts of the world, including seroprevalence of 4% in the Canary Islands of Spain, 14% in South Korea, and 17% in Zambia (Bolaños-Rivero et al., 2011; Jang et al., 2005; Okabayashi et al., 1999).

R. rickettsii also occurs throughout Central and South America, where it has been described by a variety of names, including São Paulo exanthematic typhus and Brazilian spotted fever. *R. rickettsii* is found in a variety of ticks, including *R. sanguineus* throughout its area of distribution, but also *Amblyomma cajennense* (the Cayenne tick) in Central and South America and *Amblyomma aureolatum* in South America (Labruna, 2009). In Mexico, urban infestations of *R. sanguineus* and large stray dog populations were responsible for a large urban outbreak of RMSF that was reported along the U.S.-Mexico border in Mexicali during 2009 (Sanchez et al., 2009).

R. parkeri rickettsiosis has also been documented recently in patients infected in the Paraná Delta near the city of Buenos Aires, Argentina (Romer et al., 2011), where the disease is transmitted by *Amblyomma triste* ticks (Nava et al., 2008). *R. parkeri* has also been found in specimens of *A. triste* collected from Uruguay and Brazil (Silveira et al., 2007; Venzal et al., 2004), and cases of an eschar-associated illness that is likely caused by *R. parkeri* have been described from Uruguay since 1990 (Conti-Díaz et al., 2009). Recently, a *Rickettsia* species closely related to *R. parkeri* has been reported as a cause of an eschar-associated rickettsiosis in Brazilian patients from the states of São Paulo and Bahia (Silva et al., 2011; Spolidorio et al., 2010), where *Amblyomma ovale* represents a potential vector (Sabatini et al., 2010).

Confirmed cases of rickettsialpox are now recognized from several countries around the world including the Ukraine, Croatia, South Korea, and Mexico, making this disease one of the few spotted fever rickettsioses with a cosmopolitan distribution (Paddock and Eremeeva, 2007; Radulovic et al., 1996; Zavala-Castro et al., 2009a). *R. felis* also has a worldwide distribution, and has been reported from Europe, Asia, Australia, Africa, and the Americas (Reif at al., 2009).

Typhus Group Rickettsioses

Although not a contemporary public health concern in the United States, louse-borne epidemic typhus continues to occur in some parts of the world, primarily associated with activities involving intense human crowding and stress. Epidemic typhus is endemic throughout sub-Saharan Africa (Parola, 2006), and a large outbreak estimated to have affected over 100,000 persons was reported in Burundi in 1997 (Bise and Coninx, 1997; Raoult et al., 1997, 1998). Sporadic cases are also reported in parts of Russia and Peru (Raoult et al., 1999; Tarasevich et al., 1998). *R. prowazekii* has a broad potential for distribution, in part because of the long-term interepidemic survival of the organism in some individuals and the potential for recrudescence in the form of Brill-Zinsser disease.

Murine typhus remains a risk in many parts of the world, as its principal mammalian reservoirs, *Rattus* spp., have a worldwide distribution. Despite the presumption of widespread endemicity in many parts of the world, especially Africa and Indonesia, most recent case reports focus on sporadic cases in returning travelers rather than outbreak-associated infections in obviously enzootic areas (Angelakis et al., 2010; Schulze et al., 2011; Silva and Papaiordanou, 2004; Takeshita et al., 2010). Seroprevalence studies, however, suggest that exposure to or infection with *R. typhi* is more common than is currently reported, with background seropositivity to *R. typhi* ranging from 3 to 9% in Spain and 5% in Zambia to 32% in South Korea and 35% in Indonesia (Bolaños-Rivero et al., 2011; Jang et al., 2005; Kaabia and Letaief, 2009; Nogueras et al., 2006; Okabayashi et al., 1999; Richards et al., 1997).

Scrub Typhus

Scrub typhus, or tsutsugamushi disease, is a mite-borne rickettsiosis endemic across a huge expanse of the Asia-Pacific Rim involving more than 5 million square miles, from Afghanistan to China, including Japan, the islands of the southwestern Pacific, and northern Australia (Kelly et al., 2009). The public health impact is greatest among agricultural laborers in the rural Asian tropics, with estimates of >1 million new cases annually and >1 billion persons at risk, making it the most globally important rickettsial disease in scope and magnitude (Silpapojakul, 1997; Rosenburg, 1997; Watt and Parola, 2003). The ecological and epidemiological dynamics of this disease across this vast area are reflected by its recent emergence in regions of northern China, where scrub typhus has never been previously described (Zhang et al., 2010); its reemergence in the Republic of the Maldives in 2002, where it was last identified in the mid-1940s; and its resurgence in many countries, including India, Japan, and South Korea, where a more than twofold increase in reported cases occurred during 2001 to 2006, especially among urban dwellers and nonfarmers, persons previously considered at lower risk for disease (Mathai et al., 2003; Lewis et al., 2003; Kweon et al., 2009).

Scrub typhus is caused by infection with *Orientia tsutsugamushi* and is transmitted to humans by the larval stages of at least 10 species of blood-feeding trombiculid mites of the genus *Leptotrombidium*. Mites transmit the infection transstadially and transovarially and are considered the principal reservoir of *O. tsutsugamushi*. Commensal and wild rodents influence the prevalence and distribution of scrub typhus by supporting populations of infective mites, but do serve as reservoirs; humans are infected when they encroach on mite-infested habitats. The various strains of *O. tsutsugamushi* demonstrate extensive phenotypic and genotypic diversity, and different strains often predominate among specific geographic regions. Strains phylogenetically similar to the Karp isolate comprise ~40% of all genotyped isolates. Three other genogroups (Kato-related, Kawasaki-like, and TA763-like) make up ~30% of the strains, while isolates genetically similar to the Gilliam strain comprise only 5% (Kelly et al., 2009). Scrub typhus is a potentially lethal infection, and in the pre-antibiotic era case fatality rates ranged from 35 to 50% (Kelly et al., 2009). Because many of the signs and symptoms of scrub typhus are nonspecific, and because diagnostic assays are generally unavailable in the rural tropics, the majority of cases go undiagnosed. Nonetheless, longitudinal studies in several countries where the disease is endemic reveal an enormous infectious burden. By example, 23% of all febrile illnesses diagnosed at one Malaysian hospital were scrub typhus, and *O. tsutsugamushi* was identified as the most common infectious cause of fever of unknown origin in a rural area of northern Thailand (Brown et al., 1976; Watt et al., 1996).

Ehrlichiosis and Anaplasmosis

Confirmed cases of ehrlichiosis have been described in Central and South America, and infections with *E. chaffeensis*, *E. canis*, and other *Ehrlichia* spp. have been reported in patients from Venezuela, Argentina, and Mexico (Calic et al., 2004; da Costa et al., 2006; Gongóra-Biachi et al., 1999; Martínez et al., 2008; Perez et al., 1996; Ripoll et al., 1999). Serologic evaluations of healthy patients from rural Brazil, Peru, and Argentina suggest a background seroprevalence of *E. chaffeensis* of 11, 13, and 14%, respectively, which is similar to background seroprevalence in highly endemic areas of the United States (da Costa et al., 2005; Moro et al., 2009; Ripoll et al., 1999). *E. chaffeensis* has been reported as a cause of undifferentiated febrile illness in patients from Cameroon (Ndip et al., 2009), and human infection with *E. muris* has also been reported from Russia (Nefedova et al., 2008). It seems likely that given the recent recognition of the EML agent in the northern United States, the recognition of new *Ehrlichia* spp. as human pathogens will continue to occur and challenge our understanding of this disease worldwide.

A. phagocytophilum has long been recognized as an important livestock pathogen in Europe, where it causes a disease of ruminants known as tick-borne fever that is spread by *Ixodes ricinus* ticks (Stuen, 2007). However, not all strains of *A. phagocytophilum* appear pathogenic in humans, and until recently, human illness attributable to infection with *A. phagocytophilum* had not been widely reported in Europe. In recent years, however, human infections with *A. phagocytophilum* have been reported in Austria and Slovakia (Novakova et al., 2010; Vogl et al., 2010). Serologic evaluation of persons from Poland, Italy, and Slovakia yielded background seroprevalence rates of 4, 6, and 7%, respectively, suggesting that human exposure to and possibly infection with *A. phagocytophilum* are more widespread than is currently appreciated (Grzeszczuk, 2006; Kalinova et al., 2009; Lillini et al., 2006).

There is also increasing interest in *A. phagocytophilum* as an emerging disease in China, which was bolstered by a 2008 publication suggesting nosocomial and person-to-person transmission of *A. phagocytophilum* (Krause and Wormser, 2008; Zhang et al., 2008). Subsequent investigations have suggested that this was more likely due to a tick-borne bunyavirus (Yu et al., 2011). Nonetheless, epidemiological evidence suggests that *A. phagocytophilum* is common in central and southeastern China, where background seroprevalence rates have been reported to be as high as 20% in persons at high risk for exposure to ticks and animals (Zhang et al., 2009).

TOOLS FOR PUBLIC HEALTH INVESTIGATIONS OF RICKETTSIAL DISEASES

Tick-borne diseases, and rickettsial infections in particular, often receive little attention against a background of competing public health interests. Endemic circulation of the organisms in local arthropod populations may result in sporadic reports of human disease that require no special action aside from education of the general public and health care providers. Furthermore, endemic disease patterns pose a diagnostic challenge for clinicians, who must interpret nonspecific clinical signs and correctly deduce and treat for the possibility of a rickettsial etiology—no easy task, given the rarity of these infections and the confusing array of differential diagnoses that must accompany a diagnosis of rickettsial infection.

Recognizing and responding to outbreaks, particularly when these occur in areas where a disease is not expected or in new human populations, poses even more of a challenge for local public health officials. Creating effective prevention and control programs in the face of outbreaks can be difficult for these diseases, as they require an understanding of not only human disease but circulation of the pathogen in animal reservoirs and arthropod vectors. These factors, combined with the potential lethality of some rickettsial infections, make rickettsial diseases particularly challenging public health issues.

Human Surveillance

Surveillance, whether passive or active, is a vital tool in the public health arsenal. The goal of surveillance is to define the extent and magnitude of disease in populations, whether they be human, domestic animal, wildlife, or arthropod vector. In the United States, passive surveillance for human infections forms an important cornerstone of the rickettsial public health effort (Fig. 18). SFGR infections, ehrlichiosis, and anaplasmosis are all considered

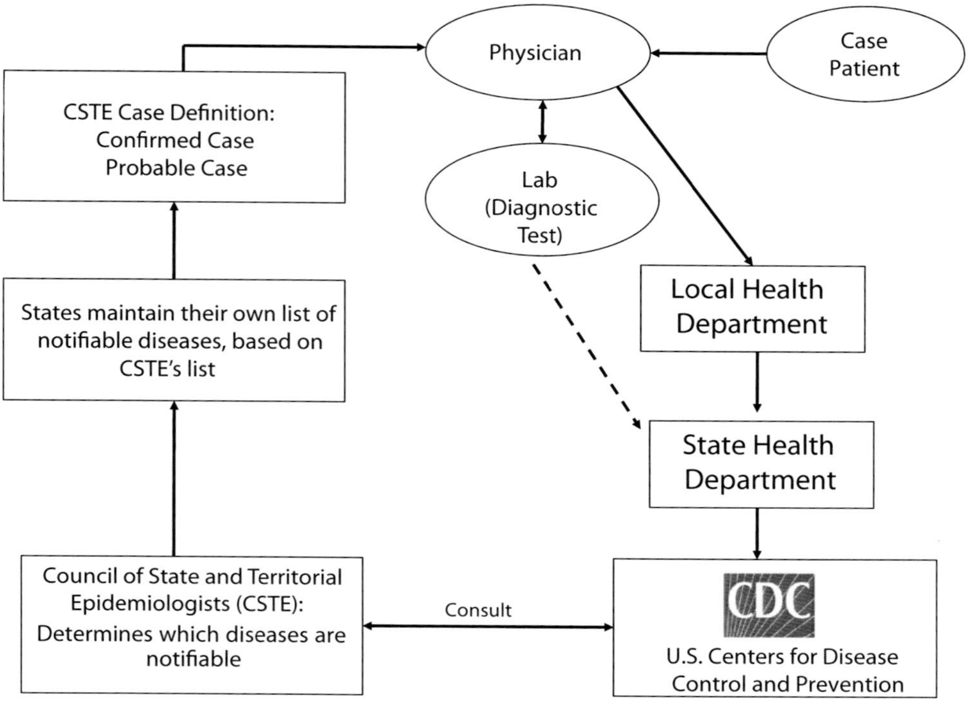

FIGURE 18 Surveillance for notifiable diseases in the United States.
doi:10.1128/9781555817336.ch2f18

notifiable conditions in the United States, as determined through the Council of State and Territorial Epidemiologists, the governing body that determines which diseases are included in the National Notifiable Diseases Surveillance System (NNDSS) (http://www.cdc.gov/osels/ph_surveillance/nndss/nndsshis.htm), and also determines standardized case definitions for case classification. Health care providers are tasked with submitting reports of lab-diagnosed cases of diseases included in the NNDSS to local public health officials, who in turn report these cases to national public health officials at the CDC.

The surveillance data collected through the NNDSS are used to determine estimates of disease incidence and the demographics of cases. Supplemental data on some cases are also collected through paper-based Case Report Forms, and these data are used to estimate case fatality rates and other pertinent demographic patterns of disease. In some states, mandatory reporting of laboratory results also contributes to surveillance data, although this requirement is not uniformly applied across all U.S. states.

Surveillance data can also be used to predict trends and to identify aberrations that may point to an outbreak or change in disease ecology. Unfortunately, as collected in the United States, these data are subject to several limitations. Data accuracy depends heavily on physician recognition and reporting of a suspected infection, and many patients are treated without pursuit of diagnostic testing that is required to confirm a rickettsial infection. There are multiple complicated steps in the reporting algorithm (Fig. 18), and a failure at any of these points affects the accuracy of the system. Conclusions regarding disease burden estimates calculated using NNDSS data must be therefore interpreted with an appropriate understanding of how data are collected and the limitations of the system.

Another factor that influences the usefulness of national surveillance is diagnostic accuracy.

Antibodies directed against many SFGR species exhibit strong cross-reactivity, and some patients diagnosed with RMSF may in some cases be infected with other pathogenic SFGR species circulating in the United States, such as *R. parkeri* and 364D (Openshaw et al., 2010; Parola et al., 2009; Raoult and Paddock, 2005). The clinical severity and likelihood of death are likely different between *R. rickettsii* and other SFGR species; thus, calculations of case fatality rates and geographic incidence, as portrayed by national surveillance systems based largely on assessment of a single serum sample, are likely to be biased (Openshaw et al., 2010).

While the accuracy of diagnostic test results, particularly those based on single serologic assays, is difficult to interpret (as discussed in chapter 1 of this text), the scope of rickettsial infections in the United States as presented by the NNDSS is nonetheless believed to underrepresent the true burden of infection. There are few published studies utilizing an active surveillance approach for assessment of rickettsial disease burden, but those that have been conducted suggest that rickettsial infections are underreported (Belongia et al., 2001; Olano et al., 2003). In a period of limited financial resources and waning support for state and local public health programs directed at vector-borne diseases, targeted active surveillance systems that more accurately capture cases may provide a more useful measure of disease burden and severity than passive surveillance systems (Childs and Paddock, 2002).

Animal and Vector Surveillance

In contrast to those diseases that only affect humans, rickettsial diseases cannot be adequately studied or controlled without an understanding of the influences and dynamics of infection in animal reservoirs and arthropod hosts. Surveillance systems utilizing animals and vectors can be useful public health tools to assess risks of human infection in a given geographic area.

Because dogs are highly susceptible to infection with *R. rickettsii* and may sustain a higher tick load than humans, they can in some circumstances be used as sentinels to evaluate human public health risks for RMSF. In eastern Arizona, where RMSF transmission occurs via *R. sanguineus*, examination of antibodies to SFGR in dogs has proven useful to evaluate regions of human risk. In the outbreak area, assessment of canine sera collected during 2003 and 2004 revealed an antibody prevalence of 70% among dogs from areas where human infections had been reported, and 57% among dogs from a community where human cases had not yet been recognized but which experienced human infections less than 1 year later (Demma et al., 2006b; CDC, unpublished). When banked sera that had been collected from dogs in this region during 1996 were evaluated, a much lower seroprevalence (5%) was found, suggesting a recent emergence (Demma et al., 2006b). Examination of canine sera from dogs collected in counties contiguous with but outside the outbreak zone during 2005 and 2006 yielded a seroprevalence of only 6%, similar to the preoutbreak findings in the outbreak zone, suggesting that human risk was still low in these areas (McQuiston et al., 2011). Indeed, no human cases have been reported from communities immediately surrounding the established outbreak region during the period from 2003 to 2011 (CDC, unpublished).

During 2009 and 2010, two human cases of RMSF residing in a single household were reported south of Phoenix, Arizona, outside of the previously recognized outbreak area region in eastern Arizona. Translocation of an infected dog from the outbreak area was suspected to be a factor in the emergence of RMSF in this new area (Baty et al., 2011). A targeted seroprevalence study of dogs from this new area suggested that the outbreak was contained to a single community, as no dogs outside that community were found to be SFGR antibody positive. Use of dogs as sentinels for human risk was very useful in this instance, because it permitted public health officials in the newly affected area to rapidly evaluate the extent of geographic spread of the organism and to prioritize prevention and control measures within the affected community (Baty et al., 2011).

Outbreak Response

Perhaps more concerning than the sporadic endemic cases of rickettsial diseases captured by national surveillance systems is the fact that, under the right circumstances, these pathogens can also cause epidemics. While only *R. prowazekii* carries a true potential for human-to-human transmission, common environmental exposures can result in efficient transmission among large groups of people. In some circumstances the concurrent appearance of signs and symptoms in family members or close contacts leads health care providers to erroneously suspect a communicable disease rather than a tick-borne illness (Centers for Disease Control and Prevention, 2004; Jones et al., 1999). Physician vigilance, therefore, is an important component of identifying outbreaks.

While rickettsial diseases like typhus have historically been associated with epidemic spread, there is a misperception that these diseases no longer pose such public health risks in contemporary settings. Epidemic typhus still has the potential to cause explosive outbreaks in vulnerable populations when conditions permit the circulation of body lice and poor public health infrastructure prevents the early identification and treatment of human cases, as was depicted in Burundi in 1998, when over 100,000 cases were suspected (Bise and Coninx, 1997; Raoult et al., 1997, 1998). The spread of murine typhus from southern Texas to central Texas was documented in 2008 with the recognition of an outbreak of murine typhus among residents of this previously nonenzootic area, resulting in over 50 suspected cases (Adjemian et al., 2010). Beginning in 2003, an outbreak of RMSF on tribal lands in eastern Arizona has resulted in over 200 human cases and 18 fatalities, in a defined geographic region of relatively limited human population density (Demma et al., 2005; McQuiston et al., 2011; CDC, unpublished). And in 2008 to 2009, a large urban outbreak of RMSF was reported in western Mexico, which reportedly caused over 1,000 cases (Sanchez et al., 2009).

Outbreaks carry the potential to quickly overwhelm available public health resources, and the interconnected roles of wildlife, domestic animals, and arthropods, as well as environmental components and the unpredictability of human behavior, pose distinct challenges for public health investigations and the eventual goal of prevention and control. Outbreak response typically focuses on identifying human cases and assessing the magnitude of the outbreak. Tools such as medical chart review and interviews of potentially exposed persons can help identify cases missed through traditional surveillance mechanisms. Public health investigations typically include descriptive epidemiology, which is characterizing the outbreak in place, time, and magnitude and determining indices of measurement such as attack rate. In addition, statistical measures of association can be explored through analytic epidemiology. A common tool is the case-control study, which includes comparing persons with the disease to persons without the disease in order to identify risk factors for infection.

In some cases, outbreak investigations should also extend to the study of disease ecology, in order to better frame appropriate public health messages. For example, during the early investigations into the emergence of RMSF in eastern Arizona in 2003 to 2004, environmental assessments were included in the initial outbreak investigation and demonstrated an absence of *Dermacentor* spp. ticks. These findings prompted a closer look at *R. sanguineus*, which was found in large numbers in the affected community, and led to the first identification of this tick species as a vector for *R. rickettsii* in the United States (Demma et al., 2005). Similarly, because *R. typhi* circulates in multiple reservoirs and with different ecological cycles, environmental assessments were conducted to assess whether rodents or opossums were the more likely animal reservoir during the 2008 murine typhus outbreak in Austin, Texas (Adjemian et al., 2010). In this outbreak, a high seroprevalence among opossums, along with a low number of rats trapped within the outbreak neighborhoods, pointed to a likely suburban cycle of *R. typhi* involving oppossums, cats, and cat fleas. The environmental investigation allowed public health officials to craft appropriate public health educational messages for local residents.

Prevention and Control

With most rickettsial diseases, education of the general public regarding health risks associated with arthropod exposure is a common public health intervention, as is educating health care providers regarding common signs and symptoms and necessary treatment with doxycycline. Education can be accomplished through health alerts sent directly to local residents or health care providers or through information distributed via Web or media. As but one example, ProMed-mail (http://www.promedmail.org/pls/apex/f?p=2400:1000) is a free-access global electronic reporting system for outbreaks that serves as both an early alert and educational resource for emerging infectious diseases. Although membership is restricted to public health professionals, Epi-X (http://www.cdc.gov/epix/) is another Web-based system that provides early notification, preliminary surveillance data, and important educational information on disease outbreaks.

In addition to education and early notification, interventions sometimes focus on control of source arthropods or animals in the environment. With *Dermacentor* spp. ticks, environmental control is difficult, given the ticks' usual location on the perimeter of woods and grasslands and the influence of wildlife as reservoirs and blood meal hosts. In eastern Arizona, however, the RMSF outbreak associated with *R. sanguineus* has challenged traditional paradigms of RMSF prevention. With *R. sanguineus*, the tick's predilection for dogs as blood meal hosts and the local concentration of ticks in the peridomestic environment has offered a unique intervention opportunity by treating dogs and households in the community with acaracides. In addition to education, intervention efforts in eastern Arizona have relied heavily on methods to reduce tick population numbers (CDC, unpublished). A missed opportunity, however, is the need to control the dog population in the affected communities, which currently lack established animal control programs. Limiting the number of stray and free-roaming dogs will reduce available blood meals for ticks and serve to also limit tick populations. In addition, limiting the number of stray and free-roaming dogs would help decrease movement of infected ticks into new areas and households. Unfortunately, local financial concerns have prevented this important step to date.

Developing effective prevention and control programs is an effort best undertaken in conjunction with partners. Close cooperation between local, state, and federal health partners is important, but so is including officials from agriculture and vector-borne disease control programs. Partners from academia are also vital to consider when investigating and responding to outbreaks, as they may bring a high degree of technical expertise on rickettsial pathogens lacking in traditional public health responses.

CONCLUSIONS

Recently observed increases in the reported incidence of rickettsial infections in the United States are a cause for renewed concern, as is the continued potential for fatal outcome associated with these infections. Remarkable scientific discoveries during the last several decades have expanded considerably the known spectrum of rickettsial diseases in the United States and around the world, and the incidence rates of these infections continue to climb (Fig. 3). Many of these agents were identified by extraordinary advances in molecular technology; however, epidemiologic tools for identifying cases and estimating incidence have not undergone similar transformative changes. Paradoxically, the discoveries of new pathogens made possible by contemporary diagnostic methods have also cast suspicion on certain aspects of the distribution, frequency, and clinical heterogeneity of some older, historically recognized tick-borne diseases (Paddock, 2009). For example, the recognition of *R. parkeri* and *Rickettsia* 364D as human pathogens has occurred only during the last decade, and there is increasing speculation that other *Rickettsia* spp. historically considered as nonpathogenic, such as *R. amblyommii*, contribute to the background of SFGR antibodies detected among the U.S. population and may also account for some proportion of mild or subclinical illness historically attributed to RMSF (Apperson et al., 2008; Stromdahl et al., 2011). In essence, the pace of pathogen discovery has eclipsed fundamental epidemiologic knowledge and the routine diagnostic capabilities for many of the diseases caused by these agents.

There are many practical reasons to explain the scarcity of surveillance systems for rickettsial diseases around the world, including perceived need, long-term interest, and funding necessary to sustain these programs. These factors affect the longevity necessary for fine-tuning of case definitions and collection instruments, and ultimately, meaningful interpretations of the collected data; however, as more unique pathogens are identified, distinctions among the diseases caused by these agents may become increasingly blurred. In general, the effort required to discover a new disease is far less than the subsequent endeavors required to establish and maintain long-lasting and accurate surveillance for that disease (Paddock and Telford, 2011). These isssues will be magnified in ecologically diverse and heavily populated regions of the globe such as India and the African continent, where rickettsioses have been poorly studied, yet the scope and magnitude of these infections are likely to be immense.

Public health officials have a duty to investigate rickettsial diseases and educate the public and medical community on contemporary best practices in the control and treatment of these infections. By example, many physicians are reluctant to prescribe doxycycline as treatment for rickettsioses in children (McElroy et al., 2010; O'Reilly et al., 2003), despite continued efforts by medical and public health experts to encourage the use of this effective and potentially life-saving drug as the antibiotic of choice for patients of all ages with these diseases (American Academy of Pediatrics, 2009; Centers for Disease Control and Prevention, 2006). Only through such diligence and a renewed appreciation of the public health impact of rickettsial diseases can further progress be achieved.

ACKNOWLEDGMENTS

Special thanks are extended to Scott Dahlgren for the language of some text, provision of several graphs and figures, and analytic assistance. Thanks are extended to Blaine Mathison for development of the life cycle images. Thanks are extended to Joanna Regan for critical review. The authors also thank the numerous state health department staff members who contribute case reports to the CDC that form the basis for national surveillance programs.

REFERENCES

Adams, J. R., E. T. Schmidtmann, and A. F. Azad. 1990. Infection of colonized cat fleas, *Ctenocephalides felis* (Bouché), with a rickettsia-like microorganism. *Am. J. Trop. Med. Hyg.* **43:**400–409.

Adjemian, J., S. Parks, K. McElroy, J. Campbell, M. E. Eremeeva, W. L. Nicholson, J. McQuiston, and J. Taylor. 2010. Murine typhus in Austin, Texas, USA, 2008. *Emerg. Infect. Dis.* **16:**412–417.

Adjemian, J. Z., J. W. Krebs, E. Mandel, and J. McQuiston. 2008. Spatial clustering by disease severity among reported Rocky Mountain spotted fever cases in the United States, 2001-2005. *Am. J. Trop. Med. Hyg.* **80:**72–77.

American Academy of Pediatrics. 2009. Rocky Mountain spotted fever, p. 573–575. *In* L. K. Pickering, C. J. Baker, D. W. Kimberlin, and S. S. Long (ed.), *Red Book: 2009 Report of the Committee on Infectious Diseases*, 28th ed. American Academy of Pediatrics, Elk Grove Village, IL.

Anderson, B. E., J. E. Dawson, S. C. Jones, and K. H. Wilson. 1991. *Ehrlichia chaffeensis*, a new species associated with human ehrlichiosis. *J. Clin. Microbiol.* **29:**2838–2842.

Angelakis, E., E. Botelho, C. Socolovschi, C. R. Sobas, C. Piketty, P. Parola, and D. Raoult. 2010. Murine typhus as a cause of fever in travelers from Tunisia and Mediterranean areas. *J. Travel Med.* **17:**310–315.

Apperson, C. S., B. Engber, W. L. Nicholson, D. G. Mead, J. Engel, M. J. Yabsley, K. Dail, J. Johnson, and D. W. Watson. 2008. Tick-borne diseases in North Carolina: is "*Rickettsia amblyommii*" a possible cause of rickettsiosis reported as Rocky Mountain spotted fever? *Vector Borne Zoonotic Dis.* **8:**597–606.

Bakken, J. S., J. S. Dumler, S. M. Chen, M. R. Eckman, L. L. Van Etta, and D. H. Walker. 1994. Human granulocytic ehrlichiosis in the upper Midwest United States. A new species emerging? *JAMA* **272:**212–218.

Baty, S. A., K. McElroy, C. Levy, W. Nicholson, and J. McQuiston. 2011. Emergence of a new focus of Rocky Mountain spotted fever—south central Arizona, 2009-2010, p. 121. *In Proc. 60th Annu. Epidemic Intelligence Serv. Conf.*, Atlanta, GA, 11 to 15 April 2011.

Bechah, Y., C. Capo, J. L. Mege, and D. Raoult. 2008. Epidemic typhus. *Lancet Infect. Dis.* **8:**417–426.

Belongia, E. A., C. M. Gale, K. D. Reed, P. D. Mitchell, M. Vandermause, M. F. Finkel, J. J. Kazmierczak, and J. P. Davis. 2001. Population-based incidence of human granulocytic ehrlichiosis in northwestern Wisconsin, 1997-1999. *J. Infect. Dis.* **184:**1470–1474.

Bise, G., and R. Coninx. 1997. Epidemic typhus in a prison in Burundi. *Trans. R. Soc. Trop. Med. Hyg.* **91:**133–134.

Bolaños-Rivero, M., E. Santana-Rodriguez, A. Angel-Moreno, M. Hernández-Cabrena, J. M. Limiñana-Canal, C. Carranza-Rodríguez, A. M. Martin-Sánchez, and J. L. Pérez-Arellano. 2011. Seroprevalence of *Rickettsia typhi* and *Rickettsia conorii* infections in the Canary Islands (Spain). *Int. J. Infect. Dis.* **15:**e481–e485.

Boostrom, A., M. S. Beier, J. A. Macaluso, K. R. Macaluso, D. Sprenger, J. Hayes, S. Radulovic, and A. F. Azad. 2002. Geographic association of *Rickettsia felis*-infected opossums with human murine typhus, Texas. *Emerg. Infect. Dis.* **8:**549–554.

Bouyer, D. H., J. Stenos, P. Croquet-Valdes, C. G. Moron, V. L. Popov, J. E. Zavala-Velasquez, L. D. Foil, D. R. Stothard, A. F. Azad, and D. H. Walker. 2001. *Rickettsia felis*: molecular characterization of a new member of the spotted fever group. *Int. J. Syst. Evol. Microbiol.* **51:**339–347.

Bozeman, F. M., S. A. Masiello, M. S. Williams, and B. L. Elisberg. 1975. Epidemic typhus rickettsiae isolated from flying squirrels. *Nature* **255:**545–547.

Bozeman, F. M., D. E. Sonenshine, M. S. Williams, D. P. Chadwick, D. M. Lauer, and B. L. Elisberg. 1981. Experimental infection of ectoparasite arthropods with *Rickettsia prowazekii* (GvF-16 strain) and transmission to flying squirrels. *Am. J. Trop. Med. Hyg.* **30:**253–263.

Breitschwerdt, E. B., B. C. Hegarty, R. G. Maggi, P. M. Lantos, D. M. Aslett, and J. M. Bradley. 2011. *Rickettsia rickettsii* transmission by a lone star tick, North Carolina. *Emerg. Infect. Dis.* **17:**873–875.

Brouqui, P., P. Parola, P. E. Fournier, and D. Raoult. 2007. Spotted fever rickettsioses in southern and eastern Europe. *FEMS Immunol. Med. Microbiol.* **49:**2–12.

Brown, G. W., D. M. Robinson, D. L. Huxsoll, T. S. Ng, and K. J. Lim. 1976. Scrub typhus: a common cause of illness in indigenous populations. *Trans. R. Soc. Trop. Med. Hyg.* **70:**444–448.

Buller, R. S., M. Arens, S. P. Hmiel, C. D. Paddock, J. W. Sumner, Y. Rikhisa, A. Unver, M. Gaudreault-Keener, F. A. Manian, A. M. Liddell, N. Schmulewitz, and G. A. Storch. 1999. *Ehrlichia ewingii*, a newly recognized agent of human ehrlichiosis. *N. Engl. J. Med.* **341:**148–155.

Calic, S. B., M. A. Galvao, F. Bacellar, C. M. Rocha, C. L. Mafra, R. C. Leite, and D. H. Walker. 2004. Human ehrlichioses in Brazil: first suspect cases. *Braz. J. Infect. Dis.* **8:**259–262.

Cascio, A., and C. Iaria. 2006. Epidemiology and clinical features of Mediterranean spotted fever in Italy. *Parassitologia* **48:**131–133.

Centers for Disease Control and Prevention. 2003. Murine typhus—Hawaii, 2002. *MMWR Morb. Mortal. Wkly. Rep.* **52:**1224–1226.

Centers for Disease Control and Prevention. 2004. Fatal cases of Rocky Mountain spotted fever in family clusters—three states, 2003. *MMWR Morb. Mortal. Wkly. Rep.* **53:**407–410.

Centers for Disease Control and Prevention. 2006. Diagnosis and management of tickborne rickettsial diseases: Rocky Mountain spotted fever, ehrlichiosis, and anaplasmosis—United States: a practical guide for physicians and other healthcare and public health professionals. *MMWR Recomm. Rep.* **55**(RR-4)**:**1–27.

Centers for Disease Control and Prevention. 2008. *Anaplasma phagocytophilum* transmitted through blood transfusion—Minnesota, 2007. *MMWR Morb. Mortal. Wkly. Rep.* **57:**1145–1148.

Centers for Disease Control and Prevention. 2009. Anaplasmosis and ehrlichiosis—Maine, 2008. *MMWR Morb. Mortal. Wkly. Rep.* **58:**1033–1036.

Chapman, A. S., S. M. Murphy, L. J. Demma, R. C. Holman, A. T. Curns, J. H. McQuiston, J. W. Krebs, and D. L. Swerdlow. 2006. Rocky Mountain spotted fever in the United States, 1997-2002. *Vector Borne Zoonotic Dis.* **6:**170–178.

Chapman, A. S., D. L. Swerdlow, V. M. Dato, A. D. Anderson, C. E. Moodie, C. Marriott, B. Amman, M. Hennessey, P. Fox, D. B. Green, E. Pegg, W. L. Nicholson, M. E. Eremeeva, and G. A. Dasch. 2009. Cluster of sylvatic epidemic typhus cases associated with flying squirrels, 2004-2006. *Emerg. Infect. Dis.* **15:**1005–1011.

Chen, S. M., J. S. Dumler, J. S. Bakken, and D. H. Walker. 1994. Identification of a granulocytotropic *Ehrlichia* species as the etiologic agent of human disease. *J. Clin. Microbiol.* **32:**589–595.

Childs, J. E., and C. D. Paddock. 2002. Passive surveillance as an instrument to identify risk factors for fatal Rocky Mountain spotted fever: is there more to learn? *Am. J. Trop. Med. Hyg.* **66:**450–457.

Choi, Y. J., W. J. Jang, J. S. Ryu, S. H. Lee, K. H. Park, H. S. Paik, Y. S. Koh, M. S. Choi, and I. S. Kim. 2005. Spotted fever group and typhus group rickettsioses in humans, South Korea. *Emerg. Infect. Dis.* **11:**237–244.

Civen, R., and V. Ngo. 2008. Murine typhus: an unrecognized suburban vectorborne disease. *Clin. Infect. Dis.* **46:**913–918.

Conti Díaz, I. A., J. Morales-Filho, R. C. Pacheco, and M. B. Labruna. 2009. Serological evidence of *Rickettsia parkeri* as the eiological agent of rickettsiosis in Uruguay. *Rev. Inst. Med. Trop. Sao Paulo* **51:**337–339.

Cragun, W. C., B. L. Bartlett, M. W. Ellis, A. Z. Hoover, S. K. Tyring, N. Mendoza, T. J. Vento, W. L. Nicholson, M. E. Eremeeva, J. P. Olano, R. P. Rapini, and C. D. Paddock. 2010. The expanding spectrum of eschar-associated rickettsioses in the United States. *Arch. Dermatol.* **146:**641–648.

da Costa, P. S., M. E. Brigatte, and D. B. Greco. 2005. Antibodies to *Rickettsia rickettsii*, *Rickettsia typhi*, *Coxiella burnetii*, *Bartonella henselae*, *Bartonella quintana*, and *Ehrlichia chaffeensis* among healthy population in Minas Gerais, Brasil. *Mem. Inst. Oswaldo Cruz* **100:**853–859.

da Costa, P. S., L. M. Valle, M. E. Brigatte, and D. B. Greco. 2006. More about human monocytic ehrlichiosis in Brazil: serological evidence of nine new cases. *Braz. J. Infect. Dis.* **10:**7–10.

Dahlgren, F. S., E. J. Mandel, J. W. Krebs, R. F. Massung, and J. H. McQuiston. 2011a. Increasing incidence of *Ehrlichia chaffeensis* and *Anaplasma phagocytophilum* in the United States: 2000 to 2007. *Am. J. Trop. Med. Hyg.* **85:**124–131.

Dahlgren, F. S., R. Moonesinghe, and J. H. McQuiston. 2011b. Race and rickettsiae: a United States perspective. *Am. J. Trop. Med. Hyg.* **85:**1124–1125.

Dalton, M. J., M. J. Clarke, R. C. Holman, J. W. Krebs, D. B. Fishbein, J. G. Olson, and J. E. Childs. 1995. National surveillance for Rocky Mountain spotted fever, 1981-1992: epidemiologic summary and evaluation of risk factors for fatal outcome. *Am. J. Trop. Med. Hyg.* **52:**405–413.

Dantas-Torres, F. 2007. Rocky Mountain spotted fever. *Lancet Infect. Dis.* **7:**724–732.

Dawson, J. E., B. E. Anderson, D. B. Fishbein, J. L. Sanchez, C. S. Goldsmith, K. H. Wilson, and C. W. Duntley. 1991. Isolation and characterization of an *Ehrlichia* sp. from a patient diagnosed with human ehrlichiosis. *J. Clin. Microbiol.* **29:**2741–2745.

Demma, L. J., R. C. Holman, C. A. Mikosz, A. T. Curns, D. L. Swerdlow, E. L. Paisano, and J. E. Cheek. 2006a. Rocky Mountain spotted fever hospitalizations among American Indians. *Am. J. Trop. Med. Hyg.* **75:**537–541.

Demma, L. J., M. Traeger, D. Blau, R. Gordon, B. Johnson, J. Dickson, R. Ethelbah, S. Piontkowski, C. Levy, W. L. Nicholson, C. Duncan, K. Heath, J. Cheek, D. L. Swerdlow, and J. H. McQuiston. 2006b. Serologic evidence for widespread exposure to *Rickettsia rickettsii* in eastern Arizona and recent emergence of Rocky Mountain spotted fever in this region. *Vector Borne Zoonotic Dis.* **6:**423–429. (Erratum, **7:**106, 2007.)

Demma, L. J., M. S. Traeger, W. L. Nicholson, C. D. Paddock, D. M. Blau, M. E. Eremeeva, G. A. Dasch, M. L. Levin, J. Singleton, Jr., S. R. Zaki, J. E. Cheek, D. L. Swerdlow, and J. H. McQuiston. 2005. Rocky Mountain spotted fever from an unexpected tick vector in Arizona. *N. Engl. J. Med.* **353:**587–594.

Dumler, J. S., J. S. Bakken, M. R. Eckman, L. L. Van Etta, S. M. Chen, and D. H. Walker. 1994. Human granulocytic ehrlichiosis: a new, potentially fatal tick-borne infection diagnosed by peripheral blood smear and PCR. *Lab. Invest.* **70:**126A.

Dumler, J. S., A. F. Barbet, C. P. Bekker, G. A. Dasch, G. H. Palmer, S. C. Ray, Y. Rikihisa, and F. R. Rurangirwa. 2001. Reorganization of genera in the families *Rickettsiaceae* and *Anaplasmataceae* in the order *Rickettsiales*: unification of some species of *Ehrlichia* with *Anaplasma*, *Cowdria* with *Ehrlichia* and *Ehrlichia* with *Neorickettsia*, descriptions of six new species combinations and designation of *Ehrlichia equi* and 'HGE agent' as subjective synonyms of *Ehrlichia phagocytophila*. *Int. J. Syst. Evol. Microbiol.* **51:**2145–2165.

Dumler, J. S., J. P. Taylor, and D. H. Walker. 1991. Clinical and laboratory features of murine typhus in south Texas, 1980 through 1987. *JAMA* **266:**1365–1370.

Elisberg, B. L., and F. M. Bozeman. 1969. Rickettsiae, p. 826–868. *In* E. H. Lennette and N. J. Schmidt (ed.), *Diagnostic Procedures for Viral and Rickettsial Infections*, 4th ed. American Public Health Association, Washington, DC.

Eremeeva, M. E., W. R. Warashina, M. M. Sturgeon, A. E. Buchholz, G. K. Olmstead, S. Y. Park, P. V. Effler, and S. E. Karpathy. 2008. *Rickettsia typhi* and *R. felis* in rat fleas (*Xenopsylla cheopis*), Oahu, Hawaii. *Emerg. Infect. Dis.* **14:**1613–1615.

Fergie, J. E., K. Purcell, and D. Wanat. 2000. Murine typhus in South Texas children. *Pediatr. Infect. Dis. J.* **19:**535–538.

Fornadel, C. M., X. Zhang, J. D. Smith, C. D. Paddock, J. R. Arias, and D. E. Norris. 2011. High rates of *Rickettsia parkeri* infection in Gulf Coast ticks (*Amblyomma maculatum*) and identification of "*Candidatus* Rickettsia andeanae" from Fairfax County, Virginia. *Vector Borne Zoonotic Dis.* doi:10.1089/vbz.2011.0654.

Galvão, M. A., J. E. Zavala-Velazquez, J. E. Zavala-Castro, C. L. Mafra, S. B. Calic, and D. H. Walker. 2006. *Rickettsia felis* in the Americas. *Ann. N. Y. Acad. Sci.* **1078:**156–158.

Gongóra-Biachi, R. A., J. Zavala-Velázquez, C. J. Castro-Sansores, and P. González-Martínez. 1999. First case of human ehrlichiosis in Mexico. *Emerg. Infect. Dis.* **5:**481.

Graf, P. C. F., J. P. Chretien, L. Ung, J. C. Gaydos, and A. L. Richards. 2008. Prevalence of seropositivity to spotted fever group rickettsiae and *Anaplasma phagocytophilum* in a large, demographically diverse U.S. sample. *Clin. Infect. Dis.* **46:**70–77.

Graves, S., and J. Stenos. 2009. Rickettsioses in Australia. *Ann. N. Y. Acad. Sci.* **1166:**151–156.

Green, C. R., D. Fishbein, and I. Gleiberman. 1990. Brill-Zinsser: still with us. *JAMA* **264:**1811–1812.

Greenberg, M., O. J. Pellitteri, and W. L. Jellison. 1947a. Rickettsialpox—a newly recognized rickettsial disease. III. Epidemiology. *Am. J. Public Health Nations Health* **37:**860–868.

Greenberg, M., O. Pellitteri, I. F. Klein, and R. J. Huebner. 1947b. Rickettsialpox—a newly recognized rickettsial disease. II. Clinical observations. *JAMA* **133:**901–906.

Grzeszczuk, A. 2006. *Anaplasma phagocyophilum* in *Ixodes ricinus* ticks and human granulocytic anaplasmosis seroprevalence among forestry rangers in Bialystok region. *Adv. Med. Sci.* **51:**282–286.

Harden, V. A. 1990. *Rocky Mountain Spotted Fever: History of a Twentieth-Century Disease.* Johns Hopkins University Press, Baltimore, MD.

Hattwick, M. A. W., A. H. Peters, M. B. Gregg, and B. Hanson. 1973. Surveillance of Rocky Mountain spotted fever. *JAMA* **225:**1338–1343.

Helmick, C. G., K. W. Bernard, and L. J. D'Angelo. 1984. Rocky Mountain spotted fever: clinical, laboratory, and epidemiological features of 262 cases. *J. Infect. Dis.* **150:**480–488.

Helmick, C. G., and W. G. Winkler. 1981. Epidemiology of Rocky Mountain spotted fever 1975-1979, p. 547–557. *In* W. Burgdorfer and R. L. Anacker (ed.), *Rickettsiae and Rickettsial Diseases.* Academic Press, New York, NY.

Hilton, E., J. DeVoti, J. L. Benach, M. L. Halluska, D. J. White, H. Paxton, and J. S. Dumler. 1999. Seroprevalence and seroconversion for tick-borne diseases in a high-risk population in the Northeast United States. *Am. J. Med.* **106:**404–409.

Jang, W. J., Y. J. Choi, J. H. Kim, K. D. Jung, J. S. Ryu, S. H. Lee, C. K. Yoo, H. S. Paik, M. S. Choi, K. H. Park, and I. S. Kim. 2005. Seroepidemiology of spotted fever group and typhus group rickettsioses in humans, South Korea. *Microbiol. Immunol.* **49:**17–24.

Jensenius, M., X. Davis, F. von Sonnenburg, E. Schwartz, J. S. Keystone, J. Leder, R. Lopéz-Véléz, E. Caumes, J. P. Cramer, L. Chen, P. Parola, and the GeoSentinel Surveillance Network. 2009. Multicenter GeoSentinel analysis of rickettsial diseases in international travelers, 1996-2008. *Emerg. Infect. Dis.* **15:**1791–1798.

Jensenius, M., P. E. Fournier, and D. Raoult. 2004. Rickettsioses and the international traveler. *Clin. Infect. Dis.* **39:**1493–1499.

Jones, T. F., A. S. Craig, C. D. Paddock, D. B. McKechnie, J. E. Childs, S. R. Zaki, and W. Schaffner. 1999. Family cluster of Rocky Mountain spotted fever. *Clin. Infect. Dis.* **28:**853–859.

Kaabia, N., and A. Letaief. 2009. Characterization of rickettsial diseases in a hospital-based population in central Tunisia. *Ann. N. Y. Acad. Sci.* **1166:**167–171.

Kalinova, Z., M. Halanova, L. Cislakova, Z. Sulinova, and P. Jarcuska. 2009. Occurrence of IgG antibodies to *Anaplasma phagocytophilum* in humans suspected of Lyme borreliosis in eastern Slovakia. *Ann. Agric. Environ. Med.* **16:**285–288.

Karpathy, S. E., E. K. Hayes, A. M. Williams, R. Hu, L. Krueger, S. Bennett, A. Tilzer, R. K. Velten, N. Kerr, W. Moore, and M. E. Eremeeva. 2009. Detection of *Rickettsia felis* and *Rickettsia typhi* in an area of California endemic for murine typhus. *Clin. Microbiol. Infect.* **15**(Suppl. 2):218–219.

Kelly, D. J., P. A. Fuerst, W. M. Ching, and A. L. Richards. 2009. Scrub typhus: the geographic distribution of phenotypic and genotypic variants of *Orientia tsutsugamushi*. *Clin. Infect. Dis.* **48**(Suppl. 3):S203–S230.

Koss, T., E. L. Carter, M. E. Grossman, D. N. Silvers, A. D. Rabinowitz, J. Singleton, Jr., S. R. Zaki, and C. D. Paddock. 2003. Increased detection of rickettsialpox in a New York City hospital following the anthrax outbreak of 2001: use of immunohistochemistry for the rapid confirmation of cases in an era of bioterrorism. *Arch. Dermatol.* **139:**1545–1552.

Krause, P. J., and G. P. Wormser. 2008. Nosocomial transmission of human granulocytic ehrlichiosis? *JAMA* **300:**2308–2309.

Kweon, S. S., J. S. Choi, H. S. Lim, J. R. Kim, K. Y. Kim, S. Y. Ryu, H. S. Yoo, and O. Park. 2009. Rapid increase of scrub typhus, South Korea, 2001-2006. *Emerg. Infect. Dis.* **15:**1127–1129.

Labruna, M. B. 2009. Ecology of *Rickettsia* in South America. *Ann. N. Y. Acad. Sci.* **1166:**156–166.

Lane, R. S., R. N. Philip, and E. A. Casper. 1981. Ecology of tick-borne agents in California.

II. Further observations on rickettsiae, p. 575–584. *In* W. Burgdorfer and R. L. Anacker (ed.), *Rickettsiae and Rickettsial Diseases*. Academic Press, New York, NY.

Lewis, M. D., A. A. Yousuf, K. Lerdthusnee, A. Razee, K. Chandranoi, and J. W. Jones. 2003. Scrub typhus reemergence in the Maldives. *Emerg. Infect. Dis.* **9:**1638–1641.

Lillini, E., G. Macri, G. Proietti, and M. Scarpulla. 2006. New findings on anaplasmosis caused by infection with *Anaplasma phagocytophilum*. *Ann. N. Y. Acad. Sci.* **1081:**360–370.

Maculaso, K. R., W. Pornwiroon, V. L. Popov, and L. D. Foil. 2008. Identification of *Rickettsia felis* in the salivary glands of cat fleas. *Vector Borne Zoonotic Dis.* **8:**391–396.

Maeda, K., N. Markowitz, R. C. Hawley, M. Ristic, D. Cox, and J. E. McDade. 1987. Human infection with *Ehrlichia canis*, a leukocytic rickettsia. *N. Engl. J. Med.* **316:**853–856.

Mahara, F. 2006. Rickettsioses in Japan and the Far East. *Ann. N. Y. Acad. Sci.* **1078:**60–73.

Manea, S. J., D. M. Sasaki, J. K. Ikeda, and P. P. Bruno. 2001. Clinical and epidemiological observations regarding the 1998 Kauai murine typhus outbreak. *Hawaii Med. J.* **60:**7–11.

Martínez, M. C., C. N. Gutiérrez, F. Monger, J. Ruiz, A. Watts, V. M. Mijares, M. G. Rojas, and F. J. Triana-Alonso. 2008. *Ehrlichia chaffeensis* in child, Venezuela. *Emerg. Infect. Dis.* **14:**519–520.

Mathai, E., J. M. Rolain, G. M. Varghese, O. C. Abraham, D. Mathai, M. Mathai, and D. Raoult. 2003. Outbreak of scrub typhus in southern India during the cooler months. *Ann. N. Y. Acad. Sci.* **990:**359–364.

McElroy, K. M., L. R. Carpenter, M. Lancaster, J. H. McQuiston, T. Ngo, F. S. Dahlgren, and J. Dunn. 2010. Practices regarding treatment of Rocky Mountain spotted fever among healthcare providers—Tennessee, 2009, p. 134. *In Proc. 59th Annu. Epidemic Intelligence Serv. Conf.*, Atlanta, GA, 19 to 23 April 2010.

McQuiston, J. H., J. E. Childs, M. E. Chamberland, and E. Tabor. 2000a. Transmission of tick-borne agents of disease by blood transfusions: a review of known and potential risks. *Transfusion* **40:**274–284.

McQuiston, J. H., M. A. Guerra, M. R. Watts, E. Lawaczek, C. Levy, W. L. Nicholson, J. Adjemian, and D. L. Swerdlow. 2011. Evidence of exposure to spotted fever group ricketttsiae among Arizona dogs outside a previously documented outbreak area. *Zoonoses Public Health* **58:**85–92.

McQuiston, J. H., R. C. Holman, A. V. Groom, S. F. Kaufman, J. E. Cheek, and J. E. Childs. 2000b. Incidence of Rocky Mountain spotted fever among American Indians in Oklahoma. *Public Health Rep.* **115:**469–475.

McQuiston, J. H., E. B. Knights, P. J. DeMartino, S. F. Paparello, W. L. Nicholson, J. Singleton, C. M. Brown, R. F. Massung, and J. C. Urbanowski. 2010. Brill-Zinsser disease in a patient following infection with sylvatic epidemic typhus associated with flying squirrels. *Clin. Infect. Dis.* **51:**712–715.

McQuiston, J. H., C. D. Paddock, R. C. Holman, and J. E. Childs. 1999. The human ehrlichioses in the United States. *Emerg. Infect. Dis.* **5:**635–642.

Mettille, F. C., K. F. Salata, K. J. Belanger, B. G. Casleton, and D. J. Kelly. 2000. Reducing the risk of transfusion-transmitted rickettsial disease by WBC filtration, using *Orientia tsutsugamushi* in a model system. *Transfusion* **40:**290–296.

Mohr, C. O., N. E. Good, and J. H. Schubert. 1953. Status of murine typhus infection in domestic rats in the United States, 1952, and relation to infestation by Oriental rat fleas. *Am. J. Public Health Nations Health* **43:**1514–1522.

Moro, P. L., J. Shah, O. Li, R. H. Gilman, N. Harris, and M. H. Moro. 2009. Short report: serologic evidence of human ehrlichiosis in Peru. *Am. J. Trop. Med. Hyg.* **80:**242–244.

Murray, E. S., G. Baehr, G. Shwartzman, R. A. Mandelbaum, N. Rosenthal, J. C. Doane, L. B. Weiss, S. Choen, and J. C. Snyder. 1950. Brill's disease. I. Clinical and laboratory diagnosis. *JAMA* **142:**1059–1066.

Nava, S., Y. Elshewany, M. E. Eremeeva, J. W. Sumner, M. Mastropaolo, and C. D. Paddock. 2008. *Rickettsia parkeri* in Argentina. *Emerg. Infect. Dis.* **14:**1894–1897.

Ndip, L. M., M. Labruna, R. N. Ndip, D. H. Walker, and J. W. McBride. 2009. Molecular and clinical evidence of *Ehrlichia chaffeensis* infection in Cameroonian patients with undifferentiated febrile illness. *Ann. Trop. Med. Parasitol.* **103:** 719–725.

Nefedova, V. V., E. I. Korenberg, I. V. Kovalevskii, N. B. Gorelova, and N. N. Vorob'eva. 2008. Microorganisms of the order Rickettsiales in taiga tick (*Ixodes persulcatus* Sch.) from the Pre-Ural region. *Vestn. Ross. Akad. Med. Nauk* **7:**47–50.

Nguyen-Hieu, T., G. Aboudharam, M. Signoli, C. Rigeade, M. Drancourt, and D. Raoult. 2010. Evidence of a louse-borne outbreak involving typhus in Douai, 1710-1712 during the war of Spanish succession. *PLoS One* **5:**e15405.

Nogueras, M. M., N. Cardenosa, I. Sanfeliu, T. Munoz, B. Font, and F. Segura. 2006. Evidence of infection in humans with *Rickettsia typhi* and *Rickettsia felis* in Catalonia in the northeast of Spain. *Ann. N. Y. Acad. Sci.* **1078:**159–161.

Novakova, M., B. Vichova, V. Majlathova, A. Lesnakova, M. Pochybova, and B. Petko. 2010. First case of granulocytic anaplasmosis from Slovakia. *Ann. Agric. Environ. Med.* **17:**173–175.

Okabayashi, T., F. Hasebe, K. L. Samui, A. S. Mweene, S. G. Pandey, T. Yanase, Y. Muramatsu, H. Ueno, and C. Morita. 1999. Prevalence of antibodies against spotted fever, murine typhus, and Q fever rickettsiae in humans living in Zambia. *Am. J. Trop. Med. Hyg.* **61:**70–72.

Olano, J. P., E. Masters, W. Hogrefe, and D. H. Walker. 2003. Human monocytotropic ehrlichiosis, Missouri. *Emerg. Infect. Dis.* **9:**1579–1586.

Openshaw, J. J., D. L. Swerdlow, J. W. Krebs, R. C. Holman, E. Mandel, A. Harvey, D. Haberling, R. F. Massung, and J. H. McQuiston. 2010. Rocky Mountain spotted fever in the United States, 2000–2007: interpreting contemporary increases in incidence. *Am. J. Trop. Med. Hyg.* **83:**174–182. (Erratum, **83:**729–730.)

O'Reilly, M., C. Paddock, B. Elchos, J. Goddard, J. Childs, and M. Currie. 2003. Physician knowledge of the diagnosis and management of Rocky Mountain spotted fever: Mississippi, 2002. *Ann. N. Y. Acad. Sci.* **990:**295–301.

Paddock, C. D. 2005. *Rickettsia parkeri* as a paradigm for multiple causes of tick-borne spotted fever in the Western Hemisphere. *Ann. N. Y. Acad. Sci.* **1063:**315–326.

Paddock, C. D. 2009. The science and fiction of emerging rickettsioses. *Ann. N. Y. Acad. Sci.* **1166:**133–143.

Paddock, C. D., and M. E. Eremeeva. 2007. Rickettsialpox, p. 63–86. *In* D. Raoult and P. Parola (ed.), *Rickettsial Diseases*. Informa Healthcare, New York, NY.

Paddock, C. D., R. W. Finley, C. S. Wright, H. N. Robinson, B. J. Schrodt, C. C. Lane, O. Ekenna, M. A. Blass, C. L. Tamminga, C. A. Ohl, S. L. McLellan, J. Goddard, R. C. Holman, J. J. Openshaw, J. W. Sumner, S. R. Zaki, and M. E. Eremeeva. 2008. *Rickettsia parkeri* rickettsiosis and its clinical distinction from Rocky Mountain spotted fever. *Clin. Infect. Dis.* **47:**1188–1196.

Paddock, C. D., P. E. Fournier, J. W. Sumner, J. Goddard, Y. Elshenawy, M. G. Metcalfe, A. D. Loftis, and A. Varela-Stokes. 2010. Isolation of *Rickettsia parkeri* and identification of a novel spotted fever group *Rickettsia* sp. from Gulf Coast ticks (*Amblyomma maculatum*) in the United States. *Appl. Environ. Microbiol.* **76:**2689–2696.

Paddock, C. D., J. W. Sumner, J. A. Comer, S. R. Zaki, C. S. Goldsmith, J. Goddard, S. L. McLellan, C. L. Tamminga, and C. A. Ohl. 2004. *Rickettsia parkeri*: a newly recognized cause of spotted fever rickettsiosis in the United States. *Clin. Infect. Dis.* **38:**805–811.

Paddock, C. D., and S. R. Telford III. 2011. Through a glass, darkly: the global incidence of tick-borne diseases, p. 221–266. *In* Institute of Medicine, *Critical Needs and Gaps in Understanding Prevention, Amelioration, and Resolution of Lyme and other Tick-Borne Diseases: the Short-Term and Long-Term Outcomes*. National Academies Press, Washington, DC.

Paddock, C. D., S. R. Zaki, T. Koss, J. Singleton, Jr., J. W. Sumner, J. A. Comer, M. E. Eremeeva, G. A. Dasch, B. Cherry, and J. E. Childs. 2003. Rickettsialpox in New York City: a persistent urban zoonosis. *Ann. N. Y. Acad. Sci.* **990:**36–44.

Parker, R. R., G. M. Kohls, G. W. Cox, and G. E. Davis. 1939. Observations on an infectious agent from *Amblyomma maculatum*. *Public Health Rep.* **54:**1482–1484.

Parola, P. 2006. Rickettsioses in sub-Saharan Africa. *Ann. N. Y. Acad. Sci.* **1078:**42–47.

Parola, P. 2011. *Rickettsia felis*: from a rare disease in the USA to a common cause of fever in sub-Saharan Africa. *Clin. Microbiol. Infect.* **17:**996–1000.

Parola, P., M. B. Labruna, and D. Raoult. 2009. Tick-borne rickettsioses in America: unanswered questions and emerging diseases. *Curr. Infect. Dis. Rep.* **11:**40–50.

Parola, P., C. D. Paddock, and D. Raoult. 2005. Tick-borne rickettsioses around the world: emerging diseases challenging old concepts. *Clin. Microbiol. Rev.* **18:**719–756.

Perez, M., M. Bodor, C. Zhang, Q. Xiong, and Y. Rikihisa. 2006. Human infection with *Ehrlichia canis* accompanied by clinical signs in Venezuela. *Ann. N. Y. Acad. Sci.* **1078:**110–117. (Erratum, **1212:**130, 2010.)

Perez, M., Y. Rikihisa, and B. Wen. 1996. *Ehrlichia canis*-like agent isolated from a man in Venezuela: antigenic and genetic characterization. *J. Clin. Microbiol.* **34:**2133–2139.

Philip, R. N., E. A. Casper, R. A. Ormsbee, M. G. Peacock, and W. Burgdorfer. 1976. Microimmunofluorescence test for the serological study of Rocky Mountain spotted fever and typhus. *J. Clin. Microbiol.* **3:**51–61.

Pier, G. B. 2008. On the greatly exaggerated reports of the death of infectious diseases. *Clin. Infect. Dis.* **47:**1113–1114.

Piranda, E. M., J. L. Faccini, A. Pinter, R. C. Pacheco, P. H. Cançado, and M. B. Labruna. 2011. Experimental infection of *Rhipicephalus sanguineus* ticks with the bacterium *Rickettsia rickettsii*, using experimentally infected dogs. *Vector Borne Zoonotic Dis.* **11:**29–36.

Portnoy, J., J. Mendelson, and B. Clecner. 1974. Brill-Zinsser disease: report of a case in Canada. *Can. Med. Assoc. J.* **111:**166.

Price, E. G. 1948. *Fighting Spotted Fever in the Rockies*. Naegele Printing Company, Helena, MT.

Price, W. H., H. Emerson, R. Blumberg, and S. Talmadge. 1958. Ecologic studies on the interepidemic survival of louse-borne epidemic typhus fever. *Am. J. Hyg.* **67:**154–178.

Pritt, B. S., L. M. Sloan, D. K. H. Johnson, U. G. Munderloh, S. M. Paskewitz, K. M. McElroy, J. D. McFadden, M. J. Binnicker, D. F. Neitzel, G. Liu, W. L. Nicholson, C. M. Nelson, J. J. Franson, S. A. Martin, S. A. Cunningham, C. R. Steward, K. Bogumill, M. E. Bjorgaard, J. P. Davis, J. H. McQuiston, D. M. Warshauer, M. P. Wilhelm, R. Patel, V. A. Trivedi, and M. E. Eremeeva. 2011. Emergence of a new pathogenic *Ehrlichia* species, Wisconsin and Minnesota, 2009. *N. Engl. J. Med.* **365:**422–429.

Radulovic, S., H. M. Feng, M. Morovic, B. Djelalija, V. Popov, P. Crocquet-Valdes, and D. H. Walker. 1996. Isolation of *Rickettsia akari* from a patient in a region where Mediterranean spotted fever is endemic. *Clin. Infect. Dis.* **22:**216–220.

Raoult, D., R. J. Birtles, M. Montoya, E. Perez, H. Tissot-Dupont, V. Roux, and H. Guerra. 1999. Survey of three bacterial louse-associated diseases among rural Andean communities in Peru: prevalence of epidemic typhus, trench fever, and relapsing fever. *Clin. Infect. Dis.* **29:**434–436.

Raoult, D., O. Dutour, L. Houhamdi, R. Jankauskas, P. E. Fournier, Y. Ardagna, M. Drancourt, M. Signoli, V. D. La, Y. Macia, and G. Aboudharam. 2006. Evidence for louse-transmitted diseases in soldiers of Napoleon's Grand Army in Vilnius. *J. Infect. Dis.* **193:**112–120.

Raoult, D., J. B. Ndihokubwayo, H. Tissot-Dupont, V. Roux, B. Faugere, R. Abegbinni, and R. J. Birtles. 1998. Outbreak of epidemic typhus associated with trench fever in Burundi. *Lancet* **352:**353–358.

Raoult, D., and C. D. Paddock. 2005. *Rickettsia parkeri* infection and other spotted fevers in the United States. *N. Engl. J. Med.* **353:**626–627.

Raoult, D., V. Roux, J. B. Ndihokubwayo, G. Bise, D. Baudon, G. Marte, and R. Birtles. 1997. Jail fever (epidemic typhus) outbreak in Burundi. *Emerg. Infect. Dis.* **3:**357–360.

Raoult, D., T. Woodward, and J. S. Dumler. 2004. The history of epidemic typhus. *Infect. Dis. Clin. North. Am.* **18:**127–140.

Reif, K. E., and K. R. Macaluso. 2009. Ecology of *Rickettsia felis*: a review. *J. Med. Entomol.* **46:**723–736.

Reilly, P. J., and R. W. Kalinske. 1980. Brill-Zinsser disease in North America. *West. J. Med.* **133:**338–340.

Reynolds, M. G., J. S. Krebs, J. A. Comer, J. W. Sumner, T. C. Rushton, C. E. Lopez, W. L. Nicholson, J. A. Rooney, S. E. Lance-Parker, J. H. McQuiston, C. D. Paddock, and J. E. Childs. 2003. Flying squirrel-associated typhus, United States. *Emerg. Infect. Dis.* **9:**1341–1343.

Rheingold, J. J., and C. L. Spurling. 1952. Chloramphenicol and aplastic anemia. *JAMA* **119:**1301–1304.

Richards, A. L., D. W. Soeatmadji, M. A. Widodo, T. W. Sardjono, B. Yanuwiadi, T. E. Hernowati, A. D. Baskoro, Roebiyoso, L. Hakim, M. Soendoro, E. Rahardjo, M. P. Putri, and J. M. Saragih. 1997. Seroepidemiologic evidence for murine and scrub typhus in Malang, Indonesia. *Am. J. Trop. Med. Hyg.* **57:**91–95.

Ripoll, C. M., C. E. Remondegui, G. Ordonez, R. Arazamendi, H. Fusaro, M. J. Hyman, C. D. Paddock, S. R. Zaki, J. G. Olson, and C. A. Santos-Buch. 1999. Evidence of rickettsial spotted fever and ehrlichial infections in a subtropical territory of Jujy, Argentina. *Am. J. Trop. Med. Hyg.* **61:**350–354.

Romer, Y., A. C. Seijo, F. Crudo, W. L. Nicholson, A. Varela-Stokes, R. R. Lash, and C. D. Paddock. 2011. *Rickettsia parkeri* rickettsiosis in Argentina. *Emerg. Infect. Dis.* **17:**1169–1173.

Rosenburg, R. 1997. Drug-resistant scrub typhus: paradigm and paradox. *Parasitol. Today* **13:**131–132.

Rucker, W. C. 1912. Rocky Mountain spotted fever. *Public Health Rep.* **27:**1465–1482.

Rudakov, N. V., S. N. Shpynov, I. E. Samoilenko, and M. A. Tankibaev. 2003. Ecology and epidemiology of spotted fever group rickettsiae and new data from their study in Russia and Kazakhstan. *Ann. N. Y. Acad. Sci.* **990:**12–24.

Sabatini, G. S., A. Pinter, F. A. Nieri-Bastos, A. Marcili, and M. B. Labruna. 2010. Survey of ticks (Acari: Ixodidae) and their *Rickettsia* in an Atlantic rain forest reserve in the State of São Paulo, Brazil. *J. Med. Entomol.* **47:**913–916.

Saini, R., J. C. Pui, and S. Burgin. 2004. Rickettsialpox: report of three cases and a review. *J. Am. Acad. Dermatol.* **51:**S65–S70.

Sanchez, R., C. Alpuche, H. Lopez-Gatell, C. Soria, J. Estrada, H. Olguin, M. A. Lezana, W. Nicholson, M. Eremeeva, CDC Infectious Disease Pathology Branch, G. Dasch, M. Fonseca, S. Montiel, S. Waterman, and J. McQuiston. 2009. *Rhipicephalus sanguineus*-associated Rocky Mountain spotted fever in Mexicali, Mexico: observations from an outbreak in 2008-2009, abstr. no. 75. *23rd Meet. Am. Soc. Rickettsiol.*, Hilton Head, SC, 15 to 18 August 2009.

Schriefer, M. E., J. B. Sacci, Jr., J. S. Dumler, M. G. Bullen, and A. F. Azad. 1994a. Identification of a novel rickettsial infection in a patient diagnosed with murine typhus. *J. Clin. Microbiol.* **32:**949–954.

Schriefer, M. E., J. B. Sacci, Jr., J. P. Taylor, J. A. Higgins, and A. F. Azad. 1994b. Murine typhus: updated roles of multiple urban components and a second typhus-like rickettsia. *J. Med. Entomol.* **31:**681–685.

Schulze, M. H., C. Keller, A. Muller, U. Ziegler, H. J. Langen, G. Hegasy, and A. Stich. 2011. *Rickettsia typhi* infection with interstitial pneumonia in a traveler treated with moxifloxacin. *J. Clin. Microbiol.* **49:**741–743.

Segura-Porta, F., B. Font-Creus, E. Espejo-Arenas, and F. Bella-Cueto. 1989. New trends in Mediterranean spotted fever. *Eur. J. Epidemiol.* **5:**438–443.

Shapiro, M. R., C. L. Fritz, K. Tait, C. D. Paddock, W. L. Nicholson, K. F. Abramowicz, S. E. Karpathy, G. A. Dasch, J. W. Sumner, P. V. Adem, J. J. Scott, K. A. Padgett, S. R. Zaki, and M. E. Eremeeva. 2010. *Rickettsia* 364D: a newly recognized cause of eschar-associated illness in California. *Clin. Infect. Dis.* **50:**541–548.

Silpapojakul, S. 1997. Scrub typhus in the western Pacific region. *Ann. Acad. Med. Singapore* **26:** 794–800.

Silva, L. J., and P. M. Papaiordanou. 2004. Murine (endemic) typhus in Brazil: case report and review. *Rev. Inst. Med. Trop. Sao Paulo* **46:**283–285.

Silva, N., M. E. Eremeeva, T. Rozental, G. S. Ribeiro, C. D. Paddock, E. A. Ramos, A. R. Favacho, M. G. Reis, G. A. Dasch, E. R. de Lemos, and A. I. Ko. 2011. Eschar-asociated spotted fever, Bahia, Brazil. *Emerg. Infect. Dis.* **17:** 275–278.

Silveira, I., R. C. Pacheco, M. P. J. Szabó, H. G. C. Ramos, and M. B. Labruna. 2007. *Rickettsia parkeri* in Brazil. *Emerg. Infect. Dis.* **13:**1111–1113.

Spolidorio, M. G., M. B. Labruna, E. Mantovani, P. E. Brandão, L. J. Richtzenhain, and N. H. Yoshinari. 2010. Novel spotted fever rickettsiosis, Brazil. *Emerg. Infect. Dis.* **16:**521–523.

Stein, A., R. Purgus, M. Olmer, and D. Raoult. 1999. Brill-Zinsser disease in France. *Lancet* **353:**1936.

Stromdahl, E. Y., J. Jiang, M. Vince, and A. L. Richards. 2011. Infrequency of *Rickettsia rickettsii* in *Dermacentor variabilis* removed from humans, with comments on the role of other human-biting ticks associated with spotted fever group rickettsiae in the United States. *Vector Borne Zoonotic Dis.* **11:**969–977.

Stuen, S. 2007. *Anaplasma phagocytophilum*—the most widespread tick-borne infection in animals in Europe. *Vet. Res. Commun.* **31**(Suppl. 1)**:**79–84.

Sumner, J. W., L. D. Durden, J. Goddard, E. Y. Stromdahl, K. L. Clark, W. K. Reeves, and C. D. Paddock. 2007. Gulf Coast ticks (*Amblyomma maculatum*) and *Rickettsia parkeri*, United States. *Emerg. Infect. Dis.* **13:**751–753.

Takeshita, N., K. Imoto, S. Ando, K. Yanagisawa, G. Ohji, Y. Kato, A. Sakata, N. Hosokawa, and T. Kishimoto. 2010. Murine typhus in two travelers returning from Bali, Indonesia: an underdiagnosed disease. *J. Travel Med.* **17:**356–358.

Tarasevich, I., E. Rydkina, and D. Raoult. 1998. Epidemic typhus in Russia. *Lancet* **352:**1151.

Telford, S. R., III, H. K. Goethert, and J. A. Cunningham. 2011. Prevalence of *Ehrlichia muris* in Wisconsin deer ticks collected during the mid 1990's. *Open Microbiol. J.* **5:**18–20.

Texas Department of State Health Services. 2008. Murine typhus in Texas. *EpiLink* **65:**1–3.

Tijsse-Klasen, E., M. Fonville, F. Gassner, A. M. Nijhof, E. K. Hovius, F. Jongejan, W. Takken, J. R. Reimerink, P. A. Overgaauw, and H. Sprong. 2011. Absence of zoonotic *Bartonella* species in questing ticks: first detection of *Bartonella clarridgeiae* and *Rickettsia felis* in cat fleas in the Netherlands. *Parasit. Vectors* **4:**61.

Treadwell, T. A., R. C. Holman, M. J. Clarke, J. W. Krebs, C. D. Paddock, and J. E. Childs. 2000. Rocky Mountain spotted fever in the United States, 1993-1996. *Am. J. Trop. Med. Hyg.* **63:** 21–26.

Varela-Stokes, A. S., C. D. Paddock, B. Engber, and M. Toliver. 2011. *Rickettsia parkeri* in*Amblyomma maculatum* ticks, North Carolina, USA, 2009–2010. *Emerg. Infect. Dis.* **17:**2350–2353.

Venzal, J. M., A. Portillo, A. Estrada-Peña, O. Castro, P. A. Cabrera, and J. A. Oteo. 2004. *Rickettsia parkeri* in *Amblyomma triste* from Uruguay. *Emerg. Infect. Dis.* **10:**1493–1495.

Virchow, R. 1985. Report on the typhus epidemic in Upper Silesia, p. 205–319. *In* L. J. Rather (ed.), *Rudolph Virchow: Collected Essays on Public Health and Epidemiology*, vol. 2. Science History Publications, Canton, MA.

Vogl, U. M., E. Presterl, G. Stanek, M. Ramharter, K. B. Gattringer, and W. Graninger. 2010. First described case of human granulocytic anaplasmosis in a patient in Eastern Austria. *Wien. Med. Wochenschr.* **160:**91–93.

Walker, D. H., H. K. Hawkins, and P. Hudson. 1983. Fulminant Rocky Mountain spotted fever. Its pathologic characteristics associated with glucose-6-phosphate dehydrogenase deficiency. *Arch. Pathol. Lab. Med.* **107:**121–125.

Watt, G., C. Chouriyagune, R. Ruangweerayud, P. Watcharapichat, D. Phulsuksombati, K. Jongsakul, P. Teja-Isavadharm, D. Bhodhidatta, K. D. Corcoran, G. A. Dasch, and D. Strickman. 1996. Scrub typhus infections poorly responsive to antibiotics in northern Thailand. *Lancet* **348:**86–89.

Watt, G., and P. Parola. 2003. Scrub typhus and tropical rickettsioses. *Curr. Opin. Infect. Dis.* **16:**429–436.

White, P. C., Jr. 1965. Murine typhus in the United States. *Mil. Med.* **130:**469–474.

Wiggers, R. J., M. C. Martin, and D. Bouyer. 2005. *Rickettsia felis* infection rates in an east Texas population. *Tex. Med.* **101:**56–58.

Wikswo, M. E., R. Hu, G. A. Dasch, L. Krueger, A. Arugay, K. Jones, B. Hess, S. Bennett, V. Kramer, and M. E. Eremeeva. 2008. Detection and identification of spotted fever group rickettsiae in *Dermacentor* species from southern California. *J. Med. Entomol.* **45:**509–516.

Williams, M., L. Izzard, S. R. Graves, J. Stenos, and J. J. Kelly. 2011. First probable Australian cases of human infection with *Rickettsia felis* (cat-flea typhus). *Med. J. Aust.* **194:**41–43.

Williams, S. G., J. B. Sacci, Jr., M. E. Schriefer, E. M. Andersen, K. K. Fujioka, F. J. Sorvillo, A. R. Barr, and A. F. Azad. 1992. Typhus and typhus-like rickettsiae associated with opossums and their fleas in Los Angeles County, California. *J. Clin. Microbiol.* **30:**1758–1762.

Wilson, L. B., and W. M. Chowning. 1904. Studies in pyroplasmosis hominis ('spotted fever' or 'tick fever' of the Rocky Mountains). *J. Infect. Dis.* **1:**31–57.

Witkop, C. J., Jr., and R. O. Wolf. 1963. Hypoplasia and intrinsic staining of enamel following tetracycline therapy. *JAMA* **185:**1008–1011.

Wright, C. L., R. M. Nadolny, J. Jiang, A. L. Richards, D. E. Sonenshine, H. D. Gaff, and W. L. Hynes. 2011. *Rickettsia parkeri* in Gulf Coast ticks, southeastern Virginia, USA. *Emerg. Infect. Dis.* **17:**896–898.

Yu, X. J., M. F. Liang, S. Y. Zhang, Y. Liu, J. D. Li, Y. L. Sun, L. Zhang, Q. F. Zhang, V. L. Popov, C. Li, J. Qu, Q. Li, Y. P. Zhang, R. Hai, W. Wu, Q. Wang, F. X. Zhan, X. J. Wang, B. Kan, S. W. Wang, K. L. Wan, H. Q. Jing, J. X. Lu, W. W. Yin, H. Zhou, X. H. Guan, J. F. Liu, Z. Q. Bi, G. H. Liu, J. Ren, H. Wang, Z. Zhao, J. D. Song, J. R. He, T. Wan, J. S. Zhang, X. P. Fu, L. N. Sun, X. P. Dong, Z. J. Feng, W. Z. Yang, T. Hong, Y. Zhang, D. H. Walker, Y. Wang, and D. X. Li. 2011. Fever with thrombocytopenia associated with a novel bunyavirus in China. *N. Engl. J. Med.* **364:**1523–1532.

Zavala-Castro, J. E., J. E. Zavala-Velázquez, G. F. Peniche-Lara, and J. E. Sulú Uicab. 2009a. Human rickettsialpox, southeastern Mexico. *Emerg. Infect. Dis.* **15:**1665–1667.

Zavala-Castro, J., J. Zavala-Velázquez, D. Walker, J. Perez-Osorio, and G. Peniche-Lara. 2009b. Severe human infection with *Rickettsia felis* associated with hepatitis in Yucatan, Mexico. *Int. J. Med. Microbiol.* **299:**529–533.

Zhang, L., Y. Liu, D. Ni, Q. Li, Y. Yu, X. J. Yu, K. Wan, D. Li, G. Liang, X. Jiang, H. Jing, J. Run, M. Luan, X. Fu, J. Zhang, W. Yang, Y. Wang, J. S. Dumler, Z. Feng, J. Ren, and J. Xu. 2008. Nosocomial transmission of human granulocytic anaplasmosis in China. *JAMA* **300:**2263–2270.

Zhang, S., H. Song, Y. Liu, Q. Li, Y. Wang, J. Wu, J. Wan, G. Li, C. Yu, X. Li, W. Yin, Z. Xu, B. Liu, Q. Zhang, K. Wan, G. Li, X. Fu, J. Zhang, J. He, R. Hai, D. Yu, D. H. Walker, J. Xu, and X. J. Yu. 2010. Scrub typhus in previously unrecognized areas of endemicity in China. *J. Clin. Microbiol.* **48:**1241–1244.

Zhang, X., R. Hai, W. Li, G. Li, G. Lin, J. He, X. Fu, J. Zhang, H. Cai, F. Ma, J. Zhang, D. Yu, and X. J. Yu. 2009. Seroprevalence of human granulocytic anaplasmosis in central and southeastern China. *Am. J. Trop. Med. Hyg.* **81:**293–295.

Zinsser, H. 1996. *Rats, Lice, and History*, p. 119. Black Dog & Leventhal Publishers, New York, NY. [Reprint of 1935 edition.]

PHYLOGENY AND COMPARATIVE GENOMICS: THE SHIFTING LANDSCAPE IN THE GENOMICS ERA

Joseph J. Gillespie, Eric K. Nordberg, Abdu F. Azad, and Bruno W. S. Sobral

3

INTRODUCTION

Bacterial species belonging to the order *Rickettsiales* are an early-branching lineage of the *Alphaproteobacteria*, forming a sister clade with all known alphaproteobacterial species (Williams et al., 2007). Excluding "*Candidatus* Pelagibacter" spp., free-living marine bacterioplankton (SAR11 clade) purportedly ancestrally related to *Rickettsiales*, all known *Rickettsiales* lineages comprise obligate intracellular species that are dependent on one or more eukaryotic hosts (Brenner et al., 1993; Amann et al., 1995; Yu and Walker, 2003; Dumler and Walker, 2005). The advent of DNA sequence-based phylogenetic analysis quickly established a link between the *Rickettsiales* and the eubacterial ancestor of the mitochondria (Olsen et al., 1994; Viale and Arakaki, 1994; Andersson et al., 1998; Roger et al., 1998; Sicheritz-Pontén et al., 1998). According to this evolutionary scenario, the symbiosis achieved between the mitochondrial ancestor and a protoeukaryote facilitated the transfer of many genes from the genome of the former to the latter, resulting in a highly reduced genome (organelle) captured within a complex higher-ordered cell (*Protoeukarya*) (Emelyanov and Sinitsyn, 1999; Emelyanov, 2001a, 2003b, 2007). Thus, it can be envisioned that the nature of the interactions of modern rickettsiae with their eukaryotic hosts is rooted in the bacterial import of proteins (as well as other molecules) that are targeted for the mitochondria (Emelyanov, 2001b). Furthermore, the secretion of various substrates into the host cell cytoplasm and organelles is likely at the forefront of established symbioses, with some species also virulent in one or more of their eukaryotic hosts and/or reservoirs.

The purpose of this contribution is to enlighten the reader on the vast treasure trove of genomic data for *Rickettsiales* that has culminated over the last 14 years. These data have fueled new discoveries in rickettsiology, including our understanding of how some of these obligate intracellular symbionts and pathogens interact with their vertebrate and invertebrate hosts. Few amendments have been made to classification, which stabilized substantially during the DNA revolution (Dumler et al., 2001; Dumler and Walker, 2005; Azad et al., 2008). We compile here statistics for 46 *Rickettsiales*

Joseph J. Gillespie, Eric K. Nordberg, and Bruno W. S. Sobral, Cyberinfrastructure Division, Virginia Bioinformatics Institute, Virginia Tech, Blacksburg, VA 24061. *Abdu F. Azad,* Department of Microbiology and Immunology, University of Maryland School of Medicine, Baltimore, MD 21201.

Intracellular Pathogens II: Rickettsiales
Edited by G. H. Palmer and A. F. Azad © 2012 ASM Press, Washington, DC
10.1128/9781555817336.ch3

genome sequences. An updated genome-based phylogeny estimation, which considers bacterioplankton species of the *Alphaproteobacteria* SAR11 clade as an ancestral lineage within the *Rickettsiales* (undefined at the family level), is provided. A phylogenomic analysis across select species from six genera (*Orientia*, *Rickettsia*, *Neorickettsia*, *Wolbachia*, *Anaplasma*, and *Ehrlichia*) with completely sequenced genomes (circularized chromosomes) highlights the genomic diversity across the *Rickettsiales*, and whole-genome alignments emphasize many of the diversifying factors shaping the blueprints of these organisms. While genomics allows for a substantial understanding of the genetic underpinnings of rickettsial biology, many questions remain. This is accentuated by the large portion of genomes that encode proteins unknown from other bacteria, especially those that are unique to individual *Rickettsiales* genera and species. We provide a look into the diversification factors of *Rickettsiales* genomes that have resulted from reductive evolution from a free-living streamlined ancestor ("*Candidatus* Pelagibacter"). Finally, in light of the age of genomics, a compilation of small-subunit rRNA-encoding genes (16S rRNA genes) from *Rickettsiales* is provided, with a phylogeny estimated from a subset of these sequences that spans the diversity of *Rickettsiales*. This analysis highlights the vast, and perhaps underappreciated, diversity that comprises the *Rickettsiales*. Understanding the phylogeny of the lineages of *Rickettsiales*, especially in light of the current information revealed by genomic data, will undoubtedly reveal more secrets about this fascinating group of eukaryotic cell denizens.

CLASSIFICATION AND THE DNA REVOLUTION

Until the implementation of DNA sequence-based systematics, species in the *Rickettsiales* were primarily distinguished from other bacteria based on chemical composition, morphology, and intracellular lifestyle. Within the *Rickettsiales*, the following five characteristics were used to support taxonomic delineations (generic groupings): (i) energy production and biosynthesis; (ii) human disease and geographic distribution; (iii) natural vertebrate and invertebrate hosts and other biological reservoirs; (iv) experimental infections and serological reactions and cross-reactions; and (v) strain cultivation, stability, and maintenance (Azad et al., 2008). By long-standing convention, obligate or facultative intracellular bacterial species from nine genera within three families comprised the *Rickettsiales*: (i) family *Rickettsiaceae* (genera *Rickettsia*, *Coxiella*, *Rochalima*, and *Ehrlichia*), (ii) family *Anaplasmataceae* (genus *Anaplasma*), and (iii) family *Bartonellaceae* (genera *Bartonella*, *Haemobartonella*, *Eperythrozoon*, and *Grahamella*). As a result of extensive revisionary work from various research groups, primarily utilizing DNA sequenced-based phylogenetic methodologies, the contemporary *Rickettsiales* classification differs greatly from the traditional classification scheme. The DNA revolution certainly shook the rickettsial tree by highlighting the tremendous degree of convergent evolution in biochemistry and cell morphology that typifies many unrelated and highly divergent bacterial species with an obligate or facultative intracellular lifestyle.

Family *Rickettsiaceae*

Of the four genera originally catalogued in the *Rickettsiaceae*, only the genus *Rickettsia* has remained. Molecular phylogenetics was used to determine that *Coxiella*, a monotypic genus (*C. burnetii*) originally treated as the sister taxon to *Rickettsia* spp., was a member of the *Gammaproteobacteria* instead of the *Alphaproteobacteria* (Weisburg et al., 1989). Specifically, it was noted that 16S rRNA gene sequences placed *C. burnetii*, as well as *Wolbachia persica*, within the vast *Gammaproteobacteria* distinct from *Enterobacteriales*, with *C. burnetii* forming a close affinity with another facultative pathogen, *Legionella*. A recent genome-based phylogeny estimate supports the monophyly of *Coxiella* and *Legionella*, as well as *Rickettsiella* (also obligate intracellular bacteria), with a basal position of this lineage within the *Gammaproteobacteria* (Williams et al., 2010). It had been suspected over decades, based on differences in cell morphology (Weiser and Zizka, 1968), reproduction mode (Götz, 1971, 1972), and a *Chlamydia*-like morphogenic cycle (Federici, 1980), that *Rickettsiella* spp. were incorrectly classified as *Rickettsiales*. Subsequent DNA-DNA hybridization (Frutos et al., 1994) and 16S rRNA gene sequence (Roux et al., 1997a) data confirmed these observations. Through comparison with DNA sequences from other diverse bacterial species, *W. persica* was suggested to be a member of *Francisella* (Forsman et al., 1994; Niebylski et al., 1997; Noda et al., 1997), another *Gammaproteobacteria* lineage forming a sister clade with the *Legionella-Coxiella-Rickettsiella* clade (Williams et al., 2010). Further studies focusing on *Wolbachia* have supported the removal of *W. persica* from the *Wolbachia* lineage (Dumler et al., 2001; Lo et al., 2007). It is noteworthy that, although not related to *Rickettsiales*, three bacterial lineages incorrectly assigned to *Rickettsiaceae* are all closely related in the distant *Gammaproteobacteria*. This epitomizes the concept that convergence in morphology arises even in diverse bacterial species with common intracellular lifestyles.

All species within the genus *Rochalima* were renamed as *Bartonella* spp. upon unification of *Rochalima* with *Bartonella* in the family *Bartonellaceae* (*Alphaproteobacteria*: *Rhizobiales*) (Brenner et al., 1993). Robust phylogeny estimation based on whole-genome sequences supports the close relationship of *Bartonellaceae* with another lineage of facultative intracellular bacteria, *Brucellaceae* (Williams et al., 2007). As the *Rhizobiales* and *Rickettsiales* are very distant lineages in the *Alphaproteobacteria* tree, the common intracellular lifestyle and host invasion strategies (e.g., translocation of effector molecules into hosts via a type IV secretion system [T4SS]) attest not only to the diversity and extreme adaptations that define these bacteria (Ettema and Andersson, 2009), but also to the phylogenetic inertia and biological limitations that ultimately result in convergent evolution of lifestyle and host invasion strategies.

The final genus previously grouped within the *Rickettsiaceae*, *Ehrlichia*, was transferred to the *Anaplasmataceae* because 16S rRNA gene sequence data suggested a close evolutionary relatedness with *Anaplasma* (Dumler et al., 2001). Significant reclassification of the species within the *Anaplasmataceae* was proposed based on similar data and is discussed below. Of significance to the *Rickettsiaceae*, molecular sequence data were also used to remove the scrub typhus group rickettsiae (STGR) agent *Rickettsia tsutsugamushi* from the genus *Rickettsia* and create a monotypic genus, *Orientia* (Tamura et al., 1995). As illustrated below, genomic characteristics (particularly differences in several metabolic pathways, especially cell wall synthesis genes) strongly support the separation of *O. tsutsugamushi* from *Rickettsia* spp.; however, the extremely proliferated mobile genetic elements (MGEs) within STGR genomes are also found in some genomes of *Rickettsia* spp. Interestingly, a recent report of a *Rickettsia*-like symbiont found in Chinese wheat pest aphid, *Sitobion miscanthi* (*Hemiptera*: *Aphididae*), suggests that *Orientia* may no longer be a monotypic genus (Li et al., 2011).

Thus, within the original *Rickettsiaceae*, only the genus *Rickettsia* remains. All species of *Rickettsia* grow freely within the host cytoplasm and have small, reductive genomes (Andersson et al., 1996, 1998; Andersson and Andersson, 1999). Molecular sequence data have illuminated a close evolutionary relatedness between *Rickettsia* spp. and the mitochondrial ancestor (Andersson et al., 1998; Müller and Martin, 1999; Emelyanov, 2001b, 2003b). Until recently, species of *Rickettsia* have been traditionally classified into either the typhus group rickettsiae (TGR) or spotted fever group rickettsiae (SFGR) (Hackstadt, 1996). Notwithstanding, it has been known for decades that *Rickettsia bellii* does not fit within either of these groups (Philip et al., 1983; Stothard et al., 1994; Stothard and Fuerst, 1995). Similarly, despite tenuous placements of *Rickettsia canadensis* within the *Rickettsia* tree in various studies based on limited molecular data, robust genome-based phylogeny estimation suggests *R. canadensis* does not belong to the TGR or SFGR, but rather to a basal lineage unique from *R. bellii* (Gillespie et al., 2008). Based on mounting molecular data, mostly from environmental studies, it is becoming clear that a large diversity of *Rickettsia* species form multiple lineages that are ancestral to the TGR and SFGR, yet still diverge from a common ancestor that forms a sister clade to *O. tsutsugamushi* (Perlman et al., 2006; Weinert et al., 2009a, 2009b). Importantly, unlike the TGR and SFGR, which are composed of many bona fide human pathogens (Azad and Radulovic, 2003; Azad, 2007; Gillespie et al., 2009a), no taxa demonstrated as belonging to this ancestral group rickettsiae (AGR) are serious vertebrate pathogens (Perlman et al., 2006). The virulence (if any) associated with both *R. bellii* and *R. canadensis* remains poorly described (Azad et al., 2008). Another lineage of *Rickettsia* species that is challenging the existing classificatory paradigm of "TGR or SFGR" has been named the transitional group rickettsiae (TRGR) and occupies a lineage that is basal to the SFGR (Gillespie et al., 2007). Some molecular phylogenetic studies have revealed this distinct clade, which comprises *Rickettsia felis*, *Rickettsia akari*, and *Rickettsia australis*, e.g. (Roux and Raoult, 1999; Sekeyova et al., 2001), as well as symbionts of wasps (*Liposcelis* spp.) and booklice (*Neochrysocharis* spp.) (Perlman et al., 2006; Braig et al., 2008). The TRGR may also include uncharacterized *Rickettsia* symbionts of the mite *Ornithonyssus bacoti* (Reeves et al., 2007). Robust phylogeny estimation and genomic characteristics support the proposal for the TRGR, along with the tendency for its members to be associated with non-tick arthropod hosts (Gillespie et al., 2008).

Family *Bartonellaceae*

DNA hybridization and molecular sequence data suggested the removal of the entire *Bartonellaceae* from the *Rickettsiales* due primarily to a close affinity of its members to *Brucellaceae* (*Alphaproteobacteria*: *Rhizobiales*) (Brenner et al., 1993). As with *Rochalima* spp. (discussed above), species of the genus *Grahamella* were transferred to the genus *Bartonella* (Birtles et al., 1995). Molecular phylogenetic analysis revealed the association of the genera *Haemobartonella* and *Eperythrozoon* with gram-positive species of the *Mycoplasmataceae* (*Mollicutes*: *Mycoplasmatales*) (Neimark and Kocan, 1997; Rikihisa et al., 1997; Neimark et al., 2001). This restructuring of *Bartonellaceae* leaves *Bartonella*, several unclassified species, and many environmental isolates as the remaining members of the family. Of note, the obligate intracellular species *Wolbachia melophagi* likely belongs to the genus *Bartonella*, as suggested by 16S rRNA gene sequence similarity. Accordingly, it has been proposed to remove *W. melophagi* from the genus *Wolbachia* (Dumler et al., 2001; Lo et al., 2007), and following a recent human infection with this organism it has been named "*Candidatus* Bartonella melophagi" (Maggi et al., 2009).

Family *Anaplasmataceae*

Molecular phylogenetic analyses from various research groups over a short time span collectively suggested that major revisions were needed in the monotypic family *Anaplasmataceae*. It was once composed solely of the genus *Anaplasma*, and amendments to membership were needed to include species within the genera *Cowdria*, *Ehrlichia*, *Neorickettsia*, and *Wolbachia* (Weisburg et al., 1989; Dame et al., 1992; van Vliet et al., 1992; Sumner et al., 1997; Zhang et al., 1997). In perhaps the most significant classificatory contribution to modern rickettsiology, Dumler et al. (2001) proceeded to arrange the majority of species within the genera *Anaplasma*, *Ehrlichia*, *Neorickettsia*, *Wolbachia*, and *Cowdria* into four of the five genera (genus *Cowdria* was dissolved under genus *Ehrlichia*). Trees based on DNA sequences generated from multiple loci (16S rRNA, *groES*, *groEL*, and surface protein-encoding genes) strongly supported these four lineages of *Anaplasmataceae*. All members within the previously recognized tribes *Ehrlichieae* and *Wolbachieae* were transferred to the *Anaplasmataceae*, with a suggestion for removal of a tribal system within the *Rickettsiaceae*. Specifically, the genus *Anaplasma* was expanded to include *Ehrlichia platys*, *Ehrlichia bovis*, and the *Ehrlichia phagocytophila* group. *E. phagocytophila*, *Ehrlichia equi*, and the human granulocytic ehrlichiosis agent were collectively renamed *Anaplasma phagocytophilum*. The heartwater disease agent *Cowdria ruminantium* was transferred to *Ehrlichia*, which decreased exceptionally in species membership upon revision. The genus *Wolbachia* was supported as a sister lineage to *Anaplasma* + *Ehrlichia*, and several species were considered misplaced in the genus (amendments discussed above). Finally, *Ehrlichia sennetsu* and *Ehrlichia risticii* were transferred to the genus *Neorickettsia*, with this lineage strongly supported as ancestral relative to all other *Anaplasmataceae*.

As with the *Rickettsiaceae*, all described species within the *Anaplasmataceae* are obligate intracellular bacteria. However, unlike *Rickettsia* spp. and *O. tsutsugamushi*, all characterized species of *Anaplasmataceae* reside within some form of a vacuole in their hosts. This trait has undoubtedly orchestrated the diversification of mechanisms by which species within these two families interact with their hosts. These differential characteristics can be observed at the genetic level. For instance, the majority of species within the *Anaplasmataceae* have lost the genes encoding enzymes in the pathways for biosynthesis of peptidoglycan (PG) and lipopolysaccharide (LPS). These major differences between cell envelope architecture have had a profound impact on the manner in which rickettsial species arrange proteins in their outer membranes (see "Age of Genomics," below). Thus, mechanisms of host immune system avoidance are likely a primary driver for genetic diversity across the *Rickettsiales*.

Mounting molecular data suggest that the genus *Wolbachia*, like the genus *Rickettsia*, is extraordinarily diverse, and this is consistent with estimates of up to 65% of arthropod species being infected with *Wolbachia* (Hilgenboecker et al., 2008). Diversity within the genera *Anaplasma*, *Ehrlichia*, and *Neorickettsia* is likely much less. The exact relationship of other poorly characterized species of *Anaplasmataceae*, such as *Aegyptianella* (Rikihisa et al., 2003), "*Candidatus* Neoehrlichia" (Fehr et al., 2010), and "*Candidatus* Neoanaplasma" (Rar et al., 2010), to the well-established lineages remains unknown. Diversity outside of these genera, including the recently described mitochondrial invader "*Candidatus* Midichloria mitochondrii" (Lo et al., 2004), various endosymbionts of protists (Vannini et al., 2010), and the agent of strawberry disease infecting rainbow trout (Lloyd et al., 2008), suggests that a second revisionary shakedown for *Anaplasmataceae* is likely impending (see "The Diversity of Rickettsiales," below).

Family *Holosporaceae*

Based on 16S rRNA gene sequence analysis, endosymbiotic bacterial species associated with (and resistant to) amoeba species (Hall and Voelz, 1985; Fritsche et al., 1993) and an alphaproteobacterial endosymbiont of *Paramecium caudatum*, named *Holospora obtusa* (Amann et al., 1991), were supported as a monophyletic lineage that forms a sister clade to *Rickettsiales* (Springer et al., 1993). This lineage has since been erected as a third family within the *Rickettsiales*, the *Holosporaceae* (Görtz and Schmidt, 2005). The lineage comprises mainly protist endosymbionts, and its monophyly and close evolutionary relationship to *Rickettsiaceae* and *Anaplasmataceae* has been supported in subsequent studies (Beier et al., 2002; Baker et al., 2003). *Caedibacter* spp., paramecia endosymbionts that form refractile bodies (R-bodies) and bestow a killer trait to their hosts, were shown by DNA sequence data to be polyphyletic, with *Caedibacter taeniospiralis* transferred to the *Gammaproteobacteria* due to its 16S rRNA gene sequence being highly similar to members of the *Francisellaceae* (Beier et al., 2002). The well-studied *Caedibacter caryophilus* (Schmidt et al., 1987) is a true member of the *Holosporaceae*. The proteinaceous R-bodies it produces, which are encoded by phage, make *C. caryophilus*-infected paramecia toxic to other uninfected paramecia enclosed together in food vacuoles (Kusch and Görtz, 2006). *C. caryophilus* can infect both the macronucleus and the cytoplasm of its paramecium host (Kusch et al., 2000). The intervening sequence within the 16S rRNA gene of *C. caryophilus* has been shown to be variable across related strains, and has been used to propose a separate species, "*Caedibacter macronucleorum*" (Schrallhammer et al., 2006).

H. obtusa Hafkine 1890 and *Holospora undulata* Hafkine 1890 were originally shown to grow in the nuclei of paramecia (Gromov and Ossipov, 1981), with the latter species serving as the type species for which the family *Holosporaceae* was named (Görtz and Schmidt, 2005). The closely related *Holospora elegans* is a bona fide parasite of *P. caudatum* and kills infected cells (Görtz and Fujishima, 1983), as well as induces a stress response in its host (Hori et al., 2008). *H. obtusa* alters its pattern of protein expression significantly upon transition from the infectious to the reproductive form inside paramecia (Görtz et al., 1990). Over 20 different *H. obtusa* proteins have been implicated in the infection process, having been characterized by N-terminal microsequencing and immunological methods (Görtz et al., 1990; Wiemann and Görtz, 1991; Dohra et al., 1997, 1998; Nakamura et al., 2004). A project to sequence the *H. obtusa* genome is currently in progress, and preliminary data (50% of the genome) indicate that the chromosome is ~25% GC and roughly 1.7 Mb in size (Lang et al., 2005), statistics that are consistent for the *Rickettsiales*. Initial highlights of the *H. obtusa* genome content include (i) a complete biotin synthesis pathway (biotin synthesis genes are variably present across *Rickettsiales*), (ii) a lack of genes involved in oxidative phosphorylation, and (iii) a high level of insertion sequences (ISs) most similar to homologs from *Legionella pneumophila* and *Francisella tularensis*. The lack of genes encoding enzymes involved in oxidative phosphorylation is consistent with life inside the host nucleus, which is oxygen limited, as well as the presence of a gene (*tlc*) encoding an ATP/ADP translocase for scavenging host ATP (Linka et al., 2003).

Molecular phylogenetic analysis of 16S rRNA gene and *groEL* sequences suggests that the mitochondrial ancestor branched off after the *Holosporaceae*, forming a sister relationship to the *Rickettsiaceae* and *Anaplasmataceae* (Emelyanov, 2001b). A preliminary genome-based phylogeny including *H. obtusa* supports the *Holosporaceae* as the basal lineage within the *Rickettsiales* (Lang et al., 2005). However, the mitochondrial sequences were not included, thus leaving a robust phylogeny estimate for the placement of the mitochondria within the *Rickettsiales* tenuous. Despite this, a growing number of species belonging to the *Holosporaceae* are emerging. The majority of these candidate species are associated with protists: "*Candidatus* Odyssella thessalonicensis" (Birtles et al., 2000), "*Candidatus* Captivus acidiprotistae" (Baker et al., 2003), endosymbiont of *Acanthamoeba* sp. KA/E23 (Xuan et al., 2007), and "*Candidatus* Paraholospora nucleivisitans" (Eschbach et al., 2009). However, the association of a *Holosporaceae* species with the prairie dog flea, *Oropsylla hirsuta*, suggests that the host range and possible reservoirs for members of this family may extend beyond protists (Jones et al., 2008).

The associations of other putative genera within the *Holosporaceae* (e.g., *Lyticum*, *Polynucleobacter*, *Pseudocaedibacter*, *Symbiotes*, and *Tectibacter*) remain untested. However, the growing diversity within the *Holosporaceae* suggests that attention to this *Rickettsiales* lineage is needed to understand the diversification of the derived families *Rickettsiaceae* and *Anaplasmataceae*. Without insight from *Holosporaceae* genomes, data will be limited for understanding (i) the evolution of vertebrate association and pathogenesis, (ii) the distinction between mechanisms underlying arthropod associations versus protist associations, (iii) the evolution of obligate intracellular species (*Holosporaceae*, *Rickettsiaceae*, and *Anaplasmataceae*) from free-living ancestral relatives of the SAR11 clade, and (iv) the phylogenetic relationship of the mitochondrial ancestor to the *Rickettsiales*.

AGE OF GENOMICS (1998 TO PRESENT)

Timeline

The agricultural and medical importance associated with many *Rickettsiales* species has resulted in over a century of research aimed at understanding the nature of pathogenesis elicited by these infectious microbes, many of which are category B Select Agents (Azad and Radulovic, 2003; Azad, 2007). In the age of genomics, the selection of rickettsial species and strains for genome sequencing has focused primarily on understanding the mechanisms associated with vertebrate pathogenicity. Over the last 14 years a plethora of genomic data has accumulated for *Rickettsiales*. Since the initial publication of the *Rickettsia prowazekii* Madrid E genome sequence in 1998 (Andersson et al., 1998), 46 genome sequences have been deposited on GenBank (Fig. 1), with additional plasmid sequences from various species and strains of *Rickettsia*. The following timeline lists the publications associated with the majority of these rickettsial genomes, illustrating the major advancements in our knowledge of the patterns and processes of reductive genome evolution and rickettsial genomic diversification. In most cases, studies relevant to these publications are also discussed.

12 NOVEMBER 1998: *R. PROWAZEKII* MADRID E (Andersson et al., 1998)

By sequencing the first genome of a species of *Rickettsiales*, Andersson et al. (1998) provided the first report of "reductive genome evolution" in obligate intracellular bacteria. The 834 predicted open reading frames (ORFs) depict a similar functional profile as mitochondria: a lack of genes required for anaerobic glycolysis, but complete pathways for the tricarboxylic acid (TCA) cycle and respiratory chain complex. The presence of five *tlc* genes suggests redundancy for ATP synthesis/utilization. Genes encoding most enzymes involved in amino acid and nucleoside synthesis are absent, implying that host reserves of these metabolites must be imported to the bacterial cytoplasm. Mechanisms of genome reduction are reflected by an unusually large percentage of noncoding DNA (24%), which comprises mainly graveyards for pseudogenes. Phylogeny estimation with other bacteria, as well as eukaryotic mitochondrial genes, suggests a *Rickettsiales* origin for modern mitochondria.

14 SEPTEMBER 2001: *RICKETTSIA CONORII* MALISH 7 (Ogata et al., 2001)

The sequencing of the *R. conorii* genome illuminated the mechanisms underlying reductive genome evolution in *Rickettsia* spp., namely gross pseudogenization. Despite its having a larger genome than *R. prowazekii* (encoding 1,374 ORFs), 804 genes are common across both genomes. Coupled with highly conserved synteny across both genomes, the 552 additional ORFs in the *R. conorii* genome anchor and identify numerous split genes and pseudogenes in the *R. prowazekii* genome. Compared with *R. prowazekii*, the *R. conorii* genome has a 10-fold increase in repetitive elements. A particular family of repetitive elements, Rickettsia palindromic elements (RPEs), which were identified prior to the completion of the *R. conorii* genome (Ogata et al., 2000), are suggested to play a role in the de novo synthesis of *Rickettsia* proteins. RPEs can insert in-frame within ORFs, adding large tracts of amino acids to segments of proteins. Thus, RPEs can

FIGURE 1 Compilation of genome statistics for 46 *Rickettsiales* taxa. Taxon codes are described in Fig. 2, with circles defining the monophyly of the major *Rickettsiales* lineages: S, SAR11 clade; O, *Orientia*; R, *Rickettsia*; N, *Neorickettsia*; W, *Wolbachia*; E, *Ehrlichia*; A, *Anaplasma*. Statistics for the outgroup *Alphaproteobacteria* taxa are not provided. Tree is a cladogram representation of the phylogram shown in Fig. 2. RefSeq data were taken from GenBank, with statistics from PATRIC provided based on the RAST automated annotation technology (Overbeek et al., 2005; Aziz et al., 2008). Pl, plasmid; HP, hypothetical protein; FA, functional annotation. doi:10.1128/9781555817336.ch3.f1

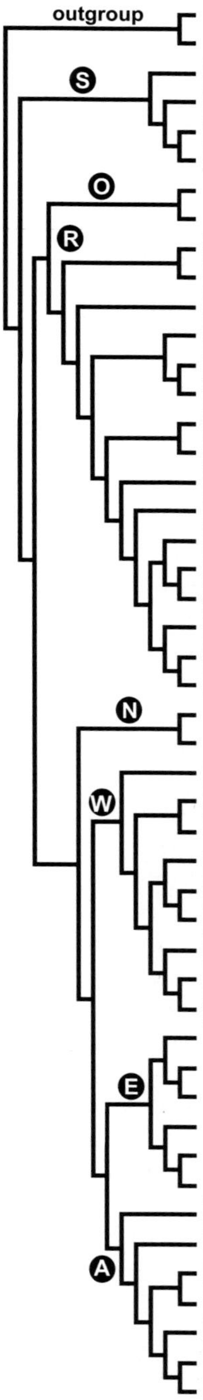

	Size (nt)	Status	%GC	PI	RefSeq %cod	RefSeq ORFs	PATRIC ORFs	PATRIC HP	PATRIC FA
Bj									
Pv									
Ap	1294706	WGS	ND	0	ND	1425	1417	319	1098
Ps	1456888	WGS	29.03	0	91	1447	1543	291	1252
Pu1	1324008	WGS	29.70	0	95	1387	1383	230	1153
Pu2	1308759	complete	29.68	0	95	1354	1364	198	1166
Bg	2127051	complete	30.53	0	48	1182	2364	718	1646
Ik	2008987	complete	30.51	0	74	1967	2197	610	1587
Bo	1528980	complete	31.63	0	80	1476	1657	575	1082
Br	1522076	complete	31.65	0	84	1429	1612	560	1052
Ca	1159772	complete	31.05	0	74	1093	1230	409	821
Ty	1111496	complete	28.92	0	75	839	892	142	750
P22	1111612	complete	29.00	0	75	952	910	146	764
Pr	1111523	complete	29.00	0	75	835	924	147	777
Ak	1231060	complete	32.33	0	75	1259	1437	431	1006
Fe	1587240	complete	32.00	1	83	1512	1810	617	1193
Is	2100092	WGS	31.00	4	78	2117	2404	1048	1356
Ma	1376184	complete	32.00	1	69	980	1721	654	1067
Pe	1314898	complete	32.00	1	67	947	1558	519	1039
Ri	1257710	complete	32.47	0	76	1345	1577	634	943
Rw	1268175	complete	32.45	0	76	1384	1595	641	954
Co	1268755	complete	32.44	0	80	1374	1578	615	963
Si	1250021	WGS	32.47	0	ND	1234	1554	589	965
Af	1276710	WGS	32.00	1	72	1041	1545	568	977
Rs	879977	complete	41.27	0	85	892	869	265	604
Se	859006	complete	41.08	0	87	932	868	271	597
Bm	1080084	complete	34.18	0	66	805	1328	583	745
Qj	1543661	WGS	ND	0	ND	1378	1545	472	1073
Qp	1482455	complete	34.19	0	81	1275	1500	465	1035
Mu	867873	WGS	ND	0	ND	827	931	271	660
Me	1267782	complete	35.23	0	80	1195	1293	387	906
Wi	1092888	WGS	ND	0	ND	899	937	328	609
Wr	1445873	complete	35.16	0	77	1150	1526	529	997
An	1440750	WGS	ND	0	ND	1802	1771	565	1206
Sm	1063100	WGS	ND	0	ND	760	951	337	614
Rg	1499920	complete	27.51	0	63	950	1003	195	808
Re	1512977	complete	27.00	0	62	958	-----	-----	-----
Ro	1516355	complete	27.00	0	63	888	-----	-----	-----
Cj	1315030	complete	28.96	0	72	925	1030	258	772
Ha	1176248	complete	30.10	0	79	1105	992	210	782
Hs	1005812	WGS	ND	0	ND	805	910	208	702
Ph	1471282	complete	41.64	0	68	1264	1506	639	867
Ce	1206806	complete	49.98	0	84	923	1153	400	753
Ms	1197687	complete	49.76	0	85	948	1171	402	769
Mi	1141520	WGS	ND	0	ND	885	1329	462	867
Mf	1202435	complete	49.77	0	85	940	1176	413	763
Mp	1158530	WGS	ND	0	ND	965	1233	435	798
Mv	1153875	WGS	ND	0	ND	961	1145	324	821

have a negative impact on enzymatic function (Abergel et al., 2006). However, an alternative hypothesis posits that RPEs are relics from the early RNA world (Dwyer, 2001), and although RPEs are subsequently described as characteristic of all *Rickettsia* (Ogata et al., 2002) and some *Wolbachia* (Ogata et al., 2005c) genomes, systematic analysis suggests that they are likely undergoing evolutionary purging rather than continual proliferation (Amiri et al., 2002; Claverie and Ogata, 2003).

10 FEBRUARY 2004: *RICKETTSIA SIBIRICA* 246 (Malek et al., 2004)

The *R. sibirica* genome sequence was generated with simultaneous construction of a bacterial two-hybrid screen, which covered ~85% of the estimated proteome (Malek et al., 2004). Genome size and ORF count (1,234) are very similar to the closely related *R. conorii*. In the bacterial two-hybrid system, several proteins of the *Rickettsiales vir* homolog (*rvh*) P-like T4SS (Gillespie et al., 2009b) were noted for their interaction with LPS-related proteins, hemolysins, protein export components, proteases, permeases, outer membrane proteins (OMPs), ABC transporters, and several uncharacterized proteins. A more recent phylogenomic study identified two large symmetrical rearrangements that, upon reindexing using the reverse complement of its circular permutation from the original position 668301, greatly improved colinearity between *R. sibirica* and closely related SFGR genomes (Gillespie et al., 2008).

APRIL 2004: *WOLBACHIA* ENDOSYMBIONT OF *DROSOPHILA MELANOGASTER* (*w*Mel) (Wu et al., 2004)

The first genome sequence from the *Anaplasmataceae* also reveals reductive genome evolution, with a genome size of 1.3 Mb and 1,195 predicted ORFs. Phylogenomics reveals that, in contrast to other obligate intracellular bacterial genomes (including *Rickettsia* spp.), *w*Mel contains high levels of repetitive DNA and MGEs. Continual population bottlenecks in *Wolbachia* spp. are attributed to a lack of selection against retention of MGEs. *w*Mel differs from *Rickettsia* spp. in metabolic capacity principally by the presence of complete pathways for glycolysis (later shown to be gluconeogenesis) and purine synthesis, a lack of *tlc* genes for ATP scavenging from the host, a lack of most genes in the LPS biosynthesis pathway, and a plethora of ORFs containing ankyrin repeat domains (ARDs). Despite the presence of a functional mobilome, no evidence was found for lateral gene transfer (LGT) between *w*Mel and its dipteran host (*D. melanogaster*). Phylogeny estimation supports the sister relationship of *w*Mel with *Rickettsia* spp., but provides limited support for the origin of the mitochondria from the *Rickettsiales*.

SEPTEMBER 2004: *RICKETTSIA TYPHI* WILMINGTON (McLeod et al., 2004)

The second genome from the TGR is highly similar to the *R. prowazekii* genome in nearly every aspect. Differences include five more predicted ORFs in *R. typhi* (877 total); more unique genes (24) than *R. prowazekii* (13); a 124-kb inversion near the origin of replication; and an increased number of pseudogenes, including a completely pseudogenized cytochrome *c* oxidase system. A comprehensive list of 43 candidate rickettsial virulence factors is provided. Comparative analysis of five *Rickettsia* genomes (the first analysis including *Rickettsia rickettsii* genomic data) illustrates the more advanced reductive evolution in TGR, along with a decreased GC content (~29%) relative to SFGR (~32.5%). A 12-kb insertion unique to the *R. prowazekii* genome was identified and occurred subsequent to the divergence of TGR. This region was later identified as a probable result of plasmid insertion into the *R. prowazekii* chromosome (Blanc et al., 2007a). In-depth bioinformatics analysis of the highly duplicated SpoT/RelA gene family, which functions in the bacterial stringent response (SR), suggests that the atypical rickettsial proteins likely contribute to a functional SR that is associated with intracellular survival.

18 JANUARY 2005: *ANAPLASMA MARGINALE* ST. MARIES

(Brayton et al., 2005)

The *A. marginale* genome sequence was one of two concomitantly published genomes from the derived *Anaplasmataceae* lineage (*Anaplasma* + *Ehrlichia*). Genome size (1.2 Mb) and predicted ORFs (949) are consistent with reductive evolution characteristic of other *Rickettsiales* genomes; however, a unique mechanism of pseudogenization is described for a portion of genes encoding proteins characteristic of the *A. marginale* surface coat. Of two superfamilies of immunodominant proteins, major surface protein (MSP) 1 and MSP2, which comprise ~80% of the predicted OMPs, functional pseudogenes within the MSP2 superfamily contribute to the well-characterized surface coat antigenic variation (Brayton et al., 2001, 2002; Meeus et al., 2003). The genome has an abnormally high GC content (49%) for an obligate intracellular bacterium, which subsequent sequencing of other closely related strains would confirm (discussed below). Despite genome plasticity in the form of gene duplication, which mostly comprises membrane-associated proteins and transporters, few pseudogenes and zero ISs were identified. Unlike *Rickettsia* spp. but similar to *w*Mel, most glycolytic enzymes are encoded and likely function in gluconeogenesis, the products of which contribute to the synthesis of cofactors, nucleotides, and phospholipids. Other metabolic profiles are also highly similar to *w*Mel, and despite the presence of minimal genes for enzymes involved in amino acid synthesis, few transporters for amino acid import are identified. A comparison with other tick-associated *Rickettsiales* species (*R. conorii* and *Ehrlichia ruminantium* Welgevonden Erwo) identified 15 ORFs present only in these genomes.

18 JANUARY 2005: *E. RUMINANTIUM* WELGEVONDEN ERWO (ERWO)

(Collins et al., 2005)

The *E. ruminantium* Welgevonden genome sequence was the second of two concomitantly published genomes from the derived *Anaplasmataceae* (Collins et al., 2005). The strain designation "Erwo" is used here to distinguish this genome sequence from that of another *E. ruminantium* Welgevonden genome sequence (Erwe), as originally suggested (Frutos et al., 2006). A larger genome size compared to other *Rickettsiales* genomes (1.52 Mb) is reported, but with an extraordinarily high percentage of noncoding DNA (38%). Of 920 predicted ORFs, 32 are pseudogenes. An unprecedented number of tandemly repeated and duplicated sequences contribute to the decrease in coding DNA as well as a very low GC content (27.5%). This low GC content has remained the lowest known for a *Rickettsiales* genome and is typical of all *E. ruminantium* genome sequences. The high similarity within several repeat families has mediated intrachromosomal recombination events that have given rise to an ongoing expanding and contracting set of duplicate genes, with the additional creation of a few novel genes. This distinct pseudogenization mechanism of repeat-mediated gene birth and death distinguishes Erwo from other *Rickettsiales* genomes.

APRIL 2005: *WOLBACHIA* ENDOSYMBIONT STRAIN TRS OF *BRUGIA MALAYI* (*w*Bm)

(Foster et al., 2005)

The *w*Bm genome was the second sequenced from the genus *Wolbachia*, and the first sequenced from a *Rickettsiales* species associated with a non-arthropod host (nematode *B. malayi*) (Foster et al., 2005). Genome size (1.08 Mb) and predicted ORF count (806) are smaller than *w*Mel, and a lack of gene synteny between these genomes is primarily due to the presence of repetitive DNA in *w*Mel. The percentage of coding DNA is much lower than in *w*Mel (66 versus 80%) and much lower than in all subsequently completed *Wolbachia* genome sequences. Unlike *w*Mel, *w*Bm does not contain prophage. Metabolic capacity is highly similar to *w*Mel, and it is suggested that *Wolbachia* spp. may provide riboflavin, flavin adenine dinucleotide, heme, and nucleotides to their hosts, in turn benefiting from a supply of host amino acids. Genomic comparison with *Rickettsia* spp. illustrates more complete pathways for nucleotide and sugar transport and metabolism, as well as coenzyme metabolism, but fewer genes involved in cell envelope biogenesis (all consistent with prior reports). A reduced pathway for PG synthesis relative to *Rickettsia* spp. is suggested to result in an atypically structured murein layer that may pose as a cytotoxin, eliciting an inflammatory response in the filarial host. Minor differences between *w*Bm and *w*Mel PG structures may correlate with a mutualistic lifestyle versus host parasitism, respectively. Similarly, the smaller accessory genome of *w*Bm may reflect a loss of genes involved in avoidance of host cell defense systems. As with *w*Mel, no evidence was found for LGT between *w*Bm and its nematode host.

APRIL 2005: *WOLBACHIA* ENDOSYMBIONTS OF *DROSOPHILA ANANASSAE* (*w*Ana), *DROSOPHILA SIMULANS* (*w*Sim), AND *DROSOPHILA MOJAVENSIS* (*w*Moj) (Salzberg et al., 2005a)

Inspection by Salzberg et al. (2005a) of disparate fruit fly genome trace file archives resulted in the identification of three novel species of *Wolbachia* endosymbionts: *D. ananassae* (*w*Ana), *D. simulans* (*w*Sim), and *D. mojavensis* (*w*Moj). The estimated genome coverage for each novel species is >95%, 75 to 80%, and 6 to 7% for *w*Ana, *w*Sim, and *w*Moj, respectively. Extensive genome rearrangement was shown across all three lineages and *w*Mel (*w*Bm was not included in the analyses). Novel genes (464 for *w*Ana, 92 for *w*Sim, and 6 for *w*Moj) unknown from the *w*Mel genome were discovered, with >100 annotated as MGEs, including 13 phage-related proteins. Conversely, ~27% of the predicted *w*Mel ORFs, including genes within two regions thought to encode important host interaction factors, were unknown from the three novel *Wolbachia* genomes, suggesting a highly variable *Wolbachia* accessory genome. Bioinformatics analysis of a large set of ARD-containing ORFs identified both a conserved set of families across the genomes as well as lineage-specific genes. Despite these contributions to dynamic *Wolbachia* genomes, no evidence was found for LGT between any of these novel species and their dipteran hosts. Subsequent reevaluation of the trace file archives determined that the *Wolbachia* sequences discovered in *D. mojavensis* were an artifact; however, 2,291 novel *Wolbachia* sequences were found in another fly genome, *Drosophila willistoni* (Salzberg et al., 2005b). This work underscores the utility of eukaryotic genomes for discovering *Rickettsiales* endosymbionts and potentially assembling partial to complete bacterial genomes from the eukaryotic reads.

AUGUST 2005: *R. FELIS* URRWXCal2, PLASMID pRF (Ogata et al., 2005a, 2005b)

The *R. felis* genome sequence revealed not only the largest *Rickettsia* (and *Rickettsiales*) genome sequenced to date (1.58 Mb), but also the first detection of a plasmid from an obligate intracellular bacterial genome (Ogata et al., 2005a, 2005b). Among genomes of *Rickettsia* spp., the *R. felis* genome is remarkable in its elevated numbers of encoded transposases (TNPs), toxin-antitoxin modules, SR proteins, and an especially large repertoire of tetratricopeptide (TPR) domain- and ARD-containing proteins. Based on the genetic profile, several phenotypic traits (e.g., β-lactamase activity, intercellular spread via actin polymerization, membrane hemolysis) are predicted and substantiated with experimental evidence. Electron microscopic images of surface appendages reminiscent of type II conjugative pili are observed, along with cell-cell mating. Surprisingly, the plasmid (pRF), referred to as a conjugative plasmid, does not encode a complete F-like T4SS for conjugation among the 68 predicted ORFs. A complete array of F-T4SS genes is also not encoded on the chromosome. Thus, the genes contributing to the F-like pili in *R. felis* remain to be identified (Darby et al., 2007; Gillespie et al., 2007, 2009b, 2010). pRF is reported as a short (39-kb) and a long (63-kb) form; however, this finding has been the subject of disagreement, as has the treatment of *R. felis* in rickettsial classification (Gillespie et al., 2007; Fournier et al., 2008).

19 AUGUST 2005: "*CANDIDATUS* PELAGIBACTER UBIQUE" HTCC1062 (Giovannoni et al., 2005)

The "*Candidatus* Pelagibacter ubique" genome sequence distinguishes "genome streamlining" from reductive genome evolution in *Rickettsiales*, as despite a small genome size (1.31 Mb) typical of other *Rickettsiales* genomes, most basic bacterial metabolic pathways are complete (e.g., TCA cycle, glycoxylate shunt, respiration, pentose phosphate cycle, fatty acid biosynthesis, cell wall biosynthesis, biosynthetic pathways for all 20 proteinogenic amino acids) (Giovannoni et al., 2005). Interestingly, the glycolysis enzymes are limited to those functioning in gluconeogenesis, as in *Anaplasmataceae* genomes, and certain pathways for cofactors are absent (pantothenate, pyridoxine phosphate, thiamine, and biotin). Although DNA uptake and competence genes are encoded by "*Candidatus* Pelagibacter ubique," zero pseudogenes, introns, transposons, extrachromosomal elements, or inteins are present, and the average length of intergenic spacers is 3 nucleotides, the shortest reported for any genome sequence. In addition, few paralogous gene families are encoded. The phylogenetic position of "*Candidatus* Pelagibacter ubique," as well as the entire SAR11 clade, is estimated to be derived in the *Alphaproteobacteria* based on analysis of RNA polymerase B sequences; however, a more robust phylogeny estimate placed all members of "*Candidatus* Pelagibacter" as the most basal lineage of *Rickettsiales*, with the origin of mitochondria estimated as a sister lineage to the remaining *Rickettsiaceae*/*Anaplasmataceae* (Williams et al., 2007). Of all *Rickettsiales* genomes, "*Candidatus* Pelagibacter ubique" has the largest metabolic capacity (including the most complete gluconeogenesis pathway), and coupled with a free-living lifestyle and lack of MGEs, its proposed ancestral position in the *Rickettsiales* tree makes intuitive sense when considering the origin of reductive genome evolution from genome streamlining (discussed below). Three subsequently sequenced genomes from the SAR11 clade ("*Candidatus* Pelagibacter" sp. HTCC7211, "*Candidatus* Pelagibacter ubique" HTCC1002, and alphaproteobacterium strain HIMB114) have confirmed the characteristics of genome streamlining in the ancestral *Rickettsiales* lineage.

DECEMBER 2005: *R. CANADENSIS* MCKIEL (Eremeeva et al., 2005)

Attributes of the *R. canadensis* genome reflect the discordant phenotypic and biochemical characteristics that have made it difficult to group this species in either the TGR or SFGR. For example, genome size (1.16 Mb) and GC content (29%) are more reflective of TGR genomes, yet both cell surface antigen (*sca*)-encoding genes *sca0* (*rompA*) and *sca5* (*rompB*) are complete (*sca0* is pseudogenized in TGR genomes) and the genome shares more common repetitive elements with SFGR genomes. Phylogenetic analyses of various datasets suggest the position of *R. canadensis* between the TGR and SFGR; however, *R. bellii* was not sampled, and subsequent studies show that without the inclusion of *R. bellii* the position of *R. canadensis* in estimated phylogenies is unstable (Vitorino et al., 2007; Gillespie et al., 2008). Aside from the genome of *R. canadensis*, this study also presented statistics on two other novel *Rickettsia* genomes: *R. akari* Hartford (1.23 Mb; 1,259 ORFs) and *R. rickettsii* Sheila Smith (1.26 Mb; 1,345 ORFs), and estimated phylogenies support the derived position of both taxa within the rickettsial tree. Additionally, Eremeeva et al. also presented preliminary data spanning an estimated 66% of the genome of the scrub typhus agent *O. tsutsugamushi* (human isolate AF PL-12 from central Thailand). Careful attention was made regarding tandem repeat sequences, which are very conserved in *Rickettsia* spp. but extremely divergent in *O. tsutsugamushi*, resulting in a total loss of genome synteny across these two genera of *Rickettsiaceae*. Comparative genomic analysis suggests that *O. tsutsugamushi* is similar to the *R. felis* genome in having an increased number of encoded TNPs, as well as TPR domain- and ARD-containing proteins. Genes unique to *O. tsutsugamushi* are reported to be involved in regulatory and transporter responses to environmental stimuli.

FEBRUARY 2006: *EHRLICHIA CHAFFEENSIS* ARKANSAS, *A. PHAGOCYTOPHILUM* HZ, AND *N. SENNETSU* MIYAYAMA (Dunning Hotopp et al., 2006)

Dunning Hotopp et al. (2006) contributed three novel genomes from the *Anaplasmataceae* (all human pathogens) and the first genome sequence from the genus *Neorickettsia*, the earliest-branching lineage of *Anaplasmataceae*. Like *w*Mel, *E. chaffeensis* and *A. phagocytophilum* contain elevated repetitive regions, and comparison with the genomes of *E. ruminantium* and *A. marginale* reveals a significant expansion of the genes (and pseudogenes) of the MSP1 and MSP2 superfamilies in all *Anaplasma* and *Ehrlichia* genomes. *N. sennetsu* has only a few repeat regions, similar to the TGR genomes. The genome size of *N. sennetsu* is the smallest for *Rickettsiales* to date (0.86 Mb), with no other subsequently sequenced genome smaller in chromosome size. However, the number of predicted ORFs for *N. sennetsu* is similar to TGR, thus resulting in a higher percentage of coding DNA and fewer pseudogenes. Genome sizes for *E. chaffeensis* (1.17 Mb) and *A. phagocytophilum* (1.47 Mb) are much smaller and larger, respectively, compared with previously sequenced species within these genera. Comparative analysis of seven *Rickettsiales* genomes revealed only three conserved syntenic regions >10 kb, as well as the conservation of an atypical split rRNA operon that was previously described for *Rickettsia* spp. and *Anaplasma* spp. (Andersson et al., 1995, 1999; Rurangirwa et al., 2002). All *Rickettsiales* genomes lack the ability to synthesize the majority of amino acids. Aside from *Rickettsia* and *Wolbachia*, the remaining genomes encode enzymes involved in the synthesis of all major vitamins and cofactors. Additionally, all genomes except *Rickettsia* spp. contain the enzymes necessary for the synthesis of purines and pyrimidines. Aside from the genomes of *Rickettsia* spp., which encode nearly all enzymes involved in the synthesis of PG, *A. marginale* and *Wolbachia* spp. encode some PG-associated enzymes; however, the ability to synthesize a murein layer (at least one of the canonical arrangement of cross-linked sugars and peptides) in these species is still unknown. Informative compilations of genes specific to certain taxa, host and vector organisms, transmission cycles, and host relationships (e.g., symbiosis versus pathogenesis) are provided. Across a generated species tree (using multiple concatenated genes), elevated rates of base substitution are characteristic of the *Anaplasmataceae* genomes relative to the genomes of *Rickettsia* spp.; however, no major differences in DNA repair systems across all genomes are reported.

APRIL 2006: *E. RUMINANTIUM* GARDEL (ERGA) AND *E. RUMINANTIUM* WELGEVONDEN ERWE (ERWE) (Frutos et al., 2006)

The genome sequences of *E. ruminantium* Gardel (Erga) and a second Welgevonden strain of *E. ruminantium* (Erwe) are highly similar to Erwo in nearly every aspect. Comparative analysis of genome synteny reveals a conserved gene order between three *Ehrlichia* (Erga, Erwe, and Erwo) and two *Anaplasma* (*A. marginale* and *A. phagocytophilum*) genomes, but a lack of synteny with the other *Rickettsiales* genomes. Conserved synteny and expansion of the *msp1* and *msp2* superfamilies are the defining characteristics of the derived lineage of *Anaplasmataceae* and suggest a very close evolutionary relatedness. Two characteristics distinguishing the three *E. ruminantium* genomes from the other *Rickettsiales* are a decreased amount of coding DNA (~63.5%) and an unusually large average length for intergenic regions (270 bp). As initially revealed from the analysis of the Erwo genome, an explosion of duplicate repeat regions comprises the majority of intergenic regions. Despite the growing number of *Rickettsiales* genome sequences, this mechanism of genome size plasticity mediated by expansion and contraction of diverse repeat families is unique to *E. ruminantium*.

MAY 2006: *R. BELLII* RML369-C (Ogata et al., 2006)

The 10th *Rickettsia* genome sequenced revealed the largest chromosome size to date (1.52 Mb; plasmid pRF added to the *R. felis* chromosome makes for the largest total genome) (Ogata et al., 2006). The ancestral phylogenetic position and questionable pathogenesis of the *R. bellii* genome make it very important for understanding the evolution of both vertebrate pathogenesis and accelerated genome reduction in derived *Rickettsia* genomes. Like *R. felis*, the *R. bellii* genome encodes an elevated number of TNPs, toxin-antitoxin modules, SR proteins, and TPR domain- and ARD-containing proteins. The genome shares limited synteny with other *Rickettsia* genomes. Several LGT events between *R. bellii* and divergent intracellular bacterial species (i.e., *Protochlamydia* spp. and *Legionella* spp.) were identified, with the most notable one being a complete F-T4SS (*tra/trb*) most similar to that encoded by *Protochlamydia amoebophila* UWE25, an obligate endosymbiont of amoebae. Consistent with the presence of a complete conjugative F-T4SS, type II pili-like cell surface appendages were visualized on *R. bellii* cells. However, unlike *R. felis*, a plasmid was not identified in the *R. bellii* genome (plasmids would later be reported from different *R. bellii* strains, discussed below). Intranuclear growth of *R. bellii* in several eukaryotic cell lines was demonstrated, as well as growth and survival in *Acanthamoeba polyphaga*. Ogata et al. conclude that ancestral protozoa likely served as melting pots for genetic exchange between intracellular bacteria, a process that equipped many species with genes involved in adaptation to life inside eukaryotic cells. A subsequent phylogenomic study including a second *R. bellii* genome (strain OSU 85-389) revealed ~100 genes unique to each genome, suggesting a variable accessory genome for *R. bellii* (Gillespie et al., 2008). A recent report of a plasmid from *R. bellii* An4 confirms this observation (Baldridge et al., 2010). As in the *R. felis* genome, the percentage of coding DNA is elevated in *R. bellii* relative to smaller *Rickettsia* genomes, suggesting that the expanded accessory genome may not be undergoing the degree of reductive evolution typical of the core *Rickettsia* genome, but rather mostly comprises LGT products (Fuxelius et al., 2007, 2008).

JUNE 2006: *EHRLICHIA CANIS* JAKE (Mavromatis et al., 2006)

The *E. canis* genome sequence was the fifth genome sequenced from the genus *Ehrlichia*, with a percentage of coding DNA (73%) similar to *E. chaffeensis*, but a genome size (1.3 Mb) and number of predicted ORFs (926) more similar to *E. ruminantium* genomes (Mavromatis et al., 2006). The metabolic profile of the *E. canis* genome is highly similar to other *Ehrlichia* genomes. *E. canis* also resembles other *Ehrlichia* genomes in the complete loss of genes encoding proteins involved in PG and LPS biosynthesis, as well as cholesterol biosynthesis and modification genes. Thus, like *E. chaffeensis* and *A. phagocytophilum* (Lin and Rikihisa, 2003), *E. canis* likely incorporates host cholesterol into its outer membrane. The *E. canis* genome also possesses a repertoire of OMPs and tandem repeat-containing proteins. All of these genus-specific traits related to cell envelope modification are suggested by Mavromatis et al. to play a key role in presenting a cell wall with continual alterations that ultimately allows the bacteria to evade host defenses, particularly Toll-like receptors 2 and 4 (Lin and Rikihisa, 2004). Compared with other *Anaplasmataceae* genomes, the *E. canis* genome contains a smaller percentage of short repetitive sequences, but is similar in the number of paralogous gene families. Bioinformatics identified a bias in serine and threonine residues in proteins predicted to interact with the host, with a particular association of this bias in glycoproteins. Other important genes and gene families were identified, including those encoding

ARD-containing ORFs, actin polymerization proteins, and proteins containing poly(G-C) tracts. Many of these genes, particularly those within paralogous families, are variably distributed across the *Ehrlichia* genomes, supporting the hypothesis for variable antigenic repertoires encoded within these genomes.

8 MAY 2007: *O. TSUTSUGAMUSHI* BORYONG (Cho et al., 2007)

The genome sequence of *O. tsutsugamushi* Boryong was the first genome sequenced from the STGR and was the largest rickettsial genome sequenced to date (2.1 Mb) (Cho et al., 2007). The genome has >100-fold-higher repeat density than the most repetitive *Rickettsia* genomes, and the substantial repetitive nature makes it the most repeated bacterial genome sequenced to date. Despite no identification of plasmids, an unprecedented 359 *tra* genes typical of F-T4SSs are present in 79 divergent sites within the genome. Associated with the amplified F-T4SSs are genes encoding proteins involved in signaling and host cell interaction, such as histidine kinases, ARD-containing proteins, and TPR domain-containing proteins. All of these F-T4SS clusters are especially characterized by various stages of pseudogenization, revealed by gene duplications and deletions, and rampant TNP integrations into several regions of the amplified segments. Over 400 TNPs, as well as genes encoding 60 phage integrases and 70 reverse transcriptases, were identified in the *O. tsutsugamushi* genome, adding to the repeat content and promoting the large-scale gene rearrangements that result in a lack of synteny with other *Rickettsiales* genomes. Cho et al. conclude that the presence of amplified conjugative elements resulted from either massive intragenomic duplications of one element or the continual invasion and integration of multiple elements. Whatever their evolutionary origin, continual population bottlenecks, which are typical of obligate intracellular bacteria, are suggested to have driven the rare fixation of these promiscuous conjugative elements.

AUGUST 2007: PLASMID pRM OF *RICKETTSIA MONACENSIS* IrR/MUNICH (Baldridge et al., 2007)

The second plasmid (pRM) from *Rickettsia* spp. and first from the SFGR was identified in *R. monacensis* (Baldridge et al., 2007). pRM was identified first by TNP-mediated insertion of coupled green fluorescent protein and chloramphenicol acetyltransferase marker genes into the plasmid, with subsequent DNA sequencing. Both circular and linear versions of pRM were detected. Despite being smaller (23.5 kb) and encoding far fewer ORFs (23) than pRF of *R. felis*, pRM contains many similar genes, with several sharing the same evolutionary history as the homologous pRF ORFs. This suggests that plasmids are more common than previously recognized in *Rickettsia* spp., and that plasmid-encoded genes have been transferred to the chromosomes of many rickettsial genomes, especially *R. felis* and AGR. Additionally, pulsed-field gel electrophoresis and Southern blot evidence is provided for a third rickettsial plasmid found in *Rickettsia amblyommii* (isolate WB-8-2), expanding the taxonomic range of rickettsial plasmids. Genes common to all three plasmids were identified, including those encoding DnaA replication initiation protein (pRM16/pRF05), alpha-crystallin-type small heat shock protein (ACD_sHsp) (pRM6/pRF51), ParA plasmid maintenance and stability protein (pRM18/pRF23), and Sca12 (pRM21/pRF24). Baldridge et al. conclude that the growing presence of rickettsial plasmids may eventually provide the basis for construction of an efficient transformation vector for *Rickettsia* spp., which remain intractable to convention genetic manipulation.

NOVEMBER 2007: *RICKETTSIA MASSILIAE* MTU5, PLASMID pRMA (Blanc et al., 2007a)

The *R. massiliae* genome was the 11th *Rickettsia* genome sequenced and third with an associated plasmid (pRMA) (Blanc et al., 2007a). The chromosome is the largest (1.38 Mb) of the sequenced SFGR genomes to date, and phylogeny estimation supports *R. massiliae* as the most basal lineage within this rickettsial group. Plasmid pRMA (15.3 kb) encodes 13 predicted ORFs and is thus the smallest identified rickettsial plasmid. Of the conserved ORFs across all known rickettsial plasmids (*dnaA*, *hsp2*, *parA*, and *sca12*), pRMA encodes only *dnaA* and *parA*. Except for a complete F-T4SS, the chromosome comprises mostly genes with counterparts in other rickettsial genomes. The F-T4SS genes and several associated components share high similarity and gene order with the *R. bellii tra* cluster, the partial *tra* cluster of pRF, and the proliferated conjugative elements present in the *O. tsutsugamushi* genome. Phylogeny estimation of common genes across these elements suggests that the partial element encoded on pRF is the most divergent. While all of these conjugative elements likely share common ancestry, their spread across rickettsial genomes occurs via LGT. Traces of similar degraded elements are reported for other *Rickettsia* genomes, suggesting that an ancestral insertion into the tRNA$^{Val\text{-}GAC}$ gene likely occurred early in *Rickettsia* evolution. Aside from this element, the chromosome is highly conserved in synteny with the other SFGR genomes. Based on the high similarity across the *tra* clusters of *R. bellii* and *R. massiliae*, the authors conclude that *recent* LGT between species of *Rickettsia* is probably more common than previously expected.

FEBRUARY 2008: *R. RICKETTSII* IOWA (Ellison et al., 2008)

The genome of the avirulent strain Iowa of *R. rickettsii* was the 12th sequenced from the genus *Rickettsia* and the fifth from the SFGR (Ellison et al., 2008). Compared with the genome of the highly virulent human pathogen *R. rickettsii* Sheila Smith (1.26 Mb; 1,345 ORFs), *R. rickettsii* Iowa encodes more ORFs (1,384) despite having roughly the same-size genome. A total of 143 genomic deletions occur across both strains, with truncation of *sca0* (*rompA*) in *R. rickettsii* Iowa being the most significant deletion. Experimental evidence demonstrates the absence of rOmpA in cells of *R. rickettsii* Iowa, supporting the deletion of this gene typically encoded within SFGR genomes. Interestingly, it was also shown that the *sca5* gene (*rompB*) is not processed to yield rOmpB, a major surface antigen encoded in all *Rickettsia* genomes sequenced to date. It is concluded that defects in these autotransporters, which deliver their passenger domains to the surface of rickettsial cells where they are abundant, are likely the major factors behind the avirulence of *R. rickettsii* Iowa. Despite an in-frame deletion within the RickA-encoding gene, the product of which is involved in actin polymerization in SFGR (Gouin et al., 2004; Jeng et al., 2004), *R. rickettsii* Iowa cells formed actin tails. This suggests that factors other than RickA and rOmpA, which also has been implicated in actin-based motility (Baldridge et al., 2004), are involved in actin tail formation-based intercellular spread in SFGR (Balraj et al., 2008; Haglund et al., 2010). Additionally, single-nucleotide polymorphism (SNP) analysis revealed 492 SNPs distinguishing *R. rickettsii* Iowa from *R. rickettsii* Sheila Smith, yet analysis of genome-wide gene expression for both strains did not reveal significant transcriptional differences, and the ORFs with different expression patterns are not noteworthy. Data were presented that show that infection with *R. rickettsii* Iowa protects guinea pigs against challenge with *R. rickettsii* Sheila Smith.

AUGUST 2008: *O. TSUTSUGAMUSHI* IKEDA (Nakayama et al., 2008)

The *O. tsutsugamushi* Ikeda genome was the second sequenced from STGR (Nakayama et al., 2008). The genome is slightly smaller than the *O. tsutsugamushi* Boryong genome (2.01 Mb) yet is predicted to encode 785 more ORFs (1,967 total predicted ORFs). However, this difference is the result of the treatment of pseudogenes, which are rampant in both genomes and are mostly associated with proliferated (yet decayed) MGEs. Adjusted numbers of predicted ORFs for both genomes, with pseudogenes included, reflect a slightly larger number of ORFs in the larger Boryong genome (2,364) versus the Ikeda genome (2,197) (Fig. 1). In addition to the integrative conjugative elements, which are highly similar in composition to the elements parasitizing the *O. tsutsugamushi* Boryong genome, 10 types of transposable elements and 7 types of short repeats are identified. As in the *O. tsutsugamushi* Boryong genome, the effects of intense insertion and pseudogenization of an assorted array of MGEs is seen predominantly in the lack of synteny with any other *Rickettsiales* genome. Even the two STGR genomes share little synteny between them. An in-depth metabolic comparison of *O. tsutsugamushi* Ikeda with *Rickettsia* spp. genomes reveals the complete loss of LPS biosynthesis genes, a reduced TCA cycle and PG biosynthesis pathway, fewer genes encoding enzymes involved in nucleotide interconversion, and the inability to synthesize malonyl coenzyme A for initiation of fatty acid biosynthesis. A pathway for simple sugar metabolism can be seen as an intermediate between reduced gluconeogenesis (*Anaplasmataceae*) and loss of most glycolytic enzymes (*Rickettsia* spp.). Such a pathway would allow for the biosynthesis of glycerolipids, but the inability to synthesize nucleotide precursors via the nonoxidative pentose phosphate pathway. Unsurprisingly, as in *Rickettsia* spp. genomes, five *tlc* genes are present within the *Orientia* genomes for nucleotide scavenging. Collectively, the STGR genomes are more metabolically dependent on their hosts than any other *Rickettsiales* genomes (Darby et al., 2007; Fuxelius et al., 2007), a likely consequence of the extreme proliferation of MGEs that have shuffled and deleted many genes.

SEPTEMBER 2008: *WOLBACHIA* ENDOSYMBIONT OF *CULEX QUINQUEFASCIATUS* Pel (*w*Pip Pel) (Klasson et al., 2008)

The *Wolbachia* endosymbiont of *C. quinquefasciatus* Pel (*w*Pip Pel) genome was the sixth sequenced from the genus *Wolbachia*, as well as the third complete genome and the first representing the B-supergroup *Wolbachia* (Klasson et al., 2008). The genome is the largest (1.48 Mb) among the *Wolbachia* genomes sequenced to date but encodes fewer ORFs (1,275) than the incomplete *w*An genome (1,802), numbers consistent across various annotation methods (Fig. 1). Comparative analysis across *w*Pip Pel, *w*Mel (A-supergroup), and *w*Bm (D-supergroup) revealed a conserved core *Wolbachia* genome, as well as a highly variable and dynamic accessory genome. Like *w*Mel, *w*Pip Pel encodes far more IS elements than *w*Bm, with 116 elements identified from eight IS families. These derived acquisitions of ISs have resulted in a substantial amount of genome rearrangement between all three genomes, with few conserved gene regions remaining. Relative to the *w*Mel and *w*Bm genomes, the *w*Pip Pel genome encodes elevated levels of prophage and genes encoding proteins with ARDs. Klasson et al. hypothesize that these factors likely play a major role in the biology of *w*Pip Pel. Adding particular complexity to the *w*Pip Pel genome are variable mosaics of the WO-B prophage, which have likely arisen through widespread insertion and duplication events that facilitate recombination across copies. A large portion of the genes encoding ARD-containing proteins, which are encoded nearly three times more in *w*Pip Pel than in *w*Mel, are located in close proximity to the WO-B prophage clusters. Collectively, these diversifying factors hint at a possible mechanism underlying reproductive manipulation of hosts by *Wolbachia* spp.

11 JANUARY 2009: *A. MARGINALE* FLORIDA, MISSISSIPPI, PUERTO RICO, AND VIRGINIA (Dark et al., 2009)

The genome sequences of strains Florida, Mississippi, Puerto Rico, and Virginia of *A. marginale* raised the number of *Anaplasma* genomes from two to six, providing a detailed look into the variation underlying the *A. marginale* pangenome (Dark et al., 2009). The complete genome sequence of *A. marginale* Florida reveals a slightly larger genome size (1.2 Mb) and ORF count (948) than the *A. marginale* St. Maries genome (Fig. 1). The nearly complete genomes of the three other strains, estimated to be greater than 96% sequenced, show similar genome sizes and numbers of predicted ORFs. Comparative genomics across diverse bacterial species (with multiple strains) suggests a highly conserved *A. marginale* pan-genome, with a comparison of SNPs between *A. marginale* St. Maries and *A. marginale* Florida revealing <1% SNPs. Of all SNPs determined across the five strains of *A. marginale*, 33.5% are present in at least two strains, with 3% present in all strains except *A. marginale* Florida. Unsurprisingly, variation in two MSP genes, *msp2* and *msp3*, was observed across strains. The *Anaplasma* appendage-associated protein family (Stich et al., 2004) also exhibited variation across strains. Despite a high level of conservation across *A. marginale* strains, comparison with both extracellular and intracellular bacterial pan-genomes indicates that *A. marginale* is intermediate in its amount of strain-specific SNPs, suggesting that differences in lifestyle across bacteria do not correlate with intrastrain variability. Dark et al. hypothesize that a highly conserved core genome in obligate intracellular bacterial species may result from pressure from the host cell environment, the process of reductive genome evolution, or both phenomena.

MARCH 2009: *WOLBACHIA* ENDOSYMBIONT OF *C. QUINQUEFASCIATUS* JHB (*w*Pip JHB) (Salzberg et al., 2009)

The *Wolbachia* endosymbiont of *C. quinquefasciatus* JHB (*w*Pip JHB) genome was the seventh genome sequence from the genus *Wolbachia* and was sequenced simultaneously with the *C. quinquefasciatus* sequencing project (Salzberg et al., 2009). After bacterial sequences were separated from nonbacterial contigs, a genome size was estimated to be 1.54 Mb, with 1,378 predicted ORFs. While the complete genome of the closely related *w*Pip Pel was used as a reference to assemble the *w*Pip JHB reads, the genomes differ by over 100 predicted ORFs (Fig. 1). This discrepancy was attributed to differences in annotation methods. However, despite high similarity across both genomes, *w*Pip JHB contains four regions not found within *w*Pip Pel but present within *w*Mel. *w*Pip Pel contains two regions not found within the *w*Pip JHB genome. These regions encode a repair enzyme, RadC, and a transcription regulator. ISs and related elements have caused 10 rearrangements between the *w*Pip JHB and *w*Pip Pel genomes, resulting mostly in inversions in one sequence relative to another. As with a previous study (Salzberg et al., 2005a, 2005b), this work by Salzberg et al. highlights the necessity to thoroughly evaluate eukaryotic (especially arthropod) genome sequencing projects for the presence of denizen obligate intracellular bacteria.

7 APRIL 2009: *WOLBACHIA* SP. *w*Ri (*w*Ri), *WOLBACHIA* ENDOSYMBIONT OF *MUSCIDIFURAX UNIRAPTOR* (*w*Uni) (Klasson et al., 2009b)

The complete genome sequence of *Wolbachia* sp. *w*Ri (*w*Ri) (1.45 Mb; 1,150 ORFs) and the partial genome sequence of the *Wolbachia* endosymbiont of *M. uniraptor* (*w*Uni) (0.87 Mb; 827 ORFs) bring the number of genome sequences from *Wolbachia* to nine (Klasson et al., 2009b). The genomic comparison was mostly limited to the A-supergroup *Wolbachia*: *w*Mel, *w*Ri, and *w*Uni. ISs and repeat regions are associated with 35 breaks in synteny between the *w*Mel and *w*Ri genomes, and the *w*Ri genome encodes 12 more ARD-containing proteins (35 total) than the *w*Mel genome. Four prophage regions were identified, as well as 114 pseudogenes, with the majority of ORFs differing between the *w*Mel and *w*Ri genomes encoding prophage, ARD-containing proteins, and TNPs. A phylogeny of all ARD-containing proteins encoded in the *w*Mel, *w*Ri, and *w*Uni genomes highlights the diversity and proliferation of these ORFs in the A-supergroup *Wolbachia*. Short tandem duplication was identified as a mechanism for the diversification of genes encoding ARD-containing proteins in A-supergroup *Wolbachia*. These sequences were used to identify likely LGT events between the A- and B-supergroup *Wolbachia*, but not between other supergroups. Remarkably, meticulous sequence analysis within the A-supergroup *Wolbachia* identified a high rate of recombination between the *w*Mel, *w*Ri, and *w*Uni genomes. A comparison with other intracellular bacteria identified the A-supergroup *Wolbachia* as the most highly recombining genomes, and phylogeny estimation revealed many different genes and gene regions that support different strain relationships. Klasson et al. suggest that the ability of multiple *Wolbachia* spp. to coinfect the same host, and hence freely participate in recombination, has resulted in mosaic genomes for A-supergroup *Wolbachia*.

20 APRIL 2009: *RICKETTSIA AFRICAE* ESF-5, PLASMID pRA (Fournier et al., 2009)

The *R. africae* genome was the 13th *Rickettsia* genome sequenced to date and revealed the third plasmid sequence associated with a complete genome sequence (pRA) (Fournier et al., 2009). The genome size is typical of SFGR (1.28 Mb), as is the ORF count (1,545, adjusted from RefSeq). Aside from an 88.5-bp inversion, the genome shares high synteny with the *R. conorii* genome. The expression profile of 18 genes unique to the *R. africae* genome identified two ORFs upregulated at 37°C, one of which encodes a putative protease-encoding gene. All 18 of these genes show varying degrees of pseudogenization across *R. conorii*, *R. rickettsii* (strain Sheila Smith), and *R. prowazekii*, and coupled with observations of an increased loss of regulatory genes in smaller *Rickettsia* genomes, prompted Fournier et al. to conclude that virulence is mostly associated with gene loss in *Rickettsia* spp. Plasmid pRA (12.3 kb; 11 ORFs) is the smallest plasmid from *Rickettsia* spp. to date, encoding several ORFs similar to those from the other three plasmids identified in *Rickettsia* spp. (pRF, pRM, and pRMA); however, only two genes appear to be common to all four plasmids: *dnaA*, encoding a replication initiator; and *parA*, encoding a protein involved in plasmid maintenance and stability. A truncated *tra* gene cluster is encoded near the $tRNA^{Val\text{-}GAC}$ gene, typical of most other *Rickettsia* genomes, illustrating the second plasmid-harboring *Rickettsia* genome without a complete F-T4SS. Finally, a robust sampling of *Amblyomma* spp. ticks revealed an infection rate of 89.6%, and generated sequences proved *R. africae* to be clonal, the first *Rickettsia* species to be identified as such.

AUGUST 2009: *N. RISTICII* ILLINOIS (Lin et al., 2009)

The genome of *N. risticii* Illinois was the second sequenced for *Neorickettsia* (Lin et al., 2009). The genome is slightly larger than the *N. sennetsu* Miyayama genome (0.88 versus 0.86 Mb) yet is predicted to encode 37 fewer ORFs (898 total predicted ORFs). Annotation by PATRIC suggests the genomes are even more similar in predicted ORFs (Fig. 1). The genomes are also similar in their metabolic capacity and, like other *Anaplasmataceae* genomes, encode enzymes responsible for the synthesis of major vitamins, cofactors, and nucleotides. Both *Neorickettsia* genomes are also highly conserved in gene order, but differ drastically from other *Rickettsiales* genomes. Despite high similarity across *N. sennetsu* and *N. risticii*, only 88.2% of the total protein-encoding genes are shared across the genomes, suggesting variability in accessory genomes at the genus level. A conserved core *Anaplasmataceae* genome of 525 ORFs is calculated, suggesting that >200 ORFs encoded in the *Neorickettsia* genomes are specific to the genus. Similar to the genomes of *Anaplasma* spp. and *Ehrlichia* spp., both *Neorickettsia* genomes encode <5 MGEs. This suggests either two independent losses of a functional mobilome relative to the *Wolbachia* genomes or a gain of diverse MGEs independently in *Wolbachia* spp. The latter is the likely scenario, given the fluidity of *Wolbachia* genomes, particularly within the A-supergroup (discussed above). A list of genes (as well as their distribution across *Anaplasmataceae* genomes) putatively involved in pathogenesis are discussed, including those encoding one- and two-component regulatory systems, the *rvh* T4SS, ARD-containing proteins, and OMPs. Expression of the duplicate major pilins (RvhB2) of the *rvh*

P-T4SS is demonstrated, as well as their bipolar localization on the surface of *N. risticii*. This is the first demonstration of localization of P-T4SS major pilin proteins on the surface of an obligate intracellular species.

21 DECEMBER 2009: *RICKETTSIA PEACOCKII* RUSTIC, PLASMID pRPR
(Felsheim et al., 2009)

The *R. peacockii* genome was the 14th *Rickettsia* genome sequenced to date and revealed the fourth plasmid sequence associated with a complete genome sequence (pRPR) (Felsheim et al., 2009). The genome size is typical of SFGR (1.31 Mb), as is the ORF count (1,558, adjusted from RefSeq). The sequencing of *R. peacockii* provided another look into the mechanisms of pathogenicity in *R. rickettsii*, as both species are closely related and infection of *Dermacentor andersoni* ticks with *R. peacockii* correlates with a reduced presence of *R. rickettsii* in tick populations. A previously described transposon, ISRPe1, which was demonstrated to interrupt the gene encoding RickA in *R. peacockii* (Simser et al., 2005), was present in 42 regions of the *R. peacockii* chromosome and 2 regions of pRPR. Phylogeny estimation of ISRPe1 and related TNPs from diverse bacteria suggest a common origin among several intracellular bacterial species. The insertion of ISRPe1, coupled with intrachromosomal recombination between similar copies, has resulted in a loss of synteny between the *R. peacockii* and the genomes of derived SFGR. ISRPe1 insertion has also contributed to pseudogenization. In comparison with the genomes of *R. rickettsii* Sheila Smith and *R. rickettsii* Iowa, several genes were shown to be entirely deleted or mutated, including *sca0*, *sca1*, *ptrB* (S9 serine protease), *dsbA*, and a gene encoding a putative phosphoethanolamine transferase. Of particular interest, a deleted gene in *R. peacockii* encoding an ARD-containing protein is mutated in *R. rickettsii* Iowa but complete in *R. rickettsii* Sheila Smith (and some other pathogenic *Rickettsia* species). Felsheim et al. conclude that all of these genomic aberrations provide areas of focus for understanding the loss of virulence in *R. peacockii*. Plasmid pRPR (26 kb) encodes 20 predicted ORFs, half of which have top BLASTP hits to *Rickettsia* sequences, which is consistent with other *Rickettsia* plasmids (Baldridge et al., 2007; Gillespie et al., 2007). Among top BLAST hits to non-*Rickettsia* sequences, a part of a flagellar glycosylation island from *Pseudomonas aeruginosa* is notable. The two genes common to plasmids pRF, pRM, pRMA, and pRA (*dnaA* and *parA*) are also encoded on pRPR. As in other *Rickettsia* genomes, a truncated *tra* gene cluster is encoded near the tRNA$^{Val\text{-}GAC}$ gene on the chromosome, illustrating the third plasmid-harboring *Rickettsia* genome without a complete F-T4SS (*R. felis* and *R. africae*). Other than ORFs encoded on pRPR and the 44 copies of ISRPe1, no evidence is shown for recent LGT in the *R. peacockii* genome.

JANUARY 2010: *ANAPLASMA CENTRALE* ISRAEL
(Herndon et al., 2010)

The genome of *A. centrale* Israel, also referred to as *A. marginale* subsp. *centrale*, was the seventh genome sequenced from the genus *Anaplasma* (Herndon et al., 2010). As a naturally attenuated close relative of *A. marginale*, and used for decades as a vaccine, its genome sequence is critical for understanding the mechanisms of virulence within the *A. marginale* sensu stricto strains. Genome size (1.2 Mb), predicted ORFs (1,153) and GC content (50%) are characteristic of other *A. marginale* genomes. Through comparison of *A. marginale* genomes, 10 genes were found to be unique to *A. centrale*, while 18 genes present in all *A. marginale* sensu stricto strains were absent in *A. centrale*. Genes of the MSP1 superfamily of *A. marginale* sensu stricto strains are not conserved in *A. centrale*, and thus are likely not associated with the immunity afforded by vaccination with this strain. The list of OMPs possibly involved in inducing protective immunity was narrowed down to four members of the MSP2 superfamily and two proteins not placed in a superfamily. Recent experimental evidence showed that only three of these proteins (along with three different OMPs) from *A. centrale* induced antibodies that recognize eight OMP orthologs in virulent *A. marginale* St. Maries (Agnes et al., 2011).

MARCH 2010: *R. PROWAZEKII* Rp22 (Bechah et al., 2010)

The *R. prowazekii* Rp22 genome was the 15th *Rickettsia* genome sequenced to date, with a genome size (1.1 Mb) and ORF count (910) highly similar to those of *R. prowazekii* Madrid E (Bechah et al., 2010). Previously, factors underlying the virulence associated with *R. prowazekii* Breinl (no genome sequence available to date) were not made readily apparent in a genome-wide transcriptional analysis comparison with *R. prowazekii* Madrid E (Ge et al., 2004). The avirulence associated with *R. prowazekii* Madrid E is considered sensitive, as cultivation of this strain in most cell types and model organisms results in restoration of virulence (with the revertant strain referred to as Evir). No differences in protein expression were previously detected between *R. prowazekii* Madrid E and Evir (Balayeva et al., 1992). Using a recent murine model for epidemic typhus (*R. prowazekii* Breinl), combined with transcriptomic, proteomic, and other approaches, Bechah et al. identified several phenotypes that differ in degree of virulence. These factors were shown to be associated with upregulation of antiapoptotic genes or the type I interferon pathway in mouse cells. In addition, several surface proteins varied in their expression profiles across strains, as did the pattern of protein methylation. *R. prowazekii* Rp22, one of the strains with a virulent phenotype, was sequenced to determine the genetic basis for the observed varying degrees of virulence. Comparative genomics revealed mutational hot spots in poly(A) and poly(T) within eight genes that resulted in their inactivation. A sequence of reactivations of these genes was observed via passage of avirulent strains in both cell lines and experimental models. The *recO* gene, encoding an enzyme involved in DNA repair and *recF* pathway recombination, was always the first gene reactivated upon reversion to virulence. Thus, if RecO functions in restoration of pseudogenes in *Rickettsia* spp., the model for gene loss accelerating gain of virulence in *Rickettsia* spp. needs reconsideration.

MAY 2010: PLASMIDS pRAM18 AND pRAM23 OF "*CANDIDATUS* RICKETTSIA AMBLYOMMII" AaR/SC (Baldridge et al., 2010)

Prior to this report, four plasmids (pRF, pRMA, pRA, and pRPR) from complete genome sequences of *R. felis*, *R. massiliae*, *R. africae*, and *R. peacockii* were published, along with the pRM plasmid of *R. monacensis*. A plasmid was also detected in *R. amblyommii* (isolate WB-8-2) (Baldridge et al., 2007), as well as in *Rickettsia helvetica*, *R. peacockii*, *R. amblyommii*, *Rickettsia hoogstraalii*, and *R. massiliae* (Baldridge et al., 2008), but sequences were not published. This study by Baldridge et al. again detected plasmids in a diverse list of *Rickettsia* species: *R. akari*, "*Candidatus* Rickettsia amblyommii," *R. bellii*, *R. rhipicephali*, and the *Rickettsia* endosymbiont of *Ixodes scapularis* (REIS). No plasmids were detected in multiple isolates of *Rickettsia parkeri* from diverse geographic locales. The suspicion of multiple divergent plasmids within individual species was confirmed via cloning and sequencing of two distinct plasmids from "*Candidatus* Rickettsia amblyommii" AaR/SC, named pRAM18 and pRAM23. A phylogeny was estimated based on the most conserved gene across these plasmids, *parA*, and suggested multiple origins for this gene and possibly plasmid incompatibility groups. Phylogenies were also estimated for two small heat shock proteins, Hsp1 and Hsp2, which are commonly encoded on *Rickettsia* plasmids. *hsp2* genes were shown to be plasmid specific and more common, whereas *hsp1* genes, unique to pRAM18, pRAM23, and plasmids from *R. felis*, *R. monacensis*, and *R. peacockii*, were suggested to have a different evolutionary origin. Real-time quantitative PCR was used to determine that plasmid copy numbers ranged from 2.4 to 9.2 in rickettsial cells. This report, coupled with previous studies, has now identified plasmids in a wide range of species in the AGR, TRGR, and SFGR, most of which are associated with blood-feeding arthropods. The significance of rickettsial plasmids remains a mystery, as there is no association of plasmid presence/absence with pathogenicity, no correlation between plasmids and a complete conjugation system (F-T4SS), and no particularly interesting plasmid genes encoding putative virulence factors. A study designed specifically to detect conjugation genes (*tra* associated with F-T4SSs) within an even broader range of *Rickettsia* species identified a prevalence of rickettsial plasmid-like genes in species primarily associated with phytophagous insects (Weinert et al., 2009a). However, results from that study cannot discern between plasmid-encoded *tra* genes and those associated with the mobile element encoded in the chromosomes of *R. massiliae* and *R. bellii tra* cluster, the partial *tra* clusters within most other sequenced *Rickettsia* genomes, and the proliferated conjugative elements present in the STGR genomes. As we discuss below regarding the REIS genome sequence, the significance of *tra* genes in the evolution of *Rickettsiaceae* is quite a fascinating story.

OCTOBER 2011: REIS, PLASMIDS pREIS1 TO -4 (Gillespie et al., 2011a)

The REIS genome was the 16th *Rickettsia* genome sequenced to date, with its four distinct plasmids (pREIS1 to -4) bringing the total number of sequenced rickettsial plasmids to 11 (Gillespie et al., 2011a). REIS is also the first genome from *Rickettsia* to be assembled from a eukaryotic sequencing project, as the bacterial reads were separated from the *I. scapularis* whole-genome shotgun sequencing data. Genome-based phylogeny estimation suggests that REIS is the most basal member of the SFGR, sharing attributes with both pathogenic and nonpathogenic rickettsiae. REIS is the largest *Rickettsia* genome sequenced to date (>2 Mb), with a chromosome of >1.8 Mb that contains 2,059 predicted ORFs. The accessory genome alone encodes almost twice as many genes as the entire *R. prowazekii* genome. Plasmids pREIS1 to -4 are novel to *Rickettsia* species, and contain 59, 83, 43, and 65 predicted ORFs, respectively. Many of these ORFs have homologs in other species of diverse obligate intracellular bacteria. The most outstanding finding within the REIS genome is the extraordinary proliferation of MGEs, which contributes to a limited synteny with other *Rickettsia* genomes and places REIS among the most repetitive bacterial genomes. In particular, a *Rickettsiales* amplified genetic element (RAGE), previously identified as a "proliferated-yet-decayed" element in STGR (*O. tsutsugamushi*) genomes (Cho et al., 2007; Nakayama et al., 2008), is present on both the REIS chromosome and plasmids (pREIS1 and pREIS3). However, unlike in *O. tsutsugamushi*, nine complete (or nearly complete) RAGEs exist within the REIS genome and encode entire F-T4SSs similar to the single-copy *tra* cluster present in the genomes of *R. bellii* and *R. massiliae* (Ogata et al., 2006; Blanc et al., 2007a). An unparalleled abundance of encoded TNPs (>650) relative to genome size, together with the RAGEs and other MGEs, comprise >35% of the total genome, making REIS one of the most dynamic bacterial genomes sequenced to date. Despite an extraordinarily plastic accessory genome, including several intriguing LGT events (e.g., a complete biotin biosynthesis operon, remnants of the WO-B prophage, and a gram-positive aminoglycoside antibiotic synthesis and resistance cluster) from other diverse intracellular bacteria, comparative analysis with 15 *Rickettsiaceae* genomes indicates that REIS does not differ in the mode and tempo of gene loss typical of rickettsial reductive genome evolution. As REIS is not known to invade vertebrate cells and has no known pathogenic effects on *I. scapularis*, its genome sequence provides an invaluable tool for deciphering the mechanisms of arthropod and vertebrate pathogenicity via comparison with the genomes of virulent and avirulent *Rickettsia* species. Of significance to *Rickettsiaceae* evolution, compelling evidence is presented suggesting that many of the genes involved in obligate intracellular lifestyle were acquired via MGEs, especially the RAGEs, through a continuum of genomic invasions by these integrative conjugative elements.

Phylogeny

A robust phylogeny based on whole-genome sequences supports the current taxonomic delineations within the *Anaplasmataceae* and *Rickettsiaceae*, with monophyly of each of the six genera strongly supported (Fig. 2). Thus, the major amendments to *Rickettsiales* classification, which were primarily based on phylogeny estimation of 16S rRNA gene sequences, suggest that this gene should still be considered a powerful estimator of lineage divergence above the genus level. However, trees estimated from single (or few) genes are not as robust as multigene-based phylogenies. For example, while individual phylogenetic analyses of the *rvh* T4SS genes of *Rickettsia* spp. failed to corroborate the species tree in virtually every case (DNA and protein sequences), the tree based on the combination of all Rvh components did agree with the genome-based phylogeny (Gillespie et al., 2009b). This doesn't necessarily imply LGT in the shaping of the *rvh* T4SS across these genomes, but rather that the phylogenetic signal within these individual genes is often not strong enough to recover the organismal phylogeny (lack of resolving power). Other factors confounding single-gene and -protein phylogeny estimation include positive selection (particularly regarding proteins interacting directly with the host), accelerated rates of evolution caused by nucleotide compositional bias, and coalescence.

While considered rare in the genomes of obligate intracellular bacteria, evidence for LGT is mounting, and *Rickettsiales* are not immune. However, depending on the evolutionary time period of the gene transfer, some products of LGT may reflect the species divergence subsequent to the gain of the gene in question. This is especially the case for the *rvh* T4SS, which was clearly acquired in the ancestor to *Anaplasmataceae*/*Rickettsiaceae* from a non-*Alphaproteobacteria* species, yet reflects the phylogeny shown in Fig. 2 (Gillespie et al., 2010). Astonishingly, the genome sequence of the obligate intracellular amoeba symbiont "*Candidatus* Amoebophilus asiaticus" (*Bacteroidetes*) revealed a plethora of genes shared with *Rickettsiaceae* genomes (Schmitz-Esser et al., 2010). Many of the phylogenies based on these genes corroborate the *Rickettsiaceae* species tree; however, the genes are clearly not of alphaproteobacterial origin (data not shown). Thus, they represent ancient LGT events that have been stabilized via long periods of vertical

FIGURE 2 Whole-genome-based phylogeny estimation for 46 *Rickettsiales* taxa. Taxon codes used in Fig. 1 are shown in parentheses after each organism name. The following pipeline was implemented to estimate *Rickettsiales* phylogeny. BLAT (refined BLAST algorithm) (Kent, 2002) searches were performed to identify similar protein sequences between all genomes, including the two outgroup taxa. To predict initial homologous protein sets, MCL (Van Dongen, 2008) was used to cluster BLAT results, with subsequent refinement of these sets using in-house hidden Markov models (Durbin et al., 1998). These protein families were then filtered to include only those with membership in >80% of the analyzed genomes (39 or more taxa included per protein family). Multiple sequence alignment of each protein family was performed using MUSCLE (default parameters) (Edgar, 2004a, 2004b), with masking of regions of poor alignment (length heterogeneous regions) done using Gblocks (default parameters) (Castresana, 2000; Talavera and Castresana, 2007). All modified alignments were then concatenated into one dataset. Tree building was performed using FastTree (Price et al., 2010). Support for generated lineages was estimated using a modified bootstrapping procedure, with 100 pseudoreplications sampling only half of the aligned protein sets per replication. (Note: Standard bootstrapping tends to produce inflated support values for very large alignments.) All branches in the illustrated tree were supported by 100%. Local refinements to tree topology were attempted in instances where highly supported nodes have subnodes with low support. This refinement is executed by running the entire pipeline on only those genomes represented by the node being refined (with additional sister taxa for rooting purposes). The refined subtree was then spliced back into the full tree. More information pertaining to this phylogeny pipeline is available at PATRIC. doi:10.1128/9781555817336.ch3.f2

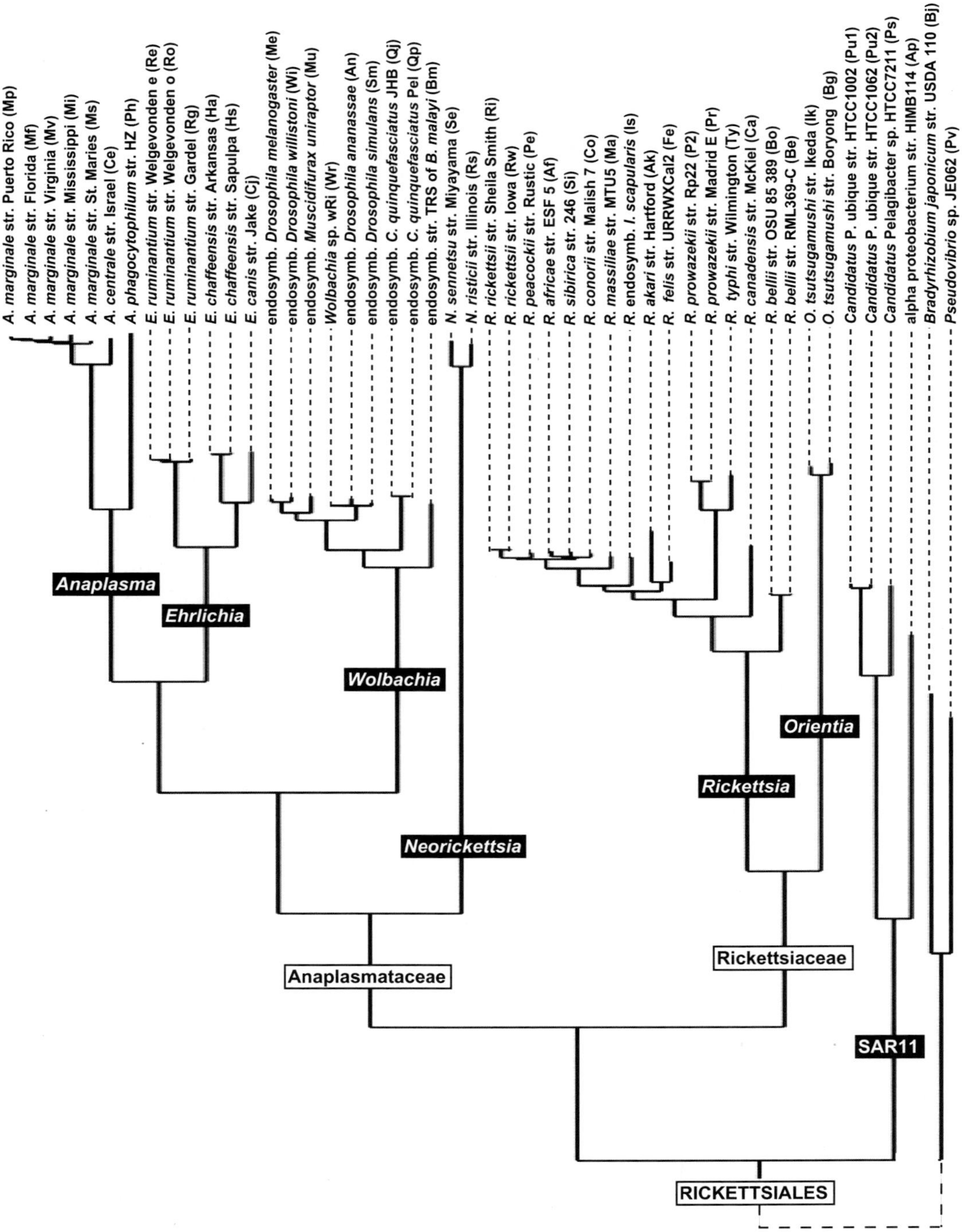
A. marginale str. Puerto Rico (Mp)
A. marginale str. Florida (Mf)
A. marginale str. Virginia (Mv)
A. marginale str. Mississippi (Mi)
A. marginale str. St. Maries (Ms)
A. centrale str. Israel (Ce)
A. phagocytophilum str. HZ (Ph)
E. ruminantium str. Welgevonden e (Re)
E. ruminantium str. Welgevonden o (Ro)
E. ruminantium str. Gardel (Rg)
E. chaffeensis str. Arkansas (Ha)
E. chaffeensis str. Sapulpa (Hs)
E. canis str. Jake (Cj)
endosymb. Drosophila melanogaster (Me)
endosymb. Drosophila willistoni (Wi)
endosymb. Muscidifurax uniraptor (Mu)
Wolbachia sp. wRi (Wr)
endosymb. Drosophila ananassae (An)
endosymb. Drosophila simulans (Sm)
endosymb. C. quinquefasciatus JHB (Qj)
endosymb. C. quinquefasciatus Pel (Qp)
endosymb. str. TRS of B. malayi (Bm)
N. sennetsu str. Miyayama (Se)
N. risticii str. Illinois (Rs)
R. rickettsii str. Sheila Smith (Ri)
R. rickettsii str. Iowa (Rw)
R. peacockii str. Rustic (Pe)
R. africae str. ESF 5 (Af)
R. sibirica str. 246 (Si)
R. conorii str. Malish 7 (Co)
R. massiliae str. MTU5 (Ma)
R. endosymb. I. scapularis (Is)
R. akari str. Hartford (Ak)
R. felis str. URRWXCal2 (Fe)
R. prowazekii str. Rp22 (P2)
R. prowazekii str. Madrid E (Pr)
R. typhi str. Wilmington (Ty)
R. canadensis str. McKiel (Ca)
R. bellii str. OSU 85 389 (Bo)
R. bellii str. RML369-C (Be)
O. tsutsugamushi str. Ikeda (Ik)
O. tsutsugamushi str. Boryong (Bg)
Candidatus P. ubique str. HTCC1002 (Pu1)
Candidatus P. ubique str. HTCC1062 (Pu2)
Candidatus Pelagibacter sp. HTCC7211 (Ps)
alpha proteobacterium str. HIMB114 (Ap)
Bradyrhizobium japonicum str. USDA 110 (Bj)
Pseudovibrio sp. JE062 (Pv)
Anaplasma
Ehrlichia
Wolbachia
Neorickettsia
Rickettsia
Orientia
Anaplasmataceae
Rickettsiaceae
SAR11
RICKETTSIALES

transmission. Recent analysis of the REIS genome identified similar instances of ancient LGT events that have stabilized in the genomes of *Rickettsia* spp. (Gillespie et al., 2011a). Many of these genes are characteristic of obligate intracellular bacteria, including those coding for metabolite transporters (*proP*, *eamA*, *glpT*, *tlc*), drug exporters (*ampG*, *mdlB*), SR regulators (*gppA*, *spoT*), and enzymes not involved in central metabolism (*ugd*, *pat1*, *rhlE*). All of these genes are associated (or once were) with MGEs, especially the RAGE that has proliferated in the genomes of STGR (Cho et al., 2007; Nakayama et al., 2008). In *Rickettsia* spp., it appears that many of the genes important for obligate intracellular survival were seeded by integrative conjugative elements, like the RAGE, that were probably spread by plasmids and are present mostly as relics across the diverse genomes (Gillespie et al., 2011a).

Recent LGT events appear to be more common within the genomes of species of *Wolbachia* and *Rickettsia* than in the genomes of other derived *Rickettsiales*. Regarding *Rickettsia* spp., it was shown that many of the genes of plasmid pRF of *R. felis* were more similar to chromosomal genes of *R. bellii* than the equivalents encoded on the *R. felis* chromosome (Gillespie et al., 2007). The genome sequence of *R. massiliae* verified this finding, demonstrating close affinities between its RAGE and that of *R. bellii* (Blanc et al., 2007a). However, analysis of the REIS genome revealed multiple RAGEs within its genome and provided substantial evidence that numerous distinct RAGE elements exist within the *Rickettsiaceae* mobilome, and that these elements are likely still capable of LGT (Gillespie et al., 2011a). Studies including phylogenetic analyses of plasmid-encoded ORFs from other species of *Rickettsia* have also determined that LGT is prominent across *Rickettsia* genomes (Baldridge et al., 2007, 2008, 2010), and that regions of plasmids are capable of insertion into chromosomes (Blanc et al., 2007a; Gillespie et al., 2011a). LGT is also widespread across *Wolbachia* genomes and is often associated with bacteriophage (Gavotte et al., 2007; Ishmael et al., 2009; Chafee et al., 2010; Kent et al., 2011). Since the initial discovery of large-scale LGT events between *Wolbachia* species and their eukaryotic hosts (Dunning Hotopp et al., 2007), subsequent studies have revealed that interdomain gene transfer between *Wolbachia* and eukaryotic genomes is prevalent (Klasson et al., 2009a; Nikoh and Nakabachi, 2009; Woolfit et al., 2009). The effect these LGT products have on the putative function and evolutionary dynamics of *Wolbachia* hosts remains an interesting area of research for rickettsiology (Blaxter, 2007; Bordenstein, 2007).

Gene duplication will also pose problems for phylogeny estimation if paralogous genes (duplicated subsequent to species divergence) are analyzed together as opposed to orthologs (duplication events occurring prior to species divergence). In *Rickettsiales*, the best example

FIGURE 3 Genus-level alignments of complete *Rickettsiales* genome sequences. (A) Sixteen *Rickettsia* genomes: Br, *R. bellii* RML369-C; Bo, *R. bellii* OSU 85-389; Ca, *R. canadensis* McKiel; Ty, *R. typhi* Wilmington; Pr, *R. prowazekii* Madrid E; P2, *R. prowazekii* Rp22; Fe, *R. felis* URRWXCal2; Ak, *R. akari* Hartford; Is, REIS; Ma, *R. massiliae* MTU5; Pe, *R. peacockii* Rustic; Ri, *R. rickettsii* Sheila Smith; Rw, *R. rickettsii* Iowa; Co, *R. conorii* Malish 7; Si, *R. sibirica* 246; Af, *R. africae* ESF-5. (B) Two *Orientia* genomes: Bg, *O. tsutsugamushi* Boryong; Ik, *O. tsutsugamushi* Ikeda. (C) Two *Neorickettsia* genomes: Rs, *N. risticii* Illinois; Se, *N. sennetsu* Miyayama. (D) Four *Wolbachia* genomes: Bm, *Wolbachia* endosymbiont strain TRS of *B. malayi*; Me, *Wolbachia* symbiont of *D. melanogaster*; Wr, *Wolbachia* endosymbiont sp. *w*Ri; Qp, *Wolbachia* symbiont of *C. quinquefasciatus* Pel. (E) Five *Ehrlichia* genomes: Cj, *E. canis* Jake; Ha, *E. chaffeensis* Arkansas; Re, *E. ruminantium* Welgevonden (Erwe); Ro, *E. ruminantium* Welgevonden (Erum); Rg, *E. ruminantium* Gardel. (F) Four *Anaplasma* genomes: Ce, *A. centrale* Israel; Mf, *A. marginale* Florida; Ms, *A. marginale* St. Maries; Ph, *A. phagocytophilum* HZ. Complete genome sequences were downloaded from PATRIC (Snyder et al., 2007; Gillespie et al., 2011b). Alignments constructed with MAUVE (progressive) v. 2.3.1 (Darling et al., 2010). Numbers above each alignment depict genome coordinates in Mb. doi:10.1128/9781555817336.ch3.f3

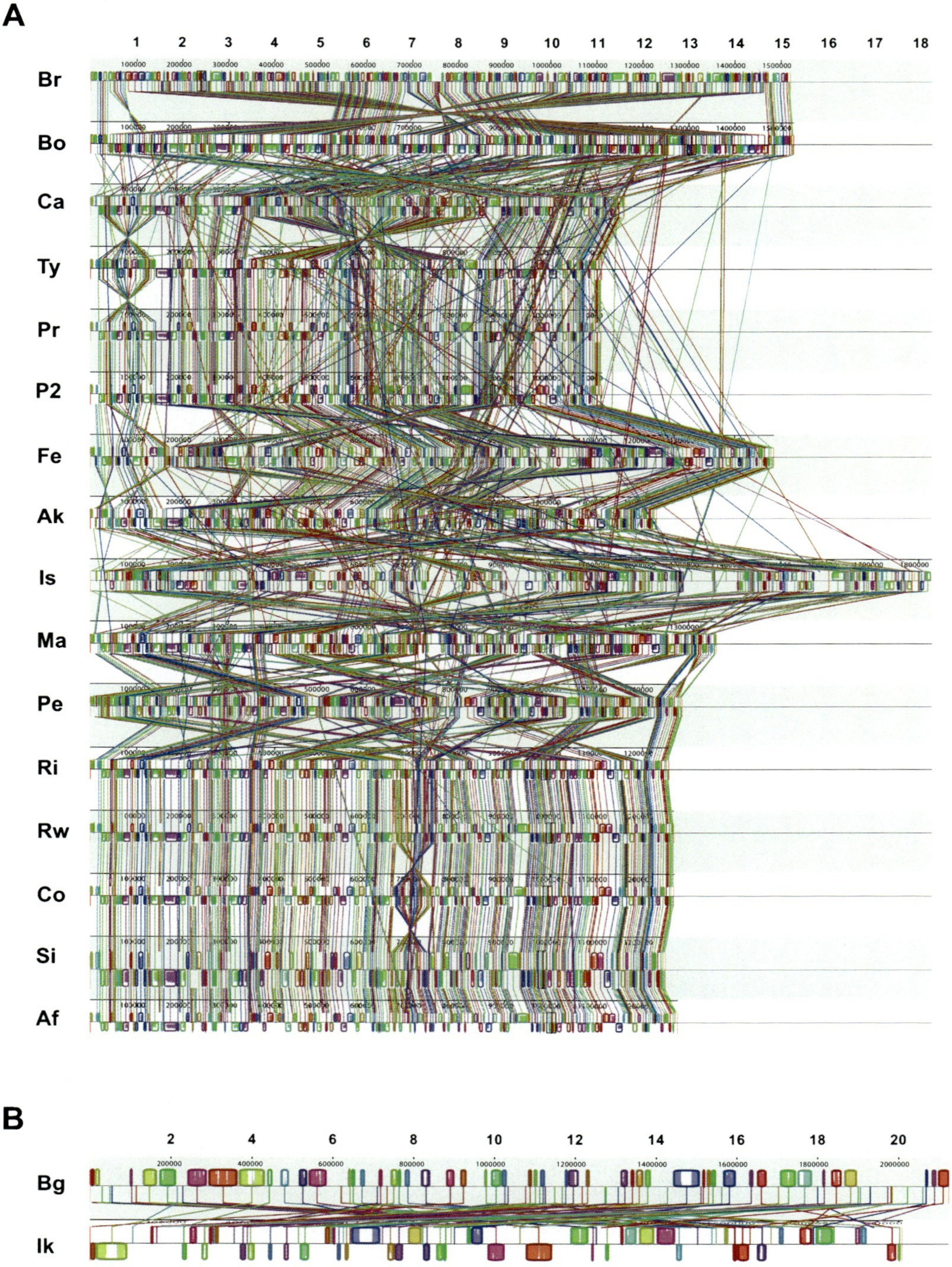
A
1 2 3 4 5 6 7 8 9 10 11 12 13 14 15 16 17 18
Br
Bo
Ca
Ty
Pr
P2
Fe
Ak
Is
Ma
Pe
Ri
Rw
Co
Si
Af
B
2 4 6 8 10 12 14 16 18 20
Bg
Ik

FIGURE 3

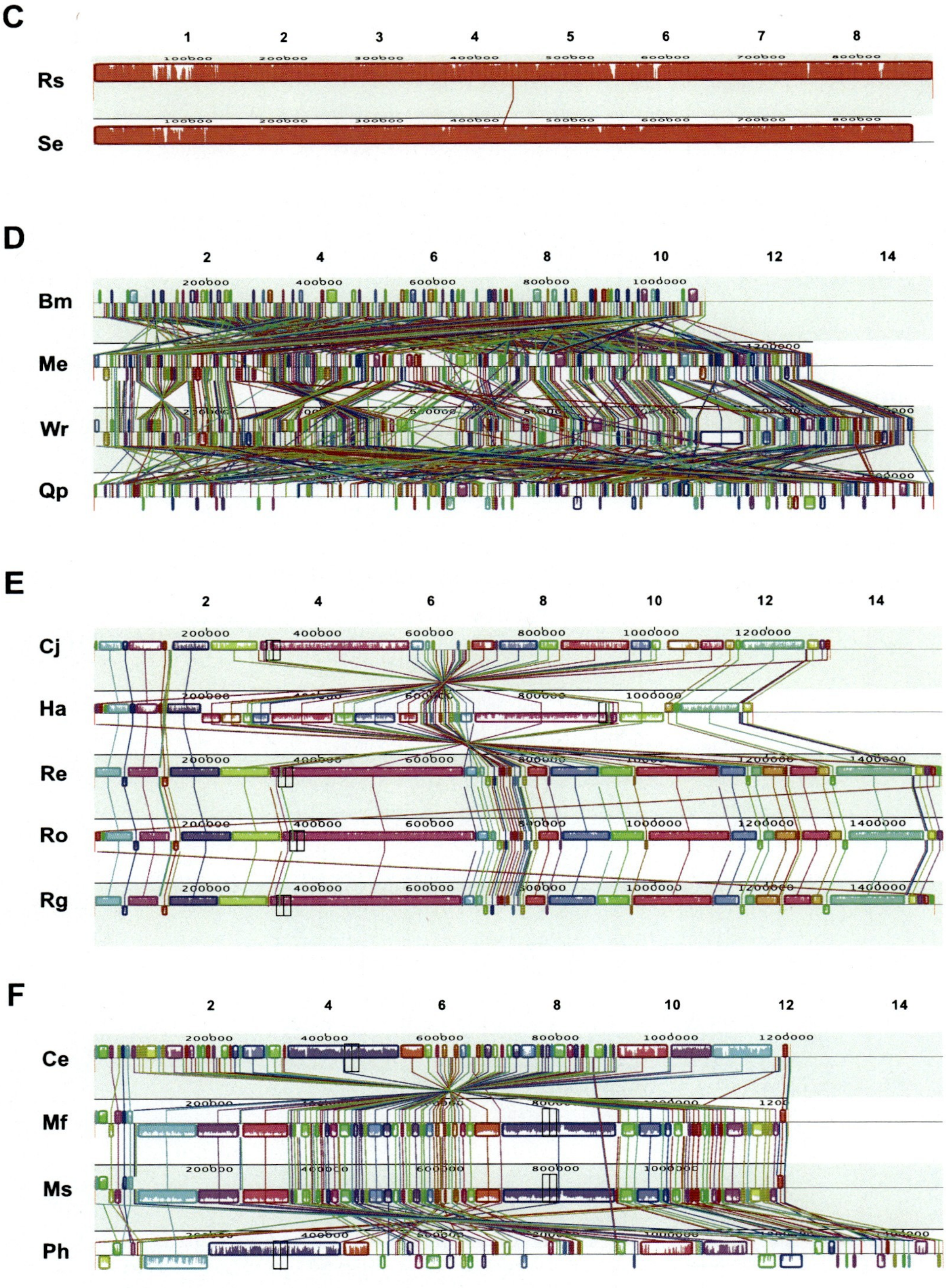

FIGURE 3 (*Continued*)

of this involves the *tlc* genes, which encode ATP/ADP translocases for host nucleotide scavenging. All genomes of the *Rickettsiaceae* encode five *tlc* orthologs (Tlc1 to -5), yet only *tlc1* appears to be functionally comparable to non-*Rickettsiaceae* transporters (Linka et al., 2003). Thus, while the individual *tlc* genes can be used for estimating *Rickettsiaceae* phylogeny, their use in higher-level phylogeny estimation should be avoided. Several duplicated families within the *rvh* T4SS share this property across the *Rickettsiales* (*rvhB4*, *rvhB6*, *rvhB8*, and *rvhB9*), as the duplication events giving rise to the divergent copies of these genes happened prior to their acquisition in the *Rickettsiaceae*-*Anaplasmataceae* ancestor (Gillespie et al., 2010). However, duplicate *rvhB2* genes are not good candidates for phylogeny estimation due to their variability across *Anaplasmataceae* genomes, being associated with the MSP2 family and contributing to surface antigenic variation (Nelson et al., 2008; Sutten et al., 2010). Regardless of the history of the duplication events, other duplicate gene families encoding surface proteins are also likely poor candidates for phylogeny estimation, given that surface-exposed regions of OMPs tend to evolve under positive selection.

Genome Synteny

The *Rickettsiales* genomes that contain high levels of ISs tend to have limited conservation in gene order (synteny) compared with closely related genomes lacking many ISs (Fig. 3). This is typified in *Rickettsia* spp., wherein the genomes of *R. bellii*, *R. felis*, and REIS are expanded and rearranged due to elevated ISs and intrachromosomal recombination across similar copies (Fig. 3A). The ISRPe1 transposon (Simser et al., 2005) has inserted 42 times in the chromosome of *R. peacockii* (Felsheim et al., 2009), with subsequent homologous recombination across similar copies breaking synteny with other highly conserved gene orders in the other derived SFGR. The proliferation of the RAGE, high repeat content, and extremely elevated number of ISs in the STGR genomes have left little synteny across these genomes (Fig. 3B). By contrast, the genomes of *Neorickettsia* spp. are extraordinarily similar despite these species being mildly divergent and having very different hosts (Fig. 3C). Accordingly, very few ISs are found within either *Neorickettsia* genome (Dunning Hotopp et al., 2006; Lin et al., 2009). The high amount of ISs in certain *Wolbachia* genomes (discussed above) has resulted in a loss of synteny across the complete genomes similar to that seen in the genus *Rickettsia* (Fig. 3D). However, unlike *Rickettsia* spp., wherein the ancestral lineage of *R. bellii* displays large amounts of genome rearrangement relative to the remaining *Rickettsia* genomes (Blanc et al., 2007b; Gillespie et al., 2008), the ancestral *w*Bm genome is lacking in ISs (particularly prophage), thus suggesting that a large amount of MGEs invaded the *Wolbachia* genomes subsequent to the divergence of the arthropod-associated species from *w*Bm. The genomes of species from both *Ehrlichia* (Fig. 3E) and *Anaplasma* (Fig. 3F) are highly conserved in synteny across the sampled genomes, consistent with a low level of ISs within these genomes (discussed above).

Intergeneric genome alignments were not possible except across *Ehrlichia* and *Anaplasma* (data not shown), implying that a lack of IS acquisition in both of these lineages has resulted in a retained gene order shared in the common ancestor (Dunning Hotopp et al., 2006). The only regions conserved in synteny across all *Rickettsiales* genomes are several stretches of genes encoding ribosomal proteins, as well as the islets forming the *rvh* T4SS archipelago (Wu et al., 2004; Gillespie et al., 2009b, 2010; Lin et al., 2009). Analysis of the REIS genome revealed rearrangements and IS disruptions even within these conserved regions, suggesting that a conservation of any specific gene order in *Rickettsiales* genomes may not be essential (Gillespie et al., 2011a).

Phylogenomics

Using the annotation of 46 *Rickettsiales* genomes available at the Pathosystems Resource Integration Center (PATRIC) (Snyder et al., 2007; Gillespie et al., 2011b), we generated protein families (FIGFams) by sampling two genomes from each of the genera within the *Anaplasmataceae* and *Rickettsiaceae* (Fig. 4). A total of 1,614 protein families contained proteins from two or more genomes, with 248 families containing a protein from every genome (Fig. 4A, skewed mostly to the top left of the heat map). This number reflects a core genome, but is lower than expected due to the subsystem annotation technology used to build the protein families, which takes into account information other than sequence similarity (Overbeek et al., 2005; Aziz et al., 2008). Thus, the core protein families can be seen as a highly conserved estimate and are probably not reflective of the actual *Rickettsiales* core genome. Previous phylogenetic studies using different protein clustering algorithms hint at a larger *Rickettsiales* core genome (~420 genes) (Dunning Hotopp et al., 2006; Mavromatis et al., 2006; Fuxelius et al., 2008), although taxon sampling was limited compared with our study here. Importantly, the FIGFams, while potentially oversplitting protein families across diverse lineages, may indicate different evolutionary origins for similar genes within analyzed genomes, and may provide information linking multiple neighboring genes in certain genomes.

Few protein families form signatures for the *Rickettsiaceae* (15%) and *Anaplasmataceae* (54%), with much more defining the six genera (70.5%) (Fig. 4B). The numbers of unique proteins per genome were quite variable across the selected taxa and hint at extraordinarily variable accessory genomes at the genus level in *Rickettsiales*. The protein families shared across multiple (but not all) genera comprised a small portion of the generated data (14.1%), as did protein families with imperfect generic distributions across the genomes (15.1%) (Fig. 4C). This hints at the conservative nature of the FIGFams, as well as the extreme genomic variation above the genus level in the *Rickettsiales*. While we provide a view at the approach to building protein families using a more conservative approach, it is important to consider that another factor determining differences in protein family estimation is input annotation. The differences between RefSeq and PATRIC annotation (Fig. 1) undoubtedly result in discrepancies in protein family clustering between this work and previous studies. We are currently investigating these differences and will present them in a future study.

Diversification

Prior phylogenomic studies have highlighted not only the defining genomic features of *Rickettsiales* but also the characteristics that accentuate each lineage. To our knowledge, our study is the first to conduct a comparative genomic analysis of all six genera of the *Rickettsiales*, and also the first to take into consideration the possibility that the SAR11 clade is the earliest-branching lineage of *Rickettsiales* (Fig. 2). Thus, we provide a brief overview of the genomic characteristics of the major lineages of *Rickettsiales*, but with comparisons to the genomic profile of "*Candidatus* Pelagibacter ubique" HTCC1062 (Giovannoni et al., 2005).

FIGURE 4 Comparative analysis of select *Rickettsiales* genome sequences. (A) Heat map representation of the distribution of proteins across 12 selected *Rickettsiales* genomes. (Note: For six genera, the two most divergent and complete genomes were selected.) O, *Orientia* genomes: Bg, *O. tsutsugamushi* Boryong; Ik, *O. tsutsugamushi* Ikeda. R, *Rickettsia* genomes: Br, *R. bellii* RML369-C; Ty, *R. typhi* Wilmington. N, *Neorickettsia* genomes: Rs, *N. risticii* Illinois; Se, *N. sennetsu* Miyayama. W, *Wolbachia* genomes: Bm, *Wolbachia* endosymbiont strain TRS of *B. malayi*; Wr, *Wolbachia* endosymbiont sp. *w*Ri. A, *Anaplasma* genomes: Ms, *A. marginale* St. Maries; Ph, *A. phagocytophilum* HZ. E, *Ehrlichia* genomes: Cj, *E. canis* Jake; Rg, *E. ruminantium* Gardel. Black depicts no representative proteins per genome, with shading spectrum (light to dark gray) illustrating one to multiple genes per genome. Heat map constructed using the Protein Family Sorter tool at PATRIC (Snyder et al., 2007; Gillespie et al., 2011b). Protein families are based on curated subsystems (FIGFams), which form the core component of the RAST automated annotation technology (Overbeek et al., 2005; Aziz et al., 2008). (B) Schema depicting the protein families shared at the order, family, and genus levels. Unique proteins (singletons) are also listed for each genome. (C) Graphic representation of the membership of 1,858 protein families and 4,289 singletons. doi:10.1128/9781555817336.ch3.f4

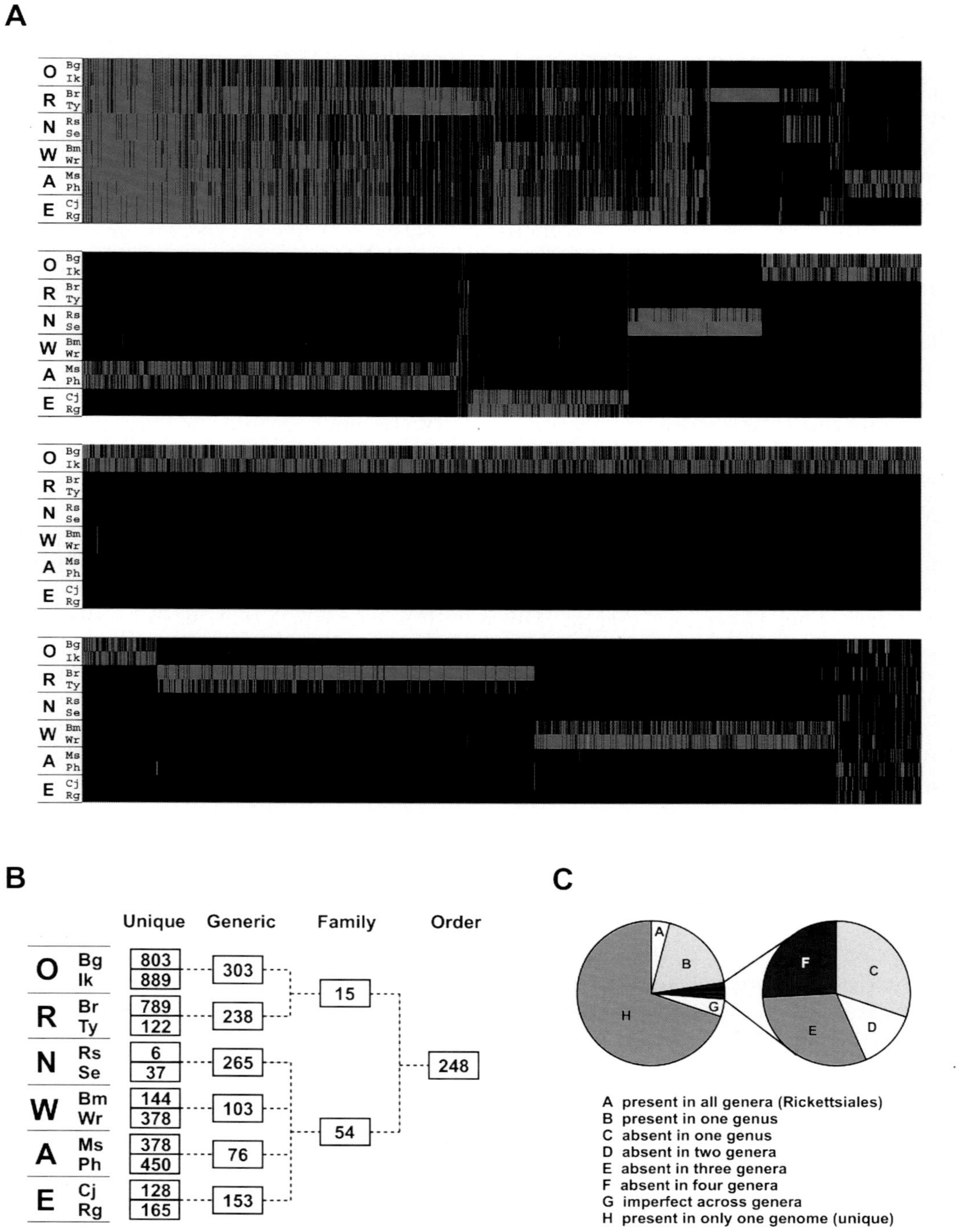
A
O Bg Ik
R Br Ty
N Rs Se
W Bm Wr
A Ms Ph
E Cj Rg
B
Unique
Generic
Family
Order
O Bg 803 Ik 889
303
R Br 789 Ty 122
238
N Rs 6 Se 37
265
W Bm 144 Wr 378
103
A Ms 378 Ph 450
76
E Cj 128 Rg 165
153
15
54
248
C
A present in all genera (Rickettsiales)
B present in one genus
C absent in one genus
D absent in two genera
E absent in three genera
F absent in four genera
G imperfect across genera
H present in only one genome (unique)

It is noteworthy that such an approach implies that the reductive genome evolution defining all derived *Rickettsiales* taxa evolved from the streamlined genome of "*Candidatus* Pelagibacter," whose genome evolved to achieve efficiency in a nutrient-depleted environment. Importantly, if the SAR11 clade is not the immediate ancestor to the *Rickettsiales*, our analysis still portrays the evolutionary scenario and selective forces that shaped the extant rickettsial lineages while transitioning from an extracellular environment to the eukaryotic cellular niche.

CELL ENVELOPE COMPOSITION

Genome comparison reveals strikingly different mechanisms for cell envelope synthesis. Among the derived *Rickettsiales*, only *Rickettsia* spp. have retained the genes coding for cell envelope structures typical of many gram-negative bacterial species, as nearly complete pathways for both PG and LPS synthesis are encoded within all genomes (Fig. 5). This is consistent with biochemical studies that support the presence of both PG and LPS in the cell envelopes of *Rickettsia* spp. (Silverman and Wisseman, 1978; Pang and Winkler, 1994). PG and LPS were previously undetected in the cell envelopes of scrub typhus agents (Silverman and Wisseman, 1978; Amano et al., 1987), and the genomes of *O. tsutsugamushi* have revealed a lack of most genes for the synthesis of both macromolecules (Cho et al., 2007; Nakayama et al., 2008). As discussed above, some species of *Wolbachia* may synthesize a modified form of PG, although a gene encoding a soluble lytic transglycosylase (*slt*), the product of which would facilitate turnover of such a murein layer, is missing. In this regard, strains of *A. marginale* may also synthesize a modified form of PG, and these genomes do encode *slt*, the product of which is necessary to remove PG subunits from the murein layer for either recycling or shedding from the bacterial cell wall. The significance of modification (*Wolbachia* spp., *A. marginale*) or deletion of the genes for the biosynthesis of PG, and complete deletion of LPS biosynthesis genes (all genera but *Rickettsia*), undoubtedly correlates with certain aspects of intracellular lifestyle. Vacuole inclusion, as opposed to a cytoplasmic lifestyle, presents different challenges for bacteria that must evade the host immune response. We recently identified a correlation with *rvh* T4SS diversification and the variable composition of cell envelope architecture across the *Rickettsiales* (Gillespie et al., 2010), and it seems that one of the most variable genomic characteristics across these diverse bacteria involves genes from various pathways (LPS synthesis, PG synthesis, OMP secretion, lipoprotein processing, metabolite import, drug export, etc.), for all of which the composition is likely influenced by coevolution with host defenses (and resources). As expected, due to its extracellular lifestyle, the pathways for LPS and PG biosynthesis in "*Candidatus* Pelagibacter" are complete (Fig. 5); however, the composition of certain transport systems is atypical for free-living bacteria (discussed below).

AMINO ACID BIOSYNTHESIS

A major defining feature of *Rickettsiales* genomes is the inability to synthesize the majority of the amino acids encoded by the universal genetic code (proteinogenic amino acids). Only three

FIGURE 5 Genomic analysis of genes involved PG synthesis, transport, degradation, and recycling in *Rickettsiales*. (A) PG pathway: brown, aminosugar metabolism; green, early-stage muropeptide synthesis; blue, lipid I and lipid II synthesis and translocation; red, anhydromuropeptide transpeptidation and transglycosylation; purple, PG degradation and turnover. Dashed pathway lines depict enzymes not encoded in any *Rickettsiales* genome. (B) Distribution of PG-associated genes across select *Rickettsiales* genomes. *Rickettsiales* taxon codes are described in the Fig. 3 legend, except for Pu ("*Candidatus* Pelagibacter ubique" HTCC1062). Letters over cladogram: P, "*Candidatus* Pelagibacter," *N, Neorickettsia*; *W, Wolbachia*; *A, Anaplasma*; *E, Ehrlichia*; *O, Orientia*; *R, Rickettsia*. Open circles depict gene absence; closed circles, gene presence; numbers, multiple gene copies; S, split ORF. doi:10.1128/9781555817336.ch3.f5

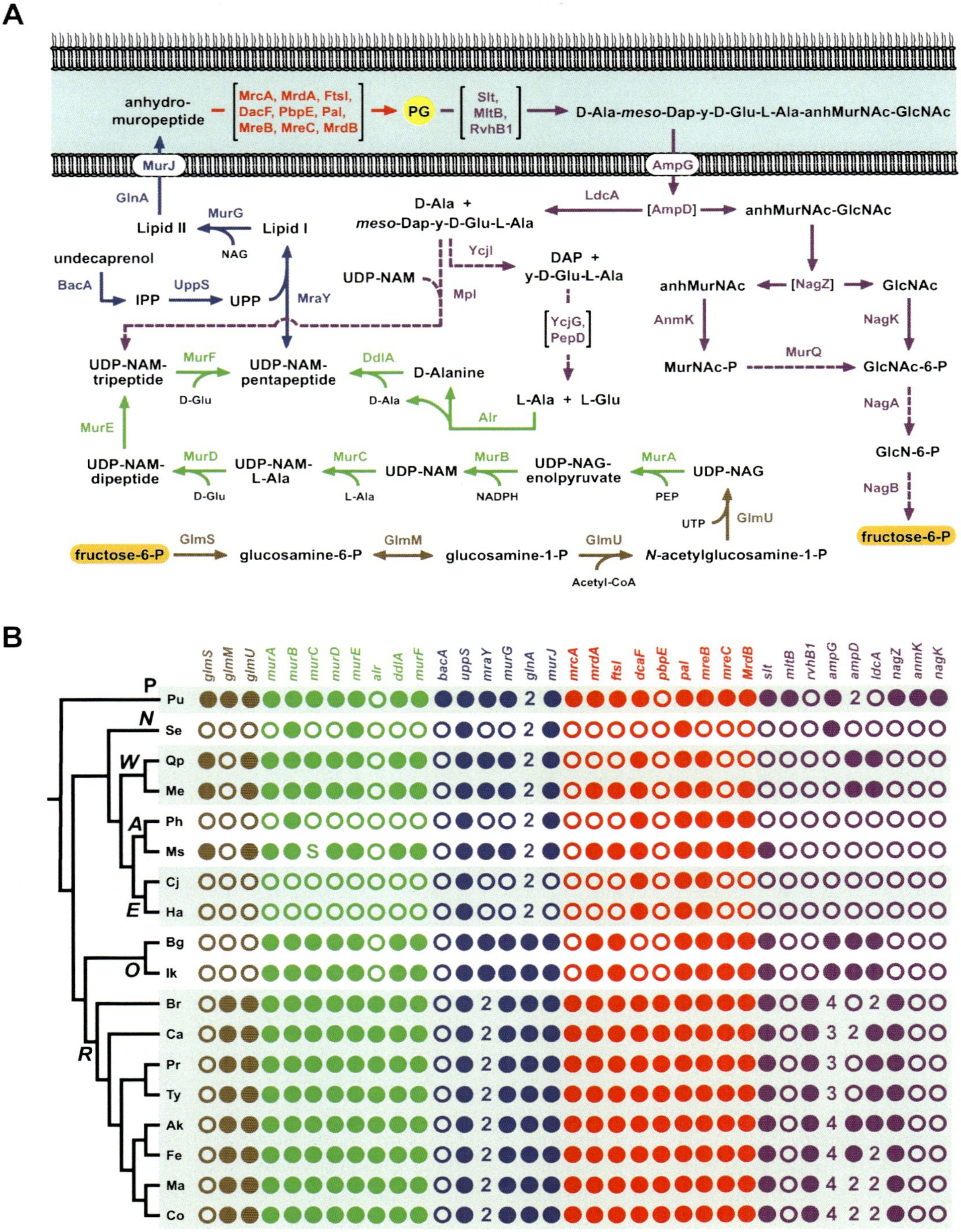
A
anhydro-muropeptide
MrcA, MrdA, FtsI, DacF, PbpE, Pal, MreB, MreC, MrdB
PG
Slt, MltB, RvhB1
D-Ala-meso-Dap-y-D-Glu-L-Ala-anhMurNAc-GlcNAc
MurJ
AmpG
GlnA
Lipid II
MurG
Lipid I
NAG
D-Ala +
meso-Dap-y-D-Glu-L-Ala
LdcA
[AmpD]
anhMurNAc-GlcNAc
undecaprenol
BacA
IPP
UppS
UPP
MraY
UDP-NAM
Mpl
YcjI
DAP +
y-D-Glu-L-Ala
YcjG, PepD
anhMurNAc
[NagZ]
GlcNAc
AnmK
NagK
MurNAc-P
MurQ
GlcNAc-6-P
UDP-NAM-tripeptide
MurF
D-Glu
UDP-NAM-pentapeptide
DdlA
D-Alanine
D-Ala
Alr
L-Ala + L-Glu
NagA
GlcN-6-P
MurE
UDP-NAM-dipeptide
MurD
D-Glu
UDP-NAM-L-Ala
MurC
L-Ala
UDP-NAM
MurB
NADPH
UDP-NAG-enolpyruvate
MurA
PEP
UDP-NAG
NagB
UTP
GlmU
fructose-6-P
fructose-6-P
GlmS
glucosamine-6-P
GlmM
glucosamine-1-P
GlmU
N-acetylglucosamine-1-P
Acetyl-CoA
B
glmS glmM glmU murA murB murC murD murE alr ddlA murF bacA uppS mraY murG glnA murJ mrcA mrdA ftsI dcaF pbpE pal mreB mreC MrdB slt mltB rvhB1 ampG ampD ldcA nagZ anmK nagK
P
N
W
A
E
O
R
Pu
Se
Qp
Me
Ph
Ms
Cj
Ha
Bg
Ik
Br
Ca
Pr
Ty
Ak
Fe
Ma
Co

proteinogenic amino acids are synthesized in all *Rickettsiales* genomes: aspartate, glycine, and glutamine, all three of which are important precursors of nucleotide synthesis. Specifically, aspartate and glycine are involved in the synthesis of purines, while glutamine plays a role in pyrimidine synthesis. As the *Rickettsiaceae* are unable to synthesize purines and pyrimidines, relying instead on nucleotide import and conversion between bases (discussed below), the strong conservation in all *Rickettsiales* genomes of enzymes involved in aspartate, glycine, and glutamine synthesis implies additional essential cellular functions for these metabolites. Some genera of the *Anaplasmataceae* can synthesize additional amino acids (arginine and lysine), but the majority of amino acids must be imported from the host cytoplasm. Regarding the proteinogenic amino acids, bioinformatic analysis predicts transporters for alanine, glycine, proline, glutamate, and aspartate. Other general amino acid transporters are predicted, but virtually none of these import systems has been experimentally characterized. In "*Candidatus* Pelagibacter," genes encoding enzymes for the biosynthesis of all 20 proteinogenic amino acids are present. Despite this, the genome encodes ABC transporters predicted to import both basic and branched amino acids, and SAR11 populations have been demonstrated to take up amino acids in high levels (Malmstrom et al., 2005). Collectively, "*Candidatus* Pelagibacter" appears to encode redundancy for amino acid synthesis/uptake, allowing it to import amino acids from the environment when available but synthesize its own when nutrients are depleted.

COFACTOR METABOLISM

Aside from *Wolbachia* spp., the *Anaplasmataceae* genomes encode genes for the synthesis of most essential cofactors (riboflavin, lipoate, pantothenate, protoheme, vitamin B_6, ubiquinone, biotin, folate, NAD, and thiamine). *Wolbachia* spp. can synthesize all but biotin, folate, NAD, and thiamine. Cofactor synthesis in the *Rickettsiaceae* is limited to lipoate, protoheme, and ubiquinone, suggesting that host reserves of the other cofactors must be imported (or converted from other metabolites). Despite an extracellular lifestyle, "*Candidatus* Pelagibacter" lacks genes for the synthesis of several essential cofactors (pantothenate, vitamin B_6, biotin, and thiamine). These cofactors must be acquired from its environment or converted from other metabolites. Interestingly, both biotin ligase (BirA) and a biotin importer (BioY) are encoded within the "*Candidatus* Pelagibacter" genome, similar to the genomes of *Wolbachia* spp. and *Rickettsia* spp., suggesting either independent losses of the biotin synthesis genes in these three lineages or a gain of the biotin biosynthesis genes in the *Anaplasmataceae* ancestor and subsequent loss in *Wolbachia* spp. Interestingly, plasmid pREIS2 of the REIS genome encodes two copies of the only known plasmid-encoded biotin operon in bacteria, and evidence strongly suggests the transfer of this operon to the genomes of *Neorickettsia* spp. and another obligate intracellular pathogen, *Lawsonia intracellularis* (*Deltaproteobacteria*) (Gillespie et al., 2011a). Thus, as is the case for biotin, genes encoding biosynthesis and transport of other cofactors may be a part of the obligate intracellular mobilome, explaining the patchy distribution of these genes across divergent genomes.

ENERGY FROM SIMPLE CARBOHYDRATES

All derived lineages of *Rickettsiales* have lost the ability to convert glucose to pyruvate (glycolysis), either through the dominant Embden-Meyerhof-Parnas pathway typical of most bacterial species or the less common Entner-Doudoroff pathway found in some aerobic bacterial species and all *Archaea*. Aside from *Rickettsia* spp., which lack genes encoding any of the typical glycolytic enzymes, *Orientia* spp. and all *Anaplasmataceae* contain at least some remnant genes for metabolism of simple sugars. For *Anaplasmataceae*, enzymes able to convert phosphoenolpyruvate (PEP) to fructose 1,6-biphosphate are well conserved (enolase, phosphoglycerate mutase, phosphoglycerate kinase, glyceraldehyde phosphate dehydrogenase, and fructose biphosphate aldolase). Thus, this set of five enzymes functions in gluconeogenesis (reverse glycolysis). Additionally, a sixth enzyme (triosephosphate isomerase) is encoded that converts the product of glyceraldehyde phosphate dehydrogenase, glyceraldehyde 3-phosphate, to dihydroxyacetone phosphate (DHAP), which is the precursor for phospholipid biosynthesis. Glyceraldehyde 3-phosphate is also used as a precursor for the synthesis of nucleotides via the nonoxidative pentose pathway (discussed below). Thus, the retention of a rudimentary glycolytic pathway (gluconeogenesis) in the *Anaplasmataceae* functions primarily to supply the bacteria with precursors for glycerolipids and nucleotides. Importantly, the ability to synthesize fructose 1,6-biphosphate supplies these bacteria with a metabolite pool that is probably essential for regulating efficient biosynthesis of both glycerolipids and nucleotides in the absence of rudimentary glycolysis.

STGR genomes encode only a subset of these gluconeogenesis enzymes (missing enolase, phosphoglycerate mutase, and fructose biphosphate aldolase). Thus, *O. tsutsugamushi* can synthesize DHAP (and hence glycerolipids) seemingly only from 3-phosphoglycerate as a precursor. This implies either unknown import mechanisms for this metabolite, unknown mechanisms for conversion of PEP to 3-phosphoglycerate, or import of host glycolytic enzymes. The absence of fructose biphosphate aldolase prevents the ability of *O. tsutsugamushi* to synthesize nucleotides, which is consistent with the presence of five *tlc* genes encoding transporters that import host cytosolic nucleotides. Recently, import of host DHAP was described for *R. prowazekii* (the transporter was not identified), explaining not only the ability of *Rickettsia* spp. to synthesize phospholipids in the absence of any glycolytic enzymes, but also the conservation of the orphan gene encoding glycerol-3-phosphate (G3P) dehydrogenase (GpsA), which converts DHAP to G3P (Frohlich et al., 2010). This pathway seemingly befits *Rickettsia* spp. with an odd metabolic redundancy, as *R. prowazekii* also encodes a transporter for G3P (GlpT). Insight into this redundancy can be gleaned via phylogenomic analysis. While the GlpT transporter is highly conserved across *Rickettsia* genomes, genes encoding GpsA in REIS and *R. canadensis* are pseudogenized (Gillespie et al., 2011a). There is strong phylogenetic support for the acquisition of *glpT* from divergent bacteria such as *Chlamydiaceae* and *Bacteroides* spp., perhaps suggesting that transport of G3P is in the process of replacing transport and subsequent conversion of DHAP for phospholipid biosynthesis (Gillespie et al., 2011a).

The glycolysis pathway encoded in the "*Candidatus* Pelagibacter" genome was described as being a modified pathway missing certain key enzymes but containing enzymes of both the nonphosphorylated Entner-Doudoroff pathway and gluconeogenesis, both of which are considered more ancient pathways than canonical glycolysis (Dandekar et al., 1999; Ronimus and Morgan, 2003). The missing key enzymes are the allosteric controllers phosphofructokinase and pyruvate kinase, also missing from the *Anaplasmataceae* pathway. Like *Anaplasmataceae*, "*Candidatus* Pelagibacter" encodes fructose 1,6-biphosphatase (*glpX*), allowing the conversion of fructose 1,6-biphosphate to fructose 6-phosphate. Unique among *Rickettsiales*, "*Candidatus* Pelagibacter" also encodes phosphoglucose isomerase (*pgi*) for synthesis of glucose 6-phosphate, which can be used to directly fuel the pentose phosphate pathway important for nucleotide and cofactor synthesis. Glucose 6-phosphate can also be directly used for LPS and PG biosynthesis. Finally, regarding pyruvate, which is needed to fuel the TCA cycle, conversion from PEP can be achieved by pyruvate phosphate dikinase, the gene for which is highly conserved in all *Rickettsiales* genomes. This enzyme completes the link between gluconeogenesis and the TCA cycle for "*Candidatus* Pelagibacter" and *Anaplasmataceae*, and presents a scenario for *Rickettsiaceae* wherein PEP (not just pyruvate) is imported into bacterial cells from the host. Otherwise, if pyruvate alone were incorporated into host cells, then pyruvate phosphate dikinase would be an orphan enzyme in the *Rickettsiaceae* genomes. Collectively, it appears that modern derived *Rickettsiales* lineages evolved from a variant of gluconeogenesis, and not glycolysis, and that specialization of this metabolic pathway has occurred across various lineages, with reduction coinciding with the gain of transporter genes for acquisition (scavenging) of key metabolic intermediates (e.g., G3P). The extreme loss of the majority of gluconeogenesis enzymes in *Rickettsia* spp. is consistent with the ability of these organisms to take up uridine 5′-diphosphoglucose (Winkler and Daugherty, 1986), which is the most abundant sugar in eukaryotic cells (Zhivkov and Tosheva, 1986).

NUCLEOTIDE PRODUCTION

The reduced version of gluconeogenesis encoded within all *Anaplasmataceae* genomes affords these species with glyceraldehyde 3-phosphate, one function of which is to feed the nonoxidative pentose phosphate pathway. The pentoses synthesized from this pathway are vital for cofactor and nucleotide synthesis. As discussed above in relation to "*Candidatus* Pelagibacter," *Anaplasmataceae* genomes do not encode a gene for phosphoglucose isomerase and thus use triosephosphate isomerase to produce glyceraldehyde 3-phosphate for the pentose phosphate pathway. Because "*Candidatus* Pelagibacter" can synthesize glucose 6-phosphate and feed it, as well as glyceraldehyde 3-phosphate, to the pentose phosphate pathway, the manner of nucleotide production in *Anaplasmatacaeae* can be considered reductive with this redundancy purged. The *Rickettsiaceae* genomes are unable to feed the pentose phosphate pathway from any entry point, thus the need for uptake of nucleotides from the host cytoplasm. Once imported, interconversion of the bases is required to obtain a full complement of pyrimidines and purines for both RNA and DNA, and these conversion pathways are more replete in *Rickettsia* spp. than in *O. tsutsugamushi* (Nakayama et al., 2008).

All *Rickettsiaceae* genomes also encode five Tlc genes (*tlc1* to *–5*), and the substrates of three have been well characterized (Audia and Winkler, 2006). Tlc1 is the sole transporter of ADP/ATP, while other ribonucleotides are transported by Tlc4 (CTP, UTP/GDP) and Tlc5 (GTP/GDP). Increasing genome sequences from obligate intracellular bacteria indicate that nucleoside triphosphate (and other metabolite) scavenging via Tlc transporters is common (Schmitz-Esser et al., 2004). At least one *tlc* gene has been identified in the genomes of *C. caryophilus* and *H. obtusa* (Linka et al., 2003), *L. intracellularis* (Schmitz-Esser et al., 2008), "*Candidatus* Liberibacter asiaticus" (*Alphaproteobacteria*) (Vahling et al., 2010), and "*Candidatus* Amoebophilus asiaticus" (Schmitz-Esser et al., 2010). The characterization of a broader host range of imported substrates for *C. caryophilus* Tlc versus *R. prowazekii* Tlc1 (Daugherty et al., 2004) suggests that *tlc* gene duplication (with subsequent diversification of imported substrates) is a likely strategy to avoid overloading the metabolic function of an individual transporter. This strategy for intracellular energy parasitism has arisen multiple times, as duplicate *tlc* genes are common in the diverse genomes of *Chlamydiales* (Schmitz-Esser et al., 2004; Gillespie et al., 2012) and *L. intracellularis* (Schmitz-Esser et al., 2008). Pathways for nucleotide production in "*Candidatus* Pelagibacter" are complete, as expected for an extracellular bacterium. Thus, the transition to an obligate intracellular lifestyle in the derived lineages resulted in two strategies for ribonucleotide import: entry into the nonoxidative pentose phosphate pathway via glyceraldehyde 3-phosphate (*Anaplasmataceae*) or complete loss of the ability to synthesize purines and pyrimidines with the acquisition (and duplication with specialization) of nucleotide transporters (*Rickettsiaceae*).

REDUCTIVE GENOME EVOLUTION FROM GENOME STREAMLINING

While initial phylogeny estimation based on a single gene did not place the SAR11 clade as a basal lineage to *Rickettsiales* (Giovannoni et al., 2005), a robust genome-based phylogeny did (Williams et al., 2007). We previously hypothesized that the acquisition of the *rvh*

T4SS was a factor in the transition from an extracellular lifestyle to an obligate intracellular lifestyle (Gillespie et al., 2010). This was rooted primarily in the fact that "*Candidatus* Pelagibacter" does not encode any *rvh* T4SS genes, which are highly conserved in all sequenced *Rickettsiales* genomes, but also based on the observation that many intracellular bacterial species use P-T4SSs to manipulate host cell functions (Alvarez-Martinez and Christie, 2009). The genome sequence of "*Candidatus* Pelagibacter" did reveal the presence of DNA uptake and competence genes (*pil*), which suggests that this organism can take up DNA from its environment (Giovannoni et al., 2005). However, we suggest that the evolution of a means to enter, survive, and replicate within protoeukaryotic cells was not initiated by any LGT events. Rather, we posit that the streamlined genome of the extracellular *Rickettsiales* ancestor ("*Candidatus* Pelagibacter"-like), designed to flourish in a nutrient-depleted marine environment, was the perfect prerequisite for the establishment of an intracellular niche that eventually evolved into strict dependence on host cell resources. This ancestor already encoded a variant form of glycolysis (gluconeogenesis), as well as multiple sugar transporters (e.g., for maltose and trehalose), and thus would thrive in an intracellular environment where sugars would be much more readily available for import. While it encoded enzymes for the biosynthesis of all 20 proteinogenic amino acids, transporters were also present for amino acid uptake; thus, an encoded redundancy would have been highly advantageous not only for protein synthesis but also for fueling metabolic pathways that use amino acids as precursors and intermediates. Measures were already in check for slow growth (typical of *Rickettsiales* species), including a single rRNA operon, as well as two-component regulatory systems for the control of phosphate limitation, osmotic stress, and nitrogen limitation. These two-component systems comprise three of the four total two-component systems encoded within the "*Candidatus* Pelagibacter" genome (Giovannoni et al., 2005), and a remarkable similarity is seen between them and the two-component systems of *Rickettsiales*, especially regarding the NtrY/NtrX (N limitation) and EnvZ/OmpR (osmotic stress) systems of *Rickettsia* spp. (data not shown). Collectively, the "*Candidatus* Pelagibacter"-like ancestor was a perfectly honed machine ready to become addicted to life inside other cells.

The genome of "*Candidatus* Pelagibacter" has the tiniest genome known from extracellular bacteria and the smallest average intergenic spacer size (3 nucleotides) of any sequenced genome (Giovannoni et al., 2005). In a sense, to evolve any further in simplicity, something would have to give. A switch to an intracellular niche would provide just the right selective forces to transition a streamlined genome to a reductive genome. Deleterious mutations in genes encoding enzymes involved in the processing of metabolites, which are readily acquired in the novel host environment, would not be selected against (Ochman and Moran, 2001). Thus, some cellular pathways would slip into disintegration, eventually leaving only those genes encoding enzymes that metabolize imported cargo. However, a second selective force would arise that would act against the maintenance of a streamlined genome. Because of the small population sizes associated with an intracellular niche, selection would not retard the retention of MGEs in the genome as it would in large population sizes (Wu et al., 2004; Cho et al., 2007). In addition, the growing number of pseudogene "graveyards" would increase the chances for MGEs to "land safely." Collectively, once an addiction to host resources was established, and hence an obligatory intracellular lifestyle forged, a steep trajectory in diversification was to follow as a consequence of tolerable LGT.

A rapid diversification among obligate intracellular species was likely propelled with the acquisition of genes enabling bacteria to secrete substrates (protein and DNA) into their novel environment. The slow but continual transfers of endosymbiont genes to resident host genomes were undoubtedly the driving factors for the downward spiral leading to capture of at least two lineages (ancestors of chloroplasts and mitochondria) and their transition to obligate organelles (Emelyanov and Sinitsyn, 1999; Emelyanov, 2001a, 2003b, 2007). The transfer of symbiont genes to host genomes may still be an active process, especially in some *Rickettsiales* taxa, as evidenced in various *Wolbachia* gene transfers to eukaryotic hosts (Dunning Hotopp et al., 2007; Klasson et al., 2009a; Nikoh and Nakabachi, 2009; Woolfit et al., 2009), as well as potential gene transfer to the paramecium macronucleus by the micronuclear-specific symbiont *H. elegans* (Hori et al., 2008). Given the evidence for lateral acquisition of many genes other than those encoding the *rvh* T4SS (Gillespie et al., 2011a), it is likely that the *Rickettsiales* ancestor, perhaps using *pil* genes like those encoded in the modern "*Candidatus* Pelagibacter," could have been parasitized by MGEs, which would contain factors important for a myriad of variations on the obligate intracellular lifestyle, resulting in colonization of a variety of eukaryotic cells. The interaction of diverse intracellular bacterial species within common eukaryotic cells, such as protists (Molmeret et al., 2005), would have given rise to the modern mobilome that is common to many of these species.

THE DIVERSITY OF *RICKETTSIALES*

The genomics era in rickettsiology has flourished over the past 13 years, culminating in 46 complete genomes, as well as a growing list of sequenced plasmids from *Rickettsia* spp. Impact on human health and agriculture has certainly influenced many of the target genomes for sequencing. Some of these organisms have no known pathogenic effect on their hosts, yet most of these are closely related to known pathogens and can be considered as naturally attenuated (e.g., *R. peacockii* Rustic relative to *R. rickettsii* [Felsheim et al., 2009]). The results from phylogenomic analyses illustrate that a large portion of *Rickettsiales* genomes at the genus level encode proteins unknown from the genomes of closely related lineages (Darby et al., 2007) (Fig. 4). The loss of genome synteny between most genera of *Rickettsiales* is evident, even in the genomes with low levels of ISs (Dunning Hotopp et al., 2006). Genome-based phylogeny estimates indicate minimal genetic divergence within genera versus between genera (Fig. 2), and metabolic profiles suggest extreme diversification from the ancestral lineage of *Rickettsiales*. All of these observations collectively suggest that a better understanding of the diversity within the *Rickettsiales* may fill gaps in the knowledge of the biology of these bacteria, especially the genomic traits associated with host cell entry, immune avoidance, intracellular growth and replication, and host cellular spreading.

Genomics aside, compilation of DNA sequence data from public databases suggests a diversity of species and strains within the *Rickettsiales*, of which completed genome sequences comprise a slight tip of a large iceberg. While protein-encoding genes have propagated on the databases, they tend to be useful mostly for intrageneric purposes, e.g., *sca* genes for *Rickettsia* spp. (Blanc et al., 2005) and *wsp* genes for *Wolbachia* (Braig et al., 1998). The citrate synthase gene, *gltA*, which has been well utilized as a phylogenetic marker (Roux et al., 1997b; Inokuma et al., 2001), has been extensively sequenced across many *Rickettsiales* species. However, given that this gene is pseudogenized in *O. tsutsugamushi* and the entire TCA cycle is deleted in *H. obtusa* (Lang et al., 2005), *gltA* is likely a poor marker for ordinal wide estimates of diversity. The chaperone *groEL* has been used to estimate large-scale phylogenies that include various lineages of *Rickettsiales* (Emelyanov, 2001b), and *groEL* sequences across the *Rickettsiales* are mounting, but they mostly represent the already well-characterized species (data not shown).

Not surprisingly, the 16S rRNA gene continues to be the dominant marker used in rickettsial phylogenetic studies, and in particular environmental samplings. We compiled 3,400 16S rRNA gene sequences from GenBank using a search strategy that allowed us to garner a rough estimate of the diversity within the *Rickettsiales* (Table 1). The results were

TABLE 1 Tabulation of complete and partial *Rickettsiales* 16S rRNA gene sequences from GenBank

Taxon[a]	16S rRNA gene sequences				
	>1,000 bp	500–999 bp	<499 bp	Total	NG[b]
Rickettsia	183	83	71	337	281 (83.4%)
Orientia	23	4	1	28	1 (3.6%)
Rickettsiaceae[c]	20	6	0	26	26 (100%)
Neorickettsia	17	37	23	77	10 (13%)
Wolbachia	169	325	138	632	599 (94.8%)
Ehrlichia	114	35	190	339	128 (37.8%)
Anaplasma[d]	118	73	164	355	307 (86.5%)
A. phagocytophilum	62	94	574	730	0
Anaplasmataceae[e]	35	6	11	52	52 (100%)
Rickettsiales[f]	260	179	385	824	820 (0)

[a]The taxon name refers to the NCBI database used in BLASTN searches.
[b]Nongenome, or subjects recovered with taxon names not matching taxa with completed genome sequences, with percentages of total subjects provided in parentheses.
[c]Taxa *Rickettsia* and *Orientia* were excluded.
[d]Taxon *A. phagocytophilum* was excluded.
[e]Taxa *Neorickettsia*, *Wolbachia*, *Rickettsia*, *Ehrlichia*, and *Anaplasma* were excluded.
[f]Taxa *Rickettsiaceae* and *Anaplasmataceae* were excluded.

unexpected, in that a plethora of poorly characterized strains, species, and isolates within nearly all of the major *Rickettsiales* lineages have associated sequence data. Intriguingly, many sequences with identified hosts made it possible to observe that *at least* protists and arthropods are prominent sources (hosts and/or reservoirs) for rickettsial species, and that a large range of higher eukaryotic hosts, vectors, and reservoirs likely exist for *Rickettsiales*. We estimated a phylogeny based on a small sampling of these sequences that illustrates the enormous diversity comprising the *Rickettsiales* (Fig. 6). Below we briefly describe this phylogeny and its impact on our current knowledge of rickettsial classification.

Holosporaceae

The ancestral relationship of the *Holosporaceae* within the *Rickettsiales*, mostly depicted by *H. obtusa* and *C. caryophila*, has been previously demonstrated (Springer et al., 1993; Emelyanov, 2001b; Lang et al., 2005; Vannini et al., 2005, 2010). A large number of 16S rRNA gene sequences were recovered that belong to members of either the *Holosporaceae* or the SAR11 clade (Table 1). We were not interested in determining the composition of the SAR11 clade; thus, 16S rRNA gene sequences from two members with complete genome sequences were used to root the tree. One of the sequences (environmental sample) we selected grouped between the SAR11 outgroup and a monophyletic lineage we refer to as *Holosporaceae*. Two sister clades formed within this group, with one containing *H. obtusa* and *C. caryophila*. This clade also contained the only specimen detected from a nonprotist source, the prairie dog flea (*O. hirsuta*) (Eschbach et al., 2009). This isolate formed a monophyletic group with *H. obtusa*, subtended by "*Candidatus* Paraholospora nucleivisitans," which lacks the complex life cycle of *H. obtusa* and resides in both the nucleus and cytoplasm of paramecia (Eschbach et al., 2009). As previously reported, an endosymbiont of *A. polyphaga* AHN-3, "*Candidatus* Caedibacter acanthamoebae," shares a close relationship with *C. caryophila* (Horn et al., 1999).

The other clade within the *Holosporaceae* comprises five protist-associated isolates and an environmental sample. The earliest-branching member of this clade is the acidophile

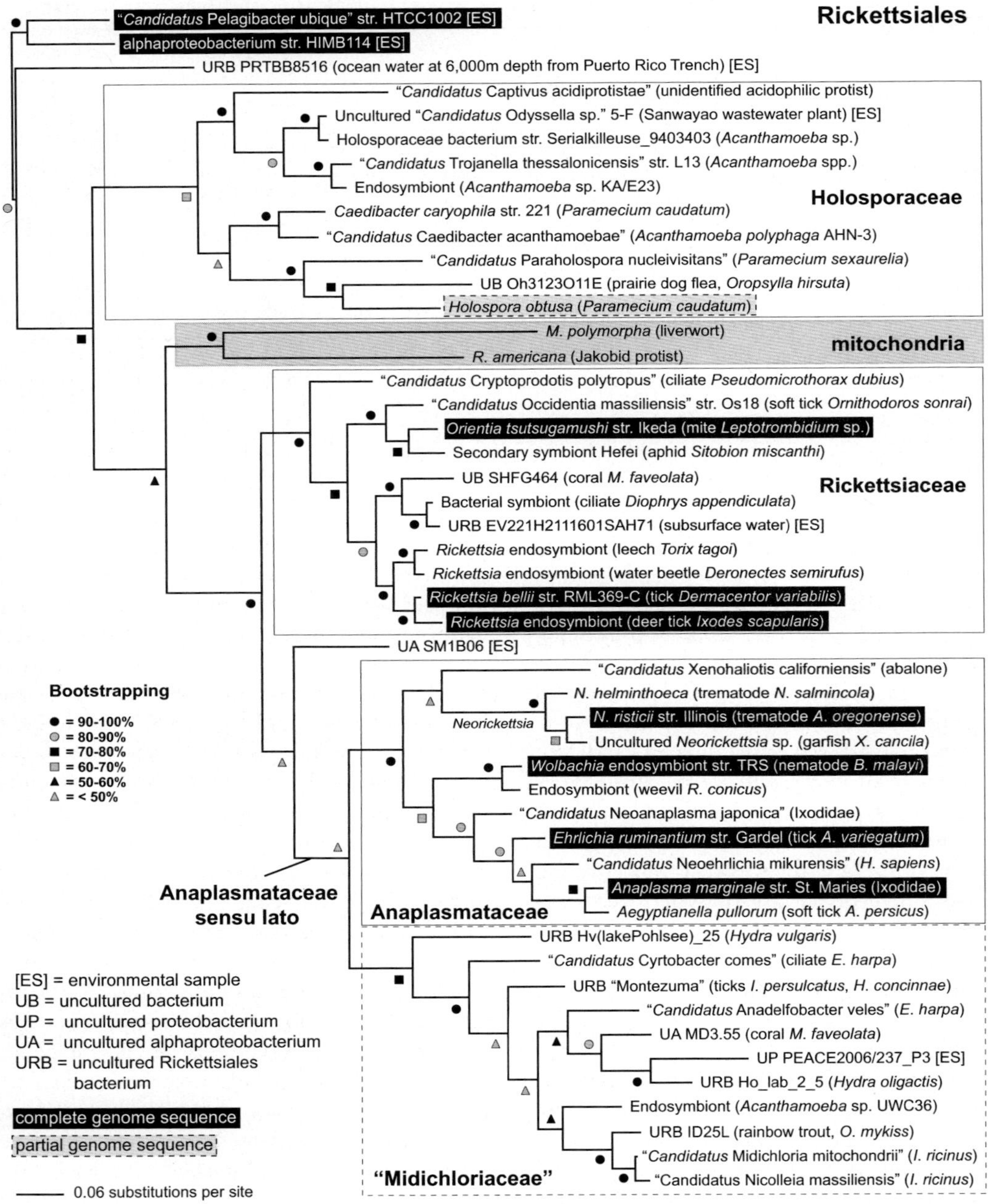
Rickettsiales
"Candidatus Pelagibacter ubique" str. HTCC1002 [ES]
alphaproteobacterium str. HIMB114 [ES]
URB PRTBB8516 (ocean water at 6,000m depth from Puerto Rico Trench) [ES]
"Candidatus Captivus acidiprotistae" (unidentified acidophilic protist)
Uncultured "Candidatus Odyssella sp." 5-F (Sanwayao wastewater plant) [ES]
Holosporaceae bacterium str. Serialkilleuse_9403403 (Acanthamoeba sp.)
"Candidatus Trojanella thessalonicensis" str. L13 (Acanthamoeba spp.)
Endosymbiont (Acanthamoeba sp. KA/E23)
Holosporaceae
Caedibacter caryophila str. 221 (Paramecium caudatum)
"Candidatus Caedibacter acanthamoebae" (Acanthamoeba polyphaga AHN-3)
"Candidatus Paraholospora nucleivisitans" (Paramecium sexaurelia)
UB Oh3123O11E (prairie dog flea, Oropsylla hirsuta)
Holospora obtusa (Paramecium caudatum)
M. polymorpha (liverwort)
R. americana (Jakobid protist)
mitochondria
"Candidatus Cryptoprodotis polytropus" (ciliate Pseudomicrothorax dubius)
"Candidatus Occidentia massiliensis" str. Os18 (soft tick Ornithodoros sonrai)
Orientia tsutsugamushi str. Ikeda (mite Leptotrombidium sp.)
Secondary symbiont Hefei (aphid Sitobion miscanthi)
UB SHFG464 (coral M. faveolata)
Rickettsiaceae
Bacterial symbiont (ciliate Diophrys appendiculata)
URB EV221H2111601SAH71 (subsurface water) [ES]
Rickettsia endosymbiont (leech Torix tagoi)
Rickettsia endosymbiont (water beetle Deronectes semirufus)
Rickettsia bellii str. RML369-C (tick Dermacentor variabilis)
Rickettsia endosymbiont (deer tick Ixodes scapularis)
UA SM1B06 [ES]
"Candidatus Xenohaliotis californiensis" (abalone)
N. helminthoeca (trematode N. salmincola)
Neorickettsia
N. risticii str. Illinois (trematode A. oregonense)
Uncultured Neorickettsia sp. (garfish X. cancila)
Wolbachia endosymbiont str. TRS (nematode B. malayi)
Endosymbiont (weevil R. conicus)
"Candidatus Neoanaplasma japonica" (Ixodidae)
Ehrlichia ruminantium str. Gardel (tick A. variegatum)
"Candidatus Neoehrlichia mikurensis" (H. sapiens)
Anaplasma marginale str. St. Maries (Ixodidae)
Aegyptianella pullorum (soft tick A. persicus)
Anaplasmataceae
Anaplasmataceae sensu lato
Bootstrapping
● = 90-100%
○ = 80-90%
■ = 70-80%
□ = 60-70%
▲ = 50-60%
△ = < 50%
URB Hv(lakePohlsee)_25 (Hydra vulgaris)
"Candidatus Cyrtobacter comes" (ciliate E. harpa)
URB "Montezuma" (ticks I. persulcatus, H. concinnae)
"Candidatus Anadelfobacter veles" (E. harpa)
UA MD3.55 (coral M. faveolata)
UP PEACE2006/237_P3 [ES]
URB Ho_lab_2_5 (Hydra oligactis)
Endosymbiont (Acanthamoeba sp. UWC36)
URB ID25L (rainbow trout, O. mykiss)
"Candidatus Midichloria mitochondrii" (I. ricinus)
"Candidatus Nicolleia massiliensis" (I. ricinus)
"Midichloriaceae"
[ES] = environmental sample
UB = uncultured bacterium
UP = uncultured proteobacterium
UA = uncultured alphaproteobacterium
URB = uncultured Rickettsiales bacterium
complete genome sequence
partial genome sequence
0.06 substitutions per site

"*Candidatus* Captivus acidiprotistae," which was sampled from acid mine drainage (Baker et al., 2003). Like *C. caryophila*, which has a large intervening sequence within variable region V3 of the 16S rRNA gene (Springer et al., 1993), "*Candidatus* Captivus acidiprotistae" contains an intervening sequence in its 16S rRNA gene, but within variable region V1 of domain I of the small-subunit rRNA. The significance of these intervening sequences in two divergent members of the *Holosporaceae* is unknown. Two derived members of this clade associated with *Acanthamoeba* spp., endosymbiont of *Acanthamoeba* sp. KA/E23 and *Holosporaceae* bacterium strain Serialkilleuse, were isolated from patients with infected corneas (keratitis) (Xuan et al., 2007). "*Candidatus* Trojanella thessalonices," also associated with *Acanthamoeba* spp., was observed to have a typical gram-negative cell wall and to induce host cell lysis when grown above 30°C (Birtles et al., 2000).

To our knowledge, this is the largest sampling of ancestral *Rickettsiales* that demonstrates the monophyly of *Holosporaceae*, and the first demonstration of a sister clade to the lineage comprising *H. obtusa* and *C. caryophila*. The diversity within the *Holosporaceae* is likely very rich, given the increasing studies detecting their presence in a wide range of hosts from an assorted list of environments. Still, we suggest caution in using the 16S rRNA gene sequences to characterize the ancestral lineages of *Rickettsiales*. We found that some of the sequences annotated as "uncultured *Rickettsiales* bacterium" (and similar names; see Fig. 6) had

FIGURE 6 Phylogeny estimation of small-subunit rRNA gene sequences from 47 diverse *Rickettsiales* taxa and 2 mitochondria. Sequences were aligned using MUSCLE v3.6 (Edgar, 2004a, 2004b) with default parameters, and subsequently analyzed under maximum likelihood using RAxML (Stamatakis et al., 2008). A gamma model of rate heterogeneity was used with estimation of the proportion of invariable sites. Brach support is from 1,000 bootstrap pseudoreplications. The term "*Candidatus*" is used as originally suggested (Murray and Stackebrandt, 1995). (Note: PATRIC accession numbers are given for rRNA gene sequences retrieved from completed genomes; all other numbers are NCBI nucleotide accession numbers.) Taxon information is as follows, moving from the top to the bottom of the tree. "*Candidatus* Pelagibacter ubique" HTCC1002 (VBICanPel113578_r025); alphaproteobacterium strain HIMB114 (VBIAlpPro140191_r032); uncultured *Rickettsiales* bacterium clone PRTBB8516 (HM798949); "*Candidatus* Captivus acidiprotistae" (AF533508); uncultured "*Candidatus* Odyssella sp." clone 5-F (EU305601); *Holosporaceae* bacterium strain Serialkilleuse_9403403 (HM138368); "*Candidatus* Trojanella thessalonicensis" L13 (AF069496); endosymbiont of *Acanthamoeba* sp. KA/E23 (EF140636); *C. caryophilus* 221 (X71837); "*Candidatus* Caedibacter acanthamoebae" (AF132138); "*Candidatus* Paraholospora nucleivisitans" (EU652696); uncultured bacterium clone Oh3123O11E (EU137369); *H. obtusa* (X58198); *Marchantia polymorpha*, mitochondria (NC_001660); *Reclinomonas americana*, mitochondria (AF007261); "*Candidatus* Cryptoprodotis polytropus" isolate PSM1 (FM201293); "*Candidatus* Occidentia massiliensis" (*Rickettsiaceae* bacterium strain Os18; GU937608); *O. tsutsugamushi* Ikeda (VBIOriTsu129072_r006); secondary symbiont of *S. miscanthi* clone Hefei (HM156647); uncultured bacterium clone SHFG464 (FJ203077); bacterial symbiont of *D. appendiculata* (AJ630204); uncultured *Rickettsiales* bacterium clone EV221H2111601SAH71 (DQ223223); *Rickettsia* endosymbiont of *T. tagoi* (AB066351); *Rickettsia* endosymbiont of *D. semirufus* (FM955311); *R. bellii* RML369-C (VBIRicBel102610_r012); REIS (VBIRicEnd40569_r031); uncultured alphaproteobacterium clone SM1B06 (AF445655); "*Candidatus* Xenohaliotis californiensis" (AF069062); *Neorickettsia helminthoeca* (U12457); *N. risticii* Illinois (VBINeoRis104330_r001); uncultured *Neorickettsia* sp. isolate 184 (EU780451); *Wolbachia* endosymbiont strain TRS of *B. malayi* (VBIWolEnd7741_r015); endosymbiont of *R. conicus* (M85267); "*Candidatus* Neoanaplasma japonica" (*Anaplasmataceae* bacterium strain IS136; AB190771); *E. ruminantium* Gardel (VBIEhrRum72196_r011); "*Candidatus* Neoehrlichia mikurensis" (uncultured "*Candidatus* Neoehrlichia sp." clone 2; GQ501090); *A. marginale* St. Maries (VBIAnaMar46146_r010); *A. pullorum* (AY125087); uncultured *Rickettsiales* bacterium clone Hv(lakePohlsee)_25 (EF667921); "*Candidatus* Cyrtobacter comes" (FN552697); uncultured *Rickettsiales* bacterium "Montezuma" (AF493952); "*Candidatus* Anadelfobacter veles" (FN552695); uncultured alphaproteobacterium clone MD3.55 (FJ425643); uncultured proteobacterium clone PEACE2006/237_P3 (EU394580); uncultured *Rickettsiales* bacterium clone Ho_lab_2_5 (EF667892); endosymbiont of *Acanthamoeba* sp. UWC36 (AF069962); uncultured *Rickettsiales* bacterium clone ID25L (EU555284); "*Candidatus* Midichloria mitochondrii" (AJ566640); "*Candidatus* Nicolleia massiliensis" France (DQ788562). doi:10.1128/9781555817336.ch3.f6 higher average bit scores per top 100 BLASTN hits to *Rhodospirillales* than to *Rickettsiales*. Two additional taxa considered to belong to *Rickettsiales*, the bacterial agent of necrotizing hepatopancreatitis (Loy et al., 1996) and a symbiont of the isopod *Porcellio scaber* (Wang et al., 2004, 2007), were removed from our analysis after showing similar properties.

Mitochondria

We included two mitochondrial small-subunit rRNA gene sequences in our phylogeny estimation, and this lineage branched after the *Holosporaceae*, but basal to the derived *Rickettsiales* (Fig. 6). Despite its usually being considered a close relative of *Rickettsiales*, the placement of the mitochondrial ancestor within the eubacterial tree has been a subject of controversy (Müller and Martin, 1999; Kurland and Andersson, 2000; Emelyanov, 2003a). The recent genome-based phylogeny estimated for *Alphaproteobacteria*, which included members of the SAR11 clade, placed the mitochondria as a sister taxon to *Rickettsiales* (Williams et al., 2007). A prior study including 16S rRNA and GroEL gene sequences from *H. obtusa* recovered the same placement of the mitochondria within the *Rickettsiales* as we demonstrate here (Emelyanov, 2001b). While not strongly supported, the results are compelling. A phylogeny estimate of *H. obtusa* based on a concatenation of multiple proteins placed *H. obtusa* as basal to the *Rickettsiaceae-Anaplasmataceae* (Lang et al., 2005), and given that the sister relationship of mitochondria to the derived *Rickettsiales* was recovered in a robust genome-based phylogeny estimate (Williams et al., 2007), we conclude that multiple lines of evidence are leading to the proposed relationships we provide here (Fig. 6). The completion and release of the genomic data for *H. obtusa* will certainly test this hypothesis.

Rickettsiaceae

The observed diversity in 16S rRNA gene sequences within the *Rickettsiaceae* was not surprising, given the growing reports of lineages that branch ancestrally to the traditional TGR and SFGR (but still form a sister clade to *O. tsutsugamushi*) (Perlman et al., 2006; Braig et al., 2008; Weinert et al., 2009b). We sampled only *R. bellii* and *R. typhi* to represent the entire genus of *Rickettsia* that contains complete genome sequences. The tendency of *R. canadensis* to group basal to *R. bellii* in 16S rRNA gene-based trees is an artifact and is incongruent with genome-based phylogenies (Gillespie et al., 2008). We sampled taxa that were demonstrated in previous studies to branch ancestrally to the clades containing *R. bellii* ("bellii group"). A sister clade to the well-characterized *Rickettsia* groups (AGR, TGR, TRGR, and SFGR) comprised endosymbionts of the leech *Torix tagoi* (Kikuchi et al., 2002) and the water beetle *Deronectes semirufus* (Kuchler et al., 2009) (Fig. 6). A diverse assemblage of *Rickettsia* species branches between these two lineages (data not shown), including pea aphid rickettsia (Chen et al., 1996), the agent of papaya bunchy top disease (Davis et al., 1998); the male-killing endosymbiont of *Adalia bipunctata* (Balayeva et al., 1995); and over a dozen other species of *Rickettsia* associated primarily with phytophagous arthropods (Perlman et al., 2006; Braig et al., 2008; Weinert et al., 2009b). Forming a sister clade to this assemblage is a group that does not include any named species of *Rickettsia*. This includes the bacterial symbionts of the ciliate *Diophrys appendiculata* (Vannini et al., 2005) and an environmental sample collected from subsurface water that is also likely associated with a protist. Subtending this clade is an uncultured bacterium (SHFG464) identified in the Caribbean coral species *Montastraea faveolata*, which may play a role in white plague disease (Sunagawa et al., 2009). The classification of this lineage in relation to *Rickettsia* spp. is currently unknown, and awaits more data and a better understanding of the biology and life histories of its diverse members.

An unexpected diversity was recovered in the sister clade to *Rickettsia* spp., which typically includes *O. tsutsugamushi*. The recently reported "*Rickettsia*-like organisms" found in Chinese wheat pest aphid, *S. miscanthi* (*Hemiptera*:

Aphididae) (Li et al., 2011), adds a new species to the long-standing monotypic lineage of scrub typhus agents. Subtending this clade is a sequence from "*Candidatus* Occidentia massiliensis" sp. recently identified in the soft tick *Ornothodoros sonrai*. It is evident from these data that the lineage comprising *O. tsutsugamushi* within the *Rickettsiaceae* comprises more diversity than previously thought. Finally, all of these lineages of *Rickettsiaceae* are subtended by "*Candidatus* Cryptoprodotis polytropus," which was identified in the ciliate *Pseudomicrothorax dubius* (Ferrantini et al., 2009). This stresses the importance of consideration of protist species as possible reservoirs, or even hosts, for other members of the *Rickettsiaceae*, particularly those only known from arthropods. As *R. bellii* was demonstrated to grow in *A. polyphaga* (Ogata et al., 2006), it stands to be reasoned that other species of *Rickettsiaceae* are likely able to do so as well.

Anaplasmataceae

The observed diversity in 16S rRNA gene sequences within the *Anaplasmataceae* is staggering in that an entire sister clade is apparent to the lineage comprising the four well-characterized genera (Fig. 6). Within the traditional *Anaplasmataceae*, which we refer to hereafter as *Anaplasmataceae* sensu stricto, a lineage comprising the diversity within *Neorickettsia* forms a sister clade to the remaining lineages, but includes the agent of withering syndrome in abalone, "*Candidatus* Xenohaliotis californiensis" (Friedman et al., 2000; Moor et al., 2001), as the most ancestral member. A diversity of rickettsia-like organisms has been reported in various species of abalone (Gardner et al., 1995; Antonio et al., 2000; Braid et al., 2005), although their placement within the *Rickettsiales* tree is unknown and awaits further analysis. Abalones, which are gastropods, suggest a similar snail association within this clade, as many of the described species of *Neorickettsia* are known from snails (Lin et al., 2009). The detection of a species of *Neorickettsia* in the garfish, *Xenentodon cancila*, expands the host range for this genus. Sister to this clade forms the remainder of the *Anaplasmataceae* sensu stricto, with *Wolbachia* (over 100 sequences; data not shown) forming a sister clade with a likely wolbachial endosymbiont of the true weevil *Rhinocyllus conicus* (Campbell et al., 1992). The remaining derived lineages of this clade contain species that are less characterized than *Ehrlichia* and *Anaplasma*, including "*Candidatus* Neoanaplasma japonica," "*Candidatus* Neoehrlichia mikurensis" (Fehr et al., 2010), and *Aegyptianella pullorum* (Rikihisa et al., 2003). The exact relationships of these taxa with species of *Ehrlichia* and *Anaplasma* remain to be validated with more data and a better understanding of the life cycle of each species.

"*Midichloriaceae*"

The emergence of a sister taxon to the traditional *Anaplasmataceae*, *Anaplasmataceae* sensu stricto, has been demonstrated in several studies (Beninati et al., 2004; Davis et al., 2009; Vannini et al., 2010), and the tree we present here is consistent with these studies, although we add a greater number of lineages to this clade (Fig. 6). The most derived lineage of this clade comprises the mitochondria-associated species "*Candidatus* Midichloria mitochondrii," which invades the mitochondria of its tick host, *Ixodes ricinus* (Lo et al., 2004; Sacchi et al., 2004; Sassera et al., 2006; Rymaszewska, 2007; Epis et al., 2008; Beninati et al., 2009). Despite destroying host mitochondria, "*Candidatus* Midichloria mitochondrii" is a widespread and abundant symbiont of ticks (Lo et al., 2006; Venzal et al., 2008) and is very closely related to "*Candidatus* Nicolleia massiliensis" (Venzal et al., 2008), which is also known from ixodid ticks. This lineage is subtended by the infectious agent of strawberry disease of rainbow trout, *Oncorhynchus mykiss* (Lloyd et al., 2008). A recent study strongly suggests that the strawberry disease agent may also be the etiological agent of red-mark syndrome in trout from the United Kingdom (Metselaar et al., 2010). These three taxa are subtended by an endosymbiont of *Acanthamoebae* sp. UWC36, which was originally suggested to be closer to *Rickettsiaceae* than *Anaplasmataceae* (Fritsche et al., 1999; Horn et al., 1999), but was subsequently shown to branch in this sister clade to *Anaplasmataceae* sensu stricto (Beninati et al., 2004), as demonstrated here. Interestingly, a *tlc* gene has been identified from this organism (NCBI protein accession no. CAE46502), and the sequence is 45% identical to a Tlc protein from "*Candidatus* Protochlamydia amoebophila" UWE25 (CAE46506) (Schmitz-Esser et al., 2004). The first report of a *tlc* gene in *Anaplasmataceae* sensu lato and its greater similarity to "*Candidatus* Protochlamydia amoebophila" than *Rickettsia* spp. *tlc* genes further suggests the LGT of these transporter genes, and accentuates the importance of protists as reservoirs for LGT among diverse obligate intracellular bacteria (Moliner et al., 2010).

A sister clade to the above-mentioned taxa comprises a protist symbiont (Vannini et al., 2010), an environmental sample, another rickettsial species associated with coral (*M. faveolata*), and a species detected in *Hydra oligactis* (*Animalia*: *Cnidaria*) (Fraune and Bosch, 2007). The protist-associated "*Candidatus* Anadelfobacter veles" was demonstrated to reside within a vacuole in its host, the ciliate *Euplotes harpa* (*Ciliophora*: *Spirotrichea*), and was the first reported ciliate-associated bacterium in the *Anaplasmataceae* sensu lato (Vannini et al., 2010). A series of three species form grades that subtend the derived lineages of the sister clade to *Anaplasmataceae* sensu stricto. The uncultured bacterium referred to as the "Montezuma" microorganism has been considered a tick-borne disease in the Far East of Russia (Mediannikov et al., 2004). Branching deeper in this clade is another symbiont of the ciliate *E. harpa*, "*Candidatus* Cyrtobacter comes," which unlike "*Candidatus* Anadelfobacter veles" was not demonstrated to reside within a vacuole in its host (Vannini et al., 2010). This trait, more typical of *Rickettsiaceae* than *Anaplasmataceae*, combined with the ancestral position of "*Candidatus* Cyrtobacter comes," coincided with its tentative association with the sister clade to the *Anaplasmataceae* sensu stricto (Vannini et al., 2010). Our demonstration that a second endosymbiont of a hydra species (*Hydra vulgaris*) (Fraune and Bosch, 2007) branches more ancestrally than "*Candidatus* Cyrtobacter comes" supports the inclusion of both of these taxa in the sister clade to *Anaplasmataceae* sensu stricto.

Based on a more limited taxon sampling than ours, it was proposed to name this sister clade to *Anaplasmataceae* sensu stricto the "*Candidatus* Midichloria" clade (Vannini et al., 2010). We feel that the information from 16S rRNA gene analyses is strongly suggesting that the *Anaplasmataceae* sensu lato be treated as two natural groups, and that assigning family status to one group based on the inclusion of better-characterized taxa is a poor reason to keep the current classification. We agree with Vannini et al. that the sister clade to the *Anaplasmataceae* sensu stricto should receive a distinct name, and we suggest that a family name be more appropriate, given that tree topology, percent nucleotide divergence, and host range diversity suggest enough evolutionary distance to propose so. Thus, following the trend to name this group after the intriguing mitochondrial invaders, we propose that the natural sister group to *Anaplasmataceae* sensu stricto be referred to hereafter as the *Midichloriaceae*. Further data are needed to support this proposal.

FUTURE DIRECTIONS

The *Rickettsiales* are a highly diverse assemblage of early-branching *Alphaproteobacteria*, with a major defining characteristic of all known species being their obligate intracellular lifestyle. The *Holosporaceae* diverged from an extracellular ancestor with a streamlined genome adapted for life in a nutrient-depleted environment (dissolved organic carbon in oceans). An ancestor to the mitochondria and the derived *Rickettsiaceae*-*Anaplasmataceae* sensu lato clade diverged from *Holosporaceae*, with all three major lineages undergoing accelerated diversification as a consequence of this transition from a free-living to an intracellular lifestyle. This diversification was likely heavily influenced by an influx of MGEs, which are afforded a lack of purifying selection in small population sizes such as intracellular niches. The rise of *Eukarya* provided an expanded array of intracellular niches that propelled further diversification, leading to the modern lineages of *Rickettsiales* as we observe them today.

At the genomic level, one of the defining features of all sequenced genomes from *Anaplasmataceae* sensu stricto and *Rickettsiaceae* is a conserved P-like T4SS, the *Rickettsiales vir* homolog (*rvh*) T4SS (Gillespie et al., 2009b, 2010; Rikihisa et al., 2009; Rikihisa and Lin, 2010). The oddity of the composition of this anomalous P-T4SS (Alvarez-Martinez and Christie, 2009), coupled with the conserved nature of its archipelago of *rvh* islets (Gillespie et al., 2009b; Gillespie et al., 2010), has increasingly attracted researchers over the past few years. Proteins from the cells of several species from the *Anaplasmataceae* sensu stricto have been characterized as translocated substrates of the *rvh* T4SS and have been shown to localize to host cell organelles (Lin et al., 2007; Niu et al., 2010). The protist symbiont *H. obtusa* also secretes proteins into the macronuclei of its host (Iwatani et al., 2005; Abamo et al., 2008), but it is not known if this is via a P-T4SS or some other mechanism. Thus, genome sequence data for *Holosporaceae* are necessary to reveal whether a common theme for protein secretion into host cytoplasm and organelles exists for derived *Rickettsiales* or if multiple strategies for host cell manipulation have evolved. This area of research is in need of continual exploration, and in particular, the characterization of additional *rvh* substrates and a possible secretion signal are likely imminent.

The F-T4SSs (*tra/trb*), which are typically associated with bacterial conjugation, appear to be specific to *Rickettsiaceae* genomes and are encoded within common (but diverse) integrative conjugative elements, termed RAGEs (Gillespie et al., 2011a). The impact RAGEs have had on the diversification of *Rickettsiaceae* genomes is staggering, ranging from total proliferation (Cho et al., 2007; Nakayama et al., 2008) to moderate parasitism (Gillespie et al., 2011a) to single maintenance (Ogata et al., 2006; Blanc et al., 2007a) and finally to total purging (remaining *Rickettsia* spp. genomes). It is evident that these MGEs introduced many genes of the obligate intracellular mobilome into at least *Rickettsiaceae* genomes, although they are likely ancient elements that may have also invaded other *Rickettsiales* genomes. Analysis of the REIS genome suggests that the RAGEs are likely still mobile, and also plasmid encoded. Thus, the correlation between plasmids in *Rickettsia* spp. and the presence of the RAGE (in some form) within nearly all *Rickettsiaceae* genomes cannot be ignored. The mechanism of intercellular transfer of rickettsial plasmids is still unknown and may not need to involve an F-T4SS. While the plasmids are diverse in number and gene composition, they seem to be specific to *Rickettsia* spp. (Baldridge et al., 2010). However, these extrachromosomal factors may provide the most promising option for the construction of a transformation vector that would benefit all of rickettsiology (Baldridge et al., 2007). We suspect that this will also be a promising area of exciting future research.

The rapid accumulation of genome sequences for *Rickettsiales* can only tell us so much, and the lack of knowledge of the status of pan-genomes (Medini et al., 2005) can keep research in the dark. The excellent work to close the pan-genome for *A. marginale* sensu stricto (Brayton et al., 2005; Dark et al., 2009) is an approach that should be followed for other *Rickettsiales* species and strain complexes, as exemplified by its utility for better understanding the use of *A. centrale* as a viable vaccine (Herndon et al., 2010). Research to unravel the underlying genomic factors involved in the attenuation of pathogens, as well as the mechanisms associated with revertant phenotypes, is also of critical importance (Bechah et al., 2010). A lot of assumptions have been made regarding the factors involved in rickettsial pathogenesis, ranging from gene loss being the primary factor involved in virulence (Fournier et al., 2009), to the acquisition of virulence factors from LGT (Ogata et al., 2005a), to initial host encounter (stress) followed by gradual wane to symbiosis (Braig et al., 2008). We suggest that it isn't that easy, or pragmatic, to blanket the factors associated with pathogenicity under one evolutionary model. Rather, it is evident from the diversification across *Rickettsiales* genomes that evolutionary continuums likely keep genome sizes and composition in flux, allowing for periods of stasis interrupted by the invasion of novel genes. Only through exhausting the pangenomes of *Rickettsiales* species can we better understand these diversifying factors.

Finally, the diversity of the *Rickettsiales* we have highlighted here presents an exciting challenge for rickettsiology. All of the major lineages (*Holosporaceae*, *Rickettsiaceae*, and *Anaplasmataceae* sensu lato) contain species, strains, and isolates identified from a wide range of eukaryotes, including protists, cnidarians, mollusks, arthropods, nematodes, annelids, and vertebrates. This highlights the importance that vectors and reservoirs may play in the life cycles of many species, and how the environment may factor in the emergence and reemergence of pathogens. From a genomics perspective, this diversity is poorly understood. Thus, we hope that future research will also extend into this uncharted area, as this would certainly shed light on the underlying factors that define the biology of obligate intracellular symbiosis and parasitism.

ACKNOWLEDGMENTS

We thank members of the Azad laboratory (University of Maryland School of Medicine) for invaluable feedback and discussion. We are grateful to Rebecca Wattam (Virginia Bioinformatics Institute) for assistance with the PATRIC Comparative Heatmap Tool. This project has been funded in whole or in part with federal funds from the National Institute of Allergy and Infectious Diseases (NIAID), National Institutes of Health, Department of Health and Human Services, under contract no. HHSN272200900040C, awarded to B. W. S. Sobral, and NIAID award numbers R01AI017828 and R01AI59118, awarded to A. Azad. The content is solely the responsibility of the authors and does not necessarily represent the official views of the NIAID or the National Institutes of Health. All data included in the analyses are publicly available at the Pathosystems Resource Integration Center (http://wwww.patricbrc.org/portal/portal/patric/Home).

ADDENDUM IN PROOF

After completion of our work, an additional 40 *Rickettsiales* genomes were made publicly available. A study focusing on the identification of candidate vaccine antigens yielded five more genomes from strains of *Anaplasma marginale* (str. Florida Relapse, str. Okeechobee, str. Oklahoma, str. South Idaho, and str. Washington Okanogan) (Dark et al., 2011). A study demonstrating complete bacteriophage transfer across co-resident *Wolbachia* strains infecting a parasitic wasp generated a novel genome sequence (*Wolbachia* endosymbiont wVitB of *Nasonia vitripennis*) (Kent et al., 2011). An additional two *Wolbachia* genomes from unpublished studies (*Wolbachia* endosymbiont of *Culex pipiens molestus* and *Wolbachia pipientis* wAlbB) brings the total number of *Anaplasmataceae* sequenced genomes to 32 as of June 2012. Regarding the *Rickettsiaceae*, an astounding 30 genomes have been published from the genus *Rickettsia*, ten of which from species previously lacking a sequenced genome: "*Candidatus* Rickettsia amblyommii str. GAT-30V," *R. australis* str. Cutlack, *R. heilongjiangensis* str. 054 (Duan et al., 2011), *R. helvetica* str. C9P9 (Dong et al, 2012), *R. japonica* str. YH, *R. montanensis* str. OSU 85-930, *R. parkeri* str. Portsmouth, *R. philipii* str. 364D, *R. rhipicephali* str. 3-7-female6-CWPP, and *R. slovaca* str. 13-B (Fournier et al., 2012). A second genome from *R. slovaca* was also released (str. D-CWPP), as well as genomes from new strains of *R. canadensis* (str. CA410), *R. massiliae* (str. AZT80) and *R. conorii* (subsp. *indica* ITTR) (Sentausa et al., 2012b). Two new genomes were also completed for *R. sibirica* (subsp. *mongolitimonae* HA-91 [Sentausa et al., 2012a] and subsp. *sibirica* BJ-90 [Sentausa et al., 2012c]) and *R. typhi* (str. B9991CWPP and str. TH1527). Finally, six genome sequences were generated for strains within *R. prowazekii* (str. BuV67-CWPP, str. Chernikova, str. Dachau, str. GvV257, str. Katsinyian, and str. RpGvF24) and *R. rickettsii* (str. Arizona, str. Brazil, str. Colombia, str. Hauke, str. Hino, and str. Hlp#2). The total number of genomes sequenced for *Anaplasmataceae* and *Rickettsiaceae*, as of June 2012, is 80.

As if perhaps answering our wish for the generation of *Rickettsiales* genome sequences outside of *Anaplasmataceae* and *Rickettsiaceae*, two studies published the valuable genomes of "*Candidatus* Midichloria mitochondrii str. IricVA" (Sassera et al., 2011) and "*Candidatus* Odyssella thessalonicensis str. L13" (Georgiades et al., 2011). The position of these organisms in estimated phylogenies differs from their placement within our 16S rRNA sequence-based phylogeny. "*Candidatus* Midichloria mitochondrii" was supported as an ancestral lineage to the *Rickettsiaceae* (Sassera et al., 2011), while the placement of "*Candidatus* Odyssella thessalonicensis str. L13" was not strongly supported as a lineage of *Rickettsiales* or any other alphaproteobacterial group in various analyses (Georgiades et al., 2011). As such, we recommend that "*Candidatus* Odyssella thessalonicensis" be treated as "*Alphaproteobacteria incertae cedis*" until a more robust phylogenomic analysis reveals its position within the alphaproteobacterial tree.

While additional SAR11 genomes have been published since the completion of our work, there is now substantial evidence, despite controversy, that this lineage is not a member of *Rickettsiales*. While some very recent studies support the placement of SAR11 within the *Rickettsiales* (Georgiades et al., 2011; Thrash et al., 2011), others have carefully selected datasets or models of DNA evolution (or both) to demonstrate that a shared base compositional bias (AT richness) between SAR11 and *Rickettsiales* results in their monophyly in analyses that do not account for this problem (Brindefalk et al., 2011; Rodriguez-Ezpeleta and Embley, 2012; Sassera et al., 2011; Viklund et al., 2012). Inclusion of mitochondrial genomes in analyses does not alleviate the base compositional problem, and spurious results have been reached claiming a SAR11-mitochondrial affiliation (Georgiades et al., 2011; Thrash et al., 2011). Modeling base compositional bias and selecting genes with unbiased (or minimally biased) base composition consistently supports the monophyly of *Rickettsiales* and mitochondria, to the exclusion of SAR11 (Brindefalk et al., 2011; Rodriguez-Ezpeleta and Embley, 2012; Sassera et al., 2011; Viklund et al., 2011). Despite this, very recent studies estimating the genetic composition of the mitochondrial ancestor have unambiguously demonstrated the chimeric nature of its genome, with genes transferred to the protoeukaryote genome having a broad range of phylogenetic signals (Abhishek et al., 2011; Thiergart et al., 2012). However, a recent study including oceanic *Alphaproteobacteria* in estimated phylogenies uncovered a bacterial lineage that is the closest free-living relative to mitochondria and shares a sister relationship with *Rickettsiales* (to the exclusion of SAR11) (Brindefalk et al., 2011). Collectively, these studies suggest that our interpretation of *Rickettsiales* reductive genomes evolving from an ancestral streamlined genome (i.e., *Pelagibacter*) needs to be interpreted with caution. However, our characterization of the modern composition of rickettsial genomes resulting from a transition from facultative to obligate intracellular lifestyle need not be dependent on the SAR11 lineage as the ancestor to *Rickettsiales*.

Accordingly, the SAR11 lineage has been removed from the *Rickettsiales* at the taxonomic level at NCBI and PATRIC. The complete list of *Rickettsiales* genomes, as well as associated genomics and post-genomics data, will continually be updated at PATRIC (http://patricbrc.org).

REFERENCES

Abamo, F., H. Dohra, and M. Fujishima. 2008. Fate of the 63-kDa periplasmic protein of the infectious form of the endonuclear symbiotic bacterium *Holospora obtusa* during the infection process. *FEMS Microbiol. Lett.* **280:**21–27.

Abergel, C., G. Blanc, V. Monchois, P. Renesto, C. Sigoillot, H. Ogata, D. Raoult, and J. M. Claverie. 2006. Impact of the excision of an ancient repeat insertion on *Rickettsia conorii* guanylate kinase activity. *Mol. Biol. Evol.* **23:**2112–2122.

Abhishek, A., A. Bavishi, and M. Choudhary. 2011. Bacterial genome chimaerism and the origin of mitochondria. *Can. J. Microbiol.* **57:**49–61.

Agnes, J. T., K. A. Brayton, M. LaFollett, J. Norimine, W. C. Brown, and G. H. Palmer. 2011. Identification of *Anaplasma marginale* outer membrane protein antigens conserved between *A. marginale* sensu stricto strains and the live *A. marginale* subsp. *centrale* vaccine. *Infect. Immun.* **79:**1311–1318.

Alvarez-Martinez, C. E., and P. J. Christie. 2009. Biological diversity of prokaryotic type IV secretion systems. *Microbiol. Mol. Biol. Rev.* **73:**775–808.

Amann, R. I., W. Ludwig, and K. H. Schleifer. 1995. Phylogenetic identification and in situ detection of individual microbial cells without cultivation. *Microbiol. Rev.* **59:**143–169.

Amann, R., N. Springer, W. Ludwig, H. D. Görtz, and K. H. Schleifer. 1991. Identification *in situ* and phylogeny of uncultured bacterial endosymbionts. *Nature* **351:**161–164.

Amano, K., A. Tamura, N. Ohashi, H. Urakami, S. Kaya, and K. Fukushi. 1987. Deficiency of peptidoglycan and lipopolysaccharide components in *Rickettsia tsutsugamushi*. *Infect. Immun.* **55:**2290–2292.

Amiri, H., C. M. Alsmark, and S. G. Andersson. 2002. Proliferation and deterioration of *Rickettsia* Palindromic Elements. *Mol. Biol. Evol.* **19:**1234–1243.

Andersson, J. O., and S. G. Andersson. 1999. Insights into the evolutionary process of genome degradation. *Curr. Opin. Genet. Dev.* **9:**664–671.

Andersson, S. G., A. S. Eriksson, A. K. Naslund, M. S. Andersen, and C. G. Kurland. 1996. The *Rickettsia prowazekii* genome: a random sequence analysis. *Microb. Comp. Genomics* **1:**293–315.

Andersson, S. G., D. R. Stothard, P. Fuerst, and C. G. Kurland. 1999. Molecular phylogeny and rearrangement of rRNA genes in *Rickettsia* species. *Mol. Biol. Evol.* **16:**987–995.

Andersson, S. G., A. Zomorodipour, J. O. Andersson, T. Sicheritz-Pontén, U. C. Alsmark, R. M. Podowski, A. K. Naslund, A. S. Eriksson, H. H. Winkler, and C. G.

Kurland. 1998. The genome sequence of *Rickettsia prowazekii* and the origin of mitochondria. *Nature* **396:**133–140.

Andersson, S. G., A. Zomorodipour, H. H. Winkler, and C. G. Kurland. 1995. Unusual organization of the rRNA genes in *Rickettsia prowazekii*. *J. Bacteriol.* **177:**4171–4175.

Antonio, D. B., K. B. Andree, J. D. Moore, C. S. Friedman, and R. P. Hedrick. 2000. Detection of Rickettsiales-like prokaryotes by *in situ* hybridization in black abalone, *Haliotis cracherodii*, with withering syndrome. *J. Invertebr. Pathol.* **75:**180–182.

Audia, J. P., and H. H. Winkler. 2006. Study of the five *Rickettsia prowazekii* proteins annotated as ATP/ADP translocases (Tlc): only Tlc1 transports ATP/ADP, while Tlc4 and Tlc5 transport other ribonucleotides. *J. Bacteriol.* **188:**6261–6268.

Azad, A. F. 2007. Pathogenic rickettsiae as bioterrorism agents. *Clin. Infect. Dis.* **45**(Suppl. 1):S52–S55.

Azad, A. F., M. S. Beier, and J. J. Gillespie. 2008. The Rickettsiaceae, p. 439–450. *In* L. Green and E. Goldman (ed.), *Practical Handbook of Microbiology*. CRC Press, Boca Raton, FL.

Azad, A. F., and S. Radulovic. 2003. Pathogenic rickettsiae as bioterrorism agents. *Ann. N. Y. Acad. Sci.* **990:**734–738.

Aziz, R. K., D. Bartels, A. A. Best, M. DeJongh, T. Disz, R. A. Edwards, K. Formsma, S. Gerdes, E. M. Glass, M. Kubal, F. Meyer, G. J. Olsen, R. Olson, A. L. Osterman, R. A. Overbeek, L. K. McNeil, D. Paarmann, T. Paczian, B. Parrello, G. D. Pusch, C. Reich, R. Stevens, O. Vassieva, V. Vonstein, A. Wilke, and O. Zagnitko. 2008. The RAST Server: rapid annotations using subsystems technology. *BMC Genomics* **9:**75.

Baker, B. J., P. Hugenholtz, S. C. Dawson, and J. F. Banfield. 2003. Extremely acidophilic protists from acid mine drainage host *Rickettsiales*-lineage endosymbionts that have intervening sequences in their 16S rRNA genes. *Appl. Environ. Microbiol.* **69:**5512–5518.

Balayeva, N. M., M. E. Eremeeva, V. F. Ignatovich, B. A. Dmitriev, E. B. Lapina, and L. S. Belousova. 1992. Protein antigens of genetically related *Rickettsia prowazekii* strains with different virulence. *Acta Virol.* **36:**52–56.

Balayeva, N. M., M. E. Eremeeva, H. Tissot-Dupont, I. A. Zakharov, and D. Raoult. 1995. Genotype characterization of the bacterium expressing the male-killing trait in the ladybird beetle *Adalia bipunctata* with specific rickettsial molecular tools. *Appl. Environ. Microbiol.* **61:**1431–1437.

Baldridge, G. D., N. Y. Burkhardt, R. F. Felsheim, T. J. Kurtti, and U. G. Munderloh. 2007. Transposon insertion reveals pRM, a plasmid of *Rickettsia monacensis*. *Appl. Environ. Microbiol.* **73:**4984–4995.

Baldridge, G. D., N. Y. Burkhardt, R. F. Felsheim, T. J. Kurtti, and U. G. Munderloh. 2008. Plasmids of the pRM/pRF family occur in diverse *Rickettsia* species. *Appl. Environ. Microbiol.* **74:**645–652.

Baldridge, G. D., N. Y. Burkhardt, M. B. Labruna, R. C. Pacheco, C. D. Paddock, P. C. Williamson, P. M. Billingsley, R. F. Felsheim, T. J. Kurtti, and U. G. Munderloh. 2010. Wide dispersal and possible multiple origins of low-copy-number plasmids in *Rickettsia* species associated with blood-feeding arthropods. *Appl. Environ. Microbiol.* **76:**1718–1731.

Baldridge, G. D., N. Y. Burkhardt, J. A. Simser, T. J. Kurtti, and U. G. Munderloh. 2004. Sequence and expression analysis of the *ompA* gene of *Rickettsia peacockii*, an endosymbiont of the Rocky Mountain wood tick, *Dermacentor andersoni*. *Appl. Environ. Microbiol.* **70:**6628–6636.

Balraj, P., K. El Karkouri, G. Vestris, L. Espinosa, D. Raoult, and P. Renesto. 2008. RickA expression is not sufficient to promote actin-based motility of *Rickettsia raoultii*. *PLoS One* **3:**e2582.

Bechah, Y., K. El Karkouri, O. Mediannikov, Q. Leroy, N. Pelletier, C. Robert, C. Medigue, J. L. Mege, and D. Raoult. 2010. Genomic, proteomic, and transcriptomic analysis of virulent and avirulent *Rickettsia prowazekii* reveals its adaptive mutation capabilities. *Genome Res.* **20:** 655–663.

Beier, C. L., M. Horn, R. Michel, M. Schweikert, H. D. Görtz, and M. Wagner. 2002. The genus *Caedibacter* comprises endosymbionts of *Paramecium* spp. related to the *Rickettsiales* (*Alphaproteobacteria*) and to *Francisella tularensis* (*Gammaproteobacteria*). *Appl. Environ. Microbiol.* **68:**6043–6050.

Beninati, T., N. Lo, L. Sacchi, C. Genchi, H. Noda, and C. Bandi. 2004. A novel alphaproteobacterium resides in the mitochondria of ovarian cells of the tick *Ixodes ricinus*. *Appl. Environ. Microbiol.* **70:**2596–2602.

Beninati, T., M. Riegler, I. M. Vilcins, L. Sacchi, R. McFadyen, M. Krockenberger, C. Bandi, S. L. O'Neill, and N. Lo. 2009. Absence of the symbiont *Candidatus* Midichloria mitochondrii in the mitochondria of the tick *Ixodes holocyclus*. *FEMS Microbiol. Lett.* **299:**241–247.

Birtles, R. J., T. G. Harrison, N. A. Saunders, and D. H. Molyneux. 1995. Proposals to unify the genera *Grahamella* and *Bartonella*, with descriptions of *Bartonella talpae* comb. nov., *Bartonella peromysci* comb. nov., and three new species, *Bartonella grahamii* sp. nov., *Bartonella taylorii* sp. nov., and

Bartonella doshiae sp. nov. *Int. J. Syst. Bacteriol.* **45:**1–8.

Birtles, R. J., T. J. Rowbotham, R. Michel, D. G. Pitcher, B. Lascola, S. Alexiou-Daniel, and D. Raoult. 2000. '*Candidatus* Odyssella thessalonicensis' gen. nov., sp. nov., an obligate intracellular parasite of *Acanthamoeba* species. *Int. J. Syst. Evol. Microbiol.* **50:**63–72.

Blanc, G., M. Ngwamidiba, H. Ogata, P. E. Fournier, J. M. Claverie, and D. Raoult. 2005. Molecular evolution of rickettsia surface antigens: evidence of positive selection. *Mol. Biol. Evol.* **22:**2073–2083.

Blanc, G., H. Ogata, C. Robert, S. Audic, J. M. Claverie, and D. Raoult. 2007a. Lateral gene transfer between obligate intracellular bacteria: evidence from the *Rickettsia massiliae* genome. *Genome Res.* **17:**1657–1664.

Blanc, G., H. Ogata, C. Robert, S. Audic, K. Suhre, G. Vestris, J. M. Claverie, and D. Raoult. 2007b. Reductive genome evolution from the mother of *Rickettsia. PLoS Genet.* **3:**e14.

Blaxter, M. 2007. Symbiont genes in host genomes: fragments with a future? *Cell Host Microbe* **2:**211–213.

Bordenstein, S. R. 2007. Evolutionary genomics: transdomain gene transfers. *Curr. Biol.* **17:**R935–R936.

Braid, B. A., J. D. Moore, T. T. Robbins, R. P. Hedrick, R. S. Tjeerdema, and C. S. Friedman. 2005. Health and survival of red abalone, *Haliotis rufescens*, under varying temperature, food supply, and exposure to the agent of withering syndrome. *J. Invertebr. Pathol.* **89:**219–231.

Braig, H. R., B. D. Turner, and M. A. Perotti. 2008. Symbiotic rickettsia, p. 221–249. *In* K. Bourtzis and T. A. Miller (ed.), *Insect Symbiosis*, vol. 3. CRC Press, Boca Raton, FL.

Braig, H. R., W. Zhou, S. L. Dobson, and S. L. O'Neill. 1998. Cloning and characterization of a gene encoding the major surface protein of the bacterial endosymbiont *Wolbachia pipientis. J. Bacteriol.* **180:**2373–2378.

Brayton, K. A., L. S. Kappmeyer, D. R. Herndon, M. J. Dark, D. L. Tibbals, G. H. Palmer, T. C. McGuire, and D. P. Knowles, Jr. 2005. Complete genome sequencing of *Anaplasma marginale* reveals that the surface is skewed to two superfamilies of outer membrane proteins. *Proc. Natl. Acad. Sci. USA* **102:**844–849.

Brayton, K. A., D. P. Knowles, T. C. McGuire, and G. H. Palmer. 2001. Efficient use of a small genome to generate antigenic diversity in tick-borne ehrlichial pathogens. *Proc. Natl. Acad. Sci. USA* **98:**4130–4135.

Brayton, K. A., G. H. Palmer, A. Lundgren, J. Yi, and A. F. Barbet. 2002. Antigenic variation of *Anaplasma marginale msp2* occurs by combinatorial gene conversion. *Mol. Microbiol.* **43:**1151–1159.

Brenner, D. J., S. P. O'Connor, H. H. Winkler, and A. G. Steigerwalt. 1993. Proposals to unify the genera *Bartonella* and *Rochalimaea*, with descriptions of *Bartonella quintana* comb. nov., *Bartonella vinsonii* comb. nov., *Bartonella henselae* comb. nov., and *Bartonella elizabethae* comb. nov., and to remove the family *Bartonellaceae* from the order *Rickettsiales. Int. J. Syst. Bacteriol.* **43:** 777–786.

Brindefalk, B., T. J. Ettema, J. Viklund, M. Thollesson, and S. G. Andersson. 2011. A phylometagenomic exploration of oceanic alphaproteobacteria reveals mitochondrial relatives unrelated to the SAR11 clade. *PLoS One* **6:**e24457.

Campbell, B. C., T. S. Bragg, and C. E. Turner. 1992. Phylogeny of symbiotic bacteria of four weevil species (Coleoptera: Curculionidae) based on analysis of 16S ribosomal DNA. *Insect Biochem. Mol. Biol.* **22:**415–421.

Castresana, J. 2000. Selection of conserved blocks from multiple alignments for their use in phylogenetic analysis. *Mol. Biol. Evol.* **17:**540–552.

Chafee, M. E., D. J. Funk, R. G. Harrison, and S. R. Bordenstein. 2010. Lateral phage transfer in obligate intracellular bacteria (*Wolbachia*): verification from natural populations. *Mol. Biol. Evol.* **27:**501–505.

Chen, D. Q., B. C. Campbell, and A. H. Purcell. 1996. A new rickettsia from a herbivorous insect, the pea aphid *Acyrthosiphon pisum* (Harris). *Curr. Microbiol.* **33:**123–128.

Cho, N. H., H. R. Kim, J. H. Lee, S. Y. Kim, J. Kim, S. Cha, A. C. Darby, H. H. Fuxelius, J. Yin, J. H. Kim, S. J. Lee, Y. S. Koh, W. J. Jang, K. H. Park, S. G. Andersson, M. S. Choi, and I. S. Kim. 2007. The *Orientia tsutsugamushi* genome reveals massive proliferation of conjugative type IV secretion system and host-cell interaction genes. *Proc. Natl. Acad. Sci. USA* **104:**7981–7986.

Claverie, J. M., and H. Ogata. 2003. The insertion of palindromic repeats in the evolution of proteins. *Trends Biochem. Sci.* **28:**75–80.

Collins, N. E., J. Liebenberg, E. P. de Villiers, K. A. Brayton, E. Louw, A. Pretorius, F. E. Faber, H. van Heerden, A. Josemans, M. van Kleef, H. C. Steyn, M. F. van Strijp, E. Zweygarth, F. Jongejan, J. C. Maillard, D. Berthier, M. Botha, F. Joubert, C. H. Corton, N. R. Thomson, M. T. Allsopp, and B. A. Allsopp. 2005. The genome of the heartwater agent *Ehrlichia ruminantium* contains multiple tandem repeats of actively variable copy number. *Proc. Natl. Acad. Sci. USA* **102:** 838–843.

Dame, J. B., S. M. Mahan, and C. A. Yowell. 1992. Phylogenetic relationship of *Cowdria ruminantium*, agent of heartwater, to *Anaplasma marginale* and other members of the order *Rickettsiales* determined on the basis of 16S rRNA sequence. *Int. J. Syst. Bacteriol.* **42:**270–274.

Dandekar, T., S. Schuster, B. Snel, M. Huynen, and P. Bork. 1999. Pathway alignment: application to the comparative analysis of glycolytic enzymes. *Biochem. J.* **343:**115–124.

Darby, A. C., N. H. Cho, H. H. Fuxelius, J. Westberg, and S. G. Andersson. 2007. Intracellular pathogens go extreme: genome evolution in the Rickettsiales. *Trends Genet.* **23:**511–520.

Dark, M. J., B. Al-Khedery, and A. F. Barbet. 2011. Multistrain genome analysis identifies candidate vaccine antigens of *Anaplasma marginale. Vaccine* **29:**4923-4932.

Dark, M. J., D. R. Herndon, L. S. Kappmeyer, M. P. Gonzales, E. Nordeen, G. H. Palmer, D. P. Knowles, Jr., and K. A. Brayton. 2009. Conservation in the face of diversity: multistrain analysis of an intracellular bacterium. *BMC Genomics* **10:**16.

Darling, A. E., B. Mau, and N. T. Perna. 2010. progressiveMauve: multiple genome alignment with gene gain, loss and rearrangement. *PLoS One* **5:**e11147.

Daugherty, R. M., N. Linka, J. P. Audia, C. Urbany, H. E. Neuhaus, and H. H. Winkler. 2004. The nucleotide transporter of *Caedibacter caryophilus* exhibits an extended substrate spectrum compared to the analogous ATP/ADP translocase of *Rickettsia prowazekii. J. Bacteriol.* **186:**3262–3265.

Davis, A. K., J. L. DeVore, J. R. Milanovich, K. Cecala, J. C. Maerz, and M. J. Yabsley. 2009. New findings from an old pathogen: intraerythrocytic bacteria (family Anaplasmatacea) in red-backed salamanders *Plethodon cinereus. Ecohealth* **6:**219–228.

Davis, M. J., Z. Ying, B. R. Brunner, A. Pantoja, and F. H. Ferwerda. 1998. Rickettsial relative associated with papaya bunchy top disease. *Curr. Microbiol.* **36:**80–84.

Dohra, H., M. Fujishima, and H. Ishikawa. 1998. Structure and expression of a GroE-homologous operon of a macronucleus-specific symbiont *Holospora obtusa* of the ciliate *Paramecium caudatum. J. Eukaryot. Microbiol.* **45:**71–79.

Dohra, H., K. Yamamoto, M. Fujishima, and H. Ishikawa. 1997. Cloning and sequencing of gene coding for a periplasmic 5.4 kDa peptide of the macronucleus-specific symbiont *Holospora obtusa* of the ciliate *Paramecium caudatum. Zoolog. Sci.* **14:**69–75.

Dong, X., K. El Karkouri, C. Robert, F. Gavory, D. Raoult, and P. E. Fournier. 2012. Genomic comparison of *Rickettsia helvetica* and other *Rickettsia species. J. Bacteriol.* **194:**2751.

Duan, C., Y. Tong, Y. Huang, X. Wang, X. Xiong, and B. Wen. 2011. Complete genome sequence of *Rickettsia heilongjiangensis*, an emerging tick-transmitted human pathogen. *J. Bacteriol.* **193:**5564-5565.

Dumler, J. S., A. F. Barbet, C. P. Bekker, G. A. Dasch, G. H. Palmer, S. C. Ray, Y. Rikihisa, and F. R. Rurangirwa. 2001. Reorganization of genera in the families *Rickettsiaceae* and *Anaplasmataceae* in the order *Rickettsiales*: unification of some species of *Ehrlichia* with *Anaplasma, Cowdria* with *Ehrlichia* and *Ehrlichia* with *Neorickettsia*, descriptions of six new species combinations and designation of *Ehrlichia equi* and 'HGE agent' as subjective synonyms of *Ehrlichia phagocytophila. Int. J. Syst. Evol. Microbiol.* **51:**2145–2165.

Dumler, J. S., and D. H. Walker. 2005. Order II. Rickettsiales Gieszczykiewicz 1939, 25AL emend. Dumler, Barbet, Bekker, Dasch, Palmer, Ray, Rikihisa and Rurangirwa 2001, 2156, p. 96–145. *In* G. M. Garrity, D. J. Brenner, N. R. Krieg, and J. T. Staley (ed.), *Bergey's Manual of Systematic Bacteriology*, 2nd ed., vol. 2. Springer-Verlag, New York, NY.

Dunning Hotopp, J. C., M. E. Clark, D. C. Oliveira, J. M. Foster, P. Fischer, M. C. Munoz Torres, J. D. Giebel, N. Kumar, N. Ishmael, S. Wang, J. Ingram, R. V. Nene, J. Shepard, J. Tomkins, S. Richards, D. J. Spiro, E. Ghedin, B. E. Slatko, H. Tettelin, and J. H. Werren. 2007. Widespread lateral gene transfer from intracellular bacteria to multicellular eukaryotes. *Science* **317:**1753–1756.

Dunning Hotopp, J. C., M. Lin, R. Madupu, J. Crabtree, S. V. Angiuoli, J. Eisen, R. Seshadri, Q. Ren, M. Wu, T. R. Utterback, S. Smith, M. Lewis, H. Khouri, C. Zhang, H. Niu, Q. Lin, N. Ohashi, N. Zhi, W. Nelson, L. M. Brinkac, R. J. Dodson, M. J. Rosovitz, J. Sundaram, S. C. Daugherty, T. Davidsen, A. S. Durkin, M. Gwinn, D. H. Haft, J. D. Selengut, S. A. Sullivan, N. Zafar, L. Zhou, F. Benahmed, H. Forberger, R. Halpin, S. Mulligan, J. Robinson, O. White, Y. Rikihisa, and H. Tettelin. 2006. Comparative genomics of emerging human ehrlichiosis agents. *PLoS Genet.* **2:**e21.

Durbin, R., S. Eddy, A. Krogh, and G. Mitchison. 1998. *Biological Sequence Analysis: Probabilistic Models of Proteins and Nucleic Acids.* Cambridge University Press, Cambridge, United Kingdom.

Dwyer, D. S. 2001. Selfish DNA and the origin of genes. *Science* **291:**252–253.

Edgar, R. C. 2004a. MUSCLE: a multiple sequence alignment method with reduced time and space complexity. *BMC Bioinformatics* **5:**113.

Edgar, R. C. 2004b. MUSCLE: multiple sequence alignment with high accuracy and high throughput. *Nucleic Acids Res.* **32:**1792–1797.

Ellison, D. W., T. R. Clark, D. E. Sturdevant, K. Virtaneva, S. F. Porcella, and T. Hackstadt. 2008. Genomic comparison of virulent *Rickettsia rickettsii* Sheila Smith and avirulent *Rickettsia rickettsii* Iowa. *Infect. Immun.* **76:**542–550.

Emelyanov, V. V. 2001a. Evolutionary relationship of Rickettsiae and mitochondria. *FEBS Lett.* **501:**11–18.

Emelyanov, V. V. 2001b. Rickettsiaceae, rickettsia-like endosymbionts, and the origin of mitochondria. *Biosci. Rep.* **21:**1–17.

Emelyanov, V. V. 2003a. Common evolutionary origin of mitochondrial and rickettsial respiratory chains. *Arch. Biochem. Biophys.* **420:**130–141.

Emelyanov, V. V. 2003b. Mitochondrial connection to the origin of the eukaryotic cell. *Eur. J. Biochem.* **270:**1599–1618.

Emelyanov, V. V. 2007. Constantin Merezhkowsky and the Endokaryotic Hypothesis, p. 201–237. *In* W. F. Martin and M. Müller (ed.), *Origin of Mitochondria and Hydrogenosomes*. Springer-Verlag, Berlin, Heidelberg.

Emelyanov, V. V., and B. V. Sinitsyn. 1999. A *groE*-based phylogenetic analysis shows the closest evolutionary relationship of mitochondria to obligate intracytoplasmic bacterium *Rickettsia prowazekii*, p. 31–37. *In* D. Raoult and P. Brouqui (ed.), *Rickettsiae and Rickettsial Diseases at the Turn of the Third Millennium*. Elsevier, Marseille, France.

Epis, S., D. Sassera, T. Beninati, N. Lo, L. Beati, J. Piesman, L. Rinaldi, K. D. McCoy, A. Torina, L. Sacchi, E. Clementi, M. Genchi, S. Magnino, and C. Bandi. 2008. *Midichloria mitochondrii* is widespread in hard ticks (Ixodidae) and resides in the mitochondria of phylogenetically diverse species. *Parasitology* **135:**485–494.

Eremeeva, M. E., A. Madan, C. D. Shaw, K. Tang, and G. A. Dasch. 2005. New perspectives on rickettsial evolution from new genome sequences of *Rickettsia*, particularly *R. canadensis*, and *Orientia tsutsugamushi*. *Ann. N. Y. Acad. Sci.* **1063:**47–63.

Eschbach, E., M. Pfannkuchen, M. Schweikert, D. Drutschmann, F. Brümmer, S. Fokin, W. Ludwig, and H. D. Görtz. 2009. "*Candidatus* Paraholospora nucleivisitans", an intracellular bacterium in *Paramecium sexaurelia* shuttles between the cytoplasm and the nucleus of its host. *Syst. Appl. Microbiol.* **32:**490–500.

Ettema, T. J., and S. G. Andersson. 2009. The α-proteobacteria: the Darwin finches of the bacterial world. *Biol. Lett.* **5:**429–432.

Federici, B. A. 1980. Reproduction and morphogenesis of *Rickettsiella chironomi*, an unusual intracellular procaryotic parasite of midge larvae. *J. Bacteriol.* **143:**995–1002.

Fehr, J. S., G. V. Bloemberg, C. Ritter, M. Hombach, T. F. Lüscher, R. Weber, and P. M. Keller. 2010. Septicemia caused by tick-borne bacterial pathogen *Candidatus* Neoehrlichia mikurensis. *Emerg. Infect. Dis.* **16:**1127–1129.

Felsheim, R. F., T. J. Kurtti, and U. G. Munderloh. 2009. Genome sequence of the endosymbiont *Rickettsia peacockii* and comparison with virulent *Rickettsia rickettsii*: identification of virulence factors. *PLoS One* **4:**e8361.

Ferrantini, F., S. I. Fokin, L. Modeo, I. Andreoli, F. Dini, H. D. Görtz, F. Verni, and G. Petroni. 2009. "*Candidatus* Cryptoprodotis polytropus," a novel *Rickettsia*-like organism in the ciliated protist *Pseudomicrothorax dubius* (Ciliophora, Nassophorea). *J. Eukaryot. Microbiol.* **56:**119–129.

Forsman, M., G. Sandström, and A. Sjöstedt. 1994. Analysis of 16S ribosomal DNA sequences of *Francisella* strains and utilization for determination of the phylogeny of the genus and for identification of strains by PCR. *Int. J. Syst. Bacteriol.* **44:**38–46.

Foster, J., M. Ganatra, I. Kamal, J. Ware, K. Makarova, N. Ivanova, A. Bhattacharyya, V. Kapatral, S. Kumar, J. Posfai, T. Vincze, J. Ingram, L. Moran, A. Lapidus, M. Omelchenko, N. Kyrpides, E. Ghedin, S. Wang, E. Goltsman, V. Joukov, O. Ostrovskaya, K. Tsukerman, M. Mazur, D. Comb, E. Koonin, and B. Slatko. 2005. The *Wolbachia* genome of *Brugia malayi*: endosymbiont evolution within a human pathogenic nematode. *PLoS Biol.* **3:**e121.

Fournier, P. E., L. Belghazi, C. Robert, K. Elkarkouri, A. L. Richards, G. Greub, F. Collyn, M. Ogawa, A. Portillo, J. A. Oteo, A. Psaroulaki, I. Bitam, and D. Raoult. 2008. Variations of plasmid content in *Rickettsia felis*. *PLoS One* **3:**e2289.

Fournier, P. E., K. El Karkouri, Q. Leroy, C. Robert, B. Giumelli, P. Renesto, C. Socolovschi, P. Parola, S. Audic, and D. Raoult. 2009. Analysis of the *Rickettsia africae* genome reveals that virulence acquisition in *Rickettsia* species may be explained by genome reduction. *BMC Genomics* **10:**166.

Fournier, P. E., K. El Karkouri, C. Robert, C. Medigue, and D. Raoult. 2012. Complete genome sequence of *Rickettsia slovaca*, the agent of tick-borne lymphadenitis. *J. Bacteriol.* **194:**1612.

Fraune, S., and T. C. Bosch. 2007. Long-term maintenance of species-specific bacterial microbiota in the basal metazoan *Hydra*. *Proc. Natl. Acad. Sci. USA* **104:**13146–13151.

Friedman, C. S., K. B. Andree, K. A. Beauchamp, J. D. Moore, T. T. Robbins, J. D. Shields, and R. P. Hedrick. 2000. '*Candidatus* Xenohaliotis californiensis', a newly described pathogen of abalone, *Haliotis* spp., along the west coast of North America. *Int. J. Syst. Evol. Microbiol.* **50:**847–855.

Fritsche, T. R., R. K. Gautom, S. Seyedirashti, D. L. Bergeron, and T. D. Lindquist. 1993. Occurrence of bacterial endosymbionts in *Acanthamoeba* spp. isolated from corneal and environmental specimens and contact lenses. *J. Clin. Microbiol.* **31:**1122–1126.

Fritsche, T. R., M. Horn, S. Seyedirashti, R. K. Gautom, K. H. Schleifer, and M. Wagner. 1999. In situ detection of novel bacterial endosymbionts of *Acanthamoeba* spp. phylogenetically related to members of the order *Rickettsiales*. *Appl. Environ. Microbiol.* **65:**206–212.

Frohlich, K. M., R. A. Roberts, N. A. Housley, and J. P. Audia. 2010. *Rickettsia prowazekii* uses an *sn*-glycerol-3-phosphate dehydrogenase and a novel dihydroxyacetone phosphate transport system to supply triose phosphate for phospholipid biosynthesis. *J. Bacteriol.* **192:**4281–4288.

Frutos, R., B. A. Federici, B. Revet, and M. Bergoin. 1994. Taxonomic studies of *Rickettsiella*, *Rickettsia*, and *Chlamydia* using genomic DNA. *J. Invertebr. Pathol.* **63:**294–300.

Frutos, R., A. Viari, C. Ferraz, A. Morgat, S. Eychenie, Y. Kandassamy, I. Chantal, A. Bensaid, E. Coissac, N. Vachiery, J. Demaille, and D. Martinez. 2006. Comparative genomic analysis of three strains of *Ehrlichia ruminantium* reveals an active process of genome size plasticity. *J. Bacteriol.* **188:**2533–2542.

Fuxelius, H. H., A. C. Darby, N. H. Cho, and S. G. Andersson. 2008. Visualization of pseudogenes in intracellular bacteria reveals the different tracks to gene destruction. *Genome Biol.* **9:**R42.

Fuxelius, H. H., A. Darby, C. K. Min, N. H. Cho, and S. G. Andersson. 2007. The genomic and metabolic diversity of *Rickettsia*. *Res. Microbiol.* **158:**745–753.

Gardner, G. R., J. C. Harshbarger, J. L. Lake, T. K. Sawyer, K. L. Price, M. D. Stephenson, P. L. Haaker, and H. A. Togstad. 1995. Association of prokaryotes with symptomatic appearance of withering syndrome in black abalone *Haliotis cracherodii*. *J. Invertebr. Pathol.* **66:**111–120.

Gavotte, L., H. Henri, R. Stouthamer, D. Charif, S. Charlat, M. Boulétreau, and F. Vavre. 2007. A Survey of the bacteriophage WO in the endosymbiotic bacteria Wolbachia. *Mol. Biol. Evol.* **24:**427–435.

Ge, H., Y. Y. Chuang, S. Zhao, M. Tong, M. H. Tsai, J. J. Temenak, A. L. Richards, and W. M. Ching. 2004. Comparative genomics of *Rickettsia prowazekii* Madrid E and Breinl strains. *J. Bacteriol.* **186:**556–565.

Georgiades, K., M. A. Madoui, P. Le, C. Robert, and D. Raoult. 2011. Phylogenomic analysis of *Odyssella thessalonicensis* fortifies the common origin of Rickettsiales, *Pelagibacter ubique* and *Reclimonas americana* mitochondrion. *PLoS One* **6:** e24857.

Gillespie, J. J., N. C. Ammerman, M. Beier-Sexton, B. S. Sobral, and A. F. Azad. 2009a. Louse- and flea-borne rickettsioses: biological and genomic analyses. *Vet. Res.* **40:**12.

Gillespie, J. J., N. C. Ammerman, S. M. Dreher-Lesnick, M. S. Rahman, M. J. Worley, J. C. Setubal, B. S. Sobral, and A. F. Azad. 2009b. An anomalous type IV secretion system in *Rickettsia* is evolutionarily conserved. *PLoS One* **4:**e4833.

Gillespie, J. J., M. S. Beier, M. S. Rahman, N. C. Ammerman, J. M. Shallom, A. Purkayastha, B. S. Sobral, and A. F. Azad. 2007. Plasmids and rickettsial evolution: insight from *Rickettsia felis*. *PLoS One* **2:**e266.

Gillespie, J. J., K. A. Brayton, K. P. Williams, M. A. Diaz, W. C. Brown, A. F. Azad, and B. W. Sobral. 2010. Phylogenomics reveals a diverse *Rickettsiales* type IV secretion system. *Infect. Immun.* **78:**1809–1823.

Gillespie, J. J., V. Joardar, K. P. Williams, T. Driscoll, J. B. Hostetler, E. Nordberg, M. Shukla, B. Wallenz, C. A. Hill, V. M. Nene, A. F. Azad, B. W. Sobral, and E. Caler. 4 November 2011a. A *Rickettsia* genome overrun by mobile genetic elements provides insight into the acquisition of genes characteristic of obligate intracellular lifestyle. *J. Bacteriol.* doi:10.1128/JB.06244-11.

Gillespie, J. J., A. R. Wattam, S. A. Cammer, J. L. Gabbard, M. P. Shukla, O. Dalay, T. Driscoll, D. Hix, S. P. Mane, C. Mao, E. K. Nordberg, M. Scott, J. R. Schulman, E. E. Snyder, D. E. Sullivan, C. Wang, A. Warren, K. P. Williams, T. Xue, H. S. Yoo, C. Zhang, Y. Zhang, R. Will, R. W. Kenyon, and B. W. Sobral. 2011b. PATRIC: the comprehensive bacterial bioinformatics resource with a focus on human pathogenic species. *Infect. Immun.* **79:**4286–4298.

Gillespie, J. J., K. Williams, M. Shukla, E. E. Snyder, E. K. Nordberg, S. M. Ceraul, C. Dharmanolla, D. Rainey, J. Soneja, J. M. Shallom, N. D. Vishnubhat, R. Wattam,

A. Purkayastha, M. Czar, O. Crasta, J. C. Setubal, A. F. Azad, and B. S. Sobral. 2008. *Rickettsia* phylogenomics: unwinding the intricacies of obligate intracellular life. *PLoS One* **3:**e2018.

Giovannoni, S. J., H. J. Tripp, S. Givan, M. Podar, K. L. Vergin, D. Baptista, L. Bibbs, J. Eads, T. H. Richardson, M. Noordewier, M. S. Rappe, J. M. Short, J. C. Carrington, and E. J. Mathur. 2005. Genome streamlining in a cosmopolitan oceanic bacterium. *Science* **309:**1242–1245.

Görtz, H. D., and M. Fujishima. 1983. Conjugation and meiosis of *Paramecium caudatum* infected with the micronucleus-specific bacterium *Holospora elegans*. *Eur. J. Cell Biol.* **32:**86–91.

Görtz, H. D., S. Lellig, O. Miosga, and M. Wiemann. 1990. Changes in fine structure and polypeptide pattern during development of *Holospora obtusa*, a bacterium infecting the macronucleus of *Paramecium caudatum*. *J. Bacteriol.* **172:**5664–5669.

Görtz, H. D., and H. J. Schmidt. 2005. Family III. Holosporaceae fam. nov., p. 146–160. *In* G. M. Garrity, D. J. Brenner, N. R. Krieg, and J. T. Staley (ed.), *Bergey's Manual of Systematic Bacteriology*, 2nd ed., vol. 2. Springer-Verlag, New York, NY.

Götz, P. 1971. "Multiple cell division" as a mode of reproduction of a cell-parasitic bacterium. *Naturwissenchaften* **58:**569–570.

Götz, P. 1972. "*Rickettsiella chironomi*": an unusual bacterial pathogen which reproduces by multiple cell division. *J. Invertebr. Pathol.* **20:**22–30.

Gouin, E., C. Egile, P. Dehoux, V. Villiers, J. Adams, F. Gertler, R. Li, and P. Cossart. 2004. The RickA protein of *Rickettsia conorii* activates the Arp2/3 complex. *Nature* **427:**457–461.

Gromov, B. V., and D. V. Ossipov. 1981. *Holospora* (ex Hafkine 1890) nom. rev., a genus of bacteria inhabiting the nuclei of paramecia. *Int. J. Syst. Bacteriol.* **31:**348–352.

Hackstadt, T. 1996. The biology of rickettsiae. *Infect. Agents Dis.* **5:**127–143.

Haglund, C. M., J. E. Choe, C. T. Skau, D. R. Kovar, and M. D. Welch. 2010. *Rickettsia* Sca2 is a bacterial formin-like mediator of actin-based motility. *Nat. Cell Biol.* **12:**1057–1063.

Hall, J., and H. Voelz. 1985. Bacterial endosymbionts of *Acanthamoeba* sp. *J. Parasitol.* **71:**89–95.

Herndon, D. R., G. H. Palmer, V. Shkap, D. P. Knowles, Jr., and K. A. Brayton. 2010. Complete genome sequence of *Anaplasma marginale* subsp. *centrale*. *J. Bacteriol.* **192:**379–380.

Hilgenboecker, K., P. Hammerstein, P. Schlattmann, A. Telschow, and J. H. Werren. 2008. How many species are infected with *Wolbachia*?—a statistical analysis of current data. *FEMS Microbiol. Lett.* **281:**215–220.

Hori, M., K. Fujii, and M. Fujishima. 2008. Micronucleus-specific bacterium *Holospora elegans* irreversibly enhances stress gene expression of the host *Paramecium caudatum*. *J. Eukaryot. Microbiol.* **55:**515–521.

Horn, M., T. R. Fritsche, R. K. Gautom, K. H. Schleifer, and M. Wagner. 1999. Novel bacterial endosymbionts of *Acanthamoeba* spp. related to the *Paramecium caudatum* symbiont *Caedibacter caryophilus*. *Environ. Microbiol.* **1:**357–367.

Inokuma, H., P. Brouqui, M. Drancourt, and D. Raoult. 2001. Citrate synthase gene sequence: a new tool for phylogenetic analysis and identification of *Ehrlichia*. *J. Clin. Microbiol.* **39:**3031–3039.

Ishmael, N., J. C. Dunning Hotopp, P. Ioannidis, S. Biber, J. Sakamoto, S. Siozios, V. Nene, J. Werren, K. Bourtzis, S. R. Bordenstein, and H. Tettelin. 2009. Extensive genomic diversity of closely related *Wolbachia* strains. *Microbiology* **155:**2211–2222.

Iwatani, K., H. Dohra, B. F. Lang, G. Burger, M. Hori, and M. Fujishima. 2005. Translocation of an 89-kDa periplasmic protein is associated with *Holospora* infection. *Biochem. Biophys. Res. Commun.* **337:**1198–1205.

Jeng, R. L., E. D. Goley, J. A. D'Alessio, O. Y. Chaga, T. M. Svitkina, G. G. Borisy, R. A. Heinzen, and M. D. Welch. 2004. A *Rickettsia* WASP-like protein activates the Arp2/3 complex and mediates actin-based motility. *Cell. Microbiol.* **6:**761–769.

Jones, R. T., K. F. McCormick, and A. P. Martin. 2008. Bacterial communities of *Bartonella*-positive fleas: diversity and community assembly patterns. *Appl. Environ. Microbiol.* **74:**1667–1670.

Kent, B. N., L. Salichos, J. G. Gibbons, A. Rokas, I. L. Newton, M. E. Clark, and S. R. Bordenstein. 2011. Complete bacteriophage transfer in a bacterial endosymbiont (*Wolbachia*) determined by targeted genome capture. *Genome Biol. Evol.* **3:**209–218.

Kent, W. J. 2002. BLAT—the BLAST-like alignment tool. *Genome Res.* **12:**656–664.

Kikuchi, Y., S. Sameshima, O. Kitade, J. Kojima, and T. Fukatsu. 2002. Novel clade of *Rickettsia* spp. from leeches. *Appl. Environ. Microbiol.* **68:** 999–1004.

Klasson, L., Z. Kambris, P. E. Cook, T. Walker, and S. P. Sinkins. 2009a. Horizontal gene transfer between *Wolbachia* and the mosquito *Aedes aegypti*. *BMC Genomics* **10:**33.

Klasson, L., T. Walker, M. Sebaihia, M. J. Sanders, M. A. Quail, A. Lord, S. Sanders, J. Earl, S. L. O'Neill, N. Thomson, S. P. Sinkins, and J. Parkhill. 2008. Genome evolution of *Wolbachia* strain *w*Pip from the *Culex pipiens* group. *Mol. Biol. Evol.* **25:**1877–1887.

Klasson, L., J. Westberg, P. Sapountzis, K. Naslund, Y. Lutnaes, A. C. Darby, Z. Veneti, L. Chen, H. R. Braig, R. Garrett, K. Bourtzis, and S. G. Andersson. 2009b. The mosaic genome structure of the *Wolbachia* *w*Ri strain infecting *Drosophila simulans*. *Proc. Natl. Acad. Sci. USA* **106:**5725–5730.

Kuchler, S. M., S. Kehl, and K. Dettner. 2009. Characterization and localization of *Rickettsia* sp. in water beetles of genus *Deronectes* (*Coleoptera*: *Dytiscidae*). *FEMS Microbiol. Ecol.* **68:** 201–211.

Kurland, C. G., and S. G. Andersson. 2000. Origin and evolution of the mitochondrial proteome. *Microbiol. Mol. Biol. Rev.* **64:**786–820.

Kusch, J., and H. D. Görtz. 2006. Towards an understanding of the killer trait: *Caedibacter* endocytobionts in *Paramecium*. *Prog. Mol. Subcell. Biol.* **41:**61–76.

Kusch, J., M. Stremmel, H. W. Breiner, V. Adams, M. Schweikert, and H. J. Schmidt. 2000. The toxic symbiont *Caedibacter caryophila* in the cytoplasm of *Paramecium novaurelia*. *Microb. Ecol.* **40:**330–335.

Lang, B. F., H. Brinkmann, L. B. Koski, M. Fujishima, H. D. Görtz, and G. Burger. 2005. On the origin of mitochondria and *Rickettsia*-related eukaryotic endosymbionts. *Jpn. J. Protozool.* **38:**171–183.

Li, T., J. H. Xiao, Z. H. Xu, R. W. Murphy, and D. W. Huang. 2011. A possibly new *Rickettsia*-like genus symbiont is found in Chinese wheat pest aphid, *Sitobion miscanthi* (Hemiptera: Aphididae). *J. Invertebr. Pathol.* **106:**418–421.

Lin, M., A. den Dulk-Ras, P. J. Hooykaas, and Y. Rikihisa. 2007. *Anaplasma phagocytophilum* AnkA secreted by type IV secretion system is tyrosine phosphorylated by Abl-1 to facilitate infection. *Cell. Microbiol.* **9:**2644–2657.

Lin, M., and Y. Rikihisa. 2003. *Ehrlichia chaffeensis* and *Anaplasma phagocytophilum* lack genes for lipid A biosynthesis and incorporate cholesterol for their survival. *Infect. Immun.* **71:**5324–5331.

Lin, M., and Y. Rikihisa. 2004. *Ehrlichia chaffeensis* downregulates surface Toll-like receptors 2/4, CD14 and transcription factors PU.1 and inhibits lipopolysaccharide activation of NF-κB, ERK 1/2 and p38 MAPK in host monocytes. *Cell. Microbiol.* **6:**175–186.

Lin, M., C. Zhang, K. Gibson, and Y. Rikihisa. 2009. Analysis of complete genome sequence of *Neorickettsia risticii*: causative agent of Potomac horse fever. *Nucleic Acids Res.* **37:**6076–6091.

Linka, N., H. Hurka, B. F. Lang, G. Burger, H. H. Winkler, C. Stamme, C. Urbany, I. Seil, J. Kusch, and H. E. Neuhaus. 2003. Phylogenetic relationships of non-mitochondrial nucleotide transport proteins in bacteria and eukaryotes. *Gene* **306:**27–35.

Lloyd, S. J., S. E. LaPatra, K. R. Snekvik, S. St-Hilaire, K. D. Cain, and D. R. Call. 2008. Strawberry disease lesions in rainbow trout from southern Idaho are associated with DNA from a *Rickettsia*-like organism. *Dis. Aquat. Organ.* **82:** 111–118.

Lo, N., T. Beninati, L. Sacchi, C. Genchi, and C. Bandi. 2004. Emerging rickettsioses. *Parassitologia* **46:**123–126.

Lo, N., T. Beninati, D. Sassera, E. A. Bouman, S. Santagati, L. Gern, V. Sambri, T. Masuzawa, J. S. Gray, T. G. Jaenson, A. Bouattour, M. J. Kenny, E. S. Guner, I. G. Kharitonenkov, I. Bitam, and C. Bandi. 2006. Widespread distribution and high prevalence of an alpha-proteobacterial symbiont in the tick *Ixodes ricinus*. *Environ. Microbiol.* **8:** 1280–1287.

Lo, N., C. Paraskevopoulos, K. Bourtzis, S. L. O'Neill, J. H. Werren, S. R. Bordenstein, and C. Bandi. 2007. Taxonomic status of the intracellular bacterium *Wolbachia pipientis*. *Int. J. Syst. Evol. Microbiol.* **57:**654–657.

Loy, J. K., F. E. Dewhirst, W. Weber, P. F. Frelier, T. L. Garbar, S. I. Tasca, and J. W. Templeton. 1996. Molecular phylogeny and in situ detection of the etiologic agent of necrotizing hepatopancreatitis in shrimp. *Appl. Environ. Microbiol.* **62:**3439–3445.

Maggi, R. G., M. Kosoy, M. Mintzer, and E. B. Breitschwerdt. 2009. Isolation of Candidatus *Bartonella melophagi* from human blood. *Emerg. Infect. Dis.* **15:**66–68.

Malek, J. A., J. M. Wierzbowski, W. Tao, S. A. Bosak, D. J. Saranga, L. Doucette-Stamm, D. R. Smith, P. J. McEwan, and K. J. McKernan. 2004. Protein interaction mapping on a functional shotgun sequence of *Rickettsia sibirica*. *Nucleic Acids Res.* **32:**1059–1064.

Malmstrom, R. R., M. T. Cottrell, H. Elifantz, and D. L. Kirchman. 2005. Biomass production and assimilation of dissolved organic matter by SAR11 bacteria in the Northwest Atlantic Ocean. *Appl. Environ. Microbiol.* **71:**2979–2986.

Mavromatis, K., C. K. Doyle, A. Lykidis, N. Ivanova, M. P. Francino, P. Chain, M. Shin, S. Malfatti, F. Larimer, A. Copeland, J. C. Detter, M. Land, P. M. Richardson, X. J. Yu, D. H. Walker, J. W. McBride, and N. C. Kyrpides. 2006. The genome of the obligately intracellular bacterium *Ehrlichia canis* reveals themes of complex membrane structure and immune evasion strategies. *J. Bacteriol.* **188:**4015–4023.

McLeod, M. P., X. Qin, S. E. Karpathy, J. Gioia, S. K. Highlander, G. E. Fox, T. Z. McNeill,

H. Jiang, D. Muzny, L. S. Jacob, A. C. Hawes, E. Sodergren, R. Gill, J. Hume, M. Morgan, G. Fan, A. G. Amin, R. A. Gibbs, C. Hong, X. J. Yu, D. H. Walker, and G. M. Weinstock. 2004. Complete genome sequence of *Rickettsia typhi* and comparison with sequences of other rickettsiae. *J. Bacteriol.* **186:**5842–5855.

Mediannikov, O. Y., L. I. Ivanov, M. Nishikawa, R. Saito, Y. N. Sidel'nikov, N. I. Zdanovskaia, E. V. Mokretsova, I. V. Tarasevich, and H. Suzuki. 2004. Microorganism "Montezuma" of the order Rickettsiales: the potential causative agent of tick-borne disease in the Far East of Russia. *Zh. Mikrobiol. Epidemiol. Immunobiol.* **2004:**7–13. (In Russian.)

Medini, D., C. Donati, H. Tettelin, V. Masignani, and R. Rappuoli. 2005. The microbial pan-genome. *Curr. Opin. Genet. Dev.* **15:**589–594.

Meeus, P. F., K. A. Brayton, G. H. Palmer, and A. F. Barbet. 2003. Conservation of a gene conversion mechanism in two distantly related paralogues of *Anaplasma marginale*. *Mol. Microbiol.* **47:**633–643.

Metselaar, M., K. D. Thompson, R. M. Gratacap, M. J. Kik, S. E. LaPatra, S. J. Lloyd, D. R. Call, P. D. Smith, and A. Adams. 2010. Association of red-mark syndrome with a *Rickettsia*-like organism and its connection with strawberry disease in the USA. *J. Fish Dis.* **33:**849–858.

Moliner, C., P. E. Fournier, and D. Raoult. 2010. Genome analysis of microorganisms living in amoebae reveals a melting pot of evolution. *FEMS Microbiol. Rev.* **34:**281–294.

Molmeret, M., M. Horn, M. Wagner, M. Santic, and Y. Abu Kwaik. 2005. Amoebae as training grounds for intracellular bacterial pathogens. *Appl. Environ. Microbiol.* **71:**20–28.

Moor, J. D., G. N. Cherr, and C. S. Friedman. 2001. Detection of *'Candidatus* Xenohaliotis californiensis' (Rickettsiales-like prokaryote) inclusions in tissue squashes of abalone (*Haliotis* spp.) gastrointestinal epithelium using a nucleic acid fluorochrome. *Dis. Aquat. Organ.* **46:**147–152.

Müller, M., and W. Martin. 1999. The genome of *Rickettsia prowazekii* and some thoughts on the origin of mitochondria and hydrogenosomes. *Bioessays* **21:**377–381.

Murray, R. G., and E. Stackebrandt. 1995. Taxonomic note: implementation of the provisional status *Candidatus* for incompletely described procaryotes. *Int. J. Syst. Bacteriol.* **45:**186–187.

Nakamura, Y., M. Aki, T. Aikawa, M. Hori, and M. Fujishima. 2004. Differences in gene expression of the ciliate *Paramecium caudatum* caused by endonuclear symbiosis with *Holospora obtusa*, revealed using differential display reverse transcribed PCR. *FEMS Microbiol. Lett.* **240:**209–213.

Nakayama, K., A. Yamashita, K. Kurokawa, T. Morimoto, M. Ogawa, M. Fukuhara, H. Urakami, M. Ohnishi, I. Uchiyama, Y. Ogura, T. Ooka, K. Oshima, A. Tamura, M. Hattori, and T. Hayashi. 2008. The whole-genome sequencing of the obligate intracellular bacterium *Orientia tsutsugamushi* revealed massive gene amplification during reductive genome evolution. *DNA Res.* **15:**185–199.

Neimark, H., K. E. Johansson, Y. Rikihisa, and J. G. Tully. 2001. Proposal to transfer some members of the genera *Haemobartonella* and *Eperythrozoon* to the genus *Mycoplasma* with descriptions of '*Candidatus* Mycoplasma haemofelis', '*Candidatus* Mycoplasma haemomuris', '*Candidatus* Mycoplasma haemosuis' and '*Candidatus* Mycoplasma wenyonii'. *Int. J. Syst. Evol. Microbiol.* **51:**891–899.

Neimark, H., and K. M. Kocan. 1997. The cell wall-less rickettsia *Eperythrozoon wenyonii* is a *Mycoplasma*. *FEMS Microbiol. Lett.* **156:**287–291.

Nelson, C. M., M. J. Herron, R. F. Felsheim, B. R. Schloeder, S. M. Grindle, A. O. Chavez, T. J. Kurtti, and U. G. Munderloh. 2008. Whole genome transcription profiling of *Anaplasma phagocytophilum* in human and tick host cells by tiling array analysis. *BMC Genomics* **9:**364.

Niebylski, M. L., M. G. Peacock, E. R. Fischer, S. F. Porcella, and T. G. Schwan. 1997. Characterization of an endosymbiont infecting wood ticks, *Dermacentor andersoni*, as a member of the genus *Francisella*. *Appl. Environ. Microbiol.* **63:** 3933–3940.

Nikoh, N., and A. Nakabachi. 2009. Aphids acquired symbiotic genes via lateral gene transfer. *BMC Biol.* **7:**12.

Niu, H., V. Kozjak-Pavlovic, T. Rudel, and Y. Rikihisa. 2010. *Anaplasma phagocytophilum* Ats-1 is imported into host cell mitochondria and interferes with apoptosis induction. *PLoS Pathog.* **6:** e1000774.

Noda, H., U. G. Munderloh, and T. J. Kurtti. 1997. Endosymbionts of ticks and their relationship to *Wolbachia* spp. and tick-borne pathogens of humans and animals. *Appl. Environ. Microbiol.* **63:**3926–3932.

Ochman, H., and N. A. Moran. 2001. Genes lost and genes found: evolution of bacterial pathogenesis and symbiosis. *Science* **292:**1096–1099.

Ogata, H., S. Audic, C. Abergel, P. E. Fournier, and J. M. Claverie. 2002. Protein coding palindromes are a unique but recurrent feature in Rickettsia. *Genome Res.* **12:**808–816.

Ogata, H., S. Audic, V. Barbe, F. Artiguenave, P. E. Fournier, D. Raoult, and J. M. Claverie. 2000. Selfish DNA in protein-coding genes of *Rickettsia*. *Science* **290:**347–350.

Ogata, H., S. Audic, P. Renesto-Audiffren, P. E. Fournier, V. Barbe, D. Samson, V. Roux, P. Cossart, J. Weissenbach, J. M. Claverie, and D. Raoult. 2001. Mechanisms of evolution in *Rickettsia conorii* and *R. prowazekii*. *Science* **293:**2093–2098.

Ogata, H., B. La Scola, S. Audic, P. Renesto, G. Blanc, C. Robert, P. E. Fournier, J. M. Claverie, and D. Raoult. 2006. Genome sequence of *Rickettsia bellii* illuminates the role of amoebae in gene exchanges between intracellular pathogens. *PLoS Genet.* **2:**e76.

Ogata, H., P. Renesto, S. Audic, C. Robert, G. Blanc, P. E. Fournier, H. Parinello, J. M. Claverie, and D. Raoult. 2005a. The genome sequence of *Rickettsia felis* identifies the first putative conjugative plasmid in an obligate intracellular parasite. *PLoS Biol.* **3:**e248.

Ogata, H., C. Robert, S. Audic, S. Robineau, G. Blanc, P. E. Fournier, P. Renesto, J. M. Claverie, and D. Raoult. 2005b. *Rickettsia felis*, from culture to genome sequencing. *Ann. N. Y. Acad. Sci.* **1063:**26–34.

Ogata, H., K. Suhre, and J. M. Claverie. 2005c. Discovery of protein-coding palindromic repeats in *Wolbachia*. *Trends Microbiol.* **13:**253–255.

Olsen, G. J., C. R. Woese, and R. Overbeek. 1994. The winds of (evolutionary) change: breathing new life into microbiology. *J. Bacteriol.* **176:**1–6.

Overbeek, R., T. Begley, R. M. Butler, J. V. Choudhuri, H. Y. Chuang, M. Cohoon, V. de Crecy-Lagard, N. Diaz, T. Disz, R. Edwards, M. Fonstein, E. D. Frank, S. Gerdes, E. M. Glass, A. Goesmann, A. Hanson, D. Iwata-Reuyl, R. Jensen, N. Jamshidi, L. Krause, M. Kubal, N. Larsen, B. Linke, A. C. McHardy, F. Meyer, H. Neuweger, G. Olsen, R. Olson, A. Osterman, V. Portnoy, G. D. Pusch, D. A. Rodionov, C. Ruckert, J. Steiner, R. Stevens, I. Thiele, O. Vassieva, Y. Ye, O. Zagnitko, and V. Vonstein. 2005. The subsystems approach to genome annotation and its use in the project to annotate 1000 genomes. *Nucleic Acids Res.* **33:**5691–5702.

Pang, H., and H. H. Winkler. 1994. Analysis of the peptidoglycan of *Rickettsia prowazekii*. *J. Bacteriol.* **176:**923–926.

Perlman, S. J., M. S. Hunter, and E. Zchori-Fein. 2006. The emerging diversity of *Rickettsia*. *Proc. Biol. Sci.* **273:**2097–2106.

Philip, R. N., E. A. Casper, R. L. Anacker, J. Cory, S. F. Hayes, W. Burgdorfer, and E. Yunker. 1983. *Rickettsia bellii* sp. nov.: a tick-borne rickettsia, widely distributed in the United States, that is distinct from the spotted fever and typhus biogroups. *Int. J. Syst. Bacteriol.* **33:** 94–106.

Price, M. N., P. S. Dehal, and A. P. Arkin. 2010. FastTree 2—approximately maximum-likelihood trees for large alignments. *PLoS One* **5:**e9490.

Rar, V. A., N. N. Livanova, V. V. Panov, E. K. Doroschenko, N. M. Pukhovskaya, N. P. Vysochina, and L. I. Ivanov. 2010. Genetic diversity of *Anaplasma* and *Ehrlichia* in the Asian part of Russia. *Ticks Tick Borne Dis.* **1:**57–65.

Reeves, W. K., A. D. Loftis, D. E. Szumlas, M. M. Abbassy, I. M. Helmy, H. A. Hanafi, and G. A. Dasch. 2007. Rickettsial pathogens in the tropical rat mite *Ornithonyssus bacoti* (Acari: Macronyssidae) from Egyptian rats (*Rattus* spp.). *Exp. Appl. Acarol.* **41:**101–107.

Rikihisa, Y., M. Kawahara, B. Wen, G. Kociba, P. Fuerst, F. Kawamori, C. Suto, S. Shibata, and M. Futohashi. 1997. Western immunoblot analysis of *Haemobartonella muris* and comparison of 16S rRNA gene sequences of *H. muris*, *H. felis*, and *Eperythrozoon suis*. *J. Clin. Microbiol.* **35:**823–829.

Rikihisa, Y., and M. Lin. 2010. *Anaplasma phagocytophilum* and *Ehrlichia chaffeensis* type IV secretion and Ank proteins. *Curr. Opin. Microbiol.* **13:** 59–66.

Rikihisa, Y., M. Lin, H. Niu, and Z. Cheng. 2009. Type IV secretion system of *Anaplasma phagocytophilum* and *Ehrlichia chaffeensis*. *Ann. N. Y. Acad. Sci.* **1166:**106–111.

Rikihisa, Y., C. Zhang, and B. M. Christensen. 2003. Molecular characterization of *Aegyptianella pullorum* (*Rickettsiales*, *Anaplasmataceae*). *J. Clin. Microbiol.* **41:**5294–5297.

Rodriguez-Ezpeleta, N., and T. M. Embley. 2012. The SAR11 Group of Alpha-Proteobacteria is not related to the origin of mitochondria. *PLoS One* **7:**e30520.

Roger, A. J., S. G. Svard, J. Tovar, C. G. Clark, M. W. Smith, F. D. Gillin, and M. L. Sogin. 1998. A mitochondrial-like chaperonin 60 gene in *Giardia lamblia*: evidence that diplomonads once harbored an endosymbiont related to the progenitor of mitochondria. *Proc. Natl. Acad. Sci. USA* **95:**229–234.

Ronimus, R. S., and H. W. Morgan. 2003. Distribution and phylogenies of enzymes of the Embden-Meyerhof-Parnas pathway from archaea and hyperthermophilic bacteria support a gluconeogenic origin of metabolism. *Archaea* **1:**199–221.

Roux, V., M. Bergoin, N. Lamaze, and D. Raoult. 1997a. Reassessment of the taxonomic position of *Rickettsiella grylli*. *Int. J. Syst. Bacteriol.* **47:**1255–1257.

Roux, V., and D. Raoult. 1999. Phylogenetic analysis and taxonomic relationships among the genus *Rickettsia*, p. 52–66. *In* D. Raoult and P. Brouqui (ed.), *Rickettsiae and Rickettsial Diseases at the Turn of the Third Millennium*. Elsevier, Marseille, France.

Roux, V., E. Rydkina, M. Eremeeva, and D. Raoult. 1997b. Citrate synthase gene comparison, a new tool for phylogenetic analysis, and its application for the rickettsiae. *Int. J. Syst. Bacteriol.* **47:**252–261.

Rurangirwa, F. R., K. A. Brayton, T. C. McGuire, D. P. Knowles, and G. H. Palmer. 2002. Conservation of the unique rickettsial rRNA gene arrangement in *Anaplasma*. *Int. J. Syst. Evol. Microbiol.* **52:**1405–1409.

Rymaszewska, A. 2007. Symbiotic bacteria in oocyte and ovarian cell mitochondria of the tick *Ixodes ricinus*: biology and phylogenetic position. *Parasitol. Res.* **100:**917–920.

Sacchi, L., E. Bigliardi, S. Corona, T. Beninati, N. Lo, and A. Franceschi. 2004. A symbiont of the tick *Ixodes ricinus* invades and consumes mitochondria in a mode similar to that of the parasitic bacterium *Bdellovibrio bacteriovorus*. *Tissue Cell* **36:**43–53.

Salzberg, S. L., J. C. Dunning Hotopp, A. L. Delcher, M. Pop, D. R. Smith, M. B. Eisen, and W. C. Nelson. 2005a. Serendipitous discovery of *Wolbachia* genomes in multiple *Drosophila* species. *Genome Biol.* **6:**R23.

Salzberg, S. L., J. C. Dunning Hotopp, A. L. Delcher, M. Pop, D. R. Smith, M. B. Eisen, and W. C. Nelson. 2005b. Correction: Serendipitous discovery of *Wolbachia* genomes in multiple *Drosophila* species. *Genome Biol.* **6:**402.

Salzberg, S. L., D. Puiu, D. D. Sommer, V. Nene, and N. H. Lee. 2009. Genome sequence of the *Wolbachia* endosymbiont of *Culex quinquefasciatus* JHB. *J. Bacteriol.* **191:**1725.

Sassera, D., T. Beninati, C. Bandi, E. A. Bouman, L. Sacchi, M. Fabbi, and N. Lo. 2006. '*Candidatus* Midichloria mitochondrii', an endosymbiont of the tick *Ixodes ricinus* with a unique intramitochondrial lifestyle. *Int. J. Syst. Evol. Microbiol.* **56:**2535–2540.

Sassera, D., N. Lo, S. Epis, G. D'Auria, M. Montagna, F. Comandatore, D. Horner, J. Pereto, A. M. Luciano, F. Franciosi, E. Ferri, E. Crotti, C. Bazzocchi, D. Daffonchio, L. Sacchi, A. Moya, A. Latorre, and C. Bandi. 2011. Phylogenomic evidence for the presence of a flagellum and cbb(3) oxidase in the free-living mitochondrial ancestor. *Mol. Biol. Evol.* **28:** 3285–3296.

Schmidt, H. J., H. D. Görtz, and R. L. Quackenbush. 1987. *Caedibacter caryophila* sp. nov., a killer symbiont inhabiting the macronucleus of *Paramecium caudatum*. *Int. J. Syst. Bacteriol.* **37:** 459–462.

Schmitz-Esser, S., I. Haferkamp, S. Knab, T. Penz, M. Ast, C. Kohl, M. Wagner, and M. Horn. 2008. *Lawsonia intracellularis* contains a gene encoding a functional rickettsia-like ATP/ADP translocase for host exploitation. *J. Bacteriol.* **190:**5746–5752.

Schmitz-Esser, S., N. Linka, A. Collingro, C. L. Beier, H. E. Neuhaus, M. Wagner, and M. Horn. 2004. ATP/ADP translocases: a common feature of obligate intracellular amoebal symbionts related to chlamydiae and rickettsiae. *J. Bacteriol.* **186:**683–691.

Schmitz-Esser, S., P. Tischler, R. Arnold, J. Montanaro, M. Wagner, T. Rattei, and M. Horn. 2010. The genome of the amoeba symbiont "*Candidatus* Amoebophilus asiaticus" reveals common mechanisms for host cell interaction among amoeba-associated bacteria. *J. Bacteriol.* **192:**1045–1057.

Schrallhammer, M., S. I. Fokin, K. H. Schleifer, and G. Petroni. 2006. Molecular characterization of the obligate endosymbiont "*Caedibacter macronucleorum*" Fokin and Görtz, 1993 and of its host *Paramecium duboscqui* strain Ku4-8. *J. Eukaryot. Microbiol.* **53:**499–506.

Sekeyova, Z., V. Roux, and D. Raoult. 2001. Phylogeny of *Rickettsia* spp. inferred by comparing sequences of 'gene D', which encodes an intracytoplasmic protein. *Int. J. Syst. Evol. Microbiol.* **51:**1353–1360.

Sentausa, E., K. El Karkouri, C. Robert, D. Raoult, and P. E. Fournier. 2012a. Genome sequence of "*Rickettsia sibirica* subsp. *mongolitimonae*." *J. Bacteriol.* **194:**2389–2390.

Sentausa, E., K. El Karkouri, C. Robert, D. Raoult, and P. E. Fournier. 2012b. Genome sequence of *Rickettsia conorii* subsp. *indica*, the agent of Indian Tick Typhus. *J. Bacteriol.* **194:** 3288–3289.

Sentausa, E., K. El Karkouri, C. Robert, D. Raoult, and P. E. Fournier. 2012c. Sequence and annotation of *Rickettsia sibirica sibirica* genome. *J. Bacteriol.* **194:**2377.

Sicheritz-Pontén, T., C. G. Kurland, and S. G. Andersson. 1998. A phylogenetic analysis of the cytochrome *b* and cytochrome *c* oxidase I genes supports an origin of mitochondria from within the Rickettsiaceae. *Biochim. Biophys. Acta* **1365:**545–551.

Silverman, D. J., and C. L. Wisseman, Jr. 1978. Comparative ultrastructural study on the cell envelopes of *Rickettsia prowazekii*, *Rickettsia rickettsii*, and *Rickettsia tsutsugamushi*. *Infect. Immun.* **21:**1020–1023.

Simser, J. A., M. S. Rahman, S. M. Dreher-Lesnick, and A. F. Azad. 2005. A novel and naturally occurring transposon, ISRpe1 in the *Rickettsia peacockii* genome disrupting the *rickA* gene involved in actin-based motility. *Mol. Microbiol.* **58:**71–79.

Snyder, E. E., N. Kampanya, J. Lu, E. K. Nordberg, H. R. Karur, M. Shukla, J. Soneja, Y. Tian, T. Xue, H. Yoo, F. Zhang, C. Dharmanolla, N. V. Dongre, J. J. Gillespie, J. Hamelius, M. Hance, K. I. Huntington, D. Jukneliene, J. Koziski, L. Mackasmiel, S. P. Mane, V. Nguyen, A. Purkayastha, J. Shallom, G. Yu, Y. Guo, J. Gabbard, D. Hix, A. F. Azad, S. C. Baker, S. M. Boyle, Y. Khudyakov, X. J. Meng, C. Rupprecht, J. Vinje, O. R. Crasta, M. J. Czar, A. Dickerman, J. D. Eckart, R. Kenyon, R. Will, J. C. Setubal, and B. W. Sobral. 2007. PATRIC: the VBI PathoSystems Resource Integration Center. *Nucleic Acids Res.* **35:**D401–D406.

Springer, N., W. Ludwig, R. Amann, H. J. Schmidt, H. D. Gortz, and K. H. Schleifer. 1993. Occurrence of fragmented 16S rRNA in an obligate bacterial endosymbiont of *Paramecium caudatum*. *Proc. Natl. Acad. Sci. USA* **90:**9892–9895.

Stamatakis, A., P. Hoover, and J. Rougemont. 2008. A rapid bootstrap algorithm for the RAxML Web servers. *Syst. Biol.* **57:**758–771.

Stich, R. W., G. A. Olah, K. A. Brayton, W. C. Brown, M. Fechheimer, K. Green-Church, S. Jittapalapong, K. M. Kocan, T. C. McGuire, F. R. Rurangirwa, and G. H. Palmer. 2004. Identification of a novel *Anaplasma marginale* appendage-associated protein that localizes with actin filaments during intraerythrocytic infection. *Infect. Immun.* **72:**7257–7264.

Stothard, D. R., J. B. Clark, and P. A. Fuerst. 1994. Ancestral divergence of *Rickettsia bellii* from the spotted fever and typhus groups of *Rickettsia* and antiquity of the genus *Rickettsia*. *Int. J. Syst. Bacteriol.* **44:**798–804.

Stothard, D. R., and P. A. Fuerst. 1995. Evolutionary analysis of the spotted fever and typhus groups of *Rickettsia* using 16S rRNA gene sequences. *Syst. Appl. Microbiol.* **18:**52–61.

Sumner, J. W., W. L. Nicholson, and R. F. Massung. 1997. PCR amplification and comparison of nucleotide sequences from the *groESL* heat shock operon of *Ehrlichia* species. *J. Clin. Microbiol.* **35:**2087–2092.

Sunagawa, S., T. Z. DeSantis, Y. M. Piceno, E. L. Brodie, M. K. DeSalvo, C. R. Voolstra, E. Weil, G. L. Andersen, and M. Medina. 2009. Bacterial diversity and White Plague Disease-associated community changes in the Caribbean coral *Montastraea faveolata*. *ISME J.* **3:**512–521.

Sutten, E. L., J. Norimine, P. A. Beare, R. A. Heinzen, J. E. Lopez, K. Morse, K. A. Brayton, J. J. Gillespie, and W. C. Brown. 2010. *Anaplasma marginale* type IV secretion system proteins VirB2, VirB7, VirB11, and VirD4 are immunogenic components of a protective bacterial membrane vaccine. *Infect. Immun.* **78:**1314–1325.

Talavera, G., and J. Castresana. 2007. Improvement of phylogenies after removing divergent and ambiguously aligned blocks from protein sequence alignments. *Syst. Biol.* **56:**564–577.

Tamura, A., N. Ohashi, H. Urakami, and S. Miyamura. 1995. Classification of *Rickettsia tsutsugamushi* in a new genus, *Orientia* gen. nov., as *Orientia tsutsugamushi* comb. nov. *Int. J. Syst. Bacteriol.* **45:**589–591.

Thiergart, T., G. Landan, M. Schenk, T. Dagan, and W. F. Martin. 2012. An evolutionary network of genes present in the eukaryote common ancestor polls genomes on eukaryotic and mitochondrial origin. *Genome Biol. Evol.* **4:**466–485.

Thrash, J. C., A. Boyd, M. J. Huggett, J. Grote, P. Carini, R. J. Yoder, B. Robbertse, J. W. Spatafora, M. S. Rappe, and S. J. Giovannoni. 2011. Phylogenomic evidence for a common ancestor of mitochondria and the SAR11 clade. *Sci. Rep.* **1:**13.

Vahling, C. M., Y. Duan, and H. Lin. 2010. Characterization of an ATP translocase identified in the destructive plant pathogen "*Candidatus* Liberibacter asiaticus." *J. Bacteriol.* **192:** 834–840.

Van Dongen, S. 2008. Graph clustering via a discrete uncoupling process. *SIAM J. Matrix Anal. Appl.* **30:**121–141.

Vannini, C., F. Ferrantini, K. H. Schleifer, W. Ludwig, F. Verni, and G. Petroni. 2010. "*Candidatus* Anadelfobacter veles" and "*Candidatus* Cyrtobacter comes," two new *Rickettsiales* species hosted by the protist ciliate *Euplotes harpa* (Ciliophora, Spirotrichea). *Appl. Environ. Microbiol.* **76:**4047–4054.

Vannini, C., G. Petroni, F. Verni, and G. Rosati. 2005. A bacterium belonging to the *Rickettsiaceae* family inhabits the cytoplasm of the marine ciliate *Diophrys appendiculata* (Ciliophora, Hypotrichia). *Microb. Ecol.* **49:**434–442.

van Vliet, A. H., F. Jongejan, and B. A. van der Zeijst. 1992. Phylogenetic position of *Cowdria ruminantium* (*Rickettsiales*) determined by analysis of amplified 16S ribosomal DNA sequences. *Int. J. Syst. Bacteriol.* **42:**494–498.

Venzal, J. M., A. Estrada-Peña, A. Portillo, A. J. Mangold, O. Castro, C. G. de Souza, M. L. Félix, L. Pérez-Martínez, S. Santibánez, and J. A. Oteo. 2008. Detection of alpha and gamma-proteobacteria in *Amblyomma triste* (Acari: Ixodidae) from Uruguay. *Exp. Appl. Acarol.* **44:**49–56.

Viale, A. M., and A. K. Arakaki. 1994. The chaperone connection to the origins of the eukaryotic organelles. *FEBS Lett.* **341:**146–151.

Viklund, J., T. J. Ettema, and S. G. Andersson. 2012. Independent genome reduction and phylogenetic reclassification of the oceanic SAR11 clade. *Mol. Biol. Evol.* **29:**599–615.

Vitorino, L., I. M. Chelo, F. Bacellar, and L. Ze-Ze. 2007. Rickettsiae phylogeny: a multigenic approach. *Microbiology* **153:**160–168.

Wang, Y., A. Brune, and M. Zimmer. 2007. Bacterial symbionts in the hepatopancreas of isopods: diversity and environmental transmission. *FEMS Microbiol. Ecol.* **61:**141–152.

Wang, Y., U. Stingl, F. Anton-Erxleben, M. Zimmer, and A. Brune. 2004. '*Candidatus* Hepatincola porcellionum' gen. nov., sp. nov., a new, stalk-forming lineage of Rickettsiales colonizing the midgut glands of a terrestrial isopod. *Arch. Microbiol.* **181:**299–304.

Weinert, L. A., J. J. Welch, and F. M. Jiggins. 2009a. Conjugation genes are common throughout the genus *Rickettsia* and are transmitted horizontally. *Proc. Biol. Sci.* **276:**3619–3627.

Weinert, L. A., J. H. Werren, A. Aebi, G. N. Stone, and F. M. Jiggins. 2009b. Evolution and diversity of *Rickettsia* bacteria. *BMC Biol.* **7:**6.

Weisburg, W. G., M. E. Dobson, J. E. Samuel, G. A. Dasch, L. P. Mallavia, O. Baca, L. Mandelco, J. E. Sechrest, E. Weiss, and C. R. Woese. 1989. Phylogenetic diversity of the rickettsiae. *J. Bacteriol.* **171:**4202–4206.

Weiser, J., and Z. Zizka. 1968. Electron microscope studies of *Rickettsiella chironomi* in the midge *Camptochironomous tentans*. *J. Invertebr. Pathol.* **12:** 222–230.

Wiemann, M., and H. D. Görtz. 1991. Identification and localization of major stage-specific polypeptides of infectious *Holospora obtusa* with monoclonal antibodies. *J. Bacteriol.* **173:**4842–4850.

Williams, K. P., J. J. Gillespie, B. W. Sobral, E. K. Nordberg, E. E. Snyder, J. M. Shallom, and A. W. Dickerman. 2010. Phylogeny of gammaproteobacteria. *J. Bacteriol.* **192:** 2305–2314.

Williams, K. P., B. W. Sobral, and A. W. Dickerman. 2007. A robust species tree for the alphaproteobacteria. *J. Bacteriol.* **189:**4578–4586.

Winkler, H. H., and R. M. Daugherty. 1986. Acquisition of glucose by *Rickettsia prowazekii* through the nucleotide intermediate uridine 5′-diphosphoglucose. *J. Bacteriol.* **167:**805–808.

Woolfit, M., I. Iturbe-Ormaetxe, E. A. McGraw, and S. L. O'Neill. 2009. An ancient horizontal gene transfer between mosquito and the endosymbiotic bacterium *Wolbachia pipientis*. *Mol. Biol. Evol.* **26:**367–374.

Wu, M., L. V. Sun, J. Vamathevan, M. Riegler, R. Deboy, J. C. Brownlie, E. A. McGraw, W. Martin, C. Esser, N. Ahmadinejad, C. Wiegand, R. Madupu, M. J. Beanan, L. M. Brinkac, S. C. Daugherty, A. S. Durkin, J. F. Kolonay, W. C. Nelson, Y. Mohamoud, P. Lee, K. Berry, M. B. Young, T. Utterback, J. Weidman, W. C. Nierman, I. T. Paulsen, K. E. Nelson, H. Tettelin, S. L. O'Neill, and J. A. Eisen. 2004. Phylogenomics of the reproductive parasite *Wolbachia pipientis w*Mel: a streamlined genome overrun by mobile genetic elements. *PLoS Biol.* **2:**E69.

Xuan, Y. H., H. S. Yu, H. J. Jeong, S. Y. Seol, D. I. Chung, and H. H. Kong. 2007. Molecular characterization of bacterial endosymbionts of *Acanthamoeba* isolates from infected corneas of Korean patients. *Korean J. Parasitol.* **45:**1–9.

Yu, X. J., and D. H. Walker. 2003. The order Rickettsiales, p. 493–528. In M. Dworkin (ed.), *The Prokaryotes: an Evolving Electronic Resource for the Microbiological Community*, 3rd ed., vol. 5, release 3.12. Springer-Verlag, New York, NY.

Zhang, Y., N. Ohashi, E. H. Lee, A. Tamura, and Y. Rikihisa. 1997. *Ehrlichia sennetsu groE* operon and antigenic properties of the GroEL homolog. *FEMS Immunol. Med. Microbiol.* **18:**39–46.

Zhivkov, V. I., and R. T. Tosheva. 1986. Uridine diphosphate sugars: concentration and rate of synthesis in tissues of vertebrates. *Int. J. Biochem.* **18:**1–6.

INVASION OF THE MAMMALIAN HOST: EARLY EVENTS AT THE CELLULAR AND MOLECULAR LEVELS

Juan J. Martinez

4

INTRODUCTION

Species belonging to the genus *Rickettsia* are small, gram-negative *Alphaproteobacteria* (0.3 to 0.5 × 0.8 to 1.0 μm) and obligate intracellular organisms. The majority of these species are divided into two groups, the spotted fever group rickettsiae (SFGR) and typhus group rickettsiae (TGR), based on the diseases that they cause, the presence of major surface antigens, and the ability to promote intracellular actin-based motility. Members of the SFGR, including *R. rickettsii* (Rocky Mountain spotted fever), *R. conorii* (Mediterranean spotted fever), and *R. japonica* (Oriental spotted fever), are organisms harbored in arthropod vectors such as ticks and are transmitted to humans during a tick blood meal. Symptoms from rickettsial disease typically manifest 2 to 14 days following inoculation and are characterized by headache, fever, and malaise. Localized replication of rickettsiae at the inoculation site and ensuing tissue damage may give rise to a necrotic lesion, or eschar. Subsequent damage to the vascular endothelium and infiltration of perivascular mononuclear cells leads to fluid leakage into the interstitial space, resulting in a dermal rash in 90% of cases. Further perturbation of vascular integrity often results in more severe manifestations of disease including encephalitis, noncardiogenic pulmonary edema, interstitial pneumonia, hypovolemia, hypotensive shock, and acute renal failure. Rocky Mountain and Mediterranean spotted fevers are responsible for causing severe morbidity and mortality in the absence of timely and appropriate antibiotic treatment (Walker, 1989a, 1989b).

Members of the TGR include *R. prowazekii* and *R. typhi*, the etiologic agents of epidemic typhus and murine typhus, respectively. TGR can be transmitted through the excrement of human body lice and are typically introduced into abraded skin by scratching. Symptoms vary, but the infections are generally characterized by headache, fever, delirium, and rash. Delayed or inappropriate antibiotic treatment can result in severe morbidity and mortality rates. In some rare cases, *R. prowazekii* establishes a latent infection which upon recurrence manifests itself as a chronic infection termed Brill-Zinsser disease (Bechah et al., 2008).

As obligate intracellular pathogens, rickettsial species have evolved mechanisms to trigger their invasion of nonprofessional phagocytes to

Juan J. Martinez, Department of Microbiology and Howard Taylor Ricketts Laboratory, University of Chicago, Chicago, IL 60637.

Intracellular Pathogens II: Rickettsiales
Edited by G. H. Palmer and A. F. Azad © 2012 ASM Press, Washington, DC
10.1128/9781555817336.ch4

survive, multiply, and ultimately spread to naive hosts. Several bacterial pathogens have developed a variety of different strategies to enter normally nonphagocytic mammalian cells. Two main pathways, termed "trigger" and "zipper," are morphologically and mechanistically distinct mechanisms by which pathogens enter cells. The trigger mechanism, utilized by invasive bacteria such as *Shigella* and *Salmonella* species, involves the translocation of bacterial proteins called effectors into the host cytosol by a secretion machinery called type III secretion. Some of these effector proteins serve to stimulate the activity of host small GTP-binding proteins, such as Rac and Cdc42, resulting in the formation of large actin-rich membrane ruffles and ultimately in bacterial uptake (Galan and Zhou, 2000; Yoshida et al., 2002). The zipper mechanism, in contrast, is independent of this specialized secretion machinery. Several invasive bacteria such as *Listeria monocytogenes*, uropathogenic *Escherichia coli*, and *Yersinia pseudotuberculosis* utilize the interactions of bacterial adhesin molecules with specific host receptors to generate signaling cascades resulting in localized host membrane rearrangements required for bacterial uptake (Cossart, 2004). An electron microscopy analysis of rickettsial entry (Teysseire et al., 1995; Gouin et al., 1999) suggested that *R. conorii* utilizes a zipperlike invasion strategy. An analysis of published rickettsial genomes failed to reveal genes that are predicted to encode a type III secretion system, further suggesting that the entry of rickettsiae into target cells is independent of this secretion system (Andersson et al., 1998; Ogata et al., 2001, 2005). This chapter focuses on the cell biology involved in the internalization of SFGR into nonphagocytic mammalian cells using *R. conorii* as a model organism and highlights the current knowledge regarding the bacterial proteins and cognate host cell receptors that are involved in initiating this process.

ACTIN DYNAMICS INVOLVED IN RICKETTSIAL INVASION

Early studies on the mechanism(s) utilized by rickettsiae to invade nonphagocytic mammalian cells identified cellular actin dynamics as playing an important role. Fluorescent microscopy studies revealed that F-actin fibers are closely associated with bacterial entry sites (Fig. 1), and perturbation of the actin cytoskeleton by pharmacological agents such as cytochalasin D diminishes the ability of *R. conorii* to enter Vero and HeLa cells. Subsequent studies further defined the complex signaling cascades that were likely involved in the reorganization of cortical actin structures required during the rickettsial entry process (Martinez and Cossart, 2004).

An investigation into host proteins likely involved in remodeling of the actin cytoskeleton during entry focused on two major regulators: the Arp2/3 complex and members of the Ras superfamily of small GTP-binding proteins,

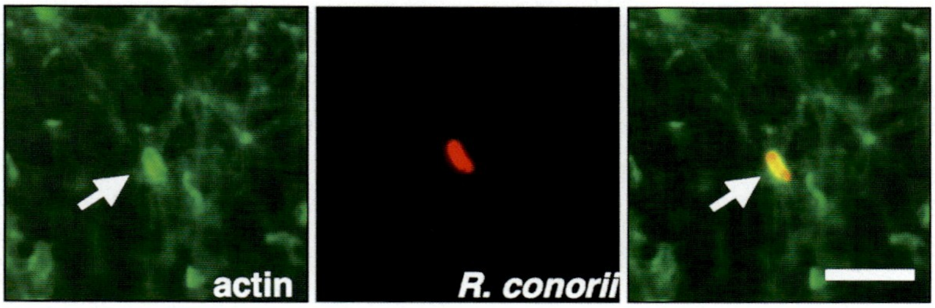

FIGURE 1 *R. conorii* recruits actin to site of entry on mammalian cells. Fluorescence microscopy of infected mammalian cells reveals colocalization of actin-rich structures (green) with invading bacteria (red). Arrows indicate areas of colocalization (yellow). White scale bar represents 2 μm. Stimulation of actin dynamics is critical to the rickettsial invasion of mammalian cells. (Reprinted from Martinez and Cossart [2004] with permission of the publisher.) doi:10.1128/9781555817336.ch4.f1

Cdc42 and Rac1. The Arp2/3 complex is a regulator of actin dynamics and has been shown to have a direct role in initiating actin nucleation and in the branching of actin filaments (Bear et al., 2001, 2002). Cdc42 and Rac1 are involved in regulating processes within host cells (Ridley and Hall, 1992; Ridley et al., 1992; Olson et al., 1995), including the reorganization of actin-rich structures in migrating cells (Hall, 1998) and the reorganization of actin during the engulfment of pathogens by nonprofessional phagocytes (Tran Van Nhieu et al., 1999; Galan and Zhou, 2000; Alrutz et al., 2001; Martinez and Hultgren, 2002). Fluorescence microscopy studies of *R. conorii*-infected Vero cells revealed that components of the Arp2/3 complex are present at *R. conorii* entry foci, strongly suggesting that the Arp2/3 complex plays a role in *R. conorii* entry. Titration of Arp2/3 protein activity by the Wiskott-Aldrich syndrome protein (WASp) family member Scar1 (Machesky et al., 1999) inhibited *R. conorii* uptake into Vero cells, demonstrating that the recruitment and activation of the Arp2/3 complex to bacterial entry foci was important to generate the observed localized recruitment of actin. These same studies also demonstrated that Cdc42, but not Rac1, colocalizes with invading *R. conorii*. Transfection of dominant-negative alleles of Cdc42, but not Rac1, perturbs the ability of *R. conorii* to enter cultured cells in vitro (Martinez and Cossart, 2004). Interestingly, Cdc42 can utilize proteins of the WASp family to activate Arp2/3 during cortical actin polymerization (Higgs and Pollard, 2001), suggesting that Cdc42 may activate Arp2/3 in this manner. These observations raised several questions. (i) What are the signaling events involved in activating the Arp2/3 complex? (ii) What are the bacterial and mammalian proteins that initiate these events?

ACTIVATION OF PROTEIN TYROSINE KINASE AND PI3-KINASE ACTIVITIES

Previous studies on *Y. pseudotuberculosis* and *L. monocytogenes* revealed that these invasive pathogens interact with specific cellular receptors to trigger tyrosine phosphorylation of proteins and the activation of specific lipid kinases such as phosphatidylinositide 3-kinase (PI3-kinase) (Finlay and Cossart, 1997; Cossart, 2004). Class I PI3-kinases are heterodimeric proteins, consisting of a noncatalytic p85 subunit and a catalytic p110 subunit, which when activated generate 3-phosphoinositides PI(3)P, PI(3,4)P2, and PI(3,4,5)P3 (also called PIP3) (Leevers et al., 1999). PI(3,4)P2 and PIP3 are important secondary messengers that function to directly or indirectly modulate actin cytoskeletal dynamics (Hartwig et al., 1995; Vanhaesebroeck and Waterfield, 1999; Dekker and Segal, 2000). The invasion of *R. conorii* into Vero cells is diminished when cells are treated with pharmacological inhibitors of protein tyrosine kinase activity (genistein and protein phosphatase 1) and inhibitors of PI3-kinases (wortmannin and LY294002). Furthermore, fluorescence microscopy revealed the colocalization of host tyrosine-phosphorylated proteins, protein tyrosine kinases such as c-Src, and cortactin, a c-Src substrate, at *R. conorii* entry foci (Martinez and Cossart, 2004). Western immunoblotting of lysates generated from infected mammalian cells revealed the induced tyrosine phosphorylation of several host proteins, including $p125^{FAK}$ (focal adhesion kinase), a protein that is involved in modulating the actin cytoskeleton during cellular migration (Richardson and Parsons, 1996) and in the entry of invasive pathogens into nonphagocytic cells (Persson et al., 1997; Martinez et al., 2000). Interestingly, PI3-kinase function can be induced following the binding of the p85 subunit to phosphotyrosine-containing proteins (Carpenter et al., 1993; Carpenter and Cantley, 1996). Indeed, the p85 subunit interacts with tyrosine-phosphorylated proteins during *R. conorii* infection of mammalian cells, suggesting that this mechanism of PI3-kinase activation may be utilized during the rickettsial invasion process.

Pharmacological inhibition of Src family tyrosine kinases by PP1 perturbed the ability of *R. conorii* to enter nonphagocytic mammalian cells in vitro. The interaction of Src with other phosphotyrosine-containing proteins is a mechanism sufficient to stimulate the activity of this nonreceptor protein tyrosine kinase (Thomas et al., 1998). *R. conorii* infection of Vero cells results in an induced increase of Src association with phosphotyrosine-containing proteins, suggesting that association with tyrosine-phosphorylated proteins could be a mechanism for Src activation during bacterial entry. Interestingly, Src, along with FAK, promotes the tyrosine phosphorylation of several focal adhesion-associated proteins including cortactin (Wu and Parsons, 1993), and Src has been shown to be involved in regulating cytoskeletal changes during the uptake of various pathogens, further highlighting their important roles in the cell biology of bacterial invasion (Fawaz et al., 1997; Dumenil et al., 2000; Agerer et al., 2003).

How different observed signals are coordinated during the *R. conorii* invasion process to stimulate actin cytoskeletal changes continues to be studied. *R. conorii* and other rickettsial species have likely adapted multiple mechanisms to trigger Arp2/3-dependent cytoskeletal changes that are required for efficient bacterial internalization. For example, active GTP-bound Cdc42 can directly interact with neural WASp (N-WASp), which then permits the acidic C-terminal (A) domain of N-WASp to recognize and potentially activate the Arp2/3 complex (Miki et al., 1998; Rohatgi et al., 1999, 2000). Moreover, N-WASp and cortactin can function cooperatively to induce Arp2/3-dependent actin polymerization as well as serving as independent activators of Arp2/3 (Weaver et al., 2002).

Ku70 IS A RECEPTOR INVOLVED IN SFGR INVASION

The ability of rickettsiae to successfully initiate an infection relies on the recognition of conserved outer membrane-associated proteins and specific cellular receptors. While rickettsial species primarily infect endothelial cells in vivo, this class of pathogen can bind to and invade a variety of cells of nonendothelial origin in vitro. Although studies revealed signaling pathways that were associated with the entry process, very little was known about the molecular details at the host-pathogen interface that were involved in triggering rickettsial invasion of target cells. An early investigation of host factors leading to rickettsial adherence identified membrane cholesterol as an essential component required for binding of *R. prowazekii* to erythrocytes, but did not reveal the identity of a putative proteinaceous receptor (Ramm and Winkler, 1976; Winkler, 1977). The identity of at least one mammalian protein receptor for *R. conorii* was revealed using a biochemical affinity approach with intact, purified rickettsiae incubated with detergent-soluble host cell lysates. Mass spectrometry of interacting proteins revealed the presence of β-actin, poly(ADP-ribose) polymerase, Ku70, Ku86, and other unidentified high-molecular-weight antigens in these samples. Further analysis revealed that at least one of these proteins, Ku70, associates with the plasma membrane and specifically interacts with *R. conorii* but not other gram-negative or gram-positive invasive pathogens tested. Fluorescence microscopy of *R. conorii*-infected Vero cells revealed that Ku70 is recruited to sites of rickettsial invasion early in the infection process. Furthermore, monoclonal antibodies against Ku70 inhibit *R. conorii* invasion, while reduction of endogenous Ku70 levels by small interfering RNAs against Ku70 and genetic deletion of *ku70* in murine embryonic fibroblasts significantly perturb the ability of *R. conorii* to invade these cells (Martinez et al., 2005).

Ku70 is ubiquitously expressed in mammalian cells as a nuclear protein; however, Ku70 expression at the plasma membrane is restricted to subsets of cell types including nontransformed endothelial cells, monocytes, macrophages, and cultured tumor cell lines such as HeLa and Vero. In monocytes and possibly other cells, Ku70 is transported via a nonclassical vesicle-mediated secretion mechanism whereby cytoplasmic pools of Ku70 are found in membrane-bound vesicles and trafficked along the actin filaments for display at the plasma membrane. Here, Ku70 is involved in heterologous and homologous cell adhesion (Koike, 2002), binding to fibronectin (Monferran et al., 2004a) and metalloprotease 9 (Monferran et al., 2004b). Subsequent localization studies determined that Ku70 at the plasma membrane is often associated with specialized cholesterol-rich microdomains termed lipid rafts (Lucero et al., 2003). As for other invasive pathogens, depletion of membrane cholesterol using methyl-β-cyclodextrin disrupts the composition of lipid rafts and inhibits *R. conorii* invasion of nonphagocytic cells, suggesting that the presence of Ku70 within these microdomains is important for efficient bacterial entry (Seveau et al., 2004; Martinez et al., 2005).

Adhesin protein interactions with host cell receptors often result in posttranslational receptor modifications that are associated with the stimulation of endocytosis and ultimately bacterial invasion (Veiga et al., 2007). Endocytosis is a process in which proteins and macromolecules are internalized via the invagination and subsequent release of membrane-bound vesicles at the plasma membrane of mammalian cells (Conner and Schmid, 2003). The ligand-dependent endocytosis of several receptor tyrosine kinases is in some cases driven by the Cbl-dependent monoubiquitination of the receptor (Haglund et al., 2003; Marmor and Yarden, 2004). Several studies have highlighted the mechanisms by which endocytic processes are exploited by viral and bacterial pathogens to enter nonphagocytic target cells (Veiga et al., 2007). Mechanistic similarities to the invasion pathways utilized by other zippering pathogens suggested that *R. conorii* also usurps these types of signaling events. Indeed, consequent to rickettsial attachment and invasion, Ku70 is ubiquitinated by c-Cbl, an E3 ubiquitin ligase. Fluorescence microscopy revealed that *R. conorii* colocalizes with c-Cbl at sites of entry, while other studies demonstrated that inhibition of endogenous c-Cbl expression by small interfering RNAs diminishes Ku70 ubiquitination and *R. conorii* invasion. These studies suggested that similar to other pathogens, *R. conorii* likely hijacks clathrin-dependent endocytosis of plasma membrane-associated receptors to gain entry into nonphagocytic mammalian cells (Martinez et al., 2005).

The discovery of a mammalian protein involved in the invasion of rickettsial species drove studies to determine the ligand(s) expressed by rickettsial species that could potentially interact with Ku70. Affinity chromatography methods using epitope-tagged recombinant Ku70 and *R. conorii* soluble protein lysates revealed the interaction of rickettsial outer membrane protein B (rOmpB/Sca5) with Ku70 (Martinez et al., 2005).

RICKETTSIAL SURFACE CELL ANTIGENS

Electron microscopic analyses of the rickettsial outer membrane fractions revealed the presence of a 7- to 16-nm-thick external layer consisting of proteins arranged in an arrayed fashion. The structure, termed an S layer, contains ~10 to 15% of the total cellular protein and is composed of immunodominant antigens. Several studies later identified at least two genes, *rompB* (originally named *spaP*) and *rompA*, whose products are components of the S layer (Palmer et al., 1974a, 1974b; Popov and Ignatovich, 1976; Silverman and Wisseman, 1978; Silverman et al., 1978; Smith and Winkler, 1979; Dasch, 1981; Dasch et al., 1981).

Bioinformatic analyses of sequenced rickettsial genomes revealed the presence of a gene family, termed *sca* for *s*urface *c*ell *a*ntigens, whose products resembled a distinct class of outer membrane proteins in gram-negative bacteria called autotransporters. Four genes in this family, namely *sca0* (*rompA*), *sca1*, *sca2*, and *sca5* (*rompB*), are present as intact open reading frames in the genomes of the majority of SFGR (Blanc et al., 2005). The proteins encoded by these genes are predicted to have modular structures, including an N-terminal signal peptide, a central passenger domain, and a C-terminal "translocation module" (β-peptide). Upon translation, autotransporter proteins are secreted across the inner membrane through

Sec-mediated recognition of an N-terminal signal sequence. The C-terminal domain, rich in β-sheet structure, is then predicted to insert into the outer membrane to form a translocation conduit through which the passenger domain is secreted. Many of these proteins can function as adhesins and proteases and are involved in the virulence of pathogens (Jacob-Dubuisson et al., 2004).

rOmpB AND ITS ROLE IN RICKETTSIAL INVASION

The *rompB* gene is present in all rickettsial species sequenced to date. rOmpB is expressed as a preprotein and then cleaved to release the passenger domain from the β-barrel translocation domain. Through a mechanism that is not well understood, the mature passenger domain does not appear to be released into the extracellular milieu; rather, the protein remains associated with the outer leaflet of the outer membrane (Hackstadt et al., 1992). Interestingly, an analysis of the predicted rOmpB protein sequences from closely and distantly related rickettsial species revealed a high degree of sequence similarity and identity, especially in regions comprising the passenger and translocation domains (Chan et al., 2010). This degree of conservation, especially within the passenger domain, suggests that rOmpB proteins may function similarly in the progression of different rickettsial diseases.

At least one role for rOmpB in mediating pathogen-host cell interactions has been elucidated. Using a heterologous expression system in *E. coli*, investigators demonstrated that rOmpB proteins from *R. conorii* and the closely related SFGR *R. japonica* are sufficient to facilitate bacterial association to and invasion of nonphagocytic cells (Uchiyama et al., 2006; Chan et al., 2009). rOmpB-mediated bacterial invasion is dependent on the expression of Ku70 in mammalian cells, confirming that rOmpB and Ku70 are a bona fide adhesin-receptor pair involved in the entry of rickettsial species. Interestingly, mammalian cells depleted of Ku70 are permissive for rOmpB-mediated adherence but not invasion, suggesting that the interactions between rOmpB and Ku70 are important to trigger bacterial entry. The importance of this protein-protein interaction was further highlighted by the ability of purified, recombinant rOmpB proteins comprising the passenger domain (amino acids 36 to 1334 in *R. conorii*) to interact directly with Ku70 and to competitively inhibit rOmpB-mediated bacterial adherence to cultured mammalian cells (Chan et al., 2009).

An analysis of the mechanism by which rOmpB-Ku70 ligation initiates bacterial entry revealed that the interaction of these two proteins is sufficient to trigger F-actin reorganization at bacterial entry sites. The signals involved in rOmpB-dependent invasion closely resemble those observed during the invasion of *R. conorii* into Vero cells and include the activation of actin and microtubule dynamics and the stimulation of protein tyrosine kinase and PI-kinase activities. Furthermore, rOmpB-induced invasion via Ku70 depends on c-Cbl ubiquitin ligase activity, suggesting c-Cbl is a signaling molecule involved in usurping host endocytic pathways (Chan et al., 2009).

While the topology of Ku70 at the plasma membrane is not known, an analysis of structural motifs present in Ku70 revealed that the N terminus of the protein shares some homology with fibronectin-binding domains present in von Willebrand factor and in certain integrins (Monferran et al., 2004a). Interestingly, *R. conorii* invasion into nonphagocytic mammalian cells involves the activation of kinases and other signals that are closely associated with the stimulation of integrin activity and with the uptake of other invasive bacteria (Isberg and Barnes, 2001). As has been demonstrated for other zippering pathogens, *R. conorii* likely trigger signals through a plasma membrane-associated receptor (Ku70) that result in localized actin polymerization and in the hijacking of the host endocytic machinery, ultimately leading to efficient rickettsial entry.

Ku70 at the plasma membrane and rOmpB play an important role in rickettsia-target cell interactions; however, inhibition of the rOmpB-Ku70 pathway disrupts *R. conorii* invasion of mammalian cells by approximately 50%, suggesting that other rickettsial adhesin-mammalian receptor pairs contribute to the invasion process (Martinez et al., 2005; Chan et al., 2010). Therefore, a closer examination of the functions of other related Sca proteins was warranted.

OTHER IMPORTANT ADHESIN AND INVASINLIKE PROTEINS

rOmpA

An intact *rompA* gene is present in the majority of SFGR species but is absent in TGR (Blanc et al., 2005). The predicted molecular weights of rOmpA proteins (224 to 247 kDa) vary from species to species due to variance in tandem repeat sequences present within the N-terminal portion of the passenger domain (Eremeeva et al., 2003; Ellison et al., 2008). Western immunoblot analyses of *R. conorii* cellular lysates with anti-rOmpA sera revealed a species with an apparent molecular weight (190 kDa) significantly lower than that of the predicted full-length protein. The difference between the predicted and observed rOmpA molecular weights can be attributed to at least one of two phenomena: either the protein migrates aberrantly on sodium dodecyl sulfate-polyacrylamide gel electrophoresis or, intriguingly, rOmpA, like other autotransporter proteins, is synthesized as a preprotein and then cleaved to release the passenger domain from the translocon module. While bioinformatic models predict that the rOmpA signal sequence from *R. conorii* is contained in the first 38 amino acids, the identity of a putative C-terminal processing motif responsible for further protein processing has yet to be experimentally defined (Chan et al., 2010).

The conservation of the *rompA* gene in SFGR but not TGR suggests that rOmpA function likely plays a prominent and important role in the progression of spotted fever infections. Interestingly, monoclonal antibodies against *R. rickettsii* rOmpA and extracted native rOmpA protein competitively inhibit rickettsial adherence to murine fibroblast cell lines (Li and Walker, 1998). In addition, rOmpA from *R. conorii*, when expressed in a surrogate nonadherent strain of *E. coli*, is sufficient to mediate adherence and invasion of cultured mammalian cells, suggesting an important role for rOmpA in mediating SFGR-host cell interactions.

Sca2

The *sca2* gene is found intact in most SFGR, but appears fragmented in many TGR species (Blanc et al., 2005). Several studies have confirmed the presence of the *sca2* mRNA transcript within rickettsial species and have localized Sca2 to the rickettsial outer membrane (Cardwell and Martinez, 2009; Haglund et al., 2010). The gene is predicted to code for an approximately 200- to 220-kDa protein in *R. conorii*; however, Western immunoblotting of whole-cell bacterial lysates reveals a reactive species of ~150 kDa, suggesting that as with other autotransporter proteins, Sca2 is processed. The molecular details of this event are currently not known.

An examination of the contribution of Sca2 to early interactions with target cells revealed that *R. conorii* Sca2 protein expressed at the outer membrane of *E. coli* is sufficient to trigger adherence and invasion of cultured mammalian cells. Interestingly, these phenotypes were also observed in experiments utilizing primary human lung microvascular endothelial cells, suggesting that Sca2 may contribute to early interactions with in vivo target cells. Competitive inhibition of *R. conorii* adherence is observed using a purified recombinant Sca2 fusion comprising an N-terminal passenger domain (amino acids 34 to 794), suggesting that the interaction domain involved in associating with plasma membrane receptors is contained within this region of the protein (Cardwell and Martinez, 2009).

Additional studies have elucidated a role for Sca2 following rickettsial invasion of host cells. An isolated *R. rickettsii sca2* mutant exhibits defects in actin-based motility, while purified recombinant Sca2 protein from the closely related SFGR *R. parkeri* is sufficient in vitro to polymerize actin monomers and to elongate filaments in a profilin-dependent manner (Haglund et al., 2010; Kleba et al., 2010). The observed adherence/invasion and actin-polymerizing activities appear to reside within the N terminus of the passenger domain. Additional analyses are required to determine the relative contribution of each protein function to rickettsial pathogenesis.

Sca1

sca1 is another member of this gene family that is nearly universally present in rickettsial genomes save for *R. prowazekii* and *Rickettsia canadensis (*Ngwamidiba et al., 2006*)*. Sca1 is expressed in *R. conorii* and localizes to the outer membrane (Riley et al., 2010). Western immunoblot analysis of whole-cell *R. conorii* lysates revealed a Sca1 reactive species that migrates significantly faster on SDS PAGE than the predicted 200-kDa protein, suggesting that like other autotransporters, Sca1 is proteolytically cleaved. Expression of Sca1 in *E. coli* was sufficient to mediate adherence to mammalian cells in vitro but was not sufficient to trigger rickettsial invasion (Riley et al., 2010). These findings demonstrated that adherence and invasion are distinct events that are governed by specific protein-protein interactions at the pathogen-host cell interface.

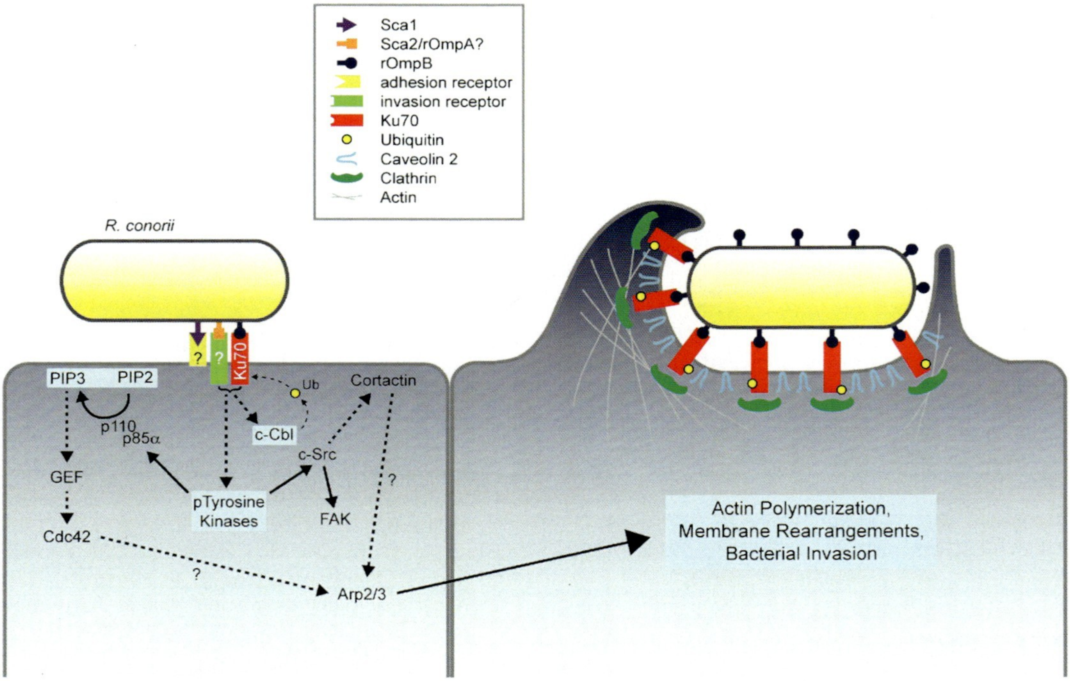

FIGURE 2 Model of *R. conorii*-induced signaling pathways in mammalian cells. The interaction of *R. conorii* with Ku70 and other mammalian receptors initiates signaling events that are coordinated to ultimately recruit actin and components of the endocytic machinery to localized areas of the membrane. These pathways include the activation of protein tyrosine kinases, PI3-kinase, Cdc42, Src, FAK, and cortactin, which are likely involved in Arp2/3-mediated actin polymerization at entry sites. c-Cbl-mediated ubiquitination (Ub) of Ku70 and the involvement of clathrin and caveolin-2 implicate the host endocytic machinery in the invasion pathway. Pathways involved in the Ku70-rOmpB invasion pathway are highlighted in light blue boxes. Putative protein-protein interactions and pathways involved in rickettsial invasion are demarcated by dashed arrows and questions marks. (Reprinted from Chan et al. [2010] with permission of the publisher.) doi:10.1128/9781555817336.ch4.f2

RC1281 (Adr1)

A biochemical approach to identify rickettsial adherence factors revealed additional gene products that may be important in early rickettsia-host cell interactions. This study revealed that in addition to the rOmpB β-peptide, rickettsial gene product RC1281 (also termed Adr1) from *R. conorii* was identified as a putative interactor with cellular membrane proteins (Renesto et al., 2006). RC1281 is conserved and ubiquitously present in the genomes of *Rickettsia* species and is paralogous to the adjacent gene RC1282 (*adr2*) in *R. conorii*. BLAST analyses of these genes show a low level of homology to other bacterial surface proteins, some which have adhesin function. Further examination is required to validate the role of these proteins as rickettsial adhesins and their importance to rickettsial pathogenesis.

CONCLUDING REMARKS

It appears that many different rickettsial proteins contribute to adherence and invasion of host cells in vitro (Li and Walker, 1998; Renesto et al., 2006; Uchiyama et al., 2006; Cardwell and Martinez, 2009; Chan et al., 2009; Riley et al., 2010). While these studies have indicated that individual gene products are sufficient to induce bacterial invasion, the internalization of rickettsial species likely results from the coordinated efforts of multiple adhesin-mammalian receptor pairs (Fig. 2). The identification of these other participants involved in pathogen-host cell interactions is of utmost importance and is the focus of current and future research endeavors.

REFERENCES

Agerer, F., A. Michel, K. Ohlsen, and C. R. Hauck. 2003. Integrin-mediated invasion of *Staphylococcus aureus* into human cells requires Src family protein-tyrosine kinases. *J. Biol. Chem.* **278:**42524–42531.

Alrutz, M. A., A. Srivastava, K. W. Wong, C. D'Souza-Schorey, M. Tang, L. E. Ch'Ng, S. B. Snapper, and R. R. Isberg. 2001. Efficient uptake of *Yersinia pseudotuberculosis* via integrin receptors involves a Rac1-Arp 2/3 pathway that bypasses N-WASP function. *Mol. Microbiol.* **42:**689–703.

Andersson, S. G., A. Zomorodipour, J. O. Andersson, T. Sicheritz-Ponten, U. C. Alsmark, R. M. Podowski, A. K. Naslund, A. S. Eriksson, H. H. Winkler, and C. G. Kurland. 1998. The genome sequence of *Rickettsia prowazekii* and the origin of mitochondria. *Nature* **396:**133–140.

Bear, J. E., M. Krause, and F. B. Gertler. 2001. Regulating cellular actin assembly. *Curr. Opin. Cell Biol.* **13:**158–166.

Bear, J. E., T. M. Svitkina, M. Krause, D. A. Schafer, J. J. Loureiro, G. A. Strasser, I. V. Maly, O. Y. Chaga, J. A. Cooper, G. G. Borisy, and F. B. Gertler. 2002. Antagonism between Ena/VASP proteins and actin filament capping regulates fibroblast motility. *Cell* **109:**509–521.

Bechah, Y., C. Capo, J. L. Mege, and D. Raoult. 2008. Epidemic typhus. *Lancet Infect. Dis.* **8:**417–426.

Blanc, G., M. Ngwamidiba, H. Ogata, P. E. Fournier, J. M. Claverie, and D. Raoult. 2005. Molecular evolution of *Rickettsia* surface antigens: evidence of positive selection. *Mol. Biol. Evol.* **20:**2073–2083.

Cardwell, M. M., and J. J. Martinez. 2009. The Sca2 autotransporter protein from *Rickettsia conorii* is sufficient to mediate adherence to and invasion of cultured mammalian cells. *Infect. Immun.* **77:**5272–5280.

Carpenter, C. L., K. R. Auger, M. Chanudhuri, M. Yoakim, B. Schaffhausen, S. Shoelson, and L. C. Cantley. 1993. Phosphoinositide 3-kinase is activated by phosphopeptides that bind to the SH2 domains of the 85-kDa subunit. *J. Biol. Chem.* **268:**9478–9483.

Carpenter, C. L., and L. C. Cantley. 1996. Phosphoinositide kinases. *Curr. Opin. Cell Biol.* **8:**153–158.

Chan, Y. G. Y., M. M. Cardwell, T. M. Hermanas, T. Uchiyama, and J. J. Martinez. 2009. Rickettsial outer-membrane protein B (rOmpB) mediates bacterial invasion through Ku70 in an actin, c-Cbl, clathrin and caveolin 2-dependent manner. *Cell. Microbiol.* **11:**629–644.

Chan, Y. G. Y., S. P. Riley, and J. J. Martinez. 2010. Adherence to and invasion of host cells by spotted fever group *Rickettsia* species. *Front. Microbiol.* **1:**1–9.

Conner, S. D., and S. L. Schmid. 2003. Regulated portals of entry into the cell. *Nature* **422:**37–44.

Cossart, P. 2004. Bacterial invasion: a new strategy to dominate cytoskeleton plasticity. *Dev. Cell* **6:**314–315.

Dasch, G. A. 1981. Isolation of species-specific protein antigens of *Rickettsia typhi* and *Rickettsia prowazekii* for immunodiagnosis and immunoprophylaxis. *J. Clin. Microbiol.* **14:**333–341.

Dasch, G. A., J. R. Samms, and J. C. Williams. 1981. Partial purification and characterization of the major species-specific protein antigens of *Rickettsia typhi* and *Rickettsia prowazekii* identified by rocket immunoelectrophoresis. *Infect. Immun.* **31:**276–288.

Dekker, L. V., and A. W. Segal. 2000. Perspectives: signal transduction. Signals to move cells. *Science* **287:**982–983.

Dumenil, G., P. Sansonetti, and G. Tran Van Nhieu. 2000. Src tyrosine kinase activity down-regulates Rho-dependent responses during *Shigella* entry into epithelial cells and stress fibre formation. *J. Cell Sci.* **113:**71–80.

Ellison, D. W., T. R. Clark, D. E. Sturdevant, K. Virtaneva, S. F. Porcella, and T. Hackstadt. 2008. Genomic comparison of virulent *Rickettsia rickettsii* Sheila Smith and avirulent *Rickettsia rickettsii* Iowa. *Infect. Immun.* **76:**542–550.

Eremeeva, M. E., R. M. Klemt, L. A. Santucci-Domotor, D. J. Silverman, and G. A. Dasch. 2003. Genetic analysis of isolates of *Rickettsia rickettsii* that differ in virulence. *Ann. N. Y. Acad. Sci.* **990:**717–722.

Fawaz, F. S., C. van Ooij, E. Homola, S. C. Mutka, and J. N. Engel. 1997. Infection with *Chlamydia trachomatis* alters the tyrosine phosphorylation and/or localization of several host cell proteins including cortactin. *Infect. Immun.* **65:**5301–5308.

Finlay, B. B., and P. Cossart. 1997. Exploitation of mammalian host cell functions by bacterial pathogens. *Science* **276:**718–725.

Galan, J. E., and D. Zhou. 2000. Striking a balance: modulation of the actin cytoskeleton by *Salmonella*. *Proc. Natl. Acad. Sci. USA* **97:**8754–8761.

Gouin, E., H. Gantelet, C. Egile, I. Lasa, H. Ohayon, V. Villiers, P. Gounon, P. J. Sansonetti, and P. Cossart. 1999. A comparative study of the actin-based motilities of the pathogenic bacteria *Listeria monocytogenes*, *Shigella flexneri* and *Rickettsia conorii*. *J. Cell Sci.* **112:**1697–1708.

Hackstadt, T., R. Messer, W. Cieplak, and M. G. Peacock. 1992. Evidence for proteolytic cleavage of the 120-kilodalton outer membrane protein of rickettsiae: identification of an avirulent mutant deficient in processing. *Infect. Immun.* **60:**159–165.

Haglund, C. M., J. E. Choe, C. T. Skau, D. R. Kovar, and M. D. Welch. 2010. *Rickettsia* Sca2 is a bacterial formin-like mediator of actin-based motility. *Nat. Cell Biol.* **12:**1057–1063.

Haglund, K., S. Sigismund, S. Polo, I. Szymkiewicz, P. P. Di Fiore, and I. Dikic. 2003. Multiple monoubiquitination of RTKs is sufficient for their endocytosis and degradation. *Nat. Cell Biol.* **5:**461–466.

Hall, A. 1998. G proteins and small GTPases: distant relatives keep in touch. *Science* **280:**2074–2075.

Hartwig, J. H., G. M. Bokoch, C. L. Carpenter, P. A. Janmey, L. A. Taylor, A. Toker, and T. P. Stossel. 1995. Thrombin receptor ligation and activated Rac uncap actin filament barbed ends through phosphoinositide synthesis in permeabilized human platelets. *Cell* **82:**643–653.

Higgs, H. N., and T. D. Pollard. 2001. Regulation of actin filament network formation through Arp2/3 complex: activation by a diverse array of proteins. *Annu. Rev. Biochem.* **70:**649–676.

Isberg, R. R., and P. Barnes. 2001. Subversion of integrins by enteropathogenic *Yersinia*. *J. Cell Sci.* **114:**21–28.

Jacob-Dubuisson, F., R. Fernandez, and L. Coutte. 2004. Protein secretion through autotransporter and two-partner pathways. *Biochim. Biophys. Acta* **1694:**235–257.

Kleba, B., T. R. Clark, E. I. Lutter, D. W. Ellison, and T. Hackstadt. 2010. Disruption of the *Rickettsia rickettsii* Sca2 autotransporter inhibits actin-based motility. *Infect. Immun.* **78:**2240–2247.

Koike, M. 2002. Dimerization, translocation and localization of Ku70 and Ku80 proteins. *J. Radiat. Res. (Tokyo)* **43:**223–236.

Leevers, S. J., B. Vanhaesebroeck, and M. D. Waterfield. 1999. Signalling through phosphoinositide 3-kinases: the lipids take centre stage. *Curr. Opin. Cell Biol.* **11:**219–225.

Li, H., and D. H. Walker. 1998. rOmpA is a critical protein for the adhesion of *Rickettsia rickettsii* to host cells. *Microb. Pathog.* **24:**289–298.

Lucero, H., D. Gae, and G. E. Taccioli. 2003. Novel localization of the DNA-PK complex in lipid rafts: a putative role in the signal transduction pathway of the ionizing radiation response. *J. Biol. Chem.* **278:**22136–22143.

Machesky, L. M., R. D. Mullins, H. N. Higgs, D. A. Kaiser, L. Blachoin, R. C. May, M. E. Hall, and T. D. Pollard. 1999. Scar, a WASp-related protein, activates nucleation of actin filaments by the Arp2/3 complex. *Proc. Natl. Acad. Sci. USA* **96:**3739–3744.

Marmor, M. D., and Y. Yarden. 2004. Role of protein ubiquitylation in regulating endocytosis of receptor tyrosine kinases. *Oncogene* **23:**2057–2070.

Martinez, J. J., and P. Cossart. 2004. Early signaling events involved in the entry of *Rickettsia conorii* into mammalian cells. *J. Cell Sci.* **117:**5097–5106.

Martinez, J. J., and S. J. Hultgren. 2002. Requirement of Rho-family GTPases in the invasion of type 1-piliated uropathogenic *Escherichia coli*. *Cell. Microbiol.* **4:**19–28.

Martinez, J. J., M. A. Mulvey, J. D. Schilling, J. S. Pinkner, and S. J. Hultgren. 2000. Type 1 pilus-mediated bacterial invasion of bladder epithelial cells. *EMBO J.* **19:**2803–2812.

Martinez, J. J., S. Seveau, E. Veiga, S. Matsuyama, and P. Cossart. 2005. Ku70, a component of DNA-dependent protein kinase, is a mammalian receptor for *Rickettsia conorii*. *Cell* **123:**1013–1023.

Miki, H., T. Sasaki, Y. Takai, and T. Takenawa. 1998. Induction of filopodium formation by a WASP-related actin-depolymerizing protein N-WASP. *Nature* **391:**93–96.

Monferran, S., C. Muller, L. Mourey, P. Frit, and B. Salles. 2004a. The membrane-associated form of the DNA repair protein Ku is involved in cell adhesion to fibronectin. *J. Mol. Biol.* **337:** 503–511.

Monferran, S., J. Paupert, S. Dauvillier, B. Salles, and C. Muller. 2004b. The membrane form of the DNA repair protein Ku interacts at the cell surface with metalloproteinase 9. *EMBO J.* **23:**3758–3768.

Ngwamidiba, M., G. Blanc, D. Raoult, and P. E. Fournier. 2006. Sca1, a previously undescribed paralog from autotransporter protein-encoding genes in *Rickettsia* species. *BMC Microbiol.* **6:**12.

Ogata, H., S. Audic, P. Renesto-Audiffren, P. E. Fournier, V. Barbe, D. Samson, V. Roux, P. Cossart, J. Weissenbach, J. M. Claverie, and D. Raoult. 2001. Mechanisms of evolution in *Rickettsia conorii* and *R. prowazekii*. *Science* **293:**2093–2098.

Ogata, H., P. Renesto, S. Audic, C. Robert, G. Blanc, P. E. Fournier, H. Parinello, J. M. Claverie, and D. Raoult. 2005. The genome sequence of *Rickettsia felis* identifies the first putative conjugative plasmid in an obligate intracellular parasite. *PLoS Biol.* **3:**e248.

Olson, M. F., A. Ashworth, and A. Hall. 1995. An essential role for Rho, Rac, and Cdc42 GTPases in cell cycle progression through G1. *Science* **269:**1270–1272.

Palmer, E. L., L. P. Mallavia, T. Tzianabos, and J. F. Obijeski. 1974a. Electron microscopy of the cell wall of *Rickettsia prowazeki*. *J. Bacteriol.* **118:**1158–1166.

Palmer, E. L., M. L. Martin, and L. Mallavia. 1974b. Ultrastucture of the surface of *Rickettsia prowazeki* and *Rickettsia akari*. *Appl. Microbiol.* **28:**713–716.

Persson, C., N. Carballeira, H. Wolf-Watz, and M. Fallman. 1997. The PTPase YopH inhibits uptake of *Yersinia*, tyrosine phosphorylation of $p130^{Cas}$ and FAK, and the associated accumulation of these proteins in peripheral focal adhesions. *EMBO J.* **16:**2307–2318.

Popov, V. L., and V. F. Ignatovich. 1976. Electron microscopy of surface structures of *Rickettsia prowazeki* stained with ruthenium red. *Acta Virologica* **20:**424–428.

Ramm, L. E., and H. H. Winkler. 1976. Identification of cholesterol in the receptor site for rickettsiae on sheep erythrocyte membranes. *Infect. Immun.* **13:**120–126.

Renesto, P., L. Samson, H. Ogata, S. Azza, P. Fourquet, J. P. Gorvel, R. A. Heinzen, and D. Raoult. 2006. Identification of two putative rickettsial adhesins by proteomic analysis. *Res. Microbiol.* **157:**605–612.

Richardson, A., and T. Parsons. 1996. A mechanism for regulation of the adhesion-associated protein tyrosine kinase $pp125^{FAK}$. *Nature* **380:** 538–540.

Ridley, A. J., and A. Hall. 1992. The small GTP-binding protein rho regulates the assembly of focal adhesions and actin stress fibers in response to growth factors. *Cell* **70:**389–399.

Ridley, A. J., H. F. Paterson, C. L. Johnston, D. Diekmann, and A. Hall. 1992. The small GTP-binding protein rac regulates growth factor-induced membrane ruffling. *Cell* **70:**401–410.

Riley, S. P., K. C. Goh, T. M. Hermanas, M. M. Cardwell, Y. G. Y. Chan, and J. J. Martinez. 2010. The *Rickettsia conorii* autotransporter protein Sca1 promotes adherence to nonphagocytic mammalian cells. *Infect. Immun.* **78:**1895–1904.

Rohatgi, R., H. Y. Ho, and M. W. Kirschner. 2000. Mechanism of N-WASP activation by CDC42 and phosphatidylinositol 4,5-bisphosphate. *J. Cell Biol.* **150:**1299–1310.

Rohatgi, R., L. Ma, H. Miki, M. Lopez, T. Kirchhausen, T. Takenawa, and M. W. Kirschner. 1999. The interaction between N-WASP and the Arp2/3 complex links Cdc42-dependent signals to actin assembly. *Cell* **97:**221–231.

Seveau, S., H. Bierne, S. Giroux, M. C. Prevost, and P. Cossart. 2004. Role of lipid rafts in E-cadherin- and HGF-R/Met-mediated entry of *Listeria monocytogenes* into host cells. *J. Cell Biol.* **166:**743–753.

Silverman, D. J., and C. L. Wisseman, Jr. 1978. Comparative ultrastructural study on the cell envelopes of *Rickettsia prowazekii*, *Rickettsia rickettsii*, and *Rickettsia tsutsugamushi*. *Infect. Immun.* **21:** 1020–1023.

Silverman, D. J., C. L. Wisseman, Jr., A. D. Waddell, and M. Jones. 1978. External layers of *Rickettsia prowazekii* and *Rickettsia rickettsii*: occurrence of a slime layer. *Infect. Immun.* **22:**233–246.

Smith, D. K., and H. H. Winkler. 1979. Separation of inner and outer membranes of *Rickettsia prowazeki* and characterization of their polypeptide compositions. *J. Bacteriol.* **137:**963–971.

Teysseire, N., J. A. Boudier, and D. Raoult. 1995. *Rickettsia conorii* entry into Vero cells. *Infect. Immun.* **63:**366–374.

Thomas, J. W., B. Ellis, R. J. Boerner, W. B. Knight, G. C. White II, and M. D. Schaller. 1998. SH2- and SH3-mediated interactions between focal adhesion kinase and Src. *J. Biol. Chem.* **273:**577–583.

Tran Van Nhieu, G., E. Caron, A. Hall, and P. J. Sansonetti. 1999. IpaC induces actin polymerization and filopodia formation during *Shigella* entry into epithelial cells. *EMBO J.* **18:**3249–3262.

Uchiyama, T., H. Kawano, and Y. Kusuhara. 2006. The major outer membrane protein rOmpB of spotted fever group rickettsiae functions in the rickettsial adherence to and invasion of Vero cells. *Microbes Infect.* **8:**801–809.

Vanhaesebroeck, B., and M. D. Waterfield. 1999. Signaling by distinct classes of phosphoinositide 3-kinases. *Exp. Cell Res.* **253:**239–254.

Veiga, E., J. A. Guttman, M. Bonazzi, E. Boucrot, A. Toledo-Arana, A. E. Lin, J. Enninga, J. Pizarro-Cerda, B. B. Finlay, T. Kirchhausen, and P. Cossart. 2007. Invasive and adherent bacterial pathogens co-opt host clathrin for infection. *Cell Host Microbe* **2:**340–351.

Walker, D. H. 1989a. Rickettsioses of the spotted fever group around the world. *J. Dermatol.* **16:**169–177.

Walker, D. H. 1989b. Rocky Mountain spotted fever: a disease in need of microbiological concern. *Clin. Microbiol. Rev.* **2:**227–240.

Weaver, A. M., J. E. Heuser, A. V. Karginov, W. L. Lee, J. T. Parsons, and J. A. Cooper. 2002. Interaction of cortactin and N-WASp with Arp2/3 complex. *Curr. Biol.* **12:**1270–1278.

Winkler, H. H. 1977. Rickettsial hemolysis: adsorption, desorption, readsorption, and hemagglutination. *Infect. Immun.* **17:**607–612.

Wu, H., and J. T. Parsons. 1993. Cortactin, an 80/85-kilodalton pp60^{src} substrate, is a filamentous actin-binding protein enriched in the cell cortex. *J. Cell Biol.* **120:**1417–1426.

Yoshida, S., E. Katayama, A. Kuwae, H. Mimuro, T. Suzuki, and C. Sasakawa. 2002. *Shigella* deliver an effector protein to trigger host microtubule destabilization, which promotes Rac1 activity and efficient bacterial internalization. *EMBO J.* **21:**2923–2935.

ESTABLISHING INTRACELLULAR INFECTION: ESCAPE FROM THE PHAGOSOME AND INTRACELLULAR COLONIZATION (*RICKETTSIACEAE*)

Matthew D. Welch, Shawna C. O. Reed, and Cat M. Haglund

5

INTRODUCTION

Almost 100 years ago, pioneering studies by Wolbach demonstrated that rickettsiae reside inside endothelial as well as other cell types (Wolbach, 1919), and that they must inhabit cells for survival and multiplication (Wolbach and Schlesinger, 1923). Since then, it has become clear that understanding how rickettsiae establish and maintain intracellular infection is key to elucidating their life cycle and mechanisms of pathogenesis.

The ability of rickettsiae to infect host cells involves a series of stages including invasion (discussed in the preceding chapter), escape from the phagosome, intracellular growth, intracellular movement for some species, and finally cell-to-cell spread. Following invasion, rickettsiae imaged by electron microscopy are free in the host cell cytosol and not present in a membrane-bound phagosome (Anderson et al., 1965), implying that they readily escape from the phagosomal compartment. The mechanism of phagosomal escape has been a matter of investigation ever since, and several bacterial phospholipases and hemolysins have been implicated in this process. However, despite many advances, the mechanism of escape remains largely mysterious.

Once free in the cytosol, rickettsiae grow by binary fission (Schaechter et al., 1957). Species in the spotted fever group rickettsiae (SFGR), including *Rickettsia rickettsii*, *Rickettsia conorii*, *Rickettsia parkeri*, *Rickettsia monacensis*, *Rickettsia montanensis*, and *Rickettsia australis*, grow as dispersed individuals in the cytosol, accumulate to lower numbers due to early release and infection of adjacent cells (Wisseman et al., 1976), and occasionally grow in the nucleus (Wolbach, 1919; Pinkerton and Hass, 1932). SFGR species also cause numerous changes in the host cell including an eventual cytopathology. In contrast, species in the typhus group rickettsiae (TGR), including *Rickettsia typhi* and *Rickettsia prowazekii*, grow in microcolonies and accumulate to high numbers in the cytosol (Wisseman and Waddell, 1975). Although TGR species are less cytotoxic than SFGR species, cells are eventually lysed when bacterial numbers reach a critical point. These observed differences between TGR and SFGR species reflect key distinctions in the behavior of the bacteria in host cells. The explanations for differences between *Rickettsia* species with respect to host cell cytopathology are unclear,

Matthew D. Welch, Shawna C. O. Reed, and Cat M. Haglund, Department of Molecular & Cell Biology, University of California, Berkeley, Berkeley, CA 94720.

Intracellular Pathogens II: Rickettsiales
Edited by G. H. Palmer and A. F. Azad © 2012 ASM Press, Washington, DC
10.1128/9781555817336.ch5

and are likely to involve differences in bacterial metabolism and interactions with the innate immune system, topics that are discussed in later chapters.

One other difference in behavior is the ability of SFGR species to undergo robust intracellular movement, a property that is not shared by TGR species. Motility is driven by actin filament polymerization and results in the formation of characteristic actin comet tails associated with bacteria (Teysseire et al., 1992; Heinzen et al., 1993). Moving bacteria can collide with the plasma membrane, resulting in the formation of short membrane protrusions. Bacteria may spread from cell to cell after being released from protrusions into the extracellular space (Schaechter et al., 1957), or alternatively, protrusions can extend into and be internalized by neighboring cells (Gouin et al., 1999). A similar mechanism of movement and spread is shared by other bacterial pathogens including *Listeria monocytogenes* (Tilney and Portnoy, 1989), *Shigella flexneri* (Bernardini et al., 1989), *Burkholderia pseudomallei* (Kespichayawattana et al., 2000), and *Mycobacterium marinum* (Stamm et al., 2003). Much progress has been made in our understanding of the mechanism of rickettsial actin-based motility, although the process of cell-to-cell spread remains poorly understood.

In this chapter, we examine historical and recent developments in our understanding of how rickettsiae establish intracellular infection. Our focus is on the three major stages mentioned above: (i) escape from the phagosome, (ii) intracellular growth, and (iii) actin-based motility, including its role in cell-to-cell spread. For each stage we provide a detailed description of the process itself, and where understood we highlight the molecular mechanisms that underlie it.

ESCAPE FROM THE PHAGOSOME

Invasion of host cells by rickettsiae was initially proposed to involve induced phagocytosis based on a requirement for active participation by both the bacteria and the host cell (Cohn et al., 1959; Walker and Winkler, 1978). These studies also established that internalization occurs rapidly and efficiently following bacterial adherence. Subsequent work has led to the identification of both rickettsial surface proteins that participate in adherence and host receptors and actin cytoskeletal proteins that mediate phagocytosis, as is discussed in the preceding chapter.

Early studies showed that, following internalization, both the TGR species *R. prowazekii* and the SFGR species *R. rickettsii* reside in the cytosol and are not enclosed in a membrane-bound phagosome, implying that bacteria escape from the phagosome after invasion (Anderson et al., 1965). However, in these and other early studies, infected cells were imaged after invasion was completed, and there was not sufficient time resolution to capture the initial steps in this process. Thus, for many years the nature of rickettsial phagocytosis and the steps that immediately follow remained a mystery.

Kinetics of Phagosome Escape

A first glimpse into the mechanism and kinetics of invasion was provided by Teysseire et al. (1995), who synchronized invasion of Vero cells by the SFGR species *R. conorii* and imaged the first 20 min of the process using electron microscopy. Their results suggested that invasion occurs in five stages: (i) adhesion to the host cell plasma membrane, (ii) engulfment, (iii) inclusion in a phagosome, (iv) phagosome lysis, and (v) release into the cytosol (Fig. 1). The kinetics of the process were observed to be very rapid. After only 3 min, 60% of bacteria were adherent to the plasma membrane and 40% were internalized. Of the internalized bacteria, two-thirds were enclosed in a phagosome, whereas one-third had already escaped into the cytosol. After 6 min, 60% of bacteria were internalized, with a similar 2:1 ratio of bacteria in the phagosome versus the cytosol. After 12 or 20 min, 90% of bacteria were internalized, with equal proportions in the phagosome and the cytosol. In a separate study, the escape of the TGR species *R. prowazekii* from the phagosome was observed

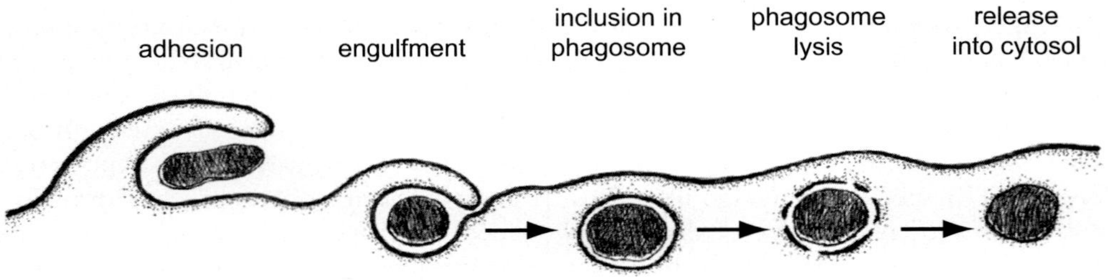

FIGURE 1 Rickettsial invasion occurs in five stages over the first 20 min of infection. These include (i) adhesion to the host cell plasma membrane, (ii) engulfment, (iii) inclusion in a phagosome, (iv) phagosome lysis, and (v) release into the cytosol. (Artwork adapted from Teysseire et al. [1995] by Taro Ohkawa.)
doi:10.1128/9781555817336.ch5.f1

following infection of Vero cells. Approximately 35% of internalized bacteria escaped from the vacuole by 30 min postinfection, while 69% escaped by 50 min (Whitworth et al., 2005). In both reports, the disrupted membranes were observed during early infection, indicating phagosomal lysis.

These observations indicated that internalization is nearly complete in the first 20 min of infection. Following phagocytosis of both species, escape from the phagosome occurs rapidly, although the kinetics of TGR escape may be slower than for the SFGR. The kinetics of rickettsial escape are faster than for other cytosolic pathogens such as *L. monocytogenes*, *S. flexneri*, and *Francisella tularensis*, which escape from the vacuole 15 to 60 min postinfection, following lysosomal associated membrane protein 1 (LAMP1) recruitment to phagosomes. For *L. monocytogenes* and *F. tularensis*, escape also occurs in response to a decrease in pH after phagolysosomal fusion (Ray et al., 2009). It has been proposed that the rapid kinetics of rickettsial escape indicate that the process may not require phagosome maturation and occurs well before phagolysosomal fusion. However, no published reports have examined the recruitment of endosomal markers to the early rickettsia-containing phagosome, so the precise stage of phagosome maturation at which rickettsiae escape remains undetermined.

Bacterial Factors Implicated in Escape

For other bacterial pathogens such as *L. monocytogenes*, escape from the phagosome involves the secretion or activation of membrane-disrupting factors including phospholipases and hemolysins in response to a drop in vacuolar pH (Ray et al., 2009). An early advance in identifying rickettsial factors that may function in phagosome escape came from the discovery that TGR species induce the hemolysis of red blood cells (Clarke and Fox, 1948). This hemolytic activity was subsequently shown to require adsorption of the bacteria to erythrocytes, membrane cholesterol, and metabolically active bacteria (Ramm and Winkler, 1973a, 1973b, 1976). It was later proposed that this contact-dependent hemolytic activity might play a role in inducing phagocytosis, mediating escape from the phagosome, and/or lysing host cells. Several bacterial activities that may function in membrane disruption have since been discovered, including phospholipase A2 (PLA2), phospholipase D (PLD), and hemolysins (TlyA and TlyC) (Table 1). Each of these activities and its potential role in phagosome escape is discussed below.

PLA2

PLA2 activity was first detected during *R. prowazekii* infection and was implicated in erythrocyte hemolysis (Winkler and Miller, 1980). The activity was shown to selectively hydrolyze lipids in erythrocyte but not bacterial membranes, and to be unusual in targeting lipids from both leaflets. It was subsequently shown that this activity was capable of hydrolyzing lipids from mouse L929 fibroblasts that are competent for bacterial infection (Winkler

TABLE 1 Bacterial factors implicated in phagosome escape

Biochemical activity	Gene names	Functional evidence for role in escape
PLA2	*R. typhi: RT0522, RT0590*	Inhibition reduced the frequency of plaque formation (Walker et al., 1983; Silverman et al., 1992)
PLD	*R. conorii: RC127* *R. prowazekii: RP819/pld*	*RP819* enabled *S.* Typhimurium phagosome escape (Whitworth et al., 2005), but *R. prowazekii* Δ*pld* mutant showed no difference from wild type in the timing of escape (Driskell et al., 2009)
Hemolysin	*R. prowazekii: RP555/tlyA, RP740/tlyC* *R. typhi: RT0725/tlyC*	*R. prowazekii tlyC* enabled *S.* Typhimurium escape from the phagosome (Whitworth et al., 2005)

and Miller, 1982). In this context, the activity required bacterial metabolism and protein synthesis and was enhanced by centrifuging bacteria onto cells, suggesting it is contact dependent (Winkler and Miller, 1982; Winkler and Daugherty, 1989). Moreover, activity was enhanced when host cell phagocytosis was inhibited, suggesting that phagocytosis can protect against extensive membrane damage. Based on the ability of the PLA2 activity to target membranes of an infection-competent cell type, it was proposed that it may play roles during infection, including inducing phagocytosis, mediating phagosome escape, and/or promoting cell lysis and spread. However, it remained unclear what role the phospholipase plays in infection, and whether the phospholipase was of bacterial or eukaryotic origin.

The first evidence that PLA2 is a bacterial enzyme came from studying *R. rickettsii*, which is also associated with phospholipase activity (Walker et al., 1983). This evidence was based on the observation that pretreatment of bacteria, but not host cells, with a PLA2 inhibitor or anti-PLA antiserum impacted bacterial infection (Silverman et al., 1992). Further support was provided by the observation that *R. prowazekii* possesses an endogenous phospholipase activity toward purified phospholipids (Ojcius et al., 1995). These studies also addressed the role of the phospholipase in infection. In particular, inhibition of phospholipase activity reduced the efficiency of bacterial internalization, but did not affect bacterial growth in cells, pointing to a specific role for this activity in phagocytosis or phagosome escape. Moreover, inhibition of phospholipase activity was shown to reduce the frequency of plaque formation (Walker et al., 1983; Silverman et al., 1992), an effect that was attributed to a role for the enzyme in invasion. Subsequent work established that *R. typhi* also possesses PLA2 activity, and that both the *R. prowazekii* and *R. typhi* PLA2 activities were robust at a range of pH values from 5 to 7 in the presence of divalent cations, suggesting they are capable of being active in the environment of the early phagosome (Ojcius et al., 1995).

Despite this extensive characterization of PLA2 activity in rickettsiae, it has been difficult to identify genes that encode PLA2 enzymes. However, a recent search for rickettsial proteins with motifs similar to known PLA2 enzymes, including patatin-like proteins and *Pseudomonas aeruginosa* ExoU, revealed potential PLA2 homologs encoded in the *R. typhi*, *R. prowazekii*, *Rickettia massiliae*, and *Rickettsia bellii* genomes (Rahman et al., 2010). In *R. typhi*, there are two such proteins, annotated as RT0522 and RT0590. Of these, only the RT0522 transcript is expressed at detectable levels in infected Vero 76 cells, with expression highest at 15 to 30 min and lowest at 8 h postinfection. Interestingly, purified recombinant RT0522 was shown to possess PLA2 enzymatic activity, but only in the presence of Vero cell cytosol, similar to ExoU. The activity of RT0522 was also 60-fold lower than that of ExoU. Mutating residues in the predicted catalytic dyad of RT0522 abolished activity, as did incubation with PLA2 chemical inhibitors, providing additional support for the notion that it is a bona fide PLA2 enzyme. Furthermore, expressing RT0522 in the yeast *Saccharomyces cerevisiae* prevented cell growth and division, and expression in mammalian cells caused lysis, indicating that RT0522 is cytotoxic. This cytotoxic effect was abolished in catalytically inactive mutants. Interestingly, RT0522 does not have a predicted signal sequence for the general secretory pathway, but in cell fractionation experiments it was found to be present in the host cell cytoplasmic fraction, suggesting it is secreted into the host cell during infection. It was postulated that secretion might occur via the rickettsial type IV secretion system, but the mechanism of secretion remains undetermined. Future work will also be needed to assess the timing of RT0522 secretion and whether the protein plays an important functional role in escape from the phagosome. Moreover, the expression and activity of PLA2 homologs from other rickettsiae have not yet been investigated.

PLD

Genes encoding PLD superfamily proteins were initially found in the *R. conorii* (*RC127*) and *R. prowazekii* (*RP819*) genomes by searching translated sequences for phospholipaselike motifs (Renesto et al., 2003). The *PLD* gene is present in all nine SFGR species tested and two TGR species based on Southern blot hybridization data. Moreover, in *R. conorii* and *R. prowazekii* the transcripts and proteins are expressed. Recombinant *R. conorii* PLD expressed in vitro was shown to possess bona fide PLD catalytic activity in hydrolyzing phosphatidylcholine, and may act as a dimer. The protein also has a general secretory pathway signal sequence, suggesting it is secreted, although this has not yet been demonstrated. In a preliminary investigation of the role of PLD in infection, it was found that pretreatment of bacteria with anti-PLD antibody inhibited cytotoxicity after 1 week, suggesting that PLD plays a role in rickettsia infection. It was suggested that PLD may play a role in phagosome escape analogous to that proposed for PLA2, although this remained untested.

A subsequent investigation of PLD from *R. prowazekii* found that the transcript was expressed at 30 min postinfection, corresponding to the time at which many bacteria are escaping from the phagosome (Whitworth et al., 2005). To investigate a potential role for PLD in escape, the enzyme was expressed in *Salmonella enterica* serovar Typhimurium, an intracellular pathogen that does not normally escape from the phagosome but instead resides in a modified phagolysosomal compartment. Interestingly, 100% of *S.* Typhimurium expressing PLD escaped into the cytosol by 4 h postinfection, indicating that PLD is sufficient to mediate bacterial escape from the phagosome. However, whether PLD mediated rickettsial escape remained uncertain.

Dissecting the functional importance of PLD in phagosome escape has historically been difficult due to a lack of tools for genetic manipulation of rickettsiae. However, a recent groundbreaking study reported a directed knockout of the *pld* gene from *R. prowazekii* using homologous recombination (Driskell et al., 2009). Although the mutation could not be complemented due to the difficulty in genetic manipulation and lack of available plasmids, the phenotype was nonetheless revealing. The Δ*pld* mutant grew as well as wild type in macrophage-like cells, indicating the gene is not important for intracellular multiplication. Surprisingly, despite the previous indication that PLD is sufficient to enable *S.* Typhimurium phagosome escape, the Δ*pld* mutant showed no significant difference from wild type in the timing of escape from the phagosome at 10, 30, 45, or 60 min postinfection. This suggests that if PLD plays a role in phagosome escape, this role is redundant with that of other bacterial factors. When introduced into guinea pigs, the Δ*pld* mutant was attenuated for virulence, did not cause fever or weight loss, and induced protective immunity against infection with the more virulent wild-type strain. Thus, PLD plays an as yet unspecified role in *R. prowazekii* virulence, may act as a redundant factor in phagosome escape for both *R. conorii* and *R. prowazekii*, and/or may function to lyse the host cell late in infection.

HEMOLYSINS

In addition to phospholipases, membrane-disrupting hemolysins are secreted by other intracellular pathogens such as *L. monocytogenes* and function in phagosome escape (Ray et al., 2009). Two genes encoding hemolysins, *tlyA* and *tlyC*, have been identified in the genome of *R. prowazekii*, and a homolog of *tlyC* was identified in *R. typhi*. The *R. typhi* TlyC protein was shown to function as a hemolysin, as it conferred hemolytic activity on normally nonhemolytic *Escherichia coli* or *Proteus mirabilis* when bacterial or culture supernatants were added to sheep erythrocytes (Radulovic et al., 1999). A second study showed that the *R. prowazekii tlyC* transcript was expressed at 30 min postinfection, during the time of phagosome escape (Whitworth et al., 2005). In addition, expression of the *R. prowazekii* protein in *S.* Typhimurium enabled a fourfold enhancement of bacterial escape from the phagosome compared with wild-type bacteria, suggesting it is sufficient to enable phagosome escape. The phenotype was, however, more subtle than for *pld* expression in *S.* Typhimurium (see above). Nevertheless, a direct role for *tlyC* in rickettsial phagosome escape has not yet been demonstrated, and the TlyC protein has not been purified nor has its biochemical activity been characterized. Given the presence of multiple membrane-disrupting factors in rickettsiae, including both phospholipases and hemolysins, it is likely that the mechanism of phagosome escape is complex and involves molecules with redundant functions.

INTRACYTOPLASMIC GROWTH AND EFFECT ON HOST CELL STRUCTURE AND PHYSIOLOGY

Early studies established that following escape from the phagosome, rickettsiae grow inside host cells (Wolbach, 1919). When observed by electron microscopy, the TGR species *R. prowazekii* was seen to grow exclusively in the cytosol, whereas the SFGR species *R. rickettsii* grew in the cytosol and occasionally in the nucleus (Anderson et al., 1965). Subsequent examinations revealed other key differences between the TGR and SFGR species with regard to the kinetics of growth, the location of growth within the host cell, and the effect on host cell structure and physiology. These differences are described below, and innate immune signaling responses by the host cell are discussed in chapter 8.

Kinetics of Growth

The first quantitative study of the growth kinetics of the TGR species *R. prowazekii* was conducted by Wisseman and colleagues, who carried out synchronous infections and compared the growth and division of both the virulent E strain and attenuated Breinl strain using light microscopy (Wisseman and Waddell, 1975). Interestingly, growth kinetics of both strains were similar, suggesting that other differences contribute to virulence. During the first growth cycle, the percentage of infected cells remained constant over time, indicating that little cell-to-cell spread occurred. Exponential growth initiated without an appreciable lag phase and continued for 36 to 48 h, with a doubling time of approximately 9 h. After this phase, the cytoplasm became so densely packed with bacteria, up to 1,000 per cell, that individuals could not be counted, and as a consequence it could not be determined whether they reached a stationary phase. Overall, *R. prowazekii* growth kinetics were similar to typical bacterial growth in culture, with the individual cell as the culture vessel. As with many bacteria grown in culture, the morphology of the rickettsia also changed over time, with smaller coccobacillary forms accumulating at later time points. Late in the exponential phase, at 36 to 48 h, cells began to lyse in an asynchronous, disordered, and prolonged process that continued for 96 h or longer. Sometimes bacteria were completely released during cell lysis, whereas at other times they remained associated with the cell carcass. Later during infection, there was an increase in the percentage of infected cells, indicating that released bacteria initiated new infection cycles. Thus, cell lysis and bacterial release mediate cell-to-cell spread of *R. prowazekii* and likely other TGR species.

Interestingly, the growth kinetics for the SFGR species *R. rickettsii* did not follow the simple kinetics observed for the TGR species *R. prowazekii*. Wisseman and colleagues observed that exponential growth of *R. rickettsii* began immediately, without an appreciable lag, and the doubling time estimated from this initial phase was 10 to 12 h (Wisseman et al., 1976). However, in contrast to *R. prowazekii*, after 10 h the percentage of infected host cells began to increase, and this continued over time until at 48 h most cells were infected and at 72 h all cells were infected with an average of 30 to 50 bacteria/cell. Overall, bacteria did not build up in large numbers in single cells and were generally dispersed throughout the cytoplasm. However, occasional concentrated growth in the nucleus of cells and the formation of compact "doughnut" colonies in the cytoplasm were observed, especially late during infection. Extracellular escaped bacteria were observed throughout infection, and their presence was also inferred by the constant increase in the number of infected cells in culture. The combination of constant release into the extracellular milieu and dense microcolony formation prevented accurate enumeration of bacteria after 72 h. Together these data indicate that during the exponential growth phase, bacteria escape rapidly and consistently into neighboring cells. Thus, the growth kinetics of SFGR species are more complicated, occurring as in a chemostat with respect to individual cells, where continuous exit enables prolonged active growth of low numbers of bacteria, while the number of bacteria increases in the culture as a whole.

Intracellular Niche and Impact on Host Cell Biology

Apart from differences in growth kinetics, TGR and SFGR species also establish distinct intracellular niches and have very different effects on host cell biology (Fig. 2). In a careful study of the ultrastructure of chick embryo fibroblast cells infected with the TGR species *R. prowazekii* (Silverman et al., 1980), bacteria were initially observed within the cytosol at 30 h postinfection, when the average number of bacteria per cell was approximately 20, and no effect on host cell ultrastructure was apparent. Even as infection progressed to 96 h, when hundreds of bacteria were observed in the cytosol, little effect on cell ultrastructure was observed beyond a limited swelling or disruption of mitochondrial structure and dissociation of ribosomes from the rough endoplasmic reticulum (RER). At 96 to 120 h, host cell lysis was seen. Interestingly, while there were few changes in host cell ultrastructure, the size and shape of the bacteria changed during the course of infection, perhaps reflecting alterations in their physiology as they grow in the closed system of the cell for an extended period of time.

In contrast to cells infected with *R. prowazekii*, initial studies showed that chicken embryo fibroblasts infected with the SFGR species *R. rickettsii* showed marked changes in their ultrastructure during the course of infection (Walker et al., 1977; Silverman and Wisseman, 1979; Walker and Cain, 1980). These changes were first noticed at 48 h postinfection, when bacteria were mostly in the cytosol. At this time the normally narrow membrane tubules of the continuous RER and outer nuclear envelope became dilated. Within this dilated space, some bacteria and other electron-dense material accumulated. At 72 h, bacteria were frequently observed within the dilated RER and nuclear envelope, and sometimes inside the nucleus. The overall architecture of the cell also was affected, with the Golgi apparatus becoming less distinct and mitochondria becoming swollen. As the infection progressed to 96 h, the degree of ER dilation increased, as did the overall number of intracellular membrane-bound compartments, suggesting a general disruption of organelle architecture. From 96 to 120 h, bacteria were often observed

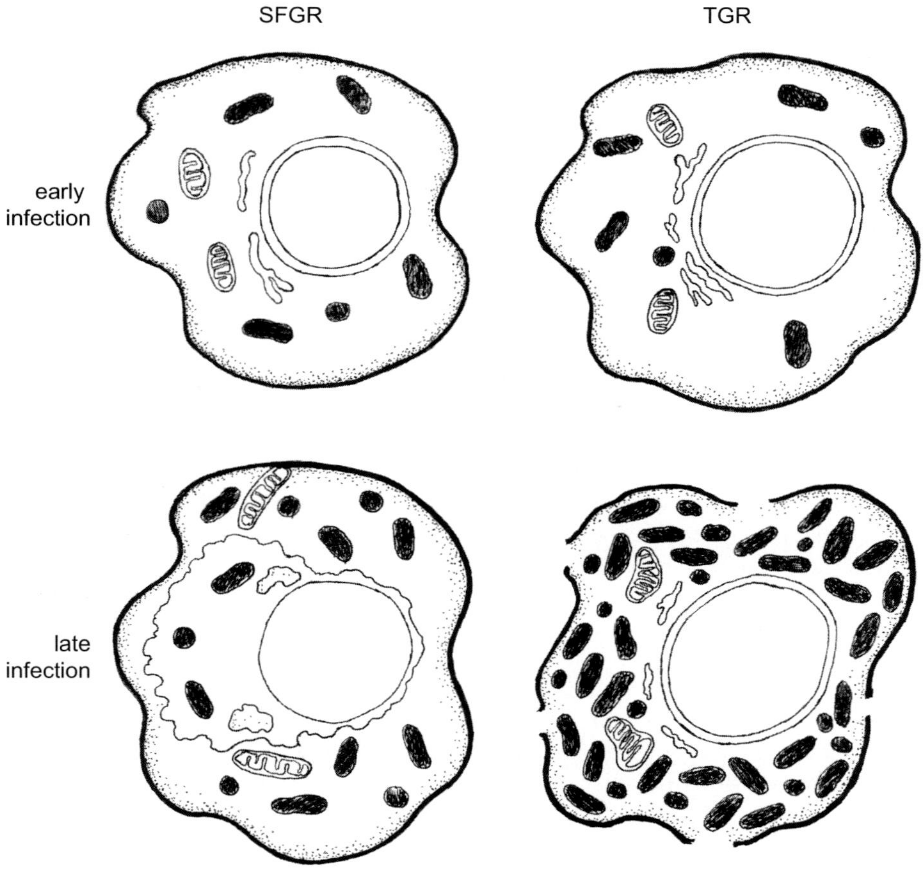

FIGURE 2 SFGR and TGR have different effects on host cell structure and physiology. (Top left) SFGR species have little effect on host cell ultrastructure at times shortly after infection (24 h). (Bottom left) However, progressive changes occur during longer infections (48 h and beyond), including dilation of the RER and outer nuclear envelope, loss of a distinct Golgi apparatus, and swollen mitochondria. Cells eventually lyse during very long infections (more than 120 h; not shown). (Top right) TGR species are initially observed in small numbers in the cytosol early in infection (30 h), and as infection progresses the number of bacteria increase with no effect on host cell ultrastructure (up to 96 h). (Bottom right) At very late times (96 to 120 h postinfection), host cell lysis and bacterial release occur. (Artwork adapted from Silverman and Wisseman [1979] and Silverman et al. [1980] by Taro Ohkawa.) doi:10.1128/9781555817336.ch5.f2

in so-called doughnut colonies that consisted of many bacteria surrounding a central vacuole. By 120 h, most cells were lysed or showed a loss of plasma membrane integrity. Later studies documented similar effects of infection by *R. rickettsii* on the ultrastructure of human umbilical vein endothelial cells (HUVECs) (Silverman, 1984) and the endothelial cell line EA.hy 926 (Eremeeva and Silverman, 1998). Interestingly, although the host cells exhibited dramatic changes, the bacterial morphology of SFGR species remained largely unchanged during the course of infection. One possibility is that the relatively lower numbers of SFGR bacteria per cell result in increased nutrient availability. Despite the profound effects of infection by *R. rickettsii* on the physiology of the host cell, exactly how the bacteria damage host cells and which host cell stress responses are induced during infection remain unclear. Dilation of the ER, disassociation of ribosomes, and swelling of mitochondria may be signs of cytoplasmic membrane damage, ATP depletion, or oxidative stress and may result from proapoptotic pathway activation (Cheville, 1994).

Given the effects on the integrity and organization of membrane-bound organelles, it has been suggested that damage may be caused by direct enzymatic targeting of membranes, perhaps by bacterial phospholipases or hemolysins (Silverman, 1984). It has also been suggested that these changes may be caused by a general depletion of cellular ATP levels as rickettsiae import and use ATP for replication (Silverman and Wisseman, 1979). Several lines of evidence also suggest that damage to host cells may be caused by a buildup of oxygen radicals, which may cause membrane modification (Silverman, 1984). For example, hydrogen peroxide has been shown to accumulate in *R. rickettsii*-infected HUVECs and EA.hy 926 cells (Silverman and Santucci, 1988; Eremeeva and Silverman, 1998). Moreover, the levels of the reducing agents glutathione and glutathione peroxidase were shown to decrease during infection (Silverman and Santucci, 1990; Eremeeva and Silverman, 1998). Treatment of cells with antioxidants reduced the cytotoxic effects of infection, inhibited cell death, and prevented profound ultrastructural changes (Silverman and Santucci, 1990; Eremeeva and Silverman, 1998). This supports the notion that buildup of oxygen radicals contributes to the cytotoxic effect of SFGR species on host cells. Why similar buildup does not occur during infection with TGR species remains unclear, but one possibility is that the bacteria specifically inhibit free-radical formation or apoptotic pathways to maintain host cell integrity.

ACTIN-BASED MOTILITY AND CELL-TO-CELL SPREAD

Most early studies of the intracellular life of rickettsiae involved the observation of bacteria in fixed cells, and as a result dynamic behaviors were overlooked. However, Schaechter and colleagues noted intracellular movements upon imaging living cells infected with the SFGR species *R. rickettsii* using phase-contrast time-lapse microscopy (Schaechter et al., 1957). They saw that bacteria were spread throughout the cytosol and that some underwent slow intracytoplasmic motion, and that bacteria in the nucleus also moved around the inner periphery of this organelle with a uniform directionality and speed. Among the cytosolic population, those near the cell periphery often entered into dynamic cell surface protrusions. If these surface structures extended outward, bacteria were carried away from the cell body, and if they retracted, bacteria either returned into the cytosol or were freed from the protrusion into the extracellular environment. It was proposed that this release of bacteria from the host cell might play a crucial role in cell-to-cell spread. However, the mechanism of intracellular movement and its precise role in spread remained enigmatic for many years.

The Actin Cytoskeleton Drives Intracellular Motility

A clue to the mechanism of rickettsial intracellular movement came from previous studies of similar behavior by the bacterial pathogens *S. flexneri* and *L. monocytogenes*, for which movement is driven by the host cell actin cytoskeleton (Bernardini et al., 1989; Tilney and Portnoy, 1989). Host cell actin exists in both monomer (globular-actin or G-actin) and filament (filamentous-actin or F-actin) forms. Actin filament assembly results from the reversible polymerization of monomers onto the ends of filaments, which are polar and have two distinct ends: the fast-growing end (barbed or plus [+] end) and the slower-growing end (pointed or minus [–] end). Polymerization occurs primarily at barbed ends in cells, and is often controlled by numerous actin-binding proteins that modulate both the nucleation of new filaments and the elongation of existing ones. In uninfected cells the force of polymerization is harnessed to drive the movement of the plasma membrane and internal membrane-bound organelles (Firat-Karalar and Welch,

2011). During infection with the above-mentioned pathogens, which like rickettsiae escape from the phagosome and reside in the cytosol, actin assembly is initiated at the bacterial surface and is harnessed for force generation to drive intracellular movement, resulting in the formation of actin comet tails that emanate from one pole and trail moving bacteria. The function of intracellular movement is to promote cell-to-cell spread, a process that is discussed below.

The first indication that rickettsiae also harness the host cell actin cytoskeleton for movement came from studies by Teysseire and coworkers as well as Heinzen and coworkers, who noted associations between the SFGR species *R. conorii* and *R. rickettsii* and host F actin (Teysseire et al., 1992; Heinzen et al., 1993). For *R. rickettsii*, association with actin was observed as early as 15 min postinfection, and the formation of actin comet tails, an indicator of movement, was observed as early as 30 min postinfection (Heinzen et al., 1993). For *R. conorii*, 20 to 50% of bacteria were associated with comet tails at 24 h postinfection, depending on the cell type, and 54 to 68% at 48 h postinfection (Teysseire et al., 1992). Cell type also affected the median length of tails; for instance, *R. conorii* tails had a median length of 3 μm in HUVECs and 5 μm in Vero cells. Within the same cell type, the length of tails varied widely, ranging from 1 to 15 μm for *R. conorii* (Teysseire et al., 1992) to up to 70 μm for *R. rickettsii* (Heinzen et al., 1993) in Vero cells.

Interestingly, association with actin tails has been observed for most SFGR species tested, including avirulent strains of *R. rickettsii*, as well as *R. montanensis*, *R. australis* (Heinzen et al., 1993), *R. monacensis* (Simser et al., 2002), *R. parkeri* (Serio et al., 2010), and *Rickettsia felis* (Ogata et al., 2005). The exception among the SFGR is the nonpathogenic species *Rickettsia peacockii*, which does not associate with actin (Simser et al., 2005). The TGR species *R. prowazekii* also does not associate with actin (Heinzen et al., 1993), but *R. typhi* does, albeit for only 1% of bacteria in HUVECs at 48 h postinfection (Teysseire et al., 1992). The actin tails formed by *R. typhi* also differ in morphology from those made by SFGR species in that they are shorter, with a median length of 1 μm and a range of 0.3 to 4 μm in HUVECs (Teysseire et al., 1992), and often have a hooked shape that contrasts with the straighter, longer tails made by SFGR species (Heinzen et al., 1993). Finally, the ancestral group rickettsiae (AGR) species *R. bellii* was recently shown to undergo actin-based motility, also forming short actin tails (Ogata et al., 2006). Together, these observations suggest that the ability to undergo actin-based motility evolved early in the genus *Rickettsia* and has been either enhanced or lost in different branches of the lineage.

Interestingly, images from both light and electron microscopy suggest that the tail structures formed by SFGR species differ from those formed by other pathogens such as *L. monocytogenes*. For example, whereas *L. monocytogenes* actin tails show a nearly uniform distribution of actin by fluorescence microscopy, *R. rickettsii* and *R. conorii* tails consist of distinct parallel filament bundles that wrap in a helical arrangement around a hollow core (Fig. 3) (Heinzen et al., 1993; Gouin et al., 1999). A similar hollow organization was also observed by electron microscopy (Gouin et al., 1999; Van Kirk et al., 2000). As with other pathogens, filaments are oriented with their fast-growing barbed or + ends facing the bacterial surface. However, while *L. monocytogenes* actin tails consist of short (0.1 to 1 μm) and densely packed filaments, SFGR tails consist of long filaments (0.3 to 3 μm) that are oriented in parallel and are much more sparsely packed (Gouin et al., 1999) (Fig. 3). How filaments are organized in the actin tails formed by the TGR species *R. typhi* or the AGR species *R. bellii* has not yet been reported, although by light microscopy the tails superficially resemble those formed by *L. monocytogenes*, suggesting that filament organization may differ from that seen for SFGR species. Together, these observations suggest that SFGR may harness a distinct mechanism for actin polymerization and

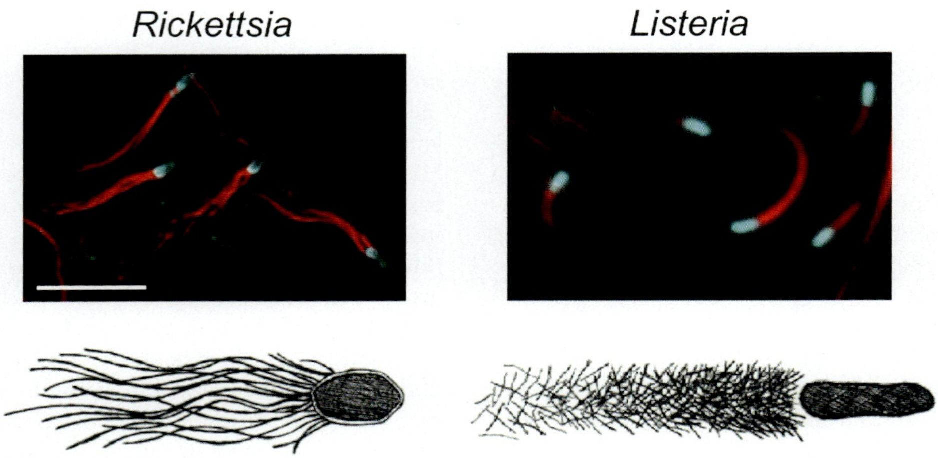

FIGURE 3 *Rickettsia* and *Listeria* actin comet tails are different in appearance and actin organization. Top panels show bacteria (blue) and actin (red) visualized by fluorescence microscopy, whereas bottom panels depict actin organization. (Left) SFGR comet tails are relatively straight, consist of a helical arrangement of actin bundles (top), and are composed of long parallel filaments (bottom). In contrast, *Listeria* comet tails are uniform in appearance (top) and are composed of short filaments organized into a dense meshwork (bottom). Bar (top panels), 5 μm. (Artwork adapted from Gouin et al. [1999] by Taro Ohkawa.) doi:10.1128/9781555817336.ch5.f3

organization compared with other pathogens, and that during the course of rickettsial evolution different molecular mechanisms may have arisen in the SFGR, TGR, and AGR species.

Motility Characteristics and Actin Dynamics

A comparison of the characteristics of SFGR actin-based motility with those of other pathogens including *L. monocytogenes* and *S. flexneri* reveals general similarities and intriguing differences. Rickettsial actin-based motility is different in overall rate, persistence of movement, and stability of the actin filaments in the comet tail. For example, rickettsiae move more slowly than their counterparts, as measured by time-lapse video microscopy. *R. rickettsii* moved at rates of 4.8 ± 0.6 μm/min, whereas *L. monocytogenes* moved at 12.0 ± 3.1 μm/min in Vero cells (Heinzen et al., 1999). Similarly, *R. conorii* moved at 8 μm/min, versus 22 μm/min for *L. monocytogenes* and 26 μm/min for *S. flexneri* (Gouin et al., 1999) in the same cell type. The fact that both *Rickettsia* species moved at ~40% of the rate of *L. monocytogenes* in parallel infections suggests that the observed difference in the motility rate of *R. rickettsii* compared to *R. conorii* is due to experimental conditions rather than an inherent difference in their motility characteristics.

The rickettsiae also differed from other pathogens in the directionality of movement. In particular, *R. rickettsii* moved in straight paths, whereas *L. monocytogenes* turned frequently (Heinzen et al., 1999). *R. rickettsii* only changed direction after colliding with intracellular structures, at which time the bacteria often reversed polarity and formed a tail from the opposite bacterial pole, a phenomenon that was not observed for *L. monocytogenes* (Heinzen, 2003). Finally, it was noted that the helical organization of rickettsial tails implies that the bacteria rotate around their long axis as they move (Heinzen et al., 1999), as does *L. monocytogenes* (Robbins and Theriot, 2003). However, *R. rickettsii* made one helical turn for every 6 μm of distance traveled, whereas *L. monocytogenes* made one turn every 30 μm, suggesting that they rotate at different rates. Although it is still unclear why rickettsiae differ in their movement characteristics, this likely reflects differences in their mechanism of actin polymerization compared with other pathogens.

In keeping with this notion, the dynamics of actin filaments also differ in rickettsial actin tails compared with those of other pathogens. Actin dynamics were observed in *R. rickettsii*-infected cells expressing green fluorescent protein-actin, where bacteria formed actin tails that resembled the corresponding structures in fixed cells (Heinzen, 2003). By measuring the decay in actin fluorescence over time, the half-life of actin filaments in *R. rickettsii* tails was measured to be 100 ± 19 s, as compared with 33 ± 8 s for actin in *L. monocytogenes* tails. The longer half-life suggests that actin filaments in rickettsial tails are more stable, which may reflect differences both in actin organization and in the activity of host proteins that function to sever and disassemble actin filaments, as discussed below.

Despite these differences, the driving force for rickettsial actin-based motility is likely to be the same as that for other pathogens. By observing the behavior of naturally occurring variations in green fluorescent protein's fluorescence along the length of *R. rickettsii* actin tails, it was noted that the actin filaments in the tail remained stationary while the bacterium moved forward (Heinzen et al., 1999), similar to what was seen for *L. monocytogenes* (Theriot et al., 1992). This indicates that new actin filaments polymerize at the bacterial surface at the same rate as bacterial movement, and implies that the force for movement is derived from actin polymerization, as it is for other pathogens.

Bacterial Proteins That May Function in Actin-Based Motility

Given the complexity of actin-based motility, it came as no surprise that the process requires a contribution of protein components from both the bacterium and its host cell. The first direct evidence for a bacterial protein requirement came from Heinzen and coworkers, who observed that treatment of infected cells with chloramphenicol, an inhibitor of bacterial protein synthesis, abolished actin polymerization by *R. rickettsii* (Heinzen et al., 1993). An effect of chloramphenicol treatment was apparent after 3 h, and actin tails disappeared after 14 h, indicating that continued synthesis of one or more bacterial proteins is required for actin polymerization by rickettsiae. However, it was many years before specific bacterial proteins that may play a role in this process were identified. These proteins include RickA and Sca2 (Table 2), and the possible role of each in actin-based motility is discussed below.

RickA

RickA was first annotated in the proteome of *R. conorii* based on its sequence similarity with host actin-nucleating proteins in the Wiskott-Aldrich syndrome protein (WASp) family (Ogata et al., 2001). It is also encoded in the genomes of other SFGR species including *R. rickettsii*, *R. montanensis*, *Rickettsia sibirica* (Jeng et al., 2004), *R. monacencis* (Baldridge et al., 2005), and *Rickettsia raoultii* (Balraj et al., 2008a), as well as the transitional group rickettsiae species *R. felis* (Ogata et al., 2005) and the AGR species *R. bellii* (Ogata et al., 2006) and

TABLE 2 Bacterial factors implicated in actin-based motility

Protein	Gene names	Functional evidence for role in motility
RickA	*R. conorii*: *rickA/RC0909* *R. rickettsii*: *rickA/RRIowa1080*	Promotes actin nucleation with host Arp2/3 complex (Gouin et al., 2004; Jeng et al., 2004); conflicting evidence against a role in motility (see text)
Sca2	*R. rickettsii*: *sca2/RRIowa0143* *R. parkeri*: *sca2* *R. typhi*: *sca2/RT0052* *R. prowazekii*: *sca2/RP0081* *R. bellii*: *sca2*	Transposon insertion prevents actin assembly and motility (Kleba et al., 2010); Sca2 nucleates and elongates actin filaments (Haglund et al., 2011)

Rickettsia canadensis (Balraj et al., 2008b), suggesting it arose early in rickettsial evolution. Additionally, RickA expression has been experimentally verified by Western blotting of crude extracts from *Rickettsia africae, R. conorii, R. felis, R. massiliae, R. montanensis, R. rickettsii, R. sibirica*, and *Rickettsia slovaca* (Balraj et al., 2008b). *R. canadensis* also expresses RickA as determined by Western blotting, but RickA was not detected in *R. bellii*, perhaps because of sequence divergence (Balraj et al., 2008b). RickA is not encoded in the genomes of the TGR species *R. prowazekii* (Ogata et al., 2001) and *R. typhi* (McLeod et al., 2004), suggesting it was lost during the evolution of the TGR lineage. The case of *R. typhi* is of particular interest because this species undergoes actin-based motility, indicating that other rickettsial proteins must play a role in this process.

RickA contains several domains that are similar to those in host WASp family proteins (Gouin et al., 2004; Jeng et al., 2004). These include a variable-length central proline-rich region that in WASp binds to the host G-actin-binding protein profilin, one or two WASp homology 2 (W2 or W) regions that bind to G actin, and a C-terminal central and acidic (CA) region that binds to the actin-nucleating Arp2/3 complex. The combined WCA domain of host WASp family proteins activates the Arp2/3 complex, which in turn nucleates actin assembly and organizes actin into Y-branched networks at the cell cortex during processes such as membrane protrusion, phagocytosis, and endocytosis (Campellone and Welch, 2010). The sequence similarity with WASp family proteins implied that RickA was a bacterial activator of the Arp2/3 complex. This was confirmed by testing the activity of purified recombinant *R. conorii* and *R. rickettsii* RickA, which were shown to promote Arp2/3-mediated actin nucleation and the formation of Y-branched filament networks in vitro (Gouin et al., 2004; Jeng et al., 2004).

Several lines of evidence suggest that RickA has the potential to be involved in rickettsial actin-based motility. First, other bacterial pathogens such as *L. monocytogenes* and *S. flexneri* activate the host Arp2/3 complex at their surface by expressing mimics of host WASp proteins (*L. monocytogenes* ActA [Welch et al., 1998; Skoble et al., 2000]) or by recruiting the host protein neural WASp (N-WASp) (*S. flexneri* via the bacterial protein IcsA/VirG [Suzuki et al., 1998; Egile et al., 1999]). This results in the formation of actin tails consisting of a Y-branched actin network that contains the Arp2/3 complex (Welch et al., 1997; Cameron et al., 2001). Second, 0.5-μm-diameter plastic beads coated with the RickA protein can undergo actin-based motility when placed into cytoplasmic extracts, resulting in the formation of actin tails that are also organized into Y-branches (Jeng et al., 2004). Thus, RickA is competent to direct actin-based motility. Third, overexpression of the dominant-active WCA domain of the WASp protein WASp-family verprolin homolog (WAVE) or of RickA inhibited actin tail formation by *R. conorii* in infected cells, likely by binding to and sequestering Arp2/3 complex (Gouin et al., 2004). Fourth, the *rickA* gene in the SFGR species *R. peacockii* is disrupted by an ISRpe1 transposon insertion mutation, and *R. peacockii* does not associate with actin or undergo actin-based motility (Simser et al., 2005). The latter two lines of evidence suggest a potential functional role for RickA in actin-based motility.

On the other hand, multiple lines of evidence call into question a role for RickA in motility. First, the unbranched organization of actin filaments in rickettsial actin tails (Gouin et al., 1999; Van Kirk et al., 2000) is inconsistent with the known propensity of RickA and the Arp2/3 complex to assemble Y-branched filaments (Gouin et al., 2004; Jeng et al., 2004). Second, the Arp2/3 complex is not generally observed in rickettsial actin tails (Gouin et al., 1999; Harlander et al., 2003; Heinzen, 2003; Serio et al., 2010), although an association of the Arp2/3 complex subunit Arp3 with *R. conorii* has been reported (Gouin et al., 2004). This contrasts with the robust and consistent localization of Arp2/3 complex to actin tails formed by *L. monocytogenes* and *S. flexneri* (Welch et al., 1997; Egile et al., 1999). Third, a different study from that cited above observed that overexpression of the dominant-active WCA domain of the WASp protein N-WASp had less of an effect on *R. rickettsii* motility than on Arp2/3-dependent *S. flexneri* motility (Harlander et al., 2003). Fourth, silencing of Arp2/3 complex expression by RNA interference (RNAi) in *Drosophila melanogaster* S2 cells had no dramatic effect on the formation of actin tails by *R. parkeri* (Serio et al., 2010). Finally, there is not a perfect correlation between the presence of the *rickA* gene in rickettsial genomes or the expression of the RickA protein and the ability of the species to undergo actin-based motility. For example, *R. canadensis* expresses RickA (Balraj et al., 2008b) yet does not undergo actin-based motility (Heinzen et al., 1993), and the SFGR species *R. raoultii* expresses RickA in diverse host cell types yet only undergoes motility in some (Balraj et al., 2008b). Conversely, the TGR species *R. typhi* does not have the *rickA* gene and yet does undergo actin-based motility, albeit with different characteristics than SFGR species. Together these findings suggest that RickA may contribute to or enhance rickettsial actin-based motility, but may not play an essential role. For example, it has been proposed that RickA and the Arp2/3 complex may function at an early stage in motility to generate a transient Y-branched actin network (Jeng et al., 2004). Nucleation by RickA and the Arp2/3 complex may then diminish, but the filaments in the network continue to elongate, resulting in the conversion of the branched network into the bundled array observed in rickettsial actin tails.

Beyond a role in actin-based motility, several questions remain about RickA function during rickettsial infection. The first concerns the mechanism of RickA secretion and surface localization. RickA lacks a canonical Sec-dependent signal sequence or transmembrane domain that would indicate a mode of secretion or tethering to the bacterial surface. It has been suggested that RickA might be secreted by the type IV secretion system (Gouin et al., 2004), but this remains speculative, and a potential mechanism of surface localization and tethering after secretion has not been proposed. Another question is whether RickA may function in processes other than motility. For example, RickA may be secreted during the entry process, which depends on Arp2/3 complex-mediated actin nucleation (Martinez and Cossart, 2004). More work is needed to understand the role of RickA during infection.

Sca2

Surface cell antigen 2 (Sca2) was first annotated as a protein encoded in the *R. prowazekii* genome (Andersson et al., 1998) that contains an autotransporter domain similar to those found in rOmpA and rOmpB, indicating that it is localized to the outer membrane. As with other autotransporter proteins, Sca2 also has a passenger domain that is predicted to be exposed on the bacterial surface. Genes encoding Sca2 have been identified in most *Rickettsia* species, indicating that it is highly conserved in their evolution (Blanc et al., 2005; Ngwamidiba et al., 2005). Sca2 has been implicated in host cell invasion (discussed in the preceding chapter), but its primary function appears to be in actin assembly during actin-based motility.

The first evidence that Sca2 functions in actin-based motility emerged from a random transposon mutagenesis screen of *R. rickettsii*, in which the *sca2* gene was disrupted by a transposon insertion, causing a small plaque phenotype (Kleba et al., 2010). The difference in plaque size was not due to a growth defect, because the *sca2* mutant replicated at the same rate as wild-type *R. rickettsii* in Vero cells. However, the *sca2* mutant did not make actin tails, as observed by light microscopy at 24 h postinfection. Moreover, the mutant bacteria were distributed in small infection foci both within each cell and within the population of host cells, in contrast to the dispersed localization of wild-type bacteria, suggesting a defect in both intra- and intercellular spread. Importantly, the *sca2* mutant caused seroconversion in guinea pigs but did not induce fever, indicating that Sca2 is a virulence factor. These experiments demonstrated that motility was disrupted in the absence of Sca2, but did not address whether Sca2 directly or indirectly affected actin assembly.

A clue to the potential role of Sca2 in actin assembly came from examining its amino acid sequence. Notably, Sca2 contains sequence motifs resembling the WH2 domains found in RickA, WASp, and other actin-nucleating proteins (Haglund et al., 2010; Kleba et al., 2010), suggesting that it might bind to actin monomers. Sca2 also contains two proline-rich regions that are similar to a domain found in a class of eukaryotic actin-nucleating proteins called formins. The proline-rich domain of formins, called a formin homology 1 (FH1) domain, facilitates actin filament assembly by binding to profilin, a small G-actin-binding protein. Finally, the N-terminal domain of Sca2, which is highly conserved in SFGR species, has a predicted secondary structure that is similar to the core formin homology 2 (FH2) domain, which is necessary and sufficient for actin nucleation and elongation. Formin proteins are distinguished by their ability to both nucleate actin filaments and then remain processively associated with barbed (or +) ends as they elongate, which can either increase or decrease the elongation rate, depending on the formin. Formins also protect barbed ends from termination by host capping proteins, enabling the growth of long, linear actin filaments. The unbranched organization and length of actin filaments in rickettsial tails are consistent with the notion that they might be assembled by a forminlike protein.

To determine whether Sca2 directly influences actin assembly, its biochemical activity was investigated using purified recombinant *R. parkeri* Sca2 in vitro (Haglund et al., 2010). Sca2 was found to be functionally similar to eukaryotic formins in its ability to nucleate unbranched actin filaments and to processively associate with growing barbed ends, protecting them from capping proteins. Moreover, Sca2 mimicked formin proteins in its requirement for profilin during filament elongation. In the presence of profilin, the elongation rate of Sca2-bound actin filaments was 40% of the rate for free actin filaments. This slower assembly rate correlates with the slower motility rate of SFGR species compared with *L. monocytogenes* in host cells. Furthermore, Sca2 is present at actin-associated surfaces of *R. parkeri* in infected cells, as detected by fluorescence microscopy. Thus, the localization and biochemical activity of Sca2 are consistent with a direct role in actin assembly to drive the motility of SFGR species.

While Sca2 is necessary for *R. rickettsii* actin-based motility and virulence, it remains unclear if Sca2 is sufficient to assemble actin during rickettsial motility, or if other bacterial factors such as RickA are also required. Moreover, the sequence of Sca2 from SFGR, TGR, and AGR species is quite divergent, suggesting potential differences in the mechanism by which it contributes to actin assembly and motility. In the TGR species *R. prowazekii*, which is unable to assemble actin, the predicted Sca2 protein is missing its entire surface-exposed passenger domain. In the TGR species *R. typhi* and the AGR species *R. bellii*, Sca2 is missing the FH2-like N-terminal domain and is divergent in the remainder of the sequence, offering a potential explanation for differences in the motility characteristics and actin tail appearance for these species. It remains unclear whether Sca2 from TGR and AGR species promotes actin assembly, and if so whether it employs a forminlike or an alternative mechanism. Future examination of the activity, function, and molecular interactions of Sca2 proteins from divergent species is likely to shed light on the evolution of actin-based motility in rickettsiae.

Host Proteins That Function in Motility

In addition to the contribution of bacterial proteins such as RickA and Sca2, it is clear that host actin cytoskeleton proteins are key contributors to the process of motility (Table 3). Initial identification of host proteins that may contribute to this process was based on a

TABLE 3 Host cell factors implicated in actin-based motility

Protein(s)	Protein function(s)	Functional evidence for role in motility
α-Actinin, filamin	Actin filament bundling and cross-linking	Localized to *Rickettsia* actin tails (Gouin et al., 1999; Van Kirk et al., 2000)
VASP	Promotes actin filament elongation	Localized to *Rickettsia* actin tails (Gouin et al., 1999; Van Kirk et al., 2000)
Vinculin	Actin binding and cell adhesion in focal adhesions	Localized to *Rickettsia* actin tails (Van Kirk et al., 2000)
Profilin	Promotes actin filament elongation, sequesters actin, and suppresses nucleation	RNAi silencing affects motility (Serio et al., 2010), localized to *Rickettsia* actin tails (Serio et al., 2010; Van Kirk et al., 2000)
CapZ	Caps actin filament barbed ends	RNAi silencing affects motility, localized to *Rickettsia* actin tails (Serio et al., 2010)
Fimbrin/T-plastin	Actin filament bundling	RNAi silencing affects motility, localized to *Rickettsia* actin tails (Serio et al., 2010)

comparison of the localization of a handful of cytoskeletal proteins in *R. conorii* and *R. rickettsii* versus *L. monocytogenes* actin tails (Gouin et al., 1999; Van Kirk et al., 2000). The actin polymerization protein vasodilator-stimulated phosphoprotein (VASP), the actin filament-bundling/cross-linking proteins α-actinin and filamin, the G-actin-binding protein profilin, and the focal adhesion protein vinculin were localized to tails from both *Rickettsia* species and *L. monocytogenes*. Other proteins such as the membrane-cytoskeleton linker ezrin, the Arp2/3 complex, the actin-severing protein cofilin, and the barbed end-capping protein CapZ were detected only in *L. monocytogenes* tails. Although these studies suggested a possible role for select cytoskeletal proteins in rickettsial motility, the functional importance of each of these proteins remained undetermined for some time.

A recent study by Serio and coworkers reported a comprehensive RNAi screening approach to identify host actin cytoskeletal proteins that function in motility (Serio et al., 2010). The approach took advantage of the ability to infect *Drosophila* S2 cells with the SFGR species *R. parkeri* and to observe robust motility and actin tail formation in these insect cells. Because S2 cells are amenable to RNAi, their use presented an advantage for the screening approach. Of the 115 known cytoskeletal genes included in the screen, targeting 20 caused a reduction in actin tail length and targeting 7 caused an increase in tail length compared with controls, indicating that numerous host proteins contribute to actin-based motility.

Of these 27 proteins, 4 were identified as a core set of proteins that have a direct and important function in motility. These included the G-actin-binding protein profilin, the F-actin-bundling protein fimbrin/T-plastin, the barbed end-capping protein CapZ, and the F-actin-severing and -depolymerizing protein cofilin. Two of these proteins, CapZ and cofilin, were among the minimal set of purified proteins required to reconstitute *L. monocytogenes* motility in vitro (Loisel et al., 1999), suggesting that there is some commonality in the cytoskeletal machinery required for pathogen actin-based motility. However, RNAi silencing of the other core proteins, profilin and fimbrin/T-plastin, in mammalian cells resulted in a specific defect in *R. parkeri* but not *L. monocytogenes* actin tail formation and motility, suggesting that these proteins play a more specific role in rickettsial motility. Of the core proteins, silencing of profilin also caused the most severe phenotype, characterized by a dramatic reduction in the motility rate. This is consistent with the biochemical function of profilin in promoting actin elongation by the Sca2 protein in vitro (Haglund et al., 2010), as discussed above.

Despite the recent identification of several host proteins that play an important functional role in rickettsial motility, numerous questions remain unanswered. First, unlike with *L. monocytogenes* and *S. flexneri*, rickettsial motility has not yet been reconstituted with purified proteins in vitro, so it remains unclear which host proteins are sufficient to enable this process. Second, it is unclear whether proteins other than those mentioned above may also be important for motility in cells. Future approaches including reconstitution of motility and genome-wide RNAi screening will be required to fully define the host cell contributions to the motility process.

Intracellular Motility in Cell-to-Cell Spread and Pathogenesis

Unlike the TGR species, which spread primarily following lysis of the host cell, SFGR species initiate the process of spread prior to host cell lysis, suggesting that early release and/or direct cell-to-cell transfer take place. The first evidence of a role for intracellular movement in cell-to-cell spread came from imaging living cells infected with the SFGR species *R. rickettsii* (Schaechter et al., 1957). Moving bacteria were observed to enter into cell surface protrusions and were occasionally released into the extracellular medium. Subsequent work established that bacterial release was mediated by actin-based motility, as release was dramatically inhibited by treatment of *R. rickettsii*-infected cells with the drug cytochalasin D, an inhibitor of actin polymerization (Heinzen et al., 1993). However, these studies did not elucidate the mechanism of cell-to-cell transfer.

Understanding the mechanism by which motility contributes to rickettsial cell-to-cell spread was aided by previous studies of other bacterial pathogens. In particular, *L. monocytogenes* was shown to move into cell surface protrusions that were then internalized by neighboring cells, a process that was followed by escape from the phagosome and another cycle of bacterial growth (Tilney and Portnoy, 1989). Careful analysis of the interaction of *R. conorii* and *R. rickettsii* with the host cell plasma membrane and adjacent cells suggests a similar picture of cell-to-cell spread by SFGR species (Fig. 4). In particular, bacteria were observed to interact with the host cell plasma membrane to form short protrusions that contacted neighboring cells (Gouin et al., 1999; Van Kirk et al., 2000). Moreover, bacteria presumed to be in the process of transfer to an adjacent cell were observed in phagosomes bounded by a double membrane, which could only result from the internalization of a bacterium surrounded by the plasma membrane of another host cell (Gouin et al., 1999). Despite this compelling evidence, actin-based movement resulting in cell-to-cell spread has not yet been observed in

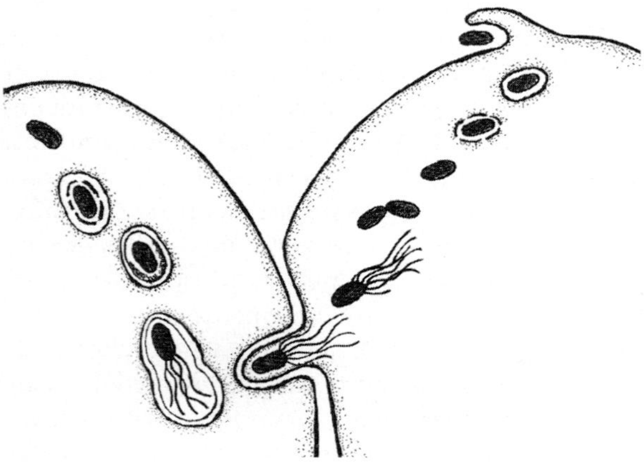

FIGURE 4 The life cycle of SFGR observed from invasion through cell-to-cell spread. From right to left, the stages include adhesion and engulfment, inclusion in a phagosome, phagosome lysis, release into the cytosol, bacterial growth, actin-based motility, movement into a protrusion, engulfment by a neighboring cell, inclusion in a double-membrane vacuole, vacuole lysis, and release into the cytosol of the second cell. (Artwork by Taro Ohkawa.)
doi:10.1128/9781555817336.ch5.f4

live cells in real time, so the precise frequency and timing of the events remains elusive.

More direct evidence for a role for actin-based motility in cell-to-cell spread came from an analysis of the phenotype of the *sca2* mutant, which is defective in motility (Kleba et al., 2010). This mutant exhibits a small plaque size and accumulates to higher numbers in individual infected cells, but exhibits no defect in bacterial replication rate, which strongly suggests that the mutant is defective in cell-to-cell spread. Moreover, the *sca2* mutant is avirulent in infected guinea pigs, suggesting that actin-based motility and cell-to-cell spread play a crucial role in pathogenesis.

Future studies will be needed to examine the precise mechanism of cell-to-cell spread and the role of bacterial and host proteins in this process. In particular, it will be important to visualize the process of spread in real time to gain insight into the frequency of events like protrusion formation, protrusion engulfment by the receiving cell, and escape from the phagosome of the receiving cell. Alternative mechanisms of spread, such as release and re-infection, might also be observed. Moreover, it will be important to determine whether other bacterial and host cell factors beyond those involved in actin-based motility play a role in cell-to-cell transmission. Such factors might include those with roles in cell-cell adhesion, exchange of cellular membranes, and phagocytosis. Because the mechanism of cell-to-cell spread is likely to be shared with other pathogens that use actin-based motility for this purpose, progress in understanding how this process works for rickettsiae may be of general importance for elucidating conserved mechanisms of pathogenesis.

CONCLUSIONS

Despite more than 100 years of study, how rickettsiae establish intracellular infection remains largely mysterious. Although numerous bacterial factors have been identified that may function in escape from the phagosome, establishment of an intracellular niche, actin-based motility, and cell-to-cell spread, the difficulty of genetically manipulating rickettsiae has remained a major barrier to defining the role of each protein in infection. Fortunately, genetic tools from transposon insertion mutagenesis (Qin et al., 2004; Liu et al., 2007; Kleba et al., 2010) to directed mutagenesis (Driskell et al., 2009) to plasmids (Baldridge et al., 2010) are now being developed (as discussed in chapter 14), which should open the floodgates for new progress in the area of rickettsial microbiology. Given the obligate intracellular growth requirement of rickettsiae, host cell factors also play a key role in rickettsial infection. Newer host genetic techniques such as RNAi screening are now being employed to dissect the host protein requirements for intracellular life (Serio et al., 2010). The combination of these newly available methods, along with the continued influx of new ideas from the related fields of immunology and cell biology, should usher in a new era of discovery in understanding the interaction of rickettsiae with their host cells.

ACKNOWLEDGMENTS

We thank Taro Ohkawa for his artistic talent and for drawing the artwork in the four figures. Work on rickettsiae in the Welch laboratory is supported by NIH/NIAID grant AI074760.

REFERENCES

Anderson, D. R., H. E. Hopps, M. F. Barile, and B. C. Bernheim. 1965. Comparison of the ultrastructure of several rickettsiae, ornithosis virus, and *Mycoplasma* in tissue culture. *J. Bacteriol.* **90:**1387–1404.

Andersson, S. G., A. Zomorodipour, J. O. Andersson, T. Sicheritz-Ponten, U. C. Alsmark, R. M. Podowski, A. K. Naslund, A. S. Eriksson, H. H. Winkler, and C. G. Kurland. 1998. The genome sequence of *Rickettsia prowazekii* and the origin of mitochondria. *Nature* **396:**133–140.

Baldridge, G. D., N. Burkhardt, M. J. Herron, T. J. Kurtti, and U. G. Munderloh. 2005. Analysis of fluorescent protein expression in transformants of *Rickettsia monacensis*, an obligate intracellular tick symbiont. *Appl. Environ. Microbiol.* **71:**2095–2105.

Baldridge, G. D., N. Y. Burkhardt, M. B. Labruna, R. C. Pacheco, C. D. Paddock, P. C. Williamson, P. M. Billingsley, R. F. Felsheim, T. J. Kurtti, and U. G. Munderloh.

2010. Wide dispersal and possible multiple origins of low-copy-number plasmids in *Rickettsia* species associated with blood-feeding arthropods. *Appl. Environ. Microbiol.* **76:**1718–1731.

Balraj, P., K. El Karkouri, G. Vestris, L. Espinosa, D. Raoult, and P. Renesto. 2008a. RickA expression is not sufficient to promote actin-based motility of *Rickettsia raoultii*. *PLoS One* **3:**e2582.

Balraj, P., C. Nappez, D. Raoult, and P. Renesto. 2008b. Western-blot detection of RickA within spotted fever group rickettsiae using a specific monoclonal antibody. *FEMS Microbiol. Lett.* **286:**257–262.

Bernardini, M. L., J. Mounier, H. d'Hauteville, M. Coquis-Rondon, and P. J. Sansonetti. 1989. Identification of *icsA*, a plasmid locus of *Shigella flexneri* that governs bacterial intra- and intercellular spread through interaction with F-actin. *Proc. Natl. Acad. Sci. USA* **86:**3867–3871.

Blanc, G., M. Ngwamidiba, H. Ogata, P. E. Fournier, J. M. Claverie, and D. Raoult. 2005. Molecular evolution of *Rickettsia* surface antigens: evidence of positive selection. *Mol. Biol. Evol.* **22**:2073–2083.

Cameron, L. A., T. M. Svitkina, D. Vignjevic, J. A. Theriot, and G. G. Borisy. 2001. Dendritic organization of actin comet tails. *Curr. Biol.* **11:**130–135.

Campellone, K. G., and M. D. Welch. 2010. A nucleator arms race: cellular control of actin assembly. *Nat. Rev. Mol. Cell Biol.* **11:**237–251.

Cheville, N. F. 1994. *Ultrastructural Pathology: an Introduction to Interpretation*. Iowa State University Press, Ames, IA.

Clarke, D. H., and J. P. Fox. 1948. The phenomenon of in vitro hemolysis produced by the rickettsiae of typhus fever, with a note on the mechanism of rickettsial toxicity in mice. *J. Exp. Med.* **88:**25–41.

Cohn, Z. A., F. M. Bozeman, J. M. Campbell, J. W. Humphries, and T. K. Sawyer. 1959. Study on growth of *Rickettsia*. V. Penetration of *Rickettsia tsutsugamushi* into mammalian cells in vitro. *J. Exp. Med.* **109:**271–292.

Driskell, L. O., X.-J. Yu, L. Zhang, Y. Liu, V. L. Popov, D. H. Walker, A. M. Tucker, and D. O. Wood. 2009. Directed mutagenesis of the *Rickettsia prowazekii pld* gene encoding phospholipase D. *Infect. Immun.* **77:**3244–3248.

Egile, C., T. P. Loisel, V. Laurent, R. Li, D. Pantaloni, P. J. Sansonetti, and M. F. Carlier. 1999. Activation of the CDC42 effector N-WASP by the *Shigella flexneri* IcsA protein promotes actin nucleation by Arp2/3 complex and bacterial actin-based motility. *J. Cell Biol.* **146:**1319–1332.

Eremeeva, M. E., and D. J. Silverman. 1998. *Rickettsia rickettsii* infection of the EA.hy 926 endothelial cell line: morphological response to infection and evidence for oxidative injury. *Microbiology* **144:**2037–2048.

Firat-Karalar, E. N., and M. D. Welch. 2011. New mechanisms and functions of actin nucleation. *Curr. Opin. Cell Biol.* **23:**4–13.

Gouin, E., C. Egile, P. Dehoux, V. Villiers, J. Adams, F. Gertler, R. Li, and P. Cossart. 2004. The RickA protein of *Rickettsia conorii* activates the Arp2/3 complex. *Nature* **427:**457–461.

Gouin, E., H. Gantelet, C. Egile, I. Lasa, H. Ohayon, V. Villiers, P. Gounon, P. J. Sansonetti, and P. Cossart. 1999. A comparative study of the actin-based motilities of the pathogenic bacteria *Listeria monocytogenes*, *Shigella flexneri* and *Rickettsia conorii*. *J. Cell Sci.* **112:**1697–1708.

Haglund, C. M., J. E. Choe, C. T. Skau, D. R. Kovar, and M. D. Welch. 2010. *Rickettsia* Sca2 is a bacterial formin-like mediator of actin-based motility. *Nat. Cell Biol.* **12:**1057–1063.

Harlander, R. S., M. Way, Q. Ren, D. Howe, S. S. Grieshaber, and R. A. Heinzen. 2003. Effects of ectopically expressed neuronal Wiskott-Aldrich syndrome protein domains on *Rickettsia rickettsii* actin-based motility. *Infect. Immun.* **71:** 1551–1556.

Heinzen, R. A. 2003. Rickettsial actin-based motility: behavior and involvement of cytoskeletal regulators. *Ann. N. Y. Acad. Sci.* **990:**535–547.

Heinzen, R. A., S. S. Grieshaber, L. S. Van Kirk, and C. J. Devin. 1999. Dynamics of actin-based movement by *Rickettsia rickettsii* in Vero cells. *Infect. Immun.* **67:**4201–4207.

Heinzen, R. A., S. F. Hayes, M. G. Peacock, and T. Hackstadt. 1993. Directional actin polymerization associated with spotted fever group *Rickettsia* infection of Vero cells. *Infect. Immun.* **61:**1926–1935.

Jeng, R. L., E. D. Goley, J. A. D'Alessio, O. Y. Chaga, T. M. Svitkina, G. G. Borisy, R. A. Heinzen, and M. D. Welch. 2004. A *Rickettsia* WASP-like protein activates the Arp2/3 complex and mediates actin-based motility. *Cell. Microbiol.* **6:**761–769.

Kespichayawattana, W., S. Rattanachetkul, T. Wanun, P. Utaisincharoen, and S. Sirisinha. 2000. *Burkholderia pseudomallei* induces cell fusion and actin-associated membrane protrusion: a possible mechanism for cell-to-cell spreading. *Infect. Immun.* **68:**5377–5384.

Kleba, B., T. R. Clark, E. I. Lutter, D. W. Ellison, and T. Hackstadt. 2010. Disruption of the *Rickettsia rickettsii* Sca2 autotransporter inhibits actin-based motility. *Infect. Immun.* **78:**2240–2247.

Liu, Z. M., A. M. Tucker, L. O. Driskell, and D. O. Wood. 2007. Mariner-based transposon mutagenesis of *Rickettsia prowazekii*. *Appl. Environ. Microbiol.* **73:**6644–6649.

Loisel, T. P., R. Boujemaa, D. Pantaloni, and M. F. Carlier. 1999. Reconstitution of actin-based motility of *Listeria* and *Shigella* using pure proteins. *Nature* **401:**613–616.

Martinez, J. J., and P. Cossart. 2004. Early signaling events involved in the entry of *Rickettsia conorii* into mammalian cells. *J. Cell Sci.* **117:**5097–5106.

McLeod, M. P., X. Qin, S. E. Karpathy, J. Gioia, S. K. Highlander, G. E. Fox, T. Z. McNeill, H. Jiang, D. Muzny, L. S. Jacob, A. C. Hawes, E. Sodergren, R. Gill, J. Hume, M. Morgan, G. Fan, A. G. Amin, R. A. Gibbs, C. Hong, X. J. Yu, D. H. Walker, and G. M. Weinstock. 2004. Complete genome sequence of *Rickettsia typhi* and comparison with sequences of other rickettsiae. *J. Bacteriol.* **186:**5842–5855.

Ngwamidiba, M., G. Blanc, H. Ogata, D. Raoult, and P. E. Fournier. 2005. Phylogenetic study of *Rickettsia* species using sequences of the autotransporter protein-encoding gene *sca2*. *Ann. N. Y. Acad. Sci.* **1063:**94–99.

Ogata, H., S. Audic, P. Renesto-Audiffren, P. E. Fournier, V. Barbe, D. Samson, V. Roux, P. Cossart, J. Weissenbach, J. M. Claverie, and D. Raoult. 2001. Mechanisms of evolution in *Rickettsia conorii* and *R. prowazekii*. *Science* **293:**2093–2098.

Ogata, H., B. La Scola, S. Audic, P. Renesto, G. Blanc, C. Robert, P. E. Fournier, J. M. Claverie, and D. Raoult. 2006. Genome sequence of *Rickettsia bellii* illuminates the role of amoebae in gene exchanges between intracellular pathogens. *PLoS Genet.* **2:**e76.

Ogata, H., P. Renesto, S. Audic, C. Robert, G. Blanc, P. E. Fournier, H. Parinello, J. M. Claverie, and D. Raoult. 2005. The genome sequence of *Rickettsia felis* identifies the first putative conjugative plasmid in an obligate intracellular parasite. *PLoS Biol.* **3:**e248.

Ojcius, D. M., M. Thibon, C. Mounier, and A. Dautry-Varsat. 1995. pH and calcium dependence of hemolysis due to *Rickettsia prowazekii*: comparison with phospholipase activity. *Infect. Immun.* **63:**3069–3072.

Pinkerton, H., and G. M. Hass. 1932. Spotted fever: I. Intranuclear rickettsiae in spotted fever studied in tissue culture. *J. Exp. Med.* **56:**151–156.

Qin, A., A. M. Tucker, A. Hines, and D. O. Wood. 2004. Transposon mutagenesis of the obligate intracellular pathogen *Rickettsia prowazekii*. *Appl. Environ. Microbiol.* **70:**2816–2822.

Radulovic, S., J. M. Troyer, M. S. Beier, A. O. Lau, and A. F. Azad. 1999. Identification and molecular analysis of the gene encoding *Rickettsia typhi* hemolysin. *Infect. Immun.* **67:**6104–6108.

Rahman, M. S., N. C. Ammerman, K. T. Sears, S. M. Ceraul, and A. F. Azad. 2010. Functional characterization of a phospholipase A_2 homolog from *Rickettsia typhi*. *J. Bacteriol.* **192:**3294–3303.

Ramm, L. E., and H. H. Winkler. 1973a. Rickettsial hemolysis: adsorption of rickettsiae to erythrocytes. *Infect. Immun.* **7:**93–99.

Ramm, L. E., and H. H. Winkler. 1973b. Rickettsial hemolysis: effect of metabolic inhibitors upon hemolysis and adsorption. *Infect. Immun.* **7:**550–555.

Ramm, L. E., and H. H. Winkler. 1976. Identification of cholesterol in the receptor site for rickettsiae on sheep erythrocyte membranes. *Infect. Immun.* **13:**120–126.

Ray, K., B. Marteyn, P. J. Sansonetti, and C. M. Tang. 2009. Life on the inside: the intracellular lifestyle of cytosolic bacteria. *Nat. Rev. Microbiol.* **7:**333–340.

Renesto, P., P. Dehoux, E. Gouin, L. Touqui, P. Cossart, and D. Raoult. 2003. Identification and characterization of a phospholipase D-superfamily gene in rickettsiae. *J. Infect. Dis.* **188:**1276–1283.

Robbins, J. R., and J. A. Theriot. 2003. *Listeria monocytogenes* rotates around its long axis during actin-based motility. *Curr. Biol.* **13:**R754–R756.

Schaechter, M., F. M. Bozeman, and J. E. Smadel. 1957. Study on the growth of rickettsiae. II. Morphologic observations of living rickettsiae in tissue culture cells. *Virology* **3:**160–172.

Serio, A. W., R. L. Jeng, C. M. Haglund, S. C. Reed, and M. D. Welch. 2010. Defining a core set of actin cytoskeletal proteins critical for actin-based motility of *Rickettsia*. *Cell Host Microbe* **7:**388–398.

Silverman, D. J. 1984. *Rickettsia rickettsii*-induced cellular injury of human vascular endothelium in vitro. *Infect. Immun.* **44:**545–553.

Silverman, D. J., and L. A. Santucci. 1988. Potential for free radical-induced lipid peroxidation as a cause of endothelial cell injury in Rocky Mountain spotted fever. *Infect. Immun.* **56:**3110–3115.

Silverman, D. J., and L. A. Santucci. 1990. A potential protective role for thiols against cell injury caused by *Rickettsia rickettsii*. *Ann. N. Y. Acad. Sci.* **590:**111–117.

Silverman, D. J., L. A. Santucci, N. Meyers, and Z. Sekeyova. 1992. Penetration of host cells by *Rickettsia rickettsii* appears to be mediated by a phospholipase of rickettsial origin. *Infect. Immun.* **60:**2733–2740.

Silverman, D. J., and C. L. Wisseman, Jr. 1979. In vitro studies of rickettsia-host cell interactions: ultrastructural changes induced by *Rickettsia rickettsii* infection of chicken embryo fibroblasts. *Infect. Immun.* **26:**714–727.

Silverman, D. J., C. L. Wisseman, Jr., and A. Waddell. 1980. In vitro studies of rickettsia-host cell interactions: ultrastructural study of *Rickettsia prowazekii*-infected chicken embryo fibroblasts. *Infect. Immun.* **29:**778–790.

Simser, J. A., A. T. Palmer, V. Fingerle, B. Wilske, T. J. Kurtti, and U. G. Munderloh. 2002. *Rickettsia monacensis* sp. nov., a spotted fever froup *Rickettsia*, from ticks (*Ixodes ricinus*) collected in a European city park. *Appl. Environ. Microbiol.* **68:**4559–4566.

Simser, J. A., M. S. Rahman, S. M. Dreher-Lesnick, and A. F. Azad. 2005. A novel and naturally occurring transposon, ISRpe1 in the *Rickettsia peacockii* genome disrupting the *rickA* gene involved in actin-based motility. *Mol. Microbiol.* **58:**71–79.

Skoble, J., D. A. Portnoy, and M. D. Welch. 2000. Three regions within ActA promote Arp2/3 complex-mediated actin nucleation and *Listeria monocytogenes* motility. *J. Cell Biol.* **150:**527–538.

Stamm, L. M., J. H. Morisaki, L. Y. Gao, R. L. Jeng, K. L. McDonald, R. Roth, S. Takeshita, J. Heuser, M. D. Welch, and E. J. Brown. 2003. *Mycobacterium marinum* escapes from phagosomes and is propelled by actin-based motility. *J. Exp. Med.* **198:**1361–1368.

Suzuki, T., H. Miki, T. Takenawa, and C. Sasakawa. 1998. Neural Wiskott-Aldrich syndrome protein is implicated in the actin-based motility of *Shigella flexneri*. *EMBO J.* **17:**2767–2776.

Teysseire, N., J. A. Boudier, and D. Raoult. 1995. *Rickettsia conorii* entry into Vero cells. *Infect. Immun.* **63:**366–374.

Teysseire, N., C. Chiche-Portiche, and D. Raoult. 1992. Intracellular movements of *Rickettsia conorii* and *R. typhi* based on actin polymerization. *Res. Microbiol.* **143:**821–829.

Theriot, J. A., T. J. Mitchison, L. G. Tilney, and D. A. Portnoy. 1992. The rate of actin-based motility of intracellular *Listeria monocytogenes* equals the rate of actin polymerization. *Nature* **357:**257–260.

Tilney, L. G., and D. A. Portnoy. 1989. Actin filaments and the growth, movement, and spread of the intracellular bacterial parasite, *Listeria monocytogenes*. *J. Cell Biol.* **109:**1597–1608.

Van Kirk, L. S., S. F. Hayes, and R. A. Heinzen. 2000. Ultrastructure of *Rickettsia rickettsii* actin tails and localization of cytoskeletal proteins. *Infect. Immun.* **68:**4706–4713.

Walker, D. H., and B. G. Cain. 1980. The rickettsial plaque. Evidence for direct cytopathic effect of *Rickettsia rickettsii*. *Lab. Invest.* **43:**388–396.

Walker, D. H., W. T. Firth, J. G. Ballard, and B. C. Hegarty. 1983. Role of phospholipase-associated penetration mechanism in cell injury by *Rickettsia rickettsii*. *Infect. Immun.* **40:**840–842.

Walker, D. H., A. Harrison, F. Henderson, and F. A. Murphy. 1977. Identification of *Rickettsia rickettsii* in a guinea pig model by immunofluorescent and electron microscopic techniques. *Am. J. Pathol.* **86:**343–358.

Walker, T. S., and H. H. Winkler. 1978. Penetration of cultured mouse fibroblasts (L cells) by *Rickettsia prowazeki*. *Infect. Immun.* **22:**200–208.

Welch, M. D., A. Iwamatsu, and T. J. Mitchison. 1997. Actin polymerization is induced by the Arp2/3 protein complex at the surface of *Listeria monocytogenes*. *Nature* **385:**265–269.

Welch, M. D., J. Rosenblatt, J. Skoble, D. Portnoy, and T. J. Mitchison. 1998. Interaction of human Arp2/3 complex and the *Listeria monocytogenes* ActA protein in actin filament nucleation. *Science* **281:**105–108.

Whitworth, T., V. L. Popov, X.-J. Yu, D. H. Walker, and D. H. Bouyer. 2005. Expression of the *Rickettsia prowazekii pld* or *tlyC* gene in *Salmonella enterica* serovar Typhimurium mediates phagosomal escape. *Infect. Immun.* **73:**6668–6673.

Winkler, H. H., and R. M. Daugherty. 1989. Phospholipase A activity associated with the growth of *Rickettsia prowazekii* in L929 cells. *Infect. Immun.* **57:**36–40.

Winkler, H. H., and E. T. Miller. 1980. Phospholipase A activity in the hemolysis of sheep and human erythrocytes by *Rickettsia prowazeki*. *Infect. Immun.* **29:**316–321.

Winkler, H. H. and E. T. Miller. 1982. Phospholipase A and the interaction of *Rickettsia prowazekii* and mouse fibroblasts (L-929 cells). *Infect. Immun.* **38:**109–113.

Wisseman, C. L., Jr., E. A. Edlinger, A. D. Waddell, and M. R. Jones. 1976. Infection cycle of *Rickettsia rickettsii* in chicken embryo and L-929 cells in culture. *Infect. Immun.* **14:**1052–1064.

Wisseman, C. L., Jr., and A. D. Waddell. 1975. In vitro studies on rickettsia-host cell interactions: intracellular growth cycle of virulent and attenuated *Rickettsia prowazeki* in chicken embryo cells in slide chamber cultures. *Infect. Immun.* **11:** 1391–1404.

Wolbach, S. B. 1919. Studies on Rocky Mountain spotted fever. *J. Med. Res.* **41:**1–198.41.

Wolbach, S. B., and M. J. Schlesinger. 1923. The cultivation of the micro-organisms of Rocky Mountain spotted fever (*Dermacentroxenus rickettsi*) and of typhus (*Rickettsia prowazeki*) in tissue cultures. *J. Med. Res.* **44:**231–256.1.

ESTABLISHING INTRACELLULAR INFECTION: MODULATION OF HOST CELL FUNCTIONS (*ANAPLASMATACEAE*)

Jason A. Carlyon

6

THE FAMILY *ANAPLASMATACEAE*

Introduction

The family *Anaplasmataceae* consists of pathogenic (parasitic) and nonpathogenic (commensal or mutualistic) obligate intracellular gram-negative *Alphaproteobacteria* that infect invertebrates. Many also infect erythrocytes, leukocytes, endothelial cells, or intestinal epithelial cells in mammals or birds. The family contains the genera *Anaplasma*, *Ehrlichia*, *Neorickettsia*, and *Wolbachia*. Together, the *Anaplasmataceae* and *Rickettsiaceae* comprise the order *Rickettsiales*. A major difference between members of these two families is that *Rickettsiaceae* members rapidly escape the host cell-derived vacuoles into which they are initially internalized to inhabit the host cell cytoplasm, whereas *Anaplasmataceae* members remain within their host cell-derived vacuoles throughout infection. The ability of these pathogens to replicate within their host cells, many of which are leukocytes that are otherwise extremely effective at destroying microbes, demonstrates that they have evolved strategies that enable them to sense their environments and modulate host cell functions to flourish and propagate infection. Studies of representative members, primarily *Anaplasma phagocytophilum* and *Ehrlichia chaffeensis*, have shed much light on the exquisite mechanisms by which *Anaplasmataceae* pathogens manipulate their host cells, and are discussed following overviews of the diseases that they cause, their notable genomic features, and their infection and developmental cycles.

Anaplasmataceae Members and Their Diseases

A. phagocytophilum, *E. chaffeensis*, *Ehrlichia ewingii*, *Ehrlichia canis*, *Neorickettsia sennetsu*, and potentially *Ehrlichia ruminantium* are known to cause disease in humans. Apart from *E. chaffeensis* and *N. sennetsu*, *Anaplasmataceae* members were initially known as veterinary pathogens (Rikihisa, 2010a). *A. phagocytophilum*, *E. chaffeensis*, and *E. ewingii* are the etiologic agents of the emerging zoonoses human granulocytic anaplasmosis (HGA; formerly human granulocytic ehrlichiosis), human monocytic ehrlichiosis (HME), and *E. ewingii* ehrlichiosis, respectively. *A. phagocytophilum* (formerly *Ehrlichia phagocytophila*, *Ehrlichia equi*, and the agent of human granulocytic ehrlichiosis) is naturally maintained in a zoonotic cycle between tick vectors of the

Jason A. Carlyon, Department of Microbiology and Immunology, Virginia Commonwealth University School of Medicine, Richmond, VA 23298-0678.

Intracellular Pathogens II: Rickettsiales
Edited by G. H. Palmer and A. F. Azad © 2012 ASM Press, Washington, DC
10.1128/9781555817336.ch6

Ixodes persulcatus complex and mammalian reservoir hosts. Humans are accidental hosts and play no role in the maintenance of the pathogen in nature. Most cases of HGA in the United States occur in the Northeast, upper Midwest, and northern California. HGA has also been reported in Europe and Asia, though less frequently than in the United States. Following inoculation through the feeding of an infected tick, HGA ranges in severity from asymptomatic seroconversion to mild or severe febrile illness. In rare instances, severe disease can result in organ failure or death (Thomas et al., 2009). Nonspecific signs include acute onset of fever, malaise, and myalgia. Less common symptoms include nausea, abdominal pain, diarrhea, and cough. Laboratory findings include leukopenia followed by rebound leukocytosis, thrombocytopenia, and elevated serum levels of hepatic enzymes (Rikihisa, 2010a; Thomas et al., 2009). *A. phagocytophilum* also infects ruminants and horses to cause tick-borne fever and equine granulocytic anaplasmosis (formerly equine granulocytic ehrlichiosis). Domestic animals can also be infected. In the mammalian host, *A. phagocytophilum* invades peripherally circulating neutrophils to reside within a host cell-derived vacuole. The resulting intravacuolar bacterial inclusion is called a morula (Latin for "mulberry") because it is stippled in appearance and stains dark blue by Romanowsky stains (Thomas et al., 2009). In addition to neutrophils and myeloid cell lines, *A. phagocytophilum* infects endothelial and megakaryocyte cell lines in vitro (Granick et al., 2008; Munderloh et al., 2004). *A. phagocytophilum* is also capable of infecting murine bone marrow-derived and human skin-derived mast cells in vitro (Ojogun et al., 2011). It is unknown whether megakaryocytes, mast cells, or endothelial cells are colonized during mammalian infection. One study detected *A. phagocytophilum* in the vascular endothelium of an experimentally infected mouse (Herron et al., 2005).

Similar to HGA, HME and *E. ewingii* ehrlichiosis present as undifferentiated febrile illnesses characterized by fever, chills, myalgia, arthralgia, and headache. The most common additional symptoms associated with HME are coughing, nausea, vomiting, diarrhea, anorexia, and abdominal pain. Complications with HME can include meningoencephalitis, acute renal failure, myocarditis, adult respiratory distress syndrome, gastrointestinal hemorrhage, and disseminated intravascular coagulation. Whereas fatalities occur in <1% of HGA cases and are usually due to opportunistic infections, 3% of HME cases are fatal, with the deaths occurring most frequently in immunocompromised patients who develop respiratory distress syndrome, hepatitis, or opportunistic nosocomial infections. No fatal cases of *E. ewingii* ehrlichiosis have been reported. *E. chaffeensis* resides primarily in the southeastern, south-central, and mid-Atlantic United States where its vector, *Amblyomma americanum*, is endemic. The primary reservoir is the white-tailed deer, though it is maintained by a diverse range of wild and domestic animals. *E. chaffeensis* DNA has been detected in Brazilian marsh deer and spotted deer in southeast Brazil and in Korea and Japan, respectively. *E. chaffeensis* replicates within monocytes and macrophages to form morulae in mammalian reservoirs and accidental human hosts. Like *A. phagocytophilum*, *E. ewingii* infects granulocytes. It is transmitted by *A. americanum* and is known to cause canine ehrlichiosis as well as human infection.

Other *Ehrlichia* spp. that infect monocytes and macrophages are *E. canis*, *E. ruminantium*, and *E. muris*. *E. canis* causes chronic and life-threatening infections of dogs and has been shown to infect humans (Thomas et al., 2009). Three fatal human infections with *E. ruminantium*, a ruminant pathogen, have been reported in Africa. However, the fatal cases were not confirmed by cytology or bacterial isolation (Allsopp et al., 2005; Rikihisa and Lin, 2010). *E. muris* causes systemic infection in mice and is used as a mouse model of monocytic ehrlichiosis (Feng and Walker, 2004).

In contrast to *A. phagocytophilum*, *Anaplasma marginale* is quite host specific, infecting only ruminants and causing disease primarily in cattle. It is the etiologic agent of bovine anaplasmosis, which occurs in tropical and subtropical areas throughout the world. In the United States, the disease is enzootic throughout the southern Atlantic states, Gulf Coast states, and several midwestern and western states. Bovine anaplasmosis is also endemic in Mexico, Central and South America, the Caribbean Islands, all Latin American countries, Mediterranean countries in Europe, and regions of Africa and Asia. *Dermacentor* spp. and *Rhipicephalus* spp. ticks are vectors for the disease. *A. marginale* is transmitted to cattle most effectively through the feeding of infected ticks, but can also be transmitted mechanically by biting flies or blood-contaminated fomites. Erythrocytes are the only known site of infection of *A. marginale* in mammals, though the bacterium can be cultivated in endothelial cell lines. Within bovine erythrocytes, *A. marginale* replicates within a membrane-bound inclusion called an initial body. Seventy percent or more of a host animal's erythrocytes may be parasitized during acute infection. Clinical signs include fever, lethargy, weight loss, abortion, icterus, and often death in animals over 2 years old. Cattle that survive acute infection develop persistent infections characterized by cyclic low-level bacteremia (Kocan et al., 2010).

Neorickettsia spp. are maintained in nature through vertical transmission in trematodes (Gibson et al., 2010; Lin et al., 2009). *N. sennetsu* (formerly *Rickettsia sennetsu* or *Ehrlichia sennetsu*) was the first bacterium in the family *Anaplasmataceae* identified as a cause of human ehrlichiosis (Rikihisa, 2010a). *N. sennetsu* infects human monocytes and macrophages to cause the disease Sennetsu neorickettsiosis, the symptoms of which are similar to infectious mononucleosis and include swelling of the lymph nodes, fever, loss of appetite, lethargy, sleeplessness, and malaise. Sennetsu neorickettsiosis is likely acquired by humans through the ingestion of gray mullet fish that are infested with *N. sennetsu*-infected metacercarial-stage trematodes (Gibson et al., 2010). *N. sennetsu* infections have been reported primarily in Japan, but also in Malaysia and Laos (Gibson et al., 2010; Rikihisa, 2010a). While *N. sennetsu* is a human pathogen that infects a trematode that presumably uses a fish as an intermediate host in Southeast Asia, *Neorickettsia risticii* (formerly *Ehrlichia risticii*) is a veterinary pathogen that uses an aquatic insect as an intermediate host in North America and South America (Lin et al., 2009; Rikihisa, 2010a). *N. risticii* causes Potomac horse fever, an acute, often severe to fatal systemic disease of horses characterized by pyrexia, depression, inappetence, dehydration, watery diarrhea, laminitis, and/or abortion. Horses and other susceptible mammals acquire *N. risticii* by ingesting aquatic insects that are infested with infected trematodes, after which the bacterium infects and replicates within monocytes, macrophages, and intestinal epithelial cells (Lin et al., 2009). Within mammalian host cells, *Neorickettsia* spp. replicate within host cell-derived vacuolar inclusions that are analogous to morulae and elementary bodies formed by *Anaplasma* spp. and *Ehrlichia* spp.

Unlike other members of the family *Anaplasmataceae*, *Wolbachia* spp. do not routinely infect vertebrates. The type species is *Wolbachia pipientis*, which was first described in the mosquito *Culex pipiens*. Based on 16S rRNA sequences and other sequence information, *Wolbachia* spp. are divided into eight supergroups, two of which are found in filarial nematodes, while the other six are found primarily in arthropods. *Wolbachia* spp. are pandemic in their distribution among their host species. *Wolbachia* spp. are fascinating microorganisms because of the manipulative effects that they have on their hosts, which include feminization of genetic males; parthenogenetic induction, which results in the development of unfertilized eggs; the killing of male progeny from infected females; and cytoplasmic incompatibility, otherwise known as sperm-egg incompatibility. Collectively, these reproductive manipulations are referred to as reproductive parasitism and aid the bacterium by increasing production of infected females (Werren et al., 2008).

NOTABLE GENOMIC FEATURES

The genomes of representative strains of most *Anaplasmataceae* members have been sequenced

(Brayton et al., 2005; Collins et al., 2005; Dunning Hotopp et al., 2006; Foster et al., 2005; Lin et al., 2009; Mavromatis et al., 2006; Wu et al., 2004). These organisms have relatively small genomes, ranging in size from 0.9 to 1.5 Mb and carrying from 859 to 1,369 open reading frames. *Anaplasma* spp., *Ehrlichia* spp., and *N. sennetsu* lack genes that are required for lipopolysaccharide (LPS) and peptidoglycan biosynthesis (Brayton et al., 2005; Collins et al., 2005; Dunning Hotopp et al., 2006; Lin et al., 2009; Mavromatis et al., 2006). *Wolbachia* endosymbiont of *Brugia malayi* lacks genes for LPS biosynthesis and is predicted to encode an unusual peptidoglycan (Foster et al., 2005). *Anaplasmataceae* members are auxotrophic for most amino acids and are unable to use glucose as a carbon or energy source (Brayton et al., 2005; Collins et al., 2005; Dunning Hotopp et al., 2006; Foster et al., 2005; Lin et al., 2009; Mavromatis et al., 2006; Wu et al., 2004). Thus, amino acids and many metabolites must be acquired from eukaryotic host cells. Accordingly, *A. phagocytophilum*, *Ehrlichia* spp., and *N. sennetsu* encode from 30 to 60 transport proteins (Collins et al., 2005; Dunning Hotopp et al., 2006; Lin et al., 2009; Mavromatis et al., 2006). Also, some *A. phagocytophilum*, *E. chaffeensis*, and *N. sennetsu* outer membrane proteins that are abundantly expressed exhibit porin activities (Gibson et al., 2010; Huang et al., 2007; Kumagai et al., 2008). *Anaplasmataceae* members have low coding capacities for central intermediary metabolism, transport, and regulatory functions. All members have pathways for aerobic respiration, including pyruvate dehydrogenase, the tricarboxylic acid cycle, and the electron transport chain. Also, *Anaplasma* spp., *Ehrlichia* spp., and *N. sennetsu*, but not *Wolbachia* spp., have the ability to synthesize all nucleotides and most vitamins and cofactors (Brayton et al., 2005; Collins et al., 2005; Dunning Hotopp et al., 2006; Foster et al., 2005; Lin et al., 2009; Mavromatis et al., 2006; Wu et al., 2004). Because this chapter primarily focuses on how *Anaplasmataceae* members manipulate their mammalian host cells to facilitate intracellular survival, *Wolbachia* spp. will not be covered further.

The outer membrane protein 1 (OMP1)-P44 superfamily (also referred to as the major surface protein 2 (Msp2) superfamily or surface antigen family (Pfam accession number PF01617) comprises proteins that are unique to the *Anaplasmataceae*. This family has expanded considerably in the genomes of *Anaplasma* spp. and *Ehrlichia* spp. There are more than 100 members (including pseudogenes) in *A. phagocytophilum*, 56 in *A. marginale*, 22 in *E. chaffeensis*, 24 in *E. canis*, and 16 in *E. ruminantium* (Dunning Hotopp et al., 2006; Rikihisa, 2010a). Such expansion has not occurred in the genomes of *N. sennetsu* or *Wolbachia* spp. (Dunning Hotopp et al., 2006; Lin et al., 2009; Rikihisa, 2010a; Wu et al., 2004). OMP1-Msp2 (P44) proteins are important for antigenic variation and have been implicated in facilitating adaptation to different host cell environments (Barbet et al., 2003; Brayton et al., 2001; IJdo et al., 2002; Jauron et al., 2001; Lin and Rikihisa, 2005; Lin et al., 2002b; Palmer and Brayton, 2007; Singu et al., 2005, 2006; Zhi et al., 1999, 2002). *Anaplasmataceae* organisms encode both a Sec-dependent and Sec-independent protein export pathway for secretion of proteins across the inner membrane (Brayton et al., 2005; Collins et al., 2005; Dunning Hotopp et al., 2006; Foster et al., 2005; Lin et al., 2009; Mavromatis et al., 2006; Wu et al., 2004). Genes encoding the type I secretion system (T1SS) for transporting effectors carrying a C-terminal secretion signal outside the bacterial cell have been documented for *A. phagocytophilum*, *E. chaffeensis*, and *N. sennetsu* (Dunning Hotopp et al., 2006). *Anaplasmataceae* bacteria encode a type IV secretion system (T4SS) that is analogous to that of *Agrobacterium tumefaciens* (Alvarez-Martinez and Christie, 2009; Brayton et al., 2005; Collins et al., 2005; Dunning Hotopp et al., 2006; Foster et al., 2005; Lin et al., 2009; Mavromatis et al., 2006; Wu et al., 2004). The *A. phagocytophilum* T4SS delivers effectors that modulate host cell functions and facilitate bacterial survival (Lin et al., 2007; Niu et al., 2010). *A. phagocytophilum*, *E. chaffeensis*, and *N. sennetsu* genomes code for multiple response regulatory two-component systems, which comprise a family of signal sensor, transduction, and response regulatory systems that enable bacteria to sense environmental signals and respond accordingly through specific gene activation or repression (Collins et al., 2005; Dunning Hotopp et al., 2006; Lin et al., 2009; Raghavan and Groisman, 2010; Rikihisa, 2010b).

LIFE CYCLE AND INTRACELLULAR INFECTION

Binding and Entry

After being inoculated into their natural mammalian reservoirs or accidental human hosts, *Anaplasma* spp., *Ehrlichia* spp., and *Neorickettsia* spp. infect peripherally circulating leukocytes or erythrocytes. *E. ruminantium* also infects endothelial cells, while *N. risticii* also infects intestinal epithelial cells (Dunning Hotopp et al., 2006; Lin et al., 2009). The most studied *Anaplasmataceae* pathogen-receptor interaction is that of *A. phagocytophilum* and P-selectin glycoprotein ligand 1 (PSGL-1). *A. phagocytophilum* binding to PSGL-1 facilitates adhesion to and invasion of human neutrophils, bone marrow progenitors, and myeloid cells lines, such as HL-60 (Goodman et al., 1999; Herron et al., 2000). *A. phagocytophilum* binding to PSGL-1 requires cooperative recognition of N-terminal primary amino acid sequence and α2,3-linked sialic acid and α1,3-linked fucose of the sialyl Lewis x tetrasaccharide (sLe^x) that modifies the N terminus of PSGL-1 (Goodman et al., 1999; Herron et al., 2000; Yago et al., 2003). This complex interaction is mediated by either multiple *A. phagocytophilum* adhesins or a single adhesin having multiple binding domains. Engagement of PSGL-1 prompts a signaling cascade that is important for *A. phagocytophilum* entry (Thomas and Fikrig, 2007). Interestingly, recognition of α1,3-fucosylated glycans and sialic acid are important for *A. phagocytophilum* infection of murine neutrophils, while PSGL-1 is not (Carlyon et al., 2003). In fact, *A. phagocytophilum* cannot even bind to murine PSGL-1 because it lacks a critical peptide sequence found in the human PSGL-1 N terminus (Yago et al., 2003). This stereospecific difference necessitates that *A. phagocytophilum* utilize a PSGL-1-independent receptor to invade murine neutrophils. *A. phagocytophilum* binding and entry also involve interactions with caveolae and glycosylphosphatidylinositol-anchored proteins (Lin and Rikihisa, 2003b). Subpopulations of naturally occurring *A. phagocytophilum* organisms that do not depend on sLe^x or, consequently, PSGL-1 can be enriched for by cultivation in sialyltransferase- and α1,3-fucosyltransferase-defective cell lines (Reneer et al., 2006, 2008; Sarkar et al., 2007). Perhaps this enriched subpopulation represents organisms exhibiting upregulated expression of one or more adhesins that are involved in binding to caveolae, glycosylphosphatidylinositol-anchored proteins, or the PSGL-1-independent murine neutrophil receptor. The human PSGL-1- and sLe^x-independent *A. phagocytophilum* organisms do not elicit the same signaling cascade that is associated with PSGL-1-mediated uptake (Reneer et al., 2008). Notably, α1,3-fucose, but not sialic acid, is necessary for *A. phagocytophilum* to colonize its *Ixodes* spp. tick vectors (Pedra et al., 2010). An elegant set of experiments performed using a flow cell examined *A. phagocytophilum* interactions with human neutrophils under shear-flow conditions that more closely reflect the high velocities under which the bacterium and neutrophils interact in a blood vessel (Schaff et al., 2010). Notably, *A. phagocytophilum* binds to neutrophils within a few seconds under flow conditions (Schaff et al., 2010), which is in contrast to the 15 to 30 minutes required for binding to neutrophils or HL-60 cells under static tissue culture conditions (Borjesson et al., 2005; Carlyon et al., 2004). Flow conditions also influence the receptors to which *A. phagocytophilum* binds. Indeed, the pathogen binds to β_2-integrin and PSGL-1 under flow (Schaff et al., 2010), but binds only to PSGL-1 under static conditions (Goodman et al., 1999; Herron et al., 2000; Reneer et al., 2006). Thus, use of the flow cell to study *A. phagocytophilum* and possibly other *Anaplasmataceae* pathogen interactions with host cells may yield more biologically relevant results.

Adhesin-receptor interactions have yet to be as thoroughly dissected for other *Anaplasmataceae* pathogens as for *A. phagocytophilum*. The *A. marginale* receptor on bovine erythrocytes consists of protein and carbohydrate components, the latter possibly being sialic acid since *A. marginale* binding is reduced following neuraminidase treatment of host cells (McGarey and Allred, 1994). *A. marginale* major surface proteins MSP1a and MSP1b facilitate adherence to bovine erythrocytes and tick cells (de la Fuente et al., 2001, 2003; McGarey et al., 1994). When expressed in recombinant form on the *Escherichia coli* surface, MSP1a binds to *Dermacentor variabilis* gut cells, cultured *Ixodes scapularis* embryonic ISE6 cells, and bovine erythrocytes, whereas MSP1b adheres only to bovine erythrocytes (de la Fuente et al., 2001, 2004; McGarey et al., 1994). When coexpressed on the *E. coli* surface, MSP1a and MSP1b provide a more than additive effect on adhesion. The adhesive domain of recombinant MSP1a has been delineated as an N-terminal stretch of a tandemly repeated peptide unit of 28 or 29 amino acids (de la Fuente et al., 2003). However, in vivo observations and genomic analyses do not support the premise that MSP1a is an adhesin (Herndon et al., 2010; Scharf et al., 2011; Ueti et al., 2007). First, naturally attenuated *A. marginale* subsp. *centrale* completely lacks the MSP1a repeat region (Herndon et al., 2010; Shkap et al., 2002), yet invades and replicates within the *Dermacentor andersonii* midgut epithelium prior to salivary gland colonization, thereby demonstrating that MSP1a N-terminal repeats are not uniformly required for *A. marginale* entry and colonization in *Dermacentor* spp. ticks (Ueti et al., 2007, 2009). Second, *A. marginale* subsp. *centrale* has the same infection cycle and achieves the same bacteremic levels in ruminant hosts (Herndon et al., 2010; Shkap et al., 2002). Third, even though the Mississippi strain of *A. marginale* has an MSP1a repeat type that is indistinguishable from that identified as being transmission competent, it fails to colonize the *D. andersonii* midgut and is consequently not transmitted. Thus, even within *A. marginale* sensu stricto strains, the MSP1a N-terminal repeat sequence is not predictive of midgut invasion of *Dermacentor* spp. ticks (Ueti et al., 2007).

Antibodies against L- and E-selectin partially block *E. chaffeensis* binding to monocytic THP-1 cells (Zhang et al., 2003). Tandem repeat-containing protein 120 (TRP120) is a multifunctional *E. chaffeensis* protein that consists of a large number of tandemly arranged direct repeats and is both secreted and associated with the ehrlichial surface. TRP120 mediates *E. coli* adherence to and uptake by HeLa cells when expressed in recombinant form on the *E. coli* surface (Popov et al., 2000). Both ehrlichial and host surface proteins are involved in *E. risticii* adhesion and invasion of host cells (Messick and Rikihisa, 1993).

FIGURE 1 *A. phagocytophilum* infection cycle in a myeloid host cell. This model is based on observations of *A. phagocytophilum*-infected HL-60 cells. (1 to 3) An *A. phagocytophilum* dense-cored bacterium binds to the host cell surface and triggers its own uptake to reside within a host cell-derived vacuole. (4) The dense-cored organism transitions to a reticulate cell. Arrowheads in corresponding electron micrograph point to two ApVs in which the bacteria are in the process of differentiating from the dense-cored to reticulate cell stage. The thick arrow denotes a vacuole harboring an *A. phagocytophilum* bacterium that is still in the dense-cored cell form. (5 and 6) The reticulate cell form divides by binary fission to fill the expanding ApV with bacteria. (7) The reticulate cell bacteria transition to the dense-cored form. (8) The mature ApV opens to release dense-cored organisms into the media. Electron micrographs 8a and 8b present ApVs at different stages of opening. Also, a host cell that is filled with several large morulae can lyse to release *A. phagocytophilum* bacteria. Similar infection/biphasic developmental cycles have been observed for *Ehrlichia* spp. and *Neorickettsia* spp. cultured in mammalian cell lines. (Electron micrograph in step 2 reprinted from Troese and Carlyon [2009] with permission of the publisher.) doi:10.1128/9781555817336.ch6.f1

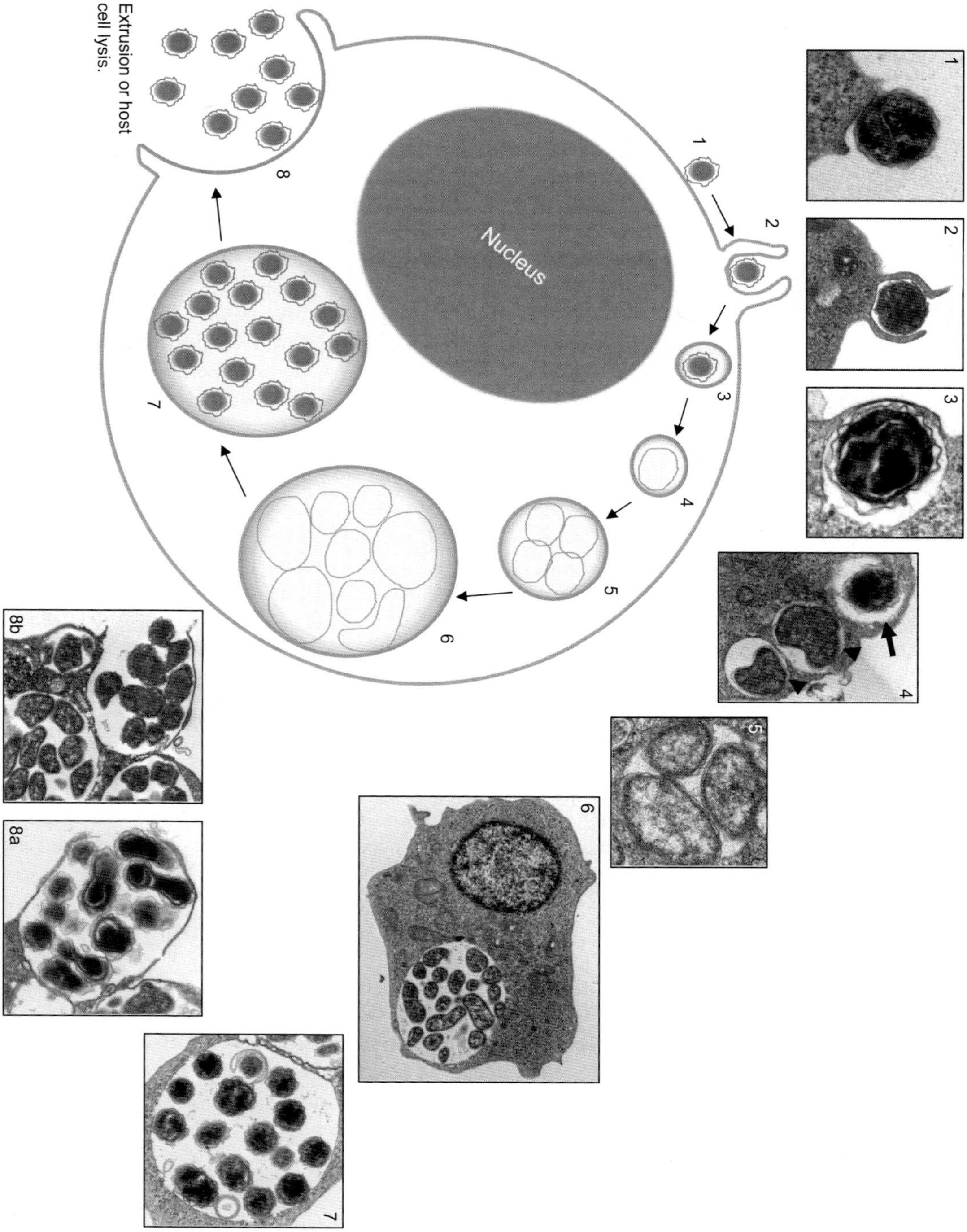
Nucleus
Extrusion or host cell lysis.
1
2
3
4
5
6
7
8
8a
8b

Infection Cycle and Biphasic Development

All *Anaplasmataceae* members are polymorphic, ranging in size from 0.2 to 2.0 µm. They exhibit a biphasic development cycle over the course of infection of cultured mammalian or invertebrate cell lines in which they transition between dense-cored cells and reticulate cells (Avakyan and Popov, 1984; Hidalgo et al., 1989; Munderloh et al., 1996, 1999, 2004; Popov et al., 1998; Thomas et al., 2010; Troese and Carlyon, 2009; Webster et al., 1998; Zhang et al., 2007). Dense-cored organisms are smaller, have dense nucleoid and ruffled outer membranes, and are spheroid. Reticulate cells are larger, have dispersed nucleoid and smooth outer membranes, and are generally pleomorphic. In cell culture, the infectious dense-cored form binds and induces its own uptake into a host cell-derived vacuole (Fig. 1) (Munderloh et al., 1999; Popov et al., 1998; Troese and Carlyon, 2009; Zhang et al., 2007). During the initial hours following entry, the dense-cored form transitions to the replicative reticulate cell form. The reticulate cell subsequently divides by binary fission to generate a morula (Fig. 2), while the vacuole enlarges to accommodate the growing bacterial population. Unlike vacuoles occupied by *Chlamydia* spp. and *Coxiella burnetii*, which are other obligate intracellular vacuole-adapted bacterial pathogens that undergo biphasic development (Abdelrahman and Belland, 2005; Heinzen et al., 1999), vacuoles harboring *A. phagocytophilum* and *Ehrlichia* spp. do not fuse with each other (Popov et al., 1998; Reneer et al., 2008; Zhang et al., 2007). After multiple rounds of binary fission, the reticulate cells mature into dense-cored cells, which are subsequently released to initiate a new wave of infection (Munderloh et al., 2004; Popov et al., 1998; Troese and Carlyon, 2009; Zhang et al., 2007). Though cellular adherence of reticulate cell organisms has been reported (Munderloh et al., 1999; Popov et al., 1998), *Anaplasmataceae* populations that are enriched in the dense-cored form exhibit much greater infectivity than those enriched in reticulate cells (Troese and Carlyon, 2009; Zhang et al., 2007). Also, *A. phagocytophilum* dense-cored organisms, but not reticulate cells, are capable of binding to the PSGL-1 receptor (Troese and Carlyon, 2009).

In cell culture, highly infected host cells that harbor multiple large vacuoles filled with *Anaplasmataceae* organisms will ultimately lyse to release bacteria (Fig. 1) (Popov et al., 1998; Troese and Carlyon, 2009; Zhang et al., 2007), but it is unknown if this is reflective of what occurs in vivo. Exit of infectious dense-cored progeny from bacteria-containing vacuoles that rupture, along with the adjacent host cell plasmalemma, without compromising the integrity of the rest of the host cell membrane has been reported for some *Anaplasmataceae* organisms, including *E. canis*, *A. phagocytophilum*, and *A. marginale* (Fig. 1) (Blouin and Kocan, 1998; Popov et al., 1998). An effective and stealthy strategy for these obligate intracellular pathogens in vivo would be to spread from cell to cell without lysing host cells and without releasing themselves into the extracellular milieu. Indeed, when neutrophils are incubated on top of *A. phagocytophilum*-infected HMEC-1 human microvascular endothelial cells, vacuoles harboring multiple bacteria can be detected within the myeloid cells within 2 hours (Herron et al., 2005). Given *A. phagocytophilum*'s slow replication cycle, this demonstrates that entire bacteria-filled vacuoles can migrate from cell to cell. Using indirect immunofluorescence microscopy and electron microscopy, we have frequently observed host cell-free vacuoles filled with *A. phagocytophilum* bound and inducing their uptake at host cell surfaces. Figure 3 presents an image from an unpublished study that demonstrated that *A. phagocytophilum* infects mast cells in vitro. The image presents a reticulate cell-filled vacuole bound at the surface and being engulfed by filopodia. *Ehrlichia* spp. induce the formation of and can spread from cell to cell by traversing through filopodia (Thomas et al., 2010).

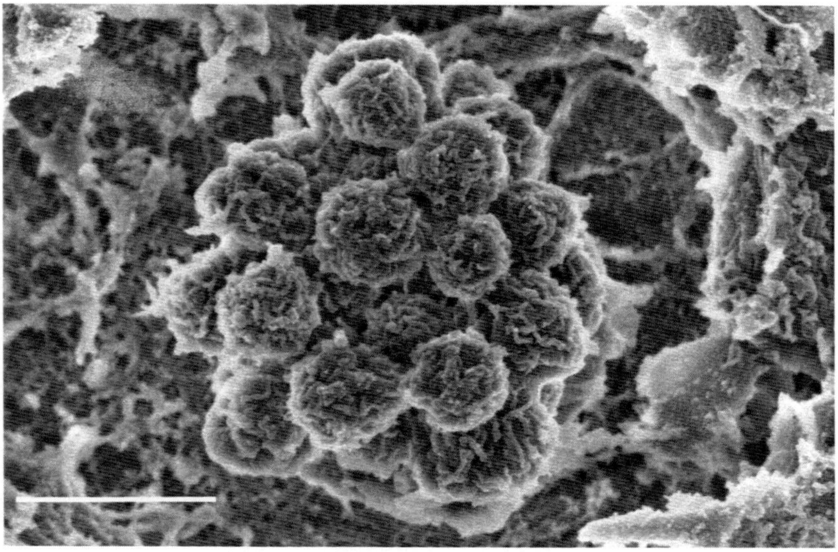

FIGURE 2 *E. chaffeensis* morulae. Scanning electron micrograph of a DH82 cell infected with *E. chaffeensis* from which the cell membrane and EVM have been removed. Bar, 1 µm. (Courtesy of Sunil Thomas, Vsevolod L. Popov, and David H. Walker, Department of Pathology and Center for Biodefense and Emerging Infectious Diseases, University of Texas Medical Branch.) doi:10.1128/9781555817336.ch6.f2

TWO-COMPONENT SYSTEM

The two-component system is a family of sensor, transduction, and response regulatory systems that enables bacteria to sense changes in their environment and rapidly respond to such stimuli via signal transduction that culminates in activation or repression of specific genes (Raghavan and Groisman, 2010). Two-component systems consist of a histidine kinase sensor and a response regulator and are critical for certain pathogenic bacteria to establish infection and survive within their hosts (Rikihisa, 2010b). *A. phagocytophilum* and *E. chaffeensis* genomes each encode three histidine kinases that are homologs of NtrX, PleC, and CckA, each of which has specific histidine-dependent autokinase activity (Cheng et al., 2006). The genomes also code for three response regulators that each contain a conserved receiver domain with an aspartate phosphorylation site. The response regulators NtrY, PleD, and CtrA pair with NtrX, PleC, and CckA, respectively. CtrA and NtrX both have a carboxy-terminal DNA-binding domain, while PleD has a C-terminal GGDEF domain. Specific aspartate-dependent in vitro phosphotransfer from NtrY to NtrX, from PleC to PleD, and from CckA to CtrA has been demonstrated using recombinant proteins (Cheng et al., 2006; Kumagai et al., 2006). The GGDEF domain is associated with production of the bacterial second messenger bis-(3′,5′)-cyclic dimeric GMP (c-di-GMP) (Römling, 2009). Recombinant PleD from *A. phagocytophilum* can produce c-di-GMP. *A. phagocytophilum* infection is inhibited by the c-di-GMP analog 2′-*O*-di(*tert*-butyldimethysilyl)-c-di-GMP (Lai et al., 2009). Infection by both *A. phagocytophilum* and *E. chaffeensis* is dependent on histidine kinase activity, as treatment with the histidine kinase inhibitor closantel or the addition of closantel to infected cells results in prevention or rapid eradication of infection, respectively (Cheng et al., 2006; Kumagai et al., 2006; Lai et al., 2009). Closantel treatment-mediated destruction of *E. chaffeensis* occurs via lysosomal fusion with *Ehrlichia*-containing vacuoles (Kumagai et al., 2006).

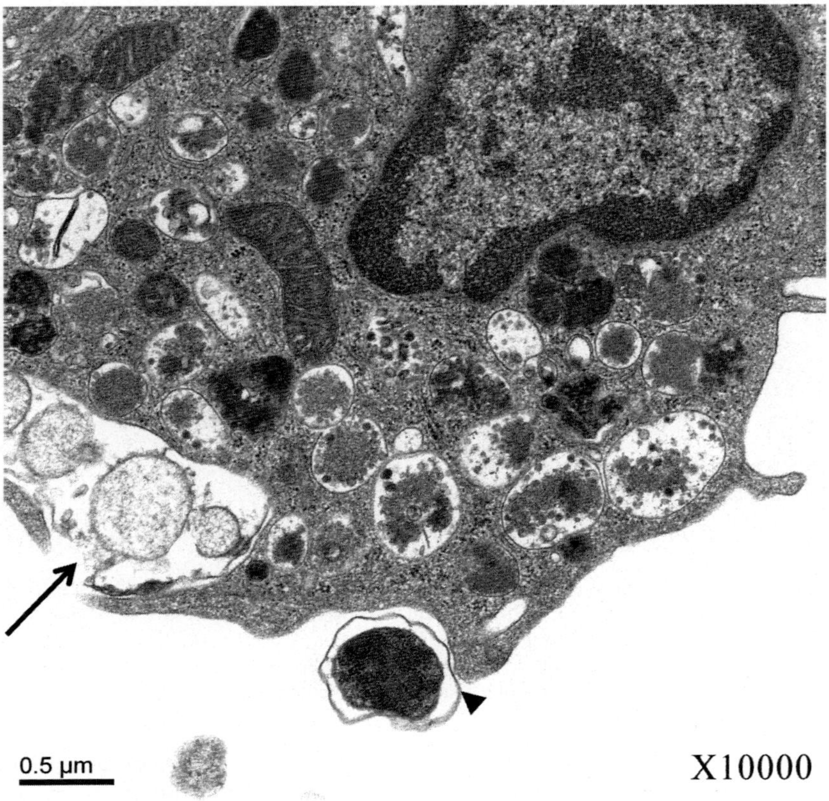

FIGURE 3 Host cell-free ApV binding to and facilitating its uptake by a host cell. Host cell-free *A. phagocytophilum* organisms and host cell-free ApVs were liberated from infected HL-60 cells and added to murine bone marrow-derived mast cells. After 40 minutes, unbound bacteria were washed off and the host cells were fixed and examined by transmission electron microscopy. The arrow denotes a host cell-free vacuole filled with *A. phagocytophilum* reticulate cell organisms that is bound to the surface of and is being internalized by a bone marrow-derived mast cell. Our laboratory has observed similar phenomena demonstrating that ApVs are also infectious for human promyelocytic HL-60 cells and monkey choroidal endothelial RF/6A cells. The arrowhead demarcates an *A. phagocytophilum* dense-cored cell bound at the host cell surface. doi:10.1128/9781555817336.ch6.f3

T4SS

T4SS Apparatus and Expression

Diverse bacterial pathogens inject effector proteins into their host cells to modulate host cell functions and generate a niche that is suitable for colonization. Vacuole-adapted pathogens secrete their effectors during binding and invasion and/or from within their host cell-derived vacuoles into the cytoplasm during intracellular residence. *Anaplasmataceae* pathogens lack type II, III, V, and VI secretion components, but do encode type I and type IV secretory apparatuses (Alvarez-Martinez and Christie, 2009; Brayton et al., 2005; Collins et al., 2005; Dunning Hotopp et al., 2006; Foster et al., 2005; Lin et al., 2009; Mavromatis et al., 2006; Rikihisa and Lin, 2010; Wu et al., 2004). The T4SS translocates protein or DNA from within the bacterial cell into host cells in an ATP-dependent manner and usually in a contact-dependent manner (Alvarez-Martinez and Christie, 2009). T4SSs are ancestrally related to the conjugation system of gram-negative bacteria. There are at least two lineages for the T4SS—the *A. tumefaciens* VirB/VirD system and the *Legionella pneumophila* Dot/Icm sytem. In *A. tumefaciens*, the *virB* operon and *virD4* encode 12 membrane-associated proteins that combine to form a transmembrane channel. Unlike the single locus of clustered *virBD* genes encoding the T4SS in other bacteria, split *virBD* loci are found in *E. chaffeensis*, *A. phagocytophilum*, and other members of the order *Rickettsiales* (Rikihisa and Lin, 2010). The *virBD* genes of *A. phagocytophilum* and *E. chaffeensis* are distributed into five clusters: *sodB–virB3–virB4–virB6-1–virB6-2–virB6-3–virB6-4*; *virB8-1–virB9-1–virB10–virB11–virD4*; *virB2-1–virB2-2–virB2-3–virB2-4–virB2-5–virB2-6–virB2-7–virB2-8–virB4-2* (in *A. phagocytophilum*) or *virB2-1–virB2-2–virB2-3–virB2-4–virB4-2* (in *E. chaffeensis*); *virB8-2*; and *virB9-2* (Rikihisa and Lin, 2010). *A. marginale* has 22 candidates of *virB2* (Gillespie et al., 2010). Thus, the split T4SS genomic islands and the duplicated *virB* genes are conserved features of the *Anaplasmataceae*, which indicates a common ancestral origin and evolutionary pressure to maintain multiple copies despite having small genomes (Rikihisa et al., 2010).

virBD genes are expressed in vivo and are environmentally regulated. *E. canis virB9* is expressed in experimentally infected dogs and infected ticks (Felek et al., 2003). *A. phagocytophilum virB9* is transcribed in peripheral blood leukocytes recovered from HGA patients and an experimentally infected horse and mouse (Ohashi et al., 2002). VirB2 is a pilus subunit that assembles to form the secretion channel of the T4SS apparatus (Alvarez-Martinez and Christie, 2009). Of the eight *virB2* paralogs carried by *A. phagocytophilum*, four (*virB2-2*, *virB2-3*, *virB2-4*, and *virB2-5*) are expressed only during infection of tick cell lines and two (*virB2-7* and *virB2-8*) are expressed only during infection of human cell lines (Nelson et al., 2008). The tick-specific paralogs are considerably more homologous to each other than to the human-specific paralogs (Fig. 4). Interestingly, the two human-specific paralogs are quite divergent. Some *Anaplasmataceae* T4SS apparatus proteins, including VirB2, are immunogenic in infected or immunized animals (Felek et al., 2003; Lopez et al., 2005, 2007; Sutten et al., 2010). Thus, having multiple nonidentical VirB2 paralogs may provide some advantage for immune evasion. Since VirB2 is a pilus component and has also been linked to adhesion of other bacterial pathogens (Backert et al., 2008; Dehio, 2008), it has also been proposed that the differentially expressed paralogs are important for initiating and/or establishing infection of different host cells (Rikihisa et al., 2010).

virBD genes are also developmentally regulated. *A. phagocytophilum virB6-1* and *virB9-1*, which are present in separate loci, are transcriptionally upregulated and VirB9 expression is upregulated during establishment of infection and growth within human neutrophils. Yet the majority of *A. phagocytophilum* organisms that are released poorly express VirB9 (Niu et al., 2006). Similarly, transcription of all five *virBD* loci is upregulated during the exponential phase of *E. chaffeensis* growth in and is downregulated prior to the release of infectious bacteria from human acute monocytic leukemia THP-1 cells. Proteomic analyses identified a unique DNA-binding protein, called EcxR (*E. chaffeensis* expression regulator), that coordinately regulates the five *virB* and *virD* loci (Cheng et al., 2008). Thus, differential expression of *Anaplasmataceae* T4SS components over the course of infection may be important to bacterial intracellular survival and proliferation and to environmental resistance upon release from host cells. In *E. chaffeensis* and *A. phagocytophilum*, the *sodB-virB6* operon is transcribed polycistronically during growth in human leukemia cell lines (Ohashi et al., 2002). Because *sodB* encodes a superoxide dismutase, it has been speculated that a protective oxidative stress response might be coupled with T4SS apparatus assembly during *Anaplasma* or *Ehrlichia* growth within host cells, where both activities are needed for survival and/or replication (Rikihisa et al., 2010).

VirB9 is exposed on the surfaces of *A. phagocytophilum* and *E. chaffeensis*, as revealed by immunogold labeling and surface labeling followed by affinity proteomic analyses, respectively (Ge

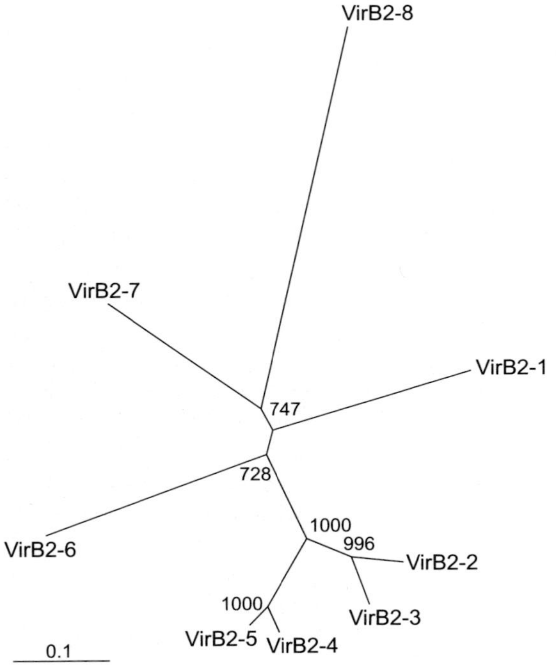

FIGURE 4 Phylogenetic tree showing the relationship, based on predicted amino acid sequences, between the eight *virB2* paralogs in the *A. phagocytophilum* HZ strain genome. Four paralogs (*virB2-1*, *virB2-2*, *virB2-3*, and *virB2-4*) are only transcribed during infection of ISE6 tick embryonic cells, while two (*virB2-7* and *virB2-8*) are expressed only during infection of human HL-60 promyelocytic leukemia and HMEC-1 human microvascular endothelial cells. This figure was generated based on data published by Nelson et al. (2008). The numbering of *virB2* paralogs is extrapolated from their relative positions to one another on the *A. phagocytophilum* chromosome and uses the numerical designations assigned by Rikihisa and Lin, 2010. Full-length VirB2 sequences were bootstrapped (N = 1,000) and an unrooted neighbor-joining tree was created (Clustal-X 2.0.8 using the Gonnet scoring matrix) (Larkin et al., 2007). The tree was visualized using the TreeView program, and bootstrap support is shown at all nodes (Page, 2002).
doi:10.1128/9781555817336.ch6.f4

and Rikihisa, 2007; Niu et al., 2006). Double immunofluorescence labeling of *N. risticii* demonstrated bipolar surface localization of VirB2, which suggests that surface distribution of the T4SS apparatus is spatially regulated (Lin et al., 2009). Though the assembly of *Anaplasmataceae* T4SS apparatuses is poorly understood, evidence suggests that they are structurally different from the prototypical *A. tumefaciens* model. *E. chaffeensis* VirB6 paralogs -1, -2, -3, and -4 are 3- to 10-fold larger than *A. tumefaciens* VirB6. During *E. chaffeensis* infection of THP-1 cells, all four VirB6 paralogs and VirB9 are expressed and VirB6-2 is cleaved into an 80-kDa fragment that associates with *Erhlichia*-occupied vacuoles. Coimmunoprecipitation and far-Western blotting reveal that the four VirB6 paralogs, the VirB6-2 80-kDa fragment, and VirB9 interact to form a unique complex (Bao et al., 2009).

T4SS Effectors and Their Roles in *Anaplasmataceae* Pathobiology

The secretion of T4SS effectors into host cells is critical for survival of many facultative and obligate intracellular bacterial pathogens. The number of confirmed T4SS effectors varies widely. For instance, *A. tumefaciens* secretes 10 T4SS effectors into the host cell (Alvarez-Martinez and Christie, 2009), whereas *C. burnetii* and *L. pneumophila* secrete over 30 and more than 100 effectors, respectively, into their host cells (Chen et al., 2010; Voth et al., 2011). There are no conserved protein sequence motifs for T4SS effectors. However, certain characteristics are shared among the C-terminal transport signals of confirmed effectors. The C-terminal 30 amino acids of *A. tumefaciens* T4SS effectors have basic amino acids and net positive charges and are hydrophilic (Vergunst et al., 2005). The T4SS effectors that have been identified in *Anaplasmataceae* pathogens thus far are ankyrin repeat-containing proteins and *Anaplasma* translocated substrate 1 (Ats-1) (Lin et al., 2007; Niu et al., 2010). These proteins exhibit C termini that share characteristics with confirmed *A. tumefaciens* T4SS effectors (Rikihisa and Lin, 2010; Scharf et al., 2011).

Ankryin is a ubiquitous motif among eukaryotic proteins that is 33 amino acids long and contains two antiparallel helices and a β-hairpin (helix-turn-helix). Ankyrin motifs may occur in combinations with other types of domains and cooperatively fold into structures that facilitate protein-protein interactions. In eukaryotes, ankyrin repeats are found in proteins that modulate many cellular functions including transcriptional regulation, cell cycle, cytoskeleton organization, developmental regulation, signal transduction, toxicity, and the inflammatory response (Zhu et al., 2009). Ankyrin repeats have been identified in confirmed or putative effector proteins of bacterial pathogens from the genera *Anaplasma*, *Ehrlichia*, *Orientia*, *Coxiella*, *Legionella*, and *Pseudomonas* (Caturegli et al., 2000; Cho et al., 2007; Collins et al., 2005; Dunning Hotopp et al., 2006; Howell et al., 2000; Klotz and Anderson, 1995; Mavromatis et al., 2006; Pan et al., 2008; Seshadri et al., 2003). The numbers of ankyrin repeat-containing proteins confirmed or predicted to be encoded by *A. phagocytophilum* HZ strain, *A. marginale* St. Maries strain, *E. chaffeensis* Arkansas strain, *E. canis* Jake strain, and *E. ruminantium* Gardel strain are four, three, four, five, and four, respectively; and their sizes range from 16.0 to 467.8 kDa (Rikihisa and Lin, 2010).

A. phagocytophilum AnkA is the most extensively studied *Anaplasmataceae* T4SS effector. It is delivered into host cells within minutes of the bacterium engaging PSGL-1 at the host cell surface and is continually delivered throughout infection, presumably, although not directly proven, from within the vacuole into the host cell (Garcia-Garcia et al., 2009b; IJdo et al., 2007; Lin et al., 2007; Reneer et al., 2008). Since genetic manipulation of any *Anaplasmataceae* bacterium is not yet possible, the Cre recombinase reporter assay of *A. tumefaciens* was used to prove that recombinant AnkA can be heterologously secreted in a VirB/D4-dependent manner (Lin et al., 2007). Interestingly, recombinant AnkA can also be heterologously secreted by the *L. pneumophila* Dot/Icm T4SS (Huang et al., 2010b). Several host cell modulation functions, which vary depending on the *A. phagocytophilum* strain examined and the laboratories in which the work was performed, have been demonstrated for AnkA (Garcia-Garcia et al., 2009b; IJdo et al., 2007; Lin et al., 2007; Park et al., 2004). Whether AnkA is a bona fide multifunctional protein or functionally varies in a strain-specific manner remains to be determined. Using the *A. phagocytophilum* HZ strain, it was demonstrated that upon delivery into HL-60 cells AnkA interacts with Abl interactor 1, which in turn interacts with Abl-1 tyrosine kinase to mediate multiple tyrosine phosphorylation of AnkA. AnkA and Abl-1 are critical for HZ infection, as infection is inhibited upon cytoplasmic delivery of an AnkA antibody, short interfering RNA-mediated knockdown of Abl-1, or pharmacologic inhibition of Abl-1 (Lin et al., 2007).

Studies performed using the NCH-1 strain demonstrated that AnkA is tyrosine-phosphorylated upon being secreted into HL-60 cells or neutrophils, *A. phagocytophilum* binding was sufficient for this phenomenon, and AnkA is phosphorylated at EPIYA motifs that are present in the C terminus of AnkA (IJdo et al., 2007). EPIYA motifs have been identified in and are critical for numerous type III secretion system and T4SS effectors to be phosphorylated by host cell kinases and interact with host cell targets (Xu et al., 2010). EPIYA motifs are recognized and phosphorylated by Src kinases. The phosphorylated EPIYA sequences can then bind Src homology 2 (SH2) domain-containing proteins. Recombinant NCH-1 AnkA expressed in COS-7 cells undergoes tyrosine phosphorylation by Src. AnkA binds to the SH2 domains of the phosphatase, SHP-1, during early infection of HL-60 cells, an event that is predicated on tyrosine phosphorylation of the EPIYA motifs (IJdo et al., 2007). Engagement of microbial ligands with host cell receptors leads to phosphorylation of signaling proteins that culminates in events that are part of the antimicrobial response, such as phagocytosis and reactive oxygen species (ROS) production. However, binding of tyrosine-phosphorylated proteins to SHP-1 activates the phosphatase activity of SHP-1, which in turn acts upon signaling proteins to negatively regulate the antimicrobial response (Poole and Jones, 2005). It has been proposed that SHP-1 recruitment and activation via T4SS-delivered AnkA at the site of *A. phagocytophilum* entry opposes host cell signals that would otherwise promote antimicrobial responses (IJdo et al., 2007).

AnkA encoded by the *A. phagocytophilum* Webster strain localizes to the HL-60 cell nucleus (Caturegli et al., 2000; Garcia-Garcia et al., 2009b). Anti-AnkA, but not isotype control antibody or antibody against the *A. phagocytophilum* major surface protein Msp2 (P44), coimmunoprecipitates HL-60 DNA. Recombinant AnkA binds to DNA fragments containing genes encoding ATPase, tyrosine phosphatase, and NADPH dehydrogenase-like functions as well as nuclear proteins in a cell-free system (Park et al., 2004). Moreover, AnkA accumulation in the host cell nucleus is coincident with transcriptional downregulation of *CYBB*, which encodes gp91^{phox}, which is critical for production of ROS (Garcia-Garcia et al., 2009b). *A. phagocytophilum* infection is associated with loss of *CYBB* transcript and loss of respiratory burst activity (Banerjee et al., 2000b; Carlyon et al., 2002; Mott and Rikihisa, 2000; Wang et al., 2002). Recombinant AnkA binds AT-rich transcriptional regulatory regions of the *CYBB* locus where eukaryotic transcriptional regulators normally bind. Histone H3 acetylation decreases pronouncedly at the *CYBB* locus, particularly around AnkA-binding sites (Garcia-Garcia et al., 2009b). Ectopically expressed YFP-AnkA downregulates not only *CYBB* expression, but also several genes, most of which encode antimicrobial factors, that are repressed during *A. phagocytophilum* infection (Carlyon et al., 2002; Garcia-Garcia et al., 2009b). Thus, Webster strain-encoded AnkA has been linked to global downregulation of host defense gene expression. Notably, Abl-1, which thus far has only been shown to interact with AnkA derived from the HZ strain (Lin et al., 2007), localizes to the cytoplasm and the nucleus, where it plays distinct roles in cell growth, cytoskeletal reorganization, cell death, and stress responses (Smith and Mayer, 2002). Accordingly, it has been suggested that AnkA may interact with both cytoplasmic and nuclear-translocated Abl-1 to differentially regulate cellular processes (Lin et al., 2007).

Analogous to the phenomenon exhibited by *A. phagocytophilum* Webster strain AnkA, *E. chaffeensis* Ank200 translocates to the nuclei of infected monocytes (Zhu et al., 2009). Chromatin immunoprecipitation coupled with DNA sequencing revealed that Ank200 interacts with host promoter and intronic *Alu-Sx* elements (Zhu et al., 2009), which are short interspersed mobile DNA elements that are distributed in a nonrandom manner and comprise approximately 5 to 10% of the human genome (Rowold and Herrera, 2000). They are believed to be involved in transcriptional regulation as a carrier of *cis*-regulatory elements and have known transcription factor-binding sites. Thus, Ank200 interaction could globally affect host cell transcription through *Alu-Sx* element-mediated transcriptional control (Wakeel et al., 2010b). Indeed, whole-genome analysis with chromatin immunoprecipitation and DNA microarray analysis revealed that genes with promoter *Alu-Sx* elements primarily related to transcription, apoptosis, ATPase activity, and structural proteins associated with the nucleus and membrane-bound organelles were targets of Ank200. Quantitative reverse transcriptase PCR revealed that three Ank200 targets that are associated with ehrlichial pathobiology—tumor necrosis factor alpha (TNF-α), signal transducer and activator of transcription 1 (STAT1), and CD48—are upregulated from approximately 13- to 37-fold during *E. chaffeensis* infection (Zhu et al., 2009). Given this striking observation, experiments analogous to those that revealed that AnkA regulates *CYBB*, which would directly prove whether or not Ank200 modulates expression of innate immune response effectors, are warranted. A perplexing feature that is shared among *E. chaffeensis* Ank200 and *A. phagocytophilum* AnkA proteins is that they each lack a classical monopartite or bipartite nuclear localization signal (Rikihisa and Lin, 2010; Zhu et al., 2009). Thus, these effectors' methods of entry into host cell nuclei are presently unknown.

A. phagocytophilum Ats-1 was identified as a T4SS effector by a bacterial two-hybrid screen using *A. phagocytophilum* VirD4 as bait (Niu et al., 2010). VirD4 is the coupling protein of the T4SS that recognizes C-terminal sequences within T4SS substrates prior to delivery into the VirB transmembrane channel (Fronzes et al., 2009). Ats-1 can be detected associated with intravacuolar bacteria at 22 hours postinfection, but cannot be detected in the host cell cytoplasm until 32 hours postinfection (Niu et al., 2010). Thus, unlike AnkA, which is secreted into host cells upon binding and throughout infection (Garcia-Garcia et al., 2009b; IJdo et al., 2007; Lin et al., 2007), Ats-1 is secreted into the host cells—at least at a suitable abundance such that it can be detected—during the latter stage of infection (Niu et al., 2010). Ats-1 carries an N-terminal eukaryoticlike mitochondrial-targeting motif that is requisite for the effector's localization to the mitochondrial matrix and inner membrane (Niu et al., 2010). The T4SS apparatus presumably attaches to the inner face of extends across the *A. phagocytophilum*-occupied vacuolar membrane (AVM) to serve as a conduit that enables secretion of Ats-1 from within the bacterium directly into the host cell cytoplasm, after which its N-terminal motif directs it to the mitochondria. Ats-1 mitochondrial association is important for preventing the loss of cytochrome *c* and blocking apoptosis (Niu et al., 2010).

SUBVERTING HOST CELLULAR PROCESSES

Evasion and Subversion of Oxidative Killing

A primary means by which neutrophils, monocytes, and macrophages kill phagocytosed microbes depends on the production of toxic reactive oxygen intermediates derived from superoxide anion (O_2^-). The multicomponent enzyme NADPH oxidase, which is unassembled in resting phagocytes, produces O_2^-. Cytochrome b_{558}, which is composed of gp91^{phox} and p22^{phox}, is integrated into the membrane of secretory vesicles and specific granules and, to a lesser extent, the cell membrane. Upon activation, the secretory vesicles and specific granules rapidly fuse with the phagosomal and cell membranes to deliver cytochrome b_{558} to the site of oxidase formation. Concurrently, the cytosolic components p40^{phox}, p47^{phox}, p67^{phox}, and Rac2 translocate to join cytochrome b_{558} and form the functional enzyme. The redox center of assembled oxidase transfers electrons from NADPH to molecular oxygen to generate O_2^- either outside the cell or within the lumen of the phagosome. The O_2^- influx triggers a pH change-dependent release of proteases into the phagosome. Within the phagosome, O_2^- is dismutated into hydrogen peroxide, which in turn is converted into hypochlorous acid. Collectively, the proteases and reactive oxygen derivatives synergistically destroy the phagocytosed microbe (Nauseef, 2007).

E. chaffeensis and *A. phagocytophilum* lack genes encoding ROS-detoxifying enzymes, such as periplasmic Cu/Zn-cofactored superoxide dismutase (SOD), Mn-cofactored SOD, and catalase, and the oxygen-sensing two-component regulatory systems OxyR and SoxRS (Rikihisa, 2010a). Both organisms are killed upon exposure to hydrogen peroxide or paraquat-derived O_2^- (Lin and Rikihisa, 2007). The abilities of *A. phagocytophilum*, *E. chaffeensis*, and other *Anaplasmataceae* members to invade and survive within the otherwise hostile confines of neutrophils and monocytes/macrophages indicate that they have evolved effective means for avoiding and/or subverting host cell ROS production. Host cells with established *A. phagocytophilum* infections are severely inhibited in their ability to generate ROS, as demonstrated ex vivo using neutrophils from HGA patients and experimentally infected mice and in vitro using long-term-infected (2 to 5 days) HL-60 cells (Banerjee et al., 2000; Carlyon et al., 2002; Wang et al., 2002). The inability of long-term-infected cells to generate a respiratory burst is attributable to the fact that *A. phagocytophilum* downregulates transcription of genes encoding two NADPH oxidase components, gp91^{phox} and Rac2 (Banerjee et al., 2000b; Carlyon et al., 2002). This phenomenon has been verified for genes in both *A. phagocytophilum*-infected HL-60 cells and HL-60 cells differentiated along the neutrophil lineage using retinoic acid and has been confirmed for *CYBB* in splenic neutrophils recovered from infected mice. The molecular mechanisms by which *A. phagocytophilum* modulates *CYBB* expression involve secreted AnkA that traverses to the nucleus to bind to the *CYBB* promoter as well as bacterial modulation of host transcription factors that regulate *CYBB* expression (Garcia-Garcia et al., 2009b; Thomas et al., 2005). These mechanisms are discussed in detail in the section "Modulation of Host Cell Gene Expression," below.

Transcriptional downregulation of *CYBB* or *rac2* has not been detected in human neutrophils infected in vitro with *A. phagocytophilum* (Borjesson et al., 2005; Lee et al., 2008; Sukumaran et al., 2005). Altering *rac2* and *CYBB* transcription is probably most relevant to infection of progenitor cells in the bone marrow. *A. phagocytophilum* infection of bone marrow progenitors has been demonstrated in vitro, in HGA patients, and in experimentally infected mice and monkeys (Anonymous, 2001; Foley et al., 1999; Hodzic et al., 2001; Klein et al., 1997; Trofe et al., 2001). Moreover, these cells are more closely related to HL-60 cells and conceivably have long enough life spans to accommodate the rate of *CYBB* downregulation. It is tempting to envision that *A. phagocytophilum* organisms initially invade peripherally circulating neutrophils and are carried to the bone marrow, where they egress to infect bone marrow progenitors. As the infected progenitors differentiate in the bone marrow, *A. phagocytophilum* inhibits *CYBB* and possibly *rac2* expression, thereby leading to the generation of an NADPH oxidase-defective neutrophil. If not properly treated, HGA can progress to a potentially fatal condition associated with increased susceptibility to opportunistic infections (Thomas et al., 2009). The detrimental effect of established *A. phagocytophilum* infection on the NADPH oxidase activity of neutrophils that were originally infected in the bone marrow as progenitor cells would be expected to be at least partially responsible for this condition. Interestingly, mice deficient in *CYBB* are unhindered in their ability to clear an *A. phagocytophilum* infection (Banerjee et al., 2000a), which indicates that although the bacterium is susceptible to ROS-mediated killing, an intact NADPH oxidase is not necessary to clear the infection.

Peripherally circulating neutrophils and monocytes/macrophages are armed with each NADPH oxidase component preformed and ready to assemble. *Anaplasmataceae* pathogens must therefore protect themselves from ROS derived from preformed NADPH oxidase. During binding and invasion, *A. phagocytophilum* does not activate NADPH oxidase, as ROS are not detected and no oxygen is consumed following the addition of the pathogen to neutrophils (Borjesson et al., 2005; Carlyon et al., 2004; IJdo and Mueller, 2004; Mott and Rikihisa, 2000). This phenomenon likely stems from *A. phagocytophilum*'s route of entry, which bypasses the phagocytic route that stimulates NADPH oxidase assembly. *A. phagocytophilum* is able to scavenge O_2^- (Carlyon et al., 2004; IJdo and Mueller, 2004), which conceivably protects it from damage and likely contributes to the inability to detect ROS in vitro. Evidence for an outer membrane-tethered cytochrome *c* oxidase or similar protein that accepts electrons from O_2^- to convert it to harmless O_2 has been presented (Herron and Goodman, 2001). Also, *A. phagocytophilum* encodes a protein containing the 12 transmembrane segments and 6 conserved histidine residues consistent with members of the heme-Cu oxidase family (Dunning Hotopp et al., 2006; Rikihisa, 2010a). Whether this protein is involved in O_2^- scavenging has yet to be investigated. Given that considerable numbers of genes in the annotated genomes of *A. phagocytophilum* and other *Anaplasmataceae* pathogens have no assigned function, it is plausible that these bacteria encode atypical proteins that confer protection from ROS.

It has been suggested that *A. phagocytophilum* and *E. chaffeensis* rapidly block preformed NADPH oxidase activity in neutrophils and monocytes (Lin and Rikihisa, 2007; Mott and Rikihisa, 2000; Mott et al., 2002), respectively, although this is a matter of dispute. For instance, while some groups observed loss of neutrophil NADPH oxidase activity after 30 minutes of incubation with *A. phagocytophilum* (Mott and Rikihisa, 2000), others have not (Carlyon et al., 2004; Choi and Dumler, 2003; IJdo and Mueller, 2004). Differences in these results may be attributable to the different assays employed to measure ROS, the timing relative to infection at which assays were performed, the fact that *A. phagocytophilum* rapidly detoxifies O_2^-, and/or possible *A. phagocytophilum* strain-specific differences. Evidence suggests that *E. chaffeensis* inhibits preformed NADPH oxidase activation by actively degrading or promoting the N-terminal cleavage of p22^{phox} (Lin and Rikihisa, 2007). Within 60 minutes, *E. chaffeensis* infection reduces the amount of p22^{phox}, but not gp91^{phox}, on the surfaces of and in membrane fractions derived from monocytes. Interestingly, this phenomenon can be partially overridden if the monocytes are primed with *E. coli* LPS prior to the addition of *E. chaffeensis*. LPS priming also induces colocalization of p22^{phox} with *E. chaffeensis*, which hints from an evolutionary standpoint one reason as to why *Anaplasmataceae* members may have lost genes for LPS biosynthesis (Lin and Rikihisa, 2007). It was also reported that *A. phagocytophilum* degrades p22^{phox} in neutrophils and HL-60 cells (Mott et al., 2002). However, our laboratory could not reproduce these results (Carlyon et al., 2004).

Following entry into its host neutrophil, *A. phagocytophilum* resides within a host cell-derived vacuole that excludes fusion with secretory vesicles and specific granules harboring gp91^{phox} and p22^{phox} (Carlyon et al., 2004; IJdo and Mueller, 2004). The inability of NADPH oxidase to assemble on the AVM does not extend to *E. coli*-containing phagosomes in the same neutrophil (IJdo and Mueller, 2004). Thus, *A. phagocytophilum* actively modifies the fusogenicity of the AVM to protect itself from NADPH oxidase-mediated killing, but does not globally inhibit NADPH oxidase assembly. Given the remarkable abilities of other *Anaplasmataceae* pathogens to reside within professional phagocytes, this may very well be a conserved intracellular survival strategy among this family.

Subversion of Autophagy

Autophagy is a highly conserved and regulated catabolic process that mediates the degradation of unwanted cytosolic constituents (Deretic, 2010). A double-layered isolation membrane forms around undesirable cytoplasmic components to yield an autophagosome, which degrades them by fusing with a lysosome. In addition to degrading proteins, large protein complexes, and damaged organelles such that their constituents can be recycled, autophagy helps to eliminate intracellular microbes and process nonself and self antigens as part of the innate and adaptive immune response (Cossart and Roy, 2010). Yet *A. phagocytophilum* subverts autophagy to aid its intracellular survival (Niu et al., 2008). Several markers of early autophagosomes are present in the *A. phagocytophilum*-occupied vacuole (ApV), including formation of a double-lipid bilayer membrane and colocalization of green fluorescent protein (GFP)-tagged forms of Beclin-1 and LC3, which are the human homologs of the *Saccharomyces cerevisiae* autophagy-related proteins Atg6 and Atg8, respectively (Deretic, 2010; Niu et al., 2008). Unlike vacuoles inhabited by other intracellular pathogens that subvert autophagy, such as *C. burnetii* and *L. pneumophila*, which recruit LC3 as early as 5 minutes and 4 hours postentry, respectively (Amer and Swanson, 2005; Romano et al., 2007), the ApV does not acquire LC3 until after 20 hours of infection (Niu et al., 2008). Another unusual feature of Beclin-1- and LC3-positive ApVs is that they are arrested in their maturation. They do not acidify or fuse with lysosomes. A functional explanation for this phenomenon is revealed upon examination of the other vacuolar markers and membrane traffic regulators that are recruited to the ApV (see below). Rapamycin-induced stimulation of autophagy favors *A. phagocytophilum* replication. Inhibition of autophagy by 3-methyladenine reversibly arrests *A. phagocytophilum* intracellular growth. Because autophagy engulfs cytosolic components, it has been speculated that the autophagosome may provide direct access to cytosolic nutrients and thereby circumvent having to transport them across the vacuolar membrane. Whether other *Anaplasmataceae* pathogens hijack autophagy is unknown.

Inhibition of Apoptosis

Apoptosis is a host cell defense mechanism that is important for killing intracellular pathogens (DeLeo, 2004). Apoptosis is initiated by enzymatic caspases, which are inactive until they are activated by apoptotic signaling pathways. Intrinsic induction of apoptosis can occur by direct activation of caspases or by intracellular stress (Lindsay et al., 2011). The intrinsic pathway is initiated by events that compromise mitochondrial membrane integrity after the insertion of proapoptotic proteins of the Bcl-2 family and culminates in activation of caspase-3. The Bcl-2 family also consists of antiapoptotic factors that maintain mitochondrial membrane integrity to block caspase activation. Extrinsic induction of apoptosis results from the engagement of surface receptors, such as Fas, by their cognate ligands (DeLeo, 2004). Numerous additional pro- and antiapoptotic factors are involved in determining whether a cell continues to live or progresses into programmed cell death. Moreover, phagocytosis of bacteria elicits specific global changes in neutrophil transcription that constitute an apoptosis differentiation program that is critical to the clearing of infection (DeLeo, 2004; Kobayashi et al., 2003; Kobayashi and DeLeo, 2004).

Peripheral neutrophils exhibit a turnover rate of approximately 10^{11} cells per day and have life spans of 1 to 3 days before undergoing apoptosis (DeLeo, 2004). Their short-lived nature and capacity to undergo apoptosis make neutrophils inhospitable for the long-term survival of intracellular pathogens. Nonetheless, *A. phagocytophilum* successfully resides within the confines of this seemingly unsuitable host cell. Given that *A. phagocytophilum* takes 24 to 30 hours to complete its biphasic life cycle and exit to initiate new infections (Troese and Carlyon, 2009), prolonging the life of its host neutrophil would be beneficial to its survival. Indeed, *A. phagocytophilum* significantly delays apoptosis of human neutrophils infected in vitro (Borjesson et al., 2005; Choi et al., 2005; Ge and Rikihisa, 2006; Ge et al., 2005; Lee and Goodman, 2006; Yoshiie et al., 2000). Similar results were observed for ex vivo cultivated neutrophils isolated from *A. phagocytophilum*-infected sheep (Gokce et al., 1999). This phenomenon depends on *Anaplasma* entry via receptor-mediated endocytosis (Yoshiie et al., 2000). It is not absolutely dependent on but occurs most effectively when executed by live, intracellularly replicating *Anaplasma* and is mediated, at least in part, by a heat-resistant bacterial surface protein (Borjesson et al., 2005; Choi et al., 2005). Live *A. phagocytophilum*-infected neutrophils and neutrophils exposed to heat-killed *A. phagocytophilum* have increased phosphorylation of p38 mitogen-activated protein kinase (MAPK) (Choi et al., 2005). Pharmacologic inhibition of p38 MAPK abrogates the delay in neutrophil apoptosis, but this pathway is bypassed after 3 to 6 hours of active *A. phagocytophilum* infection. Thus, while interaction of a heat-stable bacterial component contributes to the initial block of apoptosis, active infection is critical to prolong the survival of *A. phagocytophilum*-infected neutrophils.

A. phagocytophilum does not induce the apoptosis gene expression program that is usually triggered by bacterial ingestion, as the expression of many apoptosis-related genes either remains unchanged or is delayed throughout 24 hours of infection (Borjesson et al., 2005). *A. phagocytophilum* prevents activation of the intrinsic pathway at multiple steps. For instance, it blocks human neutrophils from reducing mRNA levels of *bfl-1* (Borjesson et al., 2005; Choi et al., 2005; Ge et al., 2005; Sukumaran et al., 2005), the encoded protein of which serves to maintain mitochondrial integrity (Danial and Korsmeyer, 2004). *A. phagocytophilum* infection inhibits the loss of mitochondrial membrane potential and activation of caspase-3 (Ge and Rikihisa, 2006; Ge et al., 2005). Microarray data from human neutrophils or NB4 cells corroborate these findings, demonstrating that *A. phagocytophilum* upregulates expression of antiapoptotic *bcl-2* family genes and downregulates expression of proapoptotic genes (Borjesson et al., 2005; Lee and Goodman, 2006; Pedra et al., 2005; Sukumaran et al., 2005). The pathogen also inhibits translocation of the proapoptotic protein Bax to mitochondria; blocks activation of caspase-9, which is the initiator caspase of the intrinsic pathway; and inhibits degradation of the potent caspase inhibitor X-chromosome-linked inhibitor of apoptosis protein (XIAP) (Ge and Rikihisa, 2006). Ats-1 is an *A. phagocytophilum* T4SS effector that is believed to contribute to the bacterium's antiapoptotic effect, as it inhibits etoposide-induced mitochondrial cytochrome *c* release, poly(ADP-ribose) polymerase cleavage, and apoptosis in mammalian cells (Niu et al., 2010). Bax is a proapoptotic Bcl-2 family protein that, when activated, integrates into the mitochondrial membrane. Bax mitochondrial membrane insertion is implicated in cytochrome *c* release. Ectopically expressed Ats-1 prevents Bax redistribution to the mitochondrial membrane in mammalian cells and prevents Bax-induced apoptosis in yeast cells.

A. phagocytophilum also blocks the extrinsic pathway at multiple steps. It prevents anti-Fas-induced apoptosis and prevents clustering of Fas on the cell surface during spontaneous neutrophil apoptosis (Borjesson et al., 2005; Ge and Rikihisa, 2006). Moreover, it inhibits cleavage of procaspase-8, activation of caspase-8, and the cleavage of Bid, which links the extrinsic and intrinsic pathways (Ge and Rikihisa, 2006).

The abilities of other *Anaplasmataceae* pathogens to delay host cell apoptosis have not been as extensively investigated as those for *A. phagocytophilum*. Yet the studies that have been performed thus far thematically point to apoptosis inhibition as a conserved intracellular survival strategy among the *Anaplasmataceae*. *E. chaffeensis* delays apoptosis of DH82 cells (Liu et al., 2011). *E. chaffeensis* infection yields an antiapoptotic gene expression profile in human monocytes, as genes encoding antiapoptotic factors such as NF-κB, BCL2A1, BIRC3, IER3, and MCL1 are induced and genes encoding proapoptotic factors such as BIK and BNIP3L are repressed (Zhang et al., 2004). In *E. chaffeensis*-infected cells, mitochondria are recruited to within close proximity of morulae, are metabolically dormant, and maintain membrane potential throughout infection (Liu et al., 2011). Based on these observations, it has been speculated that *E. chaffeensis* may be modulating mitochondria by keeping them metabolically dormant and stabilized so as not to release cytochrome *c* and thereby prevent intrinsic induction of apoptosis. *E. ewingii* also delays apoptosis of neutrophils by stabilizing mitochondrial integrity and maintaining mitochondrial membrane potential (Xiong et al., 2008).

Exploitation of the Actin Cytoskeleton To Disseminate Infection

Intracellular microbes can disseminate after host cell lysis following necrotic or apoptotic cell death or by spreading from cell to cell. When propagated in tissue culture cells, many *Anaplasmataceae* species ultimately rupture their host cells to release the bacteria into the culture medium. Yet this arguably fails to reflect how these organisms disseminate in vivo, as lytic release would deliver the bacteria and host cellular debris into the extracellular milieu, both of which would alert the host to the presence of an infection. When DH82 cells or murine macrophages that are infected with *E. chaffeensis*, *E. muris*, or the *Ixodes ovatus Ehrlichia* strain are seeded such that they are separated from one another, long filopodia that extend from the host cells and terminate in a flattened fan-shaped structure can be observed (Fig. 5). The filopodia are prominent in ehrlichiae-infected cells, but are rarely detectable in uninfected cells. The fans are frequently observed extending from an infected cell to make contact with a neighboring uninfected cell. The extensions and their terminal fan-shaped structures are filled with ehrlichiae. Formation of ehrlichiae-containing filopodia is actin dependent but not microtubule dependent. Ehrlichiae display a high affinity for filopodia of neighboring host cells. Scanning electron microscopy of contact points between ehrlichiae-filled filopodia and naïve cells reveals instances where ehrlichiae make contact with and deform the cell membrane from within, which suggests that the filopodia provide a conduit for intercellular ehrlichial transport. These data indicate that *Ehrlichia* species exploit the actin cytoskeleton to spread between host cells. This strategy is important for dissemination during the early stages of in vitro infection, as ehrlichia-filled filopodia are detectable as early as 4 hours postinfection and continue to form and lengthen over the first 24 hours of infection. By 60 hours, however, continual reinfection leads to a substantial bacterial burden that ruptures the host cells to release infectious ehrlichiae (Thomas et al., 2010).

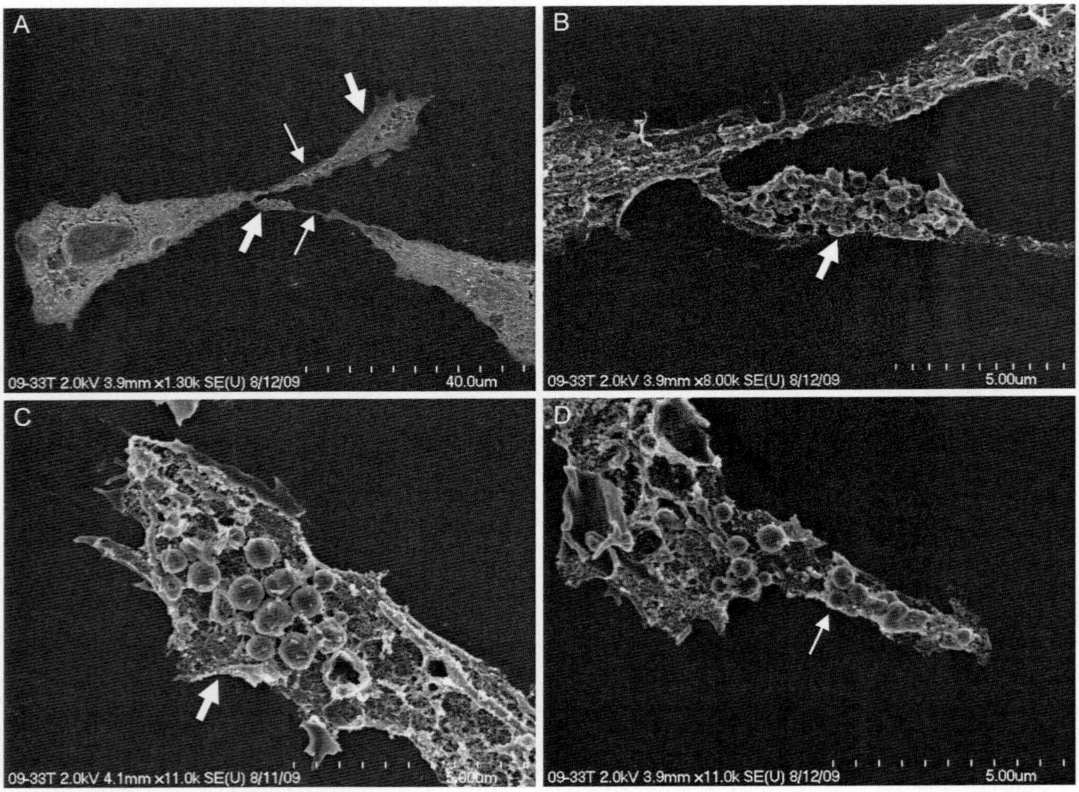

FIGURE 5 *E. chaffeensis* bacteria induce the formation of and are transported through the filopodia in DH82 cells. Shown are scanning electron micrographs of *E. chaffeensis*-infected DH82 cells. Thin arrows indicate filopodia. Thick arrows denote flattened fan-shaped structures at the terminal ends of the filopodia. (A) *E. chaffeensis* infection promotes the formation of filopodia by DH82 cells. (B to D) Filopodia and the terminal flattened fan-shaped structures are filled with *E. chaffeensis* bacteria, as revealed when cell membranes are removed from filopodia. Similar results have been reported for *E. muris* (Thomas et al., 2010). (Courtesy of Sunil Thomas, Vsevolod L. Popov, and David H. Walker, Department of Pathology and Center for Biodefense and Emerging Infectious Diseases, University of Texas Medical Branch.) doi:10.1128/9781555817336.ch6.f5

Formation of ehrlichiae-filled filopodia is likely relevant in vivo, where it would enable ehrlichiae dissemination while avoiding extracellular immune surveillance. The molecular basis for this strategy remains to be defined. It will be important to determine if ehrlichiae bind actin directly to mediate their export or whether ehrlichial proteins presented on the cytoplasmic face of the morula membrane interact with the actin cytoskeleton to facilitate transfer of entire membrane-bound colonies en masse from cell to cell. The mechanisms by which other *Anaplasmataceae* members direct their export from host cells and whether such mechanisms are influenced in vitro by host tissue culture cell density are unknown.

MODULATION OF HOST CELL GENE EXPRESSION

Several microarray analyses have revealed that infection by *Anaplasmataceae* pathogens results in global changes in host cell transcriptional profiles that promote bacterial survival. While the exact lists of differentially expressed genes vary from study to study due to whether primary cells infected in vitro, leukocytes isolated from infected animals, or a variety of immortalized host cell lines were used; the postinfection time points examined; and the gene chips used and their complement of genes, these analyses confirm that these intracellular pathogens have evolved to downregulate expression of host defense genes and proapoptotic genes and upregulate expression of prosurvival and antiapoptotic genes (Borjesson et al., 2005; Carlyon et al., 2002; de la Fuente et al., 2005; Galindo et al., 2008; Lee et al., 2010; Pedra et al., 2005; Sukumaran et al., 2005; Zhang et al., 2004; Zhu et al., 2009). Other categories of genes that are differentially expressed include transcriptional regulation, signal transduction, iron metabolism, ATPase activity, cytoskeletal remodeling, vesicular transport, cytokines, and cellular adhesion (Barnewall et al., 1999; Carlyon et al., 2002, 2005; Galindo et al., 2010; Lee et al., 2008; Pedra et al., 2005; Sukumaran et al., 2005; Zhang et al., 2004; Zhu et al., 2009).

The molecular mechanisms by which *Anaplasmataceae* organisms globally alter host cell gene expression are beginning to be elucidated. *A. phagocytophilum* uses epigenetic silencing of host defense genes to facilitate its intracellular survival. Whereas signaling pathways and transcriptional regulators act on limited subsets of genes, epigenetic regulators typically control gene expression on a global scale and impact major cellular processes to yield dramatic phenotype changes. Reversible histone acetylation, which is mediated by histone-modifying enzymes such as histone deacetylases (HDACs), is a key epigenetic regulator of chromatin structure and gene expression. *A. phagocytophilum* infection significantly increases the expression, activity, and binding of the transcriptional repressor HDAC1 to the promoters of defense genes that are known to be downregulated during infection. This results in an increase in deacetylated histones in infected cells, which in turn likely affects gene expression by leading to a highly compact chromatin conformation with limited access to transcriptional activators. Defense genes are often organized in chromosomal clusters that enable coordinate regulation by changes in chromatin organization. Indeed, many of the defense genes that are downregulated during infection form three gene clusters. Pharmacologic inhibition or silencing of HDAC1 significantly reduces *A. phagocytophilum* load and reverses the bacterium's inhibition of transcription of several defense genes. Conversely, transfecting host cells with a plasmid that overexpresses HDAC1 enhances *A. phagocytophilum* infection, which in turn results in downregulation of defense gene expression (Garcia-Garcia et al., 2009a).

Ankyrin-repeat containing proteins, which intercept host cell signaling pathways (IJdo et al., 2007; Lin et al., 2007), also modulate host cell gene expression (Garcia-Garcia et al., 2009b; Zhu et al., 2009). *A. phagocytophilum* Webster strain AnkA accumulation in the host cell nucleus is coincident with *CYBB* downregulation. Ectopic expression of AnkA in host cells is sufficient to downregulate *CYBB* expression and to alter the expression of other host genes known to be affected by *A. phagocytophilum* infection. AnkA interacts at multiple regions of high AT content in the *CYBB* proximal promoter. It also binds to several AT-rich regions in host genomic DNA, none of which are detected more than once, which suggests that the critical binding feature may not be directly related to primary nucleotide sequence. Interestingly, histone H3 acetylation decreases around AnkA-binding sites, which likely results in chromatin condensation around the *CYBB* locus and silencing of *CYBB* expression. Moreover, AnkA-binding regions overlap with transcriptional regulator-binding sites (Garcia-Garcia et al., 2009b). Downregulation of *CYBB* expression in *A. phagocytophilum*-infected cells is also associated with decreased binding of the transcriptional activators interferon regulatory factor-1 (IRF-1) and PU.1, which results from inhibition of expression of the genes that encode IRF-1 and PU.1. The lack of activator binding is coincident with increased binding of the CCAAT displacement protein repressor (Thomas et al., 2005). Thus, *A. phagocytophilum* utilizes epigenetic alteration of chromatin structure, AnkA effector silencing, and modulation of host transcription factor expression and binding to the *CYBB* promoter in a three-pronged approach that effectively silences *CYBB* expression.

E. chaffeensis Ank200 binds to adenine-rich *Alu-Sx* repeats that lie in the promoters of genes related to transcription, apoptosis, ATPase activity, and structural proteins associated with the nucleus and membrane-bound organelles. TNF-α, STAT1, and CD48, which are pronouncedly upregulated during infection of THP-1 cells, are Ank200 targets and are associated with *E. chaffeensis* pathobiology. Beginning at day 3 postinfection, TNF-α gene expression is dramatically upregulated until it reaches a 27-fold increase in expression on day 7. The upregulation of TNF-α later in infection suggests that this response is not mediated by innate immune receptors and supports that ehrlichial factors modulate TNF-α expression (Zhu et al., 2009). As elevated TNF-α levels are linked to fatal ehrlichiosis (Bitsaktsis and Winslow, 2006; Ismail et al., 2004; Stevenson et al., 2008), these observations indicate that *E. chaffeensis*-infected monocytes/macrophages may contribute to the overproduction of TNF-α and that Ank200 regulation of the TNF-α promoter may be responsible for this phenomenon (Zhu et al., 2009). The STAT1 gene is upregulated as much as 13-fold over the course of infection, and STAT1 is important for ehrlichial intracellular survival. CD48 is a glycosylphosphatidylinositol-anchored protein that is associated with caveolae and is a receptor for the uptake of multiple bacteria (Shin and Abraham, 2001a, 2001b). While CD48 has yet to be identified as an ehrlichial receptor for entry, *E. chaffeensis* invasion involves caveolae and glycosylphosphatidylinositol-anchored proteins (Lin and Rikihisa, 2003b). That *E. chaffeensis*-infected THP-1 cells exhibit as much as a 37-fold increase in CD48 gene expression over the course of infection is striking and offers the possibility that Ank200 may play a role in upregulating expression of a potential ehrlichial receptor (Zhu et al., 2009). Infection by both *E. chaffeensis* and *E. sennetsu* is associated with a considerable rise in transferrin receptor (TfR) mRNA expression, a phenomenon that has been speculated to be part of a larger strategy aimed at acquiring iron from the host cell. *E. chaffeensis* and *E. sennetsu* each upregulates TfR mRNA levels by increasing the affinity of iron-responsive protein-1 (IRP-1) for the *cis*-acting regulatory iron-responsive element that is in the 3′ untranslated region of TfR mRNA (Barnewall et al., 1999).

Anaplasmataceae pathogens also alter host cell gene expression within their arthropod vectors to enhance colonization and facilitate their survival. Indeed, *A. phagocytophilum* uses α1,3-fucose to colonize ticks. *A. phagocytophilum* infection upregulates α1,3-fucose gene expression, which likely enhances vector colonization (Pedra et al., 2010). *A. phagocytophilum* also induces *I. scapularis* expression of a gene that encodes an antifreeze protein, *I. scapularis* antifreeze glycoprotein, which enhances tick tolerance to and survival in cold temperatures (Neelakanta et al., 2010).

EXPLOITATION OF HOST SIGNALING PATHWAYS

Anaplasmataceae members alter host cell signaling pathways to facilitate invasion. For example, *A. phagocytophilum* engagement of the N terminus of PSGL-1 promotes activation of spleen tyrosine kinase (Syk), which results in tyrosine phosphorylation of Rho-associated, coiled-coiled kinase 1 (ROCK1), an effector kinase of the RhoA GTPase (Reneer et al., 2008; Thomas and Fikrig, 2007). Knockdown of Syk or ROCK1 or pharmacologic inhibition of Syk markedly reduces *A. phagocytophilum* infection (Thomas and Fikrig, 2007). When *A. phagocytophilum* interacts with neutrophils under shear-flow conditions that mimic the high-velocity conditions under which the bacterium and neutrophils interact in a blood vessel, the pathogen binds to neutrophils within seconds. Bacterial binding to neutrophils prompts a rise in intracellular calcium concentration, activates β_2-integrin, and pronouncedly inhibits p38 MAPK phosphorylation. *Anaplasma*-induced calcium flux is dependent on bacterial binding to PSGL-1 and results in neutrophil arrest via interaction with E-selectin presented on human umbilical vein endothelial cell surfaces (Schaff et al., 2010). β_2-Integrin activation is also dependent on activation of Syk, which in turn is activated by *A. phagocytophilum* engagement of PSGL-1 (Schaff et al., 2010; Thomas and Fikrig, 2007; Zarbock et al., 2007). Thus, PSGL-1 not only serves as a receptor for anchoring *A. phagocytophilum*, but also stimulates signaling events that mediate pathogen entry and promote neutrophil arrest on vascular endothelial cells. Bacterial inhibition of p38 MAPK activation, along with additional uncharacterized mechanisms, prevents neutrophil polarization and transmigration through a human umbilical vein endothelial cell monolayer and results in the release of infected neutrophils (Schaff et al., 2010). The mechanisms by which *A. phagocytophilum* transfers between host cells in vivo are unknown. One postulation is that *Anaplasma*-induced upregulation of interleukin-8 attracts naïve neutrophils to an infected neutrophil by chemotaxis (Akkoyunlu et al., 2001). It has also been proposed that infected endothelial cells may be a reservoir for *A. phagocytophilum* and serve to directly infect marginated neutrophils in the circulation (Herron et al., 2005). Both scenarios would be promoted by initial adhesion of neutrophils to inflamed endothelium (Schaff et al., 2010). Also, inhibition of transmigration and eventual demargination of infected neutrophils is likely a mechanism by which *A. phagocytophilum* ensures that its host cells remains in circulation, which would optimize its chances of being acquired by a feeding tick to continue its mammalian reservoir-tick vector life cycle. In further support of this hypothesis, it has been reported that *A. phagocytophilum*-infected neutrophils exhibit diminished PSGL-1 and L-selectin expression and, consequently, reduced adhesion to endothelial cells (Choi et al., 2003).

There are conflicting findings as to whether or not p38 MAPK activation is associated with *A. phagocytophilum* infection under static cell culture conditions. One study observed that during the first 30 minutes of neutrophil exposure to *A. phagocytophilum* in vitro, a time point that corresponds to bacterial binding and the early stages of entry, p38 MAPK is phosphorylated. Antagonism of p38 MAPK activity reverses *A. phagocytophilum*-mediated apoptosis inhibition during the first 30 minutes. However, pharmacologic inhibition of p38 MAPK at 3 hours postexposure, a postentry time point, does not reverse apoptosis inhibition. This suggests that *A. phagocytophilum* initiates p38 MAPK signaling to inhibit the apoptotic response during invasion, but that p38 MAPK is not involved in the long-term inhibition of apoptosis that is associated with bacterial differentiation and intracellular replication (Choi et al., 2005). Another study did not observe p38 MAPK activation during the first 30 minutes of or any other time point during infection (Kim and Rikihisa, 2002).

E. chaffeensis invasion is associated with host protein tyrosine phosphorylation, phospholipase C-γ2 (PLC-γ2) activation, inositol 1,4,5-triphosphate production, and an increase in intracellular calcium concentration. Treatment with inhibitors of protein tyrosine kinase, PLC, calcium mobilization inhibitors, and transglutaminase each severely antagonizes *E. chaffeensis* entry (Lin et al., 2002a). *E. chaffeensis* activates extracellular signal-related kinase 1/2 (ERK1/2) and p38 MAPK during the early stage of monocyte infection in vitro. However, monocytes with established *E. chaffeensis* infection are less responsive to *E. coli* LPS-mediated activations of ERK1/2 and p38 MAPK and exhibit reduced expression of several pattern recognition receptors, such as CD14, Toll-like receptor 2 (TLR2), and TLR4. Since treatment of monocytes with a p38 MAPK-specific inhibitor results in analogous downregulation of CD14, TLR2, and TLR4 as observed with *E. chaffeensis* infection, it is speculated that the bacterium inhibits p38 MAPK to inhibit critical monocyte signaling pathways linked to pattern recognition receptors (Lin and Rikihisa, 2004).

Anaplasmataceae pathogens exploit host cell signaling pathways to facilitate their intracellular survival. Whereas *E. chaffeensis* activates ERK1/2 only during early infection (Lin and Rikihisa, 2004), *A. phagocytophilum* only activates ERK1/2 during the intracellular replication stage (Sukumaran et al., 2011; Xiong et al., 2009). A secreted effector that is preferentially expressed during infection of mammalian versus tick cells, *A. phagocytophilum* toxin A (AptA; APH_0233), associates with the pathogen-occupied vacuole and activates ERK1/2 through MEK. AptA also interacts with the intermediate filament protein vimentin. Vimentin is required for ERK1/2 activation and maintains ERK1/2 in a phosphorylated state. Thus, through concerted actions, AptA and vimentin maintain ERK1/2 activation. Doing so is critical to *A. phagocytophilum* survival, as pharmacologic inhibition of ERK1/2 or knockdown of vimentin reduces bacterial loads (Sukumaran et al., 2011). Also, in *A. phagocytophilum*-infected cells, vimentin is reorganized as a network that surrounds the *Anaplasma*-occupied vacuole (Sukumaran et al., 2011), a phenomenon that is analogous to the actin and intermediate filament scaffolds that form around and stabilize the *Chlamydia trachomatis* vacuole (Kumar and Valdivia, 2008). The role that ERK1/2 plays in *A. phagocytophilum* infection remains unexplained, but has been speculated to be important for pathogen acquisition, since a similar strategy has been documented for *C. trachomatis* (Su et al., 2004).

Gamma interferon (IFN-γ) is elicited as part of the immune response to and plays an important role in clearing infections caused by intracellular pathogens. Indeed, IFN-γ dominates the murine cytokine response to *A. phagocytophilum* and helps to control the degree of early infection (Akkoyunlu and Fikrig, 2000). Moreover, if IFN-γ is added prior to or at the early stage of *E. chaffeensis* infection of monocytes in vitro, the infection is cleared (Barnewall and Rikihisa, 1994). However, *E. chaffeensis* and *A. phagocytophilum* have evolved strategies to inhibit host cell IFN-γ responses. If IFN-γ is added 24 hours postinfection, the infection is not cleared (Barnewall and Rikihisa, 1994). *E. chaffeensis* activates protein kinase A 25-fold to block IFN-γ-induced tyrosine phosphorylation of Janus-activated kinase 1 (JAK1), JAK2, and STAT1 and thereby prevent a host cell ehrlichiacidal response (Lee and Rikihisa, 1998). Similarly, *A. phagocytophilum*-infected neutrophils exhibit diminished STAT1 phosphorylation and reduced expression of the IFN-γ receptor α-chain CD119 (Bussmeyer et al., 2010). *A. phagocytophilum*-infected neutrophils also exhibit decreased secretion of IFN-γ-inducible chemokines CXCL10/IP-10 and CXCL9/MIG (Bussmeyer et al., 2010), which otherwise favor the recruitment of NK cells and Th1 cells to the sites of infection by intracellular pathogens (Khan et al., 2000). *A. phagocytophilum* also induces neutrophil mRNA and protein expression of suppressor of cytokine signaling 1 (SOCS1) and SOCS3, which are normally produced in a negative-feedback mechanism to inhibit IFN-γ-mediated inflammatory responses (Bussmeyer et al., 2010).

In vivo studies performed with *E. muris* provide evidence that another effective strategy that ehrlichiae use to survive within the host to maintain chronic infection is to interfere with immune-activating signals produced by T cells rather than by inhibiting antigen presentation or T-cell activation (Nandi et al., 2009). Yet further means by which *E. chaffeensis* intercepts host cell signaling pathways is through one of its many TRPs, TRP47. TRP47 is a multifunctional protein that is secreted to localize to the *Ehrlichia*-occupied vacuolar membrane. Yeast two-hybrid and ectopic expression studies reveal that TRP47 interacts with host proteins involved in transcriptional regulation, vesicular trafficking, and cell signaling (Wakeel et al., 2009). While tyrosine kinases are important for *E. chaffeensis* entry, the specific kinases are unknown (Lin et al., 2002a). TRP47 is phosphorylated during infection (Wakeel et al., 2010a). TRP47 associates with FYN and is associated with the intracellular dense-cored ehrlichial form. FYN specifically phosphorylates caveolin-1, a protein that is implicated in ehrlichial entry and is found on the *E. chaffeensis*-occupied vacuolar membrane (EVM) (Lin and Rikihisa, 2003b). Taken together, these data suggest that TRP47 interaction with FYN may be involved in caveolin-1-mediated entry of *E. chaffeensis*. TRP47 also interacts with PTPN2 (Wakeel et al., 2009), which is a protein tyrosine phosphatase that dephosphorylates several host proteins, including JAK1 and STAT1 (Stuible et al., 2008). Accordingly, the TRP47-PTPN2 interaction has been speculated to be important for the inhibition of IFN-γ-mediated JAK1 and STAT1 activation associated with *E. chaffeensis* infection (Wakeel et al., 2010b).

Little is known regarding the *Anaplasmataceae* pathogen modulation of arthropod vector signaling pathways. One remarkable study demonstrates that *A. phagocytophilum* induces actin phosphorylation in *I. scapularis* ticks to selectively enhance transcription of *salp16*, which encodes a tick salivary protein that in turn is critical for *A. phagocytophilum* survival in ticks (Sukumaran et al., 2006; Sultana et al., 2010). *A. phagocytophilum* infection increases phosphorylation of actin in an *I. ricinus* tick cell line and *I. scapularis* ticks to alter the ratio of monomeric to filamentous actin, which leads to translocation of phosphorylated/monomeric actin into the nuclei of infected tick salivary glands. *A. phagocytophilum*-induced actin phosphorylation is dependent on Gbg stimulation involving the activation of *Ixodes* phosphoinositide 3-kinase and *Ixodes* p21-activated kinase. Silencing of *pi3k*, *gbg*, or *pak1* reduces actin phosphorylation and bacterial acquisition by ticks. Phosphorylated/monomeric actin associates with RNA polymerase II (RNAPII) and enhances binding of TATA box-binding protein to RNAPII and selectively promotes transcription of *salp16* (Sultana et al., 2010).

THE *ANAPLASMATACEAE* PATHOGEN-OCCUPIED VACUOLE

Properties of the *Anaplasmataceae* Pathogen-Occupied Vacuole

Anaplasmataceae members have evolved strategies for remodeling the host cell-derived vacuoles in which they reside into permissive organelles that enable nutrient acquisition and bacterial replication while sequestering the organisms from detection, from their host cells' armamentariums of microbicidal compounds, and from delivery to lysosomes. Avoidance of fusion with lysosomes is a conserved theme among *Anaplasmataceae* members, as this phenomenon has been observed for vacuoles inhabited by *A. phagocytophilum*, *E. chaffeensis*, *E. risticii*, and *N. sennetsu* (Barnewall et al., 1997; Mott et al., 1999; Wells and Rikihisa, 1988). The diverse manners by which different *Anaplasmataceae* members remodel their host organelles are just beginning to be deciphered. Much of what we know derives from studies of vacuolar markers on the EVM and the AVM. Within minutes of their uptake, *A. phagocytophilum* and *E. chaffeensis* actively remodel the properties of their vacuoles in a bacterial protein synthesis-dependent manner. For instance, 30 minutes following the addition of tetracycline, the ApV and *E. risticii*-occupied vacuole each fuses with lysosomes and the pathogens are destroyed (Gokce et al., 1999; Huang et al., 2010a; Wells and Rikihisa, 1988).

Organelle Markers and Fusogenic Properties of the ApV

The ApV does not resemble early endosomes because the AVM lacks TfR, early endosomal antigen 1 (EEA1), Rab5, and annexins I, II, IV, and VI (Mott et al., 1999; Webster et al., 1998). The AVM lacks clathrin heavy chain, which indicates that it does not intercept clathrin-dependent endocytic vesicles. The ApV does not acidify; does not acquire the late endosomal/lysosomal markers myeloperoxidase, CD63, lysosomal-associated membrane protein-1 (LAMP-1), and vacuolar (V)-type H^+ ATPase; and avoids lysosomal fusion. Tyrosine-phosphorylated proteins and PLC-γ are found on early ApVs (Lin and Rikihisa, 2003b), which may be due to their retention following bacterial entry. The AVM carries the v-SNARE protein vesicle-associated membrane protein (VAMP2), which helps control vesicular targeting, docking, and fusion (Brumell et al., 1995). The caveolae marker protein caveolin-1 is also found on the AVM (Lin and Rikihisa, 2003b). Presence of VAMP2 on the AVM may aid in bacterial acquisition of lipid membrane and/or nutrients. The ApV does not acquire the Golgi markers β-COP or C_6-7-nitrobenzo-2-oxa-1,3-diazole (NBD)-ceramide. *A. phagocytophilum* growth is insensitive to brefeldin A, which inhibits anterograde traffic from the Golgi complex. Thus, intercepting Golgi traffic is likely inessential for *A. phagocytophilum* survival. As covered elsewhere in this chapter, the ApV excludes fusion with secretory vesicles and specific granules harboring NADPH oxidase and proteolytic enzymes (Carlyon et al., 2004; IJdo and Mueller, 2004; Mott et al., 2002).

While the ApV is negative for early endosomal, late endosomal, lysosomal, and Golgi markers, it is not an inert compartment that is completely sequestered from all membrane traffic. Endocytosed bovine serum albumin conjugated to gold, which is a useful marker for monitoring endocytic traffic because it is readily endocytosed and traffics through the endocytic pathway with defined kinetics (Alvarez-Dominguez et al., 1997; Mobius et al., 2003; Punnonen et al., 1998), traffics through the ApV (Mott et al., 1999). Incubation at 13°C, a temperature known to inhibit early-late endosomal fusion, considerably reduces the number of bovine serum albumin-gold–positive ApVs. Thus, the ApV intercepts some arm of endocytic traffic. Indeed, evidence suggests that the ApV intercepts recycling endosomes. Major histocompatibility complex class I and class II molecules, which are recycled to the plasma membrane via clathrin-independent recycling endosomes, are found on the AVM (Mott et al., 1999; Radhakrishna and Donaldson, 1997). As discussed below, the ApV selectively recruits several Rab GTPases that predominantly associate with recycling endosomes.

Organelle Markers and Fusogenic Properties of the EcV

E. chaffeensis occupies a host cell-derived vacuole that is distinct from the organelle inhabited by *A. phagocytophilum*. In fact, when both pathogens coinfect an individual HL-60 cell, each resides within organelles that are separate from one another and never fuse (Mott et al., 1999). The EVM shares some traits with the AVM in that it is negative for late endosomal, lysosomal, and Golgi markers and *E. chaffeensis* growth is unaffected by brefeldin A. Also like the AVM, PLC-γ and tyrosine-phosphorylated proteins are found on early *E. chaffeensis*-occupied vacuoles (EcVs) and caveolin-1 associates with the EVM throughout infection (Lin and Rikihisa, 2003b). Unlike the AVM, which displays no characteristics of early endosomes, the EVM is strongly positive for Rab5, EEA1, and TfR. Also unlike the ApV, to which mitochondria and endoplasmic reticulum do not associate in close proximity, both organelles accumulate around the ECV and the *E. canis*-occupied vacuole (Mott et al., 1999; Popov et al., 1998). Acquisition and/or retention of endosomal markers is selective, as the EVM is negative for clathrin heavy chain, α-adaptin, and annexins I, II, IV, and VI. Whether the EcV intercepts recycling endosomes remains to be determined. However, since major histocompatibility complex class I and class II molecules are found on less than 10% of EcVs, and Rab5, which is found on early endosomes but is absent from recycling endosomes (Stenmark, 2009), is heavily present on the EVM (Mott et al., 1999), it is unlikely that the EcV hijacks recycling endosomes. Overall, the EcV can be described as a modified early endosome that accumulates TfR throughout the course of infection.

Implications of the Different Fusogenic Traits of the ApV and EcV for Iron Acquisition

The EcV hijacks early endosomes and accumulates TfR, while the ApV does not. This difference is related, at least in part, to the disparate means by which *E. chaffeensis* and *A. phagocytophilum* obtain iron. *E. chaffeensis* likely acquires iron by hijacking TfR. *E. chaffeensis* upregulates TfR mRNA levels, which translates to a higher abundance of TfR in *E. chaffeensis*-infected cells and an accumulation of TfR on the EVM (Barnewall et al., 1999). Iron-transferrin-TfR complexes are conceivably delivered to the EcV by its interception of early endosomes. Iron dissociates from transferrin in mildly acidic environments, and acidification of endosomes delays TfR recycling back to the plasma membrane (Johnson et al., 1993). The EVM is weakly positive for V-type ATPase, and 3-(2,4-dinitroanilino)-3′-amino-*N*-methyldipropylamide, which accumulates in acidic compartments in direct proportionality to acidity (Anderson et al., 1984), weakly accumulates in the EcV (Barnewall et al., 1997, 1999). Thus, presence of the V-type ATPase on the EVM may promote an intravacuolar environment that favors release of iron to the bacteria and retention of the TfR. Support for this premise comes from the observation that deferoxamine, which is a cell-permeable iron chelator that removes iron from the labile pool that would otherwise be accessed by transferrin, prevents *E. chaffeensis* and *N. sennetsu* intracellular survival (Barnewall et al., 1999). Notably, *A. phagocytophilum* does not alter host cell TfR levels or the binding affinity of IRP-1, recruits neither TfR nor V-type ATPase to its vacuolar membrane, does not intercept early endosomes, and is only partially affected by deferoxamine (Barnewall et al., 1999; Huang et al., 2010a; Mott et al., 1999). Rather, it has been proposed that *A. phagocytophilum* degrades or promotes the degradation of ferritin to release iron for its use (Carlyon et al., 2005). A precedent for this hypothesis has been set by the observation that *Neisseria meningitidis* accelerates ferritin degradation to yield a usable iron source (Larson et al., 2004).

Anaplasmataceae-Derived PVM Proteins

Anaplasmataceae pathogen-occupied vacuoles are developmentally arrested and sequestered outside the normal endocytic continuum and selectively interact with membrane-trafficking pathways to serve as optimal niches for intracellular survival. Much of the biology of *Anaplasmataceae* pathogen-occupied vacuoles likely stems from molecular interactions between pathogen-encoded proteins that associate with/integrate into the vacuolar membrane and host cell proteins. Indeed, many proteins encoded by other vacuole-adapted bacterial pathogens play critical pathobiological roles, which include providing structural integrity to the pathogen-occupied vacuolar membrane (PVM), hijacking vesicular traffic, and modulating host cell signaling pathways. Several *Anaplasmataceae*-encoded PVM proteins have been identified and characterized, but a functional role has only been verified for one thus far. *A. phagocytophilum* AptA is a highly positive-charged, 33.0-kDa protein with an isoelectric point (pI) of 10.3 that localizes to the AVM in human neutrophils, HL-60 cells, and monkey choroidal endothelial RF/6A cells. AptA is not expressed in ticks or a tick-derived cell line. It is poorly expressed during the first 12 hours of infection, but is abundantly expressed by 24 hours. AptA interacts with ERK1/2 and vimentin. AptA and vimentin act concertedly to phosphorylate ERK1/2, a phenomenon that is integral to *A. phagocytophilum* intracellular survival. AptA recruits vimentin to the AVM, where it forms a scaffold around the ApV (Sukumaran et al., 2011).

A. phagocytophilum APH_0032 (also referred to as P130) and APH_1387 (P100) and *E. chaffeensis* TRP32 (ECH_0170; variable-length tandem repeat protein), TRP47 (ECH_0166), and TRP120 (ECH_0039) are TRPs that localize to the PVM (Doyle et al., 2006; Huang et al., 2010b, 2010c; Luo et al., 2008; Popov et al., 2000). TRPs of pathogenic bacteria have been implicated in hijacking host signaling pathways, adhesion, immune evasion, and other host-pathogen interactions (Clifton et al., 2004; Futse et al., 2009; Hussain et al., 2008; IJdo et al., 2007; Lin et al., 2007; Wakeel et al., 2009; Wistedt et al., 1995; Zhu et al., 2009). While their sequences are not homologous to each other, considerable portions (46.0 to 60.0%) of these *Anaplasmataceae* TRP proteins each consist of tandem direct repeats (Doyle et al., 2006; Luo et al., 2008; Storey et al., 1998; Wakeel et al., 2009; Yu et al., 2000). Despite amino acid sequence variation among the tandem repeats of these proteins, conservation of amino acid usage is consistent. A total of 10 amino acids are used in all of the repeats, with high frequencies of serine threonine, alanine, proline, valine, aspartate, and glutamate (Wakeel et al., 2010b). All are acidic (pI, 3.6 to 4.2) due to the predominance of acidic amino acid residues primarily in the tandem repeat region, which causes them to migrate anomalously when resolved by sodium dodecyl sulfate-polyacrylamide gel electrophoresis (Huang et al., 2010b, 2010c; Luo et al., 2008, 2009; Storey et al., 1998). TRP47, TRP120, and APH_0032 are differentially expressed late during infection by dense-cored and mature reticulate cell organisms that are differentiating to the dense-cored form (Doyle et al., 2006; Huang et al., 2010b; Popov et al., 2000). TRP47 and TRP120 have been implicated in mediating adhesion and/or entry (Doyle et al., 2006; Popov et al., 2000). APH_0032 is not detected on dense-cored bacteria bound to the host cell surface, but is only detected on intravacuolar *A. phagocytophilum* organisms and the AVM (Huang et al., 2010b). TRP32 and APH_1387 are expressed by both reticulate cell and dense-cored organisms throughout the course of infection (Huang et al., 2010c; Luo et al., 2008). Like AptA (Sukumaran et al., 2011), APH_1387 and APH_0032 are expressed and localize to the AVM during in vitro infection of HL-60 cells (Fig. 6 and 7), endothelial cells, and tick embryonic ISE6 cells, as well as in vivo in murine neutrophils (Huang et al., 2010b, 2010c). Both proteins are also expressed in *I. scapularis* salivary glands (Huang et al., 2010b, 2010c). The roles of TRP proteins are unknown. A yeast two-hybrid screen followed by confirmatory colocalization studies using ectopically expressed proteins suggests that TRP47 is a multifunctional protein that interacts with several host proteins involved in vesicular traffic, signal transduction, and transcriptional regulation (Wakeel et al., 2009). *E. chaffeensis* TRP120, TRP47, and TRP32 have orthologs in *E. canis* termed TRP140 (ECA_0017), TRP36 (ECA_0109), and TRP19 (ECAJ_0113), respectively (Doyle et al., 2006; Luo et al., 2009; McBride et al., 2007). TRP19 is expressed by both dense-cored and reticulate cell organisms and localizes to the *E. canis*-occupied vacuolar membrane (McBride et al., 2007). Whether TRP140 or TRP36 localizes to the PVM has yet to be studied. The only other confirmed *Anaplasmataceae* pathogen-derived inclusion membrane protein that has been identified is *E. canis* (Jake strain) morula membrane protein A (MmpA; ECAJ_0851), which encodes a slightly basic (pI = 8.49), 24.0-kDa protein of unknown function that lacks tandem repeats but does carry five nonlobed hydrophobic domains that are predicted to traverse the PVM (Teng et al., 2003).

All of the known *Anaplasmataceae*-derived PVM proteins are predicted to be highly antigenic (Fig. 8). Antibodies against all except AptA, against which the immune response has not yet been investigated, have been detected in the sera of naturally infected humans and animals and experimentally infected animals (Chen et al., 1994; Doyle et al., 2006; Dumler et al., 1995; Huang et al., 2010b, 2010c;

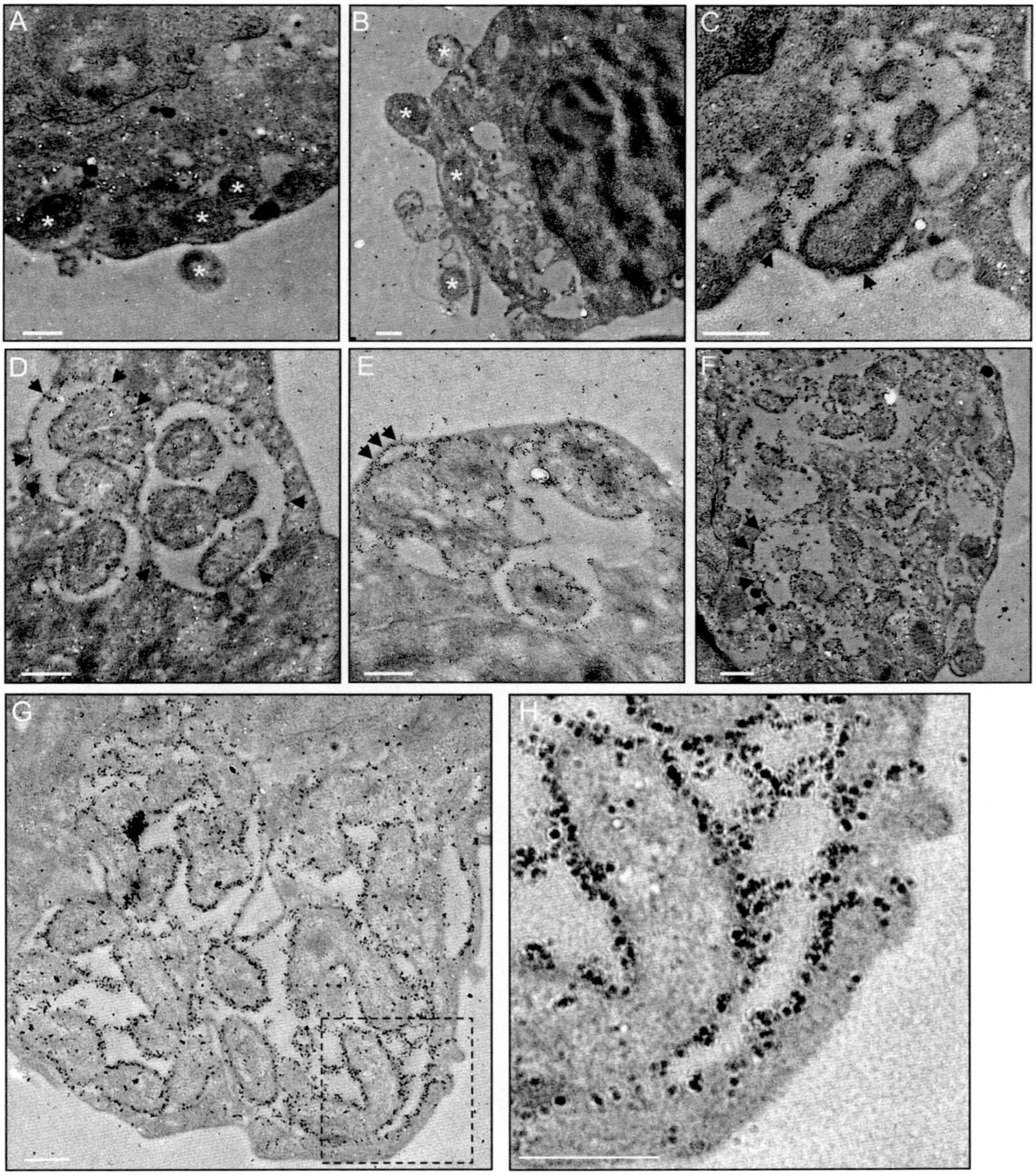

FIGURE 6 APH_1387 is expressed and localizes to the AVM throughout the course of infection. HL-60 cells were synchronously infected with *A. phagocytophilum*. At 0.7 (A), 4 (B), 8 (C), 12 (D), 18 (E), 24 (F), and 48 h (G and H) post-bacterial addition, samples were fixed and screened with anti-APH_1387 followed by goat anti-rabbit immunoglobulin G conjugated to 6-nm gold particles and examined by electron microscopy. (A and B) Asterisks denote bound or newly internalized *A. phagocytophilum* dense-cored organisms. (C to F) Arrowheads denote representative portions of the AVM that are labeled with gold particles. (H) Magnified view of the region in panel G that is demarcated by a hatched box. Bars, 0.5 μm. (Reprinted from Huang et al. [2010c] with permission of the publisher.) doi:10.1128/9781555817336.ch6.f6

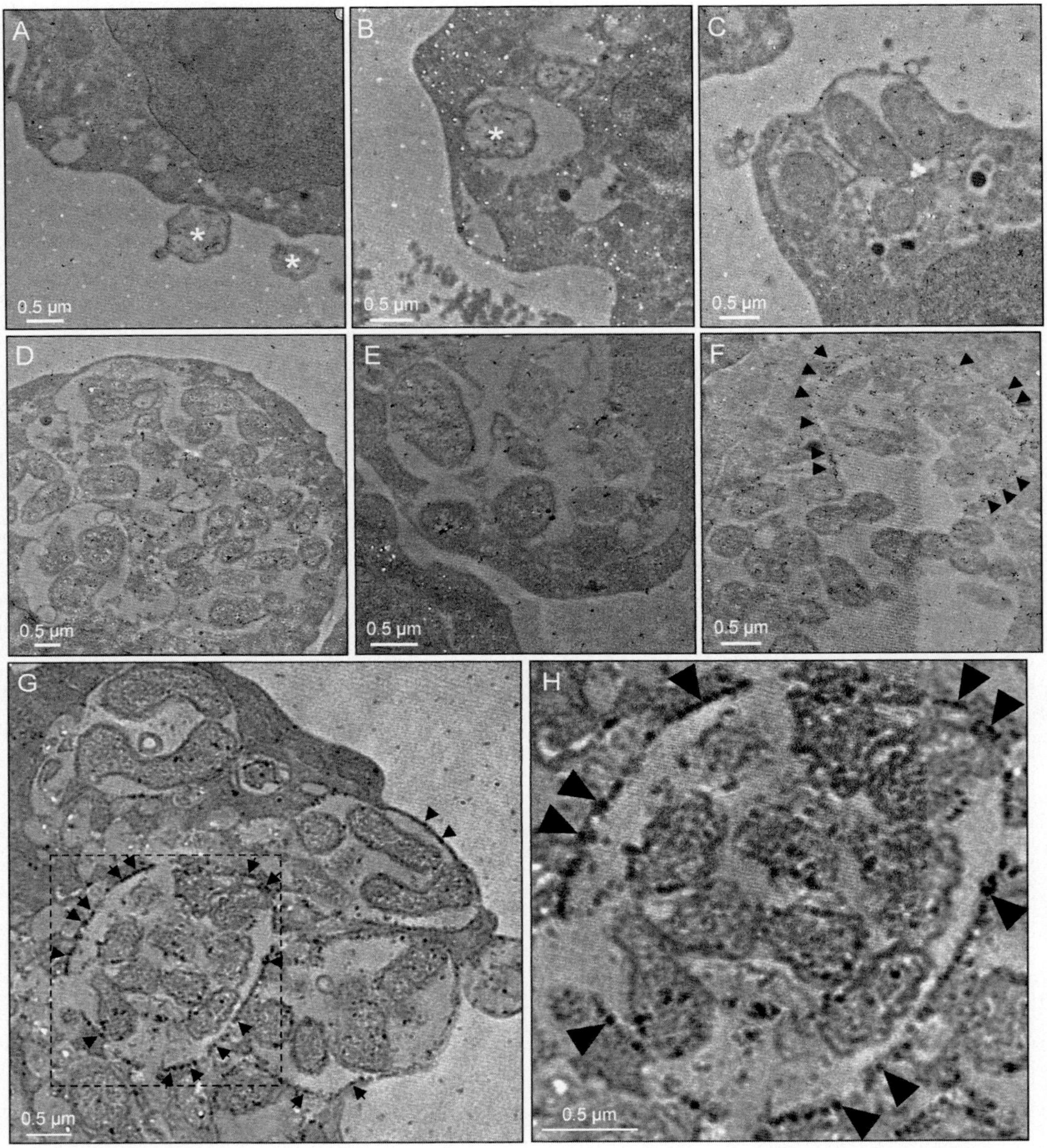

FIGURE 7 APH_0032 is expressed and localizes to the AVM late during infection. HL-60 cells were synchronously infected with *A. phagocytophilum*. At 0.7 (A), 4 (B), 8 (C), 12 (D), 18 (E), 24 (F), and 48 h (G) post-bacterial addition, samples were fixed and screened with anti-APH_0032 followed by goat anti-rabbit immunoglobulin G conjugated to 6-nm gold particles and examined by transmission electron microscopy. (A and B) Asterisks denote bound or newly internalized *A. phagocytophilum* dense-cored organisms. (F to H) Arrowheads denote representative portions of the AVM that are labeled with gold particles. (H) Magnified view of the region in panel G that is demarcated by a hatched box. (Reprinted from Huang et al. [2010b] with permission of the publisher.) doi:10.1128/9781555817336.ch6.f7

Luo et al., 2008, 2009; McBride et al., 2007; Storey et al., 1998; Teng et al., 2003; Yu et al., 1996). Each PVM protein exhibits one to several predicted hydrophobic stretches that putatively facilitate their insertion into/association with the PVM (Fig. 8). In order for these proteins to associate with the PVM, they must first be secreted from the bacterial cell. *Anaplasmataceae* members encode a T4SS and a T1SS (Dunning Hotopp et al., 2006). Though there are no conserved protein sequence motifs of type IV substrates, certain C-terminal sequences are requisite for secretion via the T4SS apparatus. In *A. tumefaciens*, three or more basic amino acids, net positive charge, and a characteristic hydrophobic hydropathy profile are important for T4SS secretion of effector proteins (Vergunst et al., 2005). *A. phagocytophilum* type IV secretion has been demonstrated for AnkA and Ats-1 (Lin et al., 2007; Niu et al., 2010), both of which have C-terminal positive-charged residues and hydropathy profiles similar to those of T4SS substrates (Rikihisa and Lin, 2010). AptA is a strong T4SS effector candidate because it carries a net charge of +25.4, is basic, and has three basic amino acids in its C terminus. MmpA is not quite as strong a candidate because it carries a net charge of +3.63, is basic, but has only two basic amino acids it its C terminus. None of the TRP proteins are T4SS candidates because they are highly acidic and have overall net negative charges. Indeed, APH_0032, APH_1387, TRP120, TRP47, and TRP32 are not secreted by heterologous T4SSs (Huang et al., 2010b; Wakeel et al.,

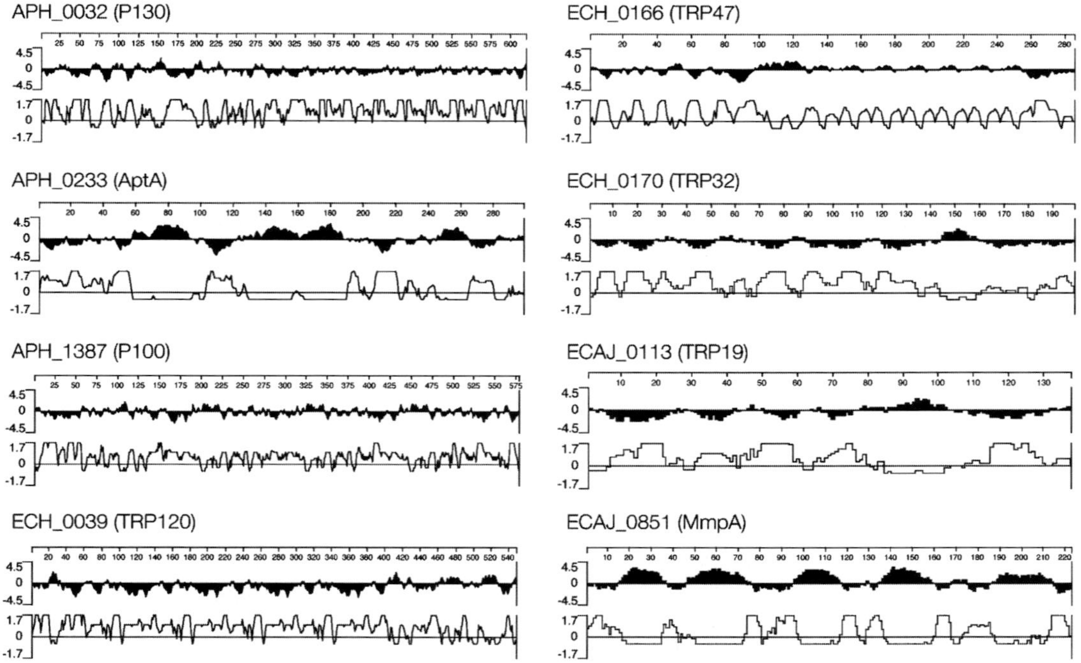

FIGURE 8 Hydropathy and antigenicity profiles of confirmed *Anaplasmataceae* PVM proteins. Numerical scales correspond to the entire amino acid sequence of each protein. Hydropathy plots were generated using the Kyte-Doolittle algorithm to denote hydrophobic (black filled histogram above the *x* axis) and hydrophilic (black filled histogram below the *x* axis) regions (Kyte and Doolittle, 1982). Antigenicity plots were generated using the Jameson-Wolf algorithm to denote regions that are predicted to be antigenic (unfilled histogram above the *x* axis) and/or nonantigenic (unfilled histogram below the *x* axis) (Jameson and Wolf, 1988). Analyses were performed using Protean, which is part of the Lasergene software package. doi:10.1128/9781555817336.ch6.f8

2010b). We hypothesize that TRP proteins are type I, or ATP-binding cassette-dependent secretion substrates. T1SSs recognize a C-terminal uncleaved secretion signal that is rich in certain amino acids (LDAVTSIF) and poor in others (KHPMWC) (Delepelaire, 2004); 72.7% of the APH_0032 C-terminal amino acids (APSTGVEIRFMDRDSDDDVLAL) are found in type I secretion signals, and 71.4% of the APH_1387 C-terminal amino acids (LVDVPTALPLKDPDDEDVLSY) are found in type I secretion signals (Huang et al., 2010b). T1SS substrates are acidic (pI, ~4) and contain very few or no cysteines (Delepelaire, 2004). Only 5 of the 619 and 4 of the 578 amino acids that compose APH_0032 and APH_1387, respectively, are cysteines (Huang et al., 2010b). Ehrlichial TRP C termini also carry few cysteines. We propose that *Anaplasmataceae* TRPs that exhibit PVM localization are delivered to the intravacuolar space via type I secretion and subsequently localize to/integrate into the PVM. We have unpublished data demonstrating that glutathione *S*-transferase N-terminal fusions of APH_0032 and APH_1387, but not glutathione *S*-transferase alone, can be heterologously secreted by type I secretion when expressed in *E. coli* and that the C termini of both proteins are requisite for their secretion.

HIJACKING HOST LIPID AND MEMBRANE-TRAFFICKING PATHWAYS

Hijacking Cholesterol

E. chaffeensis and *A. phagocytophilum* lack all genes for the biosynthesis of lipid A and most genes for the biosynthesis of peptidoglycan (Dunning Hotopp et al., 2006; Lin and Rikihisa, 2003a), both of which confer structural integrity to gram-negative bacterial membranes. This trend can be seen among other sequenced *Anaplasmataceae* members (Rikihisa, 2010a). While this likely contributes to these pathogens' ability to enter eukaryotic host cells undetected, it results in fragility of their outer membranes. As a compensatory mechanism, *E. chaffeensis* and *A. phagocytophilum* acquire and incorporate considerable amounts of host-derived cholesterol into their outer membranes (Lin and Rikihisa, 2003a). Both bacteria lack genes for cholesterol biosynthesis and require exogenous cholesterol for their survival. Indeed, treatment of the bacteria with a cholesterol-extracting agent, methyl-β-cyclodextrin, produces ultrastructural changes. Pretreatment of *E. chaffeensis* and *A. phagocytophilum* with methyl-β-cyclodextrin or a cholesterol derivative, NBD-cholesterol, prevented the bacteria from infecting naïve host cells. *A. phagocytophilum* does not require de novo synthesized cholesterol, but instead acquires cholesterol from the low-density lipoprotein (LDL) uptake pathway (Xiong et al., 2009). LDL uptake is upregulated in infected cells, as is LDL receptor (LDLR) mRNA and protein synthesis. *A. phagocytophilum* replication is significantly inhibited by depleting growth medium of cholesterol-containing lipoproteins, by blocking LDL uptake with an antibody against the LDLR, and by treating host cells with pharmacologic inhibitors that block LDL-derived cholesterol release from late endosomes and lysosomes. Though many vacuole-adapted intracellular pathogens hijack host cholesterol, *A. phagocytophilum*'s strategy of hijacking the LDL uptake pathway and enhancing LDLR expression is unique. How *A. phagocytophilum* specifically acquires LDL uptake pathway-derived cholesterol and the route from which *E. chaffeensis* obtains cholesterol remain to be deciphered. The relevance of these findings has been extended in vivo. High blood cholesterol levels correlate with increased *A. phagocytophilum* loads in the spleen, liver, and blood of infected mice (Xiong et al., 2007), which hints that HGA patients suffering from hypercholesterolemia may be at risk for developing higher *A. phagocytophilum* burdens.

The ApV Selectively Recruits Rab GTPases To Intercept Recycling Endosomes

A functional basis for the ApV's altered fusogenicity becomes evident when one examines the unique complement of Rab GTPases that it recruits. The Rab GTPase family (>60 members) is the largest member of the Ras superfamily of small guanosine triphosphatases. Rab GTPases are highly conserved among eukaryotes and coordinate many aspects of endocytic and exocytic cargo delivery. Rab GTPases cycle between a cytosolic GDP-bound inactive state and a membrane-associated GTP-bound active state to regulate membrane traffic at multiple steps, including formation of transport vesicles at donor membranes, vesicle motility, transport and docking of vesicles at acceptor membranes, and fusion of transport vesicles with acceptor membranes. Each of these steps is carried out by a diverse collection of effector molecules that bind to specific Rab GTPases in their GTP-bound states. Different Rab GTPases localize to distinct organelles and thereby dictate organelle identity (Stenmark, 2009).

Given the crucial roles that Rab GTPases play in regulating membrane traffic, it comes as no surprise that diverse vacuole-adapted bacterial pathogens selectively hijack these host proteins to their benefit. Confocal microscopic analyses of *A. phagocytophilum*-infected HL-60 cells transiently tranfected using nucleofection to express N-terminally fluorescent protein-tagged Rab GTPases revealed that a subset of Rabs is specifically recruited to the ApV. From a panel of 20 fluorescent protein-tagged Rab GTPases, GFP-Rab10, GFP-Rab11, GFP-Rab14, red fluorescent protein-Rab22, and GFP-Rab35, each of which regulates endocytic recycling, are pronouncedly recruited to the ApV. GFP-Rab4 and GFP-Rab1, which control endocytic recycling and mediate endoplasmic reticulum-to-Golgi apparatus trafficking, respectively, localize to low percentages of ApVs. Rab GTPases are recruited upon formation of and remain associated with the ApV throughout infection (Huang et al., 2010a). Rab10 and Rab14 are found on recycling endosomes and the *trans*-Golgi network (Stenmark, 2009). However, recruitment of GFP-Rab10 and GFP-Rab14 is unhindered by brefeldin A (Huang et al., 2010a), which suggests that Rab10 and Rab14 are exclusively recruited to the ApV via slow recycling endosomes. Indirect immunofluorescence microscopy using a Rab14 antibody revealed that endogenous Rab14 localizes to the AVM. The paucity of commercially available Rab GTPase antibodies that work well in immunofluorescence microscopy has thus far prevented detection of the endogenous forms of other Rab GTPases of interest on the AVM.

Endosomal recycling pathways return much of the membrane proteins and lipids that are internalized during endocytosis to the plasma membrane and can be classified as clathrin dependent or clathrin independent (Grant and Donaldson, 2009). Several Rab GTPases regulate traffic and confer vesicle identities along recycling endosome pathways. The Rab GTPases that exhibit the highest degrees of localization to the ApV—Rab10, Rab11, Rab14, Rab22, and Rab35—are found on a subset of recycling endosomes called clathrin-independent slow recycling endosomes (Fig. 9). Preferentially recruiting Rab GTPases that are predominantly found on slow recycling endosomes potentially provides *A. phagocytophilum* with four key intracellular survival advantages. First, *A. phagocytophilum* is auxotrophic for 16 amino acids (Dunning Hotopp et al., 2006). As endogenous host proteins that are critical for amino acid uptake are brought into host cells via the slow clathrin-independent endocytic pathway (Eyster et al., 2009), intercepting this pathway is a possible means by which the bacterium acquires amino acids. Second, the mechanism by which *A. phagocytophilum* obtains LDL endocytic pathway-derived cholesterol for incorporation into its cell wall is unknown (Lin and Rikihisa, 2003b; Xiong et al., 2009). Since cholesterol traffics through clathrin-independent recycling endosomes (Mayor and Pagano, 2007; Sandvig et al., 2008), the bacterium may acquire cholesterol by hijacking recycling endosomes. Third, continual delivery of recycling endosomes to the ApV would conceivably provide an unlimited supply of host membrane material to allow for expansion of the AVM, which would be necessary to accommodate growing

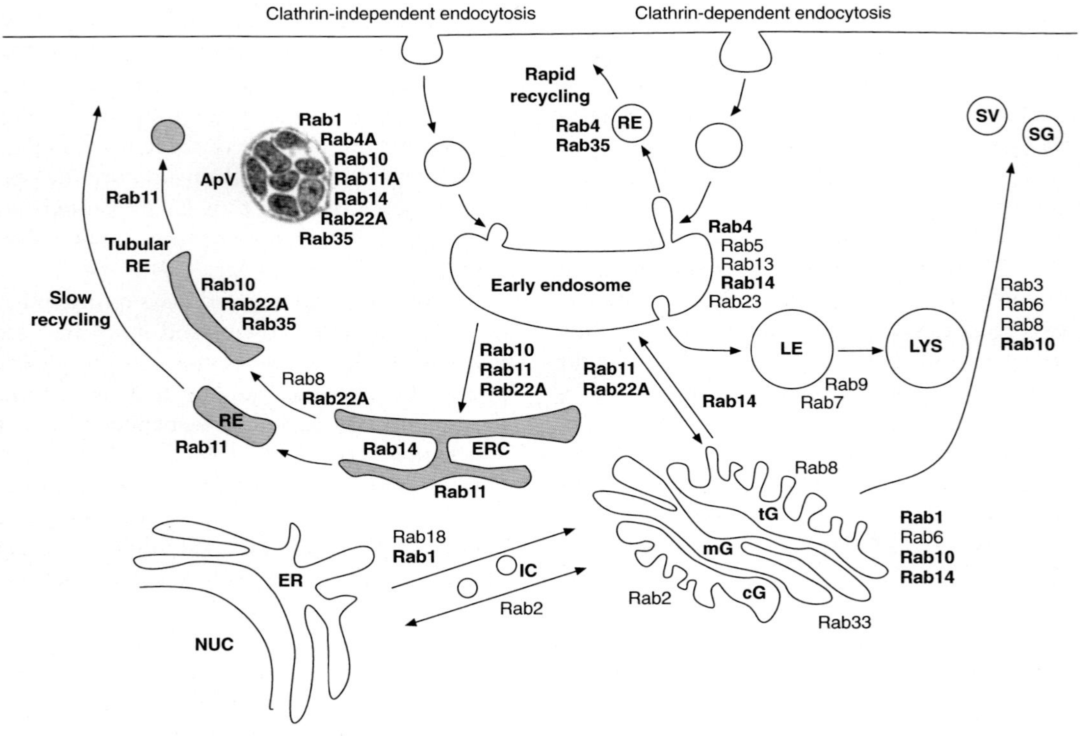

FIGURE 9 The ApV hijacks Rab GTPases. Rab GTPases that are selectively recruited to the ApV are in bold text. Recycling endosomes that are inferred as being intercepted by the ApV are shaded gray. ER, endoplasmic reticulum; cG, *cis*-Golgi; ERC, endocytic recycling center; IC, pre-Golgi intermediate compartment; LE, late endosome; LYS, lysosome; mG, *medial*-Golgi; NUC, nucleus; RE, recycling endosome; SG, secretory granule; SV, synaptic vesicle; tG, *trans*-Golgi. (Reprinted and modified from Huang et al. [2010a] with permission of the publisher.) doi:10.1128/9781555817336.ch6.f9

intravacuolar bacterial populations. Fourth, by coating the AVM with recycling endosome-associated Rab GTPases, the ApV camouflages itself as a recycling endosome, which is likely a means by which it protects itself from fusing with lysosomes. Support for this premise comes from the observation that 30 minutes post-tetracycline addition, fluorescent protein-tagged Rab10, Rab11A, Rab14, Rab22, and Rab35 dissociate from and GFP-Rab7 (localizes to late endosomes) and LAMP-1 (lysosomal marker) localize to the ApV. This latter observation not only indicates that decorating the AVM with Rab GTPases is a means by which *A. phagocytophilum* facilitates its intracellular survival, but also demonstrates that this strategy is dependent on de novo *Anaplasma* protein synthesis.

Given that the EcV is altered in its fusogenicity and it and other *Anaplasmataceae* members are able to remain undetected within their host cells, it is likely that hijacking Rab GTPases and other membrane traffic regulators is a conserved theme among this family. Extending these studies to other *Anaplasmataceae* members will lead to a more thorough understanding of how they successfully reside within their host cells. *A. phagocytophilum*'s ability to selectively hijack Rab GTPases has only been analyzed for the NCH-1 strain and only in the context of its infection of promyelocytic HL-60 cells (Huang et al., 2010a). As chlamydiae recruit Rab

GTPases to the inclusion membrane via both species-dependent and species-independent manners (Rzomp et al., 2003), it will be important to determine if the complement of Rab GTPases that are recruited to the NCH-1-occupied vacuolar membrane are conserved across all *A. phagocytophilum* strains. It will also be important to determine the mechanisms by which Rab GTPases are recruited to ApV and, presumably, other *Anaplasmataceae* organism-occupied vacuoles. One possibility is that they are directly transported to the AVM by means of *Anaplasma*-encoded proteins. Another possibility is that the ApV recruits host Rab effector molecules that interact with their cognate Rab GTPases at the AVM, as has been demonstrated for multiple *Chlamydia* species (Moorhead et al., 2007, 2010). Last, it will be critical to investigate these phenomena in the context of infection of these pathogens' natural mammalian host cells—neutrophils and macrophages—as opposed to solely relying on immortalized cell lines. This undertaking will become more feasible once better antibodies for detecting endogenous Rab GTPases using immunofluorescent methods are developed.

The ApV Recruits Rab10 in a Guanine Nucleotide-Independent Manner

Rabs localize to eukaryotic membranes in a guanine nucleotide-dependent manner (Stenmark, 2009). This switch between the active GTP-bound and inactive GDP-bound forms of Rab GTPases is controlled by guanine nucleotide exchange factors (GEFs), which trigger the binding of GTP; and GTPase-activating proteins (GAPs), which accelerate hydrolysis of the bound GTP to GDP. In their GDP-bound state, Rabs are soluble and bound to guanine nucleotide dissociation inhibitor (GDI). At the acceptor membrane, Rab-GDI interacts with GDI displacement factor (GDF), which removes GDI to enable Rab membrane insertion. A GEF converts the membrane-bound Rab to its GTP-bound state, which allows it to interact with its downstream effector proteins to control membrane traffic. After inactivation by their GAPs, the GDP-bound Rabs return to the cytosol (Brumell and Scidmore, 2007; Stenmark, 2009). Consistent with the dogma of Rab membrane cycling, GTP-bound but not GDP-bound forms of GFP-Rab1, -Rab4A, -Rab11A, -Rab14, -Rab22A, and -Rab35 localize to the AVM (Huang et al., 2010a; B. Huang and J. A. Carlyon, unpublished observation). Notably, Rab10 localization to the AVM is guanine nucleotide independent. GTP-locked, GDP-locked, and nucleotide-free (incapable of binding GTP or GDP) isoforms of GFP-Rab10 localize to the AVM with comparable efficiencies (Huang et al., 2010a). This observation suggests that an *A. phagocytophilum*-derived AVM protein may serve as a eukaryoticlike GEF and/or a GDF of Rab10 to control its cycling on the AVM. Precedents for this phenomenon are *L. pneumophila* DrrA/SidM and LepB, which are two of the only four bacterial Rab ligands whose functional roles are known. These effectors control Rab1 cycling on/off the *Legionella*-containing vacuolar membrane (LCVM) (Ingmundson et al., 2007; Machner and Isberg, 2006; Müller et al., 2010; Murata et al., 2006). DrrA/SidM functions as a GDF to displace Rab-GDI such that Rab1 can be inserted into the LCVM and as a GEF to stimulate Rab1 activation (Ingmundson et al., 2007; Machner and Isberg, 2006; Murata et al., 2006). LepB inactivates and removes Rab1 from the LCVM by stimulating GTP hydrolysis (Ingmundson et al., 2007). The *A. phagocytophilum* Rab10 ligand mimics a GEF/GDI in functionality only, as homology searches of known GEFs and GDIs against the *A. phagocytophilum* proteome revealed no primary amino acid sequence homology to any *A. phagocytophilum* protein (Huang et al., 2010a). This is not surprising since *Legionella* and *Chlamydia* Rab ligands exhibit no homology to known eukaryotic Rab regulatory proteins (Cortes et al., 2007; Ingmundson et al., 2007; Machner and Isberg, 2006; Murata et al., 2006; Rzomp et al., 2006; Schlumberger and Hardt, 2005).

FUTURE DIRECTIONS

Although many exciting advances in understanding how *Anaplasmataceae* pathogens sense environmental changes and modulate host cell functions to enable their intracellular survival have been made, many questions remain unanswered. Most of the bacterial factors responsible for manipulating host cell functions are unidentified or are insufficiently characterized. Identifying them and confirming their bona fide functional roles will be essential to fully understand the sophisticated means by which these pathogens exploit host cell processes. Development of genetic manipulation tools and/or antisense RNA-mediated silencing techniques for these obligate intracellular bacteria is sorely needed and would greatly augment such studies. Finally, much of what we know regarding *Anaplasmataceae* pathogen manipulation of host cell functions is derived from studies of *A. phagocytophilum* and *E. chaffeensis*. Thus, extending investigations into how other *Anaplasmataceae* members modulate their host cells is important.

ACKNOWLEDGMENTS

This work was supported in part by National Institutes of Health/National Institute of Allergy and Infectious Diseases grants R01AI072683 and R21AI090170. I am grateful for the generosity of Sunil Thomas, Vsevolod L. Popov, and David H. Walker for providing electron micrograph images of *Ehrlichia* spp.-infected cells. I thank Matthew J. Troese and Christopher Earnhart for preparing Fig. 4 and Bernice Huang for preparing Fig. 9.

REFERENCES

Abdelrahman, Y. M., and R. J. Belland. 2005. The chlamydial developmental cycle. *FEMS Microbiol. Rev.* **29:**949–959.

Akkoyunlu, M., and E. Fikrig. 2000. Gamma interferon dominates the murine cytokine response to the agent of human granulocytic ehrlichiosis and helps to control the degree of early rickettsemia. *Infect. Immun.* **68:**1827–1833.

Akkoyunlu, M., S. E. Malawista, J. Anguita, and E. Fikrig. 2001. Exploitation of interleukin-8-induced neutrophil chemotaxis by the agent of human granulocytic ehrlichiosis. *Infect. Immun.* **69:** 5577–5588.

Allsopp, M. T., M. Louw, and E. C. Meyer. 2005. *Ehrlichia ruminantium*: an emerging human pathogen? *Ann. N. Y. Acad. Sci.* **1063:**358–360.

Alvarez-Dominguez, C., R. Roberts, and P. D. Stahl. 1997. Internalized *Listeria monocytogenes* modulates intracellular trafficking and delays maturation of the phagosome. *J. Cell Sci.* **110:**731–743.

Alvarez-Martinez, C. E., and P. J. Christie. 2009. Biological diversity of prokaryotic type IV secretion systems. *Microbiol. Mol. Biol. Rev.* **73:**775–808.

Amer, A. O., and M. S. Swanson. 2005. Autophagy is an immediate macrophage response to *Legionella pneumophila. Cell. Microbiol.* **7:**765–778.

Anderson, R. G., J. R. Falck, J. L. Goldstein, and M. S. Brown. 1984. Visualization of acidic organelles in intact cells by electron microscopy. *Proc. Natl. Acad. Sci. USA* **81:**4838–4842.

Anonymous. 2001. Case records of the Massachusetts General Hospital. Weekly clinicopathological exercises. Case 37-2001. A 76-year-old man with fever, dyspnea, pulmonary infiltrates, pleural effusions, and confusion. *N. Engl. J. Med.* **345:** 1627–1634.

Avakyan, A. A., and V. L. Popov. 1984. Rickettsiaceae and Chlamydiaceae: comparative electron microscopic studies. *Acta Virol.* **28:**159–173.

Backert, S., R. Fronzes, and G. Waksman. 2008. VirB2 and VirB5 proteins: specialized adhesins in bacterial type-IV secretion systems? *Trends Microbiol.* **16:**409–413.

Banerjee, R., J. Anguita, and E. Fikrig. 2000a. Granulocytic ehrlichiosis in mice deficient in phagocyte oxidase or inducible nitric oxide synthase. *Infect. Immun.* **68:**4361–4362.

Banerjee, R., J. Anguita, D. Roos, and E. Fikrig. 2000b. Infection by the agent of human granulocytic ehrlichiosis prevents the respiratory burst by down-regulating *gp91^{phox}*. *J. Immunol.* **164:** 3946–3949.

Bao, W., Y. Kumagai, H. Niu, M. Yamaguchi, K. Miura, and Y. Rikihisa. 2009. Four VirB6 paralogs and VirB9 are expressed and interact in *Ehrlichia chaffeensis*-containing vacuoles. *J. Bacteriol.* **191:**278–286.

Barbet, A. F., P. F. Meeus, M. Belanger, M. V. Bowie, J. Yi, A. M. Lundgren, A. R. Alleman, S. J. Wong, F. K. Chu, U. G. Munderloh, and S. D. Jauron. 2003. Expression of multiple outer membrane protein sequence variants from a single genomic locus of *Anaplasma phagocytophilum. Infect. Immun.* **71:**1706–1718.

Barnewall, R. E., N. Ohashi, and Y. Rikihisa. 1999. *Ehrlichia chaffeensis* and *E. sennetsu*, but not the human granulocytic ehrlichiosis agent, colocalize with transferrin receptor and up-regulate transferrin receptor mRNA by activating iron-responsive protein 1. *Infect. Immun.* **67:**2258–2265.

Barnewall, R. E., and Y. Rikihisa. 1994. Abrogation of gamma interferon-induced inhibition of *Ehrlichia chaffeensis* infection in human monocytes with iron transferrin. *Infect. Immun.* **62:** 4804–4810.

Barnewall, R. E., Y. Rikihisa, and E. H. Lee. 1997. *Ehrlichia chaffeensis* inclusions are early endosomes which selectively accumulate transferrin receptor. *Infect. Immun.* **65:**1455–1461.

Bitsaktsis, C., and G. Winslow. 2006. Fatal recall responses mediated by CD8 T cells during intracellular bacterial challenge infection. *J. Immunol.* **177:**4644–4651.

Blouin, E. F., and K. M. Kocan. 1998. Morphology and development of *Anaplasma marginale* (Rickettsiales: Anaplasmataceae) in cultured *Ixodes scapularis* (Acari: Ixodidae) cells. *J. Med. Entomol.* **35:**788–797.

Borjesson, D. L., S. D. Kobayashi, A. R. Whitney, J. M. Voyich, C. M. Argue, and F. R. DeLeo. 2005. Insights into pathogen immune evasion mechanisms: *Anaplasma phagocytophilum* fails to induce an apoptosis differentiation program in human neutrophils. *J. Immunol.* **174:**6364–6372.

Brayton, K. A., L. S. Kappmeyer, D. R. Herndon, M. J. Dark, D. L. Tibbals, G. H. Palmer, T. C. McGuire, and D. P. Knowles, Jr. 2005. Complete genome sequencing of *Anaplasma marginale* reveals that the surface is skewed to two superfamilies of outer membrane proteins. *Proc. Natl. Acad. Sci. USA* **102:**844–849.

Brayton, K. A., D. P. Knowles, T. C. McGuire, and G. H. Palmer. 2001. Efficient use of a small genome to generate antigenic diversity in tick-borne ehrlichial pathogens. *Proc. Natl. Acad. Sci. USA* **98:**4130–4135.

Brumell, J. H., and M. A. Scidmore. 2007. Manipulation of Rab GTPase function by intracellular bacterial pathogens. *Microbiol. Mol. Biol. Rev.* **71:**636–652.

Brumell, J. H., A. Volchuk, H. Sengelov, N. Borregaard, A. M. Cieutat, D. F. Bainton, S. Grinstein, and A. Klip. 1995. Subcellular distribution of docking/fusion proteins in neutrophils, secretory cells with multiple exocytic compartments. *J. Immunol.* **155:**5750–5759.

Bussmeyer, U., A. Sarkar, K. Broszat, T. Lüdemann, S. Moller, G. van Zandbergen, C. Bogdan, M. Behnen, J. S. Dumler, F. D. von Loewenich, W. Solbach, and T. Laskay. 2010. Impairment of gamma interferon signaling in human neutrophils infected with *Anaplasma phagocytophilum*. *Infect. Immun.* **78:**358–363.

Carlyon, J. A., D. Abdel-Latif, M. Pypaert, P. Lacy, and E. Fikrig. 2004. *Anaplasma phagocytophilum* utilizes multiple host evasion mechanisms to thwart NADPH oxidase-mediated killing during neutrophil infection. *Infect. Immun.* **72:** 4772–4783.

Carlyon, J. A., M. Akkoyunlu, L. Xia, T. Yago, T. Wang, R. D. Cummings, R. P. McEver, and E. Fikrig. 2003. Murine neutrophils require α1,3-fucosylation but not PSGL-1 for productive infection with *Anaplasma phagocytophilum*. *Blood* **102:**3387–3395.

Carlyon, J. A., W. T. Chan, J. Galan, D. Roos, and E. Fikrig. 2002. Repression of *rac2* mRNA expression by *Anaplasma phagocytophila* is essential to the inhibition of superoxide production and bacterial proliferation. *J. Immunol.* **169:**7009–7018.

Carlyon, J. A., D. Ryan, K. Archer, and E. Fikrig. 2005. Effects of *Anaplasma phagocytophilum* on host cell ferritin mRNA and protein levels. *Infect. Immun.* **73:**7629–7636.

Caturegli, P., K. M. Asanovich, J. J. Walls, J. S. Bakken, J. E. Madigan, V. L. Popov, and J. S. Dumler. 2000. *ankA*: an *Ehrlichia phagocytophila* group gene encoding a cytoplasmic protein antigen with ankyrin repeats. *Infect. Immun.* **68:**5277–5283.

Chen, C., S. Banga, K. Mertens, M. M. Weber, I. Gorbaslieva, Y. Tan, Z. Q. Luo, and J. E. Samuel. 2010. Large-scale identification and translocation of type IV secretion substrates by *Coxiella burnetii*. *Proc. Natl. Acad. Sci. USA* **107:**21755–21760.

Chen, S. M., J. S. Dumler, H. M. Feng, and D. H. Walker. 1994. Identification of the antigenic constituents of *Ehrlichia chaffeensis*. *Am. J. Trop. Med. Hyg.* **50:**52–58.

Cheng, Z., Y. Kumagai, M. Lin, C. Zhang, and Y. Rikihisa. 2006. Intra-leukocyte expression of two-component systems in *Ehrlichia chaffeensis* and *Anaplasma phagocytophilum* and effects of the histidine kinase inhibitor closantel. *Cell. Microbiol.* **8:**1241–1252.

Cheng, Z., X. Wang, and Y. Rikihisa. 2008. Regulation of type IV secretion apparatus genes during *Ehrlichia chaffeensis* intracellular development by a previously unidentified protein. *J. Bacteriol.* **190:**2096–2105.

Cho, N. H., H. R. Kim, J. H. Lee, S. Y. Kim, J. Kim, S. Cha, A. C. Darby, H. H. Fuxelius, J. Yin, J. H. Kim, S. J. Lee, Y. S. Koh, W. J. Jang, K. H. Park, S. G. Andersson, M. S. Choi, and I. S. Kim. 2007. The *Orientia tsutsugamushi* genome reveals massive proliferation of conjugative type IV secretion system and host-cell interaction genes. *Proc. Natl. Acad. Sci. USA* **104:**7981–7986.

Choi, K. S., and J. S. Dumler. 2003. Early induction and late abrogation of respiratory burst in *A. phagocytophilum*-infected neutrophils. *Ann. N. Y. Acad. Sci.* **990:**488–493.

Choi, K. S., J. Garyu, J. Park, and J. S. Dumler. 2003. Diminished adhesion of *Anaplasma phagocytophilum*-infected neutrophils to endothelial cells is associated with reduced expression of leukocyte surface selectin. *Infect. Immun.* **71:**4586–4594.

Choi, K. S., J. T. Park, and J. S. Dumler. 2005. *Anaplasma phagocytophilum* delay of neutrophil apoptosis through the p38 mitogen-activated protein kinase signal pathway. *Infect. Immun.* **73:** 8209–8218.

Clifton, D. R., K. A. Fields, S. S. Grieshaber, C. A. Dooley, E. R. Fischer, D. J. Mead, R. A. Carabeo, and T. Hackstadt. 2004. A chlamydial type III translocated protein is tyrosine-phosphorylated at the site of entry and associated with recruitment of actin. *Proc. Natl. Acad. Sci. USA* **101:**10166–10171.

Collins, N. E., J. Liebenberg, E. P. de Villiers, K. A. Brayton, E. Louw, A. Pretorius, F. E. Faber, H. van Heerden, A. Josemans, M. van Kleef, H. C. Steyn, M. F. van Strijp, E. Zweygarth, F. Jongejan, J. C. Maillard, D. Berthier, M. Botha, F. Joubert, C. H. Corton, N. R. Thomson, M. T. Allsopp, and B. A. Allsopp. 2005. The genome of the heartwater agent *Ehrlichia ruminantium* contains multiple tandem repeats of actively variable copy number. *Proc. Natl. Acad. Sci. USA* **102:**838–843.

Cortes, C., K. A. Rzomp, A. Tvinnereim, M. A. Scidmore, and B. Wizel. 2007. *Chlamydia pneumoniae* inclusion membrane protein Cpn0585 interacts with multiple Rab GTPases. *Infect. Immun.* **75:**5586–5596.

Cossart, P., and C. R. Roy. 2010. Manipulation of host membrane machinery by bacterial pathogens. *Curr. Opin. Cell Biol.* **22:**547–554.

Danial, N. N., and S. J. Korsmeyer. 2004. Cell death: critical control points. *Cell* **116:**205–219.

Dehio, C. 2008. Infection-associated type IV secretion systems of *Bartonella* and their diverse roles in host cell interaction. *Cell. Microbiol.* **10:** 1591–1598.

de la Fuente, J., P. Ayoubi, E. F. Blouin, C. Almazán, V. Naranjo, and K. M. Kocan. 2005. Gene expression profiling of human promyelocytic cells in response to infection with *Anaplasma phagocytophilum*. *Cell. Microbiol.* **7:**549–559.

de la Fuente, J., J. C. Garcia-Garcia, A. F. Barbet, E. F. Blouin, and K. M. Kocan. 2004. Adhesion of outer membrane proteins containing tandem repeats of *Anaplasma* and *Ehrlichia* species (Rickettsiales: Anaplasmataceae) to tick cells. *Vet. Microbiol.* **98:**313–322.

de la Fuente, J., J. C. Garcia-Garcia, E. F. Blouin, and K. M. Kocan. 2001. Differential adhesion of major surface proteins 1a and 1b of the ehrlichial cattle pathogen *Anaplasma marginale* to bovine erythrocytes and tick cells. *Int. J. Parasitol.* **31:**145–153.

de la Fuente, J., J. C. Garcia-Garcia, E. F. Blouin, and K. M. Kocan. 2003. Characterization of the functional domain of major surface protein 1a involved in adhesion of the rickettsia *Anaplasma marginale* to host cells. *Vet. Microbiol.* **91:**265–283.

DeLeo, F. R. 2004. Modulation of phagocyte apoptosis by bacterial pathogens. *Apoptosis* **9:**399–413.

Delepelaire, P. 2004. Type I secretion in gram-negative bacteria. *Biochim. Biophys. Acta* **1694:** 149–161.

Deretic, V. 2010. Autophagy in infection. *Curr. Opin. Cell Biol.* **22:**252–262.

Doyle, C. K., K. A. Nethery, V. L. Popov, and J. W. McBride. 2006. Differentially expressed and secreted major immunoreactive protein orthologs of *Ehrlichia canis* and *E. chaffeensis* elicit early antibody responses to epitopes on glycosylated tandem repeats. *Infect. Immun.* **74:**711–720.

Dumler, J. S., S. M. Chen, K. Asanovich, E. Trigiani, V. L. Popov, and D. H. Walker. 1995. Isolation and characterization of a new strain of *Ehrlichia chaffeensis* from a patient with nearly fatal monocytic ehrlichiosis. *J. Clin. Microbiol.* **33:**1704–1711.

Dunning Hotopp, J. C., M. Lin, R. Madupu, J. Crabtree, S. V. Angiuoli, J. Eisen, R.Seshadri, Q. Ren, M. Wu, T. R. Utterback, S. Smith, M. Lewis, H. Khouri, C. Zhang, H. Niu, Q. Lin, N. Ohashi, N. Zhi, W. Nelson, L. M. Brinkac, R. J. Dodson, M. J. Rosovitz, J. Sundaram, S. C. Daugherty, T. Davidsen, A. S. Durkin, M. Gwinn, D. H. Haft, J. D. Selengut, S. A. Sullivan, N. Zafar, L. Zhou, F. Benahmed, H. Forberger, R. Halpin, S. Mulligan, J. Robinson, O. White, Y. Rikihisa, and H. Tettelin. 2006. Comparative genomics of emerging human ehrlichiosis agents. *PLoS Genet.* **2:**e21.

Eyster, C. A., J. D. Higginson, R. Huebner, N. Porat-Shliom, R. Weigert, W. W. Wu, R. F. Shen, and J. G. Donaldson. 2009. Discovery of new cargo proteins that enter cells through clathrin-independent endocytosis. *Traffic* **10:**590–599.

Felek, S., H. Huang, and Y. Rikihisa. 2003. Sequence and expression analysis of *virB9* of the type IV secretion system of *Ehrlichia canis* strains in ticks, dogs, and cultured cells. *Infect. Immun.* **71:** 6063–6067.

Feng, H. M., and D. H. Walker. 2004. Mechanisms of immunity to *Ehrlichia muris*: a model of monocytotropic ehrlichiosis. *Infect. Immun.* **72:**966–971.

Foley, J. E., N. W. Lerche, J. S. Dumler, and J. E. Madigan. 1999. A simian model of human granulocytic ehrlichiosis. *Am. J. Trop. Med. Hyg.* **60:**987–993.

Foster, J., M. Ganatra, I. Kamal, J. Ware, K. Makarova, N. Ivanova, A. Bhattacharyya, V. Kapatral, S. Kumar, J. Posfai, T. Vincze, J. Ingram, L. Moran, A. Lapidus, M. Omelchenko, N. Kyrpides, E. Ghedin, S. Wang, E. Golts-

man, V. Joukov, O. Ostrovskaya, K. Tsukerman, M. Mazur, D. Comb, E. Koonin, and B. Slatko. 2005. The *Wolbachia* genome of *Brugia malayi*: endosymbiont evolution within a human pathogenic nematode. *PLoS Biol.* **3:**e121.

Fronzes, R., P. J. Christie, and G. Waksman. 2009. The structural biology of type IV secretion systems. *Nat. Rev. Microbiol.* **7:**703–714.

Futse, J. E., K. A. Brayton, S. D. Nydam, and G. H. Palmer. 2009. Generation of antigenic variants via gene conversion: evidence for recombination fitness selection at the locus level in *Anaplasma marginale*. *Infect. Immun.* **77:**3181–3187.

Galindo, R. C., N. Ayllon, T. Carta, J. Vicente, K. M. Kocan, C. Gortazar, and J. de la Fuente. 2010. Characterization of pathogen-specific expression of host immune response genes in *Anaplasma* and *Mycobacterium* species infected ruminants. *Comp. Immunol. Microbiol. Infect. Dis.* **33:**e133–e142.

Galindo, R. C., P. Ayoubi, A. L. García-Pérez, V. Naranjo, K. M. Kocan, C. Gortazar, and J. de la Fuente. 2008. Differential expression of inflammatory and immune response genes in sheep infected with *Anaplasma phagocytophilum*. *Vet. Immunol. Immunopathol.* **126:**27–34.

Garcia-Garcia, J. C., N. C. Barat, S. J. Trembley, and J. S. Dumler. 2009a. Epigenetic silencing of host cell defense genes enhances intracellular survival of the rickettsial pathogen *Anaplasma phagocytophilum*. *PLoS Pathog.* **5:**e1000488.

Garcia-Garcia, J. C., K. E. Rennoll-Bankert, S. Pelly, A. M. Milstone, and J. S. Dumler. 2009b. Silencing of host cell *CYBB* gene expression by the nuclear effector AnkA of the intracellular pathogen *Anaplasma phagocytophilum*. *Infect. Immun.* **77:**2385–2391.

Ge, Y., and Y. Rikihisa. 2006. *Anaplasma phagocytophilum* delays spontaneous human neutrophil apoptosis by modulation of multiple apoptotic pathways. *Cell. Microbiol.* **8:**1406–1416.

Ge, Y., and Y. Rikihisa. 2007. Surface-exposed proteins of *Ehrlichia chaffeensis*. *Infect. Immun.* **75:** 3833–3841.

Ge, Y., K. Yoshiie, F. Kuribayashi, M. Lin, and Y. Rikihisa. 2005. *Anaplasma phagocytophilum* inhibits human neutrophil apoptosis via upregulation of *bfl-1*, maintenance of mitochondrial membrane potential and prevention of caspase 3 activation. *Cell. Microbiol.* **7:**29–38.

Gibson, K., Y. Kumagai, and Y. Rikihisa. 2010. Proteomic analysis of *Neorickettsia sennetsu* surface-exposed proteins and porin activity of the major surface protein P51. *J. Bacteriol.* **192:**5898–5905.

Gillespie, J. J., K. A. Brayton, K. P. Williams, M. A. Diaz, W. C. Brown, A. F. Azad, and B. W. Sobral. 2010. Phylogenomics reveals a diverse *Rickettsiales* type IV secretion system. *Infect. Immun.* **78:**1809–1823.

Gokce, H. I., G. Ross, and Z. Woldehiwet. 1999. Inhibition of phagosome-lysosome fusion in ovine polymorphonuclear leucocytes by *Ehrlichia (Cytoecetes) phagocytophila*. *J. Comp. Pathol.* **120:**369–381.

Goodman, J. L., C. M. Nelson, M. B. Klein, S. F. Hayes, and B. W. Weston. 1999. Leukocyte infection by the granulocytic ehrlichiosis agent is linked to expression of a selectin ligand. *J. Clin. Invest.* **103:**407–412.

Granick, J. L., D. V. Reneer, J. A. Carlyon, and D. L. Borjesson. 2008. *Anaplasma phagocytophilum* infects cells of the megakaryocytic lineage through sialylated ligands but fails to alter platelet production. *J. Med. Microbiol.* **57:**416–423.

Grant, B. D., and J. G. Donaldson. 2009. Pathways and mechanisms of endocytic recycling. *Nat. Rev. Mol. Cell Biol.* **10:**597–608.

Heinzen, R. A., T. Hackstadt, and J. E. Samuel. 1999. Developmental biology of *Coxiella burnettii*. *Trends Microbiol.* **7:**149–154.

Herndon, D. R., G. H. Palmer, V. Shkap, D. P. Knowles, Jr., and K. A. Brayton. 2010. Complete genome sequence of *Anaplasma marginale* subsp. *centrale*. *J. Bacteriol.* **192:**379–380.

Herron, M. J., M. E. Ericson, T. J. Kurtti, and U. G. Munderloh. 2005. The interactions of *Anaplasma phagocytophilum*, endothelial cells, and human neutrophils. *Ann. N. Y. Acad. Sci.* **1063:**374–382.

Herron, M. J., and J. L. Goodman. 2001. Evidence that the human granulocytic ehrlichiosis agent utilizes a novel oxygen radical detoxification system: cytochrome C re-oxidation, abstr. 50. *Proc. Am. Soc. Rickettsiol. Bartonella Emerg. Pathog. Group 2001 Joint Conf.*, Big Sky, MT, 17 to 22 August 2001.

Herron, M. J., C. M. Nelson, J. Larson, K. R. Snapp, G. S. Kansas, and J. L. Goodman. 2000. Intracellular parasitism by the human granulocytic ehrlichiosis bacterium through the P-selectin ligand, PSGL-1. *Science* **288:**1653–1656.

Hidalgo, R. J., E. W. Jones, J. E. Brown, and A. J. Ainsworth. 1989. *Anaplasma marginale* in tick cell culture. *Am. J. Vet. Res.* **50:**2028–2032.

Hodzic, E., S. Feng, D. Fish, C. M. Leutenegger, K. J. Freet, and S. W. Barthold. 2001. Infection of mice with the agent of human granulocytic ehrlichiosis after different routes of inoculation. *J. Infect. Dis.* **183:**1781–1786.

Howell, M. L., E. Alsabbagh, J. F. Ma, U. A. Ochsner, M. G. Klotz, T. J. Beveridge, K. M. Blumenthal, E. C. Niederhoffer, R. E. Morris, D. Needham, G. E. Dean, M. A. Wani, and D. J. Hassett. 2000. AnkB, a periplasmic ankyrin-like protein in *Pseudomonas aeruginosa*, is required for optimal catalase B (KatB) activity and resistance to hydrogen peroxide. *J. Bacteriol.* **182:**4545–4556.

Huang, B., A. Hubber, J. A. McDonough, C. R. Roy, M. A. Scidmore, and J. A. Carlyon. 2010a. The *Anaplasma phagocytophilum*-occupied vacuole selectively recruits Rab-GTPases that are predominantly associated with recycling endosomes. *Cell. Microbiol.* **12:**1292–1307.

Huang, B., M. J. Troese, D. Howe, S. Ye, J. T. Sims, R. A. Heinzen, D. L. Borjesson, and J. A. Carlyon. 2010b. *Anaplasma phagocytophilum* APH_0032 is expressed late during infection and localizes to the pathogen-occupied vacuolar membrane. *Microb. Pathog.* **49:**273–284.

Huang, B., M. J. Troese, S. Ye, J. T. Sims, N. L. Galloway, D. L. Borjesson, and J. A. Carlyon. 2010c. *Anaplasma phagocytophilum* APH_1387 is expressed throughout bacterial intracellular development and localizes to the pathogen-occupied vacuolar membrane. *Infect. Immun.* **78:** 1864–1873.

Huang, H., X. Wang, T. Kikuchi, Y. Kumagai, and Y. Rikihisa. 2007. Porin activity of *Anaplasma phagocytophilum* outer membrane fraction and purified P44. *J. Bacteriol.* **189:**1998–2006.

Hussain, M., A. Haggar, G. Peters, G. S. Chhatwal, M. Herrmann, J. I. Flock, and B. Sinha. 2008. More than one tandem repeat domain of the extracellular adherence protein of *Staphylococcus aureus* is required for aggregation, adherence, and host cell invasion but not for leukocyte activation. *Infect. Immun.* **76:**5615–5623.

IJdo, J., A. C. Carlson, and E. L. Kennedy. 2007. *Anaplasma phagocytophilum* AnkA is tyrosine-phosphorylated at EPIYA motifs and recruits SHP-1 during early infection. *Cell. Microbiol.* **9:**1284–1296.

IJdo, J. W., and A. C. Mueller. 2004. Neutrophil NADPH oxidase is reduced at the *Anaplasma phagocytophilum* phagosome. *Infect. Immun.* **72:** 5392–5401.

IJdo, J. W., C. Wu, S. R. Telford III, and E. Fikrig. 2002. Differential expression of the p44 gene family in the agent of human granulocytic ehrlichiosis. *Infect. Immun.* **70:**5295–5298.

Ingmundson, A., A. Delprato, D. G. Lambright, and C. R. Roy. 2007. *Legionella pneumophila* proteins that regulate Rab1 membrane cycling. *Nature* **450:**365–369.

Ismail, N., L. Soong, J. W. McBride, G. Valbuena, J. P. Olano, H. M. Feng, and D. H. Walker. 2004. Overproduction of TNF-α by CD8$^+$ type 1 cells and down-regulation of IFN-γ production by CD4$^+$ Th1 cells contribute to toxic shock-like syndrome in an animal model of fatal monocytotropic ehrlichiosis. *J. Immunol.* **172:**1786–1800.

Jameson, B. A., and H. Wolf. 1988. The antigenic index: a novel algorithm for predicting antigenic determinants. *Comput. Appl. Biosci.* **4:**181–186.

Jauron, S. D., C. M. Nelson, V. Fingerle, M. D. Ravyn, J. L. Goodman, R. C. Johnson, R. Lobentanzer, B. Wilske, and U. G. Munderloh. 2001. Host cell-specific expression of a p44 epitope by the human granulocytic ehrlichiosis agent. *J. Infect. Dis.* **184:** 1445–1450.

Johnson, L. S., K. W. Dunn, B. Pytowski, and T. E. McGraw. 1993. Endosome acidification and receptor trafficking: bafilomycin A1 slows receptor externalization by a mechanism involving the receptor's internalization motif. *Mol. Biol. Cell* **4:**1251–1266.

Khan, I. A., J. A. MacLean, F. S. Lee, L. Casciotti, E. DeHaan, J. D. Schwartzman, and A. D. Luster. 2000. IP-10 is critical for effector T cell trafficking and host survival in *Toxoplasma gondii* infection. *Immunity* **12:**483–494.

Kim, H. Y., and Y. Rikihisa. 2002. Roles of p38 mitogen-activated protein kinase, NF-κB, and protein kinase C in proinflammatory cytokine mRNA expression by human peripheral blood leukocytes, monocytes, and neutrophils in response to *Anaplasma phagocytophila. Infect. Immun.* **70:**4132–4141.

Klein, M. B., J. S. Miller, C. M. Nelson, and J. L. Goodman. 1997. Primary bone marrow progenitors of both granulocytic and monocytic lineages are susceptible to infection with the agent of human granulocytic ehrlichiosis. *J. Infect. Dis.* **176:**1405–1409.

Klotz, M. G., and A. J. Anderson. 1995. Sequence of a gene encoding periplasmic *Pseudomonas syringae* ankyrin. *Gene* **164:**187–188.

Kobayashi, S. D., K. R. Braughton, A. R. Whitney, J. M. Voyich, T. G. Schwan, J. M. Musser, and F. R. DeLeo. 2003. Bacterial pathogens modulate an apoptosis differentiation program in human neutrophils. *Proc. Natl. Acad. Sci. USA* **100:**10948–10953.

Kobayashi, S. D., and F. R. DeLeo. 2004. An apoptosis differentiation programme in human polymorphonuclear leucocytes. *Biochem. Soc. Trans.* **32:**474–476.

Kocan, K. M., J. de la Fuente, E. F. Blouin, J. F. Coetzee, and S. A. Ewing. 2010. The natural history of *Anaplasma marginale. Vet. Parasitol.* **167:**95–107.

Kumagai, Y., Z. Cheng, M. Lin, and Y. Rikihisa. 2006. Biochemical activities of three pairs of *Ehrlichia chaffeensis* two-component regulatory system proteins involved in inhibition of lysosomal fusion. *Infect. Immun.* **74:**5014–5022.

Kumagai, Y., H. Huang, and Y. Rikihisa. 2008. Expression and porin activity of P28 and OMP-1F during intracellular *Ehrlichia chaffeensis* development. *J. Bacteriol.* **190:**3597–3605.

Kumar, Y., and R. H. Valdivia. 2008. Actin and intermediate filaments stabilize the *Chlamydia trachomatis* vacuole by forming dynamic structural scaffolds. *Cell Host Microbe* **4:**159–169.

Kyte, J., and R. F. Doolittle. 1982. A simple method for displaying the hydropathic character of a protein. *J. Mol. Biol.* **4:**105-132.

Lai, T. H., Y. Kumagai, M. Hyodo, Y. Hayakawa, and Y. Rikihisa. 2009. The *Anaplasma phagocytophilum* PleC histidine kinase and PleD diguanylate cyclase two-component system and role of cyclic Di-GMP in host cell infection. *J. Bacteriol.* **191:**693–700.

Larkin, M. A., G. Blackshields, N. P. Brown, R. Chenna, P. A. McGettigan, H. McWilliam, F. Valentin, I. M. Wallace, A. Wilm, R. Lopez, J. D. Thompson, T. J. Gibson, and D. G. Higgins. 2007. Clustal W and Clustal X version 2.0. *Bioinformatics* **23:**2947-2948.

Larson, J. A., H. L. Howie, and M. So. 2004. *Neisseria meningitidis* accelerates ferritin degradation in host epithelial cells to yield an essential iron source. *Mol. Microbiol.* **53:**807–820.

Lee, E. H., and Y. Rikihisa. 1998. Protein kinase A-mediated inhibition of gamma interferon-induced tyrosine phosphorylation of Janus kinases and latent cytoplasmic transcription factors in human monocytes by *Ehrlichia chaffeensis*. *Infect. Immun.* **66:**2514–2520.

Lee, H. C., and J. L. Goodman. 2006. *Anaplasma phagocytophilum* causes global induction of antiapoptosis in human neutrophils. *Genomics* **88:**496–503.

Lee, H. C., M. Kioi, J. Han, R. K. Puri, and J. L. Goodman. 2008. *Anaplasma phagocytophilum*-induced gene expression in both human neutrophils and HL-60 cells. *Genomics* **92:**144–151.

Lee, M. J., L. Chien-Liang, J. Y. Tsai, W. T. Sue, W. S. Hsia, and H. Huang. 2010. Identification and biochemical characterization of a unique Mn^{2+}-dependent UMP kinase from *Helicobacter pylori*. *Arch. Microbiol.* **192:**739–746.

Lin, M., A. den Dulk-Ras, P. J. Hooykaas, and Y. Rikihisa. 2007. *Anaplasma phagocytophilum* AnkA secreted by type IV secretion system is tyrosine phosphorylated by Abl-1 to facilitate infection. *Cell. Microbiol.* **9:**2644–2657.

Lin, M., and Y. Rikihisa. 2003a. *Ehrlichia chaffeensis* and *Anaplasma phagocytophilum* lack genes for lipid A biosynthesis and incorporate cholesterol for their survival. *Infect. Immun.* **71:**5324–5331.

Lin, M., and Y. Rikihisa. 2003b. Obligatory intracellular parasitism by *Ehrlichia chaffeensis* and *Anaplasma phagocytophilum* involves caveolae and glycosylphosphatidylinositol-anchored proteins. *Cell. Microbiol.* **5:**809–820.

Lin, M., and Y. Rikihisa. 2004. *Ehrlichia chaffeensis* downregulates surface Toll-like receptors 2/4, CD14 and transcription factors PU.1 and inhibits lipopolysaccharide activation of NF-κB, ERK 1/2 and p38 MAPK in host monocytes. *Cell. Microbiol.* **6:**175–186.

Lin, M., and Y. Rikihisa. 2007. Degradation of $p22^{phox}$ and inhibition of superoxide generation by *Ehrlichia chaffeensis* in human monocytes. *Cell. Microbiol.* **9:**861–874.

Lin, M., C. Zhang, K. Gibson, and Y. Rikihisa. 2009. Analysis of complete genome sequence of *Neorickettsia risticii*: causative agent of Potomac horse fever. *Nucleic Acids Res.* **37:**6076–6091.

Lin, M., M. X. Zhu, and Y. Rikihisa. 2002a. Rapid activation of protein tyrosine kinase and phospholipase C-γ2 and increase in cytosolic free calcium are required by *Ehrlichia chaffeensis* for internalization and growth in THP-1 cells. *Infect. Immun.* **70:**889–898.

Lin, Q., and Y. Rikihisa. 2005. Establishment of cloned *Anaplasma phagocytophilum* and analysis of *p44* gene conversion within an infected horse and infected SCID mice. *Infect. Immun.* **73:** 5106–5114.

Lin, Q., N. Zhi, N. Ohashi, H. W. Horowitz, M. E. Aguero-Rosenfeld, J. Raffalli, G. P. Wormser, and Y. Rikihisa. 2002b. Analysis of sequences and loci of *p44* homologs expressed by *Anaplasma phagocytophila* in acutely infected patients. *J. Clin. Microbiol.* **40:**2981–2988.

Lindsay, J., M. D. Esposti, and A. P. Gilmore. 2011. Bcl-2 proteins and mitochondria—specificity in membrane targeting for death. *Biochim. Biophys. Acta* **1813:**532–539.

Liu, Y., Z. Zhang, Y. Jiang, L. Zhang, V. L. Popov, J. Zhang, D. H. Walker, and X. J. Yu. 2011. Obligate intracellular bacterium *Ehrlichia* inhibiting mitochondrial activity. *Microbes Infect.* **13:**232–238.

Lopez, J. E., G. H. Palmer, K. A. Brayton, M. J. Dark, S. E. Leach, and W. C. Brown. 2007. Immunogenicity of *Anaplasma marginale* type IV secretion system proteins in a protective outer membrane vaccine. *Infect. Immun.* **75:**2333–2342.

Lopez, J. E., W. F. Siems, G. H. Palmer, K. A. Brayton, T. C. McGuire, J. Norimine, and W. C. Brown. 2005. Identification of novel antigenic proteins in a complex *Anaplasma marginale* outer membrane immunogen by mass spectrometry and genomic mapping. *Infect. Immun.* **73:**8109–8118.

Luo, T., X. Zhang, and J. W. McBride. 2009. Major species-specific antibody epitopes of the *Ehrlichia chaffeensis* p120 and *E. canis* p140 orthologs in surface-exposed tandem repeat regions. *Clin. Vaccine Immunol.* **16:**982–990.

Luo, T., X. Zhang, A. Wakeel, V. L. Popov, and J. W. McBride. 2008. A variable-length PCR target protein of *Ehrlichia chaffeensis* contains major species-specific antibody epitopes in acidic serine-rich tandem repeats. *Infect. Immun.* **76:**1572–1580.

Machner, M. P., and R. R. Isberg. 2006. Targeting of host Rab GTPase function by the intravacuolar pathogen *Legionella pneumophila. Dev. Cell* **11:**47–56.

Mavromatis, K., C. K. Doyle, A. Lykidis, N. Ivanova, M. P. Francino, P. Chain, M. Shin, S. Malfatti, F. Larimer, A. Copeland, J. C. Detter, M. Land, P. M. Richardson, X. J. Yu, D. H. Walker, J. W. McBride, and N. C. Kyrpides. 2006. The genome of the obligately intracellular bacterium *Ehrlichia canis* reveals themes of complex membrane structure and immune evasion strategies. *J. Bacteriol.* **188:**4015–4023.

Mayor, S., and R. E. Pagano. 2007. Pathways of clathrin-independent endocytosis. *Nat. Rev. Mol. Cell Biol.* **8:**603–612.

McBride, J. W., C. K. Doyle, X. Zhang, A. M. Cardenas, V. L. Popov, K. A. Nethery, and M. E. Woods. 2007. Identification of a glycosylated *Ehrlichia canis* 19-kilodalton major immunoreactive protein with a species-specific serine-rich glycopeptide epitope. *Infect. Immun.* **75:**74–82.

McGarey, D. J., and D. R. Allred. 1994. Characterization of hemagglutinating components on the *Anaplasma marginale* initial body surface and identification of possible adhesins. *Infect. Immun.* **62:**4587–4593.

McGarey, D. J., A. F. Barbet, G. H. Palmer, T. C. McGuire, and D. R. Allred. 1994. Putative adhesins of *Anaplasma marginale*: major surface polypeptides 1a and 1b. *Infect. Immun.* **62:** 4594–4601.

Messick, J. B., and Y. Rikihisa. 1993. Characterization of *Ehrlichia risticii* binding, internalization, and proliferation in host cells by flow cytometry. *Infect. Immun.* **61:**3803–3810.

Mobius, W., E. van Donselaar, Y. Ohno-Iwashita, Y. Shimada, H. F. Heijnen, J. W. Slot, and H. J. Geuze. 2003. Recycling compartments and the internal vesicles of multivesicular bodies harbor most of the cholesterol found in the endocytic pathway. *Traffic* **4:**222–231.

Moorhead, A. M., J. Y. Jung, A. Smirnov, S. Kaufer, and M. A. Scidmore. 2010. Multiple host proteins that function in phosphatidylinositol-4-phosphate metabolism are recruited to the chlamydial inclusion. *Infect. Immun.* **78:**1990–2007.

Moorhead, A. R., K. A. Rzomp, and M. A. Scidmore. 2007. The Rab6 effector Bicaudal D1 associates with *Chlamydia trachomatis* inclusions in a biovar-specific manner. *Infect. Immun.* **75:** 781–791.

Mott, J., R. E. Barnewall, and Y. Rikihisa. 1999. Human granulocytic ehrlichiosis agent and *Ehrlichia chaffeensis* reside in different cytoplasmic compartments in HL-60 cells. *Infect. Immun.* **67:**1368–1378.

Mott, J., and Y. Rikihisa. 2000. Human granulocytic ehrlichiosis agent inhibits superoxide anion generation by human neutrophils. *Infect. Immun.* **68:**6697–6703.

Mott, J., Y. Rikihisa, and S. Tsunawaki. 2002. Effects of *Anaplasma phagocytophila* on NADPH oxidase components in human neutrophils and HL-60 cells. *Infect. Immun.* **70:**1359–1366.

Müller, M. P., H. Peters, J. Blümer, W. Blankenfeldt, R. S. Goody, and A. Itzen. 2010. The *Legionella* effector protein DrrA AMPylates the membrane traffic regulator Rab1b. *Science* **329:**946–949.

Munderloh, U. G., E. F. Blouin, K. M. Kocan, N. L. Ge, W. L. Edwards, and T. J. Kurtti. 1996. Establishment of the tick (Acari:Ixodidae)-borne cattle pathogen *Anaplasma marginale* (Rickettsiales:Anaplasmataceae) in tick cell culture. *J. Med. Entomol.* **33:**656–664.

Munderloh, U. G., S. D. Jauron, V. Fingerle, L. Leitritz, S. F. Hayes, J. M. Hautman, C. M. Nelson, B. W. Huberty, T. J. Kurtti, G. G. Ahlstrand, B. Greig, M. A. Mellencamp, and J. L. Goodman. 1999. Invasion and intracellular development of the human granulocytic ehrlichiosis agent in tick cell culture. *J. Clin. Microbiol.* **37:**2518–2524.

Munderloh, U. G., M. J. Lynch, M. J. Herron, A. T. Palmer, T. J. Kurtti, R. D. Nelson, and J. L. Goodman. 2004. Infection of endothelial cells with *Anaplasma marginale* and *A. phagocytophilum. Vet. Microbiol.* **101:**53–64.

Murata, T., A. Delprato, A. Ingmundson, D. K. Toomre, D. G. Lambright, and C. R. Roy. 2006. The *Legionella pneumophila* effector protein DrrA is a Rab1 guanine nucleotide-exchange factor. *Nat. Cell Biol.* **8:**971–977.

Nandi, B., M. Chatterjee, K. Hogle, M. McLaughlin, K. MacNamara, R. Racine, and G. M. Winslow. 2009. Antigen display, T-cell activation, and immune evasion during acute and chronic ehrlichiosis. *Infect. Immun.* **77:**4643–4653.

Nauseef, W. M. 2007. How human neutrophils kill and degrade microbes: an integrated view. *Immunol. Rev.* **219:**88–102.

Neelakanta, G., H. Sultana, D. Fish, J. F. Anderson, and E. Fikrig. 2010. *Anaplasma* *phagocytophilum* induces *Ixodes scapularis* ticks to express an antifreeze glycoprotein gene that enhances their survival in the cold. *J. Clin. Invest.* **120:** 3179–3190.

Nelson, C. M., M. J. Herron, R. F. Felsheim, B. R. Schloeder, S. M. Grindle, A. O. Chavez, T. J. Kurtti, and U. G. Munderloh. 2008. Whole genome transcription profiling of *Anaplasma phagocytophilum* in human and tick host cells by tiling array analysis. *BMC Genomics* **9:**364.

Niu, H., V. Kozjak-Pavlovic, T. Rudel, and Y. Rikihisa. 2010. *Anaplasma phagocytophilum* Ats-1 is imported into host cell mitochondria and interferes with apoptosis induction. *PLoS Pathog.* **6:** e1000774.

Niu, H., Y. Rikihisa, M. Yamaguchi, and N. Ohashi. 2006. Differential expression of VirB9 and VirB6 during the life cycle of *Anaplasma phagocytophilum* in human leucocytes is associated with differential binding and avoidance of lysosome pathway. *Cell. Microbiol.* **8:**523–534.

Niu, H., M. Yamaguchi, and Y. Rikihisa. 2008. Subversion of cellular autophagy by *Anaplasma phagocytophilum*. *Cell. Microbiol.* **10:**593–605.

Ohashi, N., N. Zhi, Q. Lin, and Y. Rikihisa. 2002. Characterization and transcriptional analysis of gene clusters for a type IV secretion machinery in human granulocytic and monocytic ehrlichiosis agents. *Infect. Immun.* **70:**2128–2138.

Ojogun, N., B. Barnstein, B. Huang, C. A. Oskeritzian, J. W. Homeister, D. Miller, J. J. Ryan, and J. A. Carlyon. 2011. *Anaplasma phagocytophilum* infects mast cells via α1,3-fucosylated but not sialylated glycans and inhibits IgE-mediated cytokine production and histamine release. *Infect. Immun.* **79:**2717–2726.

Page, R. D. 2002. Visualizing phylogenetic trees using TreeView. *Curr. Protocols Bioinformatics.* **6:**6.2. http://cda.currentprotocols.com/WileyCDA/CurPro3Title/isbn-0471250953,descCd-tableOfContents.html.

Palmer, G. H., and K. A. Brayton. 2007. Gene conversion is a convergent strategy for pathogen antigenic variation. *Trends Parasitol.* **23:**408–413.

Pan, X., A. Luhrmann, A. Satoh, M. A. Laskowski-Arce, and C. R. Roy. 2008. Ankyrin repeat proteins comprise a diverse family of bacterial type IV effectors. *Science* **320:**1651–1654.

Park, J., K. J. Kim, K. S. Choi, D. J. Grab, and J. S. Dumler. 2004. *Anaplasma phagocytophilum* AnkA binds to granulocyte DNA and nuclear proteins. *Cell. Microbiol.* **6:**743–751.

Pedra, J. H., S. Narasimhan, D. Rendic, K. DePonte, L. Bell-Sakyi, I. B. Wilson, and E. Fikrig. 2010. Fucosylation enhances colonization of ticks by *Anaplasma phagocytophilum*. *Cell. Microbiol.* **12:**1222–1234.

Pedra, J. H., B. Sukumaran, J. A. Carlyon, N. Berliner, and E. Fikrig. 2005. Modulation of NB4 promyelocytic leukemic cell machinery by *Anaplasma phagocytophilum*. *Genomics* **86:** 365–377.

Poole, A. W., and M. L. Jones. 2005. A SHPing tale: perspectives on the regulation of SHP-1 and SHP-2 tyrosine phosphatases by the C-terminal tail. *Cell. Signal.* **17:**1323–1332.

Popov, V. L., V. C. Han, S. M. Chen, J. S. Dumler, H. M. Feng, T. G. Andreadis, R. B. Tesh, and D. H. Walker. 1998. Ultrastructural differentiation of the genogroups in the genus *Ehrlichia*. *J. Med. Microbiol.* **47:**235–251.

Popov, V. L., X. Yu, and D. H. Walker. 2000. The 120 kDa outer membrane protein of *Ehrlichia chaffeensis*: preferential expression on dense-core cells and gene expression in *Escherichia coli* associated with attachment and entry. *Microb. Pathog.* **28:**71–80.

Punnonen, E. L., K. Ryhanen, and V. S. Marjomaki. 1998. At reduced temperature, endocytic membrane traffic is blocked in multivesicular carrier endosomes in rat cardiac myocytes. *Eur. J. Cell Biol.* **75:**344–352.

Radhakrishna, H., and J. G. Donaldson. 1997. ADP-ribosylation factor 6 regulates a novel plasma membrane recycling pathway. *J. Cell Biol.* **139:** 49–61.

Raghavan, V., and E. A. Groisman. 2010. Orphan and hybrid two-component system proteins in health and disease. *Curr. Opin. Microbiol.* **13:**226–231.

Reneer, D. V., S. A. Kearns, T. Yago, J. Sims, R. D. Cummings, R. P. McEver, and J. A. Carlyon. 2006. Characterization of a sialic acid- and P-selectin glycoprotein ligand-1-independent adhesin activity in the granulocytotropic bacterium *Anaplasma phagocytophilum*. *Cell. Microbiol.* **8:** 1972–1984.

Reneer, D. V., M. J. Troese, B. Huang, S. A. Kearns, and J. A. Carlyon. 2008. *Anaplasma phagocytophilum* PSGL-1-independent infection does not require Syk and leads to less efficient AnkA delivery. *Cell Microbiol* **10:**1827–1838.

Rikihisa, Y. 2010a. *Anaplasma phagocytophilum* and *Ehrlichia chaffeensis*: subversive manipulators of host cells. *Nat. Rev. Microbiol.* **8:**328–339.

Rikihisa, Y. 2010b. Molecular events involved in cellular invasion by *Ehrlichia chaffeensis* and *Anaplasma phagocytophilum*. *Vet. Parasitol.* **167:**155–166.

Rikihisa, Y., and M. Lin. 2010. *Anaplasma phagocytophilum* and *Ehrlichia chaffeensis* type IV secretion and Ank proteins. *Curr. Opin. Microbiol.* **13:**59–66.

Rikihisa, Y., M. Lin, and H. Niu. 2010. Type IV secretion in the obligatory intracellular bacterium *Anaplasma phagocytophilum*. *Cell. Microbiol.* **12:**1213–1221.

Romano, P. S., M. G. Gutierrez, W. Beron, M. Rabinovitch, and M. I. Colombo. 2007. The autophagic pathway is actively modulated by phase

II *Coxiella burnetii* to efficiently replicate in the host cell. *Cell. Microbiol.* **9:**891–909.

Römling, U. 2009. Cyclic Di-GMP (c-Di-GMP) goes into host cells—c-Di-GMP signaling in the obligate intracellular pathogen *Anaplasma phagocytophilum*. *J. Bacteriol.* **191:**683–686.

Rowold, D. J., and R. J. Herrera. 2000. *Alu* elements and the human genome. *Genetica* **108:** 57–72.

Rzomp, K. A., A. R. Moorhead, and M. A. Scidmore. 2006. The GTPase Rab4 interacts with *Chlamydia trachomatis* inclusion membrane protein CT229. *Infect. Immun.* **74:**5362–5373.

Rzomp, K. A., L. D. Scholtes, B. J. Briggs, G. R. Whittaker, and M. A. Scidmore. 2003. Rab GTPases are recruited to chlamydial inclusions in both a species-dependent and species-independent manner. *Infect. Immun.* **71:**5855–5870.

Sandvig, K., M. L. Torgersen, H. A. Raa, and B. van Deurs. 2008. Clathrin-independent endocytosis: from nonexisting to an extreme degree of complexity. *Histochem. Cell Biol.* **129:**267–276.

Sarkar, M., D. V. Reneer, and J. A. Carlyon. 2007. Sialyl-Lewis x-independent infection of human myeloid cells by *Anaplasma phagocytophilum* strains HZ and HGE1. *Infect. Immun.* **75:** 5720–5725.

Schaff, U. Y., K. A. Trott, S. Chase, K. Tam, J. L. Johns, J. A. Carlyon, D. C. Genetos, N. J. Walker, S. I. Simon, and D. L. Borjesson. 2010. Neutrophils exposed to *A. phagocytophilum* under shear stress fail to fully activate, polarize, and transmigrate across inflamed endothelium. *Am. J. Physiol. Cell Physiol.* **299:**C87–C96.

Scharf, W., S. Schauer, F. Freyburger, M. Petrovec, D. Schaarschmidt-Kiener, G. Liebisch, M. Runge, M. Ganter, A. Kehl, J. S. Dumler, A. L. Garcia-Perez, J. Jensen, V. Fingerle, M. L. Meli, A. Ensser, S. Stuen, and F. D. von Loewenich. 2011. Distinct host species correlate with *Anaplasma phagocytophilum ankA* gene clusters. *J. Clin. Microbiol.* **49:**790–796.

Schlumberger, M. C., and W. D. Hardt. 2005. Triggered phagocytosis by *Salmonella*: bacterial molecular mimicry of RhoGTPase activation/deactivation. *Curr. Top. Microbiol. Immunol.* **291:** 29–42.

Seshadri, R., I. T. Paulsen, J. A. Eisen, T. D. Read, K. E. Nelson, W. C. Nelson, N. L. Ward, H. Tettelin, T. M. Davidsen, M. J. Beanan, R. T. Deboy, S. C. Daugherty, L. M. Brinkac, R. Madupu, R. J. Dodson, H. M. Khouri, K. H. Lee, H. A. Carty, D. Scanlan, R. A. Heinzen, H. A. Thompson, J. E. Samuel, C. M. Fraser, and J. F. Heidelberg. 2003. Complete genome sequence of the Q-fever pathogen *Coxiella burnetii*. *Proc. Natl. Acad. Sci. USA* **100:**5455–5460.

Shin, J. S., and S. N. Abraham. 2001a. Caveolae as portals of entry for microbes. *Microbes Infect.* **3:**755–761.

Shin, J. S., and S. N. Abraham. 2001b. Glycosylphosphatidylinositol-anchored receptor-mediated bacterial endocytosis. *FEMS Microbiol. Lett.* **197:**131–138.

Shkap, V., T. Molad, L. Fish, and G. H. Palmer. 2002. Detection of the *Anaplasma centrale* vaccine strain and specific differentiation from *Anaplasma marginale* in vaccinated and infected cattle. *Parasitol. Res.* **88:**546–552.

Singu, V., H. Liu, C. Cheng, and R. R. Ganta. 2005. *Ehrlichia chaffeensis* expresses macrophage- and tick cell-specific 28-kilodalton outer membrane proteins. *Infect. Immun.* **73:**79–87.

Singu, V., L. Peddireddi, K. R. Sirigireddy, C. Cheng, U. Munderloh, and R. R. Ganta. 2006. Unique macrophage and tick cell-specific protein expression from the p28/p30-outer membrane protein multigene locus in *Ehrlichia chaffeensis* and *Ehrlichia canis*. *Cell. Microbiol.* **8:**1475–1487.

Smith, J. M., and B. J. Mayer. 2002. Abl: mechanisms of regulation and activation. *Front. Biosci.* **7:** d31–d42.

Stenmark, H. 2009. Rab GTPases as coordinators of vesicle traffic. *Nat. Rev. Mol. Cell Biol.* **10:** 513–525.

Stevenson, H. L., E. C. Crossley, N. Thirumalapura, D. H. Walker, and N. Ismail. 2008. Regulatory roles of CD1d-restricted NKT cells in the induction of toxic shock-like syndrome in an animal model of fatal ehrlichiosis. *Infect. Immun.* **76:**1434–1444.

Storey, J. R., L. A. Doros-Richert, C. Gingrich-Baker, K. Munroe, T. N. Mather, R. T. Coughlin, G. A. Beltz, and C. I. Murphy. 1998. Molecular cloning and sequencing of three granulocytic *Ehrlichia* genes encoding high-molecular-weight immunoreactive proteins. *Infect. Immun.* **66:**1356–1363.

Stuible, M., K. M. Doody, and M. L. Tremblay. 2008. PTP1B and TC-PTP: regulators of transformation and tumorigenesis. *Cancer Metastasis Rev.* **27:**215–230.

Su, H., G. McClarty, F. Dong, G. M. Hatch, Z. K. Pan, and G. Zhong. 2004. Activation of Raf/MEK/ERK/cPLA2 signaling pathway is essential for chlamydial acquisition of host glycerophospholipids. *J. Biol. Chem.* **279:**9409–9416.

Sukumaran, B., J. A. Carlyon, J. L. Cai, N. Berliner, and E. Fikrig. 2005. Early transcriptional response of human neutrophils to *Anaplasma phagocytophilum* infection. *Infect. Immun.* **73:**8089–8099.

Sukumaran, B., J. E. Mastronunzio, S. Narasimhan, S. Fankhauser, P. D. Uchil, R. Levy, M. Graham, T. M. Colpitts, C. F. Lesser, and E.

Fikrig. 2011. *Anaplasma phagocytophilum* AptA modulates Erk1/2 signalling. *Cell. Microbiol.* **13:**47–61.

Sukumaran, B., S. Narasimhan, J. F. Anderson, K. DePonte, N. Marcantonio, M. N. Krishnan, D. Fish, S. R. Telford, F. S. Kantor, and E. Fikrig. 2006. An *Ixodes scapularis* protein required for survival of *Anaplasma phagocytophilum* in tick salivary glands. *J. Exp. Med.* **203:** 1507–1517.

Sultana, H., G. Neelakanta, F. S. Kantor, S. E. Malawista, D. Fish, R. R. Montgomery, and E. Fikrig. 2010. *Anaplasma phagocytophilum* induces actin phosphorylation to selectively regulate gene transcription in *Ixodes scapularis* ticks. *J. Exp. Med.* **207:**1727–1743.

Sutten, E. L., J. Norimine, P. A. Beare, R. A. Heinzen, J. E. Lopez, K. Morse, K. A. Brayton, J. J. Gillespie, and W. C. Brown. 2010. *Anaplasma marginale* type IV secretion system proteins VirB2, VirB7, VirB11, and VirD4 are immunogenic components of a protective bacterial membrane vaccine. *Infect. Immun.* **78:**1314–1325.

Teng, C. H., R. U. Palaniappan, and Y. F. Chang. 2003. Cloning and characterization of an *Ehrlichia canis* gene encoding a protein localized to the morula membrane. *Infect. Immun.* **71:** 2218–2225.

Thomas, R. J., J. S. Dumler, and J. A. Carlyon. 2009. Current management of human granulocytic anaplasmosis, human monocytic ehrlichiosis and *Ehrlichia ewingii* ehrlichiosis. *Expert Rev. Anti Infect. Ther.* **7:**709–722.

Thomas, S., V. L. Popov, and D. H. Walker. 2010. Exit mechanisms of the intracellular bacterium *Ehrlichia*. *PLoS One* **5:**e15775.

Thomas, V., and E. Fikrig. 2007. *Anaplasma phagocytophilum* specifically induces tyrosine phosphorylation of ROCK1 during infection. *Cell. Microbiol.* **9:**1730–1737.

Thomas, V., S. Samanta, C. Wu, N. Berliner, and E. Fikrig. 2005. *Anaplasma phagocytophilum* modulates gp91^{phox} gene expression through altered interferon regulatory factor 1 and PU.1 levels and binding of CCAAT displacement protein. *Infect. Immun.* **73:**208–218.

Troese, M. J., and J. A. Carlyon. 2009. *Anaplasma phagocytophilum* dense-cored organisms mediate cellular adherence through recognition of human P-selectin glycoprotein ligand 1. *Infect. Immun.* **77:**4018–4027.

Trofe, J., K. S. Reddy, R. J. Stratta, S. D. Flax, K. T. Somerville, R. R. Alloway, M. F. Egidi, M. H. Shokouh-Amiri, and A. O. Gaber. 2001. Human granulocytic ehrlichiosis in pancreas transplant recipients. *Transpl. Infect. Dis.* **3:**34–39.

Ueti, M. W., D. P. Knowles, C. M. Davitt, G. A. Scoles, T. V. Baszler, and G. H. Palmer. 2009. Quantitative differences in salivary pathogen load during tick transmission underlie strain-specific variation in transmission efficiency of *Anaplasma marginale*. *Infect. Immun.* **77:**70–75.

Ueti, M. W., J. O. Reagan, Jr., D. P. Knowles, Jr., G. A. Scoles, V. Shkap, and G. H. Palmer. 2007. Identification of midgut and salivary glands as specific and distinct barriers to efficient tick-borne transmission of *Anaplasma marginale*. *Infect. Immun.* **75:**2959–2964.

Vergunst, A. C., M. C. van Lier, A. den Dulk-Ras, T. A. Stuve, A. Ouwehand, and P. J. Hooykaas. 2005. Positive charge is an important feature of the C-terminal transport signal of the VirB/D4-translocated proteins of *Agrobacterium*. *Proc. Natl. Acad. Sci. USA* **102:**832–837.

Voth, D. E., P. A. Beare, D. Howe, U. M. Sharma, G. Samoilis, D. C. Cockrell, A. Omsland, and R. A. Heinzen. 2011. The *Coxiella burnetii* cryptic plasmid is enriched in genes encoding type IV secretion system substrates. *J. Bacteriol.* **193:**1493–1503.

Wakeel, A., J. A. Kuriakose, and J. W. McBride. 2009. An *Ehrlichia chaffeensis* tandem repeat protein interacts with multiple host targets involved in cell signaling, transcriptional regulation, and vesicle trafficking. *Infect. Immun.* **77:**1734–1745.

Wakeel, A., X. Zhang, and J. W. McBride. 2010a. Mass spectrometric analysis of *Ehrlichia chaffeensis* tandem repeat proteins reveals evidence of phosphorylation and absence of glycosylation. *PLoS One* **5:**e9552.

Wakeel, A., B. Zhu, X. J. Yu, and J. W. McBride. 2010b. New insights into molecular *Ehrlichia chaffeensis*-host interactions. *Microbes Infect.* **12:**337–345.

Wang, T., S. E. Malawista, U. Pal, M. Grey, J. Meek, M. Akkoyunlu, V. Thomas, and E. Fikrig. 2002. Superoxide anion production during *Anaplasma phagocytophila* infection. *J. Infect. Dis.* **186:**274–280.

Webster, P., J. W. IJdo, L. M. Chicoine, and E. Fikrig. 1998. The agent of human granulocytic ehrlichiosis resides in an endosomal compartment. *J. Clin. Invest.* **101:**1932–1941.

Wells, M. Y., and Y. Rikihisa. 1988. Lack of lysosomal fusion with phagosomes containing *Ehrlichia risticii* in P388D1 cells: abrogation of inhibition with oxytetracycline. *Infect. Immun.* **56:**3209–3215.

Werren, J. H., L. Baldo, and M. E. Clark. 2008. *Wolbachia*: master manipulators of invertebrate biology. *Nat. Rev. Microbiol.* **6:**741–751.

Wistedt, A. C., U. Ringdahl, W. Muller-Esterl, and U. Sjobring. 1995. Identification of a plasminogen-binding motif in PAM, a bacterial surface protein. *Mol. Microbiol.* **18:**569–578.

Wu, M., L. V. Sun, J. Vamathevan, M. Riegler, R. Deboy, J. C. Brownlie, E. A. McGraw, W.

Martin, C. Esser, N. Ahmadinejad, C. Wiegand, R. Madupu, M. J. Beanan, L. M. Brinkac, S. C. Daugherty, A. S. Durkin, J. F. Kolonay, W. C. Nelson, Y. Mohamoud, P. Lee, K. Berry, M. B. Young, T. Utterback, J. Weidman, W. C. Nierman, I. T. Paulsen, K. E. Nelson, H. Tettelin, S. L. O'Neill, and J. A. Eisen. 2004. Phylogenomics of the reproductive parasite *Wolbachia pipientis* wMel: a streamlined genome overrun by mobile genetic elements. *PLoS Biol.* **2:**E69.

Xiong, Q., W. Bao, Y. Ge, and Y. Rikihisa. 2008. *Ehrlichia ewingii* infection delays spontaneous neutrophil apoptosis through stabilization of mitochondria. *J. Infect. Dis.* **197:**1110–1118.

Xiong, Q., M. Lin, and Y. Rikihisa. 2009. Cholesterol-dependent *Anaplasma phagocytophilum* exploits the low-density lipoprotein uptake pathway. *PLoS Pathog.* **5:**e1000329.

Xiong, Q., X. Wang, and Y. Rikihisa. 2007. High-cholesterol diet facilitates *Anaplasma phagocytophilum* infection and up-regulates macrophage inflammatory protein-2 and CXCR2 expression in apolipoprotein E-deficient mice. *J. Infect. Dis.* **195:**1497–1503.

Xu, S., C. Zhang, Y. Miao, J. Gao, and D. Xu. 2010. Effector prediction in host-pathogen interaction based on a Markov model of a ubiquitous EPIYA motif. *BMC Genomics* **11**(Suppl. 3)**:**S1.

Yago, T., A. Leppanen, J. A. Carlyon, M. Akkoyunlu, S. Karmakar, E. Fikrig, R. D. Cummings, and R. P. McEver. 2003. Structurally distinct requirements for binding of P-selectin glycoprotein ligand-1 and sialyl Lewis x to *Anaplasma phagocytophilum* and P-selectin. *J. Biol. Chem.* **278:**37987–37997.

Yoshiie, K., H. Y. Kim, J. Mott, and Y. Rikihisa. 2000. Intracellular infection by the human granulocytic ehrlichiosis agent inhibits human neutrophil apoptosis. *Infect. Immun.* **68:**1125–1133.

Yu, X. J., P. Crocquet-Valdes, L. C. Cullman, and D. H. Walker. 1996. The recombinant 120-kilodalton protein of *Ehrlichia chaffeensis*, a potential diagnostic tool. *J. Clin. Microbiol.* **34:**2853–2855.

Yu, X. J., J. W. McBride, C. M. Diaz, and D. H. Walker. 2000. Molecular cloning and characterization of the 120-kilodalton protein gene of *Ehrlichia canis* and application of the recombinant 120-kilodalton protein for serodiagnosis of canine ehrlichiosis. *J. Clin. Microbiol.* **38:**369–374.

Zarbock, A., C. A. Lowell, and K. Ley. 2007. Spleen tyrosine kinase Syk is necessary for E-selectin-induced $\alpha_L\beta_2$ integrin-mediated rolling on intercellular adhesion molecule-1. *Immunity* **26:**773–783.

Zhang, J. Z., J. W. McBride, and X. J. Yu. 2003. L-selectin and E-selectin expressed on monocytes mediating *Ehrlichia chaffeensis* attachment onto host cells. *FEMS Microbiol. Lett.* **227:**303–309.

Zhang, J. Z., V. L. Popov, S. Gao, D. H. Walker, and X. J. Yu. 2007. The developmental cycle of *Ehrlichia chaffeensis* in vertebrate cells. *Cell. Microbiol.* **9:**610–618.

Zhang, J. Z., M. Sinha, B. A. Luxon, and X. J. Yu. 2004. Survival strategy of obligately intracellular *Ehrlichia chaffeensis*: novel modulation of immune response and host cell cycles. *Infect. Immun.* **72:**498–507.

Zhi, N., N. Ohashi, and Y. Rikihisa. 1999. Multiple *p44* genes encoding major outer membrane proteins are expressed in the human granulocytic ehrlichiosis agent. *J. Biol. Chem.* **274:** 17828–17836.

Zhi, N., N. Ohashi, T. Tajima, J. Mott, R. W. Stich, D. Grover, S. R. Telford III, Q. Lin, and Y. Rikihisa. 2002. Transcript heterogeneity of the *p44* multigene family in a human granulocytic ehrlichiosis agent transmitted by ticks. *Infect. Immun.* **70:**1175–1184.

Zhu, B., K. A. Nethery, J. A. Kuriakose, A. Wakeel, X. Zhang, and J. W. McBride. 2009. Nuclear translocated *Ehrlichia chaffeensis* ankyrin protein interacts with a specific adenine-rich motif of host promoter and intronic *Alu* elements. *Infect. Immun.* **77:**4243–4255.

RICKETTSIAL PHYSIOLOGY AND METABOLISM IN THE FACE OF REDUCTIVE EVOLUTION

Jonathon P. Audia

7

INTRODUCTION

> "Imagine a microbe that is a combination *E. coli*-mitochondrion which, unlike the mitochondrion, has never traded its soul for symbiotic safety, but rather has evolved special membrane functions—a unique biology—that allow it to exploit the host cell's cytoplasm!" (Winkler, 1982)

A distinguishing feature of the *Rickettsiales* as a group is the intracellular growth niche to which they have committed. The *Rickettsiaceae* induce their own phagocytosis and quickly escape the endosome to gain direct access to the cytoplasm, and sometimes the nucleus, of the eukaryotic host cell. The *Anaplasmataceae* also gain entry but reside within a host-derived membranous compartment that the pathogen actively modifies to prevent lysosomal fusion and permit replication. This intimate association with a eukaryotic host cell has played an important role in the evolution of rickettsial physiology and the nature of their genomes. Our understanding of the interplay between host and the obligate intracellular rickettsia has been significantly influenced by the advent of the genomics era and the availability of genome sequences. One of the major emphases of my research program is to understand how obligate intracytoplasmic growth has affected the physiology of *Rickettsia prowazekii*, and accordingly, this chapter will discuss metabolism and reductive evolution from the pathogenic rickettsia's point of view.

Rapid advances in sequencing technologies have contributed to the ever-expanding availability of genome sequence information. This has significantly augmented our understanding of the factors that influence virulence and shape pathogen evolution at the genome level. The most familiar factor that affects the "dynamic pangenome" of a pathogen is horizontal gene transfer (HGT). Through the process of HGT, new genes/traits are readily acquired via genetic exchange with other organisms encountered in the environment. If the acquired genes/traits confer a selective advantage that allows the modified organism to outcompete others in the environment, they will become fixed within the modified organism's genome. Conversely, genes/traits that are no longer essential for survival in the environment can accumulate mutations and are eventually lost from the genome through a process known as reductive evolution—the so-called "use it or lose it" axiom (Pallen and Wren, 2007).

Jonathon P. Audia, Department of Microbiology and Immunology, University of South Alabama College of Medicine, Mobile, AL 36688.

Intracellular Pathogens II: Rickettsiales
Edited by G. H. Palmer and A. F. Azad © 2012 ASM Press, Washington, DC
10.1128/9781555817336.ch7

While it is obvious that HGT influences the biology and virulence of free-living bacteria that share common environmental niches, its potential to influence "modern-day" obligate intracellular bacteria is somewhat less clear. It is postulated that coinfection of the ancestral amoebalike protozoa by ancient facultative and obligate intracellular bacteria created a confined environment conducive to gene exchange by HGT. The isolation of members of the *Rickettsiales* from present-day protozoa and the identification of conserved conjugal-like plasmid elements in some *Rickettsia* species lend credence to this idea (Greub and Raoult, 2003; Ogata et al., 2006; Moliner et al., 2010). However, it remains unknown as to whether the lack of plasmids in the more virulent *Rickettsia* species is a function of their evolutionary loss or perhaps trivial loss due to serial laboratory passage and removal of the appropriate selection pressure(s) (Baldridge et al., 2008; Fournier et al., 2009).

The host-adapted obligate intracellular rickettsial pathogens that infect higher-order eukaryotes likely find themselves in a cloistered environment rarely cocolonized by other microbes. Thus, while the identification of rickettsial plasmids is suggestive of the potential for HGT, the intracellular growth niche may furnish little or no opportunity to do so. In support of this idea, reductive genome evolution of host-adapted obligate intracellular pathogens appears to be more heavily influenced by gene duplication/amplification, gene adaptation by mutation, and gene loss (Andersson et al., 1998; Andersson and Andersson, 1999, 2001; Fournier et al., 2009). How these fascinating organisms have evolved to exploit nutrient-rich eukaryotic cytoplasm as a sole growth niche has been the subject of much research that is discussed herein.

A BRIEF HISTORY OF RICKETTSIAL PHYSIOLOGY IN THE PRE-GENOMICS ERA

The following section provides a summary of the studies describing rickettsial physiology and metabolism before 1998, when the first rickettsial genome sequence became available (Andersson et al., 1998). This summary is presented to provide the reader with insight into some of the key experiments that guided the field during a productive period in rickettsial research. It is not meant to be all-inclusive, and the author apologizes to those rickettsiologists whose work is not discussed.

An Introduction to Rickettsial Obligate Intracellular Growth

Inevitably, any student of microbiology becomes intimately acquainted with the enteric microbe *Escherichia coli* because we know more about its physiology, biochemistry, and genetics than any other organism on Earth. This vast amount of information is owing to the ease of growing large amounts of bacteria in relatively simple culture media for the purposes of analyzing the physiology of the intact bacterial cell or the biochemical properties of its proteins, either purified or in lysed bacterial cell extracts.

Within the simplicity of its culturing lies *E. coli*'s most intriguing property—when given simple sources of carbon, nitrogen, phosphorus, sulfur, trace metals, and water, this organism possesses the metabolic capacity for the generation of energy and for the de novo synthesis of every metabolite and macromolecule required to replicate an exact duplicate daughter cell. This is in stark contrast to the obligate intracellular rickettsia, metabolically compelled to grow within a eukaryotic host cell. In fact, this property led to the early characterization of rickettsiae as viruses. This misnomer was quickly dispelled when rickettsiae were shown to possess membranes, cell wall, ribosomes, and other characteristics of bacteria.

An enduring goal of the rickettsiologist has been to understand why these organisms are confined to growth within the eukaryotic cell. The underpinning hypothesis is that rickettsiae do not synthesize their own metabolites but avail themselves of host cell metabolites. For the purposes of this discussion, the term "metabolites" refers to small molecules and cofactors such as amino acids, nucleotides, sugars, coenzymes, etc., that are required for macromolecular synthesis pathways and the term "isolated rickettsiae" refers to rickettsiae purified and isolated away from the cytoplasm of a eukaryotic host cell for assay in the test tube.

Biochemical Properties of Isolated Rickettsiae

The early pioneering studies of rickettsial physiology and biochemistry performed by Bovarnick, Snyder, Weiss, Williams, Wisseman, Winkler, and others demonstrated that rickettsiae possess a limited metabolic repertoire compared to the free-living *E. coli*. This included the inability of early studies to detect any of the enzymatic activities central to glycolysis and gluconeogenesis (reviewed in Austin and Winkler, 1988b). Isolated rickettsiae were also energetically fragile. Upon prolonged incubation at either physiological or cold temperatures, isolated rickettsiae quickly lost their infectivity for host cells and lethality for mice and chicken embryos (Bovarnick and Allen, 1957a, 1957b). The classic reactivation experiments of Bovarnick and coworkers demonstrated that the addition of ATP and/ or NAD^+ could partially restore infectivity of the fragile rickettsiae (Bovarnick et al., 1953; Bovarnick and Schneider, 1960). These early studies indicated that while unstable outside of the host cell, isolated rickettsiae were capable of metabolic activity and two molecules at the center of bioenergetics could complement their exhaustion.

Experiments to test the limits of rickettsial metabolism outside of the host cell were also quite enlightening. Incubation of isolated rickettsiae with [^{14}C]glutamate resulted in the release of $^{14}CO_2$, which is indicative of tricarboxylic acid (TCA) cycle metabolism (Bovarnick and Miller, 1949; Williams and Weiss, 1978). Amino acids and intermediates of the TCA cycle were metabolized to various extents, and glutamate proved to be a preferred energy source of the isolated *Rickettsia typhi* (Williams and Weiss, 1978). The activities of TCA cycle enzymes were readily detectable in *R. prowazekii* whole-bacterial-cell extracts, and citrate synthase was partially purified and characterized (Phibbs and Winkler, 1981, 1982).

These studies demonstrated that *R. prowazekii* citrate synthase exhibited allosteric regulation by ATP, which was unusual considering that most bacterial citrate synthase enzymes are allosterically regulated by NAD^+ (Phibbs and Winkler, 1982). In addition, the *R. prowazekii* pyruvate dehydrogenase complex was partially purified from whole-rickettsial extracts and most likely fulfills its classical role in metabolism to supply acetyl coenzyme A (acetyl-CoA) to the TCA cycle (Phibbs and Winkler, 1981). Isolated rickettsiae were shown to generate a proton motive force measured as a charge difference across the membrane ($\Delta\psi$) (Zahorchak and Winkler, 1983). Fueled by glutamate metabolism, isolated rickettsiae were shown to phosphorylate adenylate nucleotides via oxidative phosphorylation (OXPHOS) and generate an adenylate energy charge defined as the ratio (ATP + 1/2 ADP)/(ATP + ADP + AMP) (Bovarnick, 1956; Williams and Weiss, 1978).

A study of *R. typhi* adenylate energy charge by Williams and Weiss (1978) raised several intriguing and still unanswered questions that merit mentioning. *R. typhi* metabolism of glutamate increased ATP levels and produced an adenylate energy charge ratio of ~0.75, which is widely regarded as the threshold required for bacterial growth (Atkinson, 1969). However, the rickettsiae were unable to sustain their adenylate energy charge over an extended time course, and when starved for glutamate quickly exhausted their ATP pool with a rapid 5-minute half-life. Intriguingly, ATP turnover was accompanied by an increase in the AMP pool with no further degradation to adenine or adenosine. The rapid ATP half-life upon starvation and the lack of adenine or adenosine metabolism distinguish rickettsiae from free-living *E. coli*. Williams and Weiss posited that these properties contribute to rickettsial instability outside of the host cell. The mechanism of rapid ATP turnover and the reasons behind the inability of rickettsiae to sustain their adenylate energy charge outside of the host cell remain unknown. Clearly, this unique adenylate metabolism is an important determinant of the rickettsial obligate intracellular lifestyle that warrants further study.

In addition to possessing metabolic pathways such as the TCA cycle and OXPHOS to supply energy in the forms of proton motive force and adenylate energy charge, isolated rickettsiae were also shown to incorporate radiolabeled nucleotides and amino acids into trichloroacetic acid-insoluble compounds as indicators of de novo DNA, RNA, and protein synthesis (Winkler, 1987; Winkler et al., 1999). Metabolism of isolated rickettsiae was also shown to be critical to their ability to enter host cells or to lyse red blood cells, as these processes were ablated by the addition of metabolic poisons/uncouplers (Ramm and Winkler, 1973; Walker and Winkler, 1978).

As noted by Hackstadt (1996) in his review of the biology of rickettsiae, the isolated rickettsia does exhibit a surprising amount of metabolic activity. However, it is likely that the instability and failure to observe replication of the isolated rickettsia in the test tube is a consequence of not knowing the appropriate medium recipe and environmental conditions that best approximate host cell cytoplasm.

Rickettsial Transport of Host Cell Metabolites

Arguably, cytoplasm is one of the most nutrient-replete growth niches an organism could inhabit. Host cell enzymatic pathways supply all of the end products of metabolism required for macromolecule synthesis and replication. Thus, an abundance of resources is separated from the rickettsia's own cytoplasm by the mere width of the bacterial outer and inner membranes. This idea led to the hypothesis that rickettsiae would directly avail themselves of the metabolites available from the host cell. As a corollary to this hypothesis, the ability to use host cell metabolites may have rendered rickettsial biosynthetic pathways for de novo metabolite synthesis obsolete, leading to their loss from the genome by reductive evolution.

Initially, it was thought that rickettsiae might possess "leaky" membranes as a means of acquiring host cell metabolites by simply allowing passive equilibration with the contents of the host cell's cytoplasm (Moulder, 1962). This notion was dispelled by Winkler and coworkers, who identified and characterized several unusual, substrate-specific, carrier-mediated transport systems in *R. prowazekii* (reviewed in Winkler, 1990). A highlight of this work was the discovery that *R. prowazekii* possesses transport systems for large, charged molecules such as ATP (Winkler, 1976), NAD^+ (Atkinson and Winkler, 1989), and UDP-glucose (Winkler and Daugherty, 1986). These molecules are not typically transported intact across membranes by any known free-living microbes or eukaryotic cells.

The *R. prowazekii* ATP/ADP translocase is the best-characterized rickettsial transport system. It is well established that the ATP/ADP translocase functions via an obligate exchange antiport mechanism and thus requires the presence of substrate on both sides of the membrane to catalyze transport. That is, for every one molecule of ATP or ADP that enters the rickettsial cytoplasm, there is a concomitant exit of one molecule of ATP or ADP (Winkler, 1976). The importance of this obligate exchange antiport mechanism is twofold. First, ATP and ADP are not lost to the external environment when the rickettsiae are outside of the host cell, thus preserving the pools of these critical metabolites within rickettsial cytoplasm until entry into a new host cell can occur. Second, and perhaps most interesting, the activity of the ATP/ADP translocase should allow the rickettsiae to equilibrate their adenylate energy charge with that of the host cell (Fig. 1). Thus, there is no net gain of adenylate but there is a net gain of high-energy phosphate if host-derived ATP is exchanged for rickettsiae-derived ADP. In essence, the rickettsiae are energy parasites when they have infected healthy host cells. However, rickettsiae are not obligate energy parasites because, as discussed above, they generate their own adenylate energy charge via TCA/OXPHOS metabolism.

This raises a potential conundrum. Presumably, the initial infection of a host cell by a single rickettsia has no overt deleterious effect on the health and metabolism of the host cell. Thus, the rickettsia can metabolize host cell glutamate to produce its own adenylate energy charge and parasitize the host cell's energy charge via the activity of the ATP/ADP translocase. But as the infection progresses and the health of the host cell diminishes, the rickettsiae now risk the deleterious exchange of rickettsiae-derived ATP in exchange for host-derived ADP, a process that would bleed off adenylate energy charge from the rickettsiae. It has been suggested that ion concentrations may play a role in regulating the activity of the ATP/ADP translocase to prevent loss of energy charge (Winkler and Daugherty, 1984; Trentmann et al., 2008).

Alternatively, the ATP/ADP translocase could endow rickettsiae with the ability to sense changes in the energy charge of the host cell and alter their own physiology accordingly. For example, *R. prowazekii* has been shown to regulate transcription of the gene encoding the ATP/ADP translocase and the gene encoding citrate synthase (a component of the TCA cycle) in response to rickettsial numbers in the host cell (Cai and Winkler, 1996); these regulatory studies are discussed later in this chapter.

There is also the intriguing possibility that the ATP/ADP translocase allows rickettsiae to behave altruistically in a metabolically compromised host cell and supply adenylate energy charge generated by rickettsial TCA/OXPHOS metabolism. This may work to extend the life of the infected cell and maximize the yield of microbes released upon lysis. Resolving the role of the ATP/ADP translocase in obligate intracellular parasitism will require detailed analyses of the effects of rickettsial infection on the adenylate and ion gradients of both the bacteria and the host cells to determine their effects on ATP/ADP translocase activity.

Rickettsial Physiology and Growth in Cultured Eukaryotic Cells

Studies examining the physiology of rickettsiae that are growing intracellularly have contributed much to the understanding of rickettsia-host interactions. Perhaps unexpectedly, rickettsiae

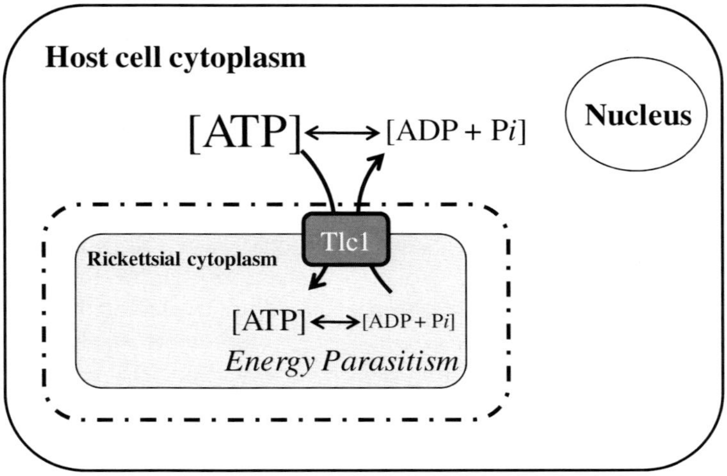

FIGURE 1 Rickettsial energy parasitism. The ATP/ADP translocase (Tlc1) functions via an obligate exchange antiport mechanism that allows the obligate intracytoplasmic rickettsia to equilibrate its adenylate energy charge with that of the host as a system to transport high-energy phosphate. The rickettsial cytoplasm is shaded gray, the solid line represents the bacterial inner membrane where the Tlc1 protein (dark gray box) resides, and the dashed line represents the bacterial outer membrane. doi:10.1128/9781555817336.ch7.f1

grow in cultured eukaryotic cells that have been enucleated (Stork and Wisseman, 1976), treated with eukaryotic protein synthesis inhibitors (Weiss et al., 1972), or gamma-irradiated (Weiss et al., 1972), indicating that de novo host transcription, protein synthesis, and cell division are not essential to support rickettsial growth. This could indicate that metabolite pools are sufficient to support viability of the host cell and rickettsiae over the time course of the experiments performed.

In contrast, there are some cultured cell model systems in which rickettsial growth *requires* the addition of eukaryotic protein synthesis inhibitors or exogenous addition of serine and glycine (Austin et al., 1987; Austin and Winkler, 1988a, 1988b). This conditional auxotrophy may suggest that fierce competition sometimes exists between rickettsiae and host for limiting nutrients. Central to this idea of competition were the observations of Stork and Wisseman (1976), which noted the presence of long filamentous rickettsiae with no obvious septa in infected host cells. Perhaps some host-derived metabolite becomes limiting during infection and affects rickettsial cell division.

This discussion points out that while host cell cytoplasm is usually considered a stable and nutrient-rich environment, it might actually be hostile toward intracellular bacteria, thus explaining the paucity of organisms that have successfully exploited it as a growth niche (Moulder, 1962). This potential hostility may be a contributing factor to the observed slow rickettsial growth rates, which have always seemed paradoxical considering the available resources (e.g., *R. prowazekii* doubles only every 8 to 12 hours [Wisseman et al., 1976; Turco and Winkler, 1989]). Studies have attempted to address the issue of slow growth as resulting from a deficit in rickettsial physiology. Some interesting considerations include the facts that (i) rickettsiae possess some of the smallest microbial genomes, encoding few open reading frames (ORFs); (ii) they can make and parasitize adenylate energy charge to fuel macromolecule synthesis; (iii) they have been shown to possess translational machinery in abundance comparable to *E. coli* despite encoding only single copies of the 16S, 23S, and 5S rRNA genes (Pang and Winkler, 1993, 1994); and (iv) their purified RNA polymerase displays processivity comparable to that of *E. coli* (Ding and Winkler, 1990, 1993). It has been postulated that slow growth helps to extend the health of the host and maximizes rickettsial numbers released to the blood as a mechanism to ensure they will be contracted by the intermittently feeding arthropod vectors that mediate transmission.

RICKETTSIAL PHYSIOLOGY IN THE GENOMICS ERA

The validity of hypotheses that drove early studies of rickettsial physiology can now be fully appreciated in the light of the genomics era. But despite this wealth of in silico information, rickettsial obligate intracellular growth and slow generation time remain poorly understood. Thus, a challenge for the modern-day rickettsiologist is to integrate the power of genomics with new, cutting-edge technologies to examine the physiology of both rickettsia and host cell to generate a better understanding of their intricate interactions.

Characteristics of Rickettsial Genomes

Rickettsial genomic analyses have correlated a striking evolutionary relationship with the eukaryotic cell's mitochondrion (Andersson et al., 1998; Sicheritz-Pontén et al., 1998). Rickettsial genomes are AT rich (~70%) and encode fewer ORFs than do free-living microbes (for example, the *R. prowazekii* genome encodes 835 predicted ORFs). Genes encoding components of glycolysis and gluconeogenesis are indeed absent, confirming the early biochemical studies. Major metabolic pathways such as the pentose phosphate pathway, amino acid biosynthetic pathways, and nucleotide metabolism pathways are also absent or are missing crucial enzymes. The missing genes are thought to have been expunged through the process of reductive evolution. The presence of large amounts of noncoding (or "junk") DNA sequences and pseudogenes within rickettsial genomes supports this idea (Andersson and Andersson, 2001).

Using Genomics To Understand Rickettsial Transport and Metabolism

Going forward, genomics should accelerate us toward a better understanding of how novel carrier-mediated transport systems allow rickettsiae to exploit metabolites available in host cell cytoplasm. For example, genomic data have confirmed the loss of rickettsial de novo amino acid biosynthetic pathways. This finding partially clarifies why rickettsial growth in some cell culture model systems requires the addition of exogenous serine and glycine to the culture medium (Austin et al., 1987; Austin and Winkler, 1988b). However, there remains no obvious explanation for why the rate of serine and glycine flux through their respective transport systems is not sufficient to compete for the levels provided endogenously by certain cultured host cells.

As a second example, pre-genomics era biochemical evidence proved that rickettsiae possess peptidoglycan (PG) and lipopolysaccharide (LPS) (Smith and Winkler, 1979). PG and LPS are unique to prokaryotes and are not found in eukaryotic cells; thus, the unusual sugars required to synthesize these critical bacterial structures cannot be transported because they are simply not present. Based on this, Winkler and coworkers postulated that rickettsiae transport sugars available in the host cytoplasm and use them for rickettsiae-mediated synthesis of the unusual components of PG and LPS (Winkler and Daugherty, 1986). Indeed, physiological studies found that rickettsiae were capable of transporting UDP-glucose and incorporating it into macromolecular compounds that sediment with the bacterial membrane fraction (Winkler and Daugherty, 1986). Genomic analyses subsequently confirmed the presence of largely conserved pathways for PG and LPS synthesis, but the identity of the UDP-glucose transport system remains unknown.

As a third example, the genome sequence of *R. prowazekii* revealed the existence of a nonsense mutation in *metK* (Andersson et al., 1998), the gene encoding methionine adenosyltransferase. Usually, such a mutation would be lethal because methionine adenosyltransferase synthesizes *S*-adenosylmethionine (AdoMet), the universal methyl donor. However, it was shown that isolated rickettsiae transport AdoMet (Tucker et al., 2003). Thus, rickettsiae presumably acquire AdoMet from host cytoplasm to support rickettsial growth. As such, *metK* is a prime example of a pseudogene in the very early stages of reductive evolution (Andersson et al., 1998). This story was given an interesting twist when other rickettsial genome sequences became available and showed the presence of intact *metK* genes. These intact rickettsial *metK* genes were subsequently shown by Wood and coworkers to functionally complement a conditional *metK* mutant of *E. coli* (Driskell et al., 2005).

Based upon these observations, it is tempting to speculate that the rickettsial AdoMet transport system is a more recent acquisition and that the *metK* gene is in the earliest stages of reductive evolution. Continuous-culture experiments to follow the fate of the rickettsial *metK* gene over time could shed light on the important aspects of mutation rates and mechanisms that underlie the process of reductive evolution. Taken together, these three examples are rather instructive—whether based on genomic or biochemical evidence, the existence of "metabolic gaps" is predictive of the existence of rickettsial transport systems.

Evolutionary Links between Transport and Obligate Intracellular Parasitism

The rickettsial paradigm of "transport what you can and synthesize only what you must" deviates from the metabolic flexibility maintained by free-living microbes. Free-living microbes transport substrates such as amino acids from the environment when available and synthesize them de novo when they are not. The economy of free-living microbial systems is apparent from the fact that the presence of a substrate in the media often induces its cognate transport systems and represses expression of the corresponding biosynthetic pathway genes. Interestingly, free-living microbes handle large, charged substrates in a different fashion. Large, charged substrates are broken down and the parts are transported from the environment and reassembled in the bacterial cytoplasm (e.g., transport nucleobases, sugars, and inorganic phosphate, not intact nucleoside triphosphates).

Why, during evolution of *Rickettsia* species, were transport systems kept and de novo biosynthetic pathways largely lost? First, consider the type of event that could establish an obligate intracellular lifestyle. As an oversimplified example, the protorickettsial ancestor likely originated as a facultative intracellular organism that evolved a novel transport system for substrate X, which is *never* found in the extracellular environment and *only* found in the ancestral host cell cytoplasm. If the protorickettsia suffers a loss-of-function mutation in its substrate X de novo biosynthetic pathway, it can no longer grow in the extracellular environment and thus obligatorily grows *only* within the host cell cytoplasm. As a consequence of this, the remainder of the genes encoding the substrate X de novo biosynthetic pathway will continue to accumulate mutations, becoming unrecognizable and eventually lost from the genome (Fig. 2).

With this example in mind, now return to the question of why transport systems were kept and de novo biosynthetic pathways lost. Consider that de novo biosynthetic pathways tend to proceed stepwise through several intermediates catalyzed by multiple proteins, whereas a transport system can be a single protein. Thus, there may simply be increased odds of suffering loss-of-function mutations in components of multistep de novo biosynthetic pathways, which eventually leads to the observed rickettsial bias toward transport systems. Perhaps there is also a hidden advantage in relying on the host to encode the multiprotein pathways to synthesize essential metabolites. The obligate intracellular bacterium would no longer be required to maintain and replicate a large genome encoding tremendous metabolic versatility.

Another selective advantage for limiting metabolic versatility may lie in a potential role of rickettsial transporters in enforcing slow growth. As discussed, even though the eukaryotic cytoplasm is an extremely nutrient-rich environment, rickettsial growth is slow, with a generation time of 8 to 12 hours (Winkler, 1990). The postulated advantage of slow growth to the rickettsia is that it delays death of the mammalian host long enough to maximize the chance that an arthropod vector will ingest and disseminate the pathogen to additional hosts. Perhaps de novo biosynthetic pathways were abandoned partly because they were too efficient and worked at a cross-purpose to slow growth. Transporters with low substrate affinity combined with limiting substrate concentrations in the host cell cytoplasm may be an efficient mechanism for obligate intracellular bacteria to ensure slow growth.

Rickettsial Phospholipid Biosynthesis: Redundancy in the Face of Reductive Evolution

In some ways, the most intriguing information generated by rickettsial genomic analyses is not about the genes and pathways lost, but about those retained. This indicates the existence of a balance between maintaining metabolic fitness and evolving toward a streamlined genome.

Prior to genomics, biochemical analyses of rickettsial phospholipids were suggestive of de novo synthesis by showing a composition comparable to that of *E. coli*, with 70% phosphatidylethanolamine (PE), 25% phosphatidylglycerol, 5% cardiolipin, and a small amount of phosphatidylserine (the precursor of PE) (Winkler and Miller, 1978; Tzianabos et al., 1981). Phosphatidylcholine, the major eukaryotic phospholipid, was also detected at low levels but was likely a contaminant in the preparation or represented a small amount of host lipid usage. As expected, various species of both saturated and unsaturated fatty acids typical of bacterial membranes were present. Genome sequence analysis has largely confirmed that rickettsiae possess de novo fatty acid and phospholipid biosynthetic pathways.

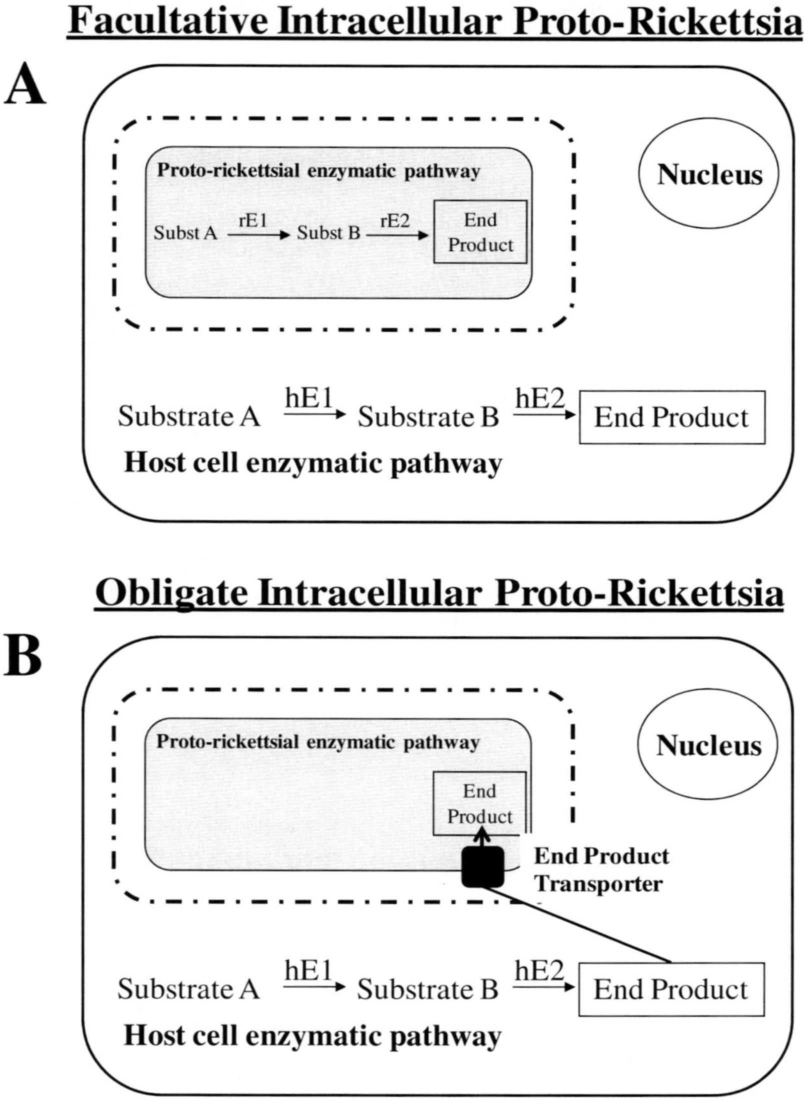

FIGURE 2 A model of reductive evolution. Panel A shows the facultative intracellular protorickettsia that possesses a typical de novo biosynthetic pathway in which rE1 denotes protorickettsial enzyme 1 and rE2 denotes protorickettsial enzyme 2. This pathway produces an essential metabolic end product. The same enzymatic pathway exists in the ancestral host cell cytoplasm, with the corresponding enzymes denoted as hE1 and hE2. This protorickettsia is not dependent on host cell metabolism for growth. Panel B shows that the obligate intracellular protorickettsia has acquired a transport system for a host cell metabolic end product (black box in the rickettsial inner membrane). The acquisition of this end product transporter has facilitated the loss of the corresponding rickettsial de novo biosynthetic pathway, as denoted by the absence of rE1 and rE2 in the protorickettsial cytoplasm.
doi:10.1128/9781555817336.ch7.f2

For instance, the *R. prowazekii fabI* gene, discovered as a result of genomic analysis, was shown to functionally complement an *E. coli fabI* temperature-sensitive mutant, confirming the genomic annotation (K. M. Frohlich and J. P. Audia, unpublished results).

There are some noteworthy features of the annotated rickettsial fatty acid and phospholipid biosynthetic pathways. The box in Fig. 3 indicates that *R. prowazekii* lacks the classical four-subunit bacterial acetyl-CoA carboxylase (ACC) complex to carry out the first committed step in fatty acid biosynthesis, which is to convert acetyl-CoA to malonyl-CoA. It is postulated that an annotated two-subunit propionyl-CoA carboxylase (PCC) may fulfill this metabolic role. While structurally different from the classical ACC complex, PCC enzymes have been shown in other microbes to catalyze the conversion of acetyl-CoA to malonyl-CoA (Chuakrut et al., 2003; Hügler et al., 2003). It is also possible that rickettsiae transport malonyl-CoA directly from the host cell. Whether one, the other, or both of these possibilities are valid remains unknown as there are no reports of malonyl-CoA transport nor has it been determined as to whether the rickettsial PCC performs the function of an ACC.

The acquisition of *sn*-glycerol-3-phosphate (G3P) for rickettsial phospholipid metabolism has been studied and presents a fascinating lesson in rickettsial physiology. As depicted in Fig. 3, rickettsiae possess an annotated GlpT homolog that could function to transport G3P from the host cell cytoplasm. The *R. prowazekii* genome also possesses an annotated G3P dehydrogenase (GpsA) protein that could provide a second source of G3P using dihydroxyacetone phosphate (DHAP) as the substrate. However, without glycolysis and gluconeogenesis, the rickettsiae have no endogenous metabolic

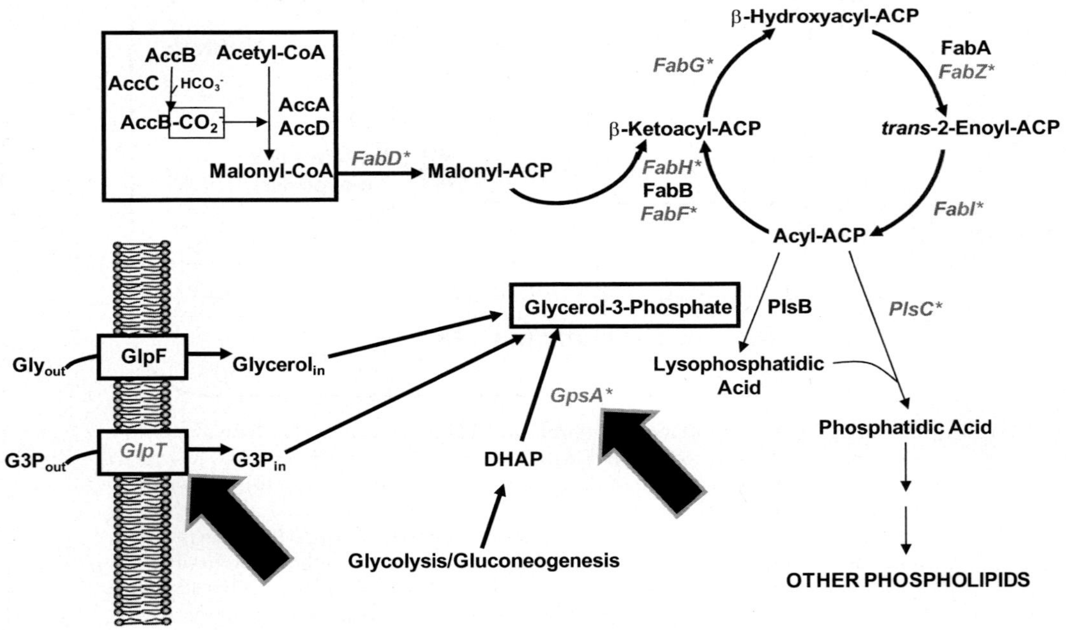

FIGURE 3 Annotated *R. prowazekii* fatty acid and phospholipid biosynthetic pathways. The components of these pathways that are annotated in the *R. prowazekii* genome are denoted in italics with an asterisk and shaded gray. The lack of a classical bacterial four-subunit ACC enzyme complex is highlighted with a box around that part of the pathway. The large black arrows highlight the annotated G3P transporter (GlpT) and the annotated G3P dehydrogenase (GpsA). doi:10.1128/9781555817336.ch7.f3

pathways to supply GpsA with DHAP as a substrate. Was the rickettsial GpsA annotation correct, or was GpsA another example of a pseudogene on its way toward expulsion from the genome?

Biochemical analyses have proven that the *R. prowazekii* GpsA protein is a functional G3P dehydrogenase. GpsA activity was measured as a purified recombinant protein using in vitro assays, and its activity was also measured in lysed cell extracts of isolated *R. prowazekii* (Frohlich et al., 2010). Furthermore, expression of the rickettsial *gpsA* gene complemented growth of an *E. coli gpsA* mutant, indicating at least enough flux through this system to support the metabolic requirements of a fast-growing, free-living microbe.

Having proven the functionality of GpsA, the next step was to determine if it played a role in rickettsial phospholipid biosynthesis. The final pieces of this puzzle included the demonstration that isolated rickettsiae were able to transport both radiolabeled DHAP and G3P across the membrane and catalyze their incorporation into organic soluble compounds, with the major radiolabeled product behaving like PE when resolved by thin-layer chromatography (Frohlich et al., 2010; Frohlich and Audia, unpublished). It is currently unknown as to whether the rickettsial GlpT functions to transport G3P, DHAP, or both (Frohlich et al., 2010). The report of DHAP transport was the first for any known bacterium and has only been previously reported for plant plastids and the glycosomes of trypanosomes (Fairlamb and Opperdoes, 1986; Borchert et al., 1993; Quick and Neuhaus, 1996).

GpsA represents an intriguing example of a functionally orphaned enzyme whose biochemical activity is intact in *R. prowazekii* but whose cognate endogenous biosynthetic pathway has been replaced by a transport system to procure its substrate from the host cell cytoplasm. Finding a role for GpsA in *R. prowazekii* physiology may serve as a model for other orphaned enzymes, and cautions against prematurely designating pseudogenes based solely on the lack of a fully intact biochemical pathway.

Another example of a functionally orphaned enzyme involves the pathway for CoA biosynthesis. The pathway has been lost from the rickettsial genome except for the last enzyme, a 3′-dephospho-CoA kinase (DPCK) that phosphorylates DPC to generate CoA. The activity of purified recombinant DPCK has been measured in vitro, and the transport of both DPC and CoA is measurable in isolated *R. prowazekii* (J. P. Audia, N. Housley, and H. H. Winkler, unpublished results). Thus, it appears rickettsiae have maintained dual acquisition pathways for some host cell substrates—that is, transporting both the end product (e.g., G3P) and its immediate precursor (e.g., DHAP) from the host cell (Fig. 4).

The selection pressure on the rickettsiae to maintain dual substrate acquisition pathways such as those described above is much less clear. Perhaps rickettsial transport of end products and their immediate precursors is merely a case of redundancy where one of the two pathways is dispensable and will eventually be lost through reductive evolution. The identification of a frameshift mutation in the *gpsA* gene in the recently sequenced *Rickettsia peacockii* may be an indicator that this is the case (Felsheim et al., 2009). It will be critical to determine whether the DHAP transporter is still intact and active in isolated *R. peacockii* and if DHAP is incorporated into phospholipids. The DPCK from *R. peacockii* still appears to be intact and could indicate that this dual substrate acquisition pathway is still functional.

Alternatively, this difference between *R. prowazekii* and *R. peacockii* could influence the host-parasite interactions between the two species. Perhaps for *R. prowazekii* the competition between the pathogen and the host cell during the natural history of the infection is fierce and the amounts of host triose phosphates are limiting. If substrates are buffered/sequestered by binding to host cell enzymes, they may be available to rickettsial transport systems at lower concentrations than expected. These dual substrate acquisition pathways could function in parallel to provide enough triose phosphate flux to support rickettsial growth.

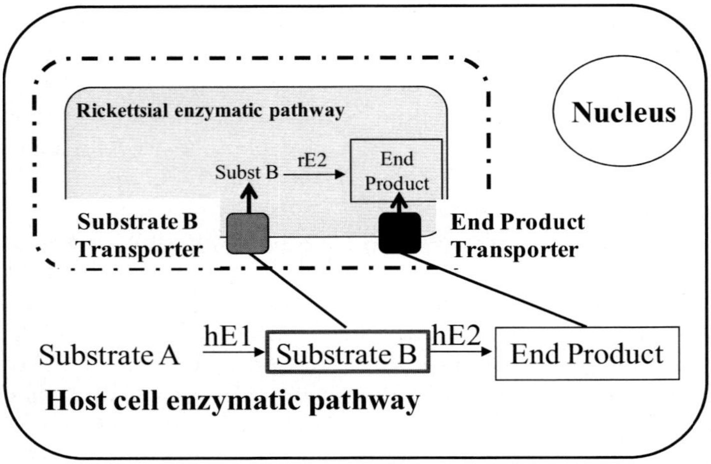

FIGURE 4 Rickettsiae possess dual G3P acquisition pathways. The rickettsiae are able to directly transport the end product of a host metabolic pathway produced by the enzymatic activities of hE1 and hE2 using the end product transporter (black box in the rickettsial inner membrane). In addition, rickettsiae are able to transport substrate B from the host cell cytoplasm using the substrate B transporter (gray box in the rickettsial inner membrane). As a consequence of evolving the capacity for substrate B transport, the rickettsiae rE2 enzyme has been conserved and functions to synthesize a second source of the end product. doi:10.1128/9781555817336.ch7.f4

Another potential advantage to maintaining dual transport requires considering the stages of *R. prowazekii* growth in the host cell. At the earliest stage of infection, host cell resources may be abundant. However, as the infection progresses and rickettsial numbers increase, there are now potentially deleterious effects on host cell metabolism as well as competition between the large numbers of rickettsiae in the cytoplasm for dwindling resources. It is known that other types of stress on the host cell can result in fluctuations in the host cell triose phosphate pools (Ralser et al., 2007, 2009). Perhaps rickettsiae have evolved differential regulation of their triose phosphate acquisition pathways to respond to changes in these host metabolite pools. A similar phenomenon has been shown for the first described dual acquisition pathways, the TCA/OXPHOS and ATP/ADP translocase pathways. The expression of *gltA* (citrate synthase) and *tlc1* (ATP/ADP translocase) genes is regulated in response to numbers of *R. prowazekii* growing in a host cell (Cai and Winkler, 1996).

Gene Amplification as a Mechanism To Generate Transporter Diversity

Further clues about rickettsial evolution come from the identification of several transporter gene families that have undergone what appear to be gene amplification events. Their discovery may partially explain how obligate intracellular bacteria living in a cloistered environment, with little opportunity for HGT, can expand their repertoire of transport systems.

An example of rickettsial gene amplification is the ProP family of putative proline/betaine transporters. The *R. prowazekii* genome possesses seven annotated ProP homologs (Andersson et al., 1998), but only one of them has been shown to transport proline when assayed by heterologous expression in an *E. coli* mutant totally deficient for endogenous proline uptake (H. H. Winkler and J. P. Audia, unpublished results). Amino acid substrates for some of the others have also been partially characterized (Winkler and Audia, unpublished). Together, these data strongly suggest that at least a subset of the ProP family members are functional and have diverged with respect to their substrate recognition to expand the repertoire of rickettsial amino acid transporters.

The *tlc1* gene encoding the ATP/ADP translocase (Tlc1) is perhaps the best-characterized example of rickettsial gene amplification. Tlc1 is thought to transport ATP from the host cytoplasm in exchange for ADP generated in the rickettsial cytoplasm as a mechanism of parasitizing the host cell's adenylate energy charge (Fig. 1). Homologs of Tlc1 have been identified in very diverse groups of organisms, including human pathogens such as rickettsiae and chlamydiae; endosymbionts of protists; and even plant plastids (reviewed in Winkler and Neuhaus, 1999). The function of many of these ATP/ADP exchange transport systems has been experimentally verified, and they represent a distinct group of nonmitochondrial nucleotide transporters (Paulsen et al., 2000; Saier et al., 2006).

Intriguingly, all sequenced rickettsial genomes possess up to four additional gene copies annotated as Tlc2, Tlc3, Tlc4, and Tlc5 based on DNA and protein sequence homology to the well-characterized Tlc1. Each Tlc homolog is encoded by its own gene, which generally do not appear to cluster together on the bacterial chromosome, and all five mRNA transcripts were detected in *R. prowazekii* at comparable levels (Audia and Winkler, 2006). The five Tlc proteins are of similar size and were all predicted to possess 11 or 12 transmembrane-spanning domains by hydrophobicity analysis, which corresponds to the 12 transmembrane domains experimentally determined for Tlc1 (Plano and Winkler, 1991; Alexeyev and Winkler, 1999).

Amplification of the *tlc* genes raised three obvious possibilities: (i) these other *tlc* genes are pseudogenes in the process of being lost from the genome, (ii) they encode functionally redundant ATP/ADP exchange transport systems, or (iii) they have diverged in function to expand the rickettsial nucleotide transporter repertoire. The first clue to discerning between these possibilities came from sequence alignments showing that the other annotated Tlc proteins shared only ~40% identity with Tlc1 and among themselves. However, sequence alignments of each Tlc protein between rickettsial species showed ~90% identity (e.g., comparing Tlc5 from *R. prowazekii* to Tlc5 from *Rickettsia conorii*). These sequence alignments were used to generate the tree diagram shown in Fig. 5, which strongly suggests that *tlc* gene amplification occurred before speciation.

It was experimentally verified that only Tlc1 was a bona fide ATP/ADP exchange transport system and that the other Tlc proteins had most likely diverged in function to fulfill other roles in rickettsial physiology (Audia and Winkler, 2006). The Tlc4 protein transports CTP, UTP, and GDP (given in order of decreasing substrate affinity), and the Tlc5 protein transports GTP and GDP (comparable substrate affinities). No substrates for Tlc2 and Tlc3 have been reported to date (summarized in Fig. 6). Thus, Tlc4 and Tlc5 appear to have diverged in substrate recognition and may function to supply nucleotides for RNA and DNA synthesis. Consistent with a role in DNA synthesis, rickettsiae possess a ribonucleotide reductase that converts transported nucleoside triphosphates to deoxynucleoside triphosphates (Cai et al., 1991).

The next logical question was to ask if these other Tlc transporters diverged from the obligate exchange antiport mechanisms known for Tlc1. Indeed, the Tlc5 guanylate transporter does not function as an obligate exchange antiport system requiring substrate on both sides of the membrane. Unlike Tlc1, Tlc5 works by active transport to catalyze concentration of guanylate nucleotides inside the bacterial cytoplasm, and the addition of metabolic poisons/uncouplers abolishes this activity (J. P. Audia, R. Roberts, and H. H. Winkler, unpublished). This observed activity of Tlc5 is reminiscent of the NTT2 nucleotide transporter from chlamydiae, another Tlc protein family member (Tjaden et al., 1999). The molecular mechanisms of action at the structural level remain unknown for the Tlc family of proteins. It will be essential to unravel these details if we are to

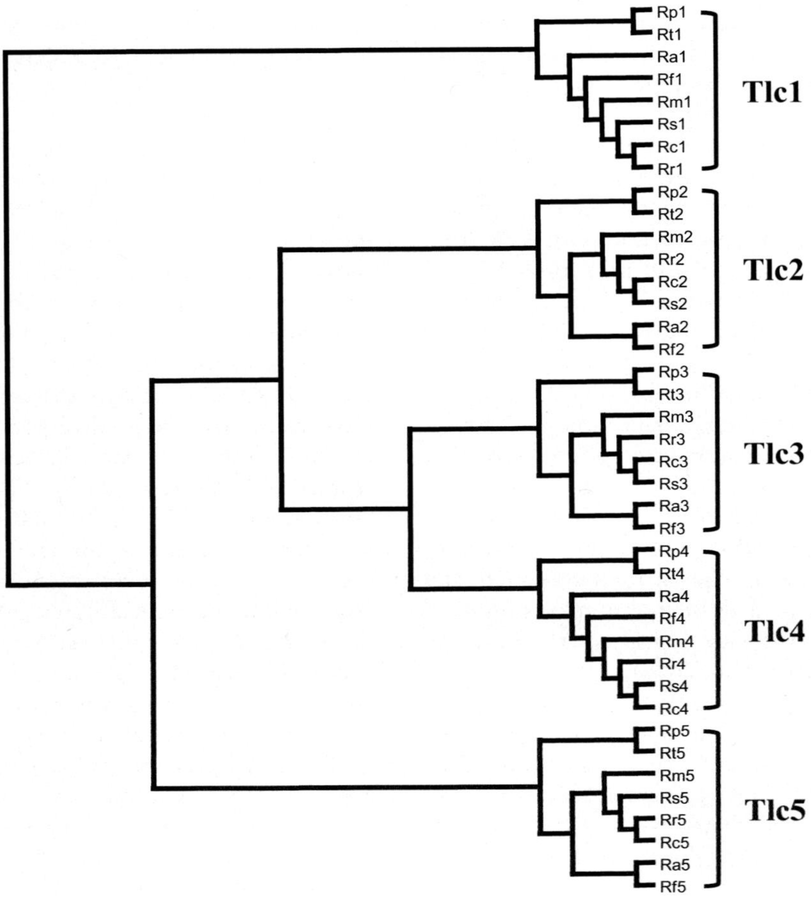

FIGURE 5 Phylogenetic analysis of the rickettsial Tlc nucleotide transporters. A partial phylogenetic analysis was performed using the Clustal X program (Larkin et al., 2007). The unrooted tree diagram shows the clustering of the five Tlc proteins and indicates that gene duplication occurred before speciation. Abbreviations for *Rickettsia* spp.: Rp, *R. prowazekii*; Rt, *R. typhi*; Ra, *R. akari*; Rf, *R. felis*; Rm, *R. massiliae*; Rs, *R. sibirica*; Rc, *R. conorii*; Rr, *R. rickettsii*. doi:10.1128/9781555817336.ch7.f5

understand the amino acid sequence changes behind the striking divergence from obligate exchange antiport to energy-dependent unidirectional transport. Equally important are the changes that govern substrate recognition.

RICKETTSIAL GENE REGULATION

Transcriptional Regulation in Rickettsiae

It is well-known that free-living microbes can coordinately alter gene regulation in response to myriad different environmental conditions. Conversely, rickettsiae have committed to a relatively narrow range of hosts and vectors in their natural environment, and grow only within the confines of another cell. Their minimal annotated capacity for gene regulation as noted by genomic analyses is consistent with the idea that rickettsiae are exposed to a limited number of environmental changes. The goal of this section is to discuss how obligate intracellular growth has affected the rickettsia's capacity for gene regulation and to highlight how little we still understand about these processes.

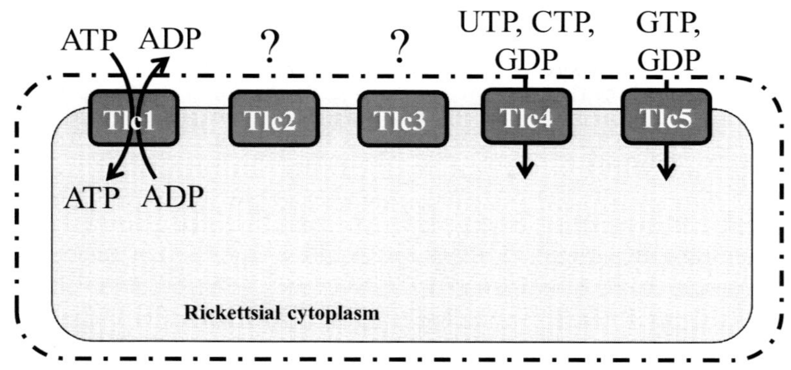

FIGURE 6 A summary of the known properties of the *R. prowazekii* Tlc family of nucleotide transporters. The Tlc transporters are shown in the rickettsial inner membrane. The question marks denote that substrates have not been identified for Tlc2 and Tlc3. Tlc1 is an obligate exchange antiporter. Tlc5 is an energy-dependent transporter. The energetics of Tlc4 are not yet known. doi:10.1128/9781555817336.ch7.f6

To date, very few rickettsial regulatory proteins have been studied. The *R. prowazekii* RNA polymerase (RNAP) has been purified and biochemically characterized (Ding and Winkler, 1990, 1993). The rickettsial RNAP was shown to have a low molar ratio of the housekeeping sigma factor σ^D subunit to core polymerase when purified from rickettsiae grown in cultured eukaryotic cells (Ding and Winkler, 1994). Interestingly, purified rickettsial RNAP formed unstable complexes with promoter DNA and demonstrated weak DNA melting capacity in vitro compared to RNAP from *E. coli*, particularly when assayed using a GC-rich template. Presumably, the DNA melting function of RNAP has diminished as rickettsiae have evolved toward an AT-rich genome. Surprisingly, the rickettsial RNAP also required negatively supercoiled template and was unable to initiate transcription in vitro using linearized plasmid DNA. This structural requirement may help govern the rickettsial RNAP's ability to discriminate specific promoter elements within its AT-rich genome.

Beyond the biochemical properties of the RNAP, there is a growing body of literature describing global gene regulation in rickettsiae. However, little is known about the mechanisms or regulatory proteins involved in coordinating gene expression. Regulatory studies have been rationally designed around genomics and around the known environments encountered by rickettsiae in nature. Rickettsiae have been shown to coexpress genes in an operon (Shaw et al., 1997), and the first gene regulation studies identified transcriptional differences between rickettsiae isolated from cultured eukaryotic cells at early and late stages of infection, akin to exponential-phase growth with few rickettsiae/cell and stationary-phase growth with many rickettsiae in a cell that is nearing lysis (Cai and Winkler, 1996, 1997).

Recall from the discussion above that rickettsiae are not strictly energy parasites and can generate adenylate energy charge via the ATP/ADP translocase (Tlc1) or TCA/OXPHOS metabolism. The mRNA levels of *tlc1* were shown to be higher early in an infection compared to late in infection. Conversely, the levels of *gltA* mRNA (citrate synthase) increased as the infection progressed (Cai and Winkler, 1996). These observations suggest that rickettsiae may attune gene regulation in response to changes in the host cell energy charge that occur as infection progresses from early to late stage. It could be beneficial to transcribe *tlc1* during early-stage infection to parasitize the healthy host cell's energy charge. However, as the infection progresses and the host cell becomes compromised, it could become necessary to increase rickettsial energy charge via TCA/OXPHOS. It is unknown whether changes in protein levels and/or activities correlate with the observed changes in mRNA levels.

Transcriptional Responses to Temperature Shift

More recent studies of rickettsial global gene regulation have employed the high-throughput microarray format. Based on the annotation of a heat shock sigma factor (σ^H) homolog and reports that *R. prowazekii* is able to adapt and grow in cultured eukaryotic cells at 40°C (Gudima, 1982), it was proposed that this pathogen would respond to temperature upshift (Audia et al., 2008). Indeed, this analysis identified 23 *R. prowazekii* mRNA transcripts that were significantly elevated (≥2-fold; $P < 0.05$) in response to a 30-minute incubation at 42°C. As one might expect, more than half of the upregulated transcripts were annotated as homologs of chaperones, proteases, and heat shock genes. Interestingly, many of the heat shock-inducible genes clustered together and, based on similar expression levels, could represent additional rickettsial gene operons driven by a common heat-regulated promoter element. It will be important for future studies to identify the promoters that control these genes to determine if they are truly operons and whether or not transcription of the *R. prowazekii* heat shock regulon is coordinated by the annotated σ^H homolog.

Other microarray-based studies of rickettsial gene expression examined the effect of temperature downshift and upshift as a model to mimic cycling of these pathogens between mammalian hosts and arthropod vectors (Dreher-Lesnick et al., 2008; Ellison et al., 2009). This rationale is based upon early studies that demonstrated that *R. rickettsii* organisms lose their ability to infect when they are harbored in starved ticks, but are reactivated when the ticks take a warm blood meal (Wike and Burgdorfer, 1972; Wike et al., 1972; Burgdorfer, 1975). When *R. typhi* was subjected to a shift from 37 to 25°C, the emerging pattern indicated downregulation of genes involved in macromolecular synthetic processes, while several putative virulence factor genes demonstrated modest levels of induction (Dreher-Lesnick et al., 2008).

Subjecting *R. rickettsii* to conditions more reminiscent of a cold shock (37 to 4°C) produced modest downregulation of macromolecular synthesis genes and, in addition, reduced expression of putative heat shock proteins such as chaperones (Ellison et al., 2009). This regulatory pattern suggests that rickettsiae respond to temperature downshift by reducing their biosynthetic capacity, perhaps in an effort to conserve resources in anticipation of a prolonged stay in an arthropod that is between blood meals. However, the modest levels of transcript downregulation in these studies could lead one to question whether the levels of the corresponding proteins decrease to a level that would affect rickettsial physiology. It may be more likely that decreases in protein catalytic efficiency in response to drastic decreases in temperature would play a larger role in decreasing rickettsial metabolic capacity to prolong any dormancy associated with extended stays in starved arthropods.

Transcriptional Regulation and Virulence

Studies of rickettsial pathogenesis have focused on identifying genomic differences between avirulent and virulent rickettsial strains and understanding the factors that influence the observed differences in virulence capacity among all identified *Rickettsia* species. Beyond genome sequence differences, the contribution of global transcriptional regulatory differences to virulence remains unknown. Microarray studies comparing the attenuated *R. prowazekii* strain Madrid E to its revertant virulent form (strain EVir) revealed no significant differences in global gene expression in rickettsiae isolated from cultured L929 murine fibroblast cells grown under standard conditions (J. P. Audia, M. Patton, and H. H. Winkler, unpublished). Furthermore, this modest capacity for gene regulation has also been observed for *R. rickettsii* in a microarray analysis comparing expression levels between the virulent Sheila Smith and avirulent Iowa strains, which revealed significant differences in the expression levels of only four genes and gave no obvious insight into the difference in virulence between the two strains (Ellison et al., 2008).

Perhaps the most promising global transcriptome analyses of virulence have come from comparisons between *R. conorii* isolated from Mediterranean spotted fever (MSF) patient skin biopsies (from the eschar) and those grown in cultured eukaryotic cells (Renesto et al., 2008). This analysis revealed that 68% of the genes identified as upregulated in patient-derived samples were specific to the spotted fever group rickettsiae, whereas 75% of the downregulated genes belonged to the core gene set shared among *Rickettsia* species genomes. Interestingly, the cadre of core genes with reduced expression paints a similar picture to temperature-downshift studies described above—conditions consistent with stress and slow growth where macromolecular synthetic capacity is downregulated. Upregulated genes putatively encoding DNA repair capacity and oxidative and osmotic stress resistance pathways are further suggestive of the unfavorable conditions encountered by the rickettsiae isolated from the MSF patient samples. The fact that the transcriptional profile was highly conserved between the different patient samples lends a high level of confidence to the study. These authors postulated that the indications of slow growth and stress may be evidence that the host response to infection is quite efficient at limiting pathogen replication, and thus MSF often manifests as a milder form of rickettsiosis.

The Stringent Response

As discussed above, several rickettsial gene families have undergone marked gene amplification events that may represent a potential mechanism for expanding metabolic capacity in the face of reductive evolution (e.g., the Tlc nucleotide transporters and the proline transporters). The rickettsial *spoT* gene is an example of a putative regulator that is undergoing reductive evolution—there are large differences in *spoT* pseudogene and/or split gene copy number among sequenced *Rickettsia* species.

In bacterial physiology it is well-known that SpoT plays an integral role in regulating the levels of two guanosine nucleotide signaling molecules, pppGpp (guanosine 3′-diphosphate 5′-triphosphate) and ppGpp (guanosine 3′-diphosphate 5′-diphosphate), that are involved in coordinating global changes in gene regulation in response to nutrient starvation—a stress response known as the stringent response (for a review, see Wu and Xie, 2009). Upon nutrient exhaustion, the (p)ppGpp signaling molecules (also referred to as alarmones) transiently accumulate and elicit downregulation of bacterial macromolecular biosynthetic processes as a mechanism of resource conservation. Typically, the levels of (p)ppGpp are balanced by synthetase and hydrolase activities that can be encoded by two different proteins or within a single protein (e.g., the *E. coli* SpoT is bifunctional [Gentry and Cashel, 1996]). Identified rickettsial SpoT paralogs are bioinformatically predicted to possess either intact synthetase or hydrolase domains (Fournier et al., 2009); however, none of these predictions have been validated in biochemical assays with purified protein.

The obvious question that arises pertains to the various rickettsial SpoT paralogs and whether they play redundant roles in response to nutrient starvation or whether they have diverged to respond to other environmental stimuli. Microarray and quantitative reverse transcriptase PCR-based studies aimed at examining the regulation of the *spoT* genes in *R. conorii* revealed that they possess the potential for differential regulation (Rovery et al., 2005; La et al., 2007). Interestingly, only a subset of the *R. conorii spoT* genes was shown to be upregulated in response to nutrient depletion, indicating possible divergence in the signaling roles for some of the *spoT* paralogs in rickettsiae. Some important questions remain unanswered. Is (p)ppGpp actually synthesized during rickettsial starvation? What are the natural environments encountered by the rickettsiae that elicit the stringent response? What is the nature of the effect of (p)ppGpp accumulation in rickettsial physiology and gene regulation (if accumulation were shown to occur)? Could (p)ppGpp be a factor in rickettsial slow growth?

Transcription Termination

The final facet of rickettsial gene regulation that we will consider in this chapter is transcriptional termination. Rickettsiae possess annotated homologs of known components of bacterial transcription elongation and termination including *nusA*, *nusB*, *nusG*, *mfd*, and *rho*; however, the sequence and structural determinants of intrinsic or Rho-dependent termination are unknown. A study of *R. prowazekii* transcriptional termination by Woodard and Wood (2011) rationalized that convergent gene pairs represented an interesting opportunity to identify rickettsial terminators because any substantial transcriptional read-through could generate deleterious antisense effects. This potential problem could necessitate the presence of discrete transcriptional terminators. Intriguingly, in the majority of convergent gene pairs examined, one gene of the pair lacked a distinct termination site and appeared to produce a series of iterated transcripts ("stuttering") that presumably extend into the adjacent gene. Furthermore, the other gene in the pair often demonstrated site-specific termination within the coding sequence of the adjacent gene. Thus it would appear, at least for the convergent gene pairs tested, that production of antisense RNA could play a role in rickettsial posttranscriptional regulation.

The picture of transcriptional termination in rickettsiae is further complicated by the observed differences between *R. prowazekii* isolated from different growth conditions. Several transcripts examined in rickettsiae isolated from cultured L929 murine fibroblasts showed a distinct termination site, whereas these same transcripts showed iterated, stuttering stop sites when the rickettsiae were isolated from embryonated hen egg yolk sacs. Based on their previous observations that rickettsiae isolated from cultured eukaryotic cells possessed higher levels of putative stress proteins (chaperones, protease, etc.) than rickettsiae isolated from yolk sacs, the authors postulated that the rickettsiae grown in cultured eukaryotic cells may be under nutrient limitation, which necessitates distinct transcriptional termination to conserve resources. The possibility of nutrient limitation in governing rickettsial transcriptional termination efficiency highlights the potential for experiments to connect a role for the stringent response discussed above.

Another interesting evolutionary question to consider is whether rickettsiae really are wasteful of their resources when abundant but capable of regulated conservation in the face of scarcity. Wastefulness would fly in the face of the *E. coli* paradigm whereby gene regulation strictly governs the hierarchical use of available resources such as carbon sources, as shown in the classic diauxy experiments. Nonetheless, it is clear that genome reduction has not completely voided the rickettsiae of complex regulatory systems that warrant closer experimental examination.

PERSPECTIVE

Unraveling the mechanisms by which *Rickettsia* species exploit eukaryotic cell cytoplasm as a sole growth medium provides a window through which to better understand the effects of rickettsial growth on the host cell's available metabolite pools and their fluxes. The impacts of environmental changes on rickettsial gene regulation are of equal importance, and while this process has been economized, its conservation at some level indicates that the rickettsia may rely on basic cues from its host cell to fully exploit this growth niche.

It is clear that rickettsiae have evolved a bias toward transport of essential metabolites from the host cell cytoplasm, leading to the loss of many de novo biosynthetic pathways through reductive evolution. This reliance on the host to encode the multiprotein pathways that synthesize essential metabolites means that the obligate intracytoplasmic rickettsia is no longer required to synthesize the proteins needed for the corresponding pathways or to maintain and replicate a large genome with tremendous metabolic versatility. It is interesting to ponder whether the effects of reductive evolution actually lead to a more virulent pathogen as regulatory and metabolic capacities are lost (Fournier et al., 2009). This ethereal question of how the virulence capacity of an intracellular bacterium evolves through reductive evolution should be justification enough to warrant the basic biological studies of host-pathogen interactions, including how they affect each other's basic physiology. In the end, who better to teach us about the biology of the eukaryotic cell than some of the most successful cell biologists—the obligate intracellular bacteria that have taken on this arduous task for millennia in order to understand what it takes to colonize and prosper in what is arguably one of the most complex and enigmatic environmental niches on Earth!

ACKNOWLEDGMENTS

I thank the editors for the opportunity to contribute to this important resource for rickettsiologists. Additional thanks go to Herb Winkler, David Wood, and John Foster for insightful discussions and for critically reading this chapter. Special thanks go to the past and present members of the Audia and Winkler laboratories for their contributions to the experimental studies described herein and to the National Institutes of Health/National Institute of Allergy and Infectious Diseases (NIH/NIAID) for financial support of my work. The content is solely the responsibility of the author and does not necessarily represent the official view of the NIAID or NIH.

REFERENCES

Alexeyev, M. F., and H. H. Winkler. 1999. Membrane topology of the *Rickettsia prowazekii* ATP/ADP translocase reported by novel dual *pho-lac* reporters. *J. Mol. Biol.* **285:**1503–1513.

Andersson, J. O., and S. G. E. Andersson. 1999. Insights into the evolutionary process of genome degradation. *Curr. Opin. Genet. Dev.* **9:**664–671.

Andersson, J. O., and S. G. E. Andersson. 2001. Pseudogenes, junk DNA, and the dynamics of *Rickettsia* genomes. *Mol. Biol. Evol.* **18:**829–839.

Andersson, S. G. E., A. Zomorodipour, J. O. Andersson, T. Sicheritz-Pontén, U. C. M. Alsmark, R. M. Podowdki, A. K. Naslund, A.-S. Eriksson, H. H. Winkler, and C. G. Kurland. 1998. The genome sequence of *Rickettsia prowazekii* and the origin of mitochondria. *Nature* **396:**133–140.

Atkinson, D. E. 1969. Regulation of enzyme function. *Annu. Rev. Microbiol.* **23:**47–68.

Atkinson, W. H., and H. H. Winkler. 1989. Permeability of *Rickettsia prowazekii* to NAD. *J. Bacteriol.* **171:**761–766.

Audia, J. P., M. C. Patton, and H. H. Winkler. 2008. DNA microarray analysis of the heat shock transcriptome of the obligate intracytoplasmic pathogen *Rickettsia prowazekii. Appl. Environ. Microbiol.* **74:**7809–7812.

Audia, J. P., and H. H. Winkler. 2006. Study of the five *Rickettsia prowazekii* proteins annotated as ATP/ADP translocases (Tlc): only Tlc1 transports ATP/ADP, while Tlc4 and Tlc5 transport other ribonucleotides. *J. Bacteriol.* **188:**6261–6268.

Austin, F. E., J. Turco, and H. H. Winkler. 1987. *Rickettsia prowazekii* requires host cell serine and glycine for growth. *Infect. Immun.* **55:** 240–244.

Austin, F. E., and H. H. Winkler. 1988a. Proline incorporation into protein by *Rickettsia prowazekii* during growth in Chinese hamster ovary (CHO-K1) cells. *Infect. Immun.* **56:**3167–3172.

Austin, F. E., and H. H. Winkler. 1988b. Relationship of rickettsial physiology and composition to the *Rickettsia*-host cell interaction, p. 29–50. *In* D. H. Walker (ed.), *Biology of Rickettsial Diseases*, vol. 2. CRC Press, Boca Raton, FL.

Baldridge, G. D., N. Y. Burkhardt, R. F. Felsheim, T. J. Kurtti, and U. G. Munderloh. 2008. Plasmids of the pRM/pRF family occur in diverse *Rickettsia* species. *Appl. Environ. Microbiol.* **74:**645–652.

Borchert, S., J. Harborth, D. Schunemann, P. Hoferichter, and H. W. Heldt. 1993. Studies of the enzymic capacities and transport properties of pea root plastids. *Plant Physiol.* **101:**303–312.

Bovarnick, M. R. 1956. Phosphorylation accompanying the oxidation of glutamate by the Madrid E strain of typhus rickettsiae. *J. Biol. Chem.* **220:**353–361.

Bovarnick, M. R., and E. G. Allen. 1957a. Reversible inactivation of the toxicity and hemolytic activity of typhus rickettsiae by starvation. *J. Bacteriol.* **74:**637–645.

Bovarnick, M. R., and E. G. Allen. 1957b. Reversible inactivation of typhus rickettsiae at 0°C. *J. Bacteriol.* **73:**56–62.

Bovarnick, M. R., E. G. Allen, and G. Pagan. 1953. The influence of diphosphopyridine nucleotide on the stability of typhus rickettsiae. *J. Bacteriol.* **66:**671–675.

Bovarnick, M. R., and J. C. Miller. 1949. Oxidation and transamination of glutamate by typhus rickettsiae. *J. Biol. Chem.* **184:**661–676.

Bovarnick, M. R., and L. Schneider. 1960. Role of adenosine triphosphate in the hemolysis of sheep erythrocytes by typhus rickettsiae. *J. Bacteriol.* **80:**344–354.

Burgdorfer, W. 1975. A review of Rocky Mountain spotted fever (tick-borne typhus), its agent, and its tick vectors in the United States. *J. Med. Entomol.* **12:**269–278.

Cai, J., R. R. Speed, and H. H. Winkler. 1991. Reduction of ribonucleotides by the obligate intracytoplasmic bacterium, *Rickettsia prowazekii*. *J. Bacteriol.* **173:**1471–1477.

Cai, J., and H. H. Winkler. 1996. Transcriptional regulation in the obligate intracytoplasmic bacterium *Rickettsia prowazekii*. *J. Bacteriol.* **178:**5543–5545.

Cai, J., and H. H. Winkler. 1997. Transcriptional regulation of the *gltA* and *tlc* genes in *Rickettsia prowazekii* growing in a respiration-deficient host cell. *Acta Virol.* **41:**285–288.

Chuakrut, S., H. Arai, M. Ishii, and Y. Igarashi. 2003. Characterization of a bifunctional archaeal acyl coenzyme A carboxylase. *J. Bacteriol.* **185:**938–947.

Ding, H.-F., and H. H. Winkler. 1990. Purification and partial characterization of the DNA-dependent RNA polymerase from *Rickettsia prowazekii*. *J. Bacteriol.* **172:**5624–5630.

Ding, H.-F., and H. H. Winkler. 1993. Characterization of the DNA-melting function of the *Rickettsia prowazekii* RNA polymerase. *J. Biol. Chem.* **268:**3897–3902.

Ding, H.-F., and H. H. Winkler. 1994. The molar ratio of σ^{73} to core polymerase in the obligate intracellular bacterium, *Rickettsia prowazekii*. *Mol. Microbiol.* **11:**869–873.

Dreher-Lesnick, S. M., S. M. Ceraul, M. S. Rahman, and A. F. Azad. 2008. Genome-wide screen for temperature-regulated genes of the obligate intracellular bacterium, *Rickettsia typhi*. *BMC Microbiol.* **8:**61.

Driskell, L. O., A. M. Tucker, H. H. Winkler, and D. O. Wood. 2005. Rickettsial *metK*-encoded methionine adenosyltransferase expression in an *Escherichia coli metK* deletion strain. *J. Bacteriol.* **187:**5719–5722.

Ellison, D. W., T. R. Clark, D. E. Sturdevant, K. Virtaneva, and T. Hackstadt. 2009. Limited transcriptional responses of *Rickettsia rickettsii* exposed to environmental stimuli. *PLoS One* **4:**e5612.

Ellison, D. W., T. R. Clark, D. E. Sturdevant, K. Virtaneva, S. F. Porcella, and T. Hackstadt. 2008. Genomic comparison of virulent *Rickettsia rickettsii* Sheila Smith and avirulent *Rickettsia rickettsii* Iowa. *Infect. Immun.* **76:**542–550.

Fairlamb, A. H., and F. R. Opperdoes. 1986. Carbohydrate metabolism in African trypanosomes, with special reference to the glycosome, p. 183–224. *In* M. J. Morgan (ed.), *Carbohydrate Metabolism in Cultured Cells*. Plenum Publishing Corporation, New York, NY.

Felsheim, R. F., T. J. Kurtti, and U. G. Munderloh. 2009. Genome sequence of the endosymbiont *Rickettsia peacockii* and comparison with virulent *Rickettsia rickettsii*: identification of virulence factors. *PLoS One* **4:**e8361.

Fournier, P. E., K. El Karkouri, Q. Leroy, C. Robert, B. Giumelli, P. Renesto, C. Socolovschi, P. Parola, S. Audic, and D. Raoult. 2009. Analysis of the *Rickettsia africae* genome reveals that virulence acquisition in *Rickettsia* species may be explained by genome reduction. *BMC Genomics* **10:**166.

Frohlich, K. M., R. A. Roberts, N. A. Housley, and J. P. Audia. 2010. *Rickettsia prowazekii* uses an *sn*-glycerol-3-phosphate dehydrogenase and a novel dihydroxyacetone phosphate transport system to supply triose phosphate for phospholipid biosynthesis. *J. Bacteriol.* **192:**4281–4288.

Gentry, D. R., and M. Cashel. 1996. Mutational analysis of the *Escherichia coli spoT* gene identifies distinct but overlapping regions involved in ppGpp synthesis and degradation. *Mol. Microbiol.* **19:** 1373–1384.

Greub, G., and D. Raoult. 2003. History of the ADP/ATP-translocase-encoding gene, a parasitism gene transferred from a *Chlamydiales* ancestor to plants 1 billion years ago. *Appl. Environ. Microbiol.* **69:**5530–5535.

Gudima, O. S. 1982. Reproduction of vaccine and virulent *Rickettsia prowazeki* strains in continuous cell lines at different temperatures. *Acta Virol.* **26:**390–394.

Hackstadt, T. 1996. The biology of rickettsiae. *Infect. Agents Dis.* **5:**127–143.

Hügler, M., R. S. Krieger, M. Jahn, and G. Fuchs. 2003. Characterization of acetyl-CoA/propionyl-CoA carboxylase in *Metallosphaera sedula*. Carboxylating enzyme in the 3-hydroxypropionate cycle for autotrophic carbon fixation. *Eur. J. Biochem.* **270:**736–744.

La, M. V., P. Francois, C. Rovery, S. Robineau, P. Barbry, J. Schrenzel, D. Raoult, and P. Renesto. 2007. Development of a method for recovering rickettsial RNA from infected cells to analyze gene expression profiling of obligate intracellular bacteria. *J. Microbiol. Methods* **71:**292–297.

Larkin, M. A., G. Blackshields, N. P. Brown, R. Chenna, P. A. McGettigan, H. McWilliam, F. Valentin, I. M. Wallace, A. Wilm, R. Lopez, J. D. Thompson, T. J. Gibson, and D. G. Higgins. 2007. Clustal W and Clustal X version 2.0. *Bioinformatics* **23:**2947–2948.

Moliner, C., P. E. Fournier, and D. Raoult. 2010. Genome analysis of microorganisms living in amoebae reveals a melting pot of evolution. *FEMS Microbiol. Rev.* **34:**281–294.

Moulder, J. W. 1962. *The Biochemistry of Intracellular Parasitism.* University of Chicago Press, Chicago, IL.

Ogata, H., B. La Scola, S. Audic, P. Renesto, G. Blanc, C. Robert, P. E. Fournier, J. M. Claverie, and D. Raoult. 2006. Genome sequence of *Rickettsia bellii* illuminates the role of amoebae in gene exchanges between intracellular pathogens. *PLoS Genet.* **2:**e76.

Pallen, M. J., and B. W. Wren. 2007. Bacterial pathogenomics. *Nature* **449:**835–842.

Pang, H., and H. H. Winkler. 1993. Copy number of the 16S ribosomal RNA gene in *Rickettsia prowazekii. J. Bacteriol.* **175:**3893–3896.

Pang, H., and H. H. Winkler. 1994. The concentrations of stable RNA and ribosomes in *Rickettsia prowazekii. Mol. Microbiol.* **12:**115–120.

Paulsen, I. T., L. Nguyen, M. K. Sliwinski, R. Rabus, and M. H. Saier, Jr. 2000. Microbial genome analyses: comparative transport capabilities in eighteen prokaryotes. *J. Mol. Biol.* **301:**75–100.

Phibbs, P. V., Jr., and H. H. Winkler. 1981. Regulatory properties of partially purified enzymes of the tricarboxylic acid cycle of *Rickettsia prowazekii*, p. 421–430. *In* W. Burgdorfer and R. Anacker (ed.), *Rickettsiae and Rickettsial Diseases.* Academic Press, New York, NY.

Phibbs, P. V., Jr., and H. H. Winkler. 1982. Regulatory properties of citrate synthase from *Rickettsia prowazekii. J. Bacteriol.* **149:**718–725.

Plano, G. V., and H. H. Winkler. 1991. Identification and initial topological analysis of the *Rickettsia prowazekii* ATP/ADP translocase. *J. Bacteriol.* **173:**3389–3396.

Quick, W. P., and H. E. Neuhaus. 1996. Evidence for two types of phosphate translocators in sweet-pepper (*Capsicum annum* L.) fruit chromoplasts. *Biochem. J.* **320**:7–10.

Ralser, M., M. M. Wamelink, A. Kowald, B. Gerisch, G. Heeren, E. A. Struys, E. Klipp, C. Jakobs, M. Breitenbach, H. Lehrach, and S. Krobitsch. 2007. Dynamic rerouting of the carbohydrate flux is key to counteracting oxidative stress. *J. Biol.* **6:**10.

Ralser, M., M. M. Wamelink, S. Latkolik, E. E. Jansen, H. Lehrach, and C. Jakobs. 2009. Metabolic reconfiguration precedes transcriptional regulation in the antioxidant response. *Nat. Biotechnol.* **27:**604–605.

Ramm, L. E., and H. H. Winkler. 1973. Rickettsial hemolysis: effect of metabolic inhibitors upon hemolysis and adsorption. *Infect. Immun.* **7:**550–555.

Renesto, P., C. Rovery, J. Schrenzel, Q. Leroy, A. Huyghe, W. Li, H. Lepidi, P. Francois, and D. Raoult. 2008. *Rickettsia conorii* transcriptional response within inoculation eschar. *PLoS One* **3:**e3681.

Rovery, C., P. Renesto, N. Crapoulet, K. Matsumoto, P. Parola, H. Ogata, and D. Raoult. 2005. Transcriptional response of *Rickettsia conorii* exposed to temperature variation and stress starvation. *Res. Microbiol.* **156:**211–218.

Saier, M. H., Jr., C. V. Tran, and R. D. Barabote. 2006. TCDB: the Transporter Classification Database for membrane transport protein analyses and information. *Nucleic Acids Res.* **34**(Database issue)**:**D181–D186.

Shaw, E. I., G. L. Marks, H. H. Winkler, and D. O. Wood. 1997. Transcriptional characterization of the *Rickettsia prowazekii* major macromolecular synthesis operon. *J. Bacteriol.* **179:**6448–6452.

Sicheritz-Pontén, T., C. G. Kurland, and S. G. E. Andersson. 1998. A phylogenetic analysis of the cytochrome *b* and cytochrome *c* oxidase I genes supports an origin of mitochondria from within the *Rickettsiaceae. Biochim. Biophys. Acta* **1365:**545–551.

Smith, D. K., and H. H. Winkler. 1979. Separation of inner and outer membranes of *Rickettsia prowazekii* and characterization of their polypeptide compositions. *J. Bacteriol.* **137:**963–971.

Stork, E., and C. L. Wisseman, Jr. 1976. Growth of *Rickettsia prowazeki* in enucleated cells. *Infect. Immun.* **13:**1743–1748.

Tjaden, J., H. H. Winkler, C. Schwöppe, M. van der Laan, T. Möhlmann, and H. E. Neuhaus. 1999. Two nucleotide transport proteins in *Chlamydia trachomatis*: one for net nucleoside triphosphate uptake and the other for transport of energy. *J. Bacteriol.* **181:**1196–1202.

Trentmann, O., B. Jung, H. E. Neuhaus, and I. Haferkamp. 2008. Nonmitochondrial ATP/ADP transporters accept phosphate as third substrate. *J. Biol. Chem.* **283:**36486–36493.

Tucker, A. M., H. H. Winkler, L. O. Driskell, and D. O. Wood. 2003. *S*-Adenosylmethionine transport in *Rickettsia prowazekii. J. Bacteriol.* **185:**3031–3035.

Turco, J., and H. H. Winkler. 1989. Isolation of *Rickettsia prowazekii* with reduced sensitivity to gamma interferon. *Infect. Immun.* **57:**1765–1772.

Tzianabos, T., C. W. Moss, and J. E. McDade. 1981. Fatty acid composition of rickettsiae. *J. Clin. Microbiol.* **13:**603–605.

Walker, T. S., and H. H. Winkler. 1978. Penetration of cultured mouse fibroblasts (L cells) by *Rickettsia prowazekii. Infect. Immun.* **22:**200–208.

Weiss, E., L. W. Newman, R. Grays, and A. E. Green. 1972. Metabolism of *Rickettsia typhi* and *Rickettsia akari* in irradiated L cells. *Infect. Immun.* **6:**50–57.

Wike, D. A., and W. Burgdorfer. 1972. Plaque formation in tissue cultures by *Rickettsia rickettsi* isolated directly from whole blood and tick hemolymph. *Infect. Immun.* **6:**736–738.

Wike, D. A., R. A. Ormsbee, G. Tallent, and M. G. Peacock. 1972. Effects of various suspending media on plaque formation by rickettsiae in tissue culture. *Infect. Immun.* **6:**550–556.

Williams, J. C., and E. Weiss. 1978. Energy metabolism of *Rickettsia typhi*: pools of adenine nucleotides and energy charge in the presence and absence of glutamate. *J. Bacteriol.* **134:** 884–892.

Winkler, H. H. 1976. Rickettsial permeability: an ADP-ATP transport system. *J. Biol. Chem.* **251:**389–396.

Winkler, H. H. 1982. Rickettsiae: intracytoplamic life. *ASM News* **48:**184–186.

Winkler, H. H. 1987. Protein and RNA synthesis by isolated *Rickettsia prowazekii. Infect. Immun.* **55:**2032–2036.

Winkler, H. H. 1990. *Rickettsia* species (as organisms). *Annu. Rev. Microbiol.* **44:**131–153.

Winkler, H. H., R. Daugherty, and F. Hu. 1999. *Rickettsia prowazekii* transports UMP and GMP, but not CMP, as building blocks for RNA synthesis. *J. Bacteriol.* **181:**3238–3241.

Winkler, H. H., and R. M. Daugherty. 1984. Regulatory role of phosphate and other anions in transport of ADP and ATP by *Rickettsia prowazekii. J. Bacteriol.* **160:**76–79.

Winkler, H. H., and R. M. Daugherty. 1986. Acquisition of glucose by *Rickettsia prowazekii* through the nucleotide intermediate uridine 5′-diphosphoglucose. *J. Bacteriol.* **167:**805–808.

Winkler, H. H., and E. T. Miller. 1978. Phospholipid composition of *Rickettsia prowazeki* grown in chicken embryo yolk sacs. *J. Bacteriol.* **136:** 175–178.

Winkler, H. H., and H. E. Neuhaus. 1999. Non-mitochondrial adenylate transport: a plant plastid to obligate intracellular bacterium connection. *Trends Biochem. Sci.* **277:**64–68.

Wisseman, C. L., Jr., E. A. Edlinger, A. D. Waddell, and M. R. Jones. 1976. Infection cycle of *Rickettsia rickettsii* in chicken embryo and L-929 cells in culture. *Infect. Immun.* **14:** 1052–1064.

Woodard, A., and D. O. Wood. 2011. Analysis of convergent gene transcripts in the obligate intracellular bacterium *Rickettsia prowazekii. PLoS One* **6:**e16537.

Wu, J., and J. Xie. 2009. Magic spot: (p) ppGpp. *J. Cell. Physiol.* **220:**297–302.

Zahorchak, R. J., and H. H. Winkler. 1983. Transmembrane electrical potential in *Rickettsia prowazekii* and its relationship to lysine transport. *J. Bacteriol.* **153:**665–671.

INNATE IMMUNE RESPONSE AND INFLAMMATION: ROLES IN PATHOGENESIS AND PROTECTION (*RICKETTSIACEAE*)

Sanjeev K. Sahni, Elena Rydkina, and Patricia J. Simpson-Haidaris

8

TARGET HOST CELLS, INTRACELLULAR BEHAVIOR, AND CELL-TO-CELL DISSEMINATION OF PATHOGENIC *RICKETTSIA* SPECIES

Rickettsiae are gram-negative bacilli characterized by strict intracellular location and replication, fastidious growth requirements, association with arthropods, and tropism for microvascular endothelium of small and medium-sized vessels in their mammalian hosts. Although spotted fever group rickettsiae (SFGR) and typhus group rickettsiae (TGR) represent two major antigenically defined and historically well-known subdivisions of pathogenic *Rickettsia* species, recent in-depth characterization by neighbor-joining phylodendrogramic analysis has distributed rickettsiae into ancestral, spotted fever, typhus, and transitional subgroups (Gillespie et al., 2008). Responsible for hundreds of cases and deaths between 1873 and early 1900s, *Rickettsia rickettsii* was first identified as "the bacillus of Rocky Mountain spotted fever" (RMSF) by Howard T. Ricketts in 1906. RMSF remains among one of the most virulent human infections, with significant morbidity and mortality, being potentially fatal even in otherwise healthy young individuals (Dantas-Torres, 2007). The highest incidence of RMSF ever reported in the United States occurred in 2004 (Demma et al., 2006; Walker et al., 2008). *Rickettsia prowazekii*, which causes epidemic typhus, also represents a severe infectious disease in humans, with a mortality rate in excess of 10 to 15%. The differential diagnosis of epidemic typhus is often difficult due to nonspecific early flulike symptoms. Of note among various rickettsial species, *R. prowazekii* is the only known rickettsial pathogen with the ability to maintain persistent subclinical infection in convalescent patients that can later manifest as recrudescent typhus, or Brill-Zinsser disease (Raoult et al., 2004). Historically, *R. rickettsii* has been described as the agent of "blue disease" or "black measles" (Woodward, 1973) and *R. prowazekii* as the "scourge of armies" (Raoult et al., 2006). However, the advent of new and improved molecular tools enabling precise detection has led to the identification of an increasing number of emerging rickettsial

Sanjeev K. Sahni, Department of Pathology, University of Texas Medical Branch, Galveston, TX 77555. *Elena Rydkina*, Department of Microbiology and Immunology, University of Rochester School of Medicine and Dentistry, Rochester, NY 14642. *Patricia J. Simpson-Haidaris*, Departments of Medicine (Hematology-Oncology Division), Microbiology and Immunology, and Pathology and Laboratory Medicine, University of Rochester School of Medicine and Dentistry, Rochester, NY 14642.

Intracellular Pathogens II: Rickettsiales
Edited by G. H. Palmer and A. F. Azad © 2012 ASM Press, Washington, DC
10.1128/9781555817336.ch8

pathogens all around the world, as evidenced by description of pathogens such as *Rickettsia africae* (Jensenius et al., 2003a). In addition, the potential of other known species such as *Rickettsia parkeri* and *Rickettsia amblyommii* to cause human rickettsioses, both of which were previously considered to be nonpathogenic, has also been realized only recently (Paddock, 2005; Apperson et al., 2008; Walker et al., 2008). Transmission of rickettsial diseases by previously unknown, unexpected tick vectors further indicates the pathogens' ability to adapt to new ecological niches and maintain virulence (Demma et al., 2005; Nicholson et al., 2010). Due to the known history and potential for exploitation as devastating bioterror agents (Walker, 2009), *R. prowazekii* and *R. rickettsii* have been classified as Select Agents and accordingly placed among National Institute of Allergy and Infectious Diseases category B or C priority pathogens for the U.S. biodefense research program.

Phylogenetically similar to *R. rickettsii, Rickettsia conorii* is the etiologic agent of Mediterranean spotted fever (MSF), or boutonneuse fever, a disease generally considered milder than RMSF. Recent data, however, suggest that MSF is much more widely distributed in Africa, Europe, and countries of the Mediterranean basin (Rovery et al., 2008). In certain instances the severity of the disease is comparable to RMSF, with mortality rates as high as 32% (de Sousa et al., 2008) and evidence of multiple inoculation eschars at the sites of tick adhesion and bite (Mouffok et al., 2009). Murine typhus, or endemic typhus, is a flea-borne disease caused by *Rickettsia typhi*, another prototypical species belonging to the TGR. With chills, fever, headache, myalgia, nausea, rash, and vomiting as the signs and symptoms, and considered to be relatively milder, murine typhus is generally characterized by the same pathophysiological and clinical features as epidemic typhus. Importantly, well-established and widely accepted in vivo models of infection closely mimicking the pathogenesis of RMSF and epidemic typhus in humans employ infection of susceptible mouse strains with *R. conorii* and *R. typhi*, respectively.

Both *R. rickettsii* and *R. prowazekii* represent major human health threats, with recent evidence for increased incidence of RMSF in the Western Hemisphere and several epidemic typhus outbreaks in different parts of the world. A majority of sequelae associated with human rickettsial diseases are attributed to widespread infection of microvascular endothelium, resulting in the damage to blood vessels, altered vascular permeability, and vascular inflammation/dysfunction, collectively defined as "rickettsial vasculitis" (Hackstadt, 1996; Walker and Ismail, 2008; Sahni and Rydkina, 2009). The clinical complications in severe, untreated cases often manifest as pulmonary edema, hypovolemic hypotension, and multiorgan failure (Walker, 2007). Rickettsiae are unique among pathogenic bacteria for several reasons: (i) their ability to escape from a transient invasion vacuole to quickly establish an intracytoplasmic niche conducive for their energy needs and strict intracellular growth requirements; (ii) their affinity to preferably infect vascular endothelial cells in mammalian hosts; and (iii) biologically relevant differences in the intracellular behavior and accumulation patterns of major SFGR and TGR species, namely *R. rickettsii* versus *R. prowazekii* (Hackstadt, 1996; Gouin et al., 2005; Walker and Ismail, 2008). The process of host cell invasion involves active participation of metabolically viable host cells and de novo protein synthesis by invading rickettsiae. Such an "induced phagocytosis" involves a zipper mechanism, wherein rearrangement of the actin cytoskeleton occurs beneath the site of rickettsial attachment, culminating in progressive apposition of host cell membrane over the invading bacterium and resultant engulfment (Olano, 2005). Evidence from transmission electron microscopy further suggests that the process of internalization into host cells occurs within a few minutes after initial contact. Genes encoding the rickettsial membranolytic functions, hemolysin C (*tlyC*) and phospholipase D (*pldA*), likely cause rapid release of organisms from phagocytic vacuoles, thereby allowing internalized rickettsiae to remain "free" in the host cytosol (Radulovic et al., 1999; Whitworth et al., 2005). Once in the intracellular niche, however, SFGR and TGR exhibit unique biological differences in terms of their growth and motility properties. While formation of a polar actin tail and its exploitation for both intracellular movements and cell-to-cell spread likely serve as a major determinant for the establishment, dissemination, and pathogenesis of SFGR, TGR either display nonlinear actin-based motility, as seen for *R. typhi*, or do not display this behavior at all, as is the case with *R. prowazekii* (Hackstadt, 1996; Stevens et al., 2006; Walker and Ismail, 2008). Evidence for the lack of motility or display of erratic movements in TGR species corresponds to the absence of a homolog of the gene governing actin-based motility, *rickA*, in the genomes of *R. prowazekii* and *R. typhi* (Heinzen et al., 1993; Andersson et al., 1998; McLeod et al., 2004; Walker and Yu, 2005). In contrast, a monoclonal antibody against recombinant *R. conorii* RickA confirms its presence in a majority of SFGR (Balraj et al., 2008). As a nucleation-promoting factor, RickA is capable of activating Arp2/3 complex and inducing actin polymerization in vitro (Gouin et al., 2004; Jeng et al., 2004). Yet the unbranched organization and absence of Arp2/3 subunits in rickettsial comet tails and display of motility in cells expressing verprolin, cofilin, and acidic domain (VCA) of neural Wiskott-Aldrich syndrome protein (N-WASp) suggest that unlike *Listeria* and *Shigella*, rickettsial actin-based motility may also encompass an Arp2/3-independent mechanism (Gouin et al., 1999; Harlander et al., 2003; Serio et al., 2010). Indeed, recent studies have implicated an autotransporter protein, surface cell antigen 2 (Sca2), in Arp2/3-independent motility of *R. rickettsii* and *R. parkeri* (Kleba et al., 2010; Haglund et al., 2010). A number of SFGR species, including *R. africae*, *Rickettsia akari*, *Rickettsia australis*, *R. conorii*, and *R. rickettsii*, encode full-length Sca2 containing an N-terminal Sec-dependent secretion signal, four putative WASp homology (WH2) domains, proline-rich regions (similar to formins), and an autotransporter domain on the C terminus. In contrast, *R. prowazekii* only expresses a truncated version of Sca2 lacking the first three WH2 domains, proline-rich segments, and secretory signal sequence. Predictably, *R. typhi* also encodes shortened Sca2 (1,483 amino acids, compared with 1,821 amino acids in SFGR) with overlapping autotransporter and last WH2 domains, but divergent formin homology 1 (FH1) domain and secretion signal (Kleba et al., 2010). Overall, the primary mechanism for intercellular dissemination of typhus species predominantly involves the buildup of bacterial load due to significant intracellular accumulation and resultant host cell lysis, which is a consequence of ongoing multiplication by binary fission with a replication time of approximately 9 to 12 h (Sporn et al., 1993).

As they are obligate intracellular parasites, propagation of rickettsiae to obtain host cell-free preparations requires their growth and purification from embryonated egg yolk sacs (similar to viruses), mammalian host cell cultures, or from infected host tissues. Pathogenic rickettsiae are able to infect and replicate in vitro in a number of different cell types, including fibroblasts, HeLa cells, and other carcinoma cell types (Clifton et al., 1998, 2005b; Martinez et al., 2005; Walker, 2006). Although rickettsiae-specific host endothelial cell receptors have yet to be identified, internalization of *R. conorii* into Vero and HeLa cells involves recruitment of Ku70, a subunit of Ku70-Ku80 heterodimers involved in double-stranded DNA break repair (Muller et al., 2005), to the sites of bacterial attachment and subsequent binding interactions between the outer membrane protein rOmpB and Ku70 (Martinez et al., 2005). A unique property displayed by rickettsiae in vivo is their affinity to preferentially infect vascular endothelial cells lining the small and medium-sized blood vessels in humans and in established animal models of infection (Walker et al., 1994; Hackstadt, 1996; Walker et al., 2000; Olano, 2005; Walker and Ismail, 2008). As a consequence, rickettsiae invade, colonize, and disseminate through the endothelium, damaging vascular networks. In a majority of patients, the resultant telltale symptoms of RMSF, MSF, and both epidemic typhus as well as endemic typhus (due to *R. typhi*) typically manifest as dermal lesions of various types such as maculopapular and petechial rash. In this regard, the well-known but recently re-emerging concept of "endothelial cell heterogeneity" dictates the existence of significant morphological and functional differences between large and small vessels and between cells derived from various microvascular endothelial beds (Dyer and Patterson, 2010). Such differences necessitate the need for comprehensive analysis of the biological basis of rickettsial affinity for vascular endothelial cells. Moreover, following dissemination to distant endothelium and disruption of its uniform monolayer, infection of and damage to underlying vascular smooth muscle cells and perivascular cells such as monocytes and macrophages may also contribute to the pathogenesis and complications of rickettsial diseases (Eremeeva et al., 1999). In this context, a notable exception to the vasculotropic nature of rickettsiae is *R. akari*, the causative agent of rickettsialpox, for which the predominant target cell type is CD68-positive macrophages (Walker et al., 1999).

ENDOTHELIAL CELL ACTIVATION BY *RICKETTSIA* INFECTION IN VITRO

Composed of a uniform monolayer of endothelial cells, the vascular endothelium constitutes a selective permeability barrier between the vessel wall and blood. Endothelial cells are unique, versatile, and multifunctional, as they are not only capable of performing a number of critical housekeeping activities under basal conditions, but can also respond to a multitude of biological triggers to perform metabolic and synthetic functions. Under normal physiological conditions, endothelial cells generally maintain a nonthrombogenic blood-tissue interface by regulating thrombosis, thrombolysis, leukocyte adhesion, and vascular tone (summarized in Fig. 1). Initial attempts to establish a relevant in vitro model to study host-pathogen interrelationships during rickettsial infections exploited infection of cultured human umbilical vein endothelial cells (HUVECs) with *R. rickettsii* and documented the course of endothelial cell injury during intracellular infection, culminating in extensive membrane damage and necrotic cell death (Silverman and Bond, 1984; Walker, 1984). Subsequent studies from a number of laboratories then went on to reveal that endothelial cells are not simply injured during infection with *R. rickettsii* and *R. conorii*, but also launch a series of adaptive cellular responses, including functional changes reflective of an "activated" phenotype. Responses that have been well characterized so far include, but are certainly not limited to, transient expression of tissue factor (Teysseire et al., 1992; Sporn et al., 1994) and E-selectin (Sporn et al., 1993), increased synthesis of plasminogen activator inhibitor-1 (Drancourt et al., 1990a; Shi et al., 2000), release of von Willebrand factor from Weibel-Palade bodies (Sporn et al., 1991, Teysseire et al., 1992), and changes in endothelial cell surface adhesiveness for platelets (Silverman, 1986). Increased expression of endothelial adhesion molecules during in vitro infection with *R. rickettsii* and *R. conorii* and higher levels of circulating intercellular adhesion molecule-1 in the blood of a patient with fulminant RMSF clearly reflect the level of activation and degree of infection-induced damage to endothelial cells (Sessler et al., 1995; Dignat-George et al., 1997). Accordingly, it is now widely accepted that the synthetic profile of endothelial cells undergoes a distinct shift from a quiescent, basal phenotype toward a procoagulant and proinflammatory state during the course of *Rickettsia* infection, a phenomenon that likely contributes to the manifestation of resultant disease symptoms. Importantly, a majority of these observations are based on the infection of primary cultures of HUVECs with purified *R. rickettsii* to achieve infection of ≥ 80% of the cells with 3 to 4 intracellular rickettsiae per infected cell.

Similar to rickettsiae causing spotted fever syndromes, the major target of *R. prowazekii*

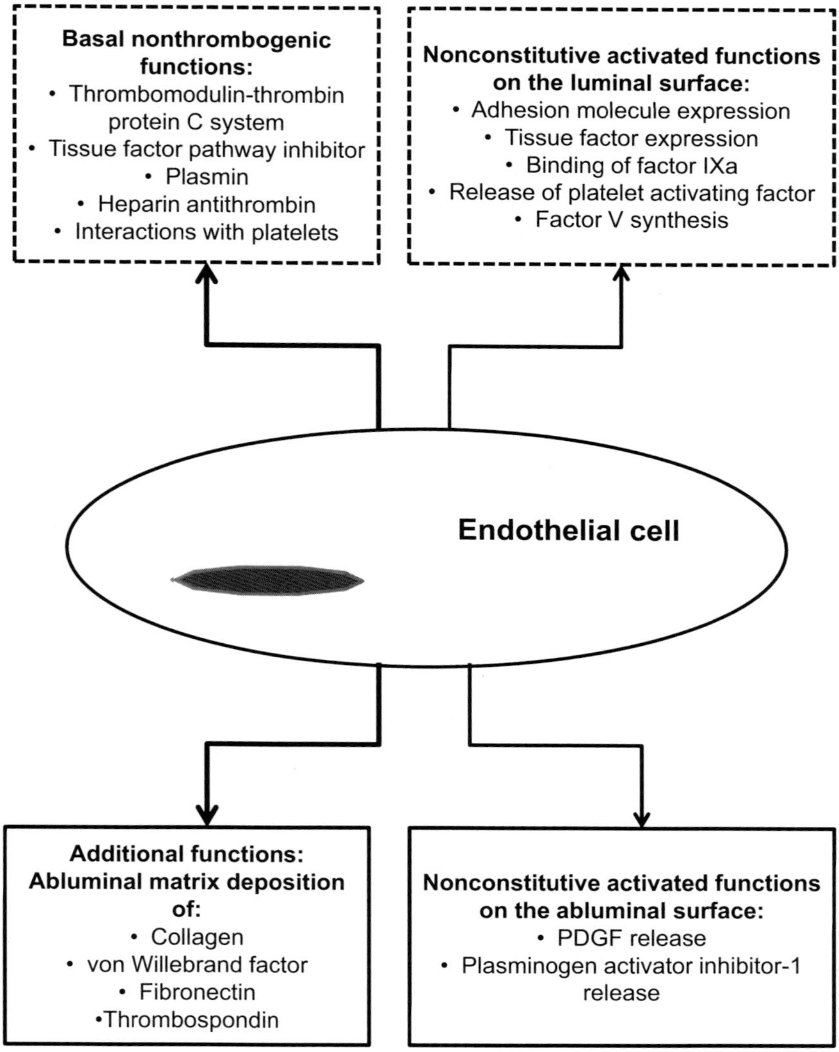

FIGURE 1 Endothelial cell functions under normal physiological conditions and in response to activation. PDGF, platelet-derived growth factor.
doi:10.1128/9781555817336.ch8.f1

during human infections is also vascular endothelium. However, very little is known about specific host cell responses and only scarce information is yet available regarding signaling mechanisms that determine the interplay between target endothelial cells and typhus rickettsiae. It is important to remember that a majority of initial, groundbreaking studies aimed at establishing intracellular growth requirements and host cell interactions with *R. prowazekii* were conducted using either L929 (fibroblasts) or Vero (epithelial) cells. As illustrated above, endothelial cells represent crucial components of the vasculature by serving to maintain vascular homeostasis. Intimately involved in the pathogenesis of epidemic typhus, endothelial cells have also emerged as key immunoreactive cells that participate in host defense and inflammation by virtue of both producing and reacting to a wide variety of mediators, including cytokines, growth factors, adhesion molecules, vasoactive substances, and chemokines, capable of exerting paracrine effects on many other different cell types (Galley and Webster, 2004). Endothelial invasion by *R. prowazekii* is an active process mediated by a rickettsial phospholipase (Walker et al., 2001a; Whitworth et al., 2005), and *R. prowazekii*-infected endothelial cells display increased platelet-activating factor synthesis and prostaglandin secretion (Walker et al., 1990, Walker and Mellott, 1993). Further, *R. prowazekii* infection of both murine and human endothelial cells in vitro results in increased transmigration of leukocytes, likely through an exacerbated inflammatory response (Bechah et al., 2008b). More recent evidence implicates engagement of $\alpha_v\beta_3$ integrins in increased adhesivity of *R. prowazekii*-infected endothelial cells to leukocytes (Bechah et al., 2009). Overall, endothelial cell responses to *R. prowazekii* and signaling mechanisms that determine the interplay between the host and this unique rickettsial pathogen, which is capable of escaping immune surveillance to cause recrudescent infections, remain critically important but neglected areas of scientific inquiry. Thus, one of the major critical gaps in the understanding of rickettsial pathogenesis is the definition of the biological basis of rickettsial affinity and consequent interactions with vascular endothelium. Enhanced understanding of rickettsial interactions with endothelial cells and manipulations of host cell signaling mechanisms may lead to the development of new and improved chemotherapeutic strategies aimed at inhibiting early intracellular growth of rickettsiae and diminishing the intensity of vascular inflammation.

MECHANISMS UNDERLYING HOST CELL TRANSCRIPTIONAL ACTIVATION BY PATHOGENIC RICKETTSIAE

Activation and nuclear translocation of NF-κB is one of the key markers of endothelial cell activation and a requisite event for many endothelial cell responses, including expression of cytokines and adhesion molecules. Although endothelial cell responses to different exogenous factors are rather complex and much remains to be understood despite intense investigations, the most thoroughly studied mediators known to stimulate inflammatory signaling in the endothelium are lipopolysaccharide (LPS), tumor necrosis factor alpha (TNF-α), and interleukin-1β (IL-1β). Signaling triggered by all three culminates in the activation of NF-κB to regulate expression of endothelial-specific genes containing κB-binding site(s) in their promoter regions, including E-selectin, plasminogen activator inhibitor-1, and tissue factor. NF-κB is a dimeric transcription factor composed of homo- and heterodimers of the proteins belonging to the Rel family, of which there are five members in mammalian cells, i.e., RelA or p65, c-Rel, RelB, NF-κB1 or p50, and NF-κB2 or p52, which share an N-terminal Rel homology domain (RHD) responsible for DNA binding and homo- and heterodimerization (Hayden and Ghosh, 2008). Importantly, the transcription activation domain (TAD) necessary for the positive regulation of gene expression is present only in p65, c-Rel, and RelB. NF-κB dimers are sequestered in the cytoplasm in an inactive form via association with the regulatory proteins called inhibitors of NF-κB (IκBs). Among the known IκBs, IκBα, IκBβ, and IκBε preferentially associate with various Rel family protein dimers (Tak and Firestein, 2001; Tak et al., 2001; Hayden and Ghosh, 2008). In endothelial cells, IκBε is associated with p65 and to a lesser extent with c-Rel, whereas IκBα and IκBβ associate with p65 but not c-Rel (Spiecker et al., 2000). Phosphorylation of IκB, an important step in the canonical pathway of NF-κB activation, is mediated by a high-molecular-weight IκB kinase (IKK) complex (DiDonato et al., 1997). The IKK consists of two catalytic kinases, IKKα (IKK1) and IKKβ (IKK2) (Mercurio et al., 1997; Zandi et al., 1997), and one regulatory subunit, IKKγ (Rothwarf et al., 1998). IKK activation initiates IκBα phosphorylation at specific amino-terminal serine residues, which signals for its ubiquitination and subsequent degradation by the 26S proteasome (Chen et al., 1995); IKK also phosphorylates IκBβ and IκBε (Tak and Firestein, 2001). Analysis of the contribution of different upstream components suggests that activation of IKKβ plays a prominent role in the primary pathway by which proinflammatory stimuli induce NF-κB function in endothelial cells, whereas the contribution of IKKα is less pronounced and differs for the specific target molecules (Denk et al., 2001). IKK, thus, represents a point of convergence for a plethora of NF-κB-activating stimuli.

R. rickettsii infection induces a biphasic pattern of NF-κB activation in cultured human endothelial cells that is characterized by an early transient activation phase and a late sustained phase; early activation occurs at 3 h and the sustained activation occurs between 18 to 24 h after infection. The activated NF-κB complex consists predominantly of p65-p50 heterodimers and p50 homodimers (Sporn et al., 1997). The transcript and protein levels of NF-κB subunits p65 and p50 and inhibitory protein IκBβ remain relatively unchanged during the course of infection; however, serine-32 phosphorylation of IκBα early during the infection is significantly increased over the basal level in uninfected cells concomitant with a significant increase in the expression of IκBα mRNA. The level of IκBα mRNA gradually declines toward the baseline, but the level of total IκBα protein remains lower than in the corresponding controls. The catalytic subunits of IKK complex are IKKα and IKKβ and can be measured by in vitro kinase assays using immunoprecipitates from uninfected and rickettsiae-infected endothelial cells. Using this assay, we showed that the activities of IKKα and IKKβ are significantly increased in correlation with the early phase of NF-κB activation in *R. rickettsii*-infected compared with uninfected endothelial cells. Activation of IKK in the early phase of NF-κB signaling is effectively attenuated by heat inactivation and completely abrogated by formalin fixation of rickettsiae. The known IKK inhibitors parthenolide (a sesquiterpene lactone) and aspirin (a nonsteroidal anti-inflammatory drug) adversely affect the activities of infection-induced IKKα and IKKβ, leading to the attenuation of nuclear translocation of NF-κB. An intriguing aspect that requires further detailed characterization, however, is the increased activity of IKKα later during the infection, coinciding with the late phase of NF-κB activation (Clifton et al., 2005a). Thus, activation of catalytic components of the IKK complex represents an important signaling event in the pathway of *R. rickettsii*-induced NF-κB response of host endothelium. Additional evidence suggests that *R. rickettsii* is capable of directly interacting with NF-κB present in its inactive from in isolated endothelial cell cytoplasm (Sahni et al., 1998). Activation of NF-κB in this "cell-free" system devoid of signaling interactions originating at the cell membrane occurs independently of the proteasome's involvement and is apparently dependent on an as yet unidentified protease activity of rickettsiae (Sahni et al., 1998, 2003). NF-κB is a critical regulator of inflammatory genes and one of the major determinants of host cell survival mechanisms during infection via both pro- as well as anti-apoptotic mechanisms that function in a context-dependent manner. Therefore, selective inhibition of IKK may provide a potential target for enhanced clearance of rickettsiae and an effective strategy to reduce inflammation-mediated damage to the host during rickettsial infections.

Since their discovery more than 15 years ago, the mitogen-activated protein kinases (MAPKs) have been implicated in generating an ever increasing, diverse array of biological effects, including inflammatory signaling cascades. Three major MAPK cascades, namely extracellular signal-regulated kinase (ERK), Jun N-terminal protein kinase (JNK), and p38 MAPK, have been identified and characterized in significant detail. These individual MAPK modules can be activated independently or concurrently to regulate and coordinate signals originating from a variety of extracellular and intracellular mediators. Specific phosphorylation and activation of enzymes in a MAPK module transmits the signal down the cascade, resulting in the phosphorylation of many proteins with substantial regulatory functions throughout the cell, including other protein kinases, cytoskeletal proteins, and transcription factors. Furthermore, MAPK cascades are ubiquitous signaling pathways that play instrumental roles in coordinating incoming signals triggered by a variety of mediators such as growth factors, cytokines, or physicochemical stress into intracellular responses, as well as in mediating host cell response(s) to infection (Bogatcheva et al., 2003; Roux and Blenis, 2004). Specifically, stress-activated MAPKs, comprising JNK and p38, play prominent roles in the innate and adaptive immune systems. A remarkable aspect of MAPK signaling is that the diversity and specificity in cellular response outcomes is achieved with an apparently simple, linear, three-tiered kinase module comprising a core of protein kinases, namely a MAPK kinase kinase, a MAPK kinase, and a MAPK, acting sequentially to achieve target protein phosphorylation. In the vasculature, MAPKs are known to regulate diverse physiological phenomena including proliferation, migration, apoptosis, and barrier function of endothelial cells. It has also been shown that proinflammatory cytokines, shear stress, and reactive oxygen species (ROS) induce signaling mechanisms resulting in the activation of selective MAPKs in a cell-type-dependent manner (Rincón and Davis, 2009). Thus, accumulating evidence for critical physiological importance in the vasculature prompted the determination of whether or not *R. rickettsii* infection of endothelial cells activates stress-activated MAPKs. Indeed, infection of cultured human endothelial cells with *R. rickettsii* results in the dose-dependent activation of p38 MAPK, as assessed by increased phosphorylation and enzymatic activity (Rydkina et al., 2005). *Rickettsia* inactivation by heat or formaldehyde results in abolished p38 kinase activation. Inhibition of cellular invasion by carrying out infection at low temperature, pretreatment of host endothelial cells with cytochalasin D, or preincubation of rickettsiae with an irreversible phospholipase inhibitor also led to diminished p38 phosphorylation, suggesting requirement of invasion by viable rickettsiae for this host cell response. Although SB 203580, a p38-specific inhibitor, exerts no inhibitory effect on infection-induced activation of the ubiquitous transcriptional regulator NF-κB, it effectively reduces the expression and secretion of the important chemoattractant cytokines IL-8 and monocyte chemoattractant protein-1 (MCP-1) by *R. rickettsii*-infected endothelial cells (Rydkina et al., 2005). Selective inhibition of p38 activity may, therefore, be exploited as an anti-inflammatory target to prevent rickettsial vasculitis and to develop new and improved chemotherapeutic agents. The possibility of a potential interrelationship between the activation of p38 MAPK and *R. rickettsii* invasion into host endothelial cells has also been explored. To this end, we employed three independent approaches, namely immunofluorescent staining of rickettsiae followed by quantitation of the proportion of infected cells and number of intracellular bacteria (Sporn et al., 1997), plaque formation assay (Rydkina et al., 2004), and citrate synthase gene (*gltA*)-based quantitative PCR (Rydkina et al., 2007). The plaque formation assay is routinely carried out using Vero cell monolayers to measure the viability of intracellular rickettsiae. On the other hand, indirect immunofluorescent staining and quantitative PCR yield quantitative measures of the intensity of infection with better accuracy but do not account for the rickettsial viability. Thus, the combination of approaches is important because the viability of intracellular rickettsiae must be determined, as well as the percentage of the endothelial cell population infected and the number of infecting organisms. Nevertheless, cumulative evidence indicates that activation of p38 MAPK also plays an important role in facilitating host cell invasion by *R. rickettsii* (Rydkina et al., 2008). A comparative evaluation of the abilities of pathogenic *Rickettsia* species to trigger phosphorylation of p38 MAPK in endothelial cells further suggests that infection of endothelial cells with both spotted fever and typhus species results in p38 MAPK activation. Although both *R. conorii* and *R. typhi* are able to stimulate phosphorylation of p38 MAPK, notable differences in the intensity but not the kinetics of its activation are also evident. In *R. conorii*- as well as *R. typhi*-infected endothelial cells, increased p38 phosphorylation is evident as early as 0.5 to 1 h postinfection, appears to peak at 3 h, remains significantly higher for up to 6 h, and then declines toward the basal level. For *R. conorii*, the average maximal activation at 3 h (relative to uninfected controls) is about 3.6 ± 0.6-fold, whereas activation of p38 in endothelial cells infected with *R. typhi* tends to be considerably weaker, as suggested by a maximum increase of 1.9 ± 0.2-fold. Thus, both TGR and SFGR species stimulate phosphorylation of p38 kinase, albeit with notable differences in the intensity of activation (Rydkina et al., 2007).

EXPRESSION AND SECRETION OF MEDIATORS OF INFLAMMATION DURING INFECTION OF CULTURED HOST CELLS WITH SFGR AND TGR

Cytokines and chemokines are major determinants of the host's systemic inflammatory responses to infection and are involved in the recruitment of inflammatory phagocytes. Therefore, expression pattern profiles of cytokines and chemokines provide a valuable insight into our understanding of innate immune responses to intracellular infections. Initial studies employing in vitro infection of cultured HUVECs with *R. conorii* revealed secretion of high levels of IL-6 and IL-8, but not IL-1α, IL-1β, or TNF-α. Interestingly, *R. conorii*-infected HUVECs also contained significant quantities of newly synthesized IL-1α, the majority of which remained cell associated, which led to enhanced secretion of IL-6 and IL-8 (Kaplanski et al., 1995). Subsequently, an identical response characterized by increased mRNA expression and synthesis of IL-1α, capable of autocrine-stimulatory mechanisms as measured by endothelial tissue factor activity, was also reported for human endothelial cells infected with *R. rickettsii* (Sporn and Marder, 1996). Contrary to these findings, *R. typhi*-infected macrophages reportedly produce and secrete IL-1β (Radulovic et al., 2002), but whether or not infected endothelial cells also respond in a similar manner remains to be determined. Intriguingly, patients diagnosed with acute MSF have significantly increased serum levels of gamma interferon (IFN-γ), TNF-α, IL-6, and IL-10, but IL-1α and IL-8 are not detectable in the blood during any phase of infection (Cillari et al., 1996; Vitale et al., 2001). A potential explanation for the absence of circulating IL-1α may be derived from in vitro findings suggesting that a majority of IL-1α in *R. conorii*- and *R. rickettsii*-infected endothelium remains cell associated and cannot be detected in the culture supernatants (Kaplanski et al., 1995; Sporn and Marder, 1996). On the other hand, IL-8 can be rapidly released from the Weibel-Palade bodies of endothelial cells and in vitro interactions between endothelial cells and *R. conorii* or *R. rickettsii* have been shown to induce the release of von Willebrand factor from Weibel-Palade bodies (Sporn et al., 1991; Teysseire et al., 1992; Knipe et al., 2010). Although a patient with fatal fulminant RMSF reportedly had an increased level of serum IL-8 despite therapy with antibiotics and vasopressors (Sessler et al., 1995), the finding of no changes in the IL-8 level in blood samples of humans with confirmed diagnoses of *R. conorii* infection is rather intriguing and dictates the need for further detailed investigation. An important role for the T-cell-targeting chemokines MIG (monokine induced by IFN-γ) and IP-10 (interferon-inducible protein of 10 kDa) in the early immune response against SFGR has been demonstrated in a mouse model and in biopsy brain stem specimens from pediatric patients with RMSF (Valbuena et al., 2003). Another striking observation of this study is the lack of MIG and IP-10 response by mouse endothelial cells infected with *R. conorii* in vitro. In addition, increased fractalkine expression coinciding with infiltration of macrophages in the tissues of experimentally infected hosts has also been documented (Valbuena and Walker, 2005a). A thorough understanding of the transcriptional control mechanisms that elicit such responses and the identity of host cell types involved in producing and secreting these chemokines, however, remains to be ascertained.

IL-8 and MCP-1 are important chemokines for activating neutrophils and monocytes, respectively, and for recruiting these circulating immune cells to the sites of inflammation (Krishnaswamy et al., 1999). Analysis of the expression and secretion of these chemokines during *R. rickettsii* infection of cultured human endothelial cells reveals increased mRNA expression of IL-8 and MCP-1 (Clifton et al., 2005b; Rydkina et al., 2005). Subsequent measurements on culture supernatants further confirm significantly enhanced secretion of both chemokines as early as 3 h postinfection and later. The presence of the peptide aldehyde compound MG132 to inhibit proteasome-mediated degradation of the inhibitor of κB protein IκBα and synthetic peptide SN-50 to inhibit the nuclear translocation of the ubiquitous transcription factor NF-κB results in significant inhibition of the chemokine response. This is further supported by expressing a superrepressor mutant of IκBα in T24 cells to render NF-κB inactive, which leads to significantly lower quantities of secreted IL-8 in response to *R. rickettsii* infection compared with mock-transfected cells. A neutralizing antibody against IL-1α or an IL-1-specific receptor antagonist has no effect on the early phase of *R. rickettsii*-induced NF-κB activation and IL-8/MCP-1 secretion early during the infection. Both of these treatments, however, partially diminish the late phase of NF-κB activation and suppress the infection-induced chemokine release later during the infection (Clifton et al., 2005b). Thus, while chemokine response early during the infection likely depends on IKK-mediated activation of NF-κB, subtle autocrine effects of newly synthesized IL-1α apparently also contribute, in part, to the control of NF-κB activation and chemokine production at later times. Taken together, these findings suggest a prominent role for host endothelial cells in recruiting immune cells to the site of *Rickettsia* infection and provide important insights into the pathogenesis of SFGR.

Comparative analysis to examine the transcriptional expression and release of regulatory chemokines by endothelial cells in response to infection with *R. typhi* and *R. conorii* reveals that at all times postinfection, both *R. typhi*- and *R. conorii*-infected endothelial cells display significant (greater than twofold) upregulation of IL-8 and MCP-1 expression compared with uninfected cells. In addition, there are also significant increases in the expression of IP-10, MCP-2, and RANTES, indicating a pattern of inflammatory/immune response similar to that for *R. rickettsii*. Further, the secretion of IL-8 and MCP-1 by infected cells is also noticeably increased. Although the magnitude of increased release of IL-8 and MCP-1 with *R. conorii* and *R. typhi* is nearly identical early, noticeable differences in the secretion pattern of both chemokines are apparent at later times (Rydkina et al., 2007). In recent experiments, we have further evaluated the induction profile of IL-8 and MCP-1 release from cultured human endothelial cells infected with *R. prowazekii* (virulent Breinl strain). Apparently more robust in intensity but similar in kinetics to *R. conorii*, *R. prowazekii* infection also triggers a time-dependent pattern of increased IL-8/MCP-1 secretion with significant differences compared with uninfected controls as early as 1.5 h postinfection (S. K. Sahni, L. C. Turpin, and E. Rydkina, unpublished results).

INHIBITION OF APOPTOSIS TO PRESERVE HOST CELL NICHE EARLY DURING THE INFECTION

The pathophysiological implications of endothelial cells' involvement in rickettsial pathogenesis warrant investigation into the mechanisms of cell death. There is increasing recognition that apoptosis, a tightly regulated process of altruistic suicide, plays a central role in complex interactions between an invading pathogen and host cell defense. Regulation of apoptosis becomes even more critical in determining the outcome of infection with intracellular pathogens that depend on their hosts to survive and propagate. While apoptosis of infected cells is an important host defense mechanism to limit the spread of infection, viruses and bacteria employ a variety of strategies to inhibit the cell's apoptotic machinery. Endothelial cells undergoing apoptosis in vivo detach from the vessel basement membrane and enter the circulation, where they are rapidly cleared by phagocytes. Such an occurrence at early stages of infection would indeed be detrimental for the growth and proliferation of intracellular rickettsiae. It is therefore possible that as for other intracellular pathogens, for example, *Chlamydia* species, successful establishment and progress of infection are contingent on the rickettsial ability to suppress apoptosis in host cells. The first evidence in support of this hypothesis was the demonstration that antiapoptotic functions of NF-κB are essential for the survival of endothelial cells infected with *R. rickettsii* (Clifton et al., 1998). Because endothelial adaptation and/or dysfunction is widely regarded as the common denominator in vascular disease, it is important to have a thorough understanding of the antiapoptotic functions of NF-κB in the unique setting of *Rickettsia*-endothelial cell interactions. Investigations focused on the potential involvement of caspase cascades and associated signaling pathways in regulation of host cell apoptosis by NF-κB suggest that infection of cultured human endothelial cells with *R. rickettsii* with simultaneous inhibition of NF-κB induces the activation of apical caspases (caspase-8 and -9) and also the executioner enzyme, caspase-3, whereas infection alone has no significant effect. Further, inhibition of either caspase-8 or caspase-9 using specific cell-permeating peptide inhibitors causes a significant decline in the extent of apoptosis, confirming the importance of caspases in NF-κB-mediated regulation of apoptosis. The peak caspase-3 activity occurs at about 12 h postinfection and leads to the cleavage of poly (ADP-ribose) polymerase, followed by DNA fragmentation and apoptosis. The activities of other important downstream executioner enzymes, namely caspase-6 and caspase-7, however, remain relatively unchanged. Caspase-9 activation is mediated through the mitochondrial pathway of apoptosis, as evidenced by loss of transmembrane potential and cytoplasmic release of cytochrome *c*. Thus, activation of NF-κB is required for the maintenance of mitochondrial integrity of host cells and protects against infection-induced apoptotic death by preventing the activation of caspase-9- and caspase-8-mediated pathways (Joshi et al., 2003). Targeted inhibition of this transcription factor may, therefore, be exploited to enhance the clearance of infections with *R. rickettsii* and other intracellular pathogens with similar survival strategies.

The Bcl-2 (B-cell lymphoma 2) family of proteins, which include both pro- and antiapoptotic factors, play a critical role in the regulation of apoptotic cell death by controlling mitochondrial permeability (Kluck et al., 1997; Scorrano and Korsmeyer, 2003). The antiapoptotic proteins Bcl-2 and Bcl-xL reside in the outer membrane of mitochondria and inhibit the release of cytochrome *c*. On the other hand, Bad, Bid, and Bax are cytosolic proapoptotic proteins and their translocation to mitochondria promotes cytochrome *c* release. The phosphorylation of Bad is necessary for its sequestration in the cytosol. The effects of Bid are dependent on the generation of tBid, the active truncated fragment capable of translocating into mitochondria (Esposti, 2002). With the goal to better understand upstream signaling mechanisms, the effects of NF-κB blockade on the status of different Bcl-2 proteins have also been examined. *R. rickettsii*-induced apoptosis of endothelial cells is associated with decreased levels of Bid and accumulation of Bad, while cytosolic levels of Bax remain relatively unchanged. In addition, cellular levels of the apoptosis antagonist Bcl-2 are downregulated and the apoptogenic mitochondrial proteins Smac (second mitochondria-derived activator of apoptosis) and cytochrome *c* are released into the cytoplasm (Joshi et al., 2004). Moreover, quantitative analysis following terminal deoxynucleotidyltransferase-mediated dUTP-biotin nick end labeling and Hoechst staining revealed that infection of endothelial cells with *R. rickettsii* in the presence of the NF-κB inhibitor MG132 results in the induction of apoptosis. Taken together, these results suggest an important regulatory role for NF-κB in controlling the intracellular levels and/or localization of pro- as well as antiapoptotic proteins of the Bcl-2 family, the intricate balance of which is a critical determinant of downstream signaling mechanisms governing cell fate during intracellular infection.

Intracellular survival of *R. rickettsii* is dependent on its ability to circumvent the crucial defense mechanism of host apoptosis. Because antiapoptotic functions of NF-κB are activated in response to *R. rickettsii* infection, we hypothesized that rickettsiae may posses the ability to modulate the host's apoptotic pathways to preserve the site of infection and prevent the clearance of infected cells early during the infection. In this regard, our available data clearly indicate that cells infected with *R. rickettsii* display a tendency to resist the effects of staurosporine, a strong pharmacologic inducer of apoptosis, and a distinct shift in the balance of certain pro- and antiapoptotic proteins to favor antiapoptosis (Bechelli et al., 2009). In conjunction with the excellent recent progress in the genomic and proteomic characterization of pathogenic *Rickettsia* species, further expansion of our existing knowledge regarding adaptations in host cells' apoptotic signaling should enable us to identify and characterize potential rickettsial antiapoptotic activity(ies) that could be targeted to develop new and improved strategies aimed at enhancing clearance of infected cells early to avoid disseminated host cell infection and associated vascular pathologies.

OXIDATIVE STRESS AND ANTIOXIDANT DEFENSE MECHANISMS DURING *RICKETTSIA*-HOST CELL INTERACTIONS

Virtually all types of cells in the vasculature generate superoxide anion (O_2^-) and hydrogen peroxide (H_2O_2), the ROS implicated in a variety of vascular diseases (Stocker and Keaney, 2004). Although ROS are known to contribute to distinct cellular stresses such as heat shock, UV irradiation, etc., it has now become apparent that ROS in the vasculature may also perform second-messenger functions. The injury to endothelial cells following infection with *R. rickettsii* culminates in extensive membrane damage. Also, cells infected with *R. rickettsii* exhibit striking changes in the structural organization of intracellular organelles and cytoskeleton, suggesting the possibility of induction of oxidative stress (Eremeeva et al., 1999). Subsequent biochemical studies confirmed the accumulation of intracellular lipid peroxides and superoxide anion (O_2^-) as well as extracellular H_2O_2; depletion of protective intracellular thiols; and reduction in the activities of antioxidant enzymes (Eremeeva and Silverman, 1998; Eremeeva et al., 2000, 2001). However, cells infected with *R. prowazekii* reportedly show no detectable morphological changes prior to undergoing lysis (Hackstadt, 1996; Eremeeva et al., 2000), and only limited information is available regarding host cell responses during TGR infection and cytopathic effects of TGR pathogens, which include etiologic agents of epidemic and endemic typhus worldwide. Severe cases of human rickettsial infections are often associated with acute renal failure and respiratory distress syndrome. The prognosis of spotted fever and typhus patients with genetic glucose-6-phosphate dehydrogenase deficiency is poor due to enhanced severity, indicating the potential importance of redox systems in combating the disease (Whelton et al., 1968; Walker and Kirkman, 1980; Drancourt et al., 1990b). Thus, oxidative stress likely plays a key role in the pathogenesis of rickettsial infections. Consequently, a promising rationale for new and improved supplemental interventions is the agents that can either quench harmful ROS or potentiate cellular protective mechanisms against oxygen radicals.

Heme oxygenase (HO), a rate-limiting enzyme, oxidatively cleaves heme, resulting in the release of ferrous iron and generation of carbon monoxide (CO) and biliverdin; the latter is subsequently converted to bilirubin. Bilirubin is a potent physiologic antioxidant capable of protecting against hydrogen peroxide-induced toxicity (Barañano et al., 2002; Kawamura et al., 2005), and CO generated through the action of HO-1 acts as an antiapoptotic molecule by preventing a series of inflammatory reactions associated with endothelial apoptosis (Brouard et al., 2000). Another study further demonstrates that upregulation of ferritin heavy chain suppresses ROS and inhibits apoptosis triggered by TNF-α (Pham et al., 2004). The most compelling evidence of HO's importance in the vasculature, however, comes from the only known human case of HO-1 deficiency, which describes a 6-year-old patient with severe, persistent endothelial damage and detachment of endothelium in vital organs (Yachie et al., 1999). Clearly, recent discoveries of diverse biological activities of HO metabolites have contributed to the paradigm shift from HO's conventional role as a "molecular wrecking ball" to a "mesmerizing trigger" of cellular events (Maines and Gibbs, 2005). Vascular endothelial cells contain two distinct HO isozymes, HO-1 and HO-2, which are encoded by different genes. Several lines of evidence suggest that the products of HO activity play an important role in the vascular endothelium by reduction of oxidative stress, diminution of vascular constriction, attenuation of inflammation, decrease in smooth muscle cell proliferation, and inhibition of apoptosis. While HO-1 has been shown to play a vital role in the pathogenesis of a number of viral infections (Yamada et al., 1997; McAllister et al., 2004), the current state of knowledge of its roles in bacterial pathogenesis is very limited. Recently, a strong association between *Helicobacter pylori* infection and HO-1 expression in carotid atherosclerotic plaques has been suggested (Ameriso et al., 2005). In addition, HO-1 induction by *Listeria monocytogenes* (Azri and Renton et al., 1991) and *Pseudomonas aeruginosa* (Zhou et al., 2004) and increased CO concentration in the exhaled air and carboxyhemoglobin levels in the blood of patients with spontaneous bacterial peritonitis have been documented (De las Heras et al., 2003). Induction of endothelial HO-1 at the cellular level by *R. rickettsii* represents the first demonstration of involvement of this versatile enzyme in response to infection with an obligate intracellular bacterium (Rydkina et al., 2002). In view of the significant role of oxidants in infection-induced cell injury and intracellular killing of rickettsiae (Feng and Walker, 2000; Walker et al., 2001b), further knowledge of the mechanisms responsible for regulation of this important antioxidant enzyme system and the physiological effects of resultant biological mediators is necessary to define the strategies these bacteria have developed in occupying their intracellular niche. It is quite likely that induction of HO-1 represents an important adaptive mechanism for mitigating the severity of cell damage and modulation of inflammatory response to infection. In this context, it is also critically important to explore the possibility of potential differences in HO-1 induction response following infection with other SFGR or TGR and to define the functional roles of biologically active products of HO-1 activity in host cell defense and survival.

The heme-HO system also functions as a cellular regulator of vascular cyclooxygenase (COX) and in so doing influences the generation of prostaglandins. Of these, prostaglandin E_2 (PGE_2) and PGI_2 are known to cause increased vascular permeability and edema, considered to be critical modulators of vascular homeostasis in both normal and pathological conditions (Mark et al., 2001). COX, the initial bifunctional enzyme in the conversion of arachidonic acid to prostaglandins, exists in two distinct isoforms: COX-1 is primarily responsible for homeostatic functions; and COX-2 is inducible by several stimuli, including growth factors, inflammatory mediators, and hormones (Smith et al., 2000; Davidge, 2001). In endothelium, HO-1 abrogates TNF-α-induced inflammation injury and angiotensin II-mediated synthesis of prostaglandins via specific inhibition of COX-2 (Kushida et al., 2002; Volti et al., 2003). Studies investigating in vitro effects of TGR suggest increased production of arachidonate-derived autacoids PGE_2 and PGI_2 by mouse leukocytes and macrophages and human endothelial cells, as well as enhanced secretion of PGE_2 by monocyte-derived macrophages infected with SFGR (Walker et al., 1990, 1991; Walker and Hoover, 1991). Our published studies using HUVECs clearly suggest increased secretion of PGE_2 and PGI_2 in response to infection with *R. rickettsii* and *R. conorii* and robust induction of endothelial COX-2 by SFGR pathogens (*R. rickettsii* and *R. conorii*). Time course studies further indicate that COX-2 induction during *R. rickettsii* and *R. conorii* infection follows a biphasic pattern with kinetics similar to that for the activation of the nuclear transcription factor NF-κB (Rydkina et al., 2006, 2009). The induction of the COX-2 system and increased release of prostaglandins in the vasculature likely contribute to increased vascular permeability observed during *Rickettsia* infection. This is particularly important in the light of experimental data suggesting that (i) hemostatic and fibrinolytic changes in patients with rickettsemia are not initial pathogenic features, and coagulopathies and thrombotic events likely represent potential complications of severe RMSF or MSF (Schmaier et al., 2001); and (ii) increased vascular permeability resulting in fluid imbalance and edema of vital organ systems is a major cardinal feature of acute inflammation (Sahni, 2007; Walker and Ismail, 2008; Valbuena and Walker, 2009). It is important to consider here that other mechanisms underlying compromised permeability of the vasculature are also beginning to be elucidated. *R. conorii* infection of human endothelial cells induces changes in the localization and staining patterns of adherens junction proteins; discontinuous adherens junctions lead to the formation of interendothelial gaps. Interestingly, such changes occur late during in vitro infection even if cells are heavily infected at earlier times and are apparently unrelated to the manipulation of host cell actin cytoskeleton by rickettsiae (Valbuena and Walker, 2005b). As suggested by a noticeable dose-dependent decrease in transendothelial electrical resistance, *R. rickettsii* infection of human cerebral endothelial cells also results in increased vascular permeability associated with the disruption of adherens junctions. Interestingly, the presence of proinflammatory cytokines IL-1β, IFN-γ, and TNF-α, but not nitric oxide, during the infection exacerbates the rickettsial effects on endothelial permeability, further suggesting that the changes in barrier properties of vascular endothelium are likely due to a combinatorial effect of the presence of intracellular rickettsiae and immune responses of the host cell (Woods and Olano, 2008).

PROTECTIVE ROLES OF IFNs AGAINST *RICKETTSIA* INFECTION

First recognized for their ability to impede viral replication, IFNs play a critical role in determining host survival in response to viral infection. During bacterial infections, IFN signaling defends the host by integrating early innate immune responses with later events governed by adaptive immunity. Type I IFNs in humans and mice include a number of IFN-α subtypes and a single species of IFN-β. IFN-γ, a type II IFN produced by activated NK and Th1 cells, not only induces antiviral function, but also activates macrophages and dendritic cells in the presence of IL-12 and IL-18 to strengthen innate immune responses to microorganisms. Initial studies on *R. prowazekii* interactions with fibroblastlike L929 cells in vitro demonstrated the production of IFN-α and -β, but not IFN-γ. Interestingly, secreted IFN-α as well as IFN-β was found to have inhibitory activity against replication of vesicular stomatitis virus and growth of *R. prowazekii* (Turco and Winkler, 1990a, 1990b). However, IFN-γ also suppresses the growth of *R. prowazekii* in mouse L929 cells, macrophagelike RAW264.7 cells, and human fibroblasts (Turco and Winkler, 1984, 1986, 1989). These observations subsequently led to the isolation and characterization of IFN-sensitive and IFN-resistant strains of *R. prowazekii* (Turco and Winkler, 1991). In addition, IFN-γ potentiates the inhibitory effect of recombinant human TNF-α on *R. conorii* growth in HEp-2 cells (Manor and Sarov, 1990). In vivo studies in animal models have also suggested an important protective role for IFN-γ in host defense during infection with *R. conorii*, *R. australis*, and *R. typhi* (Li et al., 1987; Feng et al., 1994; Walker et al., 2000, 2001b).

Clinically, plasma levels of IFN-γ in serologically confirmed cases of MSF are significantly higher during the acute phase in comparison with the convalescent phase of the disease (Vitale et al., 2001). Although a similar pattern of increased serum IFN-γ is also seen during the acute phase of fulminant Japanese spotted fever caused by *Rickettsia japonica* (Mahara, 1997; Iwasaki et al., 2001), its level did not exhibit any changes during the course of African tick bite fever caused by *R. africae*, a newly identified SFGR species (Jensenius et al., 2003b). Recently, examination of mediators of inflammation or the immune response at the site of infection shows elevated intralesional expression of IFN-γ mRNA, particularly in patients with mild to moderate MSF. Further, there is a positive correlation between high levels of IFN-γ mRNA, the absence of rickettsiae in the blood, and the production of enzymes, such as inducible nitric oxide synthase and indoleamine-2,3-dioxygenase, known to be involved in limiting the growth of rickettsiae (de Sousa et al., 2007). Taken together, available evidence suggests a protective role for IFN-γ in rickettsial diseases in humans, likely via activation of intracellular bactericidal mechanisms.

Although both type I (α, β) and type II (γ) IFNs modulate the innate immune response through Janus-activated kinase (JAK)/signal transducer and activator of transcription (STAT)-mediated transcriptional gene regulation, the activation state of JAK/STAT pathways during infection with *Rickettsia* species remains to be defined. Also, several mechanisms of negative regulation of JAK/STAT pathways have been identified. The suppressor of cytokine signaling (SOCS) proteins are a family of Src homology 2 (SH2) domain-containing cytoplasmic proteins that complete a negative-feedback loop to attenuate signal transduction from cytokines acting through JAK/STAT pathways (Baetz et al., 2007). SOCS1 is a negative regulator of LPS-induced macrophage activation and other responses (Nakagawa et al., 2002; Kimura et al., 2005). In addition, gene delivery of SOCS3 protects mice from lethal endotoxemia by decreasing serum levels of TNF-α (Fang et al., 2005). These results identify SOCS1 and SOCS3 as new targets for the treatment of endotoxin-induced shock syndrome. Recent studies indicate that SOCS proteins are implicated in a variety of immune and inflammatory disorders. Infection of mice with virulent clinical isolates of *Mycobacterium tuberculosis* causes extensive induction of the negative regulators of the JAK/STAT pathway, which may lead to a reduction of Th1-type cytokines, such as IL-12 and TNF-α, suggesting a potential association between the negative regulation of the JAK/STAT pathway and its relationship to hypervirulence of *M. tuberculosis* (Manca et al., 2005). As an explanation for its role in *Salmonella* virulence, the involvement of *Salmonella* pathogenicity island 2 in the expression of SOCS3, which is involved in the inhibition of cytokine signaling via the JAK/STAT signaling pathway, has also been recently described (Uchiya and Nikai, 2005). Our preliminary findings demonstrate increased phosphorylation of STAT3 early during the infection, followed by phosphorylation of both α and β isoforms of STAT1 at later times (Sahni et al., 2009), but no changes in the phosphorylation status of STATs 2, 5, and 6 during *R. rickettsii* and *R. conorii* infection of cultured human endothelial cells. Direct activation of the JAK/STAT system in response to *Rickettsia* infection may also serve as an important mechanism for endothelial cell migration and/or proliferation after vascular injury. The innate immune response is intimately linked to the acquired immune response, and SOCS proteins are key physiological regulators of both innate and adaptive immunity. Augmenting the innate responses by regulating STATs/SOCSs not only has the potential to promote immediate effects, but can also deliver desirable effects to the most beneficial downstream adaptive responses for defense against rickettsial pathogens. Therapeutic efforts aimed at regulating the STAT and SOCS proteins may therefore represent a new intervention approach for patients with rickettsioses.

MECHANISMS FOR INTRACELLULAR KILLING OF RICKETTSIAE WITHIN HOST CELLS

The ability to effectively control rickettsiae proliferating in the cytoplasm of nonphagocytic host cells, where they are protected from host defense mechanisms such as antibodies and exposure to phagocytes, is another important question. Studies in an established "endothelial-target" mouse model of RMSF have suggested a crucial role for synergistic, paracrine effects of IFN-γ (secreted by T lymphocytes and NK cells) and TNF-α (secreted by macrophages) in stimulating infected endothelial cells and other cell types such as hepatocytes and macrophages to produce nitric oxide, which has antirickettsial activity (Feng et al., 1994). In macrophages, limited availability of tryptophan due to enhanced degradation by indoleamine-2,3-dioxygenase also mediates cytotoxic effects against rickettsiae, whereas hydrogen peroxide-dependent killing of intracellular rickettsiae also occurs in cultured human endothelial cells, macrophages, and peripheral blood monocytes primed with cytokines prior to infection (Feng and Walker, 2000). Major mechanisms employed by Fc-dependent antibody-mediated protection against infection of target host cells also include prevention of rickettsial escape from the phagosome and phagolysosomal killing involving nitric oxide, ROS, and starvation for L-tryptophan (Feng et al., 2004).

ROLE OF DENDRITIC CELLS, NK CELLS, AND PATTERN RECOGNITION RECEPTORS

Animal models of *R. australis*, *R. conorii*, and *R. typhi* infection have been exploited for systematic investigations of pathological and biochemical aspects of human rickettsial diseases because they closely mimic disseminated infection of the vasculature and other salient features of the disease in humans. Two direct models of epidemic typhus describe the infection of BALB/c mice and cynomolgus monkeys with the virulent Breinl strain of *R. prowazekii* (Gonder et al., 1980; Bechah et al., 2008a). To enhance our understanding of the relationship between the host and pathogen, use of in vivo models (infection of selective, susceptible strains of laboratory mice) represents an attractive and useful approach to compensate for the lack of well-characterized mutants of any rickettsial species, with the exception of a recently reported phospholipase D mutant of *R. prowazekii* (Driskell et al., 2009). Conducting in vivo comparative studies using specific strains with varying degrees of virulence within a particular species (e.g., virulent Breinl versus attenuated Madrid E strains of *R. prowazekii* and highly virulent Sheila Smith versus avirulent Iowa B strains of *R. rickettsii*) may yield important new insights into rickettsial pathogenesis and virulence as well as host innate immune mechanisms. Such model systems, which yield disseminated endothelial-target infection and closely parallel the pathological features of human disease, have been extensively used to investigate immune responses during spotted fever and typhus group rickettsioses. Depletion or adoptive transfer of CD4 and CD8 T-lymphocyte subsets during in vivo infection with *R. conorii* has revealed a crucial role for immune CD8 T lymphocytes in the clearance of rickettsiae from infected endothelial cells. Major histocompatibility complex (MHC) class I knockout mice are extremely susceptible (>50,000-fold) to a lethal infection with *R. australis* compared with the wild-type C57BL/6 mice. Mice lacking genes for IFN-γ and perforin are >100-fold more susceptible, further indicating that the activity of cytotoxic T lymphocytes is apparently more critical; IFN-γ- and perforin-mediated activities contribute only partly to the antirickettsial effect of cytotoxic T lymphocytes (Valbuena et al., 2002). Expression of CXCL9 (MIG) and CXCL10 (IP-10), known CXCR3 ligands and chemokines targeting $CD8^+$ T cells and $CD4^+$ T-helper cells, is increased in tissues of experimentally infected mice and in biopsy specimens of patients with RMSF (Valbuena et al., 2003). Yet neutralization of CXCL9/10 or CXCR3 by antibodies has no effect on survival of infected hosts or on rickettsial titers in target tissues (Valbuena and Walker, 2004). Available evidence from established animal models further suggests the occurrence of reciprocal immunological cross-protection between SFGR and TGR, which is also mediated by T cells. In addition, NK cells are implicated in the early antirickettsial immune response, likely via an IFN-γ-dependent mechanism (Billings et al., 2001, Valbuena et al., 2004). Some intriguing aspects of rickettsial biology that are starting to emerge from in vitro infection of bone marrow-derived dendritic cells include localization of intracellular rickettsiae in both the cytosol and phagosomes. Differential interactions with dendritic cells during in vivo infection with rickettsiae may stimulate a protective response in resistant C57BL/6 mice and a suppressive adaptive immune response in susceptible C3H mice. It has been proposed that the presence of rickettsiae in both vacuolar and cytosolic compartments may serve to enhance the T-cell-priming function of dendritic cells through MHC class II and class I pathway-mediated antigen presentation to $CD4^+$ and $CD8^+$ T cells, respectively. Furthermore, it has been suggested that *R. conorii* infection induces maturation of dendritic cells in vitro by robust upregulation of expression of cell surface molecules such as CD40, CD80, CD86, and MHC-II, as well as production of proinflammatory cytokines. Adoptive transfer of such stimulated dendritic cells augments immune response through increased $CD4^+$-, $CD8^+$-, and NK-cell activities and confers protection against lethal in vivo challenge by inhibiting proliferation and dissemination of rickettsiae into target tissues (Fang et al., 2007; Jordan et al., 2007).

Toll-like receptors (TLRs) are pattern recognition receptors that play a vital role in the induction of antimicrobial genes in some instances and control of adaptive immune responses in others. Indeed, it is now well established that TLRs interact with different combinations of adaptor proteins to activate MAPK and NF-κB pathways in a cell-type-specific and context-dependent manner to drive specific immune responses (Fan and Cook, 2004; Kawai and Akira, 2006; Sumbayev and Yasinka, 2006). The contributions of different TLRs in bacterial infections studied so far, however, depend on the site of infection and the characteristics of invading pathogens. In general, TLR4 recognizes LPS, whereas TLR2 interacts with many different microbial products, including peptidoglycan and cell wall lipoproteins (Kawai and Akira, 2006). *Anaplasma phagocytophilum*, an obligate intracellular bacterium closely related to rickettsiae, activates NF-κB in macrophages via TLR2 (Choi et al., 2004), but in vivo infection is controlled by T- and/or B-cell immunity without the requirement for TLRs and adaptor protein MyD88 (von Loewenich et al., 2004). Recent clinical evidence indicates a potential role for TLR4 receptor polymorphisms in determining disease severity during human infections with *R. conorii* (Balistreri et al., 2005). In vitro evidence further documents selective involvement of TLR2 and TLR4 in the activation of platelets and endothelial cells by *R. africae*, the etiologic agent of African tick bite fever (Damas et al., 2006). Recent studies have further implicated TLR4-specific immune responses in rickettsial clearance from the infected host (Jordan et al., 2007, 2008, 2009).

Another important microbial detection system that plays a critical role in innate immune responses is the nucleotide oligomerization domain (NOD)-like receptors. NOD1 and NOD2 are predominantly located in the cytoplasm and also act to govern signaling mechanisms regulating NF-κB and MAPK pathways. NOD1 and NOD2 recognize peptidoglycan fragments containing *meso*-diaminopimelic acid and muramyl dipeptide, respectively (Carneiro et al., 2008). Although rickettsiae occupy an intracytoplasmic niche for their growth and replication within the host cell, making their recognition and potential interactions with cytosolic NOD proteins highly likely, nothing is known about NOD signaling during rickettsial infection of endothelial cells. Because rickettsiae contain genes for LPS and peptidoglycan biosynthesis and actually possess both in their membrane structures (Pang and Winkler, 1994; Hackstadt, 1996), and vascular endothelial cells constitutively express TLR4 and NOD1, we hypothesize that TLRs and NODs play an important role in determining the extent of host cell transcriptional activation. Accordingly, targeted comprehensive comparative analyses of the roles of TLR2, TLR4, and NOD proteins in the interplay between host endothelium and pathogenic rickettsiae of varying virulence should be conducted in future studies.

CONCLUSIONS AND FUTURE PERSPECTIVES

To identify novel chemotherapeutic targets, it is critical to define cell signaling mechanisms such as NF-κB, MAPK, JAK/STAT, and other as yet unexplored pathways. As players in the innate immune response, these pathways function as regulatory master switches to initiate, sustain, and control host cell responses to infection and likely act in a concerted manner. It is equally important to consider the potential roles of antioxidant defense mechanisms such as the inducible HO system in rickettsial pathogenesis since products of its activity have diverse physiological functions, including defense against oxidative stress and regulation of cell survival. HO-1 can also exert anti-inflammatory effects by altering expression of COX-2 and adhesion molecules (Haider et al., 2002; Soares et al., 2004) and inhibiting leukocyte infiltration (Bussolati et al., 2004). Detailed studies to address significant gaps in the knowledge of rickettsial pathogenesis by defining specialized role(s) of endothelial cells in inflammatory response, host defense, and control of vascular permeability should not only yield important insights into novel, supplemental chemotherapies, but will also enhance our understanding of the biology of vascular endothelium in general. Continuation of this line of research using infection of endothelial cells from small and medium-sized vessels (the preferred target during human infections) and those from other types of vascular beds to account for endothelial cell heterogeneity and disseminated in vivo endothelial infection of a mammalian host will provide a comprehensive understanding of the signaling pathways governing apoptotic, inflammatory, and antioxidant responses in the host cell. This information will aid in the identification of regulatory molecules with vasoprotective properties such as biological antioxidants, reactive iron chelators, and selective modulators of specific cell signaling pathways for the development of suitable supplemental interventions, and should be particularly important considering the following epidemiologic, clinical, and pathophysiological aspects of human rickettsial infections: (i) continued prevalence of SFGR and distribution of TGR agents for epidemic/recrudescent typhus and murine/endemic typhus worldwide (Raoult et al., 2004; Nicholson et al., 2010); (ii) identification of emerging strains of pathogenic rickettsiae (Raoult et al., 2004; Raoult and Paddock, 2005; Nicholson et al., 2010) and increased incidence of rickettsial diseases in international travelers (Jensenius et al., 2004); (iii) difficulties in early, accurate diagnosis and potentially fatal outcome in cases of severe infection or delayed antibiotic therapy (Raoult et al., 2004; Dumler and Walker, 2005); and (iv) potential for malicious use of virulent *Rickettsia* species as bioweapons and classification of *R. prowazekii* as a category B and other *Rickettsia* species as category C bioterror agents (Walker, 2003, 2009).

The TLR and NOD families represent important pattern recognition receptors through which the innate immune system recognizes invasive microbes and recruits phagocytes to infected tissue for subsequent microbial elimination. Since SFGR and TGR exhibit unique tropism for vascular endothelium, but distinct differences in their actin-based motility and intracellular accumulation patterns, interactions between invading rickettsiae and endothelial cells and elucidation of mechanisms underlying host cell transcriptional

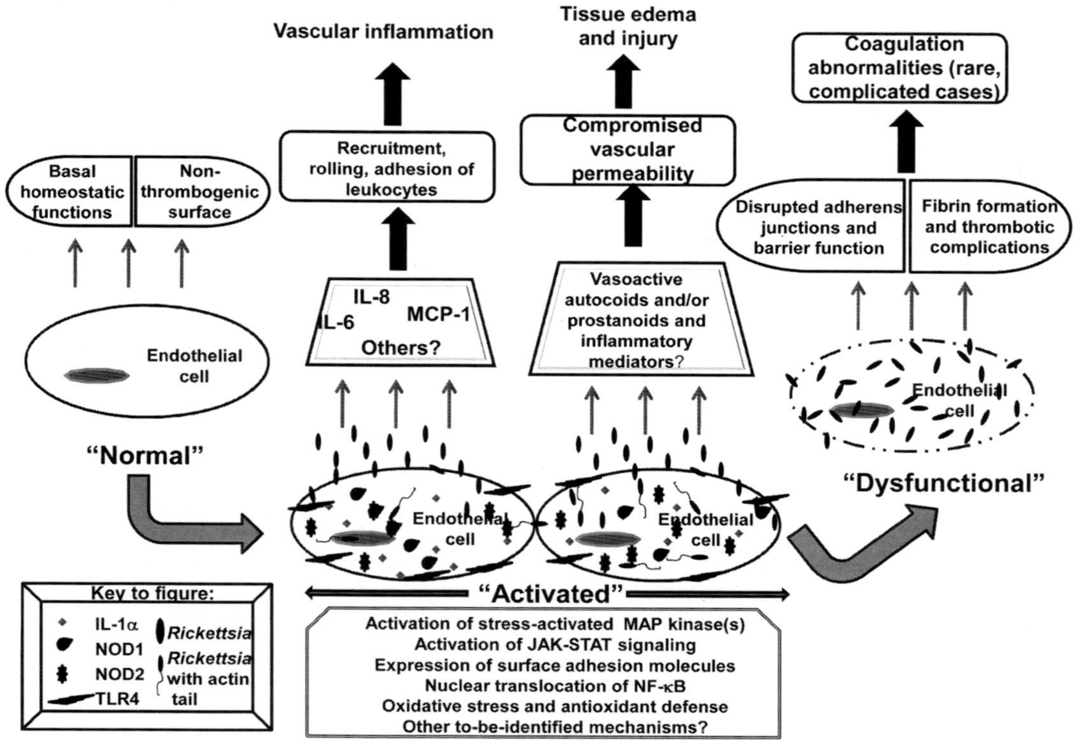

FIGURE 2 Summary of the current state of knowledge and potential future directions for investigations focused on interactions between pathogenic *Rickettsia* species and their preferred host niche, the vascular endothelium. doi:10.1128/9781555817336.ch8.f2

activation represent important avenues for expanding our knowledge of both rickettsial pathogenesis and innate immune protection mechanisms (Fig. 2). Therefore, it is critically important to decipher the potential roles of upstream signaling via TLRs and NOD proteins and downstream effects of transcriptional regulators in determining the host cell's inflammatory and apoptotic responses to infection with virulent strains of *R. rickettsii* and *R. prowazekii* and strains of varying virulence for both species. Completion of these studies should unravel critical insights into intracellular mechanisms for pattern recognition receptor-mediated activation of target host cells, thereby defining novel postulates for immunopathologic mechanisms and intervention strategies for infection-induced vascular inflammation.

ACKNOWLEDGMENTS

We thankfully acknowledge the financial support provided by the National Institute of Allergy and Infectious Diseases and National Heart, Lung, and Blood Institute of the National Institutes of Health and the American Heart Association for completed as well as ongoing research projects in their research laboratories.

REFERENCES

Ameriso, S. F., A. R. Villamil, C. Zedda, J. C. Parodi, S. Garrido, M. I. Sarchi, M. Schultz, J. Boczkowski, and G. E. Sevlever. 2005. Heme oxygenase-1 is expressed in carotid atherosclerotic plaques infected by *Helicobacter pylori* and is more prevalent in asymptomatic subjects. *Stroke* **36:**1896–1900.

Andersson, S. G., A. Zomorodipour, J. O. Andersson, T. Sicheritz-Pontén, U. C. Alsmark, R. M. Podowski, A. K. Näslund, A. S. Eriksson, H. H. Winkler, and C. G. Kurland. 1998.

The genome sequence of *Rickettsia prowazekii* and the origin of mitochondria. *Nature* **396:**133–140.

Apperson, C. S., B. Engber, W. L. Nicholson, D. G. Mead, J. Engel, M. J. Yabsley, K. Dail, J. Johnson, and D. W. Watson. 2008. Tick-borne diseases in North Carolina: is "*Rickettsia amblyommii*" a possible cause of rickettsiosis reported as Rocky Mountain spotted fever? *Vector Borne Zoonotic Dis.* **8:**597–606.

Azri, S., and K. W. Renton. 1991. Factors involved in the depression of hepatic mixed function oxidase during infections with *Listeria monocytogenes*. *Int. J. Immunopharmacol.* **13:**197–204.

Baetz, A., S. Zimmermann, and A. H. Dalpke. 2007. Microbial immune evasion employing suppressor of cytokine signaling (SOCS) proteins. *Inflamm. Allergy Drug Targets* **6:**160–167.

Balistreri, C. R., G. Candore, D. Lio, G. Colonna-Romano, G. Di Lorenzo, P. Mansueto, G. Rini, S. Mansueto, E. Cillari, C. Franceschi, and C. Caruso. 2005. Role of TLR4 receptor polymorphisms in Boutonneuse fever. *Int. J. Immunopathol. Pharmacol.* **18:**655–660.

Balraj, P., C. Nappez, D. Raoult, and P. Renesto. 2008. Western-blot detection of RickA within spotted fever group rickettsiae using a specific monoclonal antibody. *FEMS Microbiol. Lett.* **286:**257–262.

Barañano, D. E., M. Rao, C. D. Ferris, and S. H. Snyder. 2002. Biliverdin reductase: a major physiologic cytoprotectant. *Proc. Natl. Acad. Sci. USA* **99:**16093–16098.

Bechah, Y., C. Capo, G. Grau, D. Raoult, and J. L. Mege. 2009. *Rickettsia prowazekii* infection of endothelial cells increases leukocyte adhesion through αvβ3 integrin engagement. *Clin. Microbiol. Infect.* **15**(Suppl. 2)**:**249–250.

Bechah, Y., C. Capo, J. L. Mege, and D. Raoult. 2008a. Rickettsial diseases: from *Rickettsia*-arthropod relationships to pathophysiology and animal models. *Future Microbiol.* **3:**223–236.

Bechah, Y., C. Capo, D. Raoult, and J. L. Mege. 2008b. Infection of endothelial cells with virulent *Rickettsia prowazekii* increases the transmigration of leukocytes. *J. Infect. Dis.* **197:**142–147.

Bechelli, J. R., E. Rydkina, P. M. Colonne, and S. K. Sahni. 2009. *Rickettsia rickettsii* infection protects human microvascular endothelial cells against staurosporine-induced apoptosis by a cIAP$_2$-independent mechanism. *J. Infect. Dis.* **199:**1389–1398.

Billings, A. N., H. M. Feng, J. P. Olano, and D. H. Walker. 2001. Rickettsial infection in murine models activates an early anti-rickettsial effect mediated by NK cells and associated with production of gamma interferon. *Am. J. Trop. Med. Hyg.* **65:**52–56.

Bogatcheva, N. V., S. M. Dudek, J. G. Garcia, and A. D. Verin. 2003. Mitogen-activated protein kinases in endothelial pathophysiology. *J. Investig. Med.* **51:**341–352.

Brouard, S., L. E. Otterbein, J. Anrather, E. Tobiasch, F. H. Bach, A. M. K. Choi, and M. P. Soares. 2000. Carbon monoxide generated by heme oxygenase 1 suppresses endothelial cell apoptosis. *J. Exp. Med.* **192:**1015–1025.

Bussolati, B., A. Ahmed, H. Pemberton, R. C. Landis, F. Di Carlo, D. O. Haskard, and J. C. Mason. 2004. Bifunctional role for VEGF-induced heme oxygenase-1 in vivo: induction of angiogenesis and inhibition of leukocytic infiltration. *Blood* **103:**761–766.

Carneiro, L. A., J. G. Magalhaes, I. Tattoli, D. J. Philpott, and L. H. Travassos. 2008. Nod-like proteins in inflammation and disease. *J. Pathol.* **214:**136–148.

Chen, Z., J. Hagler, V. J. Palombella, F. Melandri, D. Scherer, D. Ballard, and T. Maniatis. 1995. Signal-induced site-specific phosphorylation targets IκBα to the ubiquitin-proteasome pathway. *Genes Dev.* **9:**1586–1597.

Choi, K. S., D. G. Scorpio, and J. S. Dumler. 2004. *Anaplasma phagocytophilum* ligation to Toll-like receptor (TLR) 2, but not to TLR4, activates macrophages for nuclear factor-κB nuclear translocation. *J. Infect. Dis.* **189:**1921–1925.

Cillari, E., S. Milano, P. D'Agostino, F. Arcoleo, G. Stassi, A. Galluzzo, P. Richiusa, C. Giordano, P. Quartararo, P. Colletti, G. Gambino, C. Mocciaro, A. Spinelli, G. Vitale, and S. Mansueto. 1996. Depression of CD4 T cell subsets and alteration in cytokine profile in boutonneuse fever. *J. Infect. Dis.* **174:**1051–1057.

Clifton, D. R., R. A. Goss, S. K. Sahni, D. van Antwerp, R. B. Baggs, V. J. Marder, D. J. Silverman, and L. A. Sporn. 1998. NF-κB-dependent inhibition of apoptosis is essential for host cell survival during *Rickettsia rickettsii* infection. *Proc. Natl. Acad. Sci. USA* **95:**4646–4651.

Clifton, D. R., E. Rydkina, R. S. Freeman, and S. K. Sahni. 2005a. NF-κB activation during *Rickettsia rickettsii* infection of endothelial cells involves the activation of catalytic IκB kinases IKKα and IKKβ and phosphorylation-proteolysis of the inhibitor protein IκBα. *Infect. Immun.* **73:** 155–165.

Clifton, D. R., E. Rydkina, H. Huyck, G. Pryhuber, R. S. Freeman, D. J. Silverman, and S. K. Sahni. 2005b. Expression and secretion of chemotactic cytokines IL-8 and MCP-1 by human endothelial cells after *Rickettsia rickettsii* infection: regulation by nuclear transcription factor NF-κB. *Int. J. Med. Microbiol.* **295:**267–278.

Damas, J. K., M. Jensenius, T. Ueland, K. Otterdal, A. Yndestad, S. S. Froland, J. M. Rolain, B. Myrvang, D. Raoult, and P. Aukrust. 2006. Increased levels of soluble CD40L in African tick bite fever: possible involvement of TLRs in the pathogenic interaction between *Rickettsia africae*, endothelial cells, and platelets. *J. Immunol.* **177:**2699–2706.

Dantas-Torres, F. 2007. Rocky Mountain spotted fever. *Lancet Infect. Dis.* **7:**724–732.

Davidge, S. T. 2001. Prostaglandin H synthase and vascular function. *Circ. Res.* **89:**650–660.

De las Heras, D., J. Fernández, P. Ginès, A. Cárdenas, R. Ortega, M. Navasa, J. A. Barberá, B. Calahorra, M. Guevara, R. Bataller, W. Jiménez, V. Arroyo, and J. Rodés. 2003. Increased carbon monoxide production in patients with cirrhosis with and without spontaneous bacterial peritonitis. *Hepatology* **38:**452–459.

Demma, L. J., M. Traeger, D. Blau, R. Gordon, B. Johnson, J. Dickson, R. Ethelbah, S. Piontkowski, C. Levy, W. L. Nicholson, C. Duncan, K. Heath, J. Cheek, D. L. Swerdlow, and J. H. McQuiston. 2006. Serologic evidence for exposure to *Rickettsia rickettsii* in eastern Arizona and recent emergence of Rocky Mountain spotted fever in this region. *Vector Borne Zoonotic Dis.* **6:**423–429. (Erratum, **7:**106, 2007.)

Demma, L. J., M. S. Traeger, W. L. Nicholson, C. D. Paddock, D. M. Blau, M. E. Eremeeva, G. A. Dasch, M. L. Levin, J. Singleton, Jr., S. R. Zaki, J. E. Cheek, D. L. Swerdlow, and J. H. McQuiston. 2005. Rocky Mountain spotted fever from an unexpected tick vector in Arizona. *N. Engl. J. Med.* **353:**587–594.

Denk, A., M. Goebeler, S. Schmid, I. Berberich, O. Ritz, D. Lindemann, S. Ludwig, and T. Wirth. 2001. Activation of NF-κB via the IκB kinase complex is both essential and sufficient for proinflammatory gene expression in primary endothelial cells. *J. Biol. Chem.* **276:**28451–28458.

de Sousa, R., A. França, S. D. Nobrega, A. Belo, M. Amaro, T. Abreu, J. Poças, P. Proença, J. Vaz, J. Torgal, F. Bacellar, N. Ismail, and D. H. Walker. 2008. Host- and microbe-related risk factors for and pathophysiology of fatal *Rickettsia conorii* infection in Portuguese patients. *J. Infect. Dis.* **198:**576–585.

de Sousa, R., N. Ismail, S. D. Nobrega, A. França, M. Amaro, M. Anes, J. Poças, R. Coelho, J. Torgal, F. Bacellar, and D. H. Walker. 2007. Intralesional expression of mRNA of interferon-γ, tumor necrosis factor-α, interleukin-10, nitric oxide synthase, indoleamine-2,3-dioxygenase, and RANTES is a major immune effector in Mediterranean spotted fever rickettsiosis. *J. Infect. Dis.* **196:**770–781.

DiDonato, J. A., M. Hayakawa, D. M. Rothwarf, E. Zandi, and M. Karin. 1997. A cytokine-responsive IκB kinase that activates the transcription factor NF-κB. *Nature* **388:**548–554.

Dignat-George, F., N. Teysseire, M. Mutin, N. Bardin, G. Lesaule, D. Raoult, and J. Sampol. 1997. *Rickettsia conorii* infection enhances vascular cell adhesion molecule-1- and intercellular adhesion molecule-1-dependent mononuclear cell adherence to endothelial cells. *J. Infect. Dis.* **175:**1142–1152.

Drancourt, M., M.-C. Allessi, P.-Y. Levy, I. Juhan-Vague, and D. Raoult. 1990a. Secretion of tissue-type plasminogen activator and plasminogen activator inhibitor by *Rickettsia conroii* and *Rickettsia rickettsii*-infected cultured endothelial cells. *Infect. Immun.* **58:**2459–2463.

Drancourt, M., D. Raoult, J. R. Harle, H. Chaudet, F. Janbon, C. Charrel, and H. Gallais. 1990b. Biological variations in 412 patients with Mediterranean spotted fever. *Ann. N. Y. Acad. Sci.* **590:**39–50.

Driskell, L. O., X.-J. Yu, L. Zhang, Y. Liu, V. L. Popov, D. H. Walker, A. M. Tucker, and D. O. Wood. 2009. Directed mutagenesis of the *Rickettsia prowazekii pld* gene encoding phospholipase D. *Infect. Immun.* **77:**3244–3248.

Dumler, J. S., and D. H. Walker. 2005. Rocky Mountain spotted fever—changing ecology and persisting virulence. *N. Engl. J. Med.* **353:**551–553.

Dyer, L. A., and C. Patterson. 2010. Development of the endothelium: an emphasis on heterogeneity. *Semin. Thromb. Hemost.* 36:227–235.

Eremeeva, M. E., G. A. Dasch, and D. J. Silverman. 2000. Interaction of rickettsiae with eukaryotic cells. Adhesion, entry, intracellular growth, and host cell responses, p. 479–516. *In* T. A. Oelschlaeger and J. Hacker (ed.), *Bacterial Invasion into Eukaryotic Cells.* Kluwer Academic/Plenum Publishers, New York, NY.

Eremeeva, M. E., G. A. Dasch, and D. J. Silverman. 2001. Quantitative analyses of variations in the injury of endothelial cells elicited by 11 isolates of *Rickettsia rickettsii. Clin. Diagn. Lab. Immunol.* **8:**788–796.

Eremeeva, M. E., L. A. Santucci, V. L. Popov, D. H. Walker, and D. J. Silverman. 1999. *Rickettsia rickettsii* infection of human endothelial cells: oxidative injury and reorganization of the cytoskeleton, p. 128–144. *In* D. Raoult and P. Brouqui (ed.), *Rickettsiae and Rickettsial Diseases at the Turn of the Third Millennium.* Elsevier Press, Paris, France.

Eremeeva, M. E., and D. J. Silverman. 1998. *Rickettsia rickettsii* infection of the EA.hy 926 endothelial cell line: morphological response to infection and evidence for oxidative injury. *Microbiology* **144:**2037–2048.

Esposti, M. D. 2002. The roles of Bid. *Apoptosis* **7:**433–440.

Fan, H., and J. A. Cook. 2004. Molecular mechanisms of endotoxin tolerance. *J. Endotoxin Res.* **10:**71–84.

Fang, M., H. Dai, G. Yu, and F. Gong. 2005. Gene delivery of SOCS3 protects mice from lethal endotoxic shock. *Cell. Mol. Immunol.* **2:**373–377.

Fang, R., N. Ismail, L. Soong, V. L. Popov, T. Whitworth, D. H. Bouyer, and D. H. Walker. 2007. Differential interaction of dendritic cells with *Rickettsia conorii*: impact on host susceptibility to murine spotted fever rickettsiosis. *Infect. Immun.* **75:**3112–3123.

Feng, H. M., V. L. Popov, and D. H. Walker. 1994. Depletion of gamma interferon and tumor necrosis factor alpha in mice with *Rickettsia conorii*-infected endothelium: impairment of rickettsicidal nitric oxide production resulting in fatal, overwhelming rickettsial disease. *Infect. Immun.* **62:**1952–1960.

Feng, H. M., and D. H. Walker. 2000. Mechanisms of intracellular killing of *Rickettsia conorii* in infected human endothelial cells, hepatocytes, and macrophages. *Infect. Immun.* **68:**6729–6736.

Feng, H. M., T. Whitworth, V. Popov, and D. H. Walker. 2004. Effect of antibody on the rickettsia-host cell interaction. *Infect. Immun.* **72:** 3524–3530.

Galley, H. F., and N. R. Webster. 2004. Physiology of the endothelium. *Br. J. Anaesth.* 93:105–113.

Gillespie, J. J., K. Williams, M. Shukla, E. E. Snyder, E. K. Nordberg, S. M. Ceraul, C. Dharmanolla, D. Rainey, J. Soneja, J. M. Shallom, N. D. Vishnubhat, R. Wattam, A. Purkayastha, M. Czar, O. Crasta, J. C. Setubal, A. F. Azad, and B. S. Sobral. 2008. *Rickettsia* phylogenomics: unwinding the intricacies of obligate intracellular life. *PLoS One* **3:**e2018.

Gonder, J. C., R. H. Kenyon, and C. E. Pedersen, Jr. 1980. Epidemic typhus infection in cynomolgus monkeys (*Macaca fascicularis*). *Infect. Immun.* **30:**219–223.

Gouin, E., C. Egile, P. Dehoux, V. Villiers, J. Adams, F. Gertler, R. Li, and P. Cossart. 2004. The RickA protein of *Rickettsia conorii* activates the Arp2/3 complex. *Nature* **427:**457–461.

Gouin, E., H. Gantelet, C. Egile, I. Lasa, H. Ohayon, V. Villiers, P. Gounon, P. J. Sansonetti, and P. Cossart. 1999. A comparative study of the actin-based motilities of the pathogenic bacteria *Listeria monocytogenes*, *Shigella flexneri* and *Rickettsia conorii*. *J. Cell Sci.* **112:**1697–1708.

Gouin, E., M. D. Welch, and P. Cossart. 2005. Actin-based motility of intracellular pathogens. *Curr. Opin. Microbiol.* **8:**35–45.

Hackstadt, T. 1996. The biology of rickettsiae. *Infect. Agents Dis.* **5:**127–143.

Haglund, C. M., J. E. Choe, C. T. Skau, D. R. Kovar, and M. D. Welch. 2010. *Rickettsia* Sca2 is a bacterial formin-like mediator of actin-based motility. *Nat. Cell Biol.* **12:**1057–1063.

Haider, A., R. Olszanecki, R. Gryglewski, M. L. Schwartzman, E. Lianos, A. Kappas, A. Nasjletti, and N. G. Abraham. 2002. Regulation of cyclooxygenase by the heme-heme oxygenase system in microvessel endothelial cells. *J. Pharmacol. Exp. Ther.* **300:**188–194.

Harlander, R. S., M. Way, Q. Ren, D. Howe, S. S. Grieshaber, and R. A. Heinzen. 2003. Effects of ectopically expressed neuronal Wiskott-Aldrich syndrome protein domains on *Rickettsia rickettsii* actin-based motility. *Infect. Immun.* **71:**1551–1556.

Hayden, M. S., and S. Ghosh. 2008. Shared principles in NF-κB signaling. *Cell* **132:**344–362.

Heinzen, R. A., S. F. Hayes, M. G. Peacock, and T. Hackstadt. 1993. Directional actin polymerization associated with spotted fever group *Rickettsia* infection of Vero cells. *Infect. Immun.* **61:**1926–1935.

Iwasaki, H., F. Mahara, N. Takada, H. Fujita, and T. Ueda. 2001. Fulminant Japanese spotted fever associated with hypercytokinemia. *J. Clin. Microbiol.* **39:**2341–2343.

Jeng, R. L., E. D. Goley, J. A. D'Alessio, O. Y. Chaga, T. M. Svitkina, G. G. Borisy, R. A. Heinzen, and M. D. Welch. 2004. A *Rickettsia* WASP-like protein activates the Arp2/3 complex and mediates actin-based motility. *Cell. Microbiol.* **6:**761–769.

Jensenius, M., P. E. Fournier, P. Kelly, B. Myrvang, and D. Raoult. 2003a. African tick bite fever. *Lancet Infect. Dis.* **3:**557–564.

Jensenius, M., P. E. Fournier, and D. Raoult. 2004. Rickettsioses and the international traveler. *Clin. Infect. Dis.* **39:**1493–1499.

Jensenius, M., T. Ueland, P. E. Fournier, F. Brosstad, E. Stylianou, S. Vene, B. Myrvang, D. Raoult, and P. Aukrust. 2003b. Systemic inflammatory responses in African tick-bite fever. *J. Infect. Dis.* **187:**1332–1336.

Jordan, J. M., M. E. Woods, H. M. Feng, L. Soong, and D. H. Walker. 2007. Rickettsiae-stimulated dendritic cells mediate protection against lethal rickettsial challenge in an animal model of spotted fever rickettsiosis. *J. Infect. Dis.* **196:**629–638.

Jordan, J. M., M. E. Woods, J. Olano, and D. H. Walker. 2008. Absence of TLR4 signaling in C3H/HeJ mice predisposes to overwhelming rickettsial infection and decreased protective Th1 responses. *Infect. Immun.* **76:**3717–3724.

Jordan, J. M., M. E. Woods, L. Soong, and D. H. Walker. 2009. Rickettsiae stimulate dendritic cells through Toll-like receptor 4, leading to enhanced NK cell activation in vivo. *J. Infect. Dis.* **199:**236–242.

Joshi, S. G., C. W. Francis, D. J. Silverman, and S. K. Sahni. 2003. Nuclear factor-κB protects against host cell apoptosis during *Rickettsia rickettsii* infection by inhibiting activation of apical and effector caspases and maintaining mitochondrial integrity. *Infect. Immun.* **71:**4127–4136.

Joshi, S. G., C. W. Francis, D. J. Silverman, and S. K. Sahni. 2004. NF-κB activation suppresses host cell apoptosis during *Rickettsia rickettsii* infection via regulatory effects on intracellular localization or levels of apoptogenic and anti-apoptotic proteins. *FEMS Microbiol. Lett.* **234:**333–341.

Kaplanski, G. N., N. Teysseire, C. Farnarier, S. Kaplanski, J.-C. Lissitzky, J.-M. Durand, J. Soubeyrand, C. A. Dinarello, and P. Bongrand. 1995. IL-6 and IL-8 production from cultured human endothelial cells stimulated by infection with *Rickettsia conorii* via a cell-associated IL-1α-dependent pathway. *J. Clin. Invest.* **96:** 2839–2844.

Kawai, T., and S. Akira. 2006. TLR signaling. *Cell Death Differ.* **13:**816–825.

Kawamura, K., K. Ishikawa, Y. Wada, S. Kimura, H. Matsumoto, T. Kohro, H. Itabe, T. Kodama, and Y. Maruyama. 2005. Bilirubin from heme oxygenase-1 attenuates vascular endothelial activation and dysfunction. *Arterioscler. Thromb. Vasc. Biol.* **25:**155–160.

Kimura, A., T. Naka, T. Muta, O. Takeuchi, S. Akira, I. Kawase, and T. Kishimoto. 2005. Suppressor of cytokine signaling-1 selectively inhibits LPS-induced IL-6 production by regulating JAK-STAT. *Proc. Natl. Acad. Sci. USA* **102:**17089–17094.

Kleba, B., T. R. Clark, E. I. Lutter, D. W. Ellison, and T. Hackstadt. 2010. Disruption of the *Rickettsia rickettsii* Sca2 autotransporter inhibits actin-based motility. *Infect. Immun.* **78:**2240–2247.

Kluck, R. M., E. Bossy-Wetzel, D. R. Green, and D. D. Newmeyer. 1997. The release of cytochrome *c* from mitochondria: a primary site for Bcl-2 regulation of apoptosis. *Science* **275:**1132–1136.

Knipe, L., A. Meli, L. Hewlett, R. Bierings, J. Dempster, P. Skehel, M. J. Hannah, and T. Carter. 2010. A revised model for the secretion of tPA and cytokines from cultured endothelial cells. *Blood* **116:**2183–2191.

Krishnaswamy, G., J. Kelley, L. Yerra, J. K. Smith, and D. S. Chi. 1999. Human endothelium as a source of multifunctional cytokines: molecular regulation and possible role in human disease. *J. Interferon Cytokine Res.* **19:**91–104.

Kushida, T., G. L. Volti, S. Quan, A. Goodman, and N. G. Abraham. 2002. Role of human heme oxygenase-1 in attenuating TNF-α-mediated inflammation injury in endothelial cells. *J. Cell. Biochem.* **87:**377–385.

Li, H., T. R. Jerrells, G. L. Spitalny, and D. H. Walker. 1987. Gamma interferon as a crucial host defense against *Rickettsia conorii* in vivo. *Infect. Immun.* **55:**1252–1255.

Mahara, F. 1997. Japanese spotted fever: report of 31 cases and review of the literature. *Emerg. Infect. Dis.* **3:**105–111.

Maines, M. D., and P. E. Gibbs. 2005. 30 some years of heme oxygenase: from a "molecular wrecking ball" to a "mesmerizing" trigger of cellular events. *Biochem. Biophys. Res. Commun.* **338:**568–577.

Manca, C., L. Tsenova, S. Freeman, A. K. Barczak, M. Tovey, P. J. Murray, C. Barry, and G. Kaplan. 2005. Hypervirulent *Mycobacterium tuberculosis* W/Beijing strains upregulate type I IFNs and increase expression of negative regulators of the Jak-Stat pathway. *J. Interferon Cytokine Res.* **25:**694–701.

Manor, E., and I. Sarov. 1990. Inhibition of *Rickettsia conorii* growth by recombinant tumor necrosis factor alpha: enhancement of inhibition by gamma interferon. *Infect. Immun.* **58:**1886–1890.

Mark, K. S., W. J. Trickler, and D. W. Miller. 2001. Tumor necrosis factor-α induces cyclooxygenase-2 expression and prostaglandin release in brain microvessel endothelial cells. *J. Pharmacol. Exp. Ther.* **297:**1051–1058.

Martinez, J. J., S. Seveau, E. Veiga, S. Matsuyama, and P. Cossart. 2005. Ku70, a component of DNA-dependent protein kinase, is a mammalian receptor for *Rickettsia conorii*. *Cell* **123:**1013–1023.

McAllister, S. C., S. G. Hansen, R. A. Ruhl, C. M. Raggo, V. R. DeFilippis, D. Greenspan, K.Fruh, and A. V. Moses. 2004. Kaposi sarcoma-associated herpesvirus (KSHV) induces heme oxygenase-1 expression and activity in KSHV-infected endothelial cells. *Blood* **103:**3465–3473.

McLeod, M. P., X. Qin, S. E. Karpathy, J. Gioia, S. K. Highlander, G. E. Fox, T. Z. McNeill, H. Jiang, D. Muzny, L. S. Jacob, A. C. Hawes, E. Sodergren, R. Gill, J. Hume, M. Morgan, G. Fan, A. G. Amin, R. A. Gibbs, C. Hong, X. J. Yu, D. H. Walker, and G. M. Weinstock. 2004. Complete genome sequence of *Rickettsia typhi* and comparison with sequences of other rickettsiae. *J. Bacteriol.* **186:**5842–5855.

Mercurio, F., H. Zhu, B. W. Murray, A. Shevchenko, B. L. Bennett, J. Li, D. B. Young, M. Barbosa, M. Mann, A. Manning, and A. Rao. 1997. IKK-1 and IKK-2: cytokine-activated

IκB kinases essential for NF-κB activation. *Science* **278:**860–866.

Mouffok, N., P. Parola, H. Lepidi, and D. Raoult. 2009. Mediterranean spotted fever in Algeria—new trends. *Int. J. Infect. Dis.* **13:**227–235.

Muller, C., J. Paupert, S. Monferran, and B. Salles. 2005. The double life of the Ku protein: facing the DNA breaks and the extracellular environment. *Cell Cycle* **4:**438–441.

Nakagawa, R., T. Naka, H. Tsutsui, M. Fujimoto, A. Kimura, T. Abe, E. Seki, S. Sato, O. Takeuchi, K. Takeda, S. Akira, K. Yamanishi, I. Kawase, K. Nakanishi, and T. Kishimoto. 2002. SOCS-1 participates in negative regulation of LPS responses. *Immunity* **17:** 677–687.

Nicholson, W. L., K. E. Allen, J. H. McQuiston, E. B. Breitschwerdt, and S. E. Little. 2010. The increasing recognition of rickettsial pathogens in dogs and people. *Trends Parasitol.* **26:**205–212.

Olano, J. P. 2005. Rickettsial infections. *Ann. N. Y. Acad. Sci.* **1063:**187–196.

Paddock, C. D. 2005. *Rickettsia parkeri* as a paradigm for multiple causes of tick-borne spotted fever in the Western Hemisphere. *Ann N. Y. Acad. Sci.* **1063:**315–326.

Pang, H., and H. H. Winkler. 1994. Analysis of the peptidoglycan of *Rickettsia prowazekii*. *J. Bacteriol.* **176:**923–926.

Pham, C. G., C. Bubici, F. Zazzeroni, S. Papa, J. Jones, K. Alvarez, S. Jayawardena, E. De Smaele, R. Cong, C. Beaumont, F. M. Torti, S. V. Torti, and G. Franzoso. 2004. Ferritin heavy chain upregulation by NF-κB inhibits TNFα-induced apoptosis by suppressing reactive oxygen species. *Cell* **119:**529–542.

Radulovic, S., P. W. Price, M. S. Beier, J. Gaywee, K. A. Macaluso, and A. Azad. 2002. *Rickettsia*-macrophage interactions: host cell responses to *Rickettsia akari* and *Rickettsia typhi*. *Infect. Immun.* **70:**2576–2582.

Radulovic, S., J. M. Troyer, M. S. Beier, A. O. Lau, and A. F. Azad. 1999. Identification and molecular analysis of the gene encoding *Rickettsia typhi* hemolysin. *Infect. Immun.* **67:**6104–6108.

Raoult, D., O. Dutour, L. Houhamdi, R. Jankauskas, P. E. Fournier, Y. Ardagna, M. Drancourt, M. Signoli, V. D. La, Y. Macia, and G. Aboudharam. 2006. Evidence for louse-transmitted diseases in soldiers of Napoleon's Grand Army in Vilnius. *J. Infect. Dis.* **193:**112–120.

Raoult, D., and C. D. Paddock. 2005. *Rickettsia parkeri* infection and other spotted fevers in the United States. *N. Engl. J. Med.* **353:**626–627.

Raoult, D., T. Woodward, and J. S. Dumler. 2004. The history of epidemic typhus. *Infect. Dis. Clin. North Am.* **18:**127–140.

Rincón, M., and R. J. Davis. 2009. Regulation of the immune response by stress-activated protein kinases. *Immunol. Rev.* **228:**212–224.

Rothwarf, D. M., E. Zandi, G. Natoli, and M. Karin. 1998. IKKγ is an essential regulatory subunit of the IκB kinase complex. *Nature* **395:**297–300.

Roux, P. P., and J. Blenis. 2004. ERK and p38 MAPK-activated protein kinases: a family of protein kinases with diverse biological functions. *Microbiol. Mol. Biol. Rev.* **68:**320–344.

Rovery, C., P. Brouqui, and D. Raoult. 2008. Questions on Mediterranean spotted fever a century after its discovery. *Emerg. Infect. Dis.* **14:**1360–1367.

Rydkina, E., A. Sahni, R. B. Baggs, D. J. Silverman, and S. K. Sahni. 2006. Infection of human endothelial cells with spotted fever group rickettsiae stimulates cyclooxygenase-2 expression and release of prostaglandins. *Infect. Immun.* **74:**5067–5074.

Rydkina, E., A. Sahni, D. J. Silverman, and S. K. Sahni. 2002. *Rickettsia rickettsii* infection of cultured human endothelial cells induces heme oxygenase 1 expression. *Infect. Immun.* **70:**4045–4052.

Rydkina, E., A. Sahni, D. J. Silverman, and S. K. Sahni. 2007. Comparative analysis of host cell signaling mechanisms activated in response to infection with *Rickettsia conorii* and *Rickettsia typhi*. *J. Med. Microbiol.* **56:**896–906.

Rydkina, E., S. K. Sahni, L. A. Santucci, L. C. Turpin, R. B. Baggs, and D. J. Silverman. 2004. Selective modulation of antioxidant enzyme activities in host tissues during *Rickettsia conorii* infection. *Microb. Pathog.* **36:**293–301.

Rydkina, E., D. J. Silverman, and S. K. Sahni. 2005. Activation of p38 stress-activated protein kinase during *Rickettsia rickettsii* infection of human endothelial cells: role in the induction of chemokine response. *Cell. Microbiol.* **7:**1519–1530.

Rydkina, E., L. C. Turpin, and S. K. Sahni. 2008. Activation of p38 MAP kinase module facilitates *in vitro* host cell invasion by *Rickettsia rickettsii*. *J. Med. Microbiol.* **57:**1172–1175.

Rydkina, E., L. C. Turpin, D. J. Silverman, and S. K. Sahni. 2009. *Rickettsia rickettsii* infection of human pulmonary microvascular endothelial cells: modulation of cyclooxygenase-2 expression. *Clin. Microbiol. Infect.* **15**(Suppl. 2)**:**300–302.

Sahni, S. K. 2007. Endothelial cell infection and hemostasis. *Thromb. Res.* **119:**531–549.

Sahni, S. K., S. Kiriakidi, P. M. Colonne, A. Sahni, and D. J. Silverman. 2009. Selective activation of signal transducer and activator of transcription (STAT) proteins STAT1 and STAT3 in human endothelial cells infected with *Rickettsia rickettsii*. *Clin. Microbiol. Infect.* **15**(Suppl. 2)**:**303–304.

Sahni, S. K., and E. Rydkina. 2009. Host-cell interactions with pathogenic *Rickettsia* species. *Future Microbiol.* **4:**323–339.

Sahni, S. K., E. Rydkina, S. G. Joshi, L. A. Sporn, and D. J. Silverman. 2003. Interactions of *Rickettsia rickettsii* with endothelial nuclear factor-κB in a "cell-free" system. *Ann. N. Y. Acad. Sci.* **990:**635–641.

Sahni, S. K., D. J. Van Antwerp, M. E. Eremeeva, D. J. Silverman, V. J. Marder, and L. A. Sporn. 1998. Proteasome-independent activation of nuclear factor κB in cytoplasmic extracts from human endothelial cells by *Rickettsia rickettsii*. *Infect. Immun.* **66:**1827–1833.

Schmaier, A. H., S. Srikanth, M. T. Elghetany, D. Normolle, S. Gokhale, H.-M. Feng, and D. H. Walker. 2001. Hemostatic/fibrinolytic protein changes in C3H/HeN mice infected with *Rickettsia conorii*. A model for Rocky Mountain spotted fever. *Thromb. Haemost.* **86:**871–879.

Scorrano, L., and S. J. Korsmeyer. 2003. Mechanisms of cytochrome *c* release by proapoptotic Bcl-2 family members. *Biochem. Biophys. Res. Commun.* **304:**437–444.

Serio, A. W., R. L. Jeng, C. M. Haglund, S. C. Reed, and M. D. Welch. 2010. Defining a core set of actin cytoskeletal proteins critical for actin-based motility of *Rickettsia*. *Cell Host Microbe* **7:**388–398.

Sessler, C. N., M. Schwartz, A. C. Windsor, and A. A. Fowler III. 1995. Increased serum cytokines and intercellular adhesion molecule-1 in fulminant Rocky Mountain spotted fever. *Crit. Care Med.* **23:**973–976.

Shi, R.-J., P. J. Simpson-Haidaris, V. J. Marder, D. J. Silverman, and L. A. Sporn. 2000. Post-transcriptional regulation of endothelial cell plasminogen activator inhibitor-1 expression during *Rickettsia rickettsii* infection. *Microb. Pathog.* **28:**127–133.

Silverman, D. J. 1986. Adherence of platelets to human endothelial cells infected by *Rickettsia rickettsii*. *J. Infect. Dis.* **153:**694–700.

Silverman, D. J., and S. B. Bond. 1984. Infection of human vascular endothelial cells by *Rickettsia rickettsii*. *J. Infect. Dis.* **149:**201–206.

Smith, W. L., D. L. DeWitt, and R. M. Garavito. 2000. Cyclooxygenases: structural, cellular, and molecular biology. *Annu. Rev. Biochem.* **69:**145–182.

Soares, M. P., M. P. Seldon, I. P. Gregoire, T. Vassilevskaia, P. O. Berberat, J. Yu, T.-Y. Tsui, and F. H. Bach. 2004. Heme-oxygenase-1 modulates the expression of adhesion molecules associated with endothelial cell activation. *J. Immunol.* **172:**3553–3563.

Spiecker, M., H. Darius, and J. K. Liao. 2000. A functional role of IκB-ε in endothelial cell activation. *J. Immunol.* **164:**3316–3322.

Sporn, L. A., P. J. Haidaris, R.-J. Shi, Y. Nemerson, D. J. Silverman, and V. J. Marder. 1994. *Rickettsia rickettsii* infection of cultured human endothelial cells induces tissue factor expression. *Blood* **83:**1527–1534.

Sporn, L. A., S. O. Lawrence, D. J. Silverman, and V. J. Marder. 1993. E-selectin-dependent neutrophil adhesion to *Rickettsia rickettsii*-infected endothelial cells. *Blood* **81:**2406–2412.

Sporn, L. A., and V. J. Marder. 1996. Interleukin-1α production during *Rickettsia rickettsii* infection of cultured endothelial cells: Potential role in autocrine cell stimulation. *Infect. Immun.* **64:**1609–1613.

Sporn, L. A., S. K. Sahni, N. B. Lerner, V. J. Marder, D. J. Silverman, L. C. Turpin, and A. L. Schwab. 1997. *Rickettsia rickettsii* infection of cultured human endothelial cells induces NF-κB activation. *Infect. Immun.* **65:**2786–2791.

Sporn, L. A., R.-J. Shi, S. O. Lawrence, D. J. Silverman, and V. J. Marder. 1991. *Rickettsia rickettsii* infection of cultured endothelial cells induces release of large von Willebrand factor multimers from Weibel-Palade bodies. *Blood* **78:**2595–2602.

Stevens, J. M., E. E. Galyov, and M. P. Stevens. 2006. Actin-dependent movement of bacterial pathogens. *Nat. Rev. Microbiol.* 4:91–101.

Stocker, R., and J. F. Keaney, Jr. 2004. Role of oxidative modifications in atherosclerosis. *Physiol. Rev.* **84:**1381–1478.

Sumbayev, V. V., and I. M. Yasinska. 2006. Role of MAP kinase-dependent apoptotic pathway in innate immune responses and viral infection. *Scand. J. Immunol.* **63:**391–400.

Tak, P. P., and G. S. Firestein. 2001. NF-κB: a key role in inflammatory diseases. *J. Clin. Invest.* **107:**7–11.

Tak, P. P., D. M. Gerlag, K. R. Aupperle, D. A. Van De Geest, M. Overbeek, B. L. Bennett, D. L. Boyle, A. M. Manning, and G. S. Firestein. 2001. Inhibitor of nuclear factor κB kinase β is a key regulator of synovial inflammation. *Arthritis Rheum.* **44:**1897–1907.

Teysseire, N., D. Arnoux, G. George, J. Sampol, and D. Raoult. 1992. von Willebrand factor release and thrombomodulin and tissue factor expression in *Rickettsia conorii*-infected endothelial cells. *Infect. Immun.* **60:**4388–4393.

Turco, J., and H. H. Winkler. 1984. Effect of mouse lymphokines and cloned mouse interferon-γ on the interaction of *Rickettsia prowazekii* with mouse macrophage-like RAW264.7 cells. *Infect. Immun.* **45:**303–308.

Turco, J., and H. H. Winkler. 1986. Gamma-interferon-induced inhibition of the growth of *Rickettsia prowazekii* in fibroblasts cannot be explained by the degradation of tryptophan or other amino acids. *Infect. Immun.* **53:**38–46.

Turco, J., and H. H. Winkler. 1989. Isolation of *Rickettsia prowazekii* with reduced sensitivity to gamma interferon. *Infect. Immun.* **57:**1765–1772.

Turco, J., and H. H. Winkler. 1990a. Interferon-α/β and *Rickettsia prowazekii*: induction and sensitivity. *Ann. N. Y. Acad. Sci.* **590:**168–186.

Turco, J., and H. H. Winkler. 1990b. Selection of alpha/beta interferon- and gamma interferon-resistant rickettsiae by passage of *Rickettsia prowazekii* in L929 cells. *Infect. Immun.* **58:**3279–3285.

Turco, J., and H. H. Winkler. 1991. Comparison of properties of virulent, avirulent, and interferon-resistant *Rickettsia prowazekii* strains. *Infect. Immun.* **59:**1647–1655.

Uchiya, K., and T. Nikai. 2005. *Salmonella* pathogenicity island 2-dependent expression of suppressor of cytokine signaling 3 in macrophages. *Infect. Immun.* **73:**5587–5594.

Valbuena, G., W. Bradford, and D. H. Walker. 2003. Expression analysis of the T-cell-targeting chemokines CXCL9 and CXCL10 in mice and humans with endothelial infections caused by rickettsiae of the spotted fever group. *Am. J. Pathol.* **163:**1357–1369.

Valbuena, G., H. M. Feng, and D. H. Walker. 2002. Mechanisms of immunity against rickettsiae. New perspectives and opportunities offered by unusual intracellular parasites. *Microbes Infect.* **4:**625–633.

Valbuena, G., J. M. Jordan, and D. H. Walker. 2004. T cells mediate cross-protective immunity between spotted fever group rickettsiae and typhus group rickettsiae. *J. Infect. Dis.* **190:**1221–1227.

Valbuena, G., and D. H. Walker. 2004. Effect of blocking the CXCL9/10-CXCR3 chemokine system in the outcome of endothelial-target rickettsial infections. *Am. J. Trop. Med. Hyg.* **71:**393–399.

Valbuena, G., and D. H. Walker. 2005a. Expression of CX3CL1 (fractalkine) in mice with endothelial-target rickettsial infection of the spotted-fever group. *Virchows Arch.* **446:**21–27.

Valbuena, G., and D. H. Walker. 2005b. Changes in the adherens junctions of human endothelial cells infected with spotted fever group rickettsiae. *Virchows Arch.* **446:**379–382.

Valbuena, G., and D. H. Walker. 2009. Infection of the endothelium by members of the order *Rickettsiales*. *Thromb. Haemost.* **102:**1071–1079.

Vitale, G., S. Mansueto, G. Gambino, C. Mocciaro, A. Spinelli, G. B. Rini, M. Affronti, N. Chifari, C. La Russa, S. Di Rosa, P. Colletti, C. Barbera, M. La Rosa, G. Di Bella, V. Ferlazzo, S. Milano, P. D'Agostino, and E. Cillari. 2001. The acute phase response in Sicilian patients with boutonneuse fever admitted to hospitals in Palermo, 1992-1997. *J. Infect.* **42:**33–39.

Volti, G. L., F. Seta, M. L. Schwartzman, A. Nasjletti, and N. G. Abraham. 2003. Heme oxygenase attenuates angiotensin II-mediated increase in cyclooxygenase-2 activity in human femoral endothelial cells. *Hypertension* **41:**715–719.

von Loewenich, F. D., D. G. Scorpio, U. Reischl, J. S. Dumler, and C. Bogdan. 2004. Control of *Anaplasma phagocytophilum*, an obligate intracellular pathogen, in the absence of inducible nitric oxide synthase, phagocyte NADPH oxidase, tumor necrosis factor, Toll-like receptor (TLR)2 and TLR4, or the TLR adaptor molecule MyD88. *Eur. J. Immunol.* **34:**1789–1797.

Walker, D. H. 2003. Principles of the malicious use of infectious agents to create terror: reasons for concern for organisms of the genus *Rickettsia*. *Ann. N. Y. Acad. Sci.* **990:**739–742.

Walker, D. H. 2006. Targeting *Rickettsia*. *N. Engl. J. Med.* **354:**1418–1420.

Walker, D. H. 2007. Rickettsiae and rickettsial infections: the current state of knowledge. *Clin. Infect. Dis.* **45**(Suppl. 1)**:**S39–S44.

Walker, D. H. 2009. The realities of biodefense vaccines against *Rickettsia*. *Vaccine* **27**(Suppl. 4)**:** D52–D55.

Walker, D. H., H. M. Feng, and V. L. Popov. 2001a. Rickettsial phospholipase A_2 as a pathogenic mechanism in a model of cell injury by typhus and spotted fever group rickettsiae. *Am. J. Trop. Med. Hyg.* **65:**936–942.

Walker, D. H., S. D. Hudnall, W. K. Szaniawski, and H. M. Feng. 1999. Monoclonal antibody-based immunohistochemical diagnosis of rickettsialpox: the macrophage is the principal target. *Mod. Pathol.* **12:**529–533.

Walker, D. H., and N. Ismail. 2008. Emerging and re-emerging rickettsioses: endothelial cell infection and early disease events. *Nat. Rev. Microbiol.* **6:**375–386.

Walker, D. H., and N. H. Kirkman. 1980. Rocky Mountain spotted fever and deficiency in glucose-6-phosphate dehydrogenase. *J. Infect. Dis.* **142:**771.

Walker, D. H., J. P. Olano, and H. M. Feng. 2001b. Critical role of cytotoxic T lymphocytes in immune clearance of rickettsial infection. *Infect. Immun.* **69:**1841–1846.

Walker, D. H., C. D. Paddock, and J. S. Dumler. 2008. Emerging and re-emerging tick-transmitted rickettsial and ehrlichial infections. *Med. Clin. North Am.* **92:**1345–1361.

Walker, D. H., V. L. Popov, and H. M. Feng. 2000. Establishment of a novel endothelial target mouse model of a typhus group rickettsiosis: evidence for critical roles for gamma interferon and CD8 T lymphocytes. *Lab. Invest.* **80:**1361–1372.

Walker, D. H., V. L. Popov, J. Wen, and H. M. Feng. 1994. *Rickettsia conorii* infection of C3H/HeN mice. A model of endothelial-target rickettsiosis. *Lab. Invest.* **70:**358–368.

Walker, D. H., and X. J. Yu. 2005. Progress in rickettsial genome analysis from pioneering of *Rickettsia prowazekii* to the recent *Rickettsia typhi*. *Ann. N. Y. Acad. Sci.* **1063:**13–25.

Walker, T. S. 1984. Rickettsial interactions with human endothelial cells *in vitro*: adherence and entry. *Infect. Immun.* **44:**205–210.

Walker, T. S., J. S. Brown, C. S. Hoover, and D. A. Morgan. 1990. Endothelial prostaglandin secretion: effects of typhus rickettsiae. *J. Infect. Dis.* **162:**1136–1144.

Walker, T. S., M. W. Dersch, and W. E. White. 1991. Effects of typhus rickettsiae on peritoneal and alveolar macrophages: rickettsiae stimulate leukotriene and prostaglandin secretion. *J. Infect. Dis.* **163:**568–573.

Walker, T. S., and C. S. Hoover. 1991. Rickettsial effects on leukotriene and prostaglandin secretion by mouse polymorphonuclear leukocytes. *Infect. Immun.* **59:**351–356.

Walker, T. S., and G. E. Mellott. 1993. Rickettsial stimulation of endothelial platelet-activating factor synthesis. *Infect. Immun.* **61:**2024–2029.

Whelton, A., J. V. Donadio, Jr., and B. L. Elisberg. 1968. Acute renal failure complicating rickettsial infections in glucose-6-phosphate dehydrogenase deficient individuals. *Ann. Intern. Med.* **69:**323–328.

Whitworth, T., V. L. Popov, X. J. Yu, D. H. Walker, and D. H. Bouyer. 2005. Expression of the *Rickettsia prowazekii pld* or *tlyC* gene in *Salmonella enterica* serovar Typhimurium mediates phagosomal escape. *Infect. Immun.* **73:**6668–6673.

Woods, M. E., and J. P. Olano. 2008. Host defenses to *Rickettsia rickettsii* infection contribute to increased microvascular permeability in human cerebral endothelial cells. *J. Clin. Immunol.* **28:** 174–185.

Woodward, T. E. 1973. A historical account of the rickettsial diseases with a discussion of unsolved problems. *J. Infect. Dis.* **127:**583–594.

Yachie, A., Y. Niida, T. Wada, N. Igarashi, H. Kaneda, T. Toma, K. Ohta, Y. Kasahara, and S. Koizumi. 1999. Oxidative stress causes enhanced endothelial cell injury in human heme oxygenase-1 deficiency. *J. Clin. Invest.* **103:** 129–135.

Yamada, N., M. Yamaya, S. Okinaga, M. Terajima, R. Lee, T. Suzuki, K. Sekizawa, H. Suzuki, and H. Sasaki. 1997. Heme oxygenase I inhibits rhinovirus type 14 (HRV-14) infection and replication by cultured human tracheal epithelium. *Am. J. Respir. Crit. Care Med.* **155:**A943.

Zandi, E., D. M. Rothwarf, M. Delhase, M. Hayakawa, and M. Karin. 1997. The IκB kinase complex (IKK) contains two kinase subunits, IKKα and IKKβ, necessary for IκB phosphorylation and NF-κB activation. *Cell* **91:**243–252.

Zhou, H., F. Lu, C. Latham, D. S. Zander, and G. A. Visner. 2004. Heme oxygenase-1 expression in human lungs with cystic fibrosis and cytoprotective effects against *Pseudomonas aeruginosa in vitro*. *Am. J. Respir. Crit. Care Med.* **170:**633–640.

INNATE IMMUNE RESPONSE AND INFLAMMATION: ROLES IN PATHOGENESIS AND PROTECTION (*ANAPLASMATACEAE*)

Nahed Ismail and Heather L. Stevenson

9

INTRODUCTION

The family *Anaplasmataceae* includes several species of obligate intracellular gram-negative bacteria (Anderson et al., 1992; Borjesson and Barthold, 2002; Ganta et al., 2002). The early immune responses to *Ehrlichia* and *Anaplasma* infections are likely to be important factors in determining the outcome of infection and clinical course of disease. Strong adaptive immune responses may follow, but are initiated too late to eliminate the infection. In this review, we discuss recent studies on the kinetics and quality of early immune responses to *Ehrlichia* and briefly to *Anaplasma*, and the implications for development of successful preventative immune-based therapeutic approaches.

The term "ehrlichiosis" or "anaplasmosis" is ascribed to infections caused by members of the family *Anaplasmataceae*, which consists of obligate intracellular bacteria that replicate within the host cell-derived vacuoles of their mammalian hosts (Anderson et al., 1991; Dawson et al., 1991; Fichtenbaum et al., 1993; Dumler and Bakken, 1995; Walker and Dumler, 1997; Dumler et al., 2001; Olano and Walker, 2002; Mavromatis et al., 2006) Five *Anaplasmataceae* members can infect humans, but only two of them—*Ehrlichia chaffeensis,* the causative agent of human monocytic ehrlichiosis (HME); and *Anaplasma phagocytophilum*, the causative agent of human granulocytic anaplasmosis (HGA)—have been thoroughly investigated. *Anaplasma marginale*, a species primarily of veterinary importance, is another *Anaplasma* species that is a tick-borne, intraerythrocytic rickettsial pathogen that causes significant morbidity and mortality in ruminants (Barbet et al., 2005; Palmer and Brayton, 2007). Cattle that survive infection with the intracellular bacterium *A. marginale* are incapable of completely eliminating the organism and remain persistently infected for life, although they are asymptomatic and otherwise immunocompetent.

The critical questions related to *Ehrlichia* and *Anaplasma* infections that are the focus of this review are as follows. How do these bacteria trigger the innate immune system? What are the early innate events that lead to initial control of bacterial replication? What are the immunoregulatory innate elements that control the induction

Nahed Ismail, Department of Pathology, University of Pittsburgh, Pittsburgh, PA 15261. *Heather L. Stevenson*, Department of Pathology, University of Texas Medical Branch, Galveston, TX 77555.

Intracellular Pathogens II: Rickettsiales
Edited by G. H. Palmer and A. F. Azad © 2012 ASM Press, Washington, DC
10.1128/9781555817336.ch9

and magnitude of specific acquired immune responses? How do these bacteria evade and manipulate the host immune system to create an effective replicative niche and promote bacterial dissemination? Finally, how does the interaction between both *Ehrlichia* and *Anaplasma* species and their hosts' immune systems cause disease? Recent studies of *A. phagocytophilum* and *E. chaffeensis* will be emphasized in order to address these questions.

BIOLOGY OF EARLY *EHRLICHIA* AND *ANAPLASMA* INFECTIONS

E. chaffeensis is the causative agent of HME, a potentially life-threatening emerging infectious disease. The main target cells for *E. chaffeensis* are monocytes and macrophages. HME was first described in 1986. The number of ehrlichiosis cases due to *E. chaffeensis* that have been reported to CDC has increased steadily since the disease became reportable, from 200 cases in 2000 to 961 cases in 2008. The incidence of HME increased similarly, from <1 case per million persons in 2000 to 3.4 cases per million persons in 2008. However, active surveillance in endemic areas has suggested the incidence of HME is 4.7 cases per 100,000 (Fichtenbaum et al., 1993; Yevich et al., 1995; Walker and Dumler, 1997; Dumler et al., 2001; Olano and Walker, 2002; Olano et al., 2003a, 2003b; Mavromatis et al., 2006). The true incidence of human infection with *E. chaffeensis* is likely to be much higher, as two-thirds of the infections are either asymptomatic or minimally symptomatic (Anderson et al., 1993; Ewing et al., 1995; Lockhart et al., 1996; Marshall et al., 2002; Dumler, 2005; Parola et al., 2005; Demma et al., 2006; Estrada-Peña et al., 2008). In addition, many cases are misdiagnosed by clinicians (Walker and Dumler, 1997; Dumler et al., 2001).

The dominant zoonotic cycle of *E. chaffeensis* involves a reservoir of persistently infected white-tailed deer (*Odocoileus virginianus*) and the tick vector *Amblyomma americanum*, prevalent throughout the southeastern and south-central United States (Anderson et al., 1993; Ewing et al., 1995; Lockhart et al., 1996; Marshall et al., 2002; Dumler, 2005; Parola et al., 2005; Demma et al., 2006; Estrada-Peña et al., 2008).

A. phagocytophilum is the causative agent of HGA. HGA was described in the early 1990s in patients from Michigan and Wisconsin with a tick bite history who experienced febrile illness similar to HME (Bakken et al., 1996; Walls et al., 1997; Bakken and Dumler, 2003; Chapman et al., 2006; Dumler et al., 2007b). These cases were distinguishable by the presence of inclusion bodies in granulocytes rather than monocytes, causing this syndrome initially to be termed "human granulocytic ehrlichiosis." The disease has been renamed HGA after phylogenetic analysis reclassified *Ehrlichia phagocytophilum* as a member of the genus *Anaplasma*. The number of HGA cases reported to CDC has increased steadily since the disease became reportable, from 348 cases in 2000 to 1,006 cases in 2008. The incidence of HGA has also increased, from 1.6 cases per million persons in 2000 to 4.2 cases per million persons in 2008. The highest annual incidence rates of HGA in the United States have been reported in Connecticut, Wisconsin, and New York State (Harkess, 1991; Everett et al., 1994; Bakken et al., 1996; Nadelman et al., 1997; Bakken and Dumler, 2003; Walker and Raoult, 2005). *A. phagocytophilum* is transmitted primarily by the tick species *Ixodes scapularis* in New England and the north-central United States, *Ixodes pacificus* in the western United States, *Ixodes ricinus* in Europe, and *Ixodes persulcatus* in Asia. *I. scapularis* is also the tick vector for *Borrelia burgdorferi*, *Babesia microti*, and tick-borne encephalitis viruses, resulting in approximately 10% of the patients with HGA having serologic evidence of coinfection with Lyme disease or babesiosis (Harkess, 1991; Everett et al., 1994; Bakken et al., 1996; Nadelman et al., 1997; Bakken and Dumler, 2003; Walker and Raoult, 2005). The reservoir for *A. phagocytophilum* is primarily small mammals such as the white-footed mouse (*Peromyscus leucopus*), dusky-footed wood rat (*Neotoma fuscipes*), and others such as *Apodemus* spp. or *Clethrionymus* spp., with humans serving as dead-end hosts. Transmission of *A. phagocytophilum* from the tick-mammalian reservoir to humans is similar to that of *E. chaffeensis*.

Most species within the family *Anaplasmataceae* have relatively small genomes (0.8 to 1.5 Mb) that have been molded through reductive evolution as they developed dependence on the host cell for necessary functions (Dumler et al., 2007a, 2007b). In addition, they all have an unresolved evolutionary relationship with the progenitor of the mitochondria.

Ehrlichia and *Anaplasma* enter the cells through a specialized lipid raft region of the plasma membrane that contains the protein caveolin and forms cholesterol-rich invaginations of the membrane (Lin and Rikihisa, 2003b, 2004; Dunning Hotopp et al., 2006; Rikihisa, 2006; Xiong et al., 2009). Caveolae-mediated endocytosis directs *A. phagocytophilum* to an intracellular compartment, or inclusion, which does not acquire components of NADPH oxidase or of late endosomes or lysosomes (Mott et al., 2002; Lin and Rikihisa, 2003a; Carlyon et al., 2004; IJdo and Mueller, 2004; Zhang et al., 2004a). *A. phagocytophilum* morulae acquire early autophagosome-like features (positive for Beclin-1) (Mott et al., 2002; Lin and Rikihisa, 2003a, 2007; Carlyon et al., 2004; IJdo and Mueller, 2004; Zhang et al., 2004b). In contrast, *E. chaffeensis* organisms reside within the early endosomes of monocytic cells, where inclusions retain characteristics of the early endosome, including the markers Rab5 and early endosomal antigen 1 (EEA1) and the vacuolar (H^+) ATPase, and fuse with endosomes containing transferrin and transferrin receptors (Lin and Rikihisa, 2003a; Zhang et al., 2004b; Xiong et al., 2009). This suggests that *A. phagocytophilum* and *E. chaffeensis* may localize into different host cell compartments to enable either their intracellular survival or specific evasion of the host immune system. The most studied host cell receptor for *A. phagocytophilum* infection is P-selectin glycoprotein ligand-1 (PSGL-1) (Herron et al., 2000; Carlyon et al., 2003; Munderloh et al., 2004; Reneer et al., 2006). However, little information is available about the host cell receptors for *E. chaffeensis*. Binding of *A. phagocytophilum* to cells of the human leukemia cell line HL-60 is dependent on the expression of both PSGL-1 and an α1,3-fucosyltransferase. Binding of *A. phagocytophilum* to mouse neutrophils requires expression of α1,3-fucosyltransferases but not PSGL-1 (Carlyon et al., 2003; Munderloh et al., 2004; Reneer et al., 2006). The cognate *A. phagocytophilum* molecules involved in PSGL-1-dependent and PSGL-1-independent binding and resulting infection are unknown (see chapter 6 for a detailed description of these early infection events).

Unlike *Rickettsia* spp. and most other gram-negative bacteria, the genomes of *A. phagocytophilum* and *E. chaffeensis* lack the genes for biosynthesis of the lipopolysaccharide (LPS) and peptidoglycan that activate host leukocytes (Lin and Rikihisa, 2004; Zhang et al., 2004a). The lack of LPS in these bacteria suggests that stimulation of innate responses via conventional pattern recognition receptors such as Toll-like receptors (TLRs) may not play a defining role in activation of the innate and acquired immune systems against these bacteria (Akira and Takeda, 2004; Munz et al., 2005; Akira et al., 2006). However, as is discussed later, studies have shown that LPS-lacking gram-negative bacteria such as *Ehrlichia* can stimulate innate immune responses through nontraditional lymphocytes such as natural killer T (NKT) cells in a TLR-independent manner. Activation of NKT cells has been observed in infections with other pathogens that are not known to activate TLRs (Munz et al., 2005).

Ehrlichia and *Anaplasma* have a multigene family encoding outer membrane proteins (OMPs) (Zhi et al., 1997; IJdo et al., 1999; Yu et al., 1999; Barbet et al., 2000, 2003; Cheng et al., 2003; Lin et al., 2003; van Heerden et al., 2004a; Zhang et al., 2004a; Singu et al., 2005; Miura and Rikihisa, 2007). In *E. chaffeensis*, some of the OMP multigene family is differentially expressed in vitro in mammalian and tick hosts, with p28-19 and -20 predominating within canine DH82 macrophages and p28-14 within ISE6 embryonic tick cells (Zhi et al., 1997; IJdo et al., 1999; Yu et al., 1999; Barbet et al., 2000, 2003; Unver et al., 2002; Cheng et al., 2003; Lin et al., 2003; van Heerden et al., 2004; Zhang et al., 2004a; Singu et al., 2005; Miura and Rikihisa, 2007). In vivo, a recent study by Ganta et al. (2007) determined that infection of C57BL/6 mice with ISE6 tick cell-derived *Ehrlichia* results in a higher bacterial burden and persistent infection compared with *Ehrlichia* derived from DH82 macrophages, suggesting that OMPs could play a critical role in evasion of the host immune system, bacterial persistence, and transmission between mammalian and vector hosts. *E. chaffeensis* as well as other *Ehrlichia* species harbor genes that encode tandem- and ankyrin repeat-containing proteins and actin polymerization proteins (Luo et al., 2008; Wakeel et al., 2009). A similar ankyrinlike domain has been discovered in *A. phagocytophilum* and *A. marginale* (Park et al., 2004; Lin and Rikihisa, 2007; Rikihisa and Lin, 2010; Ramabu et al., 2011). Tandem repeats within *Ehrlichia* are associated with regulation of host gene expression (IJdo et al., 2007; Lin et al., 2007; Bao et al., 2009; Garcia-Garcia et al., 2009a, 2009b; Rikihisa and Lin, 2010). *Ehrlichia* and *Anaplasma* also contain many of the known type IV secretion system (T4SS) components that may possibly mediate secretion of effector proteins into the host cell. Among the T4SS components, VirB (which is common among all of the *Rickettsiales*) has been associated with secretion of effector molecules such as AnkA (ankyrin repeat domain-containing protein A) and Ats-1 (*Anaplasma* translocated substrate 1) (Lin et al., 2007; Bao et al., 2009; Rikihisa and Lin, 2010). The secretion of microbial toxins via VirB may be a mechanism leading to the toxic shock-like syndrome that is a disease manifestation observed in fatal human *Ehrlichia* infections, a possibility that warrants further investigation (Bao et al., 2009). Using the T4SS, *A. phagocytophilum* AnkA was shown to be secreted into the host cell cytoplasm (IJdo et al., 2007; Lin et al., 2007; Garcia-Garcia et al., 2009a, 2009b). AnkA is crucial for *A. phagocytophilum* infection, as infection is inhibited by delivery of AnkA-specific antibody to the host cytoplasm or by treatment with a specific pharmacological inhibitor of AnkA-binding adaptor protein(s). Similarly, it has been reported that *E. chaffeensis* AnkA translocates into the host cell nucleus and binds to *Alu* elements, short interspersed DNA elements that are abundant in the human genome. However, it is not yet clear whether ehrlichial AnkA is delivered to the cytosol via the T4SS. Other bacterial repeat domain-containing proteins such as TRP47 are secreted by the T4SS and serve as effectors for bacterial infection of their hosts (Wakeel et al., 2009). Thus, expression of the T4SS is regulated in *A. phagocytophilum* and *E. chaffeensis*, and the secreted effector proteins may influence the activity of host cells, which in turn facilitate intracellular bacterial survival and immune evasion.

Several studies have highlighted immune evasion mechanisms exhibited by *E. chaffeensis* infection in vitro. These include inhibition of phagolysosomal fusion, which is essential for intracellular killing of phagocytosed bacteria by lysosomes; downregulation of surface expression of TLR2, TLR4, and CD14; and inhibition of activation of several transcription factors that are involved in induction of proinflammatory innate immune responses (Lin and Rikihisa, 2004; Rikihisa, 2006). The mechanism by which downregulation of TLR2 and -4 benefits survival within the macrophage is not understood, as the traditional ligands for these receptors (e.g., peptidoglycan and LPS, which activate TLR2 and -4, respectively) are not present in the bacterium, as mentioned above (Lin and Rikihisa, 2003b; Xiong et al., 2009). However, some of the genes encoding traditional pathogen-associated molecular patterns may have been lost during reductive evolution of these bacteria, allowing them to persist within their tick vectors, as ticks are known to have strong innate defenses toward these structures (Brayton et al., 2005; Foster et al., 2005; Dunning Hotopp et al., 2006).

Microarray analysis of human THP-1 cells infected with *E. chaffeensis* reveals suppression of Th1-promoting cytokines such as interleukin-12 (IL-12), IL-15, and IL-18 (Zhang et al., 2004b). These cytokines are also known to stimulate activation and cytotoxic function of NK and $CD8^+$ T cells in infectious as well as noninfectious model systems. Furthermore, the same study demonstrated that in vitro *E. chaffeensis* infection upregulates transcription of apoptosis inhibitors such as IER3, BIRC3, and Bcl-2, but inhibits apoptosis inducers such as BIK and BNIP3L during the early stage of infection, thus impairing host cell apoptosis and allowing prolonged bacterial intracellular survival within target cells (Zhang et al., 2004b). Although these in vitro studies have provided insight into host-microbial interaction within ehrlichial target cells, these data should be interpreted with caution, as other in vivo factors involved in host defenses cannot be replicated in cell cultures and were not present within these model systems.

A. phagocytophilum also inhibits spontaneous apoptosis of human neutrophils, which allows intracellular survival and replication. This phenomenon has been confirmed by several *in vitro* studies on human neutrophils as well as by an *ex vivo* study on ovine neutrophils infected *in vivo* with a sheep isolate (Scaife et al., 2003; Borjesson et al., 2005; Choi et al., 2005; Ge and Rikihisa, 2006; Lee and Goodman, 2006). *A. phagocytophilum* prevents the loss of mitochondrial membrane potential and inhibits the activation of caspase-3. Microarray data derived from human neutrophils showed that *A. phagocytophilum* infection upregulates the expression of members of the Bcl-2 family (Scaife et al., 2003; Borjesson et al., 2005; Choi et al., 2005; Ge and Rikihisa, 2006; Lee and Goodman, 2006). In addition, *A. phagocytophilum* infection inhibits translocation of the proapoptotic protein Bax to mitochondria, as well as inhibits the activation of caspase-9 (the initiator caspase in the intrinsic pathway) during spontaneous neutrophil apoptosis. Finally, it has been suggested that *A. phagocytophilum* subverts autophagy to facilitate infection (Amano et al., 2006; Schmid et al., 2006; Niu et al., 2008). Autophagy is mediated by autophagosomes that help clear intracellular infections and process nonself and self antigens in the host cytoplasm as part of the innate and adaptive immune responses (Amano et al., 2006; Schmid et al., 2006; Niu et al., 2008). Several markers of early autophagosomes are present in *A. phagocytophilum* morulae, including a double-lipid bilayer and colocalization with Beclin-1 and LC3, the human homologs of the *Saccharomyces cerevisiae* autophagy-related proteins Atg6 and Atg8, respectively (Amano et al., 2006; Schmid et al., 2006).

CLINICAL MANIFESTATIONS AND PATHOLOGY OF HME

HME is a more severe disease than HGA, with 42% of cases requiring hospitalization and a case fatality rate of 3%. The median age of patients with either infection is approximately 50 years, and slightly more males (57 to 61%) are infected than females. Patients with HME present with nonspecific flulike symptoms at early stages of the disease (summarized in Table 1) (Fichtenbaum et al., 1993; Walker and Dumler, 1997; Dumler et al., 2001; Olano and Walker, 2002; Bakken and Dumler, 2004; Chapman et al., 2006; Mavromatis et al., 2006; Thomas et al., 2007). Up to 17% of patients develop life-threatening complications, which may be fatal even in immunocompetent patients, and HME may manifest as a multisystem disease resembling toxic shock-like or septic shock syndrome (Fichtenbaum et al., 1993; Thomas et al., 2007). Disease manifestations are similar to those of Rocky Mountain spotted fever, except that the characteristic rash is rarely observed. Other life-threatening manifestations include cardiovascular failure, aseptic meningitis, hemorrhage, hepatic insufficiency or failure, interstitial pneumonia, and adult respiratory distress syndrome (Fichtenbaum et al., 1993; Walker and Dumler, 1997; Dumler et al., 2001; Olano and Walker, 2002; Bakken and Dumler, 2004; Chapman et al., 2006; Mavromatis et al., 2006; Thomas et al., 2007).

Early diagnosis of HME and HGA may be difficult due to a lack of history of tick exposure and the nonspecific manifestations of the disease. However, patients infected with HME may have some unique laboratory findings that should be important clues to the diagnosis, including thrombocytopenia, elevated liver enzymes, and lymphopenia (Table 1). Increased awareness, particularly in areas with relatively high ehrlichial prevalence, should aid in the diagnosis of HME. Specific laboratory tests employed in the laboratory diagnosis of HME and HGA that are able to confirm the clinical diagnosis are serological detection of specific antibodies using indirect immunofluorescence assays or Western blots, detection of bacterial DNA in peripheral blood using real-time PCR, staining of blood smears and detection of bacterial morulae within phagocytic cells using light microscopy, and finally isolation of *Ehrlichia* or *Anaplasma* in cell culture (Table 1) (Paddock et al., 1993; Bakken et al., 1996; Walker and Dumler, 1997; Childs et al., 1999; Standaert et al., 2000; Doyle et al., 2005).

In spite of the susceptibility of *E. chaffeensis* to doxycycline, this antibiotic is frequently ineffective in prevention of disease progression if given at late stages of disease. In a recent report, patients with HME who had doxycycline initiated within the first 24 hours of hospital admission were compared with patients who did not have empiric doxycycline therapy due to lack of definitive diagnosis of *Ehrlichia* infection. Patients not started on doxycycline treatment at hospital admission had a significantly increased rate of transfer to the intensive care unit, more commonly required mechanical ventilation, and had longer hospital stays and longer lengths of illness compared with treated patients (Hamburg et al., 2008). Overwhelming ehrlichial infections with high bacterial burdens are mainly observed in immunocompromised patients (Thomas et al., 2007). In contrast, immunocompetent patients who develop fatal toxic shock-like syndrome following *Ehrlichia* infection (Fichtenbaum et al., 1993) have very few ehrlichial organisms in their blood and organs during autopsy. These characteristics support the notion that dysregulation of the host immune response leads to tissue damage and eventually toxic shock-like syndrome with multisystem organ failure.

MURINE MODELS OF *EHRLICHIA* AND *ANAPLASMA* INFECTION

Animal Models of HME

There is only limited information on the mechanisms of pathogenesis of *Ehrlichia* and *Anaplasma*

TABLE 1 Symptoms, clinical characteristics, laboratory abnormalities, and diagnostic tests for HME[a]

Symptom or clinical characteristic	(%) (n = 237)[b]	Selected laboratory abnormalities	(%) (n = 237)[c]	Laboratory tests available for diagnosis
Fever	97	Elevated AST	90	IFA
Headache	81	Elevated ALT	84	Western blot
Myalgia	68	Thrombocytopenia	73	Real-time PCR
Nausea	48	Leukopenia	72	Light microscopy[d]
Vomiting	37	Anemia	55	Bacterial isolation/cell culture
Rash	36			
Cough	26			
Pharyngitis	26			
Diarrhea	25			
Lymphadenopathy	25			
Abdominal pain	22			
Confusion	20			

[a]Adapted with permission from H. L. Stevenson et al., 2006, 2008. Abbreviations: AST, aspartate aminotransferase; ALT, alanine aminotransferase; IFA, indirect immunofluorescence assay; PCR, polymerase chain reaction.

[b]Adapted from Fishbein et al., 1994.

[c]Adapted from Eng et al., 1990.

[d]For detection of bacterial morulae within monocytes and neutrophils for *E. chaffeensis* and *Anaplasma* spp., respectively.

and the role of host innate immune responses in contributing to disease virulence in humans; therefore, animal models are necessary to identify bacterial and host characteristics of disease. Multiple murine models have been developed to study differences in host immunity and disease susceptibility (Winslow et al., 1998; Li et al., 2001; Okada et al., 2001; Sotomayor et al., 2001; Olano et al., 2004). SCID mice infected with *E. chaffeensis* develop fatal disease; however, these mice exhibit extensive hepatic injury and pathology that does not mimic human ehrlichiosis (Winslow et al., 1998; Li et al., 2001). As mentioned above, since most patients who present with severe HME are immunocompetent, and since the pathology in SCID mice does not adequately represent the disease in patients with fatal HME, SCID mice were not considered as suitable models by some investigators.

Immunocompetent mice infected with *E. chaffeensis* clear the infection without development of pathology with signs of abortive infection (Winslow et al., 1998; Li et al., 2001). We and other investigators have employed alternative immunocompetent murine models of HME that more closely mimic the various forms of human disease using two species of *Ehrlichia* (i.e., mildly virulent *Ehrlichia muris* and highly virulent *Ixodes ovatus Ehrlichia* [IOE]) (Okada et al., 2001; Sotomayor et al., 2001; Ismail et al., 2004; Olano et al., 2004; Stevenson et al., 2006). Using these closely related surrogate ehrlichial species, we have determined that the outcome of *Ehrlichia* infection varies depending on the infectious dose and/or route of inoculation of the bacteria (Bitsaktsis et al., 2004; Ismail et al., 2004; Bitsaktsis and Winslow, 2006; Stevenson et al., 2006). *E. muris* and IOE species used in these studies are genetically and antigenically related to *E. chaffeensis*, as demonstrated by analysis of 16S rRNA, GroEL, and p28 sequences, and they have strong serological cross-reactivity (Shibata et al., 2000; Okada et al., 2001; Sotomayor et al., 2001; Olano et al., 2004). Infection of immuncompetent C57BL/6 mice with a high dose of mildly virulent *E. muris* via the intraperitoneal (i.p.) route causes mild and self-limited illness, which is associated with lymphohistiocytic infiltrates mainly in the liver and lungs and formation of well-defined granulomas (Ismail et al., 2004; Olano et al., 2004). *E. muris* was originally described as solely a murine pathogen, but was recently isolated from patients with HME in the United States and Russia (Prittet et al., 2011; Eremeeva et al., 2007). In contrast, i.p. infection of C57BL/6 mice with a high dose of IOE results in multiorgan dysfunction and eventually fatal disease (Sotomayor et al., 2001; Bitsaktsis et al., 2004; Ismail et al., 2004; Bitsaktsis and Winslow, 2006; Stevenson et al., 2006). In the murine model of fatal ehrlichiosis, mice develop substantial weight loss; lethargy; hypothermia; marked leukopenia, mainly lymphopenia; focal liver necrosis and apoptosis; severe pulmonary congestion and pneumonia; and a cytokine and chemokine storm, manifestations that are consistent with toxic shock-like syndrome (Ismail et al., 2004). The observations in this fatal murine model of HME are similar to the manifestations of severe and fatal illness in HME patients, that is, those who are not treated immediately with doxycycline at hospital admission.

Although the development of the *E. muris* and IOE models represented a significant advancement in the field, allowing us to study the differences between mild and severe disease, the peritoneal route of infection and the difference in the antigenic and genetic composition of these ehrlichial species are thought to be confounding factors when interpreting data. Recently, another model developed by our laboratory has shown that intradermal (i.d.) infection with a dose of IOE that would ordinarily cause fatal disease when administered i.p. resulted in mild and self-limited disease similar to that developed in *E. muris* infection (Stevenson et al., 2006). The i.d. route of infection also better mimics the natural route of ehrlichial transmission by ticks to humans following an infected tick bite. Analysis of host defense following the i.d. infection indicated that inoculation of virulent IOE i.d. resulted in a lower bacterial burden in different organs and much less hepatic inflammation and apoptosis compared with i.p. inoculation of mice with the same dose of organisms. However, regardless of the route of infection, the infectious/inoculum dose appears to be the most critical factor that determines the outcome of infection. I.d. infection of C57BL/6 mice with higher doses of IOE also resulted in fatal ehrlichiosis and toxic shock-like syndrome, with pathological and immunological findings consistent with those detected in mice infected with lower doses of IOE via the i.p. route (Stevenson et al., 2006).

Animal Models of HGA

Many different animal models have been used to study *Anaplasma* infection, including dogs, horses, nonhuman primates, and mice. Although the histopathological observations among these species are similar to each other and HGA, the clinical disease manifestations in humans are recapitulated only in dogs and horses (Borjesson et al., 2005; Scorpio et al., 2011). Intravenous inoculation of dogs with the neutrophil-propagated 98E4 strain resulted in clinical symptoms most closely related to HGA. On the other hand, mice do not develop clinical manifestations similar to those of human anaplasmosis (Borjesson et al., 2005). Studies have manipulated the mouse model of HGA by varying the number of passages at which the bacteria were propagated in vitro prior to inoculation, allowing comparison of relatively mild (i.e., that induced by inoculation of high-passage *A. phagocytophilum*) versus severe (i.e., that induced by inoculation of low-passage *A. phagocytophilum*) disease (Scorpio et al., 2008). A similar effect of the number of in vitro passages of *Anaplasma* on host defense and outcome of infection is also observed in horses (Pusterla et al., 2000). Interestingly, low-passage inoculation of mice results in a more robust innate immune response compared with infection with high-passaged organisms. These animal models of HGA have enabled more complex studies including comparison of NK-, NKT-, and $CD8^+$-T-cell activation that will be discussed in more detail later.

ACQUIRED IMMUNE RESPONSES TO *EHRLICHIA* AND IMMUNE-MEDIATED PATHOLOGY

Role of Antibody Responses against Infection with *Ehrlichia* and *Anaplasma*

C57BL/6 SCID mice, which are deficient in T and B cells, succumb to *E. chaffeensis* infection within 24 days, indicating that adaptive immunity is essential for protection against ehrlichiosis (Winslow et al., 1998; Li et al., 2001). It was previously thought that antibodies play a minor role in host defense against obligately intracellular pathogens because of their intracellular localization. However, passive transfer of immune sera containing *Ehrlichia*-specific immunoglobulin G (IgG) antibodies to *E. chaffeensis*-infected SCID mice before infection or at weekly intervals starting at 10 days postinfection resulted in effective elimination of ehrlichiae. However, cessation of serum treatment resulted in the recovery of bacterial numbers and development of liver pathology, which ultimately resulted in death of all infected SCID mice. These studies thus indicate that humoral immunity plays a role in protective immunity against *Ehrlichia* (Winslow et al., 2000; Li et al., 2001). The role of humoral immunity was further characterized by using mice deficient in mature B cells such as $JHD^{-/-}$ and $\mu T^{-/-}$. $JHD^{-/-}$ mice that are unable to express IgM and IgG are highly susceptible to *Ehrlichia* infection and succumbed to an ordinarily sublethal IOE infection, unlike wild-type mice (Yager et al., 2005). Our laboratory has also demonstrated that complete protection of *E. muris*-primed mice against lethal IOE challenge is positively associated with production of high levels of *Ehrlichia*-specific antibodies (IgG2a) (Ismail et al., 2004; Thirumalapura et al., 2008, 2009). Furthermore, adoptive transfer of *E. muris*-specific antibodies given in combination with gamma interferon (IFN-γ)-producing $CD4^+$ and $CD8^+$ T cells, but not when given alone, significantly increased the survival of naïve mice challenged with high-dose IOE (Ismail et al., 2004). The mechanism by which IgG2a mediates protection is most likely via FcR-enhanced opsonization

(Ravetch and Bolland, 2001; Rivera et al., 2005), because presence of high-titer antigen-specific IgM, which is an effective component of complement-mediated lysis, did not protect mice from lethal IOE infection (Yager et al., 2005). Alternatively, antibodies may also enhance the production of tumor necrosis factor alpha (TNF-α) by infected phagocytes, which may in turn enhance activation of intracellular microbicidal mechanisms and bacterial elimination, which is discussed in more detail later.

Humoral immunity is also critical for control of infections caused by *A. phagocytophilum* and persistent infections caused by *A. marginale*. *A. marginale* continually undergoes antigenic variation in major surface protein 2 (MSP2) and MSP3 during infection, and variant-specific IgG2 is produced in response to each emerging variant, which suggests that IgG2 responses control the older emerging variants, but fails to completely eliminate the pathogen because new variants are continually being created and are able to escape early detection by the immune response (French et al., 1998, 1999; Brown et al., 2003; Abbott et al., 2004, 2005) (see also chapter 12 for additional details). As with *Ehrlichia*, it is also postulated that antigen-specific IgG2 possibly promotes opsonization of *A. marginale* and/or infected erythrocytes and activates macrophages for enhanced phagocytosis, cytokine release, and nitric oxide production, which are mechanisms that aid in the elimination of intracellular bacteria (Abbott et al., 2004, 2005).

Role of CD4$^+$ T Cells against Infection with *Ehrlichia* and *Anaplasma*

The importance of CD4$^+$ T cells in protection against *E. chaffeensis* was examined using different knockout mice and adoptive transfer approaches. Resistance to fatal ehrlichiosis in immunocompetent mice infected with *E. muris* or with IOE via the i.d. route (i.e., murine models of mild/self-limited ehrlichiosis) is accompanied by strong CD4$^+$ Th1-mediated immune responses producing high concentrations of IFN-γ appropriately at the sites of infection (Bitsaktsis et al., 2004; Ismail et al., 2004; Stevenson et al., 2006). Lack of CD4$^+$ T cells in *E. chaffeensis*-infected major histocompatibility complex (MHC) class II$^{-/-}$ mice results in initial reduction in bacteremia but allows persistent infection (Ganta et al., 2004, 2007). Adoptive transfer of CD4$^+$ Th1 cells from *E. muris*-primed mice led to partial cross-protection of mice against lethal IOE challenge and complete protection when transferred in combination with *Ehrlichia*-specific antibodies (Ismail et al., 2004). Similarly, adoptive transfer of IFN-γ-producing CD4$^+$ T cells from sublethally IOE-inoculated mice protected IFN-γ$^{-/-}$ mice infected with an ordinarily lethal dose of IOE from fatal disease (Bitsaktsis et al., 2004). These observations indicate that IFN-γ produced by CD4$^+$ Th1 cells is sufficient to mediate intracellular bacterial elimination and thus protection against *Ehrlichia*.

In contrast, susceptibility to fatal *Ehrlichia*-induced toxic shock in mice is accompanied by an early suppression of CD4$^+$-T-cell proliferation and a low frequency of IFN-γ-producing CD4$^+$ Th1 cells (Ismail et al., 2004). CD4$^+$ Th1 hyporesponsiveness during severe and fatal murine ehrlichiosis is associated with lower overall levels of IL-12p70 (a Th1-promoting cytokine) but higher levels of IL-10 (an immunosuppressive cytokine) in the serum and spleen of lethally infected mice, compared with non-lethally-infected mice (Fig.1). These data suggest that lethal ehrlichial infection may influence the induction and expansion of CD4$^+$ Th1 cells via altered production of IL-12 and IL-10. Furthermore, CD4$^+$ T cells undergo apoptosis at late stages of lethal ehrlichiosis, which is associated with systemic overproduction of TNF-α, higher expression of apoptotic receptors such as Fas, and expansion of TNF-α-producing CD8$^+$ T cells compared with mild, nonlethal ehrlichiosis (Ismail et al., 2006, 2007; Stevenson et al., 2008, 2010) (Fig.1). Using depletion approaches and knockout mice, we have demonstrated that TNF-α- and IFN-γ-producing CD8$^+$ T cells are key mediators of the observed CD4$^+$-T-cell apoptosis and tissue injury during fatal ehrlichiosis via a mechanism that involves signaling through the TNF receptor (TNFR) (Ismail et al., 2006, 2007). In order to identify the ligand that binds to TNFR and mediates $CD4^+$-T-cell apoptosis and fatal disease, we next examined the effect of TNF-α neutralization on host defenses and immune responses following lethal ehrlichial infection. Unexpectedly, in constrast to the partial protective responses observed in mice deficient in TNFR, neutralization of TNF-α did not protect mice from fatal disease and did not restore the numbers of $CD4^+$ T cells (Ismail et al., 2006). These data suggest that a mediator other than TNF-α produced during IOE infection could be the ligand that binds to TNFR and mediates $CD4^+$-T-cell apoptosis and fatal disease during ehrlichial infection. In addition to TNF-α, it is also possible that a closely related ligand such as homodimeric lymphotoxin (LTα) could potentially bind to TNFR (Sacca et al., 1997; Mordue et al., 2001; Ehlers et al., 2003). However, it is not yet clear whether TNFR-LTα interaction mediates apoptotic cell death of host cells including $CD4^+$ T cells during fatal ehrlichiosis or whether signals mediated by TNF-TNFR interaction are complemented or compensated by signals mediated by binding of LTβ to an alternative receptor such as LTβR. Using different doses of virulent IOE administered via the systemic i.p. route, we have shown that expansion of $CD4^+$ Th1 cells following ehrlichial infection is dependent on inoculum dose. One of the main reasons that infection with higher bacterial doses results in more severe disease is because it causes marked early suppression of antigen-specific $CD4^+$ Th1 responses, followed by significant apoptosis of $CD4^+$ T cells (Ismail et al., 2004, 2006, 2007; Stevenson et al., 2006, 2008).

In *A. phogocytophilum* infection, $CD4^+$ T cells are also indispensible for effective bacterial elimination in infected mice, and mice defective in mounting an MHC class II-restricted $CD4^+$-T-cell response failed to eliminate *A. phagocytophilum* (Pedra et al., 2007; Birkner et al., 2008). In contrast, MHC class I-restricted $CD8^+$ T cells and B cells were not essential for control of *Anaplasma* infection. Studies have shown that clearance of *A. phagocytophilum* is achieved in the absence of perforin, Fas/FasL, and major Th1 cytokines such as IL-12, IFN-γ, and monocyte chemoattractant protein-1 (MCP-1), suggesting that control of these bacteria may not be solely dependent on conventional $CD4^+$ Th1 cells, and that other cells of innate or acquired immune responses could mediate effective elimination of bacteria (Birkner et al., 2008).

In *A. marginale* infection in cattle, the number of organisms in blood during acute and persistent infections influences the memory $CD4^+$-T-cell responses (Han et al., 2008, 2010). *A. marginale* replicates to levels of 10^9 bacteria/ml of blood during acute infection and maintains mean bacteremia levels of 10^6/ml during long-term persistent infection. Proliferation of antigen-specific $CD4^+$ T cells and IFN-γ responses in response to immunization of cattle with MSP2 disappear at the peak of bacteremia and remain undetectable for up to 1 year (Han et al., 2008, 2010). Thus, high antigen load during *A. marginale* infection may alter or modulate specific host immunity, and that may represent an important mechanism to facilitate persistent infection by *A. marginale*. Using a bovine MHC class II tetramer linked to the F2-5B peptide epitope, the antigen-specific $CD4^+$ T lymphocytes in calves immunized with a conserved T-cell epitope from *A. marginale* MSP1a were tracked, before and after homologous strain challenge (Han et al., 2008). The results from this study showed that *A. marginale* induced rapid death of antigen-specific $CD4^+$ T cells and resulted in a loss of functional memory, which may have contributed to persistence of this pathogen in the immunocompetent host.

Role of $CD8^+$ T Cells against Infection with *Ehrlichia* and *Anaplasma*

The role of $CD8^+$ T cells in the pathogenesis of HME has been examined in animal models of both mild and fatal ehrlichiosis using *E. muris* and IOE, respectively. As mentioned above, fatal ehrlichiosis in mice is associated with expansion of antigen-specific TNF-α- and IFN-γ-producing $CD8^+$ T cells (Ismail et al., 2004) (Fig.1). Lack of $CD8^+$ T cells in beta 2 microglobulin ($\beta_2m^{-/-}$) and transporter associated with antigen presentation ($TAP^{-/-}$) mice protected mice from fatal toxic shock-like syndrome, suggesting that $CD8^+$ T cells play a pathogenic role during fatal monocytotropic ehrlichiosis (Ismail et al., 2007). Enhanced resistance in $CD8^+$-T-cell-deficient mice following virulent *Ehrlichia* infection was associated with amelioration of many of the pathogenic components of the immune response observed during fatal disease and generation of protective responses, including (i) decreased serum levels of proinflammatory and immunosuppressive cytokines (TNF-α and IL-10), (ii) enhanced bacterial elimination, (iii) restoration of the number of $CD4^+$ T cells, (iv) increased frequency of antigen-specific IFN-γ-producing $CD4^+$ Th1 cells, and (v) marked attenuation of liver pathology compared with infected wild-type mice. $CD8^+$-T-cell-mediated liver injury was perforin independent since perforin knockout mice succumbed to IOE infection and developed liver injury similar to that observed in wild-type mice. In patients with fatal HME, Dierberg and Dumler (2006) reported a significantly increased number of $CD8^+$ T cells in the lymph nodes of autopsy patients who died of the disease. The increase in $CD8^+$-T-cell infiltration in liver tissues from these patients with fatal HME was associated with the presence of several lymphohistiocytic foci, centrilobular and/or coagulation necrosis, Kupffer cell hyperplasia, marked monocytic infiltration, and an increased amount of hemophagocytosis (indicative of macrophage activation). The correlation between the presence of severe tissue injury, extensive inflammatory responses in different organs, and the presence of few *Ehrlichia*-infected cells in the blood, lymphoid organs, and liver of these patients (Dumler et al., 1993, 2007b) reinforces the conclusion that severe ehrlichiosis is an immune-mediated disease. Unfortunately, most pathological evidence of HME is derived from autopsy cases; therefore, little is known about the organ pathology that occurs in acute disease that is followed by complete recovery.

The pathogenic role of $CD8^+$ T cells is not only of importance during primary ehrlichial infection, but Bitsaktsis and Winslow (2006) have shown that they also mediate a fatal recall immune response following secondary homologous *Ehrlichia* infection. IOE-primed immunocompetent mice that survived primary infection succumbed to a second sublethal challenge with a low dose of IOE. Mice deficient in $CD8^+$ T cells survived the sublethal secondary IOE challenge, and adoptive transfer of purified $CD44^{hi}$ $CD8^+$ T cells into naïve mice followed by inoculation with an ordinarily sublethal dose of IOE resulted in fatality. These observations further support the role of $CD8^+$ T cells as one of the key initiators of immune-mediated pathology during both primary and recall responses against monocytotropic *Ehrlichia*. Further analysis of the molecular mechanism(s) by which $CD8^+$ T cells mediate these fatal recall responses against *Ehrlichia* indicated that perforin and FasL are not required, while TNF-α and/or the chemokine CCL2 (i.e., MCP-1) is likely involved in the mechanism by which $CD8^+$ T cells mediate these pathogenic fatal responses following secondary ehrlichial infection (Bitsaktsis and Winslow, 2006). In addition, $CD8^+$ T cells cocultured with peritoneal cells from sublethally IOE-infected mice produced significantly higher levels of TNF-α and CCL2 compared with uninfected controls.

INNATE IMMUNE RESPONSES AGAINST *EHRLICHIA* AND *ANAPLASMA*

Proinflammatory Cytokines and Chemokines

The production of several cytokines and chemokines has been associated with mild and fatal ehrlichiosis (summarized in Table 2). Studies by us and others have suggested that the effector functions of these cytokines and chemokines depend on the cellular sources and the spatial and temporal kinetics of their

TABLE 2 Cytokines identified in monocytotropic ehrlichiosis

Cytokine(s)	Known principal source[a]	Known primary activity[a]	Described in ehrlichiosis
IL1-α and -β	Macrophages, other APCs	Costimulation of APCs and T cells, inflammation and fever, acute-phase response, hematopoiesis	Lee and Rikihisa, 1996; Lee and Rikihisa, 1997
IL-2	Activated Th1 cells, NK cells	Proliferation of B cells and activated T cells, NK functions (ADD)	Not tested
IL-4	Th2 and mast cells	B cell proliferation and mast cell growth, IgE class switching, increases MHC class II expression on B cells	IL-4 is absent in ehrlichial infection: Bitsaktsis et al., 2004; Ismail et al., 2004
IL-6	Activated Th2 cells, APCs	Acute phase response, B cell proliferation, thrombopoiesis, synergistic with IL-1 and TNF on T cells	Lee and Rikihisa, 1996; Lee and Rikihisa, 1997.
IL-10	Activated Th2 cells, $CD8^+$ T cells, B cells, macrophages	Inhibits cytokine production, promotes B cell proliferation and Ab production, suppression of cellular immunity; may also have proinflammatory functions	Lee and Rikihisa, 1996; Ismail and Walker, 2005; Winslow et al., 2005; Ismail et al., 2006; Stevenson et al., 2006
IL-12	B cells, macrophages, DCs	Proliferation of NK cells, IFN-γ production, promotes Th1 (cellular) immune responses	Zhang et al., 2004; Ismail et al., 2004; Winslow et al., 2005
IL-18	Activated macrophages, immature DCs	Activates CTLs, Kupffer cells, NK cells, enhances cytotoxicity by up-regulating the expression of perforin, granzyme, and FasL on CTLs and NKT and NK cells	Zhang et al., 2004; Stevenson et al., 2010
IFN-γ	Activated Th1 cells, NK cells	Induces expression of MHC class I on somatic cells, induces expression of MHC class II on APCs; activates macrophages, neutrophils, NK cells; promotes cellular immunity	Ismail et al., 2004; Bitsaktsis et al., 2004; Winslow et al., 2005; Ismail et al., 2006; Stevenson et al., 2006
TNF-α	Mainly activated macrophages, T cells	Promotes inflammation, enhances growth factor responses and cell proliferation, activates monocytes and neutrophils, able to activate the extrinsic apoptotic pathway	Lee and Rikihisa, 1996; Lee and Rikihisa, 1997; Ismail et al., 2004; Bitsaktsis et al., 2004; Winslow et al., 2005; Ismail et al., 2006; Stevenson et al., 2006; Winslow et al., 2006; Ismail et al., 2007

[a]Known sources and activities for a particular cytokine: A. K. Abbas, A. H. Lichtman, and S. Pillai, *Cellular and Molecular Immunology*, 7th ed., Elsevier/Saunders, Philadelphia, PA, 2011, and www.copewithcytokines.deithcytokines.de/cope.cgi.

production during the course of the infection. For example, in the murine model of fatal monocytotropic ehrlichiosis, IFN-γ production by $CD8^+$ T cells, but not by $CD4^+$ Th1 cells, is associated with severe and fatal *Ehrlichia* infection (Ismail et al., 2004). In addition, local production of pro- and anti-inflammatory cytokines in the secondary lymphoid organs is likely to be protective, while systemic production or presence of these molecules is usually associated with severe and potentially fatal ehrlichiosis (Ismail et al., 2004; Stevenson et al., 2006). In addition, many of the cytokines portray dual functionality in the host defense against *Ehrlichia* and *Anaplasma* and the subsequent pathogenesis of the diseases. For example, due to bacterial intracellular localization and existence within phagosomes, IFN-γ production, mainly by $CD4^+$ Th1 cells, is a critical requirement for the protective cell-mediated immune responses to be initiated (Akkoyunlu and Fikrig, 2000; Ismail et al., 2004). On the other hand, absence of IFN-γ decreased the degree of histopathology following *A. phagocytophilum* infection, suggesting that IFN-γ also has a pathogenic role in the murine models of HGA, possibly by promoting overactivation of macrophages, resulting in production of toxic molecules that can lead to inflammation and tissue injury even in the absence of significant bacterial loads (Akkoyunlu and Fikrig, 2000, Dumler et al., 2000; Lepidi et al., 2000; Martin et al., 2000, 2001). Higher levels of several proinflammatory and Th1 cytokines and chemokines, including IL-1β, TNF-α, IL-12p40, IL-18, CCL3 (macrophage inflammatory protein-1α [MIP-1α]), CCL4 (MIP-2), CCL2 (MCP-1), IL-8, and CCL5 (RANTES), were detected at later time points postinfection in the sera of IOE/lethally infected mice compared with *E. muris*/non-lethally-infected mice (Stevenson et al., 2008, 2010) (Fig.1). The production of these chemokines was associated with differential migration or expansion of NK and $CD8^+$ T cells in both the spleen and the liver of the lethally infected mice compared with non-lethally-infected mice. Increased production of several chemokines including CCL2 and CCL5, which are potent NK- and T-cell chemoattractants (Loetscher et al., 1996; Ingjerdingen et al., 2001), suggests that enhanced trafficking of NK cells to the liver may contribute to severe tissue injury and host cell death in lethal disease. Higher levels of CCL5 are also observed in patients with fatal Mediterranean spotted fever rickettsiosis compared with patients with mild disease (de Sousa et al., 2007). Higher serum levels of CCL2 are detected not only in fatal murine primary infection, but also during secondary ehrlichiosis (Bitsaktsis and Winslow, 2006; Bitsaktsis et al., 2007), as mentioned previously. In other infectious disease models such as those induced by influenza virus, TNF-α regulates the production of CCL2 and possibly other chemokines by host cells (Xu et al., 2004). Thus, it is possible that during fatal primary ehrlichiosis the production of TNF-α by $CD8^+$ T cells promotes increased chemokine production and chemokine-dependent inflammation, resulting in liver pathology. In addition to the chemotactic effects of those chemokines, elevated levels of CCL2 and CCL5 may contribute to the observed lymphopenia in fatal murine ehrlichiosis, possibly by reducing $CD4^+$-T-cell proliferation. This hypothesis is also supported by studies using *A. phagocytophilum* that show that increased production of MCP-1 and CCL5 during infection may be mechanisms leading to bone marrow suppression and the subsequent cytopenia observed in human anaplasmosis (Klein et al., 2000). Other studies indicate that high concentrations of RANTES could also enhance T-cell apoptosis (Anders et al., 2003), which may contribute to observed lymphopenia and $CD4^+$-T-cell apoptosis in HME patients with severe disease and animal models of fatal HME. Nevertheless, further studies are required in order to examine the roles of these chemokines in the pathogenesis of severe and potentially fatal HME and HGA.

IL-12p70 is a proinflammatory cytokine produced mainly by antigen-presenting cells (APCs) that consists of two subunits, p35 and p40. It is the prototypical cytokine that stimulates cell-mediated immunity and drives the differentiation of Th0 cells into the Th1 lineage, leading to the production of IFN-γ (Puccetti et al., 2002). The majority of IL-12 is produced by monocytes/macrophages and dendritic cells (DCs). IL-12 together with IL-18 stimulates IFN-γ production by NK, NKT, and $CD4^+$ and $CD8^+$ T cells (Berg et al., 2005; French et al., 2006). IFN-γ together with TNF-α activates the antimicrobial functions of phagocytes, leading to intracellular bacterial elimination. Decreased production of IL-12p70 (but not IL-12p40) has been observed during in vivo and in vitro ehrlichial infection (Ismail et al., 2004; Stevenson et al., 2010).

In contrast to the observed decrease in the systemic and local production of IL-12p70 in infected organs during severe ehrlichiosis, IL-18 production by hepatic mononuclear cells is increased and high concentrations are present in the sera, particularly at later time points postinfection (Stevenson et al., 2010) (Fig.1). These increased levels of IL-18 are observed simultaneously with severe tissue injury and remain elevated throughout the course of the disease. IL-18 enhances the cytotoxicity of NK and $CD8^+$ T cells by upregulating the expression of perforin and granzyme, as well as FasL (CD95) (Tsutsui et al., 2003; Berg et al., 2005; French et al., 2006) (Fig.1). The IL-18 receptor (IL-18R) is constitutively expressed on many cell types including T, NKT, and NK cells, while it is inducible on macrophages, DCs, and B cells (Lauwerys et al., 1999; Tsutsui et al., 2003). IL-18 also synergizes with IL-12 to promote IFN-γ production by T and NK cells (Lauwerys et al., 1999; O'Garra and Vieira, 2007). IL-15 and IL-18 act synergistically on NK cells to increase their proliferation and activation (Tupin et al., 2007). IL-18 has been observed to play a pathogenic role in diseases mediated by *Cryptococcus neoformans*, *Leishmania* spp., and *Plasmodium* spp., and also in pathological inflammatory diseases such as multiple sclerosis, rheumatoid arthritis, inflammatory bowel disease, atherosclerosis, and multiorgan failure caused by LPS-induced sepsis (Lauwerys et al., 1999; Tsutsui et al., 2003; French et al., 2006; O'Garra and Vieira, 2007; Tupin et al., 2007). Serum levels of IL-18 correlate with the severity of liver disease in patients with chronic liver disease and fulminant hepatitis. One of our recent studies suggested that the presence or absence of IL-18R signals governs the pathogenic versus protective immunity in an animal model of *Ehrlichia*-induced toxic shock (Ghose et al., 2011). IOE infection of mice deficient in IL-18R resulted in significantly less host cell apoptosis, decreased hepatic leukocyte recruitment, enhanced bacterial clearance, and prolonged survival compared with infected wild-type mice, suggesting a pathogenic role of IL-18/IL-18Rα in *Ehrlichia*-induced toxic shock (Fig.1). Interestingly, although lack of the IL-18R decreases the magnitude of the IFN-γ-producing type 1 immune response,enhanced resistance of the IL-$18R\alpha^{-/-}$ mice against *Ehrlichia* correlated with decreased systemic IL-10 production, increased frequency of protective NKT cells producing IFN-γ, and decreased frequency of pathogenic, TNF-α-producing $CD8^+$ T cells. Adoptive transfer of immune wild-type $CD8^+$ T cells increased the bacterial burden in IL-$18R\alpha^{-/-}$ mice following IOE infection, suggesting that IL-18 is indeed responsible for the generation of pathogenic $CD8^+$ T cells, one of the main mediators of fatal disease. Further analysis indicated that IL-18 possibly promotes the pathogenic adaptive immune responses against *Ehrlichia* via influencing the T-cell-priming functions of DCs (Ghose et al., 2011).

The Role of IL-10 in Ehrlichial Infection and Inflammation

Analysis of immune responses in an animal model of fatal monocytic ehrlichiosis indicated that IL-10 is associated with immunopathology and mortality (Ismail et al., 2004, 2006, 2007; Stevenson et al., 2008, 2010). IL-10 is considered a potent anti-inflammatory cytokine; however, the association between overproduction of IL-10 during late stages of lethal ehrlichial infection and development of severe tissue injury suggests that IL-10 may enhance inflammation rather than mediating an anti-inflammatory effect. Although direct proof of a proinflammatory effect of IL-10 has not been obtained in this study, several observations made from our studies and reports by other investigators support this conclusion. (i) Lethal ehrlichial infection enhances IL-18 production, which has been shown to enhance activation and cytotoxic function of $CD8^+$ T lymphocytes and NK cells, particularly when present with IL-10 (Cai et al., 1999). In fatal ehrlichiosis, IL-10 and IL-18 are excessively produced and this is associated with expansion of $CD8^+$ T and NK cells, known mediators of organ injury (Ismail et al., 2006; Stevenson et al., 2010). (ii) High-dose IL-10 therapy in patients with inflammatory disorders is ineffective in preventing disease progression (Ejrnaes et al., 2006). (iii) *Ehrlichia*-mediated toxic shock-like syndrome is associated with simultaneous overproduction of IL-10 and TNF-α (Ismail et al., 2006, 2007) (Fig. 1). In contrast, sepsis caused by LPS-positive gram-negative bacteria is associated with a high level of TNF-α but low levels of IL-10.

IL-10 is also a well-known immunosuppressive cytokine. Thus, it is possible that IL-10 suppresses protective immunity against *Ehrlichia*, which could enhance bacterial replication and activation of macrophages. This could lead to apoptosis of target cells and/or activation of macrophages and release of oxidative molecules that cause tissue injury. Although it is not clearly understood how IL-10 possesses immunosuppressive as well as proinflammatory functions during systemic infections, studies have shown that variability in IL-10 function is regulated by the cytokine environment (Cai et al., 1999; Ejrnaes et al., 2006).

IL-10 plays an anti-inflammatory role in *A. phagocytophilum* infection, which is consistent with its role during infections with other intracellular pathogens. Mice deficient in IL-10 developed a significantly greater degree of hepatic pathology than wild-type or IFN-γ-deficient mice, suggesting that IL-10 inhibits inflammation during HGA (Martin et al., 2000, 2001). In humans with HGA, the presence of mild clinical manifestations and recovery from infection is temporally associated with an IFN-γ-dominated cell-mediated immunity with moderate IL-10 levels (Bakken and Dumler, 2003; Dumler et al., 2007a, 2007b). Thus, murine and human studies suggest that IL-10 attenuates the pathology during *Anaplasma* infections, possibly via downregulatory effects on IFN-γ or other proinflammatory cytokines such as TNF-α, which is consistent with other infectious models (Gazzinelli et al., 1996; Wynn et al., 1998).

Phagocytic Cells and Intracellular Microbicidal Mechanisms

Although neutrophils are the main target cells for *A. phagocytophilum*, neutrophils are not efficient killer cells in *A. phagocytophilum* infection. Mice deficient in several of the antimicrobial products of neutrophils, such as myeloperoxidase, granulocyte elastase, and cathepsin G, are fully competent in pathogen elimination (Birkner et al., 2008). Other studies showed that inducible nitric oxide synthase (iNOS) and NADPH phagocyte oxidase (phox) are not required for elimination of *A. phagocytophilum* in vivo (Mott et al., 2002; Carlyon et al., 2004; von Loewenich et al., 2004). In vitro gene expression analysis of infected human neutrophils revealed that gene transcription of phox components is not repressed (Mott et al., 2002; Carlyon et al., 2004; IJdo and Mueller, 2004; Lin and Rikihisa, 2007). Earlier studies using HL-60 cells suggested the transcriptional repression of phox components as an immune evasion mechanism by *A. phagocytophilum*. In the murine model of HGA, deficiency of TNF-α, iNOS2, or phox has little impact on *A. phagocytophilum* control. In addition, innate immune responses, mainly those involving activation of infected macrophages by IFN-γ, were postulated to be a major contributing factor to tissue injury and the pathogenesis of severe HGA, although unable to restrict bacterial propagation in vivo (Choi et al., 2007).

Pattern Recognition Receptors: Molecular Recognition of Bacterial Ligands by Innate Cells

Considerable emphasis has also been placed on the identification and characterization of receptors expressed on or in host phagocytes responsible for the initial recognition of pathogens, and on the molecular structure of the pathogens recognized by host cells. Among these innate host cell receptors, TLRs, a family of evolutionarily conserved receptors, have a crucial role in early host defense against invading pathogens (Akira and Takeda, 2004; Akira et al., 2006; West et al., 2006; Chamorro et al., 2009). TLRs are known to recognize different molecular patterns on microbes, such as LPS recognized by TLR4 and CD14, lipoproteins recognized by TLR2, flagellin recognized by TLR5, and CpG-containing bacterial DNA recognized by TLR9. In addition, various endogenous ligands for TLR4, one of which is the high-mobility group protein-1 (HMGB1), an important mediator of inflammatory responses during the late phase of sepsis, are released extracellularly during acute inflammatory responses and cause tissue damage. After ligand binding, TLRs dimerize with other membrane glycoproteins, including the IL-1 receptors (IL-1Rs), and undergo the conformational change required for the recruitment of downstream signaling molecules. These include the adaptor molecule myeloid differentiation primary-response protein 88 (MyD88), IL-1R-associated kinases (IRAKs), transforming growth factor β-activated kinase (TAK1), TAK1-binding protein 1 (TAB1), TAB2, and TNFR-associated factor 6 (TRAF6) (West et al., 2006; Kawai and Akira, 2005, 2007a, 2007b). Due to the prominent roles of TLRs in the initiation of the inflammatory responses, several TLRs are potential therapeutic targets for controlling infection-induced toxic shock or sepsis.

The mechanisms underlying innate recognition of pathogens including *Ehrlichia* and *Anaplasma* are currently an area of significant interest. *Ehrlichia* and *Anaplasma* do not possess LPS and peptidoglycan, as previously mentioned, but are still able to induce inflammation, apoptosis, and necrosis, indicating that the mammalian innate immune system somehow senses the presence of the pathogens. *Ehrlichia* and *Anaplasma* cell walls are equipped with several glycoproteins, lipoproteins, and even host cell cholesterol, possible pattern recognition receptors. In addition to microbial ligands, *Ehrlichia* and *Anaplasma* infections cause severe tissue damage, which can release degraded cellular host proteins that could potentially act as endogenous ligands for stimulation of innate immune responses via TLR4. However, these interesting possibilities need further investigation, as the initial molecule(s) responsible for induction of the observed robust innate immune response is not known.

The role of TLRs in stimulation of innate responses during *Ehrlichia* infection has been emphasized in a recent study suggesting that MyD88 is required for host protection against *E. muris* infection (Koh et al., 2010). *E. muris*-infected, MyD88-deficient mice suffer from increased bacterial burdens in both the peripheral blood and spleen. Furthermore, mononuclear and $CD4^+$ T cells isolated from these splenocytes produced low levels of IL-12 and IFN-γ, respectively, suggesting that MyD88 is critical for induction of protective Th1-cell-mediated immunity following *Ehrlichia* infection. Surprisingly, despite the role of MyD88 in protective immunity against ehrlichiae, the identity of the TLR responsible for this recognition remains unclear. DCs that are deficient in individual TLRs (3 to 11) respond normally to *E. muris* infection, suggesting that *E. muris* is not detected by them (Koh et al., 2010). It is possible that synergistic signals mediated by ligation of multiple TLRs may be required for innate stimulation by *Ehrlichia* and that absence of one TLR may be compensated by another.

Considerable effort has also been made to identify the TLR or other pattern recognition receptors that recognize molecules derived from *A. phagocytophilum*. In vitro studies have demonstrated that *A. phagocytophilum* triggers NF-κB activation, which results in secretion of several cytokines and chemokines via TLR2 (von Loewenich et al., 2004). However, *A. phagocytophilum*-infected mice that are deficient in TLR2, TLR4, or MyD88 have bacterial burdens similar to those of wild-type control mice. These data suggested that signaling via these TLRs is not also critical for early stimulation of the innate response against *Anaplasma*, similar to that observed in *Ehrlichia* infection.

The lack of typical pattern recognition receptors such as LPS in *Ehrlichia* and *Anaplasma* stimulated investigations to discover alternative receptors and molecular ligands by which ehrlichiae stimulate innate and acquired immune responses. Mattner et al. (2005) demonstrated that *E. muris*-infected DCs from $MyD88^{-/-}$ mice are able to stimulate IFN-γ production by innate invariant NKT (iNKT) lymphocytes. Such responses were dependent on direct recognition of unidentified ehrlichial ligands by CD1d (an MHC class I-like molecule that presents antigens to NKT cells), but were not dependent on TLR signals, which is discussed in more detail later. Nevertheless, it is possible that the lack of conventional TLR engagement by *Ehrlichia* and *Anaplasma* may be responsible for ineffective or late activation of innate immune responses. Delayed stimulation of innate responses could in turn lead to ineffective control of bacterial replication at early stages of infection and dysregulated acquired immune responses at late stages of infection, as discussed above.

Nucleotide oligomerization domain (NOD)-like receptor (NLR) proteins are other pattern recognition receptors that have also recently been the focus of several studies. NLRs comprise a large receptor family that is characterized by the presence of a conserved NOD motif (Inohara et al., 2005; Ting and Davis, 2005; Meylan et al., 2006). The general domain organization of NLRs includes an amino-terminal effector-binding region that consists of protein-protein interaction domains such as the caspase recruitment domain (CARD), the pyrin domain (PYD), and the baculovirus inhibitor repeat (BIR) domain (an intermediary NOD that is needed for nucleotide binding and self-oligomerization), and an array of carboxy-terminal leucine-rich repeat motifs, all of which are presumed to detect conserved microbial patterns and to modulate NLR activity (Kurtz and Franz, 2003; Ting and Davis, 2005; Lara-Tejero et al., 2006; Meylan et al., 2006; Sutterwala et al., 2006; Faustin et al., 2007; Mariathasan and Monack, 2007). Moreover, the NLR family members, including PYD-containing 1 (NLRP1), NLRP3, and NLR family CARD-containing 4 (NLRC4), assemble large protein complexes known as inflammasomes that activate inflammatory caspase-1, which cleaves pro-IL-1b and pro-IL-18, leading to the production of IL-1β and IL-18. Finally, NLR family member X1 (NLRX1) and NLRC5 have been shown to inhibit NF-κB-mediated signaling pathways, indicating that these NLRs have important roles as negative regulators for the homeostatic control of innate immunity (Ting and Davis, 2005; Meylan et al., 2006; Sutterwala et al., 2006; Faustin et al., 2007; Mariathasan and Monack, 2007). Thus, NLRs and inflammasomes participate in a diverse set of innate immune signaling pathways.

Several reports have also demonstrated that intracellular pathogens can be detected by different proteins of the NLRs. For example, *Listeria monocytogenes* infection activates NALP3 and NOD2, and *Salmonella enterica* serovar Typhi infection activates NALP3, which is also called cryopyrin or NLRP3, and interleukin-1 converting enzyme protease activating factor (IPAF) (Kurtz and Franz, 2003; Lara-Tejero et al., 2006). NALP3 is a member of a family of cytoplasmic Nod-like receptors that control the activity of inflammatory caspase-1 by forming the inflammasomes. IPAF is an adaptor/protease activating factor that regulates caspase-1 (also called interleukin [IL]-1 converting enzyme), which is activated within the inflammasomes. To our knowledge, there is no clear evidence that suggests that *E. chaffeensis* or *E. muris* can stimulate innate responses via NLP pathways. However, a recent study suggested that *E. muris* is unable to activate known NLRs (NALP3 and IPAF) and *E. muris* infection did not induce IL-1β secretion by macrophages, suggesting that the bacterium does not activate the inflammasome pathway. Additionally, *E. muris* infection in NOD1- and NOD2-deficient DCs results in IL-12p40 production comparable to that produced by wild-type cells, indicating that NOD1 and NOD2 do not play major roles in *E. muris* recognition (Koh et al., 2010). On the other hand, data from our laboratory demonstrated that lethal IOE infection is able to increase IL-18, which requires caspase-1 activation for its production. This suggests that highly virulent, but not mildly virulent, ehrlichial species may stimulate innate responses through one of the inflammasome pathways (Stevenson et al., 2010).

IL-18 and IL-1β are also upregulated when *A. phagocytophilum* infects neutrophils and promyelocytic cells. Studies by Pedra et al. (2007) showed that ASC (apoptosis-associated speck-like protein containing a CARD) and caspase-1 are critical for production of IL-18 in response to *A. phagocytophilum*. In turn, IL-18 is required for IFN-γ production and clearance of *A. phagocytophilum* infection. Further, deficiency of IL-18 in *A. phagocytophilum*-infected mice resulted in defective IFN-γ production, which was paralleled by increased susceptibility to severe infection.

Role of NK Cells in Monocytotropic Ehrlichiosis

NK cells are considered a type of lymphocyte based on their morphology, expression of several lymphoid markers, and origin from a common lymphoid progenitor cell in the bone marrow. However, NK cells are generally classified as components of innate immune defense as they lack antigen-specific cell surface receptors (Lanier, 2005). NK cells are known to be major effector, cytotoxic cells that can directly induce the death of infected cells, as well as producers of high concentrations of several chemokines (e.g., CCL2, CCL3, CCL4, CCL5, XCL1 (lymphotactin), and CXCL8/IL-8) and cytokines such as IFN-γ, TNF-α and IL-10 during both physiological and pathological conditions (Yokoyama et al., 2004; Lanier, 2005; Sun et al., 2009). Thus, NK cells can negatively or positively influence downstream adaptive immune responses depending on the nature of the pathogen encountered. Production of IFN-γ by NK cells can shape acquired T-cell responses either through direct NKT cell interaction or via enhancing the T-cell-priming functions of APCs such as DCs as it promotes the induction of Th1 cells and cytotoxic $CD8^+$ T cells (Yokoyama et al., 2004; Lanier, 2005; Sun et al., 2009). NK cells require priming or activation by various factors, such as IL-12, IL-15, and IL-18, produced by APCs (including DCs) and macrophages (Lantz and Bendelac, 1994; Yoshimoto et al., 1995; Matsuda et al., 2000). NK cells express several receptors that can either stimulate (via activating receptors, e.g., NKG2D) NK cells or inhibit (via inhibitory receptors, e.g., Ly49) NK-cell activation (Yokoyama et al., 2004; Kim et al., 2005; Sun et al., 2009). The main mechanism of cell-mediated killing by NK cells is through the granule exocytosis pathway and involves perforin and granzymes (Liu et al., 2000). In virus-induced hepatitis, NK cells cause hepatocyte apoptosis, elevate serum transaminases, and stimulate the induction of antigen-specific T-cell immunity by releasing IFN-γ (Liu et al., 2000; Yokoyama et al., 2004; Kim et al., 2005; Sun et al., 2009).

We have investigated the contribution of NK cells to protective immunity against *Ehrlichia* (Stevenson et al., 2010). To this end, we examined the effect of lethal (i.p.) and nonlethal (i.d.) infection with the same high dose of IOE on NK-cell activation and function. Compared with nonlethal infection, lethal infection enhanced cytotoxicity and cytokine production by NK cells. NK cells are also the major source of IFN-γ, TNF-α, and IL-10 produced during fatal ehrlichiosis (Fig.1). Furthermore, compared with nonlethal ehrlichial infection, lethal infection enhances the expansion of NK cells in the liver, but not in the spleen. Such spatial expansion was associated with increased production of IL-18 and chemokines such as CCL2 and CCL5 in the liver and sera of infected animals. Unexpectedly, depletion of NK cells in lethally infected mice improved protective immunity against virulent *Ehrlichia*, decreased levels of proinflammatory cytokines, and prevented the development of hepatic injury, providing evidence for a major pathogenic role of NK cells in fatal ehrlichiosis (Stevenson et al., 2010). The mechanisms by which NK cells potentiate tissue injury after lethal ehrlichial infection are not completely understood. However, NK cells may mediate fatal ehrlichiosis by direct cytotoxic killing of host cells or indirectly by enhancing local and systemic inflammation (Liu et al., 2000; Sun et al., 2009) (Fig.1). In addition, NK cells may enhance the generation of cytotoxic, pathogenic $CD8^+$ T cells that promote tissue damage in fatal ehrlichiosis and may also increase death receptors on nonimmune cells, which is discussed in further detail later. Supporting this hypothesis, we observed a strong link between the expansion of NK cells, enhanced lymphohistiocytic infiltration in the liver, and an increased frequency of cytotoxic $CD8^+$ T cells at the sites of infection (Stevenson et al., 2010) (Fig. 1).

The mechanism by which NK cells negatively affect protective immunity against *Ehrlichia* is not yet completely understood. However, it is possible that NK cells mediate suppression of anti-*Ehrlichia* protective $CD4^+$ Th1 cells during fatal ehrlichiosis through production of immunosuppressive IL-10, suggesting that IL-10 may have both pro- and anti-inflammatory functions during severe ehrlichial disease. This hypothesis is supported by our studies (Ismail et al., 2004, 2006, 2007; Stevenson et al., 2008, 2010) showing that suppressed $CD4^+$-T-cell proliferation and apoptosis of $CD4^+$ T cells in lethally infected wild-type sham control mice is concomitantly associated with a significant increase in the number of intrahepatic NK cells and IL-10 overproduction. In addition, the observed decrease in IL-10 levels in NK-cell-depleted mice further supports the conclusion that NK cells mediate IL-10 production, which in turn may suppress $CD4^+$ Th1 responses in fatal ehrlichiosis. Alternatively or concurrently, NK cells may induce cytotoxicity of pathogenic $CD8^+$ T cells, resulting in apoptosis of protective $CD4^+$ T cells (Fig. 1).

Role of NKT Cells in Monocytotropic Ehrlichiosis

NKT cells are a small population of thymus-derived T cells that are found in mice and humans and are restricted by the nonclassical MHC class I molecule CD1d (Lantz and Bendelac, 1994; Yoshimoto et al., 1995; Matsuda et al., 2000; Zhou et al., 2004; Van Kaer, 2007; Van Kaer and Joyce, 2010; Wu and Van Kaer, 2011). Classical (invariant) NKT cells can be activated by self or microbial antigens in the context of CD1d (a glycolipid-binding, nonpolymorphic, MHC class I-like glycoprotein), express an $\alpha\beta$TCR (T-cell receptor) with Vα14 and Jα18, and are abundant in the liver. The best-known NKT-cell antigen is α-galactosylceramide, which is used as the prototypical antigen for stimulating NKT cells and for identifying these cells in vivo using α-galactosylceramide-loaded CD1d tetramers. Mice deficient in CD1d lack iNKT cells (Stober et al., 2003). NKT cells are known for their potent cytokine production and immunoregulatory potential. Activated NKT cells release large amounts of cytokines that can alter the magnitude and phenotype of the immune responses through cross talk with DCs, neutrophils, lymphocytes, and myeloid-derived suppressor cells, and via altering cytokine responses to Th1, Th2, and Th17 T-cell phenotypes (Lantz and Bendelac, 1994; Yoshimoto et al., 1995; Matsuda et al., 2000; Zhou et al., 2004). Similarly, NKT cells can provide $CD4^+$-T-cell-independent help to $CD8^+$ T cells by priming DCs (Stober et al., 2003). NKT cells can differentiate into NKT type 1 and type 2 phenotypes based on their production of Th1 and Th2 cytokines. The factors that control the differentiation of NKT cells into specific phenotype(s) are not yet completely understood. However, it is postulated that alteration of the antigenic structures of NKT ligands, degree or strength of binding of NKT ligand with NKT TCR, and the cytokine environment can influence the cytokine responses of NKT cells via selective stimulation of particular NKT-cell subsets or expansion of an already differentiated subset, a situation similar to that of Th1/Th2 differentiation by naïve $CD4^+$ conventional T cells (Lantz and Bendelac, 1994; Yoshimoto et al., 1995; Van Kaer and Joyce, 2010).

Recent studies provide evidence that CD1d-restricted NKT cells contribute to host defense against various microbial pathogens including *Ehrlichia* and *Anaplasma*. NKT-deficient mice have impaired antimicrobial defenses against several pathogens due in part to defective early IFN-γ secretion (Nagarajan and Kronenberg, 2007). Several studies have shown (Fujii et al., 2004) that NKT cells are activated by APCs during infection via two complementary pathways: the first one is mediated by TLR ligation and production of cytokines, and the second one involves TCR stimulation by endogenous self antigen that is presented on the surface of the APC in the context of the CD1d molecule. However, the study by Mattner et al. (2005) provided evidence for TLR-independent but CD1d-dependent activation of NKT cells against *E. muris*, as well as other gram-negative, LPS-negative α-proteobacteria such as *Sphingomonas capsulata*. Infection of DCs from wild-type or MyD88-deficient mice, but not from CD1d-deficient mice, can activate IFN-γ production by human and murine NKT cells, suggesting that *Ehrlichia*-mediated activation of NKT cells is dependent on CD1d, but not on TLR stimulation. Interestingly, stimulation of NKT-cell responses by *Ehrlichia* or *Sphingomonas* was independent of endogenous lysosomal glycosphingolipids presented by DCs (Mattner et al., 2005), suggesting that these α-proteobacteria have an exogenous microbial NKT ligand that directly activates NKT cells in the context of the CD1d restriction molecule (Fig.1). In contrast, other gram-negative bacteria that are LPS positive, such as *S. enterica* serovar Typhimurium, were found to activate NKT cells through the recognition of an endogenous lysosomal glycosphingolipid in the context of CD1d, presented by LPS-activated DCs from wild-type but not from MyD88$^{-/-}$ mice. The latter suggests that activation of NKT cells by these gram-negative, LPS-positive bacteria is dependent on signals transmitted via costimulation of TLRs (Mattner et al., 2005).

NKT cells also appear to play a role in host defense against *Anaplasma*. NKT cells are activated early during infection with *A. phagocytophilum* (Choi et al., 2007). Interestingly, infection with a low-passage *A. phagocytophilum*, which causes more severe hepatic lesions than high-passage *Anaplasma*, induces early expansion of NK1.1^{+} cells followed by a decline in NKT cells. Similar to *Anaplasma* infection, lethal ehrlichial infection is also associated with temporal and spatial changes in NKT cells compared with nonlethal ehrlichial infection (Stevenson et al., 2010). Lethal infection induced early expansion of NKT cells in the spleen and liver, followed by a decline in their numbers at later time points. In contrast, nonlethal ehrlichial infection was associated with a lower, but steadily increasing, frequency of NKT cells in both the spleen and liver of infected mice. Although the mechanism(s) resulting in a decreased frequency of NKT-cell populations at later stages of severe anaplasmosis or ehrlichiosis is not yet known, it is possible that they undergo activation-induced cell death, which has been suggested by other studies (Van Kaer, 2007; Van Kaer and Joyce, 2010). Interestingly, the numbers of NKT cells in the spleen and liver of *Ehrlichia*-infected mice are restored upon doxycycline treatment and subsequent clearance of ehrlichiae, suggesting that the observed decline in NKT cells during fatal disease could be due to continuous and direct activation of NKT cells by microbial ligand(s). Taken together, these data suggest that NKT cells are major components in the early pathogenesis of anaplasmosis and ehrlichiosis (Choi et al., 2007; Stevenson et al., 2010).

As mentioned above, NKT cells are critical for elimination of intracellular pathogens, possibly via IFN-γ production and activation of target cells. In ehrlichiosis, NKT cells appear to play dual roles in the pathogenesis of HME. Thus, absence of NKT cells in CD1d$^{-/-}$ mice increased the bacterial burden following both lethal and nonlethal ehrlichial infections compared with wild-type mice, suggesting that NKT cells are critical for protective immunity against *Ehrlichia* (Fig.1). On the other hand, lack of NKT cells in CD1d$^{-/-}$ mice infected with virulent IOE abrogated systemic overproduction of TNF-α (a known mediator of toxic shock-like syndrome) (Stevenson et al., 2008) and decreased hepatic inflammation and tissue injury, suggesting that NKT cells may mediate immunopathology in immunocompetent hosts during fatal ehrlichiosis. Of note, although lack of NKT cells in IOE-lethally-infected mice prevented the development of toxic shock, all mice succumbed to the disease, although most likely via different pathologic processes. In wild-type mice mortality is caused by immunopathology, as described above, while in NKT-cell-deficient mice mortality is most likely mediated by the observed overwhelming ehrlichial infection, which then leads to direct or indirect tissue damage. The mechanism by which NKT cells contribute to the pathogenesis of *Ehrlichia*-induced toxic shock-like syndrome is not yet understood. As mentioned above, IFN-γ-producing $CD4^+$ Th1 cells are the major protective cells during ehrlichial infection; however, as mentioned, these cells undergo apoptosis at late stages of disease. Thus, it is possible that NKT cells play a pathogenic role in ehrlichiosis via promoting the induction of pathogenic, cytotoxic $CD8^+$-T-cell responses, which then cause apoptosis of protective $CD4^+$ Th1 cells, as well as tissue injury. In support of this conclusion, IOE infection in NKT-cell-deficient mice decreases expression of Fas death receptors on splenic $CD4^+$ T cells and decreases $CD8^+$-T-cell cytotoxicity, as shown by their decreased expression of granzyme B (Stevenson et al., 2008). Increased numbers of $CD4^+$ and $CD8^+$ T cells in the spleen of IOE-infected, NKT-cell-deficient mice compared with infected wild-type mice further support this conclusion.

The mechanism by which NKT cells influence $CD8^+$-T-cell responses during ehrlichiosis has not yet been examined. However, we and others have shown that NKT cells are major producers of cytokines and that they enhance the T-cell-priming functions of several APCs, including macrophages, DCs, and B cells (Ko et al., 2007; Stevenson et al., 2008, 2010) (Fig.1). This conclusion is supported by data that showed that IOE-infected $CD1d^{-/-}$ mice lacking iNKT cells had significantly lower expression of CD40 on DCs, macrophages, and B cells. This suggests that NKT cells are essential for activation or maturation of professional APCs. NKT-cell-mediated DC maturation has been shown to activate $CD8^+$ T cells via CD40/CD40L interaction, and is able to initiate a cytotoxic-T-cell response independent of $CD4^+$-T-cell help (den Haan et al., 2000; Houde et al., 2003; Schaible et al., 2003; Wu et al., 2003). This scenario is of particular importance during severe ehrlichiosis, where infection induces substantial expansion of pathogenic $CD8^+$ T cells in the absence of strong $CD4^+$-T-cell proliferation and IL-2 production.

The ligand(s) that activates iNKT cells in response to *E. muris* or *Anaplasma* is unknown. However, several exogenous bacterial ligands identified in other pathogens, such as phosphatidylinositol mannoside in *Mycobacterium* spp. (Fischer et al., 2005), glycosylceramides in *Sphingomonas* spp. (Mattner et al., 2005), and α-galactosyl-diacylglycerols of *B. burgdorferi* (Kinjo et al., 2006), may also exist in *Ehrlichia*. Nevertheless, activation of NKT cells during infections with gram-negative bacteria that lack LPS through recognition of specific microbial ligand might be an alternative, TLR-independent pathway by which these organisms stimulate innate and acquired immune responses.

Apoptosis, Inflammation, and Hepatic Injury during Fatal Ehrlichiosis

Apoptosis is normally a rare and rapidly completed event, where few apoptotic cells in a histological section indicate the presence of extensive apoptosis in vivo (Higuchi and Gores, 2003). The key characteristics of apoptosis, or programmed cell death, are membrane blebbing, chromatin condensation, and nuclear fragmentation (Higuchi and Gores, 2003). Our data from the animal models of fatal ehrlichiosis show a link between an increased number of apoptotic cells, mainly Kupffer cells, at early stages of infection and the expansion of $CD8^+$ T cells (Fig.1). At later stages of fatal disease, many hepatic parenchymal cells also undergo apoptosis and some of these cells are associated with the presence of ehrlichial antigen, as shown by immunohistochemistry. Apoptotic blebs are able to activate cytotoxic T lymphocytes via cross-presentation by DCs (den Haan et al., 2000; Higuchi and Gores, 2003; Houde et al., 2003; Schaible et al., 2003; Wu et al., 2003; Hildebrand et al., 2006). Lymphoid DCs ($CD8^+$ $CD11c^+$ DCs) are professional APCs with a high capacity to cross-present antigens and cross-prime $CD8^+$ T cells and cytotoxic-T-lymphocyte responses. Unlike viruses or bacteria that escape into the cytoplasm (e.g., *Rickettsia*, *Shigella*, and *Listeria*), *Ehrlichia* organisms reside in the phagosomal compartment of their target cells and therefore do not access the endogenous MHC class I antigen presentation pathway. Thus, cross-presentation of ehrlichial antigens that are contained within apoptotic cells may be presented in the context of MHC class I, to naïve $CD8^+$ T cells. This may be an alternative pathway by which *Ehrlichia* infection induces substantial expansion of cytotoxic, antigen-specific $CD8^+$ T cells.

The liver is the main site of pathology during HME in humans and mice (Fig. 1). This could be partly due to marked ehrlichial tropism to the liver as well as the presence of sinusoidal lining cells such as Kupffer, stellate, and endothelial cells that makes the liver a primary target for hepatoxicity and injury following systemic infection (Chen et al., 2003). Lethal ehrlichial infection causes an early focal apoptosis of primary hepatic parenchymal cells, followed by necrosis (Ismail et al., 2004; Stevenson et al., 2008, 2010) (Fig.1). As suggested by other studies, pathogens entering the liver may be phagocytosed by Kupffer cells, which leads to their activation and release of proinflammatory cytokines such as IL-6, IFN-γ, and TNF-α (Szabo et al., 2002); chemokines such as MCP-1; and reactive oxygen and nitrogen species (Trobonjaca et al., 2001; Chen et al., 2003). These events are then followed by recruitment of inflammatory and immune cells to the liver (Higuchi and Gores, 2003; Hildebrand et al., 2006). Thus, it is likely that the onset of hepatic injury during severe ehrlichial infection proceeds from an initial noninflammatory event, such as apoptosis and ineffective clearance of apoptotic bodies by neighboring hepatocytes/phagocytes, to inflammatory events due to an overwhelming inflammatory response (Szabo et al., 2002; Chen et al., 2003; Higuchi and Gores, 2003; Hildebrand et al., 2006). If apoptotic bodies are not rapidly removed, they usually undergo autolysis and stimulate production of proinflammatory cytokines, which is followed by their phagocytosis by resident hepatic Kupffer cells or DCs. This phagocytic process results in an increase in oxidative damage through release of nitric oxide and other oxidative molecules (Trobonjaca et al., 2001; Szabo et al., 2002; Chen et al., 2003; Guicciardi and Gores, 2005; Hildebrand et al., 2006). These inflammatory processes may also increase expression of death receptors such as Fas/FasL or TNF/TNFR on hepatocytes, thus increasing their susceptibility to apoptotic cell death following their interaction with immune cells (Fig.1). As mentioned before, lethal ehrlichial infection stimulates excess production of IL-18 by hepatic mononuclear cells, which enhances the expression of death receptors such as Fas (Ismail et al., 2006, 2007; Stevenson et al., 2008, 2010). In support of this conclusion, lack of IL-18/IL-18R interaction decreases the expression of Fas on $CD4^+$ T cells in mice infected with a lethal dose of IOE (Ghose et al., 2011).

Similarly to HME, *A. phagocytophilum* causes tissue injury and hepatic apoptosis. In the animal model of severe HGA, the histopathologic lesions that correlate with disease peak are detected in infected mice at early stages of infection, before the induction of adaptive immunity. In addition, the correlation between an early peak for IFN-γ and the peak of histopathologic injury and disease in HGA in infected mice suggests that immune-mediated injury is caused by innate responses rather than acquired immune responses, similar to that observed in animal models of HME. It is postulated that IFN-γ production is a major contributor to hepatic inflammatory tissue injury via overactivation of the immune system, mainly macrophages.

CONCLUDING REMARKS AND FUTURE DIRECTIONS

Studies of the pathogenesis of severe and fatal ehrlichiosis and anaplasmosis, or protective immunity against *Ehrlichia* and *Anaplasma* infection, have largely focused on the analysis of effector $CD4^+$ T cells and their control by $CD8^+$ T cells. However, there is increasing evidence that innate immune cells have crucial roles in HME and HGA pathogenesis.

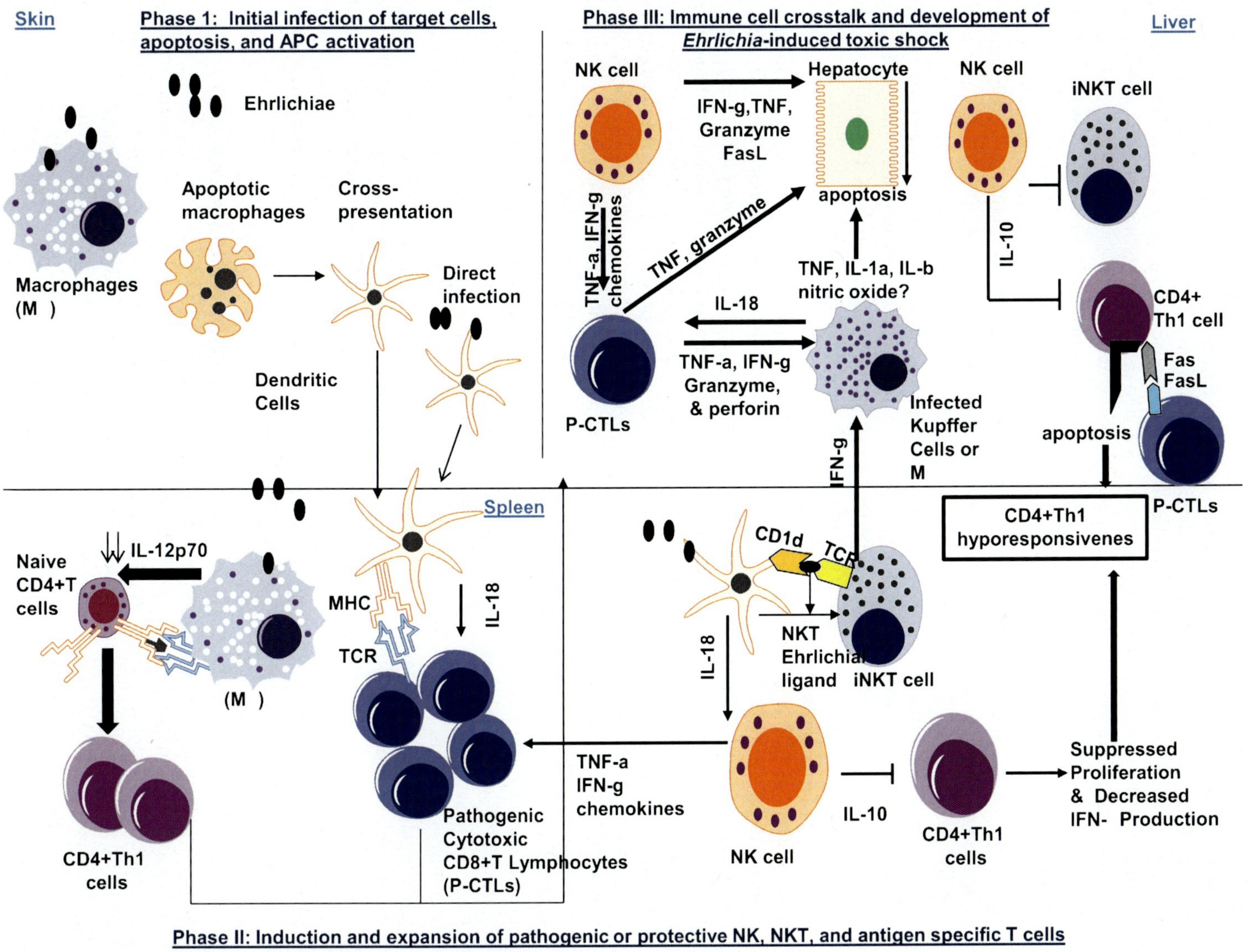

Skin
Phase 1: Initial infection of target cells, apoptosis, and APC activation
Ehrlichiae
Macrophages (M)
Apoptotic macrophages
Cross-presentation
Direct infection
Dendritic Cells
Phase III: Immune cell crosstalk and development of Ehrlichia-induced toxic shock
Liver
NK cell
Hepatocyte
IFN-g,TNF, Granzyme FasL
apoptosis
NK cell
iNKT cell
IL-10
CD4+ Th1 cell
Fas
FasL
TNF-a, IFN-g chemokines
TNF, granzyme
TNF, IL-1a, IL-b nitric oxide?
IL-18
TNF-a, IFN-g Granzyme, & perforin
P-CTLs
Infected Kupffer Cells or M
apoptosis
P-CTLs
IFN-g
Spleen
CD4+Th1 hyporesponsivenes
Naive CD4+T cells
IL-12p70
MHC
TCR
IL-18
(M)
CD1d
TCR
NKT Ehrlichial ligand
iNKT cell
IL-18
CD4+Th1 cells
Pathogenic Cytotoxic CD8+T Lymphocytes (P-CTLs)
TNF-a IFN-g chemokines
NK cell
IL-10
CD4+Th1 cells
Suppressed Proliferation & Decreased IFN- Production
Phase II: Induction and expansion of pathogenic or protective NK, NKT, and antigen specific T cells

In various animal models, innate cells such as macrophages, DCs, NK cells, and iNKT cells, as well as cytokines such as IFN-γ and TNF-α, have been shown to be required for the development of tissue injury and inflammatory responses during *Ehrlichia* and *Anaplasma* infections. These cells and cytokines have been detected not only in the secondary lymphoid organs but also at main peripheral sites of infection, such as in the liver. Innate cells such as iNKT cells and cytokines such as IFN-γ and TNF-α have also been shown to be involved in protection against this disease, although as shown in this review, some of the mediators such as TNF-α and NKT cells may play both protective and pathogenic roles, highlighting the bimodal function of many immune mechanisms in HME and possibly in HGA. NK cells mediate immunopathology and inhibit protective immunity during fatal ehrlichiosis, while NKT cells are essential for effective bacterial elimination. On the other hand, NKT cells appear to promote the induction of pathogenic $CD8^+$ T cells via enhancing the T-cell-priming function of APCs during fatal ehrlichiosis. Thus, one of the key questions that remain to be addressed is why innate immune cells are inefficient in preventing severe systemic infection during HME and HGA, but instead appear to participate in the development of the pathogenic cell-mediated immune responses. One hypothesis is that severe and lethal ehrlichial infection could be associated with early immune evasion mechanisms, including (i) inefficient recognition of bacteria by innate cells secondary to downregulation of pattern recognition receptors or lack of TLR ligands, (ii) suppression of very early inflammatory cytokines and chemokines, and/or (iii) suppression of intracellular microbicidal functions of their phagocytic host cells. Bacteria-induced immune evasion mechanisms render the innate immune cells unable to control early bacterial replication, which may lead to overactivation of the primary immune response and subsequently activation of pathogenic NK and $CD8^+$ T cells.

The knowledge gained from analysis of the immunopathogenic and protective mechanisms responsible for susceptibility and resistance, respectively, to fatal HME or HGA suggests two main pathways for targeting when attempting to develop new therapeutic approaches against these diseases. The first involves enhancing protective immunity, either by targeting bacterial components that are protective in mouse models of HME or HGA or by specifically triggering innate or acquired immune cells, such as NKT or $CD4^+$ T cells, that will exert protective roles against HME during different

FIGURE 1 Early interactions between ehrlichiae and target cells: a model of *Ehrlichia*-induced toxic shock. The induction phase of immune responses against *Ehrlichia* or *Anaplasma* infections occurs in the skin, followed by systemic dissemination through the reticuloendothelial system, where conventional DCs and macrophages (the main target cells for *Ehrlichia*) capture and process ehrlicial antigens. Infected macrophages can undergo apoptosis following intracellular bacterial replication. Apoptotic cells containing ehrlichial antigen can be phagocytosed by immature DCs and presented to naïve $CD8^+$ T cells via direct or cross-presentation pathways in the context of MHC class I molecule (i.e., cross-priming). Activated DCs prime pathogenic $CD8^+$ T cells as well as protective $CD4^+$ Th1 cells (not shown) after migration to the draining lymphoid organs (lymph nodes and spleen), and macrophages also promote this activation, possibly via IL-12p40 secretion. iNKT cells and $CD4^+$ Th1 cells control bacterial replication via IFN-γ production, preventing subsequent inflammation and severe HME (not shown). The expansion of *Ehrlichia*-specific $CD4^+$ T cells can be inhibited by NK cells through various mechanisms, including IL-10 production, but other surface molecules can be involved in this process (not shown). NK cells can promote the recruitment of pathogenic $CD8^+$ T cells to peripheral sites of infection such as the liver, and also recruit different subsets of DCs that further enhance the expansion of pathogenic $CD8^+$ T cells in this organ. In the liver, hepatocytes as well as infected target cells can be killed by pathogenic NK and $CD8^+$ T cells, possibly through the release of IFN-γ, TNF-α, granzymes, and perforin. IL-18 produced by DCs and macrophages sustains the effector functions of activated pathogenic $CD8^+$ T cells and NK cells in the liver (not shown). NK cells producing IL-10 may further suppress expansion of $CD4^+$ Th1 cells in the liver, while pathogenic $CD8^+$ T cells induce apoptosis of these cells. This complex cross talk between innate and adaptive immune cells results in the development or the prevention of severe and potentially fatal HME and HGA. doi:10.1128/9781555817336.ch9.f1 stages of infection. The second should focus on inhibiting the pathogenic effects of NK cells in HME, as these cells have been shown to have deleterious effects, similar to $CD8^+$ T cells. Regarding the exogenous activation of iNKT cells, many studies are currently under way to generate and screen new agonists in hopes of obtaining new agents that preferentially activate and direct NKT cells toward protective pathways. However, the dual role played by NKT cells during fatal ehrlichiosis needs to be considered in these immunotherapeutic approaches.

The range of clinical manifestations seen in immunocompetent hosts proves that differences in host factors cannot be ignored when studying these diseases. Additionally, the ability of organisms that lack conventional pathogen-associated molecular patterns for gram-negative bacteria, such as LPS and peptidoglycan, to cause toxic shock-like syndrome will continue to provide valuable information about alternative novel mechanisms of innate immune recognition and activation. Human studies investigating these immune phenomena are much needed and will help to confirm many of the intriguing mechanisms elucidated using animal models. Further analysis of the relevance of apoptosis/necrosis to induction of pathogenic T-cell responses against *Ehrlichia* and characterization of the mechanisms by which several cytokines and chemokines lead to tissue damage, inflammatory cell trafficking, and inflammation will enhance our understanding of the pathogenesis of acute fatal ehrlichiosis and the development of novel vaccines or immunotherapeutic strategies for prevention or treatment of these important diseases.

REFERENCES

Abbott, J. R., G. H. Palmer, C. J. Howard, J. C. Hope, and W. C. Brown. 2004. *Anaplasma marginale* major surface protein 2 $CD4^+$-T-cell epitopes are evenly distributed in conserved and hypervariable regions (HVR), whereas linear B-cell epitopes are predominantly located in the HVR. *Infect. Immun.* **72:**7360–7366.

Abbott, J. R., G. H. Palmer, K. A. Kegerreis, P. F. Hetrick, C. J. Howard, J. C. Hope, and W. C. Brown. 2005. Rapid and long-term disappearance of $CD4^+$ T lymphocyte responses specific for *Anaplasma marginale* major surface protein-2 (MSP2) in MSP2 vaccinates following challenge with live *A. marginale*. *J. Immunol.* **174:**6702–6715.

Akira, S., and K. Takeda. 2004. Toll-like receptor signaling. *Nat. Rev. Immunol.* **4:**499–511.

Akira, S., S. Uematsu, and O. Takeuchi. 2006. Pathogen recognition and innate immunity. *Cell* **124:**783–801.

Akkoyunlu, M., and E. Fikrig. 2000. Gamma interferon dominates the murine cytokine response to the agent of human granulocytic ehrlichiosis and helps to control the degree of early rickettsemia. *Infect. Immun.* **68:**1827–1833.

Amano, A., I. Nakagawa, and T. Yoshimori. 2006. Autophagy in innate immunity against intracellular bacteria. *J. Biochem.* **140:**161–166.

Anders, H. J., M. Frink, Y. Linde, B. Banas, M. Wornle, C. D. Cohen, V. Vielhauer, P. J. Nelson, H. J. Grone, and D. Schlondorff. 2003. CC chemokine ligand 5/RANTES chemokine antagonists aggravate glomerulonephritis despite reduction of glomerular leukocyte infiltration. *J. Immunol.* **170:**5658–5666.

Anderson, B. E., J. E. Dawson, D. C. Jones, and K. H. Wilson. 1991. *Ehrlichia chaffeensis*, a new species associated with human ehrlichiosis. *J. Clin. Microbiol.* **29:**2838–2842.

Anderson, B. E., K. G. Sims, J. G. Olson, J. E. Childs, J. F. Piesman, C. M. Happ, G. O. Maupin, and B. J. Johnson. 1993. *Amblyomma americanum*: a potential vector of human ehrlichiosis. *Am. J. Trop. Med. Hyg.* **49:**239–244.

Anderson, B. E., J. W. Sumner, J. E. Dawson, T. Tzianabos, C. R. Greene, J. G. Olson, D. B. Fishbein, M. Olsen-Rasmussen, B. P. Holloway, and E. H. George. 1992. Detection of the etiologic agent of human ehrlichiosis by polymerase chain reaction. *J. Clin. Microbiol.* **30:**775–780.

Bakken, J. S., and J. S. Dumler. 2003. Human granulocytic ehrlichiosis. *Clin. Infect. Dis.* **31:**554–560.

Bakken, J. S., and J. S. Dumler. 2004. Ehrlichiosis and anaplasmosis. *Infect. Med.* **21:**433–451.

Bakken, J. S., J. Krueth, C. Wilson-Nordskog, K. Asanovich, and J. S. Dumler. 1996. Clinical and laboratory characteristics of human granulocytic ehrlichiosis. *JAMA* **275:**199–205.

Bao, W., Y. Kumagai, H. Niu, M. Yamaguchi, K. Miura, and Y. Rikihisa. 2009. Four VirB6 paralogs and VirB9 are expressed and interact in *Ehrlichia chaffeensis*-containing vacuoles. *J. Bacteriol.* **191:**278–286.

Barbet, A. F., J. T. Agnes, A. L. Moreland, A. M. Lundgren, A. R. Alleman, S. M. Noh, K. A. Brayton, U. G. Munderloh, and G. H. Palmer. 2005. Identification of functional promoters in the *msp2* expression loci of *Anaplasma marginale* and *Anaplasma phagocytophilum*. *Gene* **353:**89–97.

Barbet, A. F., A. Lundgren, J. Yi, F. R. Rurangirwa, and G. H. Palmer. 2000. Antigenic variation of *Anaplasma marginale* by expression of MSP2 mosaics. *Infect. Immun.* **68:**6133–6138.

Barbet, A. F., P. F. Meeus, M. Bélanger, M. V. Bowie, J. Yi, A. M. Lundgren, A. R. Alleman, S. J. Wong, F. K. Chu, U. G. Munderloh, and S. D. Jauron. 2003. Expression of multiple outer membrane protein sequence variants from a single genomic locus of *Anaplasma phagocytophilum*. *Infect. Immun.* **71:**1706–1718.

Berg, R. E., E. Crossley, S. Murray, and J. Forman. 2005. Relative contributions of NK and CD8 T cells to IFN-g mediated innate immune protection against *Listeria monocytogenes*. *J. Immunol.* **175:**1751–1757.

Birkner, K., B. Steiner, C. Rinkler, Y. Kern, P. Aichele, C. Bogdan, and F. D. von Loewenich. 2008. The elimination of *Anaplasma phagocytophilum* requires CD4$^+$ T cells, but is independent of Th1 cytokines and a wide spectrum of effector mechanisms. *Eur. J. Immunol.* **38:**3395–3410.

Bitsaktsis, C., J. Huntington, and G. M. Winslow. 2004. Production of IFN-g by CD4 T cells is essential for resolving *Ehrlichia* infection. *J. Immunol.* **172:**6894–6901.

Bitsaktsis, C., B. Nandi, R. Racine, K. C. Macnamara, and G. Winslow. 2007. T cell-independent humoral immunity is sufficient for protection against fatal intracellular *Ehrlichia* infection. *Infect. Immun.* **75:**4933-4941.

Bitsaktsis, C., and G. Winslow. 2006. Fatal recall responses mediated by CD8 T cells during intracellular bacterial challenge infection. *J. Immunol.* **177:**4644–4651.

Borjesson, D. L., and S. W. Barthold. 2002. The mouse as a model for investigation of human granulocytic ehrlichiosis: current knowledge and future directions. *Comp. Med.* **52:**403–413.

Borjesson, D. L., S. D. Kobayashi, A. R. Whitney, J. M. Voyich, C. M. Argue, and F. R. Deleo. 2005. Insights into pathogen immune evasion mechanisms: *Anaplasma phagocytophilum* fails to induce an apoptosis differentiation program in human neutrophils. *J. Immunol.* **174:**6364–6372.

Brayton, K. A., L. S. Kappmeyer, D. R. Herndon, M. J. Dark, D. L. Tibbals, G. H. Palmer, T. C. McGuire, and D. P. Knowles, Jr. 2005. Complete genome sequencing of *Anaplasma marginale* reveals that the surface is skewed to two superfamilies of outer membrane proteins. *Proc. Natl. Acad. Sci. USA* **102:**844–849.

Brown, W. C., K. A. Brayton, C. M. Styer, and G. H. Palmer. 2003. The hypervariable region of *Anaplasma marginale* major surface protein 2 (MSP2) contains multiple immunodominant CD4$^+$ T lymphocyte epitopes that elicit variant-specific proliferative and IFN-g responses in MSP2 vaccinates. *J. Immunol.* **170:**3790–3798.

Cai, G., R. A. Kastelein, and C. A. Hunter. 1999. IL-10 enhances NK cell proliferation, cytotoxicity and production of IFN-g when combined with IL-18. *Eur. J. Immunol.* **29:**2658–2665.

Carlyon, J. A., D. Abdel-Latif, M. Pypaert, P. Lacy, and E. Fikrig. 2004. *Anaplasma phagocytophilum* utilizes multiple host evasion mechanisms to thwart NADPH oxidase-mediated killing during neutrophil infection. *Infect. Immun.* **72:** 4772–4783.

Carlyon, J. A., M. Akkoyunlu, L. Xia, T. Yago, T. Wang, R. D. Cummings, R. P. McEver, and E. Fikrig. 2003. Murine neutrophils require a1,3-fucosylation but not PSGL-1 for productive infection with *Anaplasma phagocytophilum*. *Blood* **102:**3387–3395.

Chamorro, S., J. J. García-Vallejo, W. W. Unger, R. J. Fernandes, S. C. Bruijns, S. Laban, B. O. Roep, B. A. 't Hart, and Y. van Kooyk. 2009. TLR triggering on tolerogenic dendritic cells results in TLR2 up-regulation and a reduced proinflammatory immune program. *J. Immunol.* **183:**2984–2994.

Chapman, A. S., J. S. Bakken, S. M. Folk, C. D. Paddock, K. C. Bloch, A. Krusell, D. J. Sexton, S. C. Buckingham, G. S. Marshall, G. A. Storch, G. A. Dasch, J. H. McQuiston, D. L. Swerdlow, S. J. Dumler, W. L. Nicholson, D. H. Walker, M. E. Eremeeva, and C. A. Ohl. 2006. Diagnosis and management of tickborne rickettsial diseases: Rocky Mountain spotted fever, ehrlichioses, and anaplasmosis—United States: a practical guide for physicians and other health-care and public health professionals. *MMWR Recomm. Rep.* **55:**1–27.

Chen, T., R. Zamora, B. Zuckerbraun, and T. R. Billiar. 2003. Role of nitric oxide in liver injury. *Curr. Mol. Med.* **3:**519–526.

Cheng, C., C. D. Paddock, and R. R. Ganta. 2003. Molecular heterogeneity of *Ehrlichia chaffeensis* isolates determined by sequence analysis of the 28-kilodalton outer membrane protein genes and other regions of the genome. *Infect. Immun.* **71:**187–195.

Childs, J. E., J. W. Sumner, W. L. Nicholson, R. F. Massung, S. M. Standaert, and C. D. Paddock. 1999. Outcome of diagnostic tests using samples from patients with culture-proven human monocytic ehrlichiosis: implications for surveillance. *J. Clin. Microbiol.* **37:**2997–3000.

Choi, K. S., J. T. Park, and J. S. Dumler. 2005. *Anaplasma phagocytophilum* delay of neutrophil apoptosis through the p38 mitogen-activated protein kinase signal pathway. *Infect. Immun.* **73:**8209–8218.

Choi, K. S., T. Webb, M. Oelke, D. G. Scorpio, and J. S. Dumler. 2007. Differential innate immune cell activation and proinflammatory response in *Anaplasma phagocytophilum* infection. *Infect. Immun.* **75:**3124–3130.

Dawson, J. E., B. E. Anderson, D. B. Fishbein, J. L. Sanchez, C. S. Goldsmith, K. H. Wilson, and C. W. Duntley. 1991. Isolation and characterization of an *Ehrlichia* sp. from a patient diagnosed with human ehrlichiosis. *J. Clin. Microbiol.* **29:**2741–2745.

Demma, L. J., R. C. Holman, J. H. McQuiston, J. W. Krebs, and D. L. Swerdlow. 2006. Human monocytic ehrlichiosis and human granulocytic anaplasmosis in the United States, 2001–2002. *Ann. N. Y. Acad. Sci.* **1078:**118–119.

den Haan, J. M., S. M. Lehar, and M. J. Bevan. 2000. $CD8^+$ but not $CD8^-$ dendritic cells cross-prime cytotoxic T cells in vivo. *J. Exp. Med.* **192:**1685–1696.

de Sousa, R., N. Ismail, S. D. Nobrega, A. França, M. Amaro, M. Anes, J. Poças, R. Coelho, J. Torgal, F. Bacellar, and D. H. Walker. 2007. Intralesional expression of mRNA of interferon-g, tumor necrosis factor-a, interleukin-10, nitric oxide synthase, indoleamine-2,3-dioxygenase, and RANTES is a major immune effector in Mediterranean spotted fever rickettsiosis. *J. Infect. Dis.* **196:**770–781.

Dierberg, K. L., and J. S. Dumler. 2006. Lymph node hemophagocytosis in rickettsial diseases: a pathogenetic role for CD8 T lymphocytes in human monocytic ehrlichiosis (HME)? *BMC Infect. Dis.* **6:**121.

Doyle, C. K., M. B. Labruna, E. B. Breitschwerdt, Y. W. Tang, R. E. Corstvet, B. C. Hegarty, K. C. Bloch, P. Li, D. H. Walker, and J. W. McBride. 2005. Detection of medically important *Ehrlichia* by quantitative multicolor TaqMan real-time polymerase chain reaction of the *dsb* gene. *J. Mol. Diagn.* **7:**504–510.

Dumler, J. S. 2005. *Anaplasma* and *Ehrlichia* infection. *Ann. N. Y. Acad. Sci.* **1063:**361–373.

Dumler, J. S., and J. S. Bakken. 1995. Ehrlichial diseases of humans: emerging tick-borne infections. *Clin. Infect. Dis.* **20:**1102–1110.

Dumler, J. S., N. C. Barat, C. E. Barat, and J. S. Bakken. 2007a. Human granulocytic anaplasmosis and macrophage activation. *Clin. Infect. Dis.* **45:**199–204.

Dumler, J. S., A. F. Barbet, C. P. Bekker, G. A. Dasch, G. H. Palmer, S. C. Ray, Y. Rikihisa, and F. R. Rurangirwa. 2001. Reorganization of genera in the families *Rickettsiaceae* and *Anaplasmataceae* in the order *Rickettsiales*: unification of some species of *Ehrlichia* with *Anaplasma*, *Cowdria* with *Ehrlichia* and *Ehrlichia* with *Neorickettsia*, descriptions of six new species combinations and designation of *Ehrlichia equi* and 'HGE agent' as subjective synonyms of *Ehrlichia phagocytophila*. *Int. J. Syst. Evol. Microbiol.* **51:**2145–2165.

Dumler, J. S., J. E. Dawson, and D. H. Walker. 1993. Human ehrlichiosis: hematopathology and immunohistologic detection of *Ehrlichia chaffeensis*. *Hum. Pathol.* **24:**391–396.

Dumler, J. S., J. E. Madigan, N. Pusterla, and J. S. Bakken. 2007b. Ehrlichioses in humans: epidemiology, clinical presentation, diagnosis, and treatment. *Clin. Infect. Dis.* **45**(Suppl. 1)**:**S45–S51.

Dumler, J. S., E. R. Trigiani, J. S. Bakken, M. E. Aguero-Rosenfeld, and G. P. Wormser. 2000. Serum cytokine responses during acute human granulocytic ehrlichiosis. *Clin. Diagn. Lab. Immunol.* **7:**6–8.

Dunning Hotopp, J. C., M. Lin, R. Madupu, J. Crabtree, S. V. Angiuoli, J. Eisen, R. Seshadri, Q. Ren, M. Wu, T. R. Utterback, S. Smith, M. Lewis, H. Khouri, C. Zhang, H. Niu, Q. Lin, N. Ohashi, N. Zhi, W. Nelson, L. M. Brinkac, R. J. Dodson, M. J. Rosovitz, J. Sundaram, S. C. Daugherty, T. Davidsen, A. S. Durkin, M. Gwinn, D. H. Haft, J. D. Selengut, S. A. Sullivan, N. Zafar, L. Zhou, F. Benahmed, H. Forberger, R. Halpin, S. Mulligan, J. Robinson, O. White, Y. Rikihisa, and H. Tettelin. 2006. Comparative genomics of emerging human ehrlichiosis agents. *PLoS Genet.* **2:**e21.

Ehlers, S., C. Holscher, S. Scheu, C. Tertilt, T. Hehlgans, J. Suwinski, R. Endres, and K. Pfeffer. 2003. The lymphotoxin b receptor is critically involved in controlling infections with the intracellular pathogens *Mycobacterium tuberculosis* and *Listeria monocytogenes*. *J. Immunol.* **170:**5210–5218.

Ejrnaes, M., C. M. Filippi, M. M. Martinic, E. M. Ling, L. M. Togher, S. Crotty, and M. G. von Herrath. 2006. Resolution of viral infection after interleukin-10 receptor blockade. *J. Exp. Med.* **203:**2461–2472.

Emmanuilidis, K., H. Weighardt, S. Maier, K. Gerauer, T. Fleischmann, X. X. Zheng, W. W. Hancock, B. Holzmann, and C. D. Heidecke. 2001. Critical role of Kupffer cell-derived IL-10 for host defense in septic peritonitis. *J. Immunol.* **167:**3919–3927.

Eng, T. R., J. R. Harkess, D. B. Fishbein, J. E. Dawson, C. N. Greene, M. A. Redus, and F. T. Satalowich. 1990. Epidemiologic, clinical, and laboratory findings of human ehrlichiosis in the United States, 1988. *JAMA* **264:**2251–2258.

Ereemeva, M. E., A. Oliveira, J. Moriarity, J. B. Robinson, N. K. Tokarevich, L. P. Antyukova, V. A. Pyanyh, O. N. Emeljanova, V. N. Ignatjeva, R. Buzinov, V. Pyankova, and G. A. Dasch. 2007. Detection and identification of bacterial agents in *Ixodes persulcatus* Schulze ticks from the north western region of Russia. *Vector-Borne Zoonotic Dis.* **7:**426–436.

Estrada-Peña, A., I. G. Horak, and T. Petney. 2008. Climate changes and suitability for the ticks *Amblyomma hebraeum* and *Amblyomma variegatum* (Ixodidae) in Zimbabwe (1974–1999). *Vet. Parasitol.* **151:**256–267.

Everett, E. D., K. A. Evans, R. B. Henry, and G. McDonald. 1994. Human ehrlichiosis in adults after tick exposure: diagnosis using polymerase chain reaction. *Ann. Intern. Med.* **120:**730–735.

Ewing, S. A., J. E. Dawson, A. A. Kocan, R. W. Barker, C. K. Warner, R. J. Panciera, J. C. Fox, K. M. Kocan, and E. F. Blouin. 1995. Experimental transmission of *Ehrlichia chaffeensis* (Rickettsiales: Ehrlichieae) among white-tailed deer by *Amblyomma americanum* (Acari: Ixodidae). *J. Med. Entomol.* **32:**368–374.

Faustin, B., L. Lartigue, J. M. Bruey, F. Luciano, E. Sergienko, B. Bailly-Maitre, N. Volkmann, D. Hanein, I. Rouiller, and J. C. Reed. 2007. Reconstituted NALP1 inflammasome reveals two-step mechanism of caspase-1 activation. *Mol. Cell* **25:**713–724.

Fichtenbaum, C. J., L. R. Peterson, and G. J. Weil. 1993. Ehrlichiosis presenting as a life-threatening illness with features of the toxic shock syndrome. *Am. J. Med.* **95:**351–357.

Fischer, K., E. Scotet, M. Niemeyer, H. Koebernick, J. Zerrahn, S. Maillet, R. Hurwitz, M. Kursar, M. Bonneville, S. H. Kaufmann, and U. E. Schaible. 2004. Mycobacterial phosphatidylinositol mannoside is a natural antigen for CD1d-restricted T cells. *Proc. Natl. Acad. Sci. USA* **101:**10685–10690.

Fishbein, D. B., J. E. Dawson, and L. E. Robinson. 1994. Human ehrlichiosis in the United States, 1985 to 1990. *Ann. Intern. Med.* **120:** 736–743.

Foster, J., M. Ganatra, I. Kamal, J. Ware, K. Makarova, N. Ivanova, A. Bhattacharyya, V. Kapatral, S. Kumar, J. Posfai, T. Vincze, J. Ingram, L. Moran, A. Lapidus, M. Omelchenko, N. Kyrpides, E. Ghedin, S. Wang, E. Goltsman, V. Joukov, O. Ostrovskaya, K. Tsukerman, M. Mazur, D. Comb, E. Koonin, and B. Slatko. 2005. The *Wolbachia* genome of *Brugia malayi*: endosymbiont evolution within a human pathogenic nematode. *PLoS Biol.* **3:**e121.

French, A. R., E. B. Holroyd, L. Yang, S. Kim, and W. M. Yokoyama. 2006. IL-18 acts synergistically with IL-15 in stimulating natural killer cell proliferation. *Cytokine* **35:**229–234.

French, D. M., W. C. Brown, and G. H. Palmer. 1999. Emergence of *Anaplasma marginale* antigenic variants during persistent rickettsemia. *Infect. Immun.* **67:**5834–5840.

French, D. M., T. F. McElwain, T. C. McGuire, and G. H. Palmer. 1998. Expression of *Anaplasma marginale* major surface protein 2 variants during persistent cyclic rickettsemia. *Infect. Immun.* **66:**1200–1207.

Fujii, S., K. Liu, C. Smith, A. J. Bonito, and R. M. Steinman. 2004. The linkage of innate to adaptive immunity via maturing dendritic cells in vivo requires CD40 ligation in addition to antigen presentation and CD80/86 costimulation. *J. Exp. Med.* **199:**1607–1618.

Ganta, R. R., C. Cheng, E. C. Miller, B. L. McGuire, L. Peddireddi, K. R. Sirigireddy, and S. K. Chapes. 2007. Differential clearance and immune responses to tick cell vs. macrophage culture-derived *Ehrlichia chaffeensis* in mice. *Infect. Immun.* **75:**135–145.

Ganta, R. R., C. Cheng, M. J. Wilkerson, and S. K. Chapes. 2004. Delayed clearance of *Ehrlichia chaffeensis* infection in $CD4^+$ T-cell knockout mice. *Infect. Immun.* **72:**159–167.

Ganta, R. R., M. J. Wilkerson, C. Cheng, A. M. Rokey, and S. K. Chapes. 2002. Persistent *Ehrlichia chaffeensis* infection occurs in the absence of functional major histocompatibility complex class II genes. *Infect. Immun.* **70:**380–388.

Garcia-Garcia, J. C., N. C. Barat, S. J. Trembley, and J. S. Dumler. 2009a. Epigenetic silencing of host cell defense genes enhances intracellular survival of the rickettsial pathogen *Anaplasma phagocytophilum*. *PLoS Pathog.* **5:**e1000488.

Garcia-Garcia, J. C., K. E. Rennoll-Bankert, S. Pelly, A. M. Milstone, and J. S. Dumler. 2009b. Silencing of host cell *CYBB* gene expression by the nuclear effector AnkA of the intracellular pathogen *Anaplasma phagocytophilum*. *Infect. Immun.* **77:** 2385–2391.

Gazzinelli, R. T., M. Wysocka, S. Hieny, T. Scharton-Kersten, A. Cheever, R. Kühn, W. Muller, G. Trincheri, and A. Sher. 1996. In the absence of endogenous IL-10, mice acutely infected with *Toxoplasma gondii* succumb to a lethal immune response dependent on $CD4^+$ T cells and accompanied by overproduction of IL-12, IFN-gamma and TNF-alpha. *J. Immunol.* **157:**798–805.

Ge, Y., and Y. Rikihisa. 2006. *Anaplasma phagocytophilum* delays spontaneous human neutrophil apoptosis by modulation of multiple apoptotic pathways. *Cell. Microbiol.* **8:**1406–1416.

Ghose, P., A. Ali, R. Fang, D. Forbes, B. Ballard, and N. Ismail. 2011. The interaction between IL-18 and IL-18R limits the magnitude of protective immunity and enhances pathogenic responses following infection with intracellular bacteria. *J. Immunol.* **187:**1333–1346.

Guicciardi, M. E., and G. J. Gores. 2005. Apoptosis: a mechanism of acute and chronic liver injury. *Gut* **54:**1024–1033.

Hamburg, B. J., G. A. Storch, S. T. Micek, and M. H. Kollef. 2008. The importance of early treatment with doxycycline in human ehrlichiosis. *Medicine (Baltimore)* **87:**53–60.

Han, S., J. Norimine, K. A. Brayton, G. H. Palmer, G. A. Scoles, and W. C. Brown. 2010. *Anaplasma marginale* infection with persistent high-load bacteremia induces a dysfunctional memory CD4$^+$ T lymphocyte response but sustained high IgG titers. *Clin. Vaccine Immunol.* **17:**1881–1890.

Han, S., J. Norimine, G. H. Palmer, W. Mwangi, K. K. Lahmers, and W. C. Brown. 2008. Rapid deletion of antigen-specific CD4+ T cells following infection represents a strategy of immune evasion and persistence for *Anaplasma marginale*. J. *Immunol.* **181**:7759-7769.

Harkess, J. R. 1991. Ehrlichiosis. *Infect. Dis. Clin. North. Am.* **5:**37–51.

Herron, M. J., C. M. Nelson, J. Larson, K. R. Snapp, G. S. Kansas, and J. L. Goodman. 2000. Intracellular parasitism by the human granulocytic ehrlichiosis bacterium through the P-selectin ligand, PSGL-1. *Science* **288:**1653–1656.

Higuchi, H., and G. J. Gores. 2003. Mechanisms of liver injury: an overview. *Curr. Mol. Med.* **3:**483–490.

Hildebrand, F., W. J. Hubbard, M. A. Choudhry, M. Frink, H. C. Pape, S. L. Kunkel, and I. H. Chaudry. 2006. Kupffer cells and their mediators: the culprits in producing distant organ damage after trauma-hemorrhage. *Am. J. Pathol.* **169:**784–794.

Houde, M., S. Bertholet, E. Gagnon, S. Brunet, G. Goyette, A. Laplante, M. F. Princiotta, P. Thibault, D. Sacks, and M. Desjardins. 2003. Phagosomes are competent organelles for antigen cross-presentation. *Nature* **425:**402–406.

IJdo, J. W., A. C. Carlson, and E. L. Kennedy. 2007. *Anaplasma phagocytophilum* AnkA is tyrosine-phosphorylated at EPIYA motifs and recruits SHP-1 during early infection. *Cell. Microbiol.* **9:**1284–1296.

IJdo, J. W., and A. C. Mueller. 2004. Neutrophil NADPH oxidase is reduced at the *Anaplasma phagocytophilum* phagosome. *Infect. Immun.* **72:**5392–5401.

IJdo, J. W., C. Wu, L. A. Magnarelli, and, E. Fikrig. 1999. Serodiagnosis of human granulocytic ehrlichiosis by a recombinant HGE-44-based enzyme-linked immunosorbent assay. *J. Clin. Microbiol.* **37:**3540–3544.

Ingjerdingen, M., B. Damaj, and A. A. Maghazachi. 2001. Expression and regulation of chemokine receptors in human natural killer cells. *Blood* **97:**367–375.

Inohara, N., M. Chamaillard, C. McDonald, and G. Nunez. 2005. NOD-LRR proteins: role in host-microbial interactions and inflammatory disease. *Annu. Rev. Biochem.* **74:**355–383.

Ismail, N., E. C. Crossley, H. L. Stevenson, and D. H. Walker. 2007. Relative importance of T-cell subsets in monocytotropic ehrlichiosis: a novel effector mechanism involved in *Ehrlichia*-induced immunopathology in murine ehrlichiosis. *Infect. Immun.* **75:**4608–4620.

Ismail, N., L. Soong, J. W. McBride, G. Valbuena, J. P. Olano, H. M. Feng, and D. H. Walker. 2004. Overproduction of TNF-α by CD8 type 1 cells and down-regulation of IFN-γ production by CD4 Th1 cells contribute to toxic shock-like syndrome in an animal model of fatal monocytotropic ehrlichiosis. *J. Immunol.* **172:**1786–1800.

Ismail, N., H. L. Stevenson, and D. H. Walker. 2006. Role of tumor necrosis factor alpha (TNF-α) and interleukin-10 in the pathogenesis of severe murine monocytotropic ehrlichiosis: increased resistance of TNF receptor p55- and p75-deficient mice to fatal ehrlichial infection. *Infect. Immun.* **74:**1846–1856.

Kawai, T., and S. Akira. 2005. Pathogen recognition with Toll-like receptors. *Curr. Opin. Immunol.* **17:**338–344.

Kawai, T., and S. Akira. 2007a. Signaling to NF-κB by Toll-like receptors. *Trends Mol. Med.* **13:**460–469.

Kawai, T., and S. Akira. 2007b. TLR signaling. *Semin. Immunol.* **19:**24–32.

Kim, S., J. Poursine-Laurent, S. M. Truscott, L. Lybarger, Y. J. Song, L. Yang, A. R. French, J. B. Sunwoo, S. Lemieux, T. H. Hansen, and W. M. Yokoyama. 2005. Licensing of natural killer cells by host major histocompatibility complex class I molecules. *Nature* **436:**709–713.

Kinjo, Y., E. Tupin, D. Wu, M. Fujio, R. Garcia-Navarro, M. R. Benhnia, D. M. Zajonc, G. Ben-Menachem, G. D. Ainge, G. F. Painter, A. Khurana, K. Hoebe, S. M. Behar, B. Beutler, I. A. Wilson, M. Tsuji, T. J. Sellati, C. H. Wong, and M. Kronenberg. 2006. Natural killer T cells recognize diacylglycerol antigens from pathogenic bacteria. *Nat. Immunol.* **7:**978–986.

Klein, M. B., S. Hu, C. C. Chao, and J. L. Goodman. 2000. The agent of human granulocytic ehrlichiosis induces the production of myelosuppressing chemokines without induction of proinflammatory cytokines. *J. Infect. Dis.* **182:**200–205.

Ko, S. Y., K. A. Lee, H. J. Youn, Y. J. Kim, H. J. Ko, T. H. Heo, M. N. Kweon, and C. Y. Kang. 2007. Mediastinal lymph node CD8a⁻ DC initiates antigen presentation following intranasal coadministration of a-GalCer. *Eur. J. Immunol.* **8:**2127–2137.

Koh, Y. S., J. E. Koo, A. Biswas, and K. S. Kobayashi. 2010. MyD88-dependent signaling contributes to host defense against ehrlichial infection. *PLoS One* **5:**e11758.

Kurtz, J., and K. Franz. 2003. Innate defence: evidence for memory in invertebrate immunity. *Nature* **425:**37–38.

Lanier, L. L. 2005. NK cell recognition. *Annu. Rev. Immunol.* **23:**225–274.

Lantz, O., and A. Bendelac. 1994. An invariant T cell receptor α chain is used by a unique subset of major histocompatibility complex class l-specific $CD4^+$ and $CD4^-8^-$ T cells in mice and humans. *J. Exp. Med.* **180:**1097–1106.

Lara-Tejero, M., F. S. Sutterwala, Y. Ogura, E. P. Grant, J. Bertin, A. J. Coyle, R. A. Flavell, and J. E. Galan. 2006. Role of the caspase-1 inflammasome in *Salmonella typhimurium* pathogenesis. *J. Exp. Med.* **203:**1407–1412.

Lauwerys, B. R., J. C. Renauld, and F. A. Houssiau. 1999. Synergistic proliferation and activation of natural killer cells by interleukin 12 and interleukin 18. *Cytokine* **11:**822–830.

Lee, H. C., and J. L. Goodman. 2006. *Anaplasma phagocytophilum* causes global induction of antiapoptosis in human neutrophils. *Genomics* **88:**496–503.

Lepidi, H., J. E. Bunnell, M. E. Martin, J. E. Madigan, S. Stuen, and J. S. Dumler. 2000. Comparative pathology and immunohistology associated with clinical illness after *Ehrlichia phagocytophila*-group infections. *Am. J. Trop. Med. Hyg.* **62:**29–37.

Li, J., E. Yager, M. Reilly, C. Freeman, G. R. Reddy, A. A. Reilly, F. K. Chu, and G. M. Winslow. 2001. Outer membrane protein-specific monoclonal antibodies protect SCID mice from fatal infection by the obligate intracellular bacterial pathogen *Ehrlichia chaffeensis*. *J. Immunol.* **166:**1855–1862.

Lin, M., A. den Dulk-Ras, P. J. Hooykaas, and Y. Rikihisa. 2007. *Anaplasma phagocytophilum* AnkA secreted by type IV secretion system is tyrosine phosphorylated by Abl-1 to facilitate infection. *Cell. Microbiol.* **9:**2644–2657.

Lin, M., and Y. Rikihisa. 2003a. Obligatory intracellular parasitism by *Ehrlichia chaffeensis* and *Anaplasma phagocytophilum* involves caveolae and glycosylphosphatidylinositol-anchored proteins. *Cell. Microbiol.* **5:**809–820.

Lin, M., and Y. Rikihisa. 2003b. *Ehrlichia chaffeensis* and *Anaplasma phagocytophilum* lack genes for lipid A biosynthesis and incorporate cholesterol for their survival. *Infect. Immun.* **71:**5324–5331.

Lin, M., and Y. Rikihisa. 2004. *Ehrlichia chaffeensis* downregulates surface Toll-like receptors 2/4, CD14 and transcription factors PU.1 and inhibits lipopolysaccharide activation of NF-kB, ERK 1/2 and p38 MAPK in host monocytes. *Cell. Microbiol.* **6:**175–186.

Lin, M., and Y. Rikihisa. 2007. Degradation of $p22^{phox}$ and inhibition of superoxide generation by *Ehrlichia chaffeensis* in human monocytes. *Cell. Microbiol.* **9:**861–874.

Lin, Q., Y. Rikihisa, N. Ohashi, and, N. Zhi. 2003. Mechanisms of variable *p44* expression by *Anaplasma phagocytophilum*. *Infect. Immun.* **71:**5650–5661.

Liu, Z. X., S. Govindarajan, S. Okamoto, and G. Dennert. 2000. NK cells cause liver injury and facilitate the induction of T cell-mediated immunity to a viral liver infection. *J. Immunol.* **164:**6480–6486.

Lockhart, J. M., W. R. Davidson, D. E. Stallknecht, and J. E. Dawson. 1996. Site-specific geographic association between *Amblyomma americanum* (Acari: Ixodidae) infestations and *Ehrlichia chaffeensis*-reactive (Rickettsiales: Ehrlichieae) antibodies in whitetailed deer. *J. Med. Entomol.* **33:**153–158.

Loetscher, P., M. Seitz, I. Clark-Lewis, M. Baggiolini, and B. Moser. 1996. Activation of NK cells by CC chemokines. Chemotaxis, Ca^{2+} mobilization, and enzyme release. *J. Immunol.* **156:**322–327.

Luo, T., X. Zhang, A. Wakeel, V. L. Popov, and J. W. McBride. 2008. A variable-length PCR target protein of *Ehrlichia chaffeensis* contains major species-specific antibody epitopes in acidic serinerich tandem repeats. *Infect. Immun.* **76:**1572–1580.

Mariathasan, S., and D. M. Monack. 2007. Inflammasome adaptors and sensors: intracellular regulators of infection and inflammation. *Nat. Rev. Immunol.* **7:**31–40.

Marshall, G. S., R. F. Jacobs, G. E. Schutze, H. Paxton, S. C. Buckingham, J. P. DeVincenzo, M. A. Jackson, V. H. San Joaquin, S. M. Standaert, C. R. Woods, and Tick-Borne Infections in Children Study Group. 2002. *Ehrlichia chaffeensis* seroprevalence among children in the southeast and south-central regions of the United States. *Arch. Pediatr. Adolesc. Med.* **156:**166–170.

Martin, M. E., J. E. Bunnell, and J. S. Dumler. 2000. Pathology, immunohistology, and cytokine responses in early phases of human granulocytic ehrlichiosis in a murine model. *J. Infect. Dis.* **181:** 374–378.

Martin, M. E., K. Caspersen, and J. S. Dumler. 2001. Immunopathology and ehrlichial propagation are regulated by interferon-g and interleukin-10 in a murine model of human granulocytic ehrlichiosis. *Am. J. Pathol.* **158:**1881–1888.

Matsuda, J. L., O. V. Naidenko, L. Gapin, T. Nakayama, M. Taniguchi, C. R. Wang, Y. Koezuka, and M. Kronenberg. 2000. Tracking the response of natural killer T cells to a glycolipid antigen using CD1d tetramers. *J. Exp. Med.* **192:**741–754.

Mattner, J., K. L. Debord, N. Ismail, R. D. Goff, C. Cantu III, D. Zhou, P. Saint-Mezard, V. Wang, Y. Gao, N. Yin, K. Hoebe, O. Schneewind, D. Walker, B. Beutler, L. Teyton, P. B. Savage, and A. Bendelac. 2005. Exogenous and endogenous glycolipid antigens activate NKT cells during microbial infections. *Nature* **434:**525–529.

Mavromatis, K., C. K. Doyle, A. Lykidis, N. Ivanova, M. P. Francino, P. Chain, M. Shin, S. Malfatti, F. Larimer, A. Copeland, J. C. Detter, M. Land, P. M. Richardson, X. J. Yu, D. H. Walker, J. W. McBride, and N. C. Kyrpides. 2006. The genome of the obligately intracellular bacterium *Ehrlichia canis* reveals themes of complex membrane structure and immune evasion strategies. *J. Bacteriol.* **188:**4015–4023.

Meylan, E., J. Tschopp, and M. Karin. 2006. Intracellular pattern recognition receptors in the host response. *Nature* **442:**39–44.

Miura, K., and Y. Rikihisa. 2007. Virulence potential of *Ehrlichia chaffeensis* strains of distinct genome sequences. *Infect. Immun.* **75:**3604–3613.

Mordue, D. G., F. Monroy, M. La Regina, C. A. Dinarello, and L. D. Sibley. 2001. Acute toxoplasmosis leads to lethal overproduction of Th1 cytokines. *J. Immunol.* **167:**4574-4584.

Mott, J., Y. Rikihisa, and S. Tsunawaki. 2002. Effects of *Anaplasma phagocytophila* on NADPH oxidase components in human neutrophils and HL-60 cells. *Infect. Immun.* **70:**1359–1366.

Munderloh, U. G., M. J. Lynch, M. J. Herron, A. T. Palmer, T. J. Kurtti, R. D. Nelson, and J. L. Goodman. 2004. Infection of endothelial cells with *Anaplasma marginale* and *A. phagocytophilum*. *Vet. Microbiol.* **101:**53–64.

Munz, C., R. M. Steinman, and S. Fujii. 2005. Dendritic cell maturation by innate lymphocytes: coordinated stimulation of innate and adaptive immunity. *J. Exp. Med.* **202:**203–207.

Nadelman, R. B., H. W. Horowitz, T-C. Hsieh, J. M. Wu, M. E. Aguero-Rosenfeld, I. Schwartz, J. Nowakowski, S. Varde, and G. P. Wormser. 1997. Simultaneous human granulocytic ehrlichiosis and Lyme borreliosis. *N. Engl. J. Med.* **337:**27–30.

Nagarajan, N. A., and M. Kronenberg. 2007. Invariant NKT cells amplify the innate immune response to lipopolysaccharide. *J. Immunol.* **178:**2706–2713.

Niu, H., M. Yamaguchi, and Y. Rikihisa. 2008. Subversion of cellular autophagy by *Anaplasma phagocytophilum*. *Cell. Microbiol.* **10:**593–605.

O'Garra, A., and P. Vieira. 2007. TH1 cells control themselves by producing interleukin-10. *Nat. Rev. Immunol.* **7:**425–428.

Okada, H., T. Tajima, M. Kawahara, and Y. Rikihisa. 2001. Ehrlichial proliferation and acute hepatocellular necrosis in immunocompetent mice experimentally infected with the HF strain of *Ehrlichia*, closely related to *Ehrlichia chaffeensis*. *J. Comp. Pathol.* **124:**165–171.

Olano, J. P., W. Hogrefe, B. Seaton, and D. H. Walker. 2003a. Clinical manifestations, epidemiology, and laboratory diagnosis of human monocytotropic ehrlichiosis in a commercial laboratory setting. *Clin. Diagn. Lab. Immunol.* **10:**891–896.

Olano, J. P., E. Masters, W. Hogrefe, and D. H. Walker. 2003b. Human monocytotropic ehrlichiosis, Missouri. *Emerg. Infect. Dis.* **9:**1579–1586.

Olano, J. P., and D. H. Walker. 2002. Human ehrlichioses. *Med. Clin. North Am.* **86:**375–392.

Olano, J. P., G. Wen, H. M. Feng, J. W. McBride, and D. H. Walker. 2004. Histologic, serologic, and molecular analysis of persistent ehrlichiosis in a murine model. *Am. J. Pathol.* **165:**997–1006.

Paddock, C. D., D. P. Suchard, K. L. Grumbach, W. K. Hadley, R. L. Kerschmann, N. W. Abbey, J. E. Dawson, B. E. Anderson, K. G. Sims, and J. S. Dumler. 1993. Brief report: fatal seronegative ehrlichiosis in a patient with HIV infection. *N. Engl. J. Med.* **329:**1164–1167.

Palmer, G. H., and K. A. Brayton. 2007. Gene conversion is a convergent strategy for pathogen antigenic variation. *Trends Parasitol.* **23:**408–413.

Park, J., K. J. Kim, K. S. Choi, D. J. Grab, and J. S. Dumler. 2004. *Anaplasma phagocytophilum* AnkA binds to granulocyte DNA and nuclear proteins. *Cell. Microbiol.* **6:**743–751.

Parola, P., B. Davoust, and D. Raoult. 2005. Tick- and flea-borne rickettsial emerging zoonoses. *Vet. Res.* **36:**469–492.

Pedra, J. H., F. S. Sutterwala, B. Sukumaran, Y. Ogura, F. Qian, R. R. Montgomery, R. A. Flavell, and E. Fikrig. 2007. ASC/PYCARD and caspase-1 regulate the IL-18/IFN-γ axis during *Anaplasma phagocytophilum* infection. *J. Immunol.* **179:**4783–4791.

Pritt, B. S., L. M. Sloan, D. K. Johnson, U. G. Munderloh, S. M. Paskewitz, K. M. McElroy, J. D. McFadden, M. J. Binnicker, D. F. Neitzel, G. Liu, W. L. Nicholson, C. M. Nelson, J. J.

Franson, S. A. Martin, S. A. Cunningham, C. R. Steward, K. Bogumill, M. E. Bjorgaard, J. P. Davis, J. H. McQuiston, D. M. Warshauer, M. P. Wilhelm, R. Patel, V. A. Trivedi, and M. E. Eremeeva. 2011. Emergence of a new pathogenic *Ehrlichia* species, Wisconsin and Minnesota, 2009. *N. Engl. J. Med.* **365:**422–429.

Puccetti, P., M. L. Belladonna, and U. Grohmann. 2002. Effects of IL-12 and IL-23 on antigen-presenting cells at the interface between innate and adaptive immunity. *Crit. Rev. Immunol.* **22:**373–390.

Pusterla, N., J. E. Madigan, K. M. Asanovich, J. S. Chae, E. Derock, C. M. Leutenegger, J. B. Pusterla, H. Lutz, and J. S. Dumler. 2000. Experimental inoculation with human granulocytic *Ehrlichia* agent derived from high- and low-passage cell culture in horses. *J. Clin. Microbiol.* **38:**1276–1278.

Ramabu, S. S., D. A. Schneider, K. A. Brayton, M. W. Ueti, T. Graça, J. E. Futse, S. M. Noh, T. V. Baszler, and G. H. Palmer. 2011. Expression of *Anaplasma marginale* ankyrin repeat-containing proteins during infection of the mammalian host and tick vector. *Infect. Immun.* **79:**2847–2855.

Ravetch, J. V., and S. Bolland. 2001. IgG Fc receptors. *Annu. Rev. Immunol.* **19:**275–290.

Reneer, D. V., S. A. Kearns, T. Yago, J. Sims, R. D. Cummings, R. P. McEver, and J. A. Carlyon. 2006. Characterization of a sialic acid- and P-selectin glycoprotein ligand-1-independent adhesin activity in the granulocytotropic bacterium *Anaplasma phagocytophilum*. *Cell. Microbiol.* **8:**1972–1984.

Rikihisa, Y. 2006. *Ehrlichia* subversion of host innate responses. *Curr. Opin. Microbiol.* **9:**95–101.

Rikihisa, Y., and M. Lin. 2010. *Anaplasma phagocytophilum* and *Ehrlichia chaffeensis* type IV secretion and Ank proteins. *Curr. Opin. Microbiol.* **13:**59–66.

Rivera, J., O. Zaragoza, and A. Casadevall. 2005. Antibody-mediated protection against *Cryptococcus neoformans* pulmonary infection is dependent on B cells. *Infect. Immun.* **73:**1141–1150.

Sacca, R., C. A. Cuff, and N. H. Ruddle. 1997. Mediators of inflammation. *Curr. Opin. Immunol.* **9:**851–857.

Scaife, H., Z. Woldehiwet, C. A. Hart, and S. W. Edwards. 2003. *Anaplasma phagocytophilum* reduces neutrophil apoptosis *in vivo*. *Infect. Immun.* **71:**1995–2001.

Schaible, U. E., F. Winau, P. A. Sieling, K. Fischer, H. L. Collins,, K. Hagens, R. L. Modlin, V. Brinkmann, and S. H. Kaufmann. 2003. Apoptosis facilitates antigen presentation to T lymphocytes through MHC-I and CD1 in tuberculosis. *Nat. Med.* **9:**1039–1046.

Schmid, D., J. Dengjel, O. Schoor, S. Stevanovic, and C. Munz. 2006. Autophagy in innate and adaptive immunity against intracellular pathogens. *J. Mol. Med.* **84:**194–202.

Scorpio, D. G., J. S. Dumler, N. C. Barat, J. A. Cook, C. E. Barat, B. A. Stillman, K. C. Debisceglie, M. J. Beall, and R. Chandrashekar. 2011. Comparative strain analysis of *Anaplasma phagocytophilum* infection and clinical outcomes in a canine model of granulocytic anaplasmosis. *Vector Borne Zoonotic Dis.* **3:**223–229.

Scorpio, D. G., C. Leutenegger, J. Berger, N. Barat, J. E. Madigan, and J. S. Dumler. 2008. Sequential analysis of *Anaplasma phagocytophilum msp2* transcription in murine and equine models of human granulocytic anaplasmosis. *Clin. Vaccine Immunol.* **15:**418–424.

Shibata, S., M. Kawahara, Y. Rikihisa, H. Fujita, Y. Watanabe, C. Suto., and T. Ito. 2000. New *Ehrlichia* species closely related to *Ehrlichia chaffeensis* isolated from *Ixodes ovatus* ticks in Japan. *J. Clin. Microbiol.* **38:**1331–1338.

Singu, V., H. Liu, C. Cheng, and R. R. Ganta. 2005. *Ehrlichia chaffeensis* expresses macrophage- and tick cell-specific 28-kilodalton outer membrane proteins. *Infect. Immun.* **73:**79–87.

Sotomayor, E. A., V. L. Popov, H. M. Feng, D. H. Walker, and J. P. Olano. 2001. Animal model of fatal human monocytotropic ehrlichiosis. *Am. J. Pathol.* **158:**757–769.

Standaert, S. M., T. Yu, M. A. Scott, J. E. Childs, C. D. Paddock, W. L. Nicholson, J. Singleton, Jr., and M. J. Blaser. 2000. Primary isolation of *Ehrlichia chaffeensis* from patients with febrile illnesses: clinical and molecular characteristics. *J. Infect. Dis.* **181:**1082–1088.

Stevenson, H. L., E. C. Crossley, N. Thirumalapura, D. H. Walker, and N. Ismail. 2008. Regulatory roles of CD1d-restricted NKT cells in the induction of toxic shock-like syndrome in an animal model of fatal ehrlichiosis. *Infect. Immun.* **76:**1434–1444.

Stevenson, H. L., D. M. Estes, R. N. Thirumalapura, D. H. Walker, and N. Ismail. 2010. Natural killer cells promote tissue injury and systemic inflammatory responses during fatal *Ehrlichia*-induced toxic shock-like syndrome. *Am. J. Pathol.* **177:**766–776.

Stevenson, H. L., J. M. Jordan, Z. Peerwani, H. Q. Wang, D. H. Walker, and N. Ismail. 2006. An intradermal environment promotes a protective type-1 response against lethal systemic monocytotropic ehrlichial infection. *Infect. Immun.* **74:**4856–4864.

Stober, D., I. Jomantaite, R. Schirmbeck, and J. Reimann. 2003. NKT cells provide help for dendritic cell-dependent priming of MHC class I-restricted $CD8^+$ T cells in vivo. *J. Immunol.* **170:**2540–2548.

Sun, J. C., J. N. Beilke, and L. L. Lanier. 2009. Adaptive immune features of natural killer cells. *Nature* **457:**557–561.

Sutterwala, F. S., Y. Ogura, M. Szczepanik, M. Lara-Tejero, S. G. Lichtenberger, E. Grant, J. Bertin, A. J. Coyle, J. E. Galán, P. W. Askenase, and R. A. Flavell. 2006. Critical role for NALP3/CIAS1/cryopyrin in innate and adaptive immunity through its regulation of caspase-1. *Immunity* **24:**317–327.

Szabo, G., L. Romics, Jr., and G. Frendl. 2002. Liver in sepsis and systemic inflammatory response syndrome. *Clin. Liver Dis.* **6:**1045–1066.

Thirumalapura, N. R., E. C. Crossley, D. H. Walker, and N. Ismail. 2009. Persistent infection contributes to heterologous protective immunity against fatal ehrlichiosis. *Infect. Immun.* **77:** 5582–5589.

Thirumalapura, N. R., H. L. Stevenson, D. H. Walker, and N. Ismail. 2008. Protective heterologous immunity against fatal ehrlichiosis and lack of protection following homologous challenge. *Infect. Immun.* **76:**1920–1930.

Thomas, L. D., I. Hongo, K. C. Bloch, Y. W. Tang, and S. Dummer. 2007. Human ehrlichiosis in transplant recipients. *Am. J. Transplant.* **7:**1641–1647.

Ting, J., and B. Davis. 2005. CATERPILLER: a novel gene family important in immunity, cell death, and diseases. *Annu. Rev. Immunol.* **23:**387–414.

Trobonjaca, Z., F. Leithauser, P. Moller, R. Schirmbeck, and J. Reimann. 2001. Activating immunity in the liver. I. Liver dendritic cells (but not hepatocytes) are potent activators of IFN-γ release by liver NKT cells. *J. Immunol.* **167:**1413–1422.

Tsutsui, H., K. Adachi, E. Seki, and K. Nakanishi. 2003. Cytokine-induced inflammatory liver injuries. *Curr. Mol. Med.* **3:**545–559.

Tupin, E., Y. Kinjo, and M. Kronenberg. 2007. The unique role of natural killer T cells in the response to microorganisms. *Nat. Rev. Microbiol.* **5:**405–417.

Unver, A., Y. Rikihisa, R. W. Stich, N. Ohashi, and S. Felek. 2002. The *omp-1* major outer membrane multigene family of *Ehrlichia chaffeensis* is differentially expressed in canine and tick hosts. *Infect. Immun.* **70:**4701–4704.

van Heerden, H., N. E. Collins, K. A. Brayton, C. Rademeyer, and B. A. Allsopp. 2004. Characterization of a major outer membrane protein multigene family in *Ehrlichia ruminantium*. *Gene* **330:**159–168.

Van Kaer, L. 2007. NKT cells T lymphocytes with innate effector functions. *Curr. Opin. Immunol.* **19:**354–364.

Van Kaer, L., and S. Joyce. 2010. The hunt for iNKT cell antigens: a-galactosidase-deficient mice to the rescue? *Immunity* **33:**143–145.

von Loewenich, F. D., D. G. Scorpio, U. Reischl, J. S. Dumler, and C. Bogdan. 2004. Control of *Anaplasma phagocytophilum*, an obligate intracellular pathogen, in the absence of inducible nitric oxide synthase, phagocyte NADPH oxidase, tumor necrosis factor, Toll-like receptor (TLR)2 and TLR4, or the TLR adaptor molecule MyD88. *Eur. J. Immunol.* **34:**1789–1797.

Wakeel, A., J. A. Kuriakose, and J. W. McBride. 2009. An *Ehrlichia chaffeensis* tandem repeat protein interacts with multiple host targets involved in cell signaling, transcriptional regulation, and vesicle trafficking. *Infect. Immun.* **77:**1734–1745.

Walker, D. H., and J. S. Dumler. 1997. Human monocytic and granulocytic ehrlichioses. Discovery and diagnosis of emerging tick-borne infections and the critical role of the pathologist. *Arch. Pathol. Lab. Med.* **121:**785–791.

Walker, D. H., and D. Raoult. 2005. *Rickettsia rickettsii* and other spotted fever group rickettsiae: Rocky Mountain spotted fever and other spotted fevers. p. 2287–2295. *In* G. L. Mandell, J. E. Bennett, and R. Dolin (ed.), *Mandell, Douglas, and Bennett's Principles and Practice of Infectious Diseases*, 6th ed. Churchill Livingstone, Philadelphia, PA.

Walls, J. J., B. Greig, D. F. Neitzel, and J. S. Dumler. 1997. Natural infection of small mammal species in Minnesota with the agent of human granulocytic ehrlichiosis. *J. Clin. Microbiol.* **35:**853–855.

West, A. P., A. A. Koblansky, and S. Ghosh. 2006. Recognition and signaling by Toll-like receptors. *Annu. Rev. Cell Dev. Biol.* **22:**409–437.

Winslow, G. M., E. Yager, K. Shilo, D. N. Collins, and F. K. Chu. 1998. Infection of the laboratory mouse with the intracellular pathogen *Ehrlichia chaffeensis*. *Infect. Immun.* **66:**3892–3899.

Winslow, G. M., E. Yager, K. Shilo, E. Volk, A. Reilly, and F. K. Chu. 2000. Antibody-mediated elimination of the obligate intracellular bacterial pathogen *Ehrlichia chaffeensis* during active infection. *Infect. Immun.* **68:**2187–2195.

Wu, D. Y., N. H. Segal, S. Sidobre, M. Kronenberg, and P. B. Chapman. 2003. Cross-presentation of disialoganglioside GD3 to natural killer T cells. *J. Exp. Med.* **1:**173–181.

Wu, L., and L. Van Kaer. 2011. Natural killer T cells in health and disease. *Front. Biosci. (Schol. Ed.)* **3:**236–251.

Wynn, T. A., A. W. Cheever, M. E. Williams, S. Hieny, P. Caspar, R. Kühn, W. Müller, and A. Sher. 1998. IL-10 regulates liver pathology in acute murine *Schistosomiasis mansoni* but is not required for immune down-modulation of chronic disease. *J. Immunol.* **160:**4473–4480.

Xiong, Q., M. Lin, and Y. Rikihisa. 2009. Cholesterol-dependent *Anaplasma phagocytophilum* exploits the low-density lipoprotein uptake pathway. *PLoS Pathog.* **5:**e1000329.

Xu, L., H. Yoon, M. Q. Zhao, J. Liu, C. V. Ramana, and R. I. Enelow. 2004. Pulmonary immunopathology mediated by antigen-specific expression of TNF-a by antiviral CD8$^+$ T cells. *J. Immunol.* **173:**721–725.

Yager, E., C. Bitsaktsis, B. Nandi, J. W. McBride, and G. Winslow. 2005. Essential role for humoral immunity during *Ehrlichia* infection in immunocompetent mice. *Infect. Immun.* **73:** 8009–8016.

Yevich, S. J., J. L. Sanchez, R. F. DeFraites, C. C. Rives, J. E. Dawson, I. J. Uhaa, B. J. Johnson, and D. B. Fishbein. 1995. Seroepidemiology of infections due to spotted fever group rickettsiae and *Ehrlichia* species in military personnel exposed in areas of the United States where such infections are endemic. *J. Infect. Dis.* **171:**1266–1273.

Yokoyama, W. M., S. Kim, and A. R. French. 2004. The dynamic life of natural killer cells. *Annu. Rev. Immunol.* **22:**405–429.

Yoshimoto, T., A. Bendelac, C. Watson, J. Hu-Li, and W. E. Paul. 1995. Role of NK1.1$^+$ T cells in a TH2 response and in immunoglobulin E production. *Science* **270:**1845–1847.

Yu, X. J., J. W. McBride, and D. H. Walker. 1999. Genetic diversity of the 28-kilodalton outer membrane protein gene in human isolates of *Ehrlichia chaffeensis*. *J. Clin. Microbiol.* **37:**1137–1143.

Zhang, J. Z., H. Guo, G. M. Winslow, and X. J. Yu. 2004a. Expression of members of the 28-kilodalton major outer membrane protein family of *Ehrlichia chaffeensis* during persistent infection. *Infect. Immun.* **72:**4336–4343.

Zhang, J. Z., M. Sinha, B. A. Luxon, and X. J. Yu. 2004b. Survival strategy of obligately intracellular *Ehrlichia chaffeensis*: novel modulation of immune response and host cell cycles. *Infect. Immun.* **72:**498–507.

Zhi, N., Y. Rikihisa, H. Y. Kim, G. P. Wormser, and H. W. Horowitz. 1997. Comparison of major antigenic proteins of six strains of the human granulocytic ehrlichiosis agent by Western immunoblot analysis. *J. Clin. Microbiol.* **35:**2606–2611.

Zhou, D., J. Mattner, C. Cantu III, L. Teyton, and A. Bendelac. 2004. Lysosomal glycosphingolipid recognition by NKT cells. *Science* **306:**1786–1789.

ADAPTIVE IMMUNE RESPONSES TO INFECTION AND OPPORTUNITIES FOR VACCINE DEVELOPMENT (*RICKETTSIACEAE*)

Gustavo Valbuena

10

INTRODUCTION

Both genera of the family *Rickettsiaceae*, *Rickettsia* and *Orientia*, include human pathogens. For the discussion herein, the diseases they cause will be referred to as rickettsioses. They share three critical factors with implications in pathogenesis and immunity: (i) they are transmitted by arthropod vectors; (ii) they are obligate intracellular bacteria that inhabit the cell cytoplasm; and (iii) the predominant target is the endothelium, a single layer of endothelial cells that line the luminal side of all blood and lymphatic vessels. The only exception is *Rickettsia akari*, the agent of rickettsialpox, which predominantly infects monocytes and macrophages. Those three factors are very important because (i) arthropod saliva can modify the immune response; (ii) a cytoplasmic niche requires special handling by the immune system; and (iii) the target cells are multifunctional cells with important regulatory functions in angiogenesis, hemostasis, permeability and solute exchange, vascular tone, and inflammation (Michiels, 2003; Danese et al., 2007; Pober et al., 2009). However, our current understanding of the implications of those three factors in pathogenesis and the development of an adaptive antirickettsial immune response is very incomplete and mostly based on inferences from other models. For example, the role of arthropod factors inoculated together with rickettsial agents during transmission has not been directly addressed. It is thought to be very important because the saliva of *Ixodes* ticks, which do not transmit *Rickettsia* in the Americas, modifies the immune response against other tick-transmitted microbes such as *Borrelia* (Wikel, 1999; Brossard and Wikel, 2004; Francischetti et al., 2009; Socolovschi et al., 2009).

The cytoplasmic location in endothelial cells is a critical determinant of pathogenesis because it implies involvement of all organs—these are true systemic infections. In fact, detached infected endothelial cells can be identified in the circulation (George et al., 1993; La Scola and Raoult, 1996); they may be the source of new foci of infection once they lodge in distal capillaries. Not surprisingly, rickettsioses can be very severe febrile diseases in individuals of any age independently of their immune status (Centers for Disease Control and Prevention, 2004; Paddock et al., 1999, 2002; Lee et al., 2008; Lee et al., 2009a). Severe manifestations such as noncardiogenic pulmonary edema,

Gustavo Valbuena, Department of Pathology, University of Texas Medical Branch, Galveston, TX 77555.

Intracellular Pathogens II: Rickettsiales
Edited by G. H. Palmer and A. F. Azad © 2012 ASM Press, Washington, DC
10.1128/9781555817336.ch10

interstitial pneumonia, adult respiratory distress syndrome, meningoencephalitis, seizures, and coma (Walker et al., 2003; Rizzo et al., 2004; Bechah et al., 2008a; Chen and Sexton, 2008; Demeester et al., 2010) are a consequence of endothelial damage in the lungs and brain; involvement of these organs explains the majority of the mortality (Fig. 1). Nevertheless, not all rickettsioses are equally severe despite sharing common pathogenic mechanisms; this is probably explained by interacting and synergistic factors that include differences in virulence of the individual species and vector-related issues.

The most important rickettsioses, as determined by factoring severity of the disease, geographic distribution, potential for transmission, and incidence, are Rocky Mountain spotted fever (caused by *Rickettsia rickettsii*), Mediterranean spotted fever (caused by *Rickettsia conorii*), scrub typhus (caused by *Orientia tsutsugamushi*), epidemic typhus (caused by *Rickettsia prowazekii*), and murine typhus (caused by *Rickettsia typhi*).

THE IMMUNE RESPONSE

Several recent reviews have summarized the most relevant literature on the pathogenesis and development of the antirickettsial immune response (Walker and Ismail, 2008; Valbuena et al., 2002). Herein we emphasize some findings and suggest a framework for a modern conceptualization of the field of rickettsiology at the interface with immunology.

The development of the adaptive immune response is conditioned by the innate immune mechanisms activated during early events of the infection. Since rickettsioses are transmitted by arthropod vectors, factors in the saliva are likely to be important determinants of the final outcome; however, this aspect has not been thoroughly examined. Typhus group rickettsiae (TGR) are transmitted mainly by insect vectors. The human body louse (*Pediculus humanus corporis*) is the vector of epidemic typhus (*R. prowazekii*). The mechanism of transmission is by inoculation of infected louse feces into small open wounds created by scratching. In this case, the vector's saliva is not coinoculated with the rickettsiae; however, they are inoculated in an area already containing the vector saliva as well as louse feces, both of which could have a modulatory effect on the immune response. This probably also applies to murine typhus (*R. typhi*), which is transmitted to humans by inoculation of infected flea feces into skin abraded by scratching (direct inoculation through the bite has only been described in the laboratory setting). Spotted fever group rickettsiae (SFGR) are transmitted by the bite of any developmental stage (larva, nymph, and adult) of hard ticks (mainly *Amblyomma*, *Dermacentor*, and *Rhipicephalus*). They feed for a long time (several days for nymphs and

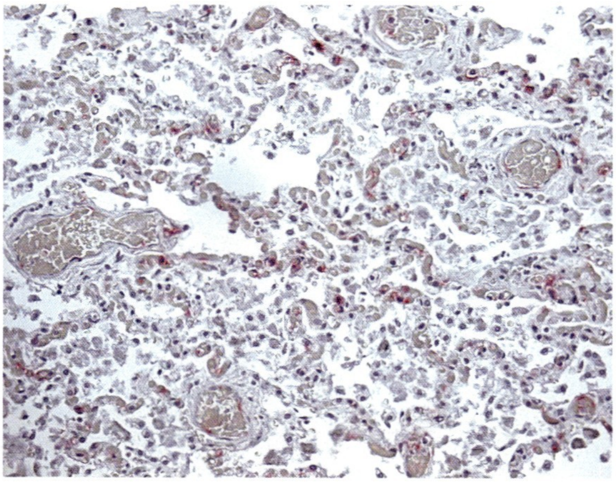

FIGURE 1 Immunohistochemical detection of *R. prowazekii* (original magnification, ×400). The red precipitate, a product of the alkaline phosphatase tag on the secondary antibody, marks the predominant location of rickettsiae in the capillaries of lung alveolar walls as well as larger vessels. The hematoxylin contrast allows the visualization of abundant macrophages in the alveolar spaces and infiltrating interstitial mononuclear leukocytes. doi:10.1128/9781555817336.ch10.f1

adults), which requires the injection of anticoagulants and immunomodulators in order to avoid a host response and feed successfully (Francischetti et al., 2009; Reck et al., 2009). Scrub typhus (*O. tsutsugamushi*) is transmitted by the bite of tribiculid mites (only the larval stage).

We do not know which cells are infected immediately after inoculation. Some rickettsiae produce an eschar (area of necrosis with a rich inflammatory infiltrate) after proliferating locally at the site of injection (Walker et al., 1988). Many rickettsioses that produce this phenomenon also produce local lymphadenitis, suggesting initial spread through lymphatics. When considering only the spotted fever group rickettsioses, it is interesting to note that the most severe one, Rocky Mountain spotted fever, does not manifest with an eschar or local lymph node reaction. It is tempting to speculate that *R. rickettsii* disseminates hematogenously faster than the less severe spotted fever rickettsioses, thus giving less time for the immune system to react. This hypothesis needs to be tested experimentally.

One of the problems for the study of early events is that there are no animal models for *Rickettsia* or *Orientia* that include transmission by appropriate vectors. This is an area of current active investigation, but the results are not yet published. Currently available animal models bypass the vector and use bacteria grown in mammalian cells in vitro or in the yolk sac of chicken embryos. For the spotted fever and typhus group rickettsioses, the best animal model is C3H/HeN mice inoculated intravenously with *R. conorii* or *R. typhi* (Walker et al., 2000; La et al., 2007). Note that the HeN mouse substrain, unlike the more frequently used HeJ substrain, does not have a defect in Toll-like receptor 4 (TLR4) activation and signaling. Infection of these animals is predominantly endothelial and disseminated, just like human infections (Fig. 2 and 3). A low number of rickettsiae produce a mild disease that results in solid immunity, which is important because humans who survive an infection with *Rickettsia* appear to develop lifelong immunity against *Rickettsia* (this is not the case for scrub typhus, as we shall see later). An inoculum with a high number of rickettsiae produces a severe disease resulting in lethality associated with a very large number of rickettsiae throughout the vasculature and prominent perivascular infiltrates with mononuclear cells. These models have been extremely valuable in understanding the pathophysiology and immune response against *Rickettsia*; however, one must keep in mind at least two caveats when considering the information derived from these mouse models.

1. The physiological state (and transcriptional profile) of rickettsiae used in the inoculum is probably different from that of rickettsiae inside the appropriate vec-

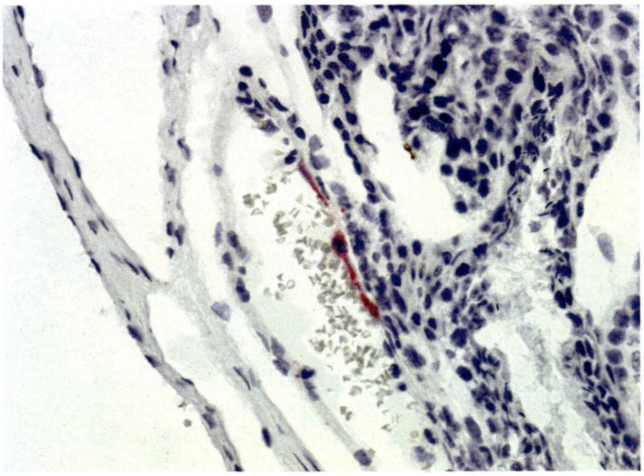

FIGURE 2 Immunohistochemical detection of *R. conorii* in the testis of a mouse infected with a sublethal inoculum (original magnification, ×400). The red precipitate, a product of the alkaline phosphatase tag on the secondary antibody, reveals the endothelial location of rickettsiae in a vein. doi:10.1128/9781555817336.ch10.f2

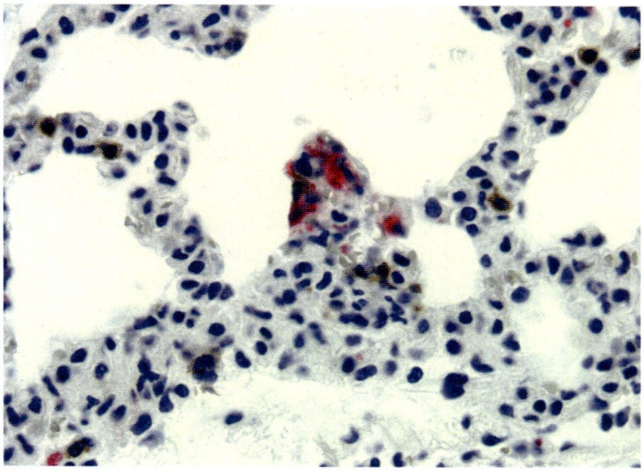

FIGURE 3 Immunohistochemical detection of *R. conorii* and $CD8^+$ T cells in the lung of a mouse infected with a sublethal inoculum (original magnification, ×400). The red precipitate, a product of the alkaline phosphatase tag on the secondary antibody, shows the rickettsiae in a capillary of an alveolar wall infiltrated by mononuclear leukocytes. The brown precipitate, a product of the horseradish peroxidase tag on a different secondary antibody, marks the location of $CD8^+$ T cells.
doi:10.1128/9781555817336.ch10.f3

tor; it certainly is different for *Rickettsia* in tissue culture versus *Rickettsia* in an infected host (Renesto et al., 2008). For instance, for rickettsiae transmitted by ticks, we know that they are in a dormant state and they reactivate once the tick begins feeding (Hayes and Burgdorfer, 1982). This might be one of the explanations for why transmission does not begin immediately upon initiation of feeding.

2. The current animal models do not take into account the immunomodulatory factors found in the vector's saliva. This is a very important consideration not only for understanding pathogenesis and development of the natural immune response but also for vaccine testing. There are examples of vaccines that provide appropriate protection in the context of direct inoculation of the pathogen but not in the context of vector-mediated transmission (Peters et al., 2009).

The animal model most frequently used for the study of scrub typhus is mice inoculated with *O. tsutsugamushi* via the intraperitoneal route. In addition to the same two problems mentioned above for *Rickettsia* models, the major limitation of this model is that disseminated endothelial infection is not a feature of the model and the predominant pathology is peritonitis, with macrophages and mesothelial cells as the main targets of infection (Kobayashi et al., 1985; Kundin et al., 1964; Koh et al., 2004); this is not a characteristic of human scrub typhus. Recently, a promising (albeit expensive) nonhuman primate model was described (Walsh et al., 2007).

Most of our current understanding of the pathogenesis and the immune response against *Rickettsia* and *Orientia* derives from the study of the mouse models that we just referred to. For *Rickettsia*, the mouse models suggest the following.

- NK cells become activated during the first 2 to 4 days of infection and are important producers of gamma interferon (IFN-γ) (Billings et al., 2001; Jordan et al., 2009). This cytokine is important because it activates the bactericidal functions of the endothelium. The experimental design of one of the studies consisted of antibody-mediated depletion of NK cells; pathogen loads were significantly higher in depleted animals.
- As expected for an intracellular bacterial parasite such as *Rickettsia*, cellular immunity of the Th1 type is fundamental in the host defenses against rickettsial infections (Mansueto et al., 2008). Nonimmune mice are protected from a lethal rickettsial challenge, even from a heterologous rickettsial group, if adoptively

transferred with immune T cells (Valbuena et al., 2004). Moreover, $CD8^+$ T lymphocytes are essential in the effective immune response against rickettsiae; an ordinarily sublethal dose is lethal or causes persistent infection in mice depleted of $CD8^+$ T lymphocytes. Adoptive transfer of immune $CD4^+$ or $CD8^+$ T lymphocytes before infection with 3.6 times the 50% lethal dose (LD_{50}) of *R. conorii* protects immunocompetent naïve mice; however, the immunity provided by the transferred cells can be overwhelmed, as they do not protect mice against a higher dose of rickettsiae (Feng et al., 1997). Other mechanisms must cooperate with immune $CD8^+$ and $CD4^+$ T lymphocytes since mice that become immune by a sublethal rickettsial infection survive a challenge with even higher doses of rickettsiae. Higher levels of IFN-γ and tumor necrosis factor alpha (TNF-α) in serum from animals depleted of either $CD8^+$ T lymphocytes or $CD8^+$ and $CD4^+$ T lymphocytes versus sham-depleted animals support this hypothesis. The origin of the cytokines under these conditions is probably from NK cells. This would be in agreement with the finding that rickettsia-infected mice depleted of NK cells have lower levels of IFN-γ (Billings et al., 2001).

- The importance of cytolytic mechanisms is suggested by *Rickettsia australis* infection of C57BL/6 perforin gene knockout mice, which are 1,000-fold more susceptible to lethal infection with rickettsiae than are wild-type mice, and by adoptive transfer of immune IFN-γ gene knockout $CD8^+$ T lymphocytes into naïve IFN-γ gene knockout mice infected with a lethal dose of *R. australis*. This adoptive transfer resulted in lower rickettsial infectivity titers in the organs analyzed (spleen, lungs, and liver), most likely resulting from cytotoxic-T-lymphocyte activity since neither the adoptively transferred cells nor the recipients' cells were capable of producing IFN-γ. Furthermore, other cytotoxic-T-lymphocyte mechanisms (e.g., Fas/Fas ligand or granulysin) may also play a role since the LD_{50} of *R. australis* for perforin gene knockout mice was 2 orders of magnitude higher than the LD_{50} of major histocompatibility complex (MHC) class I gene knockout mice. In these mice, an infectious dose of *R. australis* as low as 0.5 PFU (probably a rickettsial body count of 1 to 2 rickettsiae) produced a lethal infection (Walker et al., 2001). On the other hand, we must be aware that cytolytic mechanisms are potentially very dangerous when acted upon endothelial cells because these cells normally prevent inappropriate activation of coagulation and regulate the equilibrium of fluid between the intravascular and interstitial spaces. Indiscriminate killing of endothelial cells could lead to widespread thrombosis and severe edema. Since these are in fact characteristics of severe rickettsioses, the question arises as to whether those phenomena are caused by direct endothelial damage produced by rickettsiae, or immune-mediated damage, or both. This is an important question that has not been addressed yet. However, it is interesting to note that endothelial cells may be equipped with anticytolysis mechanisms; in vitro, ATP produced by human umbilical vein endothelial cells can inhibit the CX3CL1-driven cytotoxicity of NK cells (Gorini et al., 2010). In addition, endothelial cells may protect themselves by not presenting immunodominant epitopes efficiently (Kummer et al., 2005).
- The necessary bias of the immune response toward cellular immunity in rickettsial infections depends on multiple factors, particularly the cytokine environment during the induction of immune T lymphocytes. One of the most important

cytokines in this respect is interleukin-12 (IL-12). IL-12 serum levels are increased early after *R. conorii* infection of C3H/HeN mice (Billings et al., 2001).

- IFN-γ and TNF-α activate the rickettsicidal effector function of endothelial cells in vivo, as indicated by the overwhelming infection resulting from IFN-γ and/or TNF-α neutralization, which is associated with decreased nitric oxide production and greater growth of rickettsiae (Feng et al., 1994; Walker et al., 1997).
- In *R. conorii*-infected C3H/HeN mice, the endothelium expresses several chemokines (Fig. 4), including CCL2, CCL3, CCL4, CCL12, CX3CL1, CXCL9, and CXCL10 (Valbuena et al., 2003; Valbuena and Walker, 2004). Many of those chemokines target T cells. Although their initial expression is focal, it becomes widespread at a time when the rickettsial infection remains focal. The significance of this finding has not been explored further.
- Regulatory T cells may participate in the development of a lethal outcome when the rickettsial load is high (Fang et al., 2009). These cells were characterized as $CD4^+CD25^+Foxp3^-$ T cells capable of promoting IL-10 production and of suppressing T-cell proliferation and IFN-γ secretion. The importance of these cells in vivo is not clear; however, it is interesting to note that endothelial cells can stimulate regulatory T cells in an allogeneic mouse model (Krupnick et al., 2005) and that true regulatory T cells can induce an inhibitory phenotype on endothelial cells in an in vitro model of atherosclerosis (He et al., 2010). Regulatory T cells can also induce indoleamine-2,3-dioxygenase (IDO) expression in endothelial cells through IFN-γ stimulation (Thebault et al., 2007), which is very interesting in light of the role of IDO as an effector mechanism of rickettsial killing (Feng and Walker, 2000). Also, endothelial cells can enhance the suppressive phenotype of regulatory T cells (Bedke et al., 2010).
- Dendritic cells from susceptible C3H/HeN mice appear to be less efficient at activation of naïve T cells than dendritic cells from the resistant C57BL/6 mice in an in vitro model of rickettsial infection (Fang et al., 2007).Nevertheless, dendritic cells from susceptible mice mature and become activated upon rickettsial infection with a phenotype capable of inducing Th1 responses, which includes the production of IL-12. More importantly, these dendritic cells, when transferred to naïve mice, protected mice from a lethal rickettsial challenge given a day later (Jordan et al., 2007).

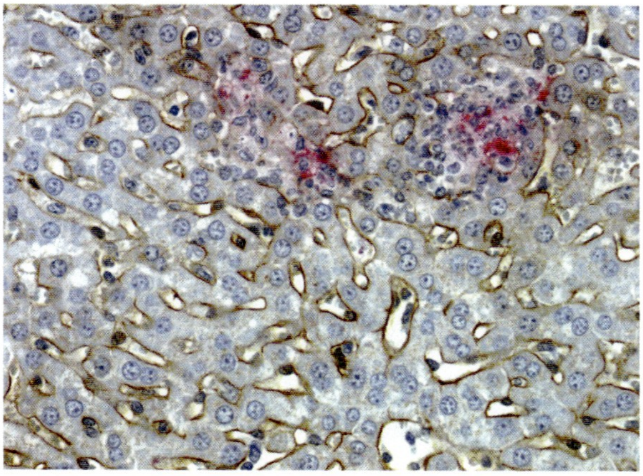

FIGURE 4 Immunohistochemical detection of *R. conorii* and the chemokine CXCL9 in the liver of a mouse infected with a sublethal inoculum 5 days earlier (original magnification, ×400). The red precipitate, a product of the alkaline phosphatase tag on the secondary antibody, shows the rickettsiae in focal areas infiltrated by mononuclear leukocytes. The brown precipitate, a product of the horseradish peroxidase tag on a different secondary antibody, marks the location of the chemokine (CXCL9); it outlines the sinusoidal endothelium.
doi:10.1128/9781555817336.ch10.f4

- Although *Rickettsia* does have lipopolysaccharide, it is not endotoxic. Yet mice deficient in TLR4 (C3H/HeJ) develop a lethal disease with inocula that are not lethal to C3H/HeN mice. Also, the TLR4-deficient mice did not develop a Th17 response or early activation of NK cells, while normal mice did (Jordan et al., 2008, 2009). Interestingly, blocking of TLR4 in human endothelial cells cultured in vitro blunts their chemokine response (monocyte chemoattractant protein-1 and IL-8) to heat-killed *Rickettsia* (Damås et al., 2009). While the in vitro responses of primary endothelial cells from either mouse strain were not different, it is noteworthy that in a mouse model that expresses TLR4 only on endothelial cells, TLR4 ligation on endothelial cells by systemic bacteria (in a bacteremia model) led to efficient clearance of the infection (with no mortality) (Andonegui et al., 2009). These findings begin to build a point that we emphasize later: the endothelium in vitro is very different from the endothelium in vivo.

The general paradigm of the importance of Th1 immunity that applies to *Rickettsia* also applies to *O. tsutsugamushi*. The studies leading to this conclusion have been reviewed (Seong et al., 2001). Human studies from clinical cases are not abundant, but the data are obviously extremely relevant. The following is some of what we know about human rickettsioses from clinical case studies.

- Both $CD4^+$ T cells and $CD8^+$ T cells infiltrate perivascular areas of infection (Valbuena et al., 2003). Of course, the role played by those cells cannot be discerned since the analysis was performed on fatal cases (Fig. 5). In the case of scrub typhus, patients present with elevated levels of granzymes A and B, which suggests the activation of cytolytic cells (de Fost et al., 2005). Also, T cells and $CD68^+$ myeloid cells are abundant in both the eschar and erythematous plaques (Lee et al., 2009b).
- Monocytes from patients with Mediterranean spotted fever produce higher levels of IL-12 than monocytes obtained during the convalescent phase; this response was abrogated by the elimination of $CD14^+$ cells from the cell culture system and was amplified by IFN-γ (Milano et al., 2000).
- Polymorphisms of the IFN-γ gene that are associated with increased production of the corresponding protein are associated with better clinical outcomes in patients with Mediterranean spotted fever (Forte et al., 2009). Higher levels of transcript from this gene (together with those for TNF-α, inducible nitric oxide synthase, and IDO)

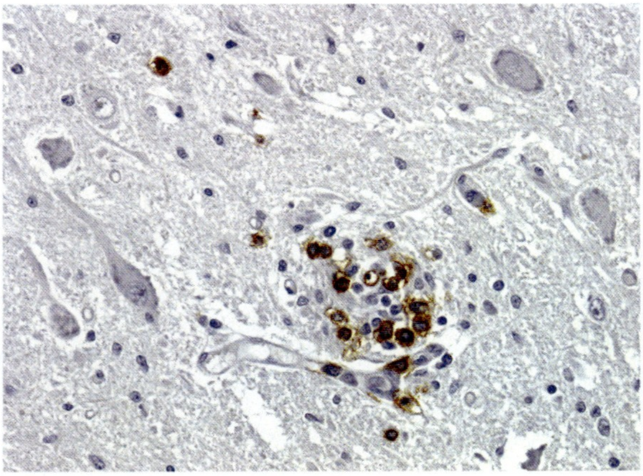

FIGURE 5 Immunohistochemical detection of $CD8^+$ T cells in the brain of a child who died of Rocky Mountain spotted fever (original magnification, ×400). The brown precipitate, a product of the horseradish peroxidase tag on the secondary antibody, marks the location of $CD8^+$ T cells around two small vessels.
doi:10.1128/9781555817336.ch10.f5

are present in patients with better outcomes than in those with worse outcomes (de Sousa et al., 2007). IFN-γ, together with IL-10, is also found in patients with scrub typhus (Kramme et al., 2009).

- When more severe rickettsioses are compared with less severe rickettsioses (e.g., Mediterranean spotted fever versus African tick bite fever) during the acute phase, the sera of these patients have higher levels of IL-8, E-selectin, and vascular cell adhesion molecule-1 (VCAM-1) (Damås et al., 2009), which probably reflects enhanced inflammation. Experiments in vitro support this concept; the transmigration of mononuclear cells across a monolayer of endothelial cells is much higher when the infection is caused by a virulent strain of *R. prowazekii* than when the infection is caused by the less virulent Madrid E strain of *R. prowazekii* (Bechah et al., 2008b).

THE ENDOTHELIUM IN THE ANTIRICKETTSIAL IMMUNE RESPONSE

Our understanding of the role of the human endothelium in rickettsial pathogenesis and immunity has mainly derived from in vitro studies (Valbuena and Walker, 2009). For *Rickettsia*, those studies have shown that endothelial cells become activated upon rickettsial infection, resulting in the secretion of cytokines and chemokines such as IL-1a, IL-6, and CXCL8 (Kaplanski et al., 1995; Sporn and Marder, 1996) and the expression of adhesion molecules such as E-selectin, VCAM-1 (Fig. 6), intercellular adhesion molecule-1 (ICAM-1; Sporn et al., 1993; Dignat-George et al., 1997; Damås et al., 2009), and $\alpha_v\beta_3$ integrin (Bechah et al., 2009). Such response may be initiated by direct rickettsia-triggered activation of the transcription factor NF-κB (Sahni et al., 1998, 2003; Sporn et al., 1997). In addition, *R. rickettsii* infection of primary human endothelial cells results in the secretion of prostanoids as a consequence of the induction of cyclooxygenase-2 via signaling through p38 mitogen-activated protein kinase (Rydkina et al., 2006, 2009). Other signaling mediators that get activated upon rickettsial infection included signal transducer and activator of transcription 1 (STAT1) and STAT3 (Sahni et al., 2009). Interestingly, the combination of activated STAT3 with IL-6 production has the potential of driving a Th17 response in a transplantation model (Taflin et al., 2011). In vitro infection of an endothelial cell line with *O. tsutsugamushi* triggers the expression of many of the same cytokines described for *Rickettsia* and also IL-32 as a consequence of activation of the intracellular pattern recognition receptor nucleotide oligomerization domain 1 (NOD1) (Cho et al., 2010). All the studies above did not analyze the possible role

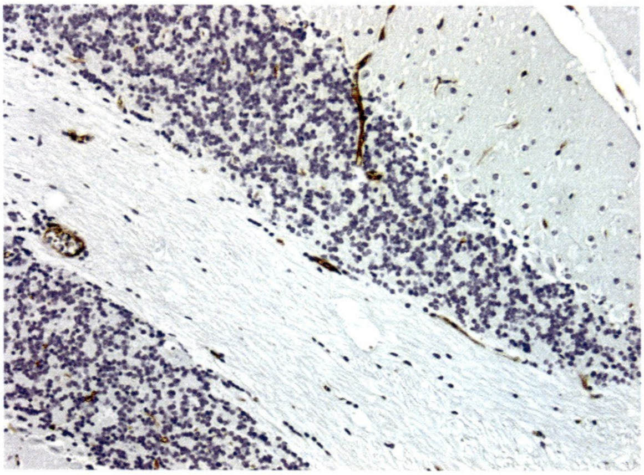

FIGURE 6 Immunohistochemical detection of VCAM-1 in the brain of a mouse infected with a lethal inoculum of *R. conorii* (original magnification, ×400). The brown precipitate, a product of the horseradish peroxidase tag on the secondary antibody, marks the endothelial location of the adhesion molecule VCAM-1. doi:10.1128/9781555817336.ch10.f6

of other cells that are naturally in contact with endothelial cells, such as platelets and leukocytes. One recent report demonstrated the role of soluble CD40L derived from platelets in enhancing the activation of the endothelium in response to rickettsial infection (Damås et al., 2006). Furthermore, there appears to be a positive-feedback loop of this response from *Rickettsia*-infected endothelial cells mediated by CX3CL1.

However, we do not yet know if the response of the endothelium to rickettsial infection in vitro is a good approximation of the response of the endothelium to rickettsial infection in vivo. The molecularly detailed data obtained thus far in regard to infection of the human endothelium with *Rickettsia* and *Orientia* have been derived from studies in vitro using endothelial cell lines and primary endothelial cells derived from large vessels (such as human umbilical vein endothelial cells) cultured under static conditions. More recently, primary human microvascular endothelial cells have also been used, albeit under static conditions (Rydkina et al., 2010). The drawback is that endothelial cells cultivated in vitro are remarkably different from endothelial cells in vivo due to dynamic microenvironment-mediated regulation (including the constantly changing environment of inflammation) (Durr et al., 2004; Lacorre et al., 2004; Roy et al., 2007; Amatschek et al., 2007; Calabria and Shusta, 2008); blood flow-mediated regulation, particularly its anti-inflammatory effect (Dai et al., 2004; Gaucher et al., 2007; Liu et al., 2005); and differences that result from the significant heterogeneity of the endothelium throughout the different anatomic portions of the vasculature (Davies, 2008; Murphy et al., 1999; St. Croix et al., 2000). Indeed, in vivo, endothelial cells develop remarkable functional and structural adaptations tailored specifically toward the physiological homeostasis of each organ (Aird, 2005) through regulatory functions in the trafficking of cells and nutrients; the formation of new vessels (angiogenesis); regulation of blood coagulation and vascular integrity (hemostasis); and control of vascular permeability, vascular tone, and inflammation (Michiels, 2003; Wagner and Frenette, 2008; Pober and Sessa, 2007; Rao et al., 2007; Pober et al., 2009; Komarova and Malik, 2010). When cultured in vitro, endothelial cells lose much of their native phenotypic characteristics. For instance, the endothelial anti-inflammatory phenotype normally triggered by laminar shear stress (Liu et al., 2005; Tsai et al., 2007; Di Francesco et al., 2009) and the endothelial glycocalyx (Chappell et al., 2009; Potter et al., 2009; Potter and Damiano, 2008) are completely lost when endothelial cells are cultured in vitro. The glycocalyx itself appears to provide an anti-inflammatory phenotype; when the synthesis of hyaluronan (an integral molecule of the glycocalyx) is inhibited, inflammation increased in an in vivo model of atherosclerosis (Nagy et al., 2010). Thus, the result of the opposing phenotypes during a rickettsial infection in vivo remains unknown.

Clearly, experimental studies with static in vitro systems, such as those performed for studies with *Rickettsia* and *Orientia*, are necessarily biased by those dramatic phenotypic changes that the endothelium undergoes when cultured in vitro. Naturally, such data should not be discounted, yet should be considered in the light of our current knowledge about the physiology of the endothelium. Given the availability of systems to study the endothelium in vitro under conditions of shear stress and, better yet, the technology to study the endothelium in vivo (Hickey and Kubes, 2009), the results mentioned above need to be confirmed and new data need to be generated in these more physiological systems. The study of the endothelium in the context of these true endothelium-target infections offers new opportunities to explore the role of the endothelium in orchestrating or modifying immune responses. It is noteworthy that of the infectious agents that have been demonstrated to infect the endothelium, only *Ehrlichia ruminantium* (Valbuena and Walker, 2009), hantaviruses, human herpesvirus 8, *O. tsutsugamushi*, and *Rickettsia* use endothelial cells as their main targets (Valbuena and Walker, 2006).

While leukocyte recruitment is one aspect of the immune response function of the endothelium that has been extensively investigated (Wagner and Frenette, 2008; Langer and Chavakis, 2009; Muller, 2009, 2011), the study of rickettsioses offers the opportunity to explore underdeveloped areas such as antigen-specific interactions between endothelial cells and T cells, which have been investigated almost exclusively from the perspective of organ transplantation (Choi et al., 2004; Valujskikh and Heeger, 2003; Al-Lamki et al., 2008). Endothelial cells at the level of the microvasculature (arterioles, capillaries, and venules) have frequent tight contacts with circulating T cells; most of them are naïve T cells. This physiological characteristic, together with normal constitutive endothelial expression of MHC class I and class II molecules (only class I for mice), makes endothelial cells uniquely positioned for the function of antigen presentation (Choi et al., 2004; Pober and Sessa, 2007). In fact, there is evidence that endothelial antigenic presentation triggers memory-T-cell activation and proliferation (Epperson and Pober, 1994). In addition, antigen presentation by the endothelium triggers antigen-specific translocation of T cells into the perivascular space (Savinov et al., 2003; Marelli-Berg et al., 2004), a mechanism that implies antigenic cross-presentation and that has been demonstrated in other models as well (Valujskikh et al., 2002).

Although the capacity of endothelial cells to activate secondary (memory) T-cell responses has been demonstrated and is well accepted (Rothermel et al., 2004; Choi et al., 2004; Pober and Sessa, 2007), the debate regarding their ability to stimulate primary immune responses has not been settled. Endothelial cells express various kinds of costimulatory molecules through which they could activate primary immune T-cell responses. They have been analyzed in various models and include LFA-3 (CD58) (Smith and Thomas, 1990), B7-2 (CD86) (Jollow et al., 1999), B7-1 (CD80) (Prat et al., 2000; Omari and Dorovini-Zis, 2001), CD2 (Ma and Pober, 1998), CD40 (Karmann et al., 1995; Omari and Dorovini-Zis, 2003), CD154 (Reul et al., 1997), CD134 (OX40) ligand (Kunitomi et al., 2000), and ICOS-L (Khayyamian et al., 2002). In addition, integrins and cytokines expressed by endothelial cells could also serve costimulatory functions (Choi et al., 2004). Moreover, the capacity of endothelial cells to activate naïve $CD8^+$ T cells is physiologically plausible since naïve T cells continually recirculate between secondary lymphoid organs through the bloodstream. This results in a proportion of circulating naïve and memory T cells that is very similar (Bisset et al., 2004). In mouse models with bone marrow and organ transplantation, there is now evidence that endothelial cells can activate naïve $CD8^+$ T cells in vivo (Kreisel et al., 2002a, 2002b). Also, liver sinusoidal endothelial cells, which normally trigger tolerance through B7-H1 and programmed cell death 1(PD1), are capable of activating cytolytic T cells when the antigen dose is high (Schurich et al., 2010). It remains to be tested whether endothelial cells can also activate naïve $CD8^+$ T cells recognizing nominal antigen such as that presented by endothelial cells during the course of endothelium-target infections. The rickettsial infection model is a very interesting one because this is a situation in which the endothelium can present the antigen in the context of an endothelial inflammatory phenotype. In some models in which the endothelium presents antigen, priming of naïve T cells does not occur (Bolinger et al., 2008); however, in those models there was no endothelial cell activation, unlike the situation with a rickettsial infection, because the expression of the antigen was driven from an endogenous endothelial promoter without any source of inflammatory stimulus.

From the perspective of immunity, rickettsiologists need to begin working with endothelial cells as integral members of the immune system (Pober and Sessa, 2007). These cells can sense pathogens through TLRs and other innate immunity pattern recognition receptors (Opitz et al., 2007). They respond through the expression of an activated phenotype that includes production of cytokines, chemokines, and adhesion molecules. They can process and present antigen to T cells; they are even capable of cross-presentation (Bagai et al., 2005).

VACCINE DEVELOPMENT

The first vaccine against epidemic typhus was reported in 1930 and consisted of a preparation produced from formalin-fixed intestines of lice that had been previously inoculated intrarectally with live *R. prowazekii*. Attenuated vaccines derived from vertebrate animals were later developed in the 1940s, but their protection was not rigorously tested, side effects were frequent, and the production techniques were very cumbersome (Woodward, 1986). Even a vaccine consisting of dried flea feces containing *R. typhi* (then known as *R. mooseri*) diluted in ox bile and saline was tried in North Africa, but only individuals who developed murine typhus became immune to epidemic typhus, a fact that was already known (Zinsser, 1937). Around the same time, purified inactivated *R. prowazekii* vaccines were shown to prevent death from epidemic typhus but did not stimulate complete protection (Woodward, 1986). An attenuated strain of *R. prowazekii*, denominated Madrid E, was also produced; however, the apparent reversion to a virulent phenotype precluded further development and testing (Balayeva and Nikolskaya, 1973; Nikolskaya and Balayeva, 1973). Although Russian scientists produced a vaccine consisting of *R. prowazekii*-extracted antigens (Nikol'skaia et al., 1982; Imamaliev et al., 1981), no further development was undertaken. Lastly, the production of the most recent epidemic typhus vaccine was stopped due to its pronounced variability in antigenicity and potency (Mason et al., 1976). Considerable effort, time, and money were also invested in the development and testing of vaccines for Rocky Mountain spotted fever, the most severe of the spotted fever group rickettsioses; however, they were not sufficiently effective (DuPont et al., 1973; Clements et al., 1983). For more detailed information, excellent accounts of the history of *Rickettsia* vaccines were recently published (Richards, 2004; Walker, 2009).

For scrub typhus, attempts to develop a vaccine have not been successful because there are many strains of *O. tsutsugamushi* and they do not trigger long-lasting cross-protective immunity (Kelly et al., 2009). Moreover, homologous immunity after natural infection does not last longer than a few years. More recent approaches to vaccine development have focused on immunodominant antigens for the humoral immune response. Those studies have not addressed the fact that this is an intracellular infection for which cellular immunity is critical (Shirai et al., 1976; Jerrells and Osterman, 1982; de Fost et al., 2005). A comprehensive review of vaccine development studies for scrub typhus is available (Chattopadhyay and Richards, 2007).

More recently, progress has been made in the identification of fragments of rickettsial proteins that may trigger protective immunity, including outer membrane protein A (OmpA) (Crocquet-Valdes et al., 2001; Sumner et al., 1995) and OmpB (Churilla et al., 1990; Li et al., 2003; Chan et al., 2011). However, although these approaches aim at the eventual design of subunit vaccines, they are limited and biased because of their focus on proteins that elicit a strong humoral response.

The production of an effective universal *Rickettsia* vaccine is a public health priority because (i) the combined incidence of all rickettsioses is very high; (ii) the initial clinical presentation is not specific, making clinical diagnosis very difficult; (iii) laboratory diagnosis is very difficult because there are no commercially available diagnostic tests that can be used during the acute stage when antibiotic intervention is helpful; (iv) no antirickettsial vaccines are currently available; (v) two rickettsiae (*R. prowazekii* and *R. rickettsii*) could potentially be used as bioweapons; and (vi) Rocky Mountain spotted fever is highly lethal not only in humans but also in dogs. At the present moment, our advance toward the development of a clinically effective universal vaccine against *Rickettsia* is impeded by the lack of comprehensive knowledge of antigens that are common to different species of *Rickettsia* and that, as a whole, can stimulate broad branches of the adaptive immunity.

The production of an effective universal *Orientia* vaccine is also a public health priority because (i) it affects a large geographic area, which extends from Afghanistan in the west to Japan, the Philippines, and New Guinea in the east, and from northern Australia in the south to Russia in the north; (ii) more than 1 billion people living in those areas are at risk of acquiring the infection; (iii) in addition to the more than 1 million cases estimated to occur every year, this disease has had a dramatic impact on our troops since World War II, and is now emerging as an important disease for ecotourism travelers; (iv) mortality can be as high as 35%; (v) the clinical presentation is not specific, making clinical diagnosis very difficult; (vi) there are no commercially available diagnostic tests that can be used in the rural setting, making on-site laboratory diagnosis generally impossible; and (vii) forms of the disease have emerged that do not respond well to currently available antibiotics (Watt et al., 1996).

Rickettsial Antigens and Cross-Protection

From the perspective of the humoral immune response, it is known that there is serological cross-reactivity among members of the same group. Although there is some evidence that occasionally the human antirickettsial immunoglobulin G response (and very rarely the immunoglobulin M response) is cross-reactive between the TGR and SFGR (Ormsbee et al., 1978), the particular antigenic preparation used might have favored these cross-reactions, in contrast with other antigenic preparations that appear to exclude this phenomenon (Shirai et al., 1975). It had been generally thought that the T-cell response against rickettsiae of different groups was usually not cross-reactive in the same manner that the humoral response is usually not cross-reactive between the SFGR and TGR. Although the rickettsial genomes are closely related, there are differences that supported the traditional view that the two groups are antigenically distinct (Vishwanath, 1991).

The concept of cross-protective immunity within the TGR and SFGR was established early (Zinsser and Castaneda, 1933; Zinsser, 1937). Remarkably, the concept of cross-protective immunity between SFGR and TGR was suggested in studies performed by the legendary R. R. Parker. For instance, one of those studies, published in 1939, indicated that in guinea pigs there is some degree of cross-immunity between endemic typhus and an SFGR later named *Rickettsia parkeri* (Parker et al., 1939). In another study published around the same time, he showed that guinea pigs immune to epidemic typhus are occasionally immune to maculatum rickettsial infection (*R. parkeri*) (Parker and Cox, 1940). In a later publication, Parker et al. (1951) showed that only two of six guinea pigs immune to *R. rickettsii* showed signs of infection upon challenge with *R. prowazekii*; however, no evidence of cross-immunity was found when *R. prowazekii*-immune guinea pigs were challenged with *R. rickettsii*. Subsequently, in the 1980s, cross-protection between the nonpathogenic *Rickettsia montanensis* and *R. rickettsii* was demonstrated in guinea pigs (Feng and Waner, 1980), and cross-stimulation of lymphocytes between several members of the SFGR was also established (Jerrells et al., 1986). Moreover, immunization of C3H/HeJ mice with various SFGR protected them from a lethal challenge with *R. conorii* (Eisemann et al., 1984), as well as *R. akari* and *Rickettsia sibirica* (Jerrells et al., 1986). More recently, cross-stimulation of T cells between two SFGR, *R. rickettsii* and *Rickettsia rhipicephali*, was demonstrated in guinea pigs. Furthermore, *R. rhipicephali*-immune guinea pigs are in fact immune to challenge with *R. rickettsii* (Gage and Jerrells, 1992). Feng and Walker (2003) also demonstrated cross-protection between the distantly related *R. australis* (a member of the transitional group) and *R. conorii*. The concept of cross-protection is not entirely surprising given the remarkable similarity between rickettsiae at the genomic level (Andersson et al., 1998; Ogata et al., 2001; McLeod et al., 2004; Gillespie et al., 2008; Weinert et al., 2009).

It appears that cross-reactivity at the level of T-cell recognition follows rules that are different from those governing the humoral immune response. For instance, the same $CD8^+$-T-cell line or clone can be activated by peptides with no obvious homology at the sequence level (Anderson et al., 1992; Kuwano et al., 1991; Kulkarni et al., 1993; Clute et al., 2010). Indeed, the precursor frequency of cytotoxic T lymphocytes against one virus may be increased by a previous infection with another virus through the mechanism of cross-reactivity at the T-cell level (Selin et al., 1994). This seems to be a natural consequence of the "rules" that peptides must follow in order to comply with the sequence motif for the presenting MHC molecule. Thus, the possibility of finding cross-reactive antigens at the T-cell level between closely related organisms, such as those within the genera *Rickettsia* and *Orientia*, is very high.

Modern Vaccinology

Vaccinology is a field that naturally converges with immunology; however, not until relatively recently have the two disciplines truly embraced each other. The long history of separation probably relates to the fact that scientists working on the microbe side led the vaccine development effort, which was very fruitful for pathogens with limited or no antigenic variation and those for which humoral immunity is critical. On the other hand, many basic immunologists were working on model antigens outside the context of infection. Today, immunology and vaccinology have clearly converged, since most vaccines that could be developed by pathogen attenuation or inactivation and trial-and-error approaches have already been produced and most immunologists are using infectious disease models that can inform vaccine development in a more relevant manner.

The contemporary development of a vaccine has two initial essential aspects, namely identification of the relevant antigens and definition of immunological correlates of protection to guide the selection of vehicles, vectors, schedules, and adjuvants. In the case of infections caused by *Rickettsia*, due to the availability of excellent murine models, relevant correlates of protective immunity can be derived from the characterization of experimental infections because animals (as well as humans) that survive the infection become solidly immune to reinfection. In the case of *Orientia*, currently available murine models (Shirai et al., 1976; Jerrells and Osterman, 1982), which do not mimic the pathogenesis of human infections (Moron et al., 2001), may not provide appropriate correlates.

In regard to immunological correlates of protection, the magnitude of a response assessed by a single parameter (e.g., IFN-γ for intracellular pathogens, as frequently reported) is not enough. Now we know that there is functional heterogeneity of the T-cell effector responses (including cytokine secretion, cytolytic activity, and development of various memory phenotypes) and that there are particular subsets of T cells, which express unique combinations of effector functions, that are more protective (Wu et al., 2002; Darrah et al., 2007; Lindenstrøm et al., 2009). We probably should approach the definition of correlates of protective immunity in a way that parallels the complexity of physiological immunity, which is a multifaceted and integrated response that includes many different cells, receptors, ligands, and signaling modules that function in a combinatorial mode. For infections in which cellular immunity plays a predominant role, there is evidence from experimental models that multifunctional T cells are the best correlate of protection described thus far. More importantly, this has been demonstrated in humans as well (Tenzer et al., 2009; Miller et al., 2008; Akondy et al., 2009). In a recent study, a new metric that integrates several parameters was introduced: the integrated MFI (iMFI) (Darrah et al., 2007). This metric multiplies the median fluorescence intensity (MFI) for a particular cytokine assessed by flow cytometry, an indicator of the magnitude of the response, by the total frequency of cells producing that cytokine. Although the metric applies to individual cytokines, when the iMFI is high for a cytokine such as IFN-γ, those cells tend to be producers of multiple cytokines.

The technologies for understanding the integrated functioning of the immune system are now available and accessible. Those are the tools of systems biology, the "-omics" methods and bioinformatics tools that permit the analysis of complex interactions in biological systems through the investigation of massively parallel data acquired from each experimental condition (Rinaudo et al., 2009). The application of systems biology to vaccinology is already identifying transcriptional signatures of protective immune responses that include subsignatures of appropriate innate and adaptive responses (Haining and Wherry, 2010). Moreover, early predictive signatures of appropriate adaptive immune responses immediately after vaccination have been defined and verified using the yellow fever (17D) vaccine as a model (Querec et al., 2009). It is expected that such knowledge will provide paradigms for the development of novel vaccines for which limited data from humans are currently available. That is certainly the case for infections caused by *Rickettsia* and *Orientia* because it is unlikely that we will be able to collect sufficient human samples from clinical cases with diverse outcomes in order to define broad signatures of protective immunity. A promising solution to this problem is to use our current understanding of well-known effective immune responses as guiding principles. The study of the response to two of the most successful human vaccines in history, the yellow fever vaccine (Gaucher et al., 2008; Akondy et al., 2009) and the smallpox vaccine (Miller et al., 2008), is likely to yield relevant paradigms that we could use as guideposts in rickettsiology.

Antigen Identification

For most of the history of vaccinology, antigen identification has been biased toward the humoral immune response because of the technical challenges associated with screening methods for antigens recognized by T cells. The use of immune sera to identify pathogens' proteins in Western blots or expression libraries was relatively straightforward. Although antibodies can mediate protection against many pathogens, true physiological immune responses always involve all branches of the adaptive immune response in various proportions. Antibody-based screening methods are not designed to identify antigens recognized by $CD4^+$ T cells or, particularly, $CD8^+$ T cells. Indeed, there is good evidence that $CD4^+$ T cells and $CD8^+$ T cells target different antigens (Moutaftsi et al., 2007). Thus, a better approach to antigen identification should use parallel strategies to screen microbial proteins for antigenicity in each one of the components of the adaptive immune response. This is especially important for intracellular pathogens such as *Rickettsia* and *Orientia*, for which there is evidence that all branches of the adaptive immune response participate in effective immunity. Since individual proteins recognized by antibodies, $CD4^+$ T cells, and $CD8^+$ T cells simultaneously are expected to be very rare, the identification of groups of antigens that can target all branches of the adaptive immune response as a whole is one of the most desirable aspects of modern vaccine development. Another one is the identification of proteins that promote an immune response that is equally effective against several strains of a pathogenic species (which clearly applies to *O. tsutsugamushi*) or several pathogenic members of a genus (such as *Rickettsia*).

Modern approaches for antigen identification for T cells include a very large number of strategies. Representative ones include the identification of microbial proteins with a high level of expression in vivo (Rollenhagen et al., 2004), screening retrovirus-based cDNA libraries (Wang et al., 1998; Aoshi et al., 2005), isolation of antigenic peptides from MHC-peptide complexes and characterization by mass spectrometry (Lemmel and Stevanović, 2003), stimulation of immune cells with fractions of sodium dodecyl sulfate-polyacrylamide gel electrophoresis-separated microbial lysates and peptide mass mapping by matrix-assisted laser desorption ionization mass spectrometry of relevant fractions (Olsen et al., 2007), stimulation of immune cells to screen random expression genomic libraries in *Escherichia coli* (Gonçalves et al., 2006), screening of large sets of peptides (Moutaftsi et al., 2007), and bioinformatics-based approaches (Sylvester-Hvid et al., 2004; Capo et al., 2005; Moriel et al., 2010). The last one belongs to the field of reverse vaccinology. This emerging field is a branch of systems biology that analyzes entire microbial genomes in order to predict immunogenic proteins based on predefined rules derived from the analysis of large empirical datasets. Success stories using reverse vaccinology are described in a recent review article (Sette and Rappuoli, 2010). One of the caveats of this strategy, particularly for antigens recognized by T cells, is that its predicting power has not been thoroughly tested by direct experimentation. From a set of potential antigens predicted by reverse vaccinology, protective antigens have always been identified; on the other hand, an entire microbial genome has not been validly screened in order to determine the proportion of truly antigenic proteins that were predicted through bioinformatics approaches. In fact, at least for bacterial proteins, known protective antigens actually have fewer predicted epitopes than randomly selected bacterial protein sets used as a control (Halling-Brown et al., 2008). Examples of modern approaches and their potential problems are summarized in Table 1.

For a very long time, immunodominance has been one of the main drivers in the identification of relevant antigenic proteins. The concept of immunodominance refers to a hierarchy in the immune response that limits the diversity of T-cell receptors and antibodies used in the immune response against a pathogen; a consequence of immunodominance is that only some antigens within a microbial pathogen drive the expansion of a relatively limited number of T-cell and B-cell clones, while other antigens (called cryptic or subdominant) do not produce identifiable T-cell or B-cell clonal responses. The factors that intervene in determining immunodominance

TABLE 1 Selected examples of current approaches to antigen identification

Strategy[a]	Selects for cross-reactivity	Discovers antigens of all branches of adaptive immunity	Knowledge of which antigens are validly screened
Identification of microbial proteins with a high level of expression	No	No	No
Screening of retrovirus-based cDNA libraries	No	No	No
Isolation of antigenic peptides from MHC-peptide complexes with characterization by mass spectrometry	No	No	No
Stimulation of immune cells with fractions of SDS-PAGE-separated microbial lysates and peptide mass mapping by MALDI-MS	No	No	No
Stimulation of immune cells to screen a random expression genomic library in *E. coli*	No	No	No
Screening of large sets of peptides	No	No	Yes
Genomic immunization	No	Yes	No
Reverse vaccinology	Yes	Yes	No

[a]MALDI-MS, matrix-assisted laser desorption ionization mass spectrometry; SDS-PAGE, sodium dodecyl sulfate-polyacrylamide gel electrophoresis.

are many and are not completely understood. They include pathogen factors such as the level of expression of different proteins and host factors such as the initial frequency of responding cells, type of innate immune response, cytosolic proteasome preferences, and the relative abundance of epitopes that they generate (Akondy et al., 2009).

However, immunodominant responses are not necessarily protective. In chronic infections, they are clearly not protective. In the case of *O. tsutsugamushi*, the immunodominant response is directed toward antigens that are very variable between strains, which probably explains why cross-protective immunity between different strains of *O. tsutsugamushi* is so poor. A possible solution to this problem may be the use of subdominant conserved antigens for vaccination. This strategy would be supported by recent evidence showing that vaccination with subdominant antigens can provide firm protection (Riedl et al., 2009; Ruckwardt et al., 2010). A modern antigen discovery strategy should take these concepts into account. For example, any screening strategy that uses immune serum or immune T cells for antigen discovery is necessarily biased toward immunodominant antigens because it relies on memory B and T cells. Those methods are necessarily blind to cryptic antigens.

The resources are now available to test the antigenicity of each open reading frame (ORF) from many *Rickettsia* and *Orientia* species. At the time of this writing, there were 23 entries for *Rickettsia* in the NCBI genome database; 14 of these entries correspond to complete genome sequences and the rest to complete sequences of rickettsial plasmids. Also, two strains of *O. tsutsugamushi* have been fully sequenced. With this information, the construction of libraries with all ORFs from several *Rickettsia* and *Orientia* species (the ORFeome) is possible through PCR-mediated amplification of each ORF or through gene synthesis. In fact, at least one of them, that of *R. prowazekii*, is available from the NIH through a contract with the Craig Venter Institute. In order to use this kind of resource for antigen identification, the rickettsial genes need to be transferred individually to several appropriate expression plasmid vectors that would allow specific screening for antigens presented through MHC class I molecules (for $CD8^+$ T cells), MHC class II molecules (for $CD4^+$ T cells), or directly to antibodies. In addition, in the case of T cells, the eukaryotic expression vectors need to be transferred (through nucleofection or electroporation, for example) to histocompatible antigen-presenting cells that can present to matched immune T cells. Activation of effector functions of anti-*Rickettsia* memory T cells could then be used as an indicator of recognition of an antigen.

Most reports of antigen identification that have used genomic data utilized native microbial sequences to express the corresponding microbial proteins in eukaryotic cells. This approach is problematic because codon preference is always different between prokaryotes and eukaryotic cells, which results in limited or even absent expression. We have observed several examples of this phenomenon with ORFs of *R. prowazekii*. One way to overcome this obstacle is to use the strategy of gene synthesis with codon optimization for expression in mammalian cells. In fact, such a strategy has been shown to enhance adaptive immune responses to DNA vaccines due to increased expression of the antigen (Ternette et al., 2007; Siegismund et al., 2009).

One strategy for antigen identification that is not biased toward immunodominant antigens is genomic immunization, or expression library immunization (Sykes, 2008). In this technique, pools of eukaryotic expression vectors with cloned pathogen genes are used to immunize animals. The animals are subsequently challenged with lethal doses of the microbial pathogen. The gene pools that trigger protection are then deconvoluted by testing one at a time. This method allows the priming of naïve T cells by the expressed cloned microbial genes regardless of whether they are subdominant or dominant during a natural infection as long as the appropriate T-cell receptors are present. Although expression library immunization has been successfully used (Stemke-Hale et al.,

2005; Li et al., 2006), it has its own problems. It relies on a DNA immunization strategy, so antigen expression is not guaranteed in all cases. Accordingly, it is not possible to know which pathogen genes were not screened validly; a negative response can be due to lack of an immunological response or to failed expression of the microbial gene. We propose that in order to validly screen all possible protein antigens in a microbial genome, it is best to use a strategy of immunizing with optimized antigen-presenting cells expressing individual microbial genes with an identifiable tag such as a fluorescent protein that can be easily verified before injection.

The development of a modern vaccine against *Rickettsia* or *Orientia* should aim at mimicking a physiological immune response in the sense that all branches of adaptive immunity should be stimulated: the purposeful harnessing of the combined power of the cellular and humoral immune responses against these intracellular bacteria. This implies the identification of a combination of antigens that together can stimulate protective responses mediated by $CD8^+$ T cells, $CD4^+$ T cells, and B cells (antibodies). The next step in the development of a modern vaccine is the identification of the best strategy to optimally stimulate $CD8^+$ T cells, $CD4^+$ T cells, and B cells. The idea here is to do better than nature by testing different adjuvants and immunization routes. Some of these concepts are not only beyond the scope of this chapter but are evolving very rapidly; thus, the reader is referred to excellent reviews that were recently published (Coffman et al., 2010; Liu, 2010). Finally, although the previous discussion emphasizes the development of subunit vaccines, attenuated vaccines also have the potential of being very effective. Indeed, the Madrid E strain of *R. prowazekii* was a successful epidemic typhus vaccine when it did not revert to the virulent phenotype. The gene with a point mutation that explains the attenuation is an *S*-adenosylmethionine-dependent methyltransferase (Zhang et al., 2006). Deletion of the entire gene would permit the production of a safer vaccine. One of the problems is that this is a difficult endeavor because the tools for genetic manipulation of *Rickettsia* are not optimal and have not been standardized.

REFERENCES

Aird, W. C. 2005. Spatial and temporal dynamics of the endothelium. *J. Thromb. Haemost.* **3:**1392–1406.

Akondy, R. S., N. D. Monson, J. D. Miller, S. Edupuganti, D. Teuwen, H. Wu, F. Quyyumi, S. Garg, J. D. Altman, C. Del Rio, H. L. Keyserling, A. Ploss, C. M. Rice, W. A. Orenstein, M. J. Mulligan, and R. Ahmed. 2009. The yellow fever virus vaccine induces a broad and polyfunctional human memory $CD8^+$ T cell response. *J. Immunol.* **183:**7919–7930.

Al-Lamki, R. S., J. R. Bradley, and J. S. Pober. 2008. Endothelial cells in allograft rejection. *Transplantation* **86:**1340–1348.

Amatschek, S., E. Kriehuber, W. Bauer, B. Reininger, P. Meraner, A. Wolpl, N. Schweifer, C. Haslinger, G. Stingl, and D. Maurer. 2007. Blood and lymphatic endothelial cell-specific differentiation programs are stringently controlled by the tissue environment. *Blood* **109:**4777–4785.

Anderson, R. W., J. R. Bennink, J. W. Yewdell, W. L. Maloy, and J. E. Coligan. 1992. Influenza basic polymerase 2 peptides are recognized by influenza nucleoprotein-specific cytotoxic T lymphocytes. *Mol. Immunol.* **29:**1089–1096.

Andersson, S. G., A. Zomorodipour, J. O. Andersson, T. Sicheritz-Pontén, U. C. Alsmark, R. M. Podowski, A. K. Näslund, A. S. Eriksson, H. H. Winkler, and C. G. Kurland. 1998. The genome sequence of *Rickettsia prowazekii* and the origin of mitochondria. *Nature* **396:**133–140.

Andonegui, G., H. Zhou, D. Bullard, M. M. Kelly, S. C. Mullaly, B. McDonald, E. M. Long, S. M. Robbins, and P. Kubes. 2009. Mice that exclusively express TLR4 on endothelial cells can efficiently clear a lethal systemic Gram-negative bacterial infection. *J. Clin. Invest.* **119:**1921–1930.

Aoshi, T., M. Suzuki, M. Uchijima, T. Nagata, and Y. Koide. 2005. Expression mapping using a retroviral vector for $CD8^+$ T cell epitopes: definition of a *Mycobacterium tuberculosis* peptide presented by H2-D^d. *J. Immunol. Methods* **298:**21–34.

Bagai, R., A. Valujskikh, D. H. Canaday, E. Bailey, P. N. Lalli, C. V. Harding, and P. S. Heeger. 2005. Mouse endothelial cells cross-present lymphocyte-derived antigen on class I MHC via a TAP1- and proteasome-dependent pathway. *J. Immunol.* **174:**7711–7715.

Balayeva, N. M., and V. N. Nikolskaya. 1973. Analysis of lung culture of *Rickettsia prowazedi* E strain with regard to its capacity of increasing virulence in passages on the lungs of white mice. *J. Hyg. Epidemiol. Microbiol. Immunol.* **17:**294–303.

Bechah, Y., C. Capo, G. Grau, D. Raoult, and J. L. Mege. 2009. *Rickettsia prowazekii* infection of endothelial cells increases leukocyte adhesion through $\alpha v\beta 3$ integrin engagement. *Clin. Microbiol. Infect.* **15**(Suppl. 2)**:**249–250.

Bechah, Y., C. Capo, J. L. Mege, and D. Raoult. 2008a. Epidemic typhus. *Lancet Infect. Dis.* **8:**417–426.

Bechah, Y., C. Capo, D. Raoult, and J. L. Mege. 2008b. Infection of endothelial cells with virulent *Rickettsia prowazekii* increases the transmigration of leukocytes. *J. Infect. Dis.* **197:**142–147.

Bedke, T., L. Pretsch, S. Karakhanova, A. H. Enk, and K. Mahnke. 2010. Endothelial cells augment the suppressive function of $CD4^+CD25^+Foxp3^+$ regulatory T cells: involvement of programmed death-1 and IL-10. *J. Immunol.* **184:**5562–5570.

Billings, A. N., H. M. Feng, J. P. Olano, and D. H. Walker. 2001. Rickettsial infection in murine models activates an early anti-rickettsial effect mediated by NK cells and associated with production of gamma interferon. *Am. J. Trop. Med. Hyg.* **65:**52–56.

Bisset, L. R., T. L. Lung, M. Kaelin, E. Ludwig, and R. W. Dubs. 2004. Reference values for peripheral blood lymphocyte phenotypes applicable to the healthy adult population in Switzerland. *Eur. J. Haematol.* **72:**203–212.

Bolinger, B., P. Krebs, Y. Tian, D. Engeler, E. Scandella, S. Miller, D. C. Palmer, N. P. Restifo, P. A. Clavien, and B. Ludewig. 2008. Immunologic ignorance of vascular endothelial cells expressing minor histocompatibility antigen. *Blood* **111:**4588–4595.

Brossard, M., and S. K. Wikel. 2004. Tick immunobiology. *Parasitology* **129**(Suppl.)**:**S161–S176.

Calabria, A. R., and E. V. Shusta. 2008. A genomic comparison of *in vivo* and *in vitro* brain microvascular endothelial cells. *J Cereb. Blood Flow Metab.* **28:**135–148.

Capo, S., S. Nuti, M. Scarselli, S. Tavarini, S. Montigiani, E. Mori, O. Finco, S. Abrignani, G. Grandi, and G. Bensi. 2005. *Chlamydia pneumoniae* genome sequence analysis and identification of HLA-A2-restricted $CD8^+$ T cell epitopes recognized by infection-primed T cells. *Vaccine* **23:**5028–5037.

Centers for Disease Control and Prevention. 2004. Fatal cases of Rocky Mountain spotted fever in family clusters—three states, 2003. *MMWR Morb. Mortal. Wkly. Rep.* **53:**407–410.

Chan, Y. G., S. P. Riley, E. Chen, and J. J. Martinez. 2011. Molecular basis of immunity to rickettsial infection conferred through outer membrane protein B. *Infect. Immun.* **79:**2303–2313.

Chappell, D., M. Jacob, O. Paul, M. Rehm, U. Welsch, M. Stoeckelhuber, P. Conzen, and B. F. Becker. 2009. The glycocalyx of the human umbilical vein endothelial cell: an impressive structure ex vivo but not in culture. *Circ. Res.* **104:**1313–1317.

Chattopadhyay, S., and A. L. Richards. 2007. Scrub typhus vaccines: past history and recent developments. *Hum. Vaccin.* **3:**73–80.

Chen, L. F., and D. J. Sexton. 2008. What's new in Rocky Mountain spotted fever? *Infect. Dis. Clin. North Am.* **22:**415–432, vii–viii.

Cho, K. A., Y. H. Jun, J. W. Suh, J. S. Kang, H. J. Choi, and S. Y. Woo. 2010. *Orientia tsutsugamushi* induced endothelial cell activation via the NOD1-IL-32 pathway. *Microb. Pathog.* **49:**95–104.

Choi, J., D. R. Enis, K. P. Koh, S. L. Shiao, and J. S. Pober. 2004. T lymphocyte-endothelial cell interactions. *Annu. Rev. Immunol.* **22:**683–709.

Churilla, A., W. M. Ching, G. A. Dasch, and M. Carl. 1990. Human T lymphocyte recognition of cyanogen bromide fragments of the surface protein of *Rickettsia typhi*. *Ann. N. Y. Acad. Sci.* **590:**215–220.

Clements, M. L., C. L. Wisseman, T. E. Woodward, P. Fiset, J. S. Dumler, W. McNamee, R. E. Black, J. Rooney, T. P. Hughes, and M. M. Levine. 1983. Reactogenicity, immunogenicity, and efficacy of a chick embryo cell-derived vaccine for Rocky Mountain spotted fever. *J. Infect. Dis.* **148:**922–930.

Clute, S. C., Y. N. Naumov, L. B. Watkin, N. Aslan, J. L. Sullivan, D. A. Thorley-Lawson, K. Luzuriaga, R. M. Welsh, R. Puzone, F. Celada, and L. K. Selin. 2010. Broad cross-reactive TCR repertoires recognizing dissimilar Epstein-Barr and influenza A virus epitopes. *J Immunol.* **185:**6753–6764.

Coffman, R. L., A. Sher, and R. A. Seder. 2010. Vaccine adjuvants: putting innate immunity to work. *Immunity* **33:**492–503.

Crocquet-Valdes, P. A., C. M. Díaz-Montero, H. M. Feng, H. Li, A. D. Barrett, and D. H. Walker. 2001. Immunization with a portion of rickettsial outer membrane protein A stimulates protective immunity against spotted fever rickettsiosis. *Vaccine* **20:**979–988.

Dai, G., M. R. Kaazempur-Mofrad, S. Natarajan, Y. Zhang, S. Vaughn, B. R. Blackman, R. D. Kamm, G. García-Cardeña, and M. A. Gimbrone. 2004. Distinct endothelial phenotypes evoked by arterial waveforms derived from atherosclerosis-susceptible and -resistant regions of human vasculature. *Proc. Natl. Acad. Sci. USA* **101:**14871–14876.

Damås, J. K., G. Davì, M. Jensenius, F. Santilli, K. Otterdal, T. Ueland, T. H. Flo, E. Lien, T. Espevik, S. S. Frøland, G. Vitale, D. Raoult, and P. Aukrust. 2009. Relative chemokine and adhesion molecule expression in Mediterranean spotted fever and African tick bite fever. *J. Infect.* **58:**68–75.

Damås, J. K., M. Jensenius, T. Ueland, K. Otterdal, A. Yndestad, S. S. Frøland, J. M. Rolain, B. Myrvang, D. Raoult, and P. Aukrust. 2006. Increased levels of soluble CD40L in African tick bite fever: possible involvement of TLRs in the pathogenic interaction between *Rickettsia africae*, endothelial cells, and platelets. *J. Immunol.* **177:**2699–2706.

Danese, S., E. Dejana, and C. Fiocchi. 2007. Immune regulation by microvascular endothelial cells: directing innate and adaptive immunity, coagulation, and inflammation. *J. Immunol.* **178:**6017–6022.

Darrah, P. A., D. T. Patel, P. M. De Luca, R. W. Lindsay, D. F. Davey, B. J. Flynn, S. T. Hoff, P. Andersen, S. G. Reed, S. L. Morris, M. Roederer, and R. A. Seder. 2007. Multifunctional T_H1 cells define a correlate of vaccine-mediated protection against *Leishmania major*. *Nat. Med.* **13:**843–850.

Davies, P. F. 2008. Endothelial transcriptome profiles *in vivo* in complex arterial flow fields. *Ann. Biomed. Eng.* **36:**563–570.

de Fost, M., W. Chierakul, K. Pimda, A. M. Dondorp, N. J. White, and T. Van der Poll. 2005. Activation of cytotoxic lymphocytes in patients with scrub typhus. *Am. J. Trop. Med. Hyg.* **72:**465–467.

Demeester, R., M. Claus, M. Hildebrand, E. Vlieghe, and E. Bottieau. 2010. Diversity of life-threatening complications due to Mediterranean spotted fever in returning travelers. *J. Travel Med.* **17:**100–104.

de Sousa, R., N. Ismail, S. D. Nobrega, A. França, M. Amaro, M. Anes, J. Poças, R. Coelho, J. Torgal, F. Bacellar, and D. H. Walker. 2007. Intralesional expression of mRNA of interferon-γ, tumor necrosis factor-α, interleukin-10, nitric oxide synthase, indoleamine-2,3-dioxygenase, and RANTES is a major immune effector in Mediterranean spotted fever rickettsiosis. *J. Infect. Dis.* **196:**770–781.

Di Francesco, L., L. Totani, M. Dovizio, A. Piccoli, A. Di Francesco, T. Salvatore, A. Pandolfi, V. Evangelista, R. A. Dercho, F. Seta, and P. Patrignani. 2009. Induction of prostacyclin by steady laminar shear stress suppresses tumor necrosis factor-α biosynthesis via heme oxygenase-1 in human endothelial cells. *Circ. Res.* **104:**506–513.

Dignat-George, F., N. Teysseire, M. Mutin, N. Bardin, G. Lesaule, D. Raoult, and J. Sampol. 1997. *Rickettsia conorii* infection enhances vascular cell adhesion molecule-1- and intercellular adhesion molecule-1-dependent mononuclear cell adherence to endothelial cells. *J. Infect. Dis.* **175:**1142–1152.

DuPont, H. L., R. B. Hornick, A. T. Dawkins, G. G. Heiner, I. B. Fabrikant, C. L. Wisseman, and T. E. Woodward. 1973. Rocky Mountain spotted fever: a comparative study of the active immunity induced by inactivated and viable pathogenic *Rickettsia rickettsii*. *J. Infect. Dis.* **128:**340–344.

Durr, E., J. Yu, K. M. Krasinska, L. A. Carver, J. R. Yates, J. E. Testa, P. Oh, and J. E. Schnitzer. 2004. Direct proteomic mapping of the lung microvascular endothelial cell surface *in vivo* and in cell culture. *Nat. Biotechnol.* **22:**985–992.

Eisemann, C. S., M. J. Nypaver, and J. V. Osterman. 1984. Susceptibility of inbred mice to rickettsiae of the spotted fever group. *Infect. Immun.* **43:**143–148.

Epperson, D. E., and J. S. Pober. 1994. Antigen-presenting function of human endothelial cells. Direct activation of resting CD8 T cells. *J. Immunol.* **153:**5402–5412.

Fang, R., N. Ismail, T. Shelite, and D. H. Walker. 2009. $CD4^+$ $CD25^+$ $Foxp3^-$ T-regulatory cells produce both gamma interferon and interleukin-10 during acute severe murine spotted fever rickettsiosis. *Infect Immun.* **77:**3838–3849.

Fang, R., N. Ismail, L. Soong, V. L. Popov, T. Whitworth, D. H. Bouyer, and D. H. Walker. 2007. Differential interaction of dendritic cells with *Rickettsia conorii*: impact on host susceptibility to murine spotted fever rickettsiosis. *Infect. Immun.* **75:**3112–3123.

Feng, H. M., V. L. Popov, and D. H. Walker. 1994. Depletion of gamma interferon and tumor necrosis factor alpha in mice with *Rickettsia conorii*-infected endothelium: impairment of rickettsicidal nitric oxide production resulting in fatal, overwhelming rickettsial disease. *Infect Immun.* **62:**1952–1960.

Feng, H., V. L. Popov, G. Yuoh, and D. H. Walker. 1997. Role of T lymphocyte subsets in immunity to spotted fever group rickettsiae. *J. Immunol.* **158:**5314–5320.

Feng, H. M., and D. H. Walker. 2000. Mechanisms of intracellular killing of *Rickettsia conorii* in infected human endothelial cells, hepatocytes, and macrophages. *Infect Immun.* **68:**6729–6736.

Feng, H. M., and D. H. Walker. 2003. Cross-protection between distantly related spotted fever group rickettsiae. *Vaccine* **21:**3901–3905.

Feng, W. C., and J. L. Waner. 1980. Serological cross-reaction and cross-protection in guinea pigs infected with *Rickettsia rickettsii* and *Rickettsia montana*. *Infect Immun.* **28:**627–629.

Forte, G. I., L. Scola, G. Misiano, S. Milano, P. Mansueto, G. Vitale, F. Bellanca, M. Sanacore, L. Vaccarino, G. B. Rini, C. Caruso, E. Cillari, D. Lio, and S. Mansueto. 2009. Relevance of gamma interferon, tumor necrosis factor alpha, and interleukin-10 gene polymorphisms to susceptibility to Mediterranean spotted fever. *Clin. Vaccine Immunol.* **16:**811–815.

Francischetti, I. M., A. Sá-Nunes, B. J. Mans, I. M. Santos, and J. M. Ribeiro. 2009. The role of saliva in tick feeding. *Front. Biosci.* **14:**2051–2088.

Gage, K. L., and T. R. Jerrells. 1992. Demonstration and partial characterization of antigens of *Rickettsia rhipicephali* that induce cross-reactive cellular and humoral immune responses to *Rickettsia rickettsii*. *Infect Immun.* **60:**5099–5106.

Gaucher, C., C. Devaux, C. Boura, P. Lacolley, J. F. Stoltz, and P. Menu. 2007. In vitro impact of physiological shear stress on endothelial cells gene expression profile. *Clin. Hemorheol. Microcirc.* **37:**99–107.

Gaucher, D., R. Therrien, N. Kettaf, B. R. Angermann, G. Boucher, A. Filali-Mouhim, J. M. Moser, R. S. Mehta, D. R. Drake III, E. Castro, R. Akondy, A. Rinfret, B. Yassine-Diab, E. A. Said, Y. Chouikh, M. J. Cameron, R. Clum, D. Kelvin, R. Somogyi, L. D. Greller, R. S. Balderas, P. Wilkinson, G. Pantaleo, J. Tartaglia, E. K. Haddad, and R. P. Sékaly. 2008. Yellow fever vaccine induces integrated multilineage and polyfunctional immune responses. *J. Exp. Med.* **205:**3119–3131.

George, F., P. Brouqui, M. C. Boffa, M. Mutin, M. Drancourt, C. Brisson, D. Raoult, and J. Sampol. 1993. Demonstration of *Rickettsia conorii*-induced endothelial injury in vivo by measuring circulating endothelial cells, thrombomodulin, and von Willebrand factor in patients with Mediterranean spotted fever. *Blood* **82:**2109–2116.

Gillespie, J. J., K. Williams, M. Shukla, E. E. Snyder, E. K. Nordberg, S. M. Ceraul, C. Dharmanolla, D. Rainey, J. Soneja, J. M. Shallom, N. D. Vishnubhat, R. Wattam, A. Purkayastha, M. Czar, O. Crasta, J. C. Setubal, A. F. Azad, and B. S. Sobral. 2008. *Rickettsia* phylogenomics: unwinding the intricacies of obligate intracellular life. *PLoS One* **3:**e2018.

Gonçalves, R. B., O. Leshem, K. Bernards, J. R. Webb, P. P. Stashenko, and A. Campos-Neto. 2006. T-cell expression cloning of *Porphyromonas gingivalis* genes coding for T helper-biased immune responses during infection. *Infect. Immun.* **74:**3958–3966.

Gorini, S., G. Callegari, G. Romagnoli, C. Mammi, D. Mavilio, G. Rosano, M. Fini, F. Di Virgilio, S. Gulinelli, S. Falzoni, A. Cavani, D. Ferrari, and A. la Sala. 2010. ATP secreted by endothelial cells blocks CX_3CL1-elicited natural killer cell chemotaxis and cytotoxicity via $P2Y_{11}$ receptor activation. *Blood* **116:** 4492–4500.

Haining, W. N., and E. J. Wherry. 2010. Integrating genomic signatures for immunologic discovery. *Immunity* **32:**152–161.

Halling-Brown, M., C. E. Sansom, M. Davies, R. W. Titball, and D. S. Moss. 2008. Are bacterial vaccine antigens T-cell epitope depleted? *Trends Immunol.* **29:**374–379.

Hayes, S. F., and W. Burgdorfer. 1982. Reactivation of *Rickettsia rickettsii* in *Dermacentor andersoni* ticks: an ultrastructural analysis. *Infect. Immun.* **37:**779–785.

He, S., M. Li, X. Ma, J. Lin, and D. Li. 2010. $CD4^+CD25^+Foxp3^+$ regulatory T cells protect the proinflammatory activation of human umbilical vein endothelial cells. *Arterioscler. Thromb. Vasc. Biol.* **30:**2621–2630.

Hickey, M. J., and P. Kubes. 2009. Intravascular immunity: the host-pathogen encounter in blood vessels. *Nat. Rev. Immunol.* **9:**364–375.

Imamaliev, O. G., A. A. Sumarokov, V. N. Nikol'skaia, V. L. Lelikov, and M. S. Vorob'eva. 1981. Results of a study to determine the optimal vaccination dose and schedule for primary immunization with chemical typhus vaccine. *Zh. Mikrobiol. Epidemiol. Immunobiol.* **1981:**88–91. (In Russian.)

Jerrells, T. R., D. L. Jarboe, and C. S. Eisemann. 1986. Cross-reactive lymphocyte responses and protective immunity against other spotted fever group rickettsiae in mice immunized with *Rickettsia conorii*. *Infect. Immun.* **51:**832–837.

Jerrells, T. R., and J. V. Osterman. 1982. Host defenses in experimental scrub typhus: delayed-type hypersensitivity responses of inbred mice. *Infect. Immun.* **35:**117–123.

Jollow, K. C., J. C. Zimring, J. B. Sundstrom, and A. A. Ansari. 1999. CD40 ligation induced phenotypic and functional expression of CD80 by human cardiac microvascular endothelial cells. *Transplantation* **68:**430–439.

Jordan, J. M., M. E. Woods, H. M. Feng, L. Soong, and D. H. Walker. 2007. Rickettsiae-stimulated dendritic cells mediate protection against lethal rickettsial challenge in an animal model of spotted fever rickettsiosis. *J. Infect. Dis.* **196:**629–638.

Jordan, J. M., M. E. Woods, J. Olano, and D. H. Walker. 2008. The absence of Toll-like receptor 4 signaling in C3H/HeJ mice predisposes them to overwhelming rickettsial infection and decreased protective Th1 responses. *Infect. Immun.* **76:**3717–3724.

Jordan, J. M., M. E. Woods, L. Soong, and D. H. Walker. 2009. Rickettsiae stimulate dendritic cells through Toll-like receptor 4, leading to enhanced NK cell activation in vivo. *J. Infect. Dis.* **199:**236–242.

Kaplanski, G., N. Teysseire, C. Farnarier, S. Kaplanski, J.-C. Lissitzky, J.-M. Durand, J. Soubeyrand, C. A. Dinarello, and P. Bongrand. 1995. IL-6 and IL-8 production from cultured human endothelial cells stimulated by infection with *Rickettsia conorii* via a cell-associated IL-1α-dependent pathway. *J. Clin. Invest.* **96:**2839–2844.

Karmann, K., C. C. Hughes, J. Schechner, W. C. Fanslow, and J. S. Pober. 1995. CD40 on human endothelial cells: inducibility by cytokines and functional regulation of adhesion molecule expression. *Proc. Natl. Acad. Sci. USA* **92:** 4342–4346.

Kelly, D. J., P. A. Fuerst, W. M. Ching, and A. L. Richards. 2009. Scrub typhus: the geographic distribution of phenotypic and genotypic variants of *Orientia tsutsugamushi*. *Clin. Infect. Dis.* **48**(Suppl. 3): S203–S230.

Khayyamian, S., A. Hutloff, K. Büchner, M. Gräfe, V. Henn, R. A. Kroczek, and H. W. Mages. 2002. ICOS-ligand, expressed on human endothelial cells, costimulates Th1 and Th2 cytokine secretion by memory $CD4^+$ T cells. *Proc. Natl. Acad. Sci. USA* **99:**6198–6203.

Kobayashi, Y., S. Kawamura, and T. Oyama. 1985. Immunological studies of experimental tsutsugamushi disease in congenitally athymic (nude) mice. *Am. J. Trop. Med. Hyg.* **34:**568–577.

Koh, Y. S., J. H. Yun, S. Y. Seong, M. S. Choi, and I. S. Kim. 2004. Chemokine and cytokine production during *Orientia tsutsugamushi* infection in mice. *Microb. Pathog.* **36:**51–57.

Komarova, Y., and A. B. Malik. 2010. Regulation of endothelial permeability via paracellular and transcellular transport pathways. *Annu Rev Physiol.***72**:463–493.

Kramme, S., l.. V. An, N. D. Khoa, l.. V. Trin, E. Tannich, J. Rybniker, B. Fleischer, C. Drosten, and M. Panning. 2009. *Orientia tsutsugamushi* bacteremia and cytokine levels in Vietnamese scrub typhus patients. *J. Clin. Microbiol.* **47:**586–589.

Kreisel, D., A. S. Krupnick, K. R. Balsara, M. Riha, A. E. Gelman, S. H. Popma, W. Y. Szeto, L. A. Turka, and B. R. Rosengard. 2002a. Mouse vascular endothelium activates $CD8^+$ T lymphocytes in a B7-dependent fashion. *J. Immunol.* **169:**6154–6161.

Kreisel, D., A. S. Krupnick, A. E. Gelman, F. H. Engels, S. H. Popma, A. M. Krasinskas, K. R. Balsara, W. Y. Szeto, L. A. Turka, and B. R. Rosengard. 2002b. Non-hematopoietic allograft cells directly activate $CD8^+$ T cells and trigger acute rejection: an alternative mechanism of allorecognition. *Nat. Med.* **8:**233–239.

Krupnick, A. S., A. E. Gelman, W. Barchet, S. Richardson, F. H. Kreisel, L. A. Turka, M. Colonna, G. A. Patterson, and D. Kreisel. 2005. Murine vascular endothelium activates and induces the generation of allogeneic $CD4^+25^+Foxp3^+$ regulatory T cells. *J. Immunol.* **175:**6265–6270.

Kulkarni, A. B., H. C. Morse, J. R. Bennink, J. W. Yewdell, and B. R. Murphy. 1993. Immunization of mice with vaccinia virus-M2 recombinant induces epitope-specific and cross-reactive Kd-restricted $CD8^+$ cytotoxic T cells. *J. Virol.* **67:**4086–4092.

Kummer, M., A. Lev, Y. Reiter, and B. C. Biedermann. 2005. Vascular endothelial cells have impaired capacity to present immunodominant, antigenic peptides: a mechanism of cell type-specific immune escape. *J. Immunol.* **174:**1947–1953.

Kundin, W. D., C. Liu, P. Harmon, and P. Rodina. 1964. Pathogenesis of scrub typhus infection (*Rickettsia tsutsugamushi*) as studied by immunofluorescence. *J. Immunol.* **93:**772–781.

Kunitomi, A., T. Hori, A. Imura, and T. Uchiyama. 2000. Vascular endothelial cells provide T cells with costimulatory signals via the OX40/gp34 system. *J. Leukoc. Biol.* **68:**111–118.

Kuwano, K., V. E. Reyes, R. E. Humphreys, and F. A. Ennis. 1991. Recognition of disparate HA and NS1 peptides by an H-2Kd-restricted, influenza specific CTL clone. *Mol. Immunol.* **28:**1–7.

La, M. V., P. François, C. Rovery, S. Robineau, P. Barbry, J. Schrenzel, D. Raoult, and P. Renesto. 2007. Development of a method for recovering rickettsial RNA from infected cells to analyze gene expression profiling of obligate intracellular bacteria. *J. Microbiol. Methods* **71:**292–297.

Lacorre, D. A., E. S. Baekkevold, I. Garrido, P. Brandtzaeg, G. Haraldsen, F. Amalric, and J. P. Girard. 2004. Plasticity of endothelial cells: rapid dedifferentiation of freshly isolated high endothelial venule endothelial cells outside the lymphoid tissue microenvironment. *Blood* **103:**4164–4172.

Langer, H. F., and T. Chavakis. 2009. Leukocyte-endothelial interactions in inflammation. *J. Cell. Mol. Med.* **13:**1211–1220.

La Scola, B., and D. Raoult. 1996. Diagnosis of Mediterranean spotted fever by cultivation of *Rickettsia conorii* from blood and skin samples using the centrifugation-shell vial technique and by detection of *R. conorii* in circulating endothelial cells: a 6-year follow-up. *J. Clin. Microbiol.* **34:**2722–2727.

Lee, C. S., J. H. Hwang, H. B. Lee, and K. S. Kwon. 2009a. Risk factors leading to fatal outcome in scrub typhus patients. *Am. J. Trop. Med. Hyg.* **81:**484–488.

Lee, J. S., M. Y. Park, Y. J. Kim, H. I. Kil, Y. H. Choi, and Y. C. Kim. 2009b. Histopathological features in both the eschar and erythematous lesions of tsutsugamushi disease: identification of $CD30^+$ cell infiltration in tsutsugamushi disease. *Am. J. Dermatopathol.* **31:**551–556.

Lee, N., M. Ip, B. Wong, G. Lui, O. T. Tsang, J. Y. Lai, K. W. Choi, R. Lam, T. K. Ng, J. Ho, Y. Y. Chan, C. S. Cockram, and S. T. Lai. 2008. Risk factors associated with life-threatening rickettsial infections. *Am. J. Trop. Med. Hyg.* **78:**973–978.

Lemmel, C., and S. Stevanović. 2003. The use of HPLC-MS in T-cell epitope identification. *Methods* **29:**248–259.

Li, D., A. Borovkov, A. Vaglenov, C. Wang, T. Kim, D. Gao, K. F. Sykes, and B. Kaltenboeck. 2006. Mouse model of respiratory *Chlamydia pneumoniae* infection for a genomic screen of subunit vaccine candidates. *Vaccine* **24:**2917–2927.

Li, Z., C. M. Díaz-Montero, G. Valbuena, X. J. Yu, J. P. Olano, H. M. Feng, and D. H. Walker. 2003. Identification of CD8 T-lymphocyte epitopes in OmpB of *Rickettsia conorii*. *Infect. Immun.* **71:**3920–3926.

Lindenstrøm, T., E. M. Agger, K. S. Korsholm, P. A. Darrah, C. Aagaard, R. A. Seder, I. Rosenkrands, and P. Andersen. 2009. Tuberculosis subunit vaccination provides long-term protective immunity characterized by multifunctional CD4 memory T cells. *J. Immunol.* **182:**8047–8055.

Liu, M. A. 2010. Immunologic basis of vaccine vectors. *Immunity* **33:**504–515.

Liu, Y., Y. Zhang, K. Schmelzer, T. S. Lee, X. Fang, Y. Zhu, A. A. Spector, S. Gill, C. Morisseau, B. D. Hammock, and J. Y. Shyy. 2005. The antiinflammatory effect of laminar flow: the role of PPARγ, epoxyeicosatrienoic acids, and soluble epoxide hydrolase. *Proc. Natl. Acad. Sci. USA* **102:**16747–16752.

Ma, W., and J. S. Pober. 1998. Human endothelial cells effectively costimulate cytokine production by, but not differentiation of, naive $CD4^+$ T cells. *J. Immunol.* **161:**2158–2167.

Mansueto, P., G. Vitale, G. Di Lorenzo, F. Arcoleo, S. Mansueto, and E. Cillari. 2008. Immunology of human rickettsial diseases. *J. Biol. Regul. Homeost. Agents* **22:**131–139.

Marelli-Berg, F. M., M. J. James, J. Dangerfield, J. Dyson, M. Millrain, D. Scott, E. Simpson, S. Nourshargh, and R. I. Lechler. 2004. Cognate recognition of the endothelium induces HY-specific $CD8^+$ T-lymphocyte transendothelial migration (diapedesis) in vivo. *Blood* **103:** 3111–3116.

Mason, R. A., R. P. Wenzel, E. B. Seligmann, and R. K. Ginn. 1976. A reference, inactivated, epidemic typhus vaccine: clinical trials in man. *J. Biol. Stand.* **4:**217–224.

McLeod, M. P., X. Qin, S. E. Karpathy, J. Gioia, S. K. Highlander, G. E. Fox, T. Z. McNeill, H. Jiang, D. Muzny, L. S. Jacob, A. C. Hawes, E. Sodergren, R. Gill, J. Hume, M. Morgan, G. Fan, A. G. Amin, R. A. Gibbs, C. Hong, X. J. Yu, D. H. Walker, and G. M. Weinstock. 2004. Complete genome sequence of *Rickettsia typhi* and comparison with sequences of other rickettsiae. *J. Bacteriol.* **186:**5842–5855.

Michiels, C. 2003. Endothelial cell functions. *J. Cell. Physiol.* **196:**430–443.

Milano, S., P. D'Agostino, G. Di Bella, M. La Rosa, C. Barbera, V. Ferlazzo, P. Mansueto, G. B. Rini, A. Barera, G. Vitale, S. Mansueto, and E. Cillari. 2000. Interleukin-12 in human boutonneuse fever caused by *Rickettsia conorii*. *Scand. J. Immunol.* **52:**91–95.

Miller, J. D., R. G. van der Most, R. S. Akondy, J. T. Glidewell, S. Albott, D. Masopust, K. Murali-Krishna, P. L. Mahar, S. Edupuganti, S. Lalor, S. Germon, C. Del Rio, M. J. Mulligan, S. I. Staprans, J. D. Altman, M. B. Feinberg, and R. Ahmed. 2008. Human effector and memory $CD8^+$ T cell responses to smallpox and yellow fever vaccines. *Immunity* **28:**710–722.

Moriel, D. G., I. Bertoldi, A. Spagnuolo, S. Marchi, R. Rosini, B. Nesta, I. Pastorello, V. A. Corea, G. Torricelli, E. Cartocci, S. Savino, M. Scarselli, U. Dobrindt, J. Hacker, H. Tettelin, L. J. Tallon, S. Sullivan, L. H. Wieler, C. Ewers, D. Pickard, G. Dougan, M. R. Fontana, R. Rappuoli, M. Pizza, and L. Serino. 2010. Identification of protective and broadly conserved vaccine antigens from the genome of extraintestinal pathogenic *Escherichia coli*. *Proc. Natl. Acad. Sci. USA* **107:**9072–9077.

Moron, C. G., V. L. Popov, H. M. Feng, D. Wear, and D. H. Walker. 2001. Identification of the target cells of *Orientia tsutsugamushi* in human cases of scrub typhus. *Mod. Pathol.* **14:**752–759.

Moutaftsi, M., H. H. Bui, B. Peters, J. Sidney, S. Salek-Ardakani, C. Oseroff, V. Pasquetto, S. Crotty, M. Croft, E. J. Lefkowitz, H. Grey, and A. Sette. 2007. Vaccinia virus-specific $CD4^+$ T cell responses target a set of antigens largely distinct from those targeted by $CD8^+$ T cell responses. *J. Immunol.* **178:**6814–6820.

Muller, W. A. 2009. Mechanisms of transendothelial migration of leukocytes. *Circ. Res.* **105:**223–230.

Muller, W. A. 2011. Mechanisms of leukocyte transendothelial migration. *Annu. Rev. Pathol.* **6:**323–344.

Murphy, T. J., G. Thurston, T. Ezaki, and D. M. McDonald. 1999. Endothelial cell heterogeneity in venules of mouse airways induced by polarized inflammatory stimulus. *Am. J. Pathol.* **155:**93–103.

Nagy, N., T. Freudenberger, A. Melchior-Becker, K. Röck, M. ter Braak, H. Jastrow, M. Kinzig, S. Lucke, T. Suvorava, G. Kojda, A. A. Weber, F. Sörgel, B. Levkau, S. Ergün, and J. W. Fischer. 2010. Inhibition of hyaluronan synthesis accelerates murine iatherosclerosis: novel Insights into the role of hyaluronan synthesis. *Circulation* **122:**2313–2322.

Nikol'skaia, V. N., O. G. Imamaliev, and N. D. Klimchuk. 1982. Characteristics of the antitoxic immunity in persons inoculated with chemical typhus vaccine. *Zh. Mikrobiol. Epidemiol. Immunobiol.* **1982:**86–89. (In Russian.)

Nikolskaya, V. N., and N. M. Balayeva. 1973. Homogeneity of *Rickettsia prowazeki* E strain egg culture as to the capacity to increase virulence in passages on white mouse lungs. *J. Hyg. Epidemiol. Microbiol. Immunol.* **17:**505–506.

Ogata, H., S. Audic, P. Renesto-Audiffren, P. E. Fournier, V. Barbe, D. Samson, V. Roux, P. Cossart, J. Weissenbach, J. M. Claverie, and D. Raoult. 2001. Mechanisms of evolution in *Rickettsia conorii* and *R. prowazekii*. *Science* **293:** 2093–2098.

Olsen, A. W., F. Follmann, P. Højrup, R. Leah, C. Sand, P. Andersen, and M. Theisen. 2007. Identification of human T cell targets recognized during *Chlamydia trachomatis* genital infection. *J. Infect. Dis.* **196:**1546–1552.

Omari, K. I., and K. Dorovini-Zis. 2001. Expression and function of the costimulatory molecules B7-1 (CD80) and B7-2 (CD86) in an in vitro model of the human blood-brain barrier. *J. Neuroimmunol.* **113:**129–141.

Omari, K. M., and K. Dorovini-Zis. 2003. CD40 expressed by human brain endothelial cells regulates $CD4^{+}$ T cell adhesion to endothelium. *J. Neuroimmunol.* **134:**166–178.

Opitz, B., S. Hippenstiel, J. Eitel, and N. Suttorp. 2007. Extra- and intracellular innate immune recognition in endothelial cells. *Thromb. Haemost.* **98:**319–326.

Ormsbee, R., M. Peacock, R. Philip, E. Casper, J. Plorde, T. Gabre-Kidan, and L. Wright. 1978. Antigenic relationships between the typhus and spotted fever groups of rickettsiae. *Am. J. Epidemiol.* **108:**53–59.

Paddock, C. D., P. W. Greer, T. L. Ferebee, J. Singleton, D. B. McKechnie, T. A. Treadwell, J. W. Krebs, M. J. Clarke, R. C. Holman, J. G. Olson, J. E. Childs, and S. R. Zaki. 1999. Hidden mortality attributable to Rocky Mountain spotted fever: immunohistochemical detection of fatal, serologically unconfirmed disease. *J. Infect. Dis.* **179:**1469–1476.

Paddock, C. D., R. C. Holman, J. W. Krebs, and J. E. Childs. 2002. Assessing the magnitude of fatal Rocky Mountain spotted fever in the United States: comparison of two national data sources. *Am. J. Trop. Med. Hyg.* **67:**349–354.

Parker, R. R., and H. R. Cox. 1940. A pathogenic rickettsia from the Gulf Coast tick, *Amblyomma maculatum*, p. 390–391. *In* Proceedings of the Third International Congress of Microbiology. International Association of Microbiologists, New York, NY.

Parker, R. R., G. M. Kohls, G. W. Cox, and G. E. Davis. 1939. Observations of an infectious agent from *Amblyomma maculatum*. *Public Health Rep.* **54:**1482–1484.

Parker, R. R., E. G. Pickens, D. B. Lackman, E. J. Belle, and F. B. Thraikill. 1951. Isolation and characterization of Rocky Mountain spotted fever rickettsiae from the rabbit tick *Haemaphysalis leporis-palustris* Packard. *Public Health Rep.* **66:**455–463.

Peters, N. C., N. Kimblin, N. Secundino, S. Kamhawi, P. Lawyer, and D. L. Sacks. 2009. Vector transmission of *Leishmania* abrogates vaccine-induced protective immunity. *PLoS Pathog.* **5:**e1000484.

Pober, J. S., W. Min, and J. R. Bradley. 2009. Mechanisms of endothelial dysfunction, injury, and death. *Annu. Rev. Pathol.* **4:**71–95.

Pober, J. S., and W. C. Sessa. 2007. Evolving functions of endothelial cells in inflammation. *Nat. Rev. Immunol.* **7:**803–815.

Potter, D. R., and E. R. Damiano. 2008. The hydrodynamically relevant endothelial cell glycocalyx observed in vivo is absent in vitro. *Circ. Res.* **102:**770–776.

Potter, D. R., J. Jiang, and E. R. Damiano. 2009. The recovery time course of the endothelial cell glycocalyx in vivo and its implications in vitro. *Circ. Res.* **104:**1318–1325.

Prat, A., K. Biernacki, B. Becher, and J. P. Antel. 2000. B7 expression and antigen presentation by human brain endothelial cells: requirement for proinflammatory cytokines. *J. Neuropathol. Exp. Neurol.* **59:**129–136.

Querec, T. D., R. S. Akondy, E. K. Lee, W. Cao, H. I. Nakaya, D. Teuwen, A. Pirani, K. Gernert, J. Deng, B. Marzolf, K. Kennedy, H. Wu, S. Bennouna, H. Oluoch, J. Miller, R. Z. Vencio, M. Mulligan, A. Aderem, R. Ahmed, and B. Pulendran. 2009. Systems biology approach predicts immunogenicity of the yellow fever vaccine in humans. *Nat. Immunol.* **10:**116–125.

Rao, R. M., L. Yang, G. Garcia-Cardena, and F. W. Luscinskas. 2007. Endothelial-dependent mechanisms of leukocyte recruitment to the vascular wall. *Circ. Res.* **101:**234–247.

Reck, J., M. Berger, F. S. Marks, R. B. Zingali, C. W. Canal, C. A. Ferreira, J. A. Guimarães, and C. Termignoni. 2009. Pharmacological action of tick saliva upon haemostasis and the neutralization ability of sera from repeatedly infested hosts. *Parasitology* **136:**1339–1349.

Renesto, P., C. Rovery, J. Schrenzel, Q. Leroy, A. Huyghe, W. Li, H. Lepidi, P. François, and D. Raoult. 2008. *Rickettsia conorii* transcriptional response within inoculation eschar. *PLoS One* **3:**e3681.

Reul, R. M., J. C. Fang, M. D. Denton, C. Geehan, C. Long, R. N. Mitchell, P. Ganz, and D. M. Briscoe. 1997. CD40 and CD40 ligand (CD154) are coexpressed on microvessels in vivo in human cardiac allograft rejection. *Transplantation* **64:**1765–1774.

Richards, A. L. 2004. Rickettsial vaccines: the old and the new. *Expert Rev. Vaccines* **3:**541–555.

Riedl, P., A. Wieland, K. Lamberth, S. Buus, F. Lemonnier, K. Reifenberg, J. Reimann, and R. Schirmbeck. 2009. Elimination of immunodominant epitopes from multispecific DNA-based vaccines allows induction of CD8 T cells that have a striking antiviral potential. *J. Immunol.* **183:**370–380.

Rinaudo, C. D., J. L. Telford, R. Rappuoli, and K. L. Seib. 2009. Vaccinology in the genome era. *J. Clin. Invest.* **119:**2515–2525.

Rizzo, M., P. Mansueto, G. Di Lorenzo, S. Morselli, S. Mansueto, and G. B. Rini. 2004. Rickettsial disease: classical and modern aspects. *New Microbiol.* **27:**87–103.

Rollenhagen, C., M. Sörensen, K. Rizos, R. Hurvitz, and D. Bumann. 2004. Antigen selection based on expression levels during infection facilitates vaccine development for an intracellular pathogen. *Proc. Natl. Acad. Sci. USA* **101:**8739–8744.

Rothermel, A. L., Y. Wang, J. Schechner, B. Mook-Kanamori, W. C. Aird, J. S. Pober, G. Tellides, and D. R. Johnson. 2004. Endothelial cells present antigens *in vivo*. *BMC Immunol.* **5:**5.

Roy, S., D. Patel, S. Khanna, G. M. Gordillo, S. Biswas, A. Friedman, and C. K. Sen. 2007. Transcriptome-wide analysis of blood vessels laser captured from human skin and chronic wound-edge tissue. *Proc. Natl. Acad. Sci. USA* **104:**14472–14477.

Ruckwardt, T. J., C. Luongo, A. M. Malloy, J. Liu, M. Chen, P. L. Collins, and B. S. Graham. 2010. Responses against a subdominant $CD8^+$ T cell epitope protect against immunopathology caused by a dominant epitope. *J. Immunol.* **185:**4673–4680.

Rydkina, E., A. Sahni, R. B. Baggs, D. J. Silverman, and S. K. Sahni. 2006. Infection of human endothelial cells with spotted fever group rickettsiae stimulates cyclooxygenase 2 expression and release of vasoactive prostaglandins. *Infect. Immun.* **74:**5067–5074.

Rydkina, E., L. C. Turpin, and S. K. Sahni. 2010. *Rickettsia rickettsii* infection of human macrovascular and microvascular endothelial cells reveals activation of both common and cell type-specific host response mechanisms. *Infect. Immun.* **78:**2599–2606.

Rydkina, E., L. C. Turpin, D. J. Silverman, and S. K. Sahni. 2009. *Rickettsia rickettsii* infection of human pulmonary microvascular endothelial cells: modulation of cyclooxygenase-2 expression. *Clin. Microbiol. Infect.* **15**(Suppl. 2):300–302.

Sahni, S. K., S. Kiriakidi, P. M. Colonne, A. Sahni, and D. J. Silverman. 2009. Selective activation of signal transducer and activator of transcription (STAT) proteins STAT1 and STAT3 in human endothelial cells infected with *Rickettsia rickettsii*. *Clin. Microbiol. Infect.* **15**(Suppl. 2):303–304.

Sahni, S. K., E. Rydkina, S. G. Joshi, L. A. Sporn, and D. J. Silverman. 2003. Interactions of *Rickettsia rickettsii* with endothelial nuclear factor-κB in a "cell-free" system. *Ann. N. Y. Acad. Sci.* **990:**635–641.

Sahni, S. K., D. J. Van Antwerp, M. E. Eremeeva, D. J. Silverman, V. J. Marder, and L. A. Sporn. 1998. Proteasome-independent activation of nuclear factor κB in cytoplasmic extracts from human endothelial cells by *Rickettsia rickettsii*. *Infect. Immun.* **66:**1827–1833.

Savinov, A. Y., F. S. Wong, A. C. Stonebraker, and A. V. Chervonsky. 2003. Presentation of antigen by endothelial cells and chemoattraction are required for homing of insulin-specific $CD8^+$ T cells. *J. Exp. Med.* **197:**643–656.

Schurich, A., M. Berg, D. Stabenow, J. Böttcher, M. Kern, H. J. Schild, C. Kurts, V. Schuette, S. Burgdorf, L. Diehl, A. Limmer, and P. A. Knolle. 2010. Dynamic regulation of CD8 T cell tolerance induction by liver sinusoidal endothelial cells. *J. Immunol.* **184:**4107–4114.

Selin, L. K., S. R. Nahill, and R. M. Welsh. 1994. Cross-reactivities in memory cytotoxic T lymphocyte recognition of heterologous viruses. *J. Exp. Med.* **179:**1933–1943.

Seong, S. Y., M. S. Choi, and I. S. Kim. 2001. *Orientia tsutsugamushi* infection: overview and immune responses. *Microbes Infect.* **3:**11–21.

Sette, A., and R. Rappuoli. 2010. Reverse vaccinology: developing vaccines in the era of genomics. *Immunity* **33:**530–541.

Shirai, A., P. J. Catanzaro, S. M. Phillips, and J. V. Osterman. 1976. Host defenses in experimental scrub typhus: role of cellular immunity in heterologous protection. *Infect. Immun.* **14**:39–46.

Shirai, A., J. W. Dietel, and J. V. Osterman. 1975. Indirect hemagglutination test for human antibody to typhus and spotted fever group rickettsiae. *J. Clin. Microbiol.* **2**:430–437.

Siegismund, C. S., O. Hohn, R. Kurth, and S. Norley. 2009. Enhanced T- and B-cell responses to simian immunodeficiency virus (SIV)agm, SIVmac and human immunodeficiency virus type 1 Gag DNA immunization and identification of novel T-cell epitopes in mice via codon optimization. *J. Gen. Virol.* **90**:2513–2518.

Smith, M. E., and J. A. Thomas. 1990. Cellular expression of lymphocyte function associated antigens and the intercellular adhesion molecule-1 in normal tissue. *J. Clin. Pathol.* **43**:893–900.

Socolovschi, C., O. Mediannikov, D. Raoult, and P. Parola. 2009. The relationship between spotted fever group *Rickettsiae* and ixodid ticks. *Vet. Res.* **40**:34.

Sporn, L. A., S. O. Lawrence, D. J. Silverman, and V. J. Marder. 1993. E-selectin-dependent neutrophil adhesion to *Rickettsia rickettsii*-infected endothelial cells. *Blood* **81**:2406–2412.

Sporn, L. A., and V. J. Marder. 1996. Interleukin-1α production during *Rickettsia rickettsii* infection of cultured endothelial cells: Potential role in autocrine cell stimulation. *Infect. Immun.* **64**:1609–1613.

Sporn, L. A., S. K. Sahni, N. B. Lerner, V. J. Marder, D. J. Silverman, L. C. Turpin, and A. L. Schwab. 1997. *Rickettsia rickettsii* infection of cultured human endothelial cells induces NF-κB activation. *Infect. Immun.* **65**:2786–2791.

St. Croix, B., C. Rago, V. Velculescu, G. Traverso, K. E. Romans, E. Montgomery, A. Lal, G. J. Riggins, C. Lengauer, B. Vogelstein, and K. W. Kinzler. 2000. Genes expressed in human tumor endothelium. *Science* **289**:1197–1202.

Stemke-Hale, K., B. Kaltenboeck, F. J. DeGraves, K. F. Sykes, J. Huang, C. H. Bu, and S. A. Johnston. 2005. Screening the whole genome of a pathogen in vivo for individual protective antigens. *Vaccine* **23**:3016–3025.

Sumner, J. W., K. G. Sims, D. C. Jones, and B. E. Anderson. 1995. Protection of guinea-pigs from experimental Rocky Mountain spotted fever by immunization with baculovirus-expressed *Rickettsia rickettsii* rOmpA protein. *Vaccine* **13**:29–35.

Sykes, K. 2008. Progress in the development of genetic immunization. *Expert Rev. Vaccines* **7**:1395–1404.

Sylvester-Hvid, C., M. Nielsen, K. Lamberth, G. Røder, S. Justesen, C. Lundegaard, P. Worning, H. Thomadsen, O. Lund, S. Brunak, and S. Buus. 2004. SARS CTL vaccine candidates; HLA supertype-, genome-wide scanning and biochemical validation. *Tissue Antigens* **63**:395–400.

Taflin, C., B. Favier, J. Baudhuin, A. Savenay, P. Hemon, A. Bensussan, D. Charron, D. Glotz, and N. Mooney. 2011. Human endothelial cells generate Th17 and regulatory T cells under inflammatory conditions. *Proc. Natl. Acad. Sci. USA* **108**:2891–2896.

Tenzer, S., E. Wee, A. Burgevin, G. Stewart-Jones, L. Friis, K. Lamberth, C. H. Chang, M. Harndahl, M. Weimershaus, J. Gerstoft, N. Akkad, P. Klenerman, L. Fugger, E. Y. Jones, A. J. McMichael, S. Buus, H. Schild, P. van Endert, and A. K. Iversen. 2009. Antigen processing influences HIV-specific cytotoxic T lymphocyte immunodominance. *Nat. Immunol.* **10**:636–646.

Ternette, N., B. Tippler, K. Uberla, and T. Grunwald. 2007. Immunogenicity and efficacy of codon optimized DNA vaccines encoding the F-protein of respiratory syncytial virus. *Vaccine* **25**:7271–7279.

Thebault, P., T. Condamine, M. Heslan, M. Hill, I. Bernard, A. Saoudi, R. Josien, I. Anegon, M. C. Cuturi, and E. Chiffoleau. 2007. Role of IFNγ in allograft tolerance mediated by $CD4^+CD25^+$ regulatory T cells by induction of IDO in endothelial cells. *Am. J. Transplant.* **7**: 2472–2482.

Tsai, Y. C., H. J. Hsieh, F. Liao, C. W. Ni, Y. J. Chao, C. Y. Hsieh, and D. L. Wang. 2007. Laminar flow attenuates interferon-induced inflammatory responses in endothelial cells. *Cardiovasc. Res.* **74**:497–505.

Valbuena, G., W. Bradford, and D. H. Walker. 2003. Expression analysis of the T-cell-targeting chemokines CXCL9 and CXCL10 in mice and humans with endothelial infections caused by rickettsiae of the spotted fever group. *Am. J. Pathol.* **163**:1357–1369.

Valbuena, G., H. M. Feng, and D. H. Walker. 2002. Mechanisms of immunity against rickettsiae. New perspectives and opportunities offered by unusual intracellular parasites. *Microbes Infect.* **4**:625–633.

Valbuena, G., J. M. Jordan, and D. H. Walker. 2004. T cells mediate cross-protective immunity between spotted fever group rickettsiae and typhus group rickettsiae. *J. Infect. Dis.* **190**:1221–1227.

Valbuena, G., and D. H. Walker. 2004. Effect of blocking the CXCL9/10-CXCR3 chemokine system in the outcome of endothelial-target rickettsial infections. *Am. J. Trop. Med. Hyg.* **71**:393–399.

Valbuena, G., and D. H. Walker. 2006. The endothelium as a target for infections. *Annu. Rev. Pathol.* **1:**171–198.

Valbuena, G., and D. H. Walker. 2009. Infection of the endothelium by members of the order Rickettsiales. *Thromb. Haemost.* **102:**1071–1079.

Valujskikh, A., and P. S. Heeger. 2003. Emerging roles of endothelial cells in transplant rejection. *Curr. Opin. Immunol.* **15:**493–498.

Valujskikh, A., O. Lantz, S. Celli, P. Matzinger, and P. S. Heeger. 2002. Cross-primed $CD8^+$ T cells mediate graft rejection via a distinct effector pathway. *Nat. Immunol.* **3:**844–851.

Vishwanath, S. 1991. Antigenic relationships among the rickettsiae of the spotted fever and typhus groups. *FEMS Microbiol Lett.* **65:**341–344.

Wagner, D. D., and P. S. Frenette. 2008. The vessel wall and its interactions. *Blood* **111:**5271–5281.

Walker, D. H. 2009. The realities of biodefense vaccines against *Rickettsia. Vaccine* **27**(Suppl. 4)**:**D52–D55.

Walker, D. H., and N. Ismail. 2008. Emerging and re-emerging rickettsioses: endothelial cell infection and early disease events. *Nat. Rev. Microbiol.* **6:**375–386.

Walker, D. H., C. Occhino, G. R. Tringali, S. Di Rosa, and S. Mansueto. 1988. Pathogenesis of rickettsial eschars: the tache noire of boutonneuse fever. *Hum. Pathol.* **19:**1449–1454.

Walker, D. H., J. P. Olano, and H. M. Feng. 2001. Critical role of cytotoxic T lymphocytes in immune clearance of rickettsial infection. *Infect. Immun.* **69:**1841–1846.

Walker, D. H., V. L. Popov, P. A. Crocquet-Valdes, C. J. Welsh, and H. M. Feng. 1997. Cytokine-induced, nitric oxide-dependent, intracellular antirickettsial activity of mouse endothelial cells. *Lab. Invest.* **76:**129–138.

Walker, D. H., V. L. Popov, and H. M. Feng. 2000. Establishment of a novel endothelial target mouse model of a typhus group rickettsiosis: evidence for critical roles for gamma interferon and CD8 T lymphocytes. *Lab. Invest.* **80:**1361–1372.

Walker, D. H., G. A. Valbuena, and J. P. Olano. 2003. Pathogenic mechanisms of diseases caused by *Rickettsia. Ann. N. Y. Acad. Sci.* **990:**1–11.

Walsh, D. S., E. C. Delacruz, R. M. Abalos, E. V. Tan, J. Jiang, A. L. Richards, C. Eamsila, W. Rodkvantook, and K. S. Myint. 2007. Clinical and histological features of inoculation site skin lesions in cynomolgus monkeys experimentally infected with *Orientia tsutsugamushi. Vector Borne Zoonotic Dis.* **7:**547–554.

Wang, R. F., X. Wang, S. L. Johnston, G. Zeng, P. F. Robbins, and S. A. Rosenberg. 1998. Development of a retrovirus-based complementary DNA expression system for the cloning of tumor antigens. *Cancer Res.* **58:**3519–3525.

Watt, G., C. Chouriyagune, R. Ruangweerayud, P. Watcharapichat, D. Phulsuksombati, K. Jongsakul, P. Teja-Isavadharm, D. Bhodhidatta, K. D. Corcoran, G. A. Dasch, and D. Strickman. 1996. Scrub typhus infections poorly responsive to antibiotics in northern Thailand. *Lancet* **348:**86–89.

Weinert, L. A., J. H. Werren, A. Aebi, G. N. Stone, and F. M. Jiggins. 2009. Evolution and diversity of *Rickettsia* bacteria. *BMC Biol.* **7:**6.

Wikel, S. K. 1999. Tick modulation of host immunity: an important factor in pathogen transmission. *Int. J. Parasitol.* **29:**851–859.

Woodward, T. E. 1986. Rickettsial vaccines with emphasis on epidemic typhus. Initial report of an old vaccine trial. *S. Afr. Med. J.* **1986**(Suppl.)**:**73–76.

Wu, C. Y., J. R. Kirman, M. J. Rotte, D. F. Davey, S. P. Perfetto, E. G. Rhee, B. L. Freidag, B. J. Hill, D. C. Douek, and R. A. Seder. 2002. Distinct lineages of T_H1 cells have differential capacities for memory cell generation *in vivo. Nat. Immunol.* **3:**852–858.

Zhang, J. Z., J. F. Hao, D. H. Walker, and X. J. Yu. 2006. A mutation inactivating the methyltransferase gene in avirulent Madrid E strain of *Rickettsia prowazekii* reverted to wild type in the virulent revertant strain Evir. *Vaccine* **24:**2317–2323.

Zinsser, H. 1937. The rickettsial diseases: variety, epidemiology and geographical distribution. *Am. J. Hyg.* **25:**430–463.

Zinsser, H., and M. R. Castaneda. 1933. On the isolation from a case of Brill's disease of a typhus strain resembling the European type. *N. Engl. J. Med.* **209:**815–819.

ADAPTIVE IMMUNE RESPONSES TO INFECTION AND OPPORTUNITIES FOR VACCINE DEVELOPMENT (*ANAPLASMATACEAE*)

Susan M. Noh and Wendy C. Brown

11

INTRODUCTION

The critical interface between the host immune system and the pathogen in large part determines the outcome of infection. We study immunity to understand how a pathogen is able to establish infection and how the immune system is able to control or clear infection. The adaptive immune response is of particular interest because under the appropriate conditions it can lead to long-lasting, pathogen-specific immunity, which is the basis for vaccine development. Vaccines are the most cost-effective and powerful tools available to prevent disease in both humans and animals.

In general, the most effective vaccines target invariant epitopes on pathogens that cause transient infection (Sallusto et al., 2010; Telford, 2008). For example, vaccines that induce protective immunity to a variety of morbilliviruses, including human measles virus, canine distemper virus, and most notable for its recent eradication, rinderpest virus, were developed in the 1950s and 1960s (Domenech et al., 2010; Greene and Vandevelde, 2012; Hilleman, 2001). Several features of these viruses provide insight into the reasons that vaccine development has proven so rewarding in some cases. First, natural infection with a morbilli virus, if not lethal, results in lifelong immunity against all members of the viral population (Domenech et al., 2010; Greene and Vandevelde, 2012). Genetic and serotypic differences in measles virus populations can be used to track the source of disease outbreaks (Rima et al., 1995a, 1995b). While this variation among viral populations exists, remarkably a single vaccine with a single attenuated live virus protects against all variants (Domenech et al., 2010; Greene and Vandevelde, 2012; Hilleman, 2001). This broadly protective immune response targets surface-exposed glycoproteins, which mediate virus attachment and entry to the host cell (Griffin, 2010). Importantly, not only are these epitopes conserved, but they are also invariant through time. A second, related feature of these pathogens is that antigenic variation leading to persistent infection does not play a role in the epidemiology of disease (Domenech et al., 2010; Greene and Vandevelde, 2012). Rather, the pathogen is maintained by transient shedding from recently infected humans or animals, and transmission relies on the

Susan M. Noh, Animal Disease Research Unit, Agriculture Research Service, U.S. Department of Agriculture, Pullman, WA 99164. *Wendy C. Brown*, Program in Vector-borne Diseases, Department of Veterinary Microbiology and Pathology, Washington State University, Pullman, WA 99164.

Intracellular Pathogens II: Rickettsiales
Edited by G. H. Palmer and A. F. Azad © 2012 ASM Press, Washington, DC
10.1128/9781555817336.ch11

continued presence of a cohort of naïve individuals. Thus, administration of the appropriate antigen to the naïve cohort results in a high level of immunity among the population (Domenech et al., 2010). The successes of vaccine development in preventing morbillivirus infection are instructive, but so are the failures, in particular with regard to the measles virus.

Morbilliviruses modulate the host immune response, which can directly affect vaccine efficacy and safety. For example, marked anergy and immune dysfunction often follow measles virus infection, particularly in infants (Aaby et al., 1990; Griffin, 2010). Administration of a formalin-killed measles virus vaccine, licensed in the United States in 1963, resulted in the induction of neutralizing antibodies, relatively rapidly waning protection from infection, and unfortunately, a predisposition to developing atypical measles upon exposure to the natural virus (Hilleman, 1992). Atypical measles was attributed to an anamnestic, nonprotective antibody response that led to the deposition of immunoglobulin G (IgG) immune complexes and complement activation, as well as an exaggerated Th2 cytokine response, prolonged increases in IgE, and eosinophilia (Griffin, 2010; Hilleman, 2001; Polack et al., 1999). This example illustrates the fact that immunization must not only be effective but, importantly, must not potentiate disease.

In contrast to invariant pathogens, development of vaccines that target more variable pathogens has met with limited success (Domenech et al., 2010; Telford, 2008). For example, immunity to the influenza virus is limited, for the most part, to the serotypes to which an individual has been exposed (Telford, 2008). The majority of neutralizing antibodies block the binding and entry function of hemagglutinin (HA) protein, which is immunogenic, under strong immune selection, and can be highly variable from season to season (Ekiert et al., 2009; Han and Marasco, 2011). A particular vaccine formulation only targets a limited number of HA variants; thus, the host population remains susceptible to variants to which it has not been exposed or which have not been included in the vaccine. Pathogens for which vaccine development has proven exceptionally challenging, such as HIV, are those that vary at the population level such that protective immunity to one strain is ineffective against a second strain (Blish et al., 2008), are able to establish persistent infection in the host, and modulate the host immune system.

Bacterial pathogens within the family *Anaplasmataceae* generally fall into this category. This family includes members of the genera *Anaplasma*, *Ehrlichia*, *Neorickettsia*, and *Wolbachia*, which cause a variety of diseases in humans and animals (Dumler et al., 2001). When considering this group of pathogens, three major themes emerge that present particular challenges in terms of vaccine development. The first is the extent and variety of ways in which many of these organisms modulate the host immune response. In the case of *Ehrlichia* spp. and possibly *Neorickettsia* spp., immune dysregulation plays a role in the development of severe disease (Rikihisa, 2010). In contrast, *Anaplasma phagocytophilum* is generally immunosuppressive, while the immune suppression caused by *Anaplasma marginale* is pathogen specific (Han et al., 2010). Interestingly, through unknown mechanisms, some populations of *Wolbachia* are able to alter the immune competence and survival of their invertebrate host (Braquart-Varnier et al., 2008; Saridaki and Bourtzis, 2010). The second theme is the ability of many of these pathogens to establish persistent infection. This ability is well established in all but the *Neorickettsia* spp. The *Wolbachia* spp. represent the extreme of persistent infection, as they are endosymbionts of many invertebrates and thus are lifelong residents of their hosts. The third and related theme is strain variation, specifically meaning that a protective immune response is effective against some but not all members of a population. In this context, the defining features of a strain have not been completely identified for any of these pathogens. Despite the challenges presented by these pathogens in terms of vaccine development, there is strong precedent for the ability to induce protection against disease, if not infection. This occurs in particular with *A. marginale*, the causative agent of bovine anaplasmosis, and with *Ehrlichia ruminantium*, the causative agent of heartwater, two pathogens that have been studied for many years. An association between the pathogen and disease was made in 1910 in the case of *A. marginale* (Theiler, 1910), and in 1925 in the case of *E. ruminantium* (Bezuidenhout, 1985). Due to the great economic impact of these diseases and the length of time for which they have been studied, more is known about these two pathogens than other members of this family. Consequently, this chapter is biased toward these two organisms. Additionally, emphasis is placed on data derived from in vivo systems using natural host species when possible.

RESPONSE TO INFECTION AND PATHOGEN IMMUNE MODULATION

There are several outcomes of infection with pathogens in the family *Anaplasmataceae*, depending on the pathogen, the host, and the adaptive immune response. In the case of members of the genera *Ehrlichia* and *Neorickettsia*, dysregulation of the immune response, rather than direct cytotoxic effects of the pathogen, leads to clinical disease. These pathogens, with the partial exception of *E. ruminantium*, are monocytotropic. Thus, the immune dysregulation is not surprising as the macrophage acts as a bridge between the innate and adaptive immune responses through phagocytosis of invading microorganisms and subsequent secretion of signaling proteins that activate and recruit immune effectors. In the early stages of infection, *E. ruminantium* is thought to replicate in the phagocytic reticuloendothelial system of the lymph nodes and spleen (Allsopp, 2010). Then, similarly to rickettsiae, it preferentially infects endothelial cells. In contrast to rickettsial disease, where endothelial cell infection results in necrotizing vasculitis and activation of both the coagulation system and fibrinolytic system (Greene et al., 2012), vasculitis is not a hallmark of heartwater. Rather, *E. ruminantium* causes increased capillary permeability, leading to hydropericardium, abdominal and thoracic effusion, and cerebral edema (Maxie and Robinson, 2007) in the absence of systemic, microscopically detectable tissue destruction. In some cases, mild vascular injury and perivascular inflammation are apparent in the central nervous system. However, the severity of the vascular permeability is disproportionate to the degree of endothelial damage and number of organisms present; thus, cytokines and immune effectors are thought to be responsible for the pathology that leads to severe disease and death (Totté et al., 1999). The mechanisms that mediate this process are unknown.

Ehrlichia chaffeensis, the causative agent of human monocytic ehrlichiosis (HME), infects cells in the monocyte/macrophage lineage and is associated with a broad spectrum of clinical disease varying from inapparent infection to a multisystemic toxic shock-like syndrome with fever, hypotension, rash, and multiorgan dysfunction (Paddock and Childs, 2003). The most consistent hematologic features of HME are leukopenia and thrombocytopenia (Olano et al., 2003). Surprisingly, the toxic shock-like syndrome occurs in the absence of lipopolysaccharide, and in the presence of an intact immune system (Dunning Hotopp et al., 2006; Lin and Rikihisa, 2003). The disproportionately low bacterial burden in association with expansion of the mononuclear phagocyte cell population in the spleen, lymph nodes, and bone marrow, and lymphohistiocytic infiltrates in multiple organs including the lung, kidney, heart, liver, brain, and meninges, suggest that immune dysregulation is responsible for severe fatal disease (Dierberg and Dumler, 2006; Rikihisa, 1999).

Immunocompetent mice are generally resistant to *E. chaffeensis* infection. Rather, experiments comparing the immune response to *Ehrlichia muris* infection, which causes mild, self-limiting disease in mice, and *Ixodes ovatus Ehrlichia* (IOE), a monocytotropic ehrlichial species isolated from *I. ovatus* ticks in Japan, which in mice mimics severe fatal HME (Sotomayor et al., 2001), have identified mechanisms responsible for immune dysregulation. Multisystemic disease is mediated by tumor necrosis factor alpha (TNF-α)- and gamma interferon (IFN-γ)-producing, antigen-specific $CD8^+$ T cells accompanied by systemic overproduction of interleukin-10 (IL-10) and TNF-α and the loss of $CD4^+$ T cells by apoptosis (Ismail et al., 2004, 2007). Similarly, in humans an increase in $CD8^+$ T cells in lymph nodes from patients with severe HME has been observed (Dierberg and Dumler, 2006; chapter 9, this volume).

Similarly to *E. chaffeensis*, *Ehrlichia canis* primarily infects monocytes, and the spectrum of disease in dogs is variable, with many infections being mild or subclinical. Immune dysregulation likely also plays a prominent role in the pathogenesis of disease. Even in mild cases, pancytopenia, including thrombocytopenia, is a hallmark of canine monocytic erhlichiosis (Neer and Harrus, 2006). Multiple mechanisms cause the thrombocytopenia, including immune-mediated destruction (Harrus et al., 1999). The loss of other cell types is attributed to progressive bone marrow aplasia, the cause of which is unknown. Fatal forms of the disease can develop during acute or chronic infection (Neer and Harrus, 2006). A prominent feature of severe canine ehrlichiosis is plasmacytic cellular infiltrates in multiple organs, including medullary sinuses of lymph nodes, bone marrow, liver, and kidney, in the absence of a heavy bacterial burden (Reardon and Pierce, 1981). These inflammatory changes occur in concert with hypergammaglobulinemia that does not correspond to anti-*E. canis* antibody titers, suggesting nonspecific Ig production (Harrus et al., 1999). Immunosuppressive therapy delayed the onset of clinical disease and hematologic abnormalities as well as decreased the severity of lesions in dogs experimentally infected with *E. canis*, further suggesting that immune dysregulation is in large part the pathologic basis for disease (Reardon and Pierce, 1981). The mechanisms by which the immune dysregulation occurs are less well-known than in *E. chaffeensis*. However, similarly to *E. chaffeensis*, $CD8^+$ T cells may play a role, as an inverted ratio of $CD4^+$ to $CD8^+$ T cells was reported in one case (Heeb et al., 2003). Additionally, a spike in TNF-α, IFN-γ, and IL-10, all of which are associated with lethal IOE infection in mice, occurred in dogs between days 18 and 30 after infection (Faria et al., 2011).

Similarly to *E. chaffeensis* and *E. canis*, *Neorickettsia helminthoeca*, *Neorickettsia risticii*, and *Neorickettsia sennetsu* primarily infect and replicate in mononuclear phagocytes, though *N. risticii* also infects enterocytes (Neer and Harrus, 2006). Little is known about the pathogenesis of disease caused by these pathogens, and particularly whether immune dysregulation is a defining component. However, some clues can be gained by study of the associated clinical features and lesions. For example, *N. sennetsu* in humans is associated with peripheral mononucleosis with atypical lymphocytes, which is strongly suggestive of immune dysregulation (Newton et al., 2009). Salmon poisoning in dogs, due to *N. helminthoeca*, is a systemic disease associated with lymphopenia and thrombocytopenia, the pathogenesis of which is unknown (Neer and Harrus, 2006). The most prominent histological features of *N. helminthoeca* infection are lysis of the mature lymphocytes in the spleen and lymph nodes throughout the body accompanied by hyperplasia of mononuclear phagocytic cells (Brown et al., 2007; Gorham et al., 2012; Headley et al., 2011). All of these features are suggestive of immune dysregulation similar to that seen in severe *E. chaffeensis* infections in humans. However, *N. helminthoeca* is little studied, and such mechanisms have not been identified.

In contrast to *N. helminthoeca*, the primary lesions associated with *N. risticii*, which causes Potomac horse fever, are confined to the gastrointestinal tract (Jones, 2004). While *N. risticii* can be found in circulating monocytes, the predominant lesions are early mononuclear cell infiltrates in the colon followed by necrosis of the colonic epithelium, leading to ulceration and severe inflammation (Cordes et al., 1986; Dutra et al., 2001). Similarly to *N. helmintheoca* and human and canine ehrlichioses, *N. risticii* is associated with lymphopenia and thrombocytopenia. The lymphopenia is accompanied by neutropenia and is attributed to endotoxemia secondary to breach of the intestinal barrier (Jones, 2004). The cause of the thrombocytopenia is unknown. Additionally, it is unknown if the events leading to enterocolitis are mediated by virulence factors produced by the pathogen, an excessive or dysregulated immune response, or a combination of these and other factors.

A. phagocytophilum and *A. marginale* also modulate the host immune response, though in a somewhat different fashion. In particular, *A. phagocytophilum* is generally immunosuppressive through subversion of the innate and adaptive immune responses, though in the case of the adaptive immune response, the mechanisms have not been well defined. *A. phagocytophilum* replicates in neutrophils and inhibits neutrophil function, which is discussed in detail in chapter 6. In humans and horses, the spectrum of disease varies from subclinical to severe (Bullock et al., 2000; Butler et al., 2008; Dumler et al., 2005; Madigan et al., 1990; Pusterla and Madigan, 2007a). Severe and fatal cases of human and equine granulocytic anaplasmosis are often associated with secondary opportunistic infections, indicative of systemic immune suppression (Dumler et al., 2007; Ismail et al., 2010; Pusterla and Madigan, 2007a). However, direct evidence of dysregulation of the adaptive immune system in infected horses and humans is lacking. As with monocytic ehrlichiosis, the hallmarks of human and equine granulocytic ehlichiosis are leukopenia and thrombocytopenia, with mononuclear phagocyte hyperplasia in the lymph nodes and spleen in humans and accumulations of lymphocytes and histiocytes in multiple organs in both humans and horses (Bakken et al., 1996; Dumler et al., 2005; Lepidi et al., 2000). Infected neutrophils are often not associated with the lesions, suggesting that immune dysregulation rather than direct bacteria-mediated injury (Dumler et al., 2005; Lepidi et al., 2000) is responsible for these lesions.

In Europe, sheep infected with *A. phagocytophilum* develop tick-borne fever, which is characterized by fever and severe leukopenia due to early lymphocytopenia, prolonged neutropenia, and thrombocytopenia, with few other clinical signs (Stuen and Longbottom, 2011; Whist et al., 2002). Proliferation of the lymphohistocytic population, as seen in severe human granulocytic ehrlichiosis, is not observed. Death, when it occurs, is due to secondary bacterial infections (Brodie et al., 1986; Stuen and Longbottom, 2011). Impaired innate and adaptive immunity accounts for the immune suppression. In the case of the adaptive immune response, there is suppression of both *A. phagocytophilum*-specific and -nonspecific responses. The lymphocytopenia is primarily due to reductions in $\gamma\delta$ T cells, $CD4^+$ T cells, and B cells (Whist et al., 2003). The reduction in numbers is accompanied by impaired antibody production, impaired lymphocyte proliferation, and decreased IFN-γ responses to mitogens and vaccines (Whist et al., 2003). Additionally, all *A. phagocytophilum*-specific antibody responses including those targeting major surface protein 2 (MSP2/p44) diminish markedly over time with infection (Granquist et al., 2010).

Unlike other pathogens in the family *Anaplasmataceae*, the majority of clinical signs of bovine anaplasmosis are associated with fever and anemia, rather than alteration in total leukocyte numbers (Palmer, 2009). The anemia is due to the removal of *A. marginale*-infected erythrocytes by the reticuloendothelial system (Palmer, 2009). Additionally, there is no evidence of an association among immune dysregulation, severity of disease, and pathology. Interestingly, modulation of the *A. marginale*-specific immune response contributes to the establishment and maintenance of persistent infection. In animals infected with *A. marginale* by needle inoculation or tick transmission, *A. marginale*-specific $CD4^+$-T-cell responses were first detected at 5 to 7 weeks postinfection, concurrent with either rising or declining bacteremia (Han et al., 2010). Thereafter, regardless of the method of inoculation, antigen-specific

T-cell responses were transient and recurred only sporadically in animals that were followed for nearly 1 year (Han et al., 2010). In comparison, T-cell responses to *Clostridium* spp. vaccine were not impaired throughout the study period, indicating that the poor *A. marginale*-specific T-cell responses were not due to generalized immune suppression. Paradoxically, high titers of antibody, including IgG2, were detected by 2.5 to 3 weeks postinoculation, concurrent with rising bacteremia, and were maintained throughout persistent infection (Han et al., 2010).

The broad spectrum of immune suppression and dysregulation mediated by these pathogens highlights the difficulty and importance of careful selection of antigens to ensure that a vaccine will elicit a protective immune response and not exacerbate disease.

KNOWN EFFECTIVE IMMUNOGENS

In the case of *A. marginale*, *A. phagocytophilum*, and *E. ruminantium*, infection in the naïve mammalian host is established through tick feeding. After a prepatent period, acute disease develops, which is characterized by variable clinical signs, depending upon the pathogen. Importantly, most animals that survive acute disease are able to control but not clear the infection. Thus, an equilibrium is reached between the host immune response and the pathogen. The only effective vaccines currently in use that target this family of pathogens are based on this observation (Table 1).

Heartwater and bovine anaplasmosis, two of the most economically significant diseases of ruminants worldwide, are caused by *E. ruminantium* and *A. marginale*, respectively. The vaccines used to control these diseases are based on the administration of live bacteria and result in the prevention of disease rather than prevention of infection (Neitz et al., 1947; Theiler, 1912). These live vaccines are variably effective, and in the case of *A. marginale* allow for the maintenance of a reservoir of infected hosts (Allsopp, 2010; Bock and de Vos, 2001; Krigel et al., 1992; Neitz et al., 1947; Pipano, 1995). However, the advantages and limitations of these vaccines are instructive and set the precedent for the development of more effective vaccines.

The *E. ruminantium* vaccination protocol was developed in South Africa and is an infection-and-treatment technique, which consists of infecting animals intravenously with cryopreserved blood containing virulent *E. ruminantium* organisms (Allsopp, 2010; Neitz and Alexander, 1941). Infection is followed by antibiotic treatment when a rise in body temperature occurs. Apart from the technical difficulties, this technique provides a high level of protection from disease when the vaccinee is challenged with the same strain as the immunogen, but not a heterologous strain (du Plessis et al., 1989). The limited heterologous protection is a major obstacle in vaccine development to prevent heartwater (du Plessis et al., 1989, 1990; Gueye et al., 1994).

Importantly, heterologous protection can be achieved and is linked to immunity to the virulent Welgevonden isolate, which is difficult to use in infection-and-treatment because the rise in temperature is rapidly followed by death; thus, the window of opportunity for treatment is limited. However, this isolate induces wide heterologous protection when used in infection-and-treatment (du Plessis et al., 1989). Interestingly, the Welgevonden isolate has been attenuated by serial passage in DH82 cells (Collins et al., 2003b). This attenuated strain induces fever and confers heterologous protection to at least four different strains (Zweygarth et al., 2005), but has not been tested in the field. Immunity to challenge with field isolates administered by tick feeding tends to be poorer than that using known isolates administered by needle inoculation (Allsopp, 2010). Regardless, these data suggest that heterologous protection may in large part be strain dependent. If this is true, specific antigens or groups of antigens displayed by the Welgevonden strain, when presented in the right context, may confer strong homologous and heterologous protection.

The second example of a live vaccine is *A. marginale* subsp. *centrale*, which is naturally

TABLE 1 Protective immunogens used commercially or experimentally

Pathogen	Host in which disease most commonly recognized	Vaccine currently or previously commercially available	Experimentally tested immunogen that confers protection (animal model used for testing)[a]
E. canis	Canine		None
E. chaffeensis	Human		None
E. muris	Murine		Live: related organism (IOE, mouse) (6)
IOE	Murine		Recombinant protein: p28-Omp19, related organism (*E. chaffeensis*, SCID mouse) (10)
E. ruminantium	Ovine, caprine, bovine	Infect-and-treat: Ball3	Infect-and-treat: Welgevonden, Ball3, Mara87/7, Gardel, Kwanyanga, Blaauwkrans (5) Attenuated: Senegal (caprine) (7), Welgevonden (ovine, caprine) (19) Inactivated: Gardel (caprine) (9), Crystal Springs (ovine) (8) Recombinant protein: pooled recombinant bacterial lysates (murine) (2) DNA vaccine: MAP1 (murine) (12); Erum2510, Erum2540, Erum2550, Erum2580, Erum2590 (ovine) (15)
N. helmintheoca	Canine		None
N. risticii	Equine		Live: related organism (*N. sennetsu*, horse) (13, 16)
N. sennetsu	Human		None
A. marginale	Bovine	Live: related organism (*A. marginale* subsp. *centrale*)	Live: related organism (*A. marginale* subsp. *centrale*) (1, 3, 14, 18) Pathogen component: outer membrane proteins (4, 17), surface exposed protein complex (bovine) (11)
A. phagocytophilum	Ovine, equine, human		Infect-and-treat: horse (13)

[a]Numbers indicate the following references: 1. Anziani et al., 1987; 2. Barbet et al., 2001; 3. Bock and de Vos, 2001; 4. Brown et al., 1998a; 5. Collins et al., 2003a; 6. Ismail et al., 2004; 7. Jongejan, 1991; 8. Mahan et al., 1995; 9. Martinez et al., 1994; 10. Nandi et al., 2007; 11. Noh et al., 2008; 12. Nyika et al., 2002; 13. Nyindo et al., 1978; 14. Potgieter and Van Rensburg, 1983; 15. Pretorius et al., 2007; 16. Rikihisa et al., 1988; 17. Tebele et al., 1991; 18. Turton et al., 1998; 19. Zweygarth et al., 2005.

attenuated compared with *A. marginale*, and was first isolated in South Africa in the early 1900s (Theiler, 1911). This subspecies, which is currently used as a vaccine in many parts of the world, is passaged in splenectomized animals (Bock and de Vos, 2001; Shkap et al., 2008). Blood from these animals is cryopreserved and used to infect susceptible animals in order to induce protection to more virulent field strains. Immunization with *A. marginale* subsp. *centrale* results in long-term persistent infection (up to 5 years) in the majority of animals (Krigel et al., 1992). Infection with this attenuated organism does not prevent infection and subsequent transmission of wild-type *A. marginale*, as 55% of *A. marginale* subsp. *centrale*-vaccinated animals were infected with both *A. marginale* subsp. *centrale* and *A. marginale* (Shkap et al., 2002, 2008). This approach results in near complete protection from clinical disease in some circumstances and poor protection in others (Anziani et al., 1987; Bock and de Vos, 2001; Potgieter and Van Rensburg, 1983; Shkap et al., 2002; Turton et al., 1998), indicating that heterologous protection from disease is achievable, but not uniform.

Apart from *A. marginale* subsp. *centrale*, a second effective immunogen composed of purified outer membranes protects against *A. marginale* (Tebele et al., 1991). In challenge experiments using a homologous strain to that from which the immunogen was derived, protection against anemia and bacteremia occurred in nearly all immunized animals and protection against infection in 40 to 70% of vaccinees (Brown et al., 1998a; Noh et al., 2008; Tebele et al., 1991). These examples demonstrate both the potential for vaccine development and the challenges in terms of achieving a high level of protection to a wide variety of strains. Both of the live preparations described above as well as the outer membrane immunogen are problematic in terms of widespread practical use. All three require live animals for their production. Additionally, they are difficult, expensive, and time-consuming to formulate. In many instances protection is inadequate, and the use of live products carries the risk of spreading blood-borne pathogens. Thus, the work started in the early 1900s to develop vaccines to prevent anaplasmosis and heartwater continues today.

THE PROTECTIVE IMMUNE RESPONSE AND RESPONSE TO IMMUNIZATION

Defining the immune response elicited by infection and immunization is useful in terms of characterization of the specific responses that correlate with immunity and identifying those that are detrimental. In general, control of intracellular pathogens occurs through cell-mediated immunity biased toward a Th1-type response. Briefly, the antigen-presenting cell (APC) secretes IFN-γ and IL-12, which biases $CD4^+$-T-cell differentiation toward a Th1 effector subset. In turn, Th1-type cells secrete IFN-γ and IL-2. In general, IL-2 is required for T-cell proliferation and differentiation and is a key cytokine in the development of an adaptive immune response. The primary function of IFN-γ is the activation of macrophages. Additionally, Th1-type cells activate infected macrophages as well as naïve B cells. The activation of macrophages results in increased opsonization and macrophage killing. The activation of B cells results in antibody production and induction of class switching to allow for production of high-affinity and opsonizing antibodies. Despite the association of the Th1-type response with protective immunity to these pathogens, the specific mechanisms required for pathogen control and elimination are largely unknown.

In the case of *A. marginale*, a model closely paralleling the Th1 paradigm has been proposed (Palmer et al., 1999). Protection in outer membrane-immunized cattle is associated with IgG2 production and induction of IFN-γ-secreting $CD4^+$ T cells (Brown et al., 1998a). In cattle, $CD4^+$ T cells that express IFN-γ provide help to B cells for the production of IgG2 antibody (Brown et al., 1999), while IFN-γ enhances IgG2 synthesis (Estes et al., 1994). Additionally, IFN-γ can activate macrophages to increase Fc receptor expression, phagocytosis, and nitric oxide-mediated killing of intracellular pathogens.

In contrast to immunization with the protective outer membranes of *A. marginale*, immunity in response to individual native or recombinant proteins is poor. One possible reason for this is the immune modulation mediated by *A. marginale*. Immunization with either the immunodominant, hypervariable MSP2 or MSP1a resulted in no protection against challenge, coupled with the loss of MSP2- or MSP1a-specific $CD4^+$ T cells upon challenge (Abbott et al., 2005; Han et al., 2008). Functional suppressor cells were not detected in the peripheral blood of nonresponding animals, the frequencies of circulating regulatory $CD25^+CD4^+$ lymphocytes during the course of infection did not vary significantly, and the production of the T-regulatory cytokines transforming growth factor β and IL-10 by these cells in response to antigen did not increase following infection (Abbott et al., 2005). These results suggest the sudden loss of T-cell responses is not simply due to an increase in regulatory T cells in peripheral blood in response to infection (Abbott et al., 2005). Additionally, antigen-specific T cells were not sequestered in the spleen or lymph nodes, as demonstrated using tetramer staining (Han et al., 2008). Thus, it is hypothesized that the loss of $CD4^+$-T-cell responses is the result of a high antigen load from a pathogen that replicates to $\geq 10^9$ bacteria per ml of blood during acute infection, but maintains an average of 10^6 bacteria per ml of blood during persistent infection, leading to T-cell apoptosis. However, it is unknown if protectively immunized animals experience a similar loss of antigen-specific $CD4^+$-T-cell responses upon challenge. If so, developing a means to overcome the loss of *A. marginale*-specific $CD4^+$ T cells may greatly improve vaccine efficacy.

Little is known about the characteristics of the immune response that results in protection against *A. phagocytophilum* in humans or animals. In sheep, primary infection associated with a low level of tick exposure (Kimberling, 1988) results in variable degrees of resistance to homologous challenge (Woldehiwet, 2006; Woldehiwet and Scott, 1982), indicating that induction of a protective immune response is possible. Additionally, protection against reinfection is associated with high antibody titers (Woldehiwet and Scott, 1982). In horses and humans (Aguero-Rosenfeld et al., 2002; Bakken et al., 1998), subclinical infection or mild disease in response to *A. phagocytophilum* infection is common (Artursson et al., 1999; Bullock et al., 2000; Madigan et al., 1990; Pusterla and Madigan, 2007a), indicating that humans and horses are able to overcome the immunosuppressive and other effects of infection and mount an effective immune response. Additionally, in horses, protective immunity to challenge can be induced through infection-and-treatment (Nyindo et al., 1978).

The characteristics of the immune response that lead to control of *A. phagocytophilum* infection are unknown, although cellular and humoral immune responses occur in response to infection (Artursson et al., 1999; Nyindo et al., 1978). High acute-phase antibody titers occur in approximately 40% of human patients and 44% of equine patients, though the role of antibody in control of infection is unknown (Artursson et al., 1999; Dumler and Bakken, 1998). In horses experimentally infected with *A. phagocytophilum* that developed mild disease, only IL-1β and TNF-α were consistently upregulated in peripheral blood leukocytes (Kim et al., 2002). IFN-γ and IL-10 were weakly detected in only one of the four horses. In contrast, in humans, in vivo cytokine responses that accompany *A. phagocytophilum* infection were dominated by IFN-γ and moderate levels of IL-10, but not TNF-α, IL-1β, or IL-4 (Dumler et al., 2000). These cytokine responses were temporally associated with clinical manifestations and recovery, but were similar in severely and mildly affected patients (Dumler et al., 2000). The source of IFN-γ and IL-10 during *A. phagocytophilum* infection has not been identified.

Mouse models of *A. phagocytophilum* are useful for studying the immune mechanisms that lead to pathogen control and clearance because experimental infection of mice results in a subclinical, transient infection with histologic lesions similar to those seen in cases of human granulocytic ehrlichiosis (Martin et al., 2000). In mice, the peak bacterial levels occur at day 7 and are followed by an IFN-γ peak at day 10 and the peak in tissue injury at day 14 (Martin et al., 2000, 2001). In IFN-γ knockout mice, increased bacterial levels occurred during early infection but were later cleared (Akkoyunlu and Fikrig, 2000; Martin et al., 2001). In the absence of IFN-γ, histopathologic lesions did not develop. In comparison, IL-10 knockout mice, which have IFN-γ levels similar to those of wild-type mice, had increased severity of histologic lesions compared with controls, but did not have increased bacteremia. These data suggest that IFN-γ mediates both pathology and early pathogen control, though is not essential for bacterial clearance. IL-10, which generally acts as an anti-inflammatory cytokine, helps to decrease tissue injury.

The immune mechanisms responsible for control of *E. ruminantium* are dominated by a Th1-type response. Protective immunity in immunized cattle and goats associates with a cellular immune response mediated by IFN-γ-producing $CD4^+$ T cells. IL-4, a hallmark of a Th2-type response, is absent, indicating a bias toward a Th1 response (Esteves et al., 2004b). Additionally, IFN-γ-producing $CD8^+$ T cells from immune animals proliferate in the presence of antigen. IFN-α and IFN-γ are thought to be of particular importance because treatment with these IFNs reduced replication of *E. ruminantium* in endothelial cells in vitro (Totté et al., 1994, 1996). Additionally, cattle that resisted a lethal challenge produced IFN-α, while cattle that succumbed to challenge did not (Totté et al., 1994). $\gamma\delta$ T cells that proliferate in response to antigen have also been identified in immune cattle, but not in immune goats (Esteves et al., 2004b; Mwangi et al., 1998).

The role of antibody in the control of *E. ruminantium* is controversial. In cattle infected by needle inoculation, 43% (3/7) of animals that survived also seroconverted (Totté et al., 1994), although one of these animals required treatment. Of the four animals that did not seroconvert, two died and two survived. One of the survivors required antibiotic treatment. This animal did not seroconvert until it was later challenged with the same strain. The other survivor was never treated, survived both the initial inoculation and subsequent challenge, and did not seroconvert. Thus, antibody is apparently not required for protective immunity in all animals. Similarly, in a mouse model, passive transfer of immune serum or gamma globulins failed to protect animals or modify the course of disease (du Plessis, 1970).

Up to two-thirds of human infections with *E. chaffeensis* are asymptomatic or minimally symptomatic (Ismail et al., 2010), suggesting that many humans mount an appropriate, protective immune response upon infection. A single model of protective immunity to heterologous challenge has been described. C57BL/6 mice first exposed to *E. muris*, which causes mild self-limiting disease, are protected from lethal infection upon challenge with the virulent IOE (described above) isolated from *I. ovatus*. The level of protection afforded by this immunization was remarkable in that IOE was eliminated from liver, spleen, and lung 7 days after challenge and the levels of *E. muris* were also lower in the immunized group compared with the group that received only *E. muris* (Ismail et al., 2004). Protection was associated with a Th1-type response, including expansion of IFN-γ-producing $CD4^+$ and $CD8^+$ T cells, a high IgG2a titer, and in contrast to severe disease, low serum levels of TNF-α. Antibody also plays a role in protective immunity to ehrlichial infection. Mice lacking B cells succumbed to infection with a low dose of IOE, whereas the control mice receiving the same dose were able to resolve the infection (Yager et al., 2005). Administration of immune serum to mice infected with a high dose of IOE reduced the number of bacteria within the spleen, though there was a high degree of variability among the mice. Additionally, passive transfer of *E. chaffeensis*-specific antibodies (mainly IgG2a) protected SCID mice, for which infection is consistently fatal, from severe infection with *E. chaffeensis* and increased the rate at which bacteria were cleared from immunocompetent mice (Winslow et al., 1998; Yager et al., 2005).

E. canis infection in dogs varies from clinically inapparent to severe (Neer and Harrus, 2006). There are many variables that influence the outcome of infection, such as the virulence of the pathogen, the dose administered, and individual and breed-specific differences in immune responses (Little, 2010). These factors are further complicated by the use of serology, which lacks specificity, as a diagnostic test (Little, 2010). Regardless, attempts to induce protective immunity have proven particularly unrewarding and serve as a cautionary tale. Infection followed by treatment with doxycycline resulted in little protection against homologous challenge. With heterologous challenge, the severity of anemia, fever, neutropenia, and thrombocytopenia were increased (Breitschwerdt et al., 1998). Similarly, antigen derived from inactivated cell culture induced an antibody response in vaccinees. However, upon challenge, clinical disease was more severe in the immunized animals than in nonimmunized controls (Ristic and Holland, 1993). Thus, overcoming the immune dysregulation mediated by *E. canis* appears particularly problematic.

Little is known about the protective immune response directed against pathogens within the genus *Neorickettsia*. Horses that recover from infection with *N. risticii* are resistant to clinical disease upon rechallenge. Similarly, inoculation with *N. sennetsu*, a human pathogen, protected ponies from disease upon challenge with *N. risticii* (Pusterla and Madigan, 2007b). In these experiments, *N. sennetsu* inoculation induced a high IgG titer, which included antibodies that bound both *N. risticii* and *N. sennetsu* (Rikihisa et al., 1988). Interestingly, antibodies from the protected *N. sennetsu*-inoculated ponies recognized a 44-kDa antigen, which was poorly recognized by antibodies induced by *N. risticii* inoculation alone. It is unknown if the antibody response to the 44-kDa antigen plays a direct role in protective immunity; however, these experiments demonstrate the potential utility of identifying epitopes shared among related pathogens as a means to finding protective antigens.

IMMUNODOMINANT, VARIABLE SURFACE PROTEINS AS PROTECTIVE ANTIGENS AND IMMUNOLOGIC STRAIN DETERMINANTS

Despite the immunomodulatory effects these pathogens have on the mammalian host, the examples above clearly demonstrate the potential for the induction of a protective immune response. A second obstacle to vaccine development is antigenic variability, in terms of the ability of the pathogen to establish persistent infection and in terms of lack of cross-protection between strains, displayed by many of these pathogens, particularly *A. marginale* and *E. ruminantium*. It is reasonable to hypothesize that understanding the source of the variation and identifying the proteins that allow for immune escape could serve as the basis of rational vaccine design.

While the ehrlichiae establish long-term persistent infection, the mechanisms and molecules involved in this process have not been identified. In the case of the *Anaplasma* spp., persistent infection is established and maintained in large part through antigenic variation (Palmer et al., 2009), which is mediated by MSP2 and MSP3 of *A. marginale* and MSP2/p44 of *A. phagocytophilum* (chapter 12, this volume). In *A. marginale*, MSP2 is expressed from a single expression site and is composed of a central hypervariable region that is flanked by highly conserved regions. The variation is generated by gene conversion in which one of multiple *msp2* donor alleles is recombined into a single, operon-linked expression site (Barbet et al., 2000; Brayton et al., 2001, 2002). The donor alleles have truncated 5′ and 3′ regions that are identical to the expression-site copy and flank a unique, allele-specific hypervariable domain (Brayton et al., 2001, 2005).

The donor alleles lack the functional elements for in situ transcription and are only expressed following recombination into the single expression site. The *A. phagocytophilum msp2/p44* expression site is similar and consists of 1 *msp2/p44* expression-site gene and approximately 100 *msp2/p44* pseudogenes with a similar structure of hypervariable and conserved region variants (IJdo et al., 1998; Murphy et al., 1998; Zhi et al., 1998). A similar mechanism of gene conversion gives rise to multiple variants (Barbet et al., 2003; Lin et al., 2003).

During infection, MSP2 is the dominant antigen recognized by sera from cattle infected with *A. marginale*, and MSP2/p44 is the dominant antigen recognized during *A. phagocytophilum* infection (IJdo et al., 1997; Murphy et al., 1998; Palmer et al., 1986; Zhi et al., 1997). In cattle, the anti-MSP2 antibody response is predominantly directed toward the hypervariable region rather than the flanking conserved regions (Abbott et al., 2004; Zhuang et al., 2007). Importantly, in *A. marginale*, the hypervariable regions of newly emergent variants are not recognized by existing antibody (French et al., 1999). Similarly, in sheep infected with *A. phagocytophilum*, the antibody response to a particular hypervariable region is detected shortly after the appearance of the variant in a rickettsemic peak (Granquist et al., 2008, 2010). Additionally, immunization of mice with a single MSP2/p44 variant failed to prevent *A. phagocytophilum* infection due to the emergence of other MSP2/p44 variants (IJdo et al., 2002). These data suggest that the generation of MSP2 or MSP2/p44 variants allows for immune escape and contributes to the ability of the pathogen to establish long-term persistent infection (French et al., 1999; Palmer et al., 2007). There is one caveat: during *A. phagocytophilum* infection, antibody responses to a particular variant may be short-lived and do not consistently associate with clearance of that variant (Granquist et al., 2010).

Not only are these pathogens able to establish persistent infection within a host, but strain variation is great, such that a protective immune response directed to one strain is not uniformly protective against other strains, indicating antigenic differences among strains in the protective outer membrane proteins (OMPs). Stated differently, at the level of the individual host, the population of bacteria infecting the host is able to evade the existing immune response in order to persist. At the level of the host population, a second strain of the bacteria is able to overcome the existing immune response in a persistently infected animal in order to establish infection (superinfection) and be transmitted within that herd. It has been proposed that the same set of proteins and mechanisms that allow for antigenic variation and persistent infection also allow for superinfection. Identification of these proteins may provide the basis for development of a broadly cross-protective vaccine.

In the case of *A. marginale*, it has been hypothesized that variation in the *msp2* donor repertoire circulating within a population allows for superinfection of a persistently infected animal by a second strain (Futse et al., 2008). As described above, superinfection of *A. marginale* subsp. *centrale*-immunized animals with circulating *A. marginale* strains is common. Similarly, though to a lesser degree, infection with one strain of *A. marginale* does not preclude infection with a second strain. For example, in one herd 5 of 75 infected animals each harbored two *A. marginale* strains with markedly distinct genotypes based on the number and identity of tandem repeats in the MSP encoded by *msp1a* (Palmer et al., 2004). These two *A. marginale* strains also had markedly different *msp2* donor repertoires (Rodriguez et al., 2005). Similarly with *A. phagocytophilum*, up to five different *msp4* genotypes were detected in blood samples collected sequentially in lambs housed on tick-infested pasture (Ladbury et al., 2008). Two different genotypes were found concurrently in four animals. Additionally, coinfection with two variants of *A. phagocytophilum* in five of seven dogs from the western United States was identified by sequencing the 16S rRNA gene (Poitout et al., 2005). Based on these findings, superinfection occurs in animals naturally infected with *A. marginale* and *A. phagocytophilum*. Due to the complexity of the *msp2/p44* pseudogene repertoire in *A. phagocytophilum*, tracking expressed variants in the original and superinfecting strains is difficult.

In *A. marginale*, the situation is somewhat simplified due to the limited number of *msp2* donor. Importantly, the diversity of the hypervariable regions of the *msp2* donors is similar when comparing the diversity within a strain with the diversity between strains (Futse et al., 2008). This suggests that the selection pressure within an animal to overcome immunity and persist is the same as the selection pressure to establish infection within an endemically infected herd. In experimental infections established by tick feeding on persistently infected animals, a single distinct *msp2* allele was shown to be sufficient to allow for superinfection (Futse et al., 2008). Importantly, the distinct allele was uniquely expressed during establishment of infection by the second strain, demonstrating use of this *msp2* allele to evade the existing immune response and establish infection.

However, in these studies, while superinfection occurred, clinical disease did not, suggesting that while the existing immune response was inadequate to prevent infection, it was adequate to prevent disease. Thus, while variation in *msp2* donor allelic repertoire may allow for reinfection in an immune animal, it is not sufficient to cause clinical disease. This conclusion is supported by field studies conducted in Israel examining animals immunized with *A. marginale* subsp. *centrale*. These animals had a high rate of coinfection with *A. marginale*, as stated above, but no history of clinical disease. *A. marginale* and *A. marginale* subsp. *centrale* have completely different *msp2* pseudogene repertoires, allowing for superinfection; however, the determinants of the development of clinical disease coincident with superinfection are unknown.

Regardless, one potentially effective approach to vaccine development is to identify the antigens that allow for immune escape and superinfection and immunize animals with all possible variants, thus inducing a broadly protective immune response. For example, if an animal was immunized with all possible expressed variants of MSP2 from a regional population of *A. marginale*, would that animal then be protected from infection by all *A. marginale* strains within that region? This question is based on the assumption that the variable antigens that allow for immune evasion are the same antigens that confer protective immunity upon challenge. While this experiment has not been done, there are data that lend insight into this question.

Using one strain of *A. marginale* (St. Maries), gel-purified native MSP2 containing a wide variety of antigenic variants was used to immunize cattle. These animals were then challenged with organisms expressing the same MSP2 variants, as determined by sequencing the *msp2* expression site within the challenge dose. No protection was afforded using this approach (Abbott et al., 2005), suggesting that the protective immune response is directed toward proteins other than the immunodominant and antigenically variable MSP2. To further determine if the protective immune response correlates with MSP2-specific immunity, the antibody response to both conserved and hypervariable region epitopes of MSP2 was compared in protectively vaccinated and nonvaccinated, infected animals (Noh et al., 2010). Both immunized and infected animals had comparable antibody repertoires to MSP2 in terms of breadth of response and titer. Among the immunized animals, there was no association between either breadth or magnitude of the anti-MSP2 response and either complete protection from infection or control of bacteremia. Together, these data argue that protection afforded with the outer membrane vaccine is due to immune responses directed at OMPs other than the immunodominant and antigenically variable MSP2. Based on these data, development of a vaccine targeting immunodominant, antigenically variable surface proteins is likely to be unrewarding, at least in the case of *A. marginale*.

IMMUNODOMINANT SURFACE PROTEINS AS VACCINE CANDIDATES

The specific antigens that induce protective immunity to pathogens in the family *Anaplasmataceae* are unknown. Bacteria in the genera *Ehrlichia* and *Anaplasma* lack genes encoding lipopolysaccharide (Brayton et al., 2005; Dunning Hotopp et al., 2006; Fenn and Blaxter, 2006; Lin and Rikihisa, 2003; Rikihisa, 2010), so the bacterial outer membrane forms the interface between these bacteria and the host. In many instances a protective immune response correlates with the antibody titer directed toward surface proteins. Additionally, these proteins are the easiest to identify; thus, most studies with *Anaplasma* and *Ehrlichia* spp. designed to identify protective antigens have focused on immunodominant surface proteins.

For these pathogens, the most immunodominant proteins are encoded by a protein superfamily, Pfam accession number PF01617, and include the predominant OMPs of each species, *A. marginale* MSP2 and MSP3 and

A. phagocytophilum MSP2/p44, which are involved in antigenic variation, as discussed above. Additionally, *E. chaffeensis* p28-OMP (Chen et al., 1994; Ohashi et al., 1998; Reddy et al., 1998; Rikihisa et al., 1994), *E. canis* p30 (Rikihisa et al., 1994), *E. ruminantium* major antigenic protein 1 (MAP1) (Reddy et al., 1996; Sulsona et al., 1999; van Heerden et al., 2004), and *Wolbachia* surface protein (WSP) (Braig et al., 1998; Dunning Hotopp et al., 2006) are members of this multigene family.

Wolbachia spp. are bacterial endosymbionts in many invertebrates, including filarial nematodes, that cause human lymphatic filariasis and river blindness and canine and feline heartworm. WSP is an abundant surface protein that displays areas of marked amino acid variability and conservation across populations of *Wolbachia* (Baldo et al., 2005). Both the innate and adaptive immune responses of the nematode-infected vertebrate to *Wolbachia* WSP play a role in filarial-associated disease (Hise et al., 2007; Porksakorn et al., 2007; Saint André et al., 2002). For example, the presence and levels of antibody directed toward WSP correlate with filaria-associated disease (Punkosdy et al., 2003; Taylor et al., 2001). Dogs infected with *Dirofilaria immitis*, which causes canine heartworm, mount an IgG response to WSP of the *Wolbachia* carried by *D. immitis* (Kramer et al., 2005; Werren et al., 2008). The role this immune response plays in development or prevention of disease is unknown.

In the ehrlichiae, the polymorphic, immunodominant outer membrane gene locus is composed of gene paralogs arranged in tandem and located downstream of a transcriptional regulator. The number of paralogs varies from 16 in *E. ruminantium* to 22 in *E. chaffeensis* and 25 in *E. canis*. When aligning the protein sequences of the paralogs, areas of amino acid conservation and areas of variability, which tend to be hydrophilic (Reddy and Streck, 1999; Reddy et al., 1998), are apparent. In the case of *E. chaffeensis*, the antibody response is often directed toward the variable regions of the proteins, suggesting that these proteins are under immune selection. Additionally, these proteins may have a role in the protective immune response.

For example, in cattle, immunization by infection with subsequent protection from challenge with *E. ruminantium* results in a strong antibody and T-cell response directed toward MAP1 and MAP2. Significant protection against homologous challenge with *E. ruminantium* was observed in a mouse model following immunization with MAP1 DNA followed by a recombinant protein boost (Nyika et al., 2002). MAP1 varies in molecular size between strains (Mahan, 1995); if this molecular size difference reflects antigenic differences, inclusion of multiple variants within a vaccine would likely be necessary. The ortholog of MAP1 is p28-OMP19 of *E. chaffeensis*, which also displays diversity among strains. In *E. chaffeensis*, p28-OMP19 and -20 are the dominant paralogs expressed in macrophages (Ganta et al., 2009; Peddireddi et al., 2009), though protein from 18 of the 22 paralogs has been identified in macrophage-derived *E. chaffeensis* (Seo et al., 2008). In SCID mice, antibody specific for the variable region of p28-OMP19 protects against fatal infection (Li et al., 2002; Winslow et al., 2000). In a second model using IOE, a strain of *Ehrlichia* that is fatal for immunocompetent mice, immunization with recombinant p28-OMP19 encoded by IOE resulted in protection from fatal infection and the development of a robust humoral and $CD4^+$-T-cell response (Nandi et al., 2007).

Additional dominant antigens in *A. phagocytophilum* and ehrlichiae were identified by antibody binding on Western blots. Most of these have been sequenced and many are secreted proteins, although there is little information on protective efficacy of these proteins. They are designated by molecular size: 100-, 130-, and 160-kDa (AnkA) antigens of *A. phagocytophilum* (Caturegli et al., 2000; Storey et al., 1998); 200-, 120-, 88-, 55-, 47-, 40-, and 23-kDa proteins of *E. chaffeensis* (Chen et al., 1994; Rikihisa et al., 1994); 200-, 140-, 95-, 75-, 47-, and 36-kDa antigens of *E. canis* (McBride et al., 2003); and 160-, 85-, 58-, 46-, 40-, 32-, and 21-kDa proteins of *E. ruminantium* (Mahan et al., 1994). Tandem repeat proteins (TRPs) of *E. chaffeensis* have also been defined, including TRP120, TRP47, and TRP32, that contain immunodominant epitopes in the repeat regions that are recognized by sera of convalescent patients (Luo et al., 2008; Wakeel et al., 2010). *E. canis* orthologs TRP140, TRP36, and TRP19 were also recognized by immune sera (Doyle et al., 2006; McBride et al., 2007). The potential for these proteins as vaccine candidates in the natural host remains to be determined.

As for *A. marginale*, MSP2 and MSP3, MSP1a, MSP1b, MSP4, and MSP5 were initially discovered using surface radiolabeling in combination with immunoprecipitation and are the dominant protein bands on Western blots probed with serum from infected animals (Palmer and McGuire, 1984; Tebele et al., 1991). MSP1a, MSP2, MSP3, and MSP4 prime naïve $CD4^+$ T cells and stimulate recall responses from outer membrane vaccinees to proliferate and secrete high levels of IFN-γ, the cytokine associated with protection (Brown et al., 1998a, 1998b). In spite of the ability of these immunodominant proteins to prime $CD4^+$-T-cell responses, when they are used as individual immunogens in vaccine trials, protection equivalent to that of outer membranes is not observed (Abbott et al., 2005; Palmer et al., 1986, 1988, 1999; Palmer and McElwain, 1995).

When considered together, the best-described protective antigens in the case of *E. chaffeensis* (p28-OMP19) and *E. ruminantium* (MAP1) are variable among strains and thus likely will not induce wide cross-protection. In the case of *A. marginale*, individual protective OMPs have not been identified. These results raise the question of what protein or proteins in the outer membrane confer broadly protective immunity. Addressing this question perhaps requires a shift in thinking. As proposed by Byron Waksman in the 1970s, and restated by Alan Sher, "It seems intuitive that for a pathogen to survive, expression of immunodominant antigens targeted by a protective immune response would be lethal for the pathogen" and "antigens that induce poor or immunosuppressed responses during natural infection should not be ignored since they may be molecules essential for parasite survival" (Sher, 1988). Indeed, many pathogens, including *Anaplasma* spp. and *Ehrlichia* spp., persist in the face of strong immune responses. Therefore, more rational choices of vaccine antigens may be those that do not naturally evoke a strong immune response. For example, the HA protein of the influenza virus mediates membrane fusion and viral entry and is the target of the protective immune response. Specifically, the vaccine in use today targets an immunodominant, highly variable epitope located on the globular head of HA. Consequently, the vaccine only protects against a limited number of variants and must be reformulated each year. Encouragingly, antibodies that target the highly conserved stem region of the influenza virus HA molecule have been identified (Ekiert et al., 2009; Kubota-Koketsu et al., 2009). These antibodies neutralize a wide variety of viral variants, but are poorly immunogenic in the context of both current immunization formulations and natural infection. The identification of this epitope will likely provide the foundation to develop a widely protective influenza vaccine. Perhaps similarly widely protective epitopes exist among the *Anaplasmataceae*.

IDENTIFICATION OF SUBDOMINANT, POTENTIALLY PROTECTIVE ANTIGENS

Subdominant antigens are of interest for vaccine development not only because they are usually antigenically conserved, but because in other disease models protective immunity does not necessarily correlate with the immunodominance hierarchy of epitopes (Gallimore et al., 1998a, 1998b; Oukka et al., 1996; van der Most et al., 1996). Using genomic and proteomic approaches, the full spectrum of surface-exposed proteins can be identified, thus allowing for rapid and comprehensive identification of vaccine candidates. The challenge then becomes identifying those antigens that are required for protective immunity. In the next two sections, approaches to identify subdominant antigens are discussed.

Unbiased Proteomic Approaches To Identify Vaccine Candidates

One comprehensive strategy to identify subdominant OMPs relies on labeling of surface-exposed proteins, separation of the labeled components from other cellular components, followed by tandem mass spectrometric analysis (Ge and Rikihisa, 2007a, 2007b; Noh et al., 2008). For example, to identify the surface-exposed proteome of *A. phagocytophilum* and *E. chaffeensis*, isolated bacteria were labeled with a membrane-impermeable, cleavable biotin reagent (Sulfo-NHS-SS-Biotin) (Ge and Rikihisa, 2007a, 2007b). The biotinylated bacterial surface proteins were affinity purified and separated by electrophoresis, and proteins in the bands were identified by liquid chromatography followed by mass spectroscopy. In *A. phagocytophilum*, in addition to P44/MSP2 proteins, predicted surface-exposed proteins identified were members of the OMP85 family (PF01103); three hypothetical proteins, APH_0441, APH_040 (Asp62), and APH_0405 (Asp55); PF01617 member OMP1A; type IV secretion system (T4SS) protein VirB8-1; and a thiol-disulfide oxidoreductase involved in protein folding and stabilization. Additional proteins, such as translation elongation factor G, chaperone proteins DnaK and GroEL, and cochaperone GrpE, not predicted to be membrane associated were also identified (Ge and Rikihisa, 2007a). Little is known about the immunogenicity of these proteins, although Asp62 and Asp55 are conserved among *A. phagocytophilum* strains and are recognized by patient sera. Additionally, antisera raised against these proteins were able to partially block in vitro infection, indicating these may be vaccine candidates (Ge and Rikihisa, 2007a).

In a similar study using *E. chaffeensis*, several proteins predicted to be surface exposed, besides the P28-OMPs, were identified, including the hypothetical protein ECH_0525 (also termed Esp73), a member of the OMP85 family (PF01103); immunodominant surface protein gp47; VirB9-1, a component of the T4SS; disulfide oxidoreductase and serine protease, two proteins involved in protein folding and processing; and dihydrodipicolinate reductase, which may be involved in biosynthesis of bacterial cell walls (Ge and Rikihisa, 2007b). Additionally, proteins predicted to localize to the cytoplasmic, periplasmic, or inner membrane were identified. The presence of these proteins in the outer membrane preparation may reflect contamination. Alternatively, some of these proteins may truly be surface exposed and have functions that are yet to be identified. While comprehensive, these studies did not link specific proteins to the protective immune response.

A similar approach was used to identify surface-exposed proteins from *A. marginale* isolated from bovine erythrocytes (Noh et al., 2008). The isolated bacteria were treated with a membrane-impermeable chemical cross-linker. The resulting protein complexes were isolated, characterized by liquid chromatography followed by tandem mass spectrometry, and used to immunize cattle. The individual components identified in the complex included MSP1a, MSP2, MSP3, MSP4, OMP1, OMP7, OMP9, OpAG2, AM854, and AM779, many of which are members of the PF01617 family. The protein complex-immunized cattle had levels of protection following challenge similar to those of animals immunized with outer membranes, thus limiting the number of possible vaccine candidates (Fig. 1).

Another unbiased proteomic approach has been employed to identify the specific antigens in a complex mixture that are immunogenic, thus linking a protective immune response directly to the antigen. In this strategy, *A. marginale* OMPs were separated using two-dimensional electrophoresis (Fig 2). The immunoreactive proteins were then identified by Western blotting using IgG2 from the vaccinees as a correlate of $CD4^+$-T-cell responses, usually required for class switching. The animals were vaccinated with either *A. marginale* outer membranes (Lopez et al., 2005) or the *A. marginale* subsp. *centrale* vaccine strain (Agnes et al., 2011). The immunoreactive proteins were excised and subjected to liquid chromatography followed by mass spectrometry for definitive identification by mapping to the annotated genome. This approach restricted the identification to antigens relevant to the induction of an immune response while permitting the identification of subdominant antigens weakly reactive with immune sera.

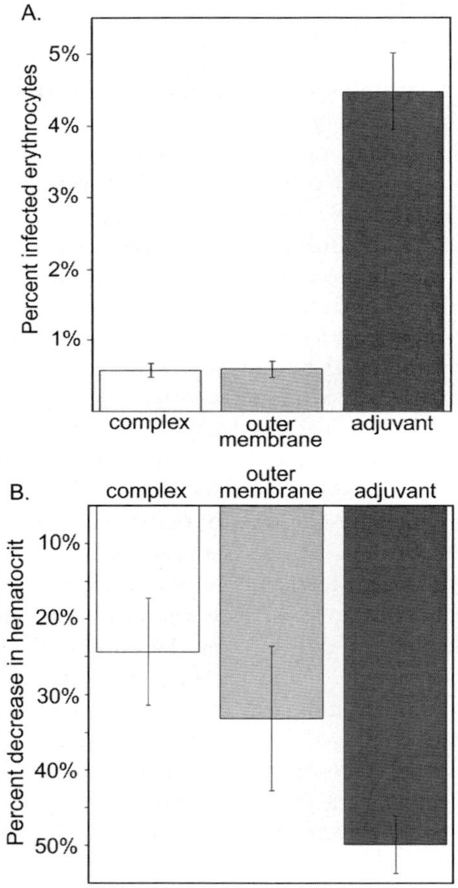

FIGURE 1 Surface complex immunization recapitulates the protection induced by whole *A. marginale* outer membrane immunization. (A) Protection against high-level bacteremia. (B) Protection against severe anemia. (Reprinted from Noh et al. [2008] with permission.) doi:10.1128/9781555817336.ch11.f1

When *A. marginale* outer membranes were used as the protective immunogen, 24 antigens were identified (Lopez et al., 2005). These included the previously characterized surface-exposed OMPs MSP2, MSP3, and MSP5. Among 21 newly described antigenic proteins, 14 were annotated in the *A. marginale* genome and included the product of the OpAG2 gene of the *msp2* operon (Löhr et al., 2002); three appendage-associated proteins thought to associate with host actin filaments within the erythrocyte (Stich et al., 2004); three T4SS structural proteins; outer membrane antigen OMA87 (PF01103); OMP4, -7, -10, and -14; EF-Tu; and PepA cytosol aminopeptidase (Lopez et al., 2005). AM854, annotated as a peptidoglycan-associated protein, was also identified in this study.

The rationale for the second set of experiments using immune serum from *A. marginale* subsp. *centrale*-immunized animals was to identify antigens in *A. marginale* that are conserved in the protective vaccine strain (Agnes et al., 2010). In addition to MSP2 and MSP3, surface proteins recognized were OMP7, -8, -9, -11, -13, and -14; AM779; and AM854. Several housekeeping genes, including EF-Tu and GroEL, were also recognized. Excluding the hypervariable MSP2 and MSP3, conservation between the antigenic surface proteins of *A. marginale* and *A. marginale* subsp. *centrale* ranged from 60 to 84%, indicating that a high degree of overall identity is not required to maintain antigenic cross-reactivity.

After excluding the MSPs, these three studies identified three OMPs in common—OMP7, OMP9, and AM854—while three OMPs—OMP8, OMP14, and AM779—were identified in common in two of the experiments. In addition, two of the studies identified EF-Tu as antigenic. EF-Tu was predicted by TMHMM analysis to be surface associated, and the association of EF-Tu with the surface of other bacterial pathogens has been previously documented (Dallo et al., 2002; Granato et al., 2004; Jacobson and Rosenbusch, 1976; Marques et al., 1998) Thus, in some bacteria,

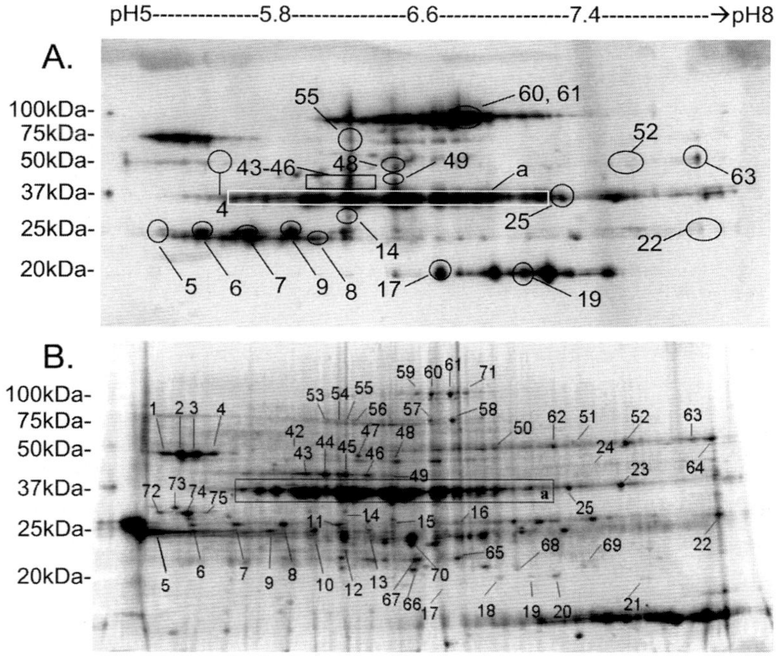

FIGURE 2 OMPs separated by two-dimensional electrophoresis. One gel (B) was stained with SYPRO Ruby for total protein detection, while proteins from the other gel (A) were transferred to a nitrocellulose membrane for immunoblotting with serum (IgG2) from immunized animals. (A) A representative immunoblot from one of three animals tested. To identify spots of interest for mass spectroscopy, images from the SYPRO Ruby-stained gel and immunoblot were overlaid. Block a represents proteins identified as MSP2. Spots 60 and 61 represent MSP3, and spot 17 represents MSP5. Spots 12 and 70 were not immunoreactive but were used as two of several reference spots to determine reproducibility of outer membrane-separated two-dimensional gels. Spots 6, 63, and 17 were used to align two-dimensional immunoblots. (Reprinted from Lopez et al. [2005] with permission.) doi:10.1128/9781555817336.ch11.f2

EF-Tu localizes to the surface, where it can function in bacterial attachment to host cells. Thus, this protein may also be of interest for vaccine development.

Rather than using two-dimensional electrophoresis to separate proteins as described above, *E. ruminantium* was fractionated using continuous-flow electrophoresis (Esteves et al., 2004a; Van Kleef et al., 2000). Fractions containing low-molecular-weight proteins of from 11 to 19 kDa and 23 to 29 kDa stimulated proliferation of blood mononuclear cells and CD4$^+$-enriched T cells that produced increased amounts of IFN-γ, which correlates with protective immunity (Esteves et al., 2004a; Van Kleef et al., 2002). At the time, the genome sequence was not available; thus, specific identification of these proteins was problematic. With the sequencing of the *E. ruminantium* genome, five open reading frames (ORFs) predicted to encode genes smaller than 20 kDa were selected for cloning and expression. All five recombinant proteins induced proliferative responses and IFN-γ production in peripheral blood mononuclear cells from immune animals (Sebatjane et al., 2010). Disappointingly, when animals were immunized with these candidates using DNA vaccines, protection against needle challenge was poor.

High-Throughput Screening of Predicted OMPs

Many OMPs and predicted OMPs, which could serve as vaccine candidates, have been identified through genome sequencing (Collins et al., 2005; Dunning Hotopp et al., 2006; Noh et al., 2006; Sebatjane et al., 2010). However, reliable high-throughput screening strategies to identify the most likely vaccine candidates are lacking. To overcome this difficulty for *E. ruminantium*, genomic DNA libraries were made using expression clones. The protein products from these clones were then screened for immunogenicity based on antibody recognition and peripheral blood mononuclear cell proliferation from cattle immunized using infection-and-treatment (Barbet et al., 2001; Brayton et al., 1998). Candidates were then chosen for an immunization and challenge trial. Mice immunized with pools of recombinant proteins from these clones were, in some cases, significantly (58 to 89% survival) protected from death. The genes comprising these pools were not reported (Barbet et al., 2001).

Individual and groups of ORFs from *E. ruminantium* have been screened for their ability to induce protective immunity when presented as DNA vaccines with variable results (Collins et al., 2003b; Pretorius et al., 2007, 2010). To date, a variety of ORFs (Table 2), predicted to encode a transported protein, integral membrane proteins, a conjugal transport protein, and several hypothetical proteins have been tested. The most promising candidates identified were Erum2540, predicted to be an exported protein, and Erum2510, a hypothetical protein, both of which induced protective immunity to needle challenge. The best protection occurred when sheep were immunized with the DNA-containing vector and boosted with recombinant protein (Pretorius et al., 2008, 2010). However, heterologous protection during field challenge has not been achieved, suggesting that the protective epitopes are variable among strains. The three different genotypes of Erum2510 that have been identified support this conclusion (Collins et al., 2003b; Pretorius et al., 2010), though Erum2540 is more conserved (Pretorius et al., 2008). Additionally, for unknown reasons, tick challenge tends to be more virulent than needle challenge (Allsopp, 2010; Collins et al., 2003b).

Another high-throughput approach to identifying vaccine candidates is in vitro transcription and translation (IVTT) (Fig. 3). This method allows rapid expression and identification of T-lymphocyte antigens for any pathogen for which the genome sequence is available. Fifty selected vaccine candidate antigens identified from the *A. marginale* genome were expressed from transcriptionally active PCR products using an *Escherichia coli*-based IVTT system (Lopez et al., 2008). IVTT-expressed antigens were affinity purified and processed and presented by APCs to T lymphocytes collected from outer membrane-immunized animals. Antigens that stimulated strong lymphocyte proliferation were members of the T4SS—VirB9-1, VirB9-2, and VirB10—and OMPs MSP2, MSP3, Ana29, OMA87, OMP4, OMP7, OMP8, OMP9, and EF-Tu. Although the IVTT-expressed antigens were not tested for antibody recognition, this has been accomplished for other pathogens (Davies et al., 2005; Gat et al., 2006).

VIRULENCE FACTORS AS VACCINE CANDIDATES

Virulence factors are targets for some of the most effective vaccines currently in use. For example, while the MSP (SlpA) of *Clostridium tetani* is variable, the major virulence factor, regardless of the host species, is a single conserved toxin (Qazi et al., 2007; Telford, 2008). Thus, induction of an antibody response that neutralizes this toxin prevents disease. Due to variation among individuals in response to infection with pathogens in the family *Anaplasmataceae*, virulence is difficult to measure. Additionally, few specific factors that confer virulence have been defined. However, the proteins that comprise the T4SS are of particular interest. In gram-negative bacteria, this system functions in invasion and transfer of DNA and proteins directly to the eukaryotic host cell (Alvarez-Martinez and

Christie, 2009; Cascales and Christie, 2003). Importantly, while there is variation in the repertoire of T4SS components among bacteria, overall the structure is highly conserved.

All sequenced members of the *Anaplasmataceae* to date have a T4SS (Gillespie et al., 2010). The function of the T4SS in most of the *Anaplasmataceae* has not been determined. In *A. phagocytophilum* it is a virulence determinant, secretes effector proteins into the host cell, and may be involved in host cell attachment and invasion (Niu et al., 2010; Rikihisa et al., 2010). Importantly, when comparing members of the T4SS among *A. marginale*, *A. phagocytophilum*, *E. chaffeensis*, and *E. canis*, there is between 53 and 74% amino acid identity, depending on the protein being evaluated (Lopez et al., 2007). Variation in amino acid identity between strains of the same pathogen is minimal. For example, the amino acid identity in comparing VirB9, VirB10, and conjugal transfer protein (CTP) of *E. ruminantium* Welgevonden and Gardel strains is 99 to 100%. Similarly, in comparing all the *A. marginale* strains as well as *A. marginale* subsp. *centrale*, there is between 91 and 99% amino acid identity for individual members of the secretion system (Lopez et al., 2007). Due to the high degree of conservation, both among species and within strains, these proteins could prove to be highly useful in developing broadly cross-protective vaccines. There is evidence that at least some of the proteins of the T4SS are surface exposed and immunogenic.

For example, VirB9, VirB10, and conjugal transfer protein were identified in the protective outer membrane fraction of *A. marginale* (Lopez et al., 2005). These proteins were recognized by IgG2 in immune serum, and stimulated memory-T-lymphocyte proliferation and IFN-γ production (Lopez et al., 2007). Additionally, VirB2, VirB7, VirB11, and VirD4 were immunogenic for cattle with distinct major histocompatibility complex (MHC) class II haplotypes (Sutten et al., 2010). In a separate study, VirB9 and VirB10 induced IgG responses in experimentally and naturally infected cattle (Araújo et al., 2008; Vidotto et al., 2008). Dogs infected with *E. canis* also mount a strong antibody response to VirB9 (Felek et al., 2003). However, neither individual nor groups of T4SS proteins have been tested for the ability to induce protective immunity.

FUTURE DIRECTIONS

To achieve maximal vaccine efficacy, antigen selection must be combined with careful formulation (Bachmann and Jennings, 2010). Live vaccine products tend to be more immunogenic than recombinant products. However, due to the relative ease and cost of production, subunit vaccines rather than whole attenuated or inactive organisms are more likely to be the primary component in vaccines targeting the *Anaplasmataceae*, particularly those that primarily affect livestock. Similarly, in humans, vaccine reactions and the possibility of reversion to virulence are likely to prohibit the use of live products. Development of vaccines that target this group of pathogens provides additional challenges, which include overcoming the immune-modulatory/suppressive effects of these pathogens while inducing the appropriate immune response to protective antigens that may be inherently poorly immunogenic. However, many tools have been developed or are in the process of development that effectively boost and modulate the immune response in a predictable fashion (Leroux-Roels, 2010; Zepp, 2010).

DNA vaccines are one potentially convenient means of delivering an antigen to the immune system in a fashion that mimics some features of a live organism. DNA vaccines carry DNA sequences that encode the protective antigens cloned into a plasmid backbone. The DNA molecules are taken up by the host cell and translated into protein. These antigens are then processed and presented on the surface of the cell in the context of MHC molecules, particularly MHC class I molecules, improving the stimulation of cytotoxic effector T cells. Additionally, antigen may be secreted or released upon lysis of the transfected cell. This cell-free antigen can then be taken up and processed by APCs and presented on both MHC class I and class II molecules, thus

TABLE 2 *E. ruminantium* ORFs that have been tested as vaccine candidates in sheep

Gene designation[a]	Annotation	PBMC proliferation (no. positive/ no. tested)	IFN-γ secretion (no. positive/ no. tested)	Method of immunization	Outcome of challenge[d]
Erum7340 (3)	Low-molecular-weight membrane protein	Pos (2/3)[b]	Pos (2/5)[b]	Pool of five ORFs; DNA prime/ protein boost	E: 2/5; D: 2/5; S: 1/5
Erum7350	Low-molecular-weight membrane protein	Pos (3/3)[b]	Pos (2/5)[b]		
Erum7360	Low-molecular-weight membrane protein	Pos (1/3)[b]	Pos (1/5)[b]		
Erum7380	Low-molecular-weight membrane protein	Pos (1/3)[b]	Pos (2.5)[b]		
Erum4360	Unknown	Pos (2/3)[b]	Pos (2/5)[b]		
Tested as cocktail					
Erum2540 (1)	Exported protein	Pos (2/6)[c]		Pool of five ORFs; DNA vaccine	Trial 1: S: 5/5
Erum2550	ATP-binding protein (ABC transporter)				Trial 2: S: 6/6
Erum2580	Periplasmic solute-binding protein				
Erum2590	ATP-binding protein (ABC transporter)				
Tested individually					
Erum2540 (1)	Exported protein	Pos (1/5)[c]		DNA vaccine	S: 5/5
Erum2550	ATP-binding protein (ABC transporter)	Pos (3/4)[c]		DNA vaccine	S: 4/4
Erum2580	Periplasmic solute-binding protein	Pos (2/4)[c]		DNA vaccine	S: 4/4
Erum2590	ATP-binding protein (ABC transporter)	Pos (2/5)[c]		DNA vaccine	S: 5/5

Tested as cocktail				
Erum7450 (1)	Probable integral membrane protein		Pool of four ORFs; DNA vaccine	E: 5/5
Erum7490	Probable inorganic polyphosphate/ ATP-NAD kinase			
Erum7510	Hypothetical protein			
Erum7530	Putative conjugal transfer protein			
Tested as cocktail				
Erum6480 (1)	Probable peptidase		Pool of three ORFs; DNA vaccine	E: 5/5
Erum6490	Hypothetical protein			
Erum6530	Hypothetical protein			
Tested individually				
Erum2510 (2)	*Cowdria* polymorphic gene 1, hypothetical protein	Pos (1/5)[c]	DNA vaccine	Trial 1: E: 1/5; T: 3/5; S: 1/5 Trial 2: T: 3/5; S: 2/5
Erum2510	*Cowdria* polymorphic gene 1, hypothetical protein	Pos (1/5)[c]	DNA prime/protein boost	S: 5/5
Tested as cocktail				
Erum2500 (2)	Hypothetical protein		Pool of three ORFs; DNA vaccine	E: 5/5
Erum2490	Hypothetical protein			
Erum2480	Hypothetical protein			

[a] Numbers indicate the following references: 1. Pretorius et al., 2007; 2. Pretorius et al., 2010; 3. Sebatjane et al., 2010.
[b] Lymphocyte proliferation and IFN-γ secretion tested in peripheral blood mononuclear cells (PBMCs) from sheep immunized with infect-and-treat protocol. Pos, positive.
[c] Lymphocyte proliferation tested in PBMCs after immunization.
[d] S, survived; T, treated; E, euthanized; D, died.

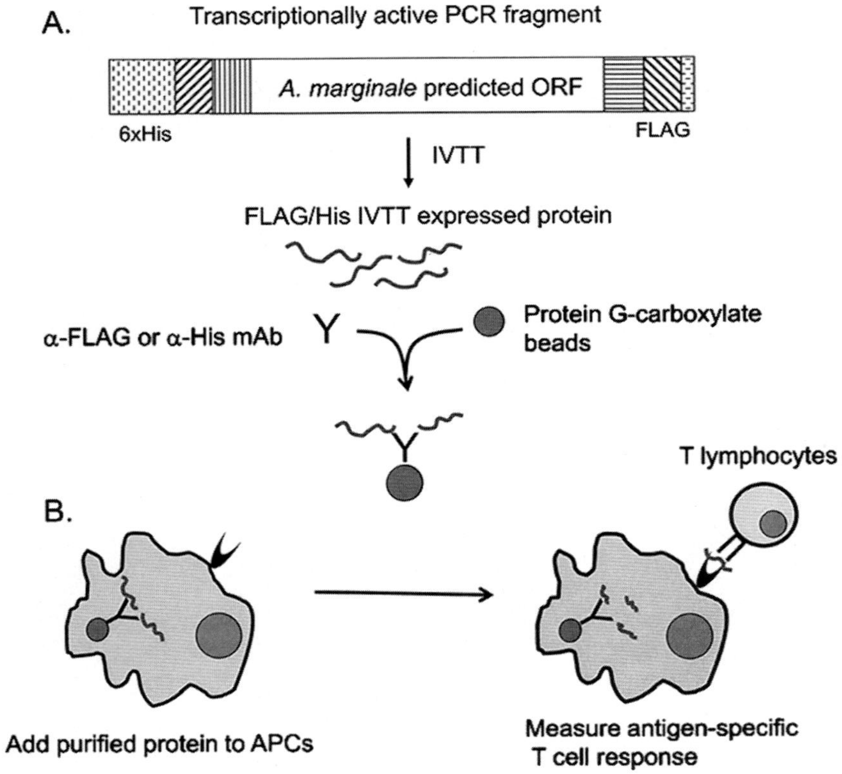

FIGURE 3 IVTT is a high-throughput method that allows rapid expression and identification of T-lymphocyte antigens for any pathogen for which the genome sequence is available. (A) Candidate antigens identified from the *A. marginale* genome were expressed from transcriptionally active PCR products, which included C-terminal FLAG and N-terminal six-His tags, using an *E. coli*-based IVTT system. Purification of the target antigen was accomplished using antibodies to six-His and FLAG epitope tags and protein G bead-affinity purification. (B) Purified IVTT-expressed, bead-bound antigens were processed and presented by APCs to T lymphocytes from immunized animals and evaluated for immunogenicity in proliferation assays. doi:10.1128/9781555817336.ch11.f3

enhancing the $CD4^+$-T-cell and antibody response. The advantages of DNA vaccines include the potential for priming and expansion of the immune response with one dose, ease of manufacturing and alteration of included antigens, and stability such that maintenance of a cold chain is not required. However, to date no DNA vaccines have been licensed for use, and first-generation vaccines have over all lacked potency, in part due to poor antigen spreading and dendritic cell transfection efficiency (Ingolotti et al., 2010; Zepp, 2010).

In terms of the *Anaplasmataceae*, DNA vaccines carrying a variety of *E. ruminantium* ORFs have produced promising results. However, multiple doses of the DNA vaccine (Pretorius et al., 2007) were used to induce immunity. Alternatively, priming the immune response with a DNA vaccine followed by boosting with recombinant protein dramatically improved the level of protective immunity to MAP1 (Nyika et al., 2002). In mice, following the DNA priming dose with recombinant protein improved protection against death from 13 to 66%, with two protein boosts being optimal. In these mice, improved survival correlated with a Th1-type response, with increased production of IFN-γ, IL-2, and MAP1-specific IgG2a. In comparison, the mice that received only recombinant protein produced a Th2-type response characterized by production of IL-4, IL-5, IL-10, and MAP1-specific IgG1. Similarly, in sheep immunized with Erum2510, two of five animals that received three doses of the DNA construct survived challenge, while all five animals that received the DNA construct followed by a recombinant protein boost survived challenge (Pretorius et al., 2010).

However, the requirement for multiple vaccine doses is limiting, particularly in extensively reared livestock. Incorporating intercellular trafficking and targeting motifs in the vaccine may improve priming and expansion of the immune response such that a single vaccine dose is sufficient. Specifically, an intracellular trafficking domain and an invariant-chain MHC class II-targeting motif capable of enhancing dendritic cell antigen uptake and presentation were fused to B- and T-cell antigens from MSP1a of *A. marginale* (Mwangi et al., 1998). The addition of these two domains enhanced antigen-specific $CD4^+$-T-cell proliferation compared with those animals that received only the B- and T-cell antigens. A single inoculation of the construct primed the immune system, resulting in proliferation of IFN-γ-secreting $CD4^+$ T cells and IgG production as well the induction of a memory response. However, in these studies the inoculation site was first enriched with dendritic cells by intradermal administration of DNA encoding fetal liver tyrosine kinase 3 ligand and granulocyte-macrophage colony-stimulating factor (Mwangi et al., 1998, 2002). These studies illustrate the potential utility of immune enhancers and trafficking motifs; however, it remains to be seen if DNA vaccines can effectively be used to prevent disease caused by the *Anaplasmataceae*.

While DNA vaccines, in theory, enhance the T-cell response, physical linkage of that B-cell epitope to a strong T-cell epitope can enhance the antibody response. This enhanced antibody response is based on the principle of linked recognition, which means that a given B cell is optimally activated by a helper T cell that responds to the same or a physically associated antigen. While the two cells need not recognize the same epitope, the epitopes must be spatially associated (Murphy et al., 2008). In the context of vaccine design (Fig. 4), a poorly immunogenic but protective B-cell epitope is linked to a T-cell epitope. Upon exposure to the conjugated antigens, dendritic cells take up, process, and present the resulting T-cell epitope on MHC class II molecules. Helper $CD4^+$ T cells that recognize and bind the peptide become primed by the APC. Concurrently, B cells bind the B-cell epitope, internalize the entire construct, and process the antigen via the MHC class II pathway. The T-cell epitope is then displayed in the context of MHC class II molecules on the surface of the B cell. The previously primed T cell can recognize and bind this antigen and thus provide help to the B cell, resulting in B cell activation, isotype switching, and production of high-affinity antibodies targeting the protective antigen. This method is used in the *Haemophilus influenzae* type b childhood vaccine to prevent meningitis (Murphy et al., 2008; Zepp, 2010). In this vaccine, the poorly immunogenic but protective capsular polysaccharide is conjugated to the strongly antigenic tetanus toxoid. $CD4^+$ T cells primed during recognition of tetanus toxoid antigen can provide help to B cells, which produce antibodies targeting the capsular polysaccharide. In the case of *A. marginale*, experimental evidence suggests that linkage between molecules enhances antibody production. For example, MSP1 occurs in the outer membrane as a heteromeric complex composed of MSP1a and MSP1b1. Infection or immunization with MSP1 results in development of antibodies to both proteins, while T-cell epitopes are predominantly located only on MSP1a (Brown et al., 2002; Macmillan et al., 2006; Vidotto et al., 1994). To test the hypothesis that physical linkage between the molecules enhances antibody production, animals were immunized

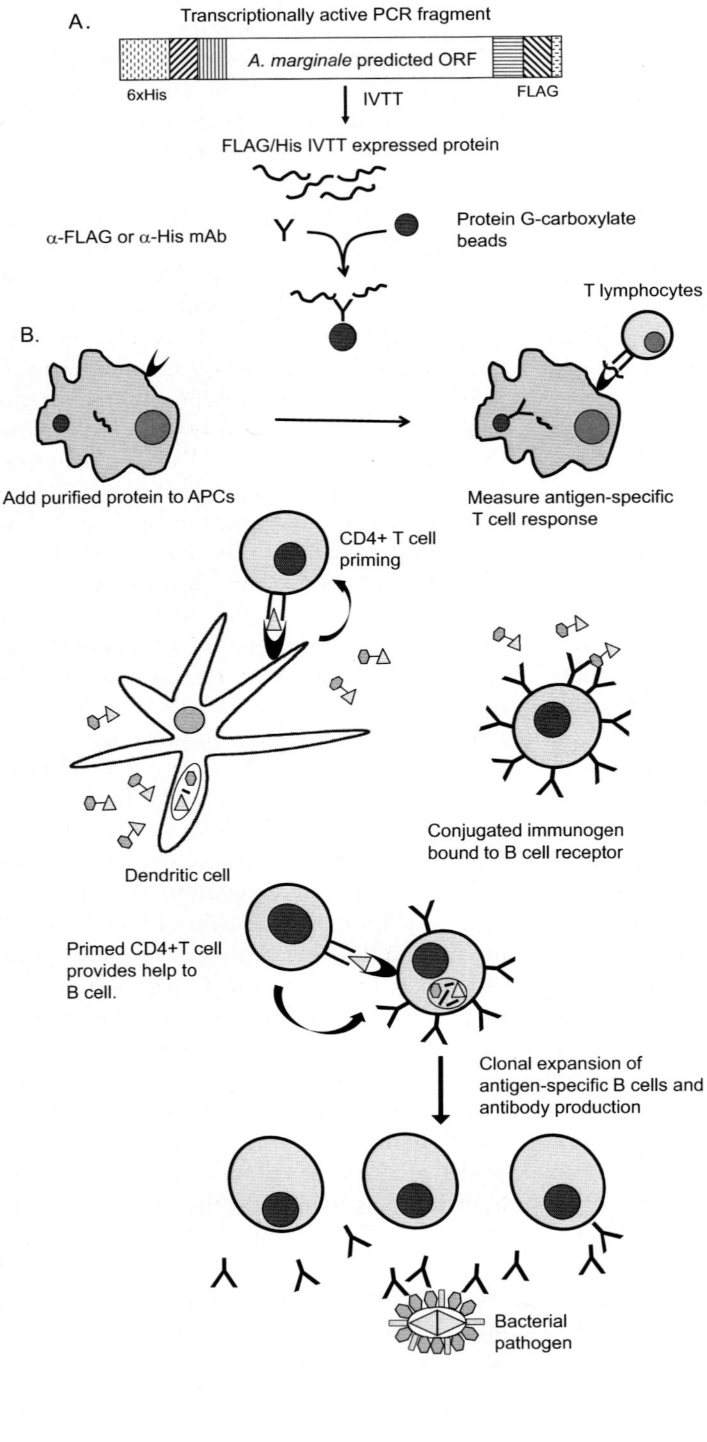

FIGURE 4 The physical association between a T-cell epitope (Δ) and B-cell epitope (⬢) can enhance the antibody response to the B-cell epitope. Upon exposure to the conjugated antigens, dendritic cells take up, process, and present the T-cell epitope on MHC class II molecules. Helper CD4$^+$ T cells that recognize and bind the peptide become primed by the APC. Concurrently, B cells bind the B-cell epitope, internalize the entire construct, and process the antigen via the MHC class II pathway. The T-cell epitope is then displayed in the context of MHC class II molecules on the surface of the B cell. The previously primed T cell can recognize and bind this antigen and thus provide help to the B cell, resulting in B-cell activation, isotype switching, and production of high-affinity antibodies targeting the B-cell epitope. doi:10.1128/9781555817336.ch11.f4

with either the $CD4^+$-T-cell epitopes from MSP1a linked to MSP1b1 or MSP1a and MSP1b1 administered separately. The animals that received the linked molecules had significantly enhanced IgG titers to MSP1b1 compared with the animals that received the antigens in separate sites (Macmillan et al., 2008). Similarly, animals immunized with a group of covalently cross-linked OMPs had significantly higher titers (≥10,000) than did animals immunized with the same variety of OMPs in which the cross-links had been reduced (≤300) such that the proteins were individualized (S. M. Noh, unpublished data).

Many other methods of vaccine delivery, as well as an increasing variety of adjuvants that can boost and bias the immune response in predictable ways, have been or are in the process of being developed. These delivery systems and adjuvants include viral vectors, viruslike particles, liposomes, emulsions, and adjuvant systems (Bachmann and Jennings, 2010; Leroux-Roels, 2010; Zepp, 2010). The promise of these technologies has yet to be fully realized; however, these technologies have been successfully used to develop new vaccines and improve the safety and efficacy of existing vaccines (Bachmann and Jennings, 2010). For example, the development of a vaccine targeting human papillomavirus (HPV) has added cervical cancer to the list of vaccine-preventable diseases (Frazer et al., 2011). The HPV vaccine is composed of viruslike particles that self-assemble into a surface coat that mimics the structure of the viral particle. In the case of HPV, these viruslike particles are able to induce systemic immunity to a virus that is typically poorly immunogenic and confined to the mucosa of the reproductive tract. These technologies have also been instrumental in the development of a malaria vaccine currently being tested in clinical trials (Cohen et al., 2010; Kester et al., 2009). While the efficacy of the malaria vaccine has yet to be fully demonstrated, the trials to date have been promising. This recombinant vaccine, RTS,S, is composed of the carboxy-terminal half of the *Plasmodium falciparum* circumsporozoite protein (designated RT) fused to a surface protein of the hepatitis B virus (S) and coexpressed with the hepatitis B surface protein. The inclusion of the viral surface protein allows for self-assembly into viruslike particles. Essential in the development of this vaccine is the induction of an adequate and appropriate immune response through the use a combination of adjuvants, termed an adjuvant system, to stimulate both innate immunity and the appropriate adaptive immune response.

These examples illustrate the potential for the rational use of these technologies to continue to lengthen the list of vaccine-preventable diseases. The increasing variety of vaccine platforms and effective adjuvants combined with comprehensive and efficient methods to identify protective antigens will serve as the basis for the development of the next generation of vaccines, particularly those that target antigenically variable pathogens that are able to modulate the host immune response and establish persistent infection.

REFERENCES

Aaby, P., J. Bukh, D. Kronborg, I. M. Lisse, and M. C. da Silva. 1990. Delayed excess mortality after exposure to measles during the first six months of life. *Am. J. Epidemiol.* **132:**211–219.

Abbott, J. R., G. H. Palmer, C. J. Howard, J. C. Hope, and W. C. Brown. 2004. *Anaplasma marginale* major surface protein 2 $CD4^+$-T-cell epitopes are evenly distributed in conserved and hypervariable regions (HVR), whereas linear B-cell epitopes are predominantly located in the HVR. *Infect. Immun.* **72:**7360–7366.

Abbott, J. R., G. H. Palmer, K. A. Kegerreis, P. F. Hetrick, C. J. Howard, J. C. Hope, and W. C. Brown. 2005. Rapid and long-term disappearance of $CD4^+$ T lymphocyte responses specific for *Anaplasma marginale* major surface protein-2 (MSP2) in MSP2 vaccinates following challenge with live *A. marginale*. *J. Immunol.* **174:**6702–6715.

Agnes, J. T., K. A. Brayton, M. Lafollett, J. Norimine, W. C. Brown, and G. H. Palmer. 2011. Identification of *Anaplasma marginale* outer membrane protein antigens conserved between *sensu stricto* strains and the live *A. marginale* ss. *centrale* vaccine. *Infect. Immun.* **79:**1311–1318.

Agnes, J. T., D. Herndon, M. W. Ueti, S. S. Ramabu, M. Evans, K. A. Brayton, and G. H. Palmer. 2010. Association of pathogen strain-specific gene transcription and transmission efficiency phenotype of *Anaplasma marginale*. *Infect. Immun.* **78:**2446–2453.

Aguero-Rosenfeld, M. E., L. Donnarumma, L. Zentmaier, J. Jacob, M. Frey, R. Noto, C. A. Carbonaro, and G. P. Wormser. 2002. Seroprevalence of antibodies that react with *Anaplasma phagocytophila*, the agent of human granulocytic ehrlichiosis, in different populations in Westchester County, New York. *J. Clin. Microbiol.* **40:**2612–2615.

Akkoyunlu, M., and E. Fikrig. 2000. Gamma interferon dominates the murine cytokine response to the agent of human granulocytic ehrlichiosis and helps to control the degree of early rickettsemia. *Infect. Immun.* **68:**1827–1833.

Allsopp, B. A. 2010. Natural history of *Ehrlichia ruminantium*. *Vet. Parasitol.* **167:**123–135.

Alvarez-Martinez, C. E., and P. J. Christie. 2009. Biological diversity of prokaryotic type IV secretion systems. *Microbiol. Mol. Biol. Rev.* **73:**775–808.

Anziani, O. S., H. D. Tarabla, C. A. Ford, and C. Galleto. 1987. Vaccination with *Anaplasma centrale*: response after an experimental challenge with *Anaplasma marginale*. *Trop. Anim. Health Prod.* **19:**83–87.

Araújo, F. R., C. M. Costa, C. A. Ramos, T. A. Farias, I. I. de Souza, E. S. Melo, C. Elisei, G. M. Rosinha, C. O. Soares, S. P. Fragoso, and A. H. Fonseca. 2008. IgG and IgG2 antibodies from cattle naturally infected with *Anaplasma marginale* recognize the recombinant vaccine candidate antigens VirB9, VirB10, and elongation factor-Tu. *Mem. Inst. Oswaldo Cruz* **103:**186–190.

Artursson, K., A. Gunnarsson, U. B. Wikstrom, and E. O. Engvall. 1999. A serological and clinical follow-up in horses with confirmed equine granulocytic ehrlichiosis. *Equine Vet. J.* **31:**473–477.

Bachmann, M. F., and G. T. Jennings. 2010. Vaccine delivery: a matter of size, geometry, kinetics and molecular patterns. *Nat. Rev. Immunol.* **10:**787–796.

Bakken, J. S., P. Goellner, M. Van Etten, D. Z. Boyle, O. L. Swonger, S. Mattson, J. Krueth, R. L. Tilden, K. Asanovich, J. Walls, and J. S. Dumler. 1998. Seroprevalence of human granulocytic ehrlichiosis among permanent residents of northwestern Wisconsin. *Clin. Infect. Dis.* **27:**1491–1496.

Bakken, J. S., J. Krueth, C. Wilson-Nordskog, R. L. Tilden, K. Asanovich, and J. S. Dumler. 1996. Clinical and laboratory characteristics of human granulocytic ehrlichiosis. *JAMA* **275:**199–205.

Baldo, L., N. Lo, and J. H. Werren. 2005. Mosaic nature of the *Wolbachia* surface protein. *J. Bacteriol.* **187:**5406–5418.

Barbet, A. F., A. Lundgren, J. Yi, F. R. Rurangirwa, and G. H. Palmer. 2000. Antigenic variation of *Anaplasma marginale* by expression of MSP2 mosaics. *Infect. Immun.* **68:**6133–6138.

Barbet, A. F., P. F. Meeus, M. Belanger, M. V. Bowie, J. Yi, A. M. Lundgren, A. R. Alleman, S. J. Wong, F. K. Chu, U. G. Munderloh, and S. D. Jauron. 2003. Expression of multiple outer membrane protein sequence variants from a single genomic locus of *Anaplasma phagocytophilum*. *Infect. Immun.* **71:**1706–1718.

Barbet, A. F., W. M. Whitmire, S. M. Kamper, B. H. Simbi, R. R. Ganta, A. L. Moreland, D. M. Mwangi, T. C. McGuire, and S. M. Mahan. 2001. A subset of *Cowdria ruminantium* genes important for immune recognition and protection. *Gene* **275:**287–298.

Bezuidenhout, J. D. 1985. The epizootiology and control of heartwater and other tick-borne diseases of cattle in South Africa. *Onderstepoort J. Vet. Res.* **52:**211–214.

Blish, C. A., O. C. Dogan, N. R. Derby, M. A. Nguyen, B. Chohan, B. A. Richardson, and J. Overbaugh. 2008. Human immunodeficiency virus type 1 superinfection occurs despite relatively robust neutralizing antibody responses. *J. Virol.* **82:**12094–12103.

Bock, R. E., and A. J. de Vos. 2001. Immunity following use of Australian tick fever vaccine: a review of the evidence. *Aust. Vet. J.* **79:**832–839.

Braig, H. R., W. Zhou, S. L. Dobson, and S. L. O'Neill. 1998. Cloning and characterization of a gene encoding the major surface protein of the bacterial endosymbiont *Wolbachia pipientis*. *J. Bacteriol.* **180:**2373–2378.

Braquart-Varnier, C., M. Lachat, J. Herbiniere, M. Johnson, Y. Caubet, D. Bouchon, and M. Sicard. 2008. *Wolbachia* mediate variation of host immunocompetence. *PLoS One* **3:**e3286.

Brayton, K. A., L. S. Kappmeyer, D. R. Herndon, M. J. Dark, D. L. Tibbals, G. H. Palmer, T. C. McGuire, and D. P. Knowles, Jr. 2005. Complete genome sequencing of *Anaplasma marginale* reveals that the surface is skewed to two superfamilies of outer membrane proteins. *Proc. Natl. Acad. Sci. USA* **102:**844–849.

Brayton, K. A., D. P. Knowles, T. C. McGuire, and G. H. Palmer. 2001. Efficient use of a small genome to generate antigenic diversity in tick-borne ehrlichial pathogens. *Proc. Natl. Acad. Sci. USA* **98:**4130–4135.

Brayton, K. A., G. H. Palmer, A. Lundgren, J. Yi, and A. F. Barbet. 2002. Antigenic variation of *Anaplasma marginale msp2* occurs by combinatorial gene conversion. *Mol. Microbiol.* **43:**1151–1159.

Brayton, K. A., S. W. Vogel, and B. A. Allsopp. 1998. Expression library immunization to identify protective antigens from *Cowdria ruminantium*. *Ann. N. Y. Acad. Sci.* **849:**369–371.

Breitschwerdt, E. B., B. C. Hegarty, and S. I. Hancock. 1998. Doxycycline hyclate treatment of experimental canine ehrlichiosis followed by challenge inoculation with two *Ehrlichia canis* strains. *Antimicrob. Agents Chemother.* **42:**362–368.

Brodie, T. A., P. H. Holmes, and G. M. Urquhart. 1986. Some aspects of tick-borne diseases of British sheep. *Vet. Rec.* **118:**415–418.

Brown, C. C., D. C. Baker, and I. K. Barker. 2007. Salmon poisoning disease, p. 258–259. *In* M. G. Maxie (ed.), *Jubb, Kennedy, and Palmer's Pathology of Domestic Animals*, 5th ed., vol. 2. Saunders, Philadelphia, PA.

Brown, W. C., T. F. McElwain, G. H. Palmer, S. E. Chantler, and D. M. Estes. 1999. Bovine CD4+ T-lymphocyte clones specific for rhoptry-associated protein 1 of *Babesia bigemina* stimulate enhanced immunoglobulin G1 (IgG1) and IgG2 synthesis. *Infect. Immun.* **67:**155–164.

Brown, W. C., T. C. McGuire, W. Mwangi, K. A. Kegerreis, H. Macmillan, H. A. Lewin, and G. H. Palmer. 2002. Major histocompatibility complex class II DR-restricted memory CD4+ T lymphocytes recognize conserved immunodominant epitopes of *Anaplasma marginale* major surface protein 1a. *Infect. Immun.* **70:**5521–5532.

Brown, W. C., V. Shkap, D. Zhu, T. C. McGuire, W. Tuo, T. F. McElwain, and G. H. Palmer. 1998a. CD4+ T-lymphocyte and immunoglobulin G2 responses in calves immunized with *Anaplasma marginale* outer membranes and protected against homologous challenge. *Infect. Immun.* **66:**5406–5413.

Brown, W. C., D. Zhu, V. Shkap, T. C. McGuire, E. F. Blouin, K. M. Kocan, and G. H. Palmer. 1998b. The repertoire of *Anaplasma marginale* antigens recognized by CD4+ T-lymphocyte clones from protectively immunized cattle is diverse and includes major surface protein 2 (MSP-2) and MSP-3. *Infect. Immun.* **66:**5414–5422.

Bullock, P. M., T. R. Ames, R. A. Robinson, B. Greig, M. A. Mellencamp, and J. S. Dumler. 2000. *Ehrlichia equi* infection of horses from Minnesota and Wisconsin: detection of seroconversion and acute disease investigation. *J. Vet. Intern. Med.* **14:**252–257.

Butler, C. M., A. M. Nijhof, F. Jongejan, and J. H. van der Kolk. 2008. *Anaplasma phagocytophilum* infection in horses in the Netherlands. *Vet. Rec.* **162:**216–217.

Cascales, E., and P. J. Christie. 2003. The versatile bacterial type IV secretion systems. *Nat. Rev. Microbiol.* **1:**137–149.

Caturegli, P., K. M. Asanovich, J. J. Walls, J. S. Bakken, J. E. Madigan, V. L. Popov, and J. S. Dumler. 2000. *ankA*: an *Ehrlichia phagocytophila* group gene encoding a cytoplasmic protein antigen with ankyrin repeats. *Infect. Immun.* **68:**5277–5283.

Chen, S. M., J. S. Dumler, H. M. Feng, and D. H. Walker. 1994. Identification of the antigenic constituents of *Ehrlichia chaffeensis*. *Am. J. Trop. Med. Hyg.* **50:**52–58.

Cohen, J., V. Nussenzweig, R. Nussenzweig, J. Vekemans, and A. Leach. 2010. From the circumsporozoite protein to the RTS,S/AS candidate vaccine. *Hum. Vaccin.* **6:**90–96.

Collins, N. E., J. Liebenberg, E. P. de Villiers, K. A. Brayton, E. Louw, A. Pretorius, F. E. Faber, H. van Heerden, A. Josemans, M. van Kleef, H. C. Steyn, M. F. van Strijp, E. Zweygarth, F. Jongejan, J. C. Maillard, D. Berthier, M. Botha, F. Joubert, C. H. Corton, N. R. Thomson, M. T. Allsopp, and B. A. Allsopp. 2005. The genome of the heartwater agent *Ehrlichia ruminantium* contains multiple tandem repeats of actively variable copy number. *Proc. Natl. Acad. Sci. USA* **102:**838–843.

Collins, N. E., A. Pretorius, M. van Kleef, K. A. Brayton, M. T. Allsopp, E. Zweygarth, and B. A. Allsopp. 2003a. Development of improved attenuated and nucleic acid vaccines for heartwater. *Dev. Biol. (Basel)* **114:**121–136.

Collins, N. E., A. Pretorius, M. van Kleef, K. A. Brayton, E. Zweygarth, and B. A. Allsopp. 2003b. Development of improved vaccines for heartwater. *Ann. N. Y. Acad. Sci.* **990:**474–484.

Cordes, D. O., B. D. Perry, Y. Rikihisa, and W. R. Chickering. 1986. Enterocolitis caused by *Ehrlichia* sp. in the horse (Potomac horse fever). *Vet. Pathol.* **23:**471–477.

Dallo, S. F., T. R. Kannan, M. W. Blaylock, and J. B. Baseman. 2002. Elongation factor Tu and E1 β subunit of pyruvate dehydrogenase complex act as fibronectin binding proteins in *Mycoplasma pneumoniae*. *Mol. Microbiol.* **46:**1041–1051.

Davies, D. H., X. Liang, J. E. Hernandez, A. Randall, S. Hirst, Y. Mu, K. M. Romero, T. T. Nguyen, M. Kalantari-Dehaghi, S. Crotty, P. Baldi, L. P. Villarreal, and P. L. Felgner. 2005. Profiling the humoral immune response to infection by using proteome microarrays: high-throughput vaccine and diagnostic antigen discovery. *Proc. Natl. Acad. Sci. USA* **102:**547–552.

Dierberg, K. L., and J. S. Dumler. 2006. Lymph node hemophagocytosis in rickettsial diseases: a pathogenetic role for CD8 T lymphocytes in human monocytic ehrlichiosis (HME)? *BMC Infect. Dis.* **6:**121.

Domenech, J., J. Lubroth, and K. Sumption. 2010. Immune protection in animals: the examples of rinderpest and foot-and-mouth disease. *J. Comp. Pathol.* **142**(Suppl. 1):S120–S124.

Doyle, C. K., K. A. Nethery, V. L. Popov, and J. W. McBride. 2006. Differentially expressed and secreted major immunoreactive protein orthologs of *Ehrlichia canis* and *E. chaffeensis* elicit early antibody responses to epitopes on glycosylated tandem repeats. *Infect. Immun.* **74:**711–720.

Dumler, J. S., and J. S. Bakken. 1998. Human ehrlichioses: newly recognized infections transmitted by ticks. *Annu. Rev. Med.* **49:**201–213.

Dumler, J. S., A. F. Barbet, C. P. Bekker, G. A. Dasch, G. H. Palmer, S. C. Ray, Y. Rikihisa, and F. R. Rurangirwa. 2001. Reorganization of genera in the families *Rickettsiaceae* and *Anaplasmataceae* in the order *Rickettsiales*: unification of some species of *Ehrlichia* with *Anaplasma*, *Cowdria* with *Ehrlichia* and *Ehrlichia* with *Neorickettsia*, descriptions of six new species combinations and designation of *Ehrlichia equi* and 'HGE agent' as subjective synonyms of *Ehrlichia phagocytophila*. *Int. J. Syst. Evol. Microbiol.* **51:**2145–2165.

Dumler, J. S., K. S. Choi, J. C. Garcia-Garcia, N. S. Barat, D. G. Scorpio, J. W. Garyu, D. J. Grab, and J. S. Bakken. 2005. Human granulocytic anaplasmosis and *Anaplasma phagocytophilum*. *Emerg. Infect. Dis.* **11:**1828–1834.

Dumler, J. S., J. E. Madigan, N. Pusterla, and J. S. Bakken. 2007. Ehrlichioses in humans: epidemiology, clinical presentation, diagnosis, and treatment. *Clin. Infect. Dis.* **45**(Suppl. 1):S45–S51.

Dumler, J. S., E. R. Trigiani, J. S. Bakken, M. E. Aguero-Rosenfeld, and G. P. Wormser. 2000. Serum cytokine responses during acute human granulocytic ehrlichiosis. *Clin. Diagn. Lab. Immunol.* **7:**6–8.

Dunning Hotopp, J. C., M. Lin, R. Madupu, J. Crabtree, S. V. Angiuoli, J. Eisen, R. Seshadri, Q. Ren, M. Wu, T. R. Utterback, S. Smith, M. Lewis, H. Khouri, C. Zhang, H. Niu, Q. Lin, N. Ohashi, N. Zhi, W. Nelson, L. M. Brinkac, R. J. Dodson, M. J. Rosovitz, J. Sundaram, S. C. Daugherty, T. Davidsen, A. S. Durkin, M. Gwinn, D. H. Haft, J. D. Selengut, S. A. Sullivan, N. Zafar, L. Zhou, F. Benahmed, H. Forberger, R. Halpin, S. Mulligan, J. Robinson, O. White, Y. Rikihisa, and H. Tettelin. 2006. Comparative genomics of emerging human ehrlichiosis agents. *PLoS Genet.* **2:** e21.

du Plessis, J. L. 1970. Immunity in heartwater. I. A preliminary note on the role of serum antibodies. *Onderstepoort J. Vet. Res.* **37:**147–149.

du Plessis, J. L., F. T. Potgieter, and L. van Gas. 1990. An attempt to improve the immunization of sheep against heart-water by using different combinations of 3 stocks of *Cowdria ruminantium*. *Onderstepoort J. Vet. Res.* **57:**205–208.

du Plessis, J. L., L. van Gas, J. A. Olivier, and J. D. Bezuidenhout. 1989. The heterogenicity of *Cowdria ruminantium* stocks: cross-immunity and serology in sheep and pathogenicity to mice. *Onderstepoort J. Vet. Res.* **56:**195–201.

Dutra, F., L. F. Schuch, E. Delucchi, B. R. Curcio, H. Coimbra, M. B. Raffi, O. Dellagostin, and F. Riet-Correa. 2001. Equine monocytic ehrlichiosis (Potomac horse fever) in horses in Uruguay and southern Brazil. *J. Vet. Diagn. Invest.* **13:**433–437.

Ekiert, D. C., G. Bhabha, M. A. Elsliger, R. H. Friesen, M. Jongeneelen, M. Throsby, J. Goudsmit, and I. A. Wilson. 2009. Antibody recognition of a highly conserved influenza virus epitope. *Science* **324:**246–251.

Estes, D. M., N. M. Closser, and G. K. Allen. 1994. IFN-γ stimulates IgG_2 production from bovine B cells costimulated with anti-μ and mitogen. *Cell. Immunol.* **154:**287–295.

Esteves, I., D. Martinez, and P. Totté. 2004a. Identification of *Ehrlichia ruminantium* (Gardel strain) IFN-γ inducing proteins after vaccination with a killed vaccine. *Vet. Microbiol.* **100:**233–240.

Esteves, I., N. Vachiéry, D. Martinez, and P. Totté. 2004b. Analysis of *Ehrlichia ruminantium*-specific T1/T2 responses during vaccination with a protective killed vaccine and challenge of goats. *Parasite Immunol.* **26:**95–103.

Faria, J. L., T. D. Munhoz, C. F. João, G. Vargas-Hernández, M. R. André, W. A. Pereira, R. Z. Machado, and M. Tinucci-Costa. 2011. *Ehrlichia canis* (Jaboticabal strain) induces the expression of TNF-α in leukocytes and splenocytes of experimentally infected dogs. *Rev. Bras. Parasitol. Vet.* **20:**71–74.

Felek, S., H. Huang, and Y. Rikihisa. 2003. Sequence and expression analysis of *virB9* of the type IV secretion system of *Ehrlichia canis* strains in ticks, dogs, and cultured cells. *Infect. Immun.* **71:**6063–6067.

Fenn, K., and M. Blaxter. 2006. *Wolbachia* genomes: revealing the biology of parasitism and mutualism. *Trends Parasitol.* **22:**60–65.

Frazer, I. H., G. R. Leggatt, and S. R. Mattarollo. 2011. Prevention and treatment of papillomavirus-related cancers through immunization. *Annu. Rev. Immunol.* **29:**111–138.

French, D. M., W. C. Brown, and G. H. Palmer. 1999. Emergence of *Anaplasma marginale* antigenic variants during persistent rickettsemia. *Infect. Immun.* **67:**5834–5840.

Futse, J. E., K. A. Brayton, M. J. Dark, D. P. Knowles, Jr., and G. H. Palmer. 2008. Superinfection as a driver of genomic diversification in antigenically variant pathogens. *Proc. Natl. Acad. Sci. USA* **105:**2123–2127.

Gallimore, A., T. Dumrese, H. Hengartner, R. M. Zinkernagel, and H. G. Rammensee. 1998a. Protective immunity does not correlate with the hierarchy of virus-specific cytotoxic T cell responses to naturally processed peptides. *J. Exp. Med.* **187:**1647–1657.

Gallimore, A., J. Hombach, T. Dumrese, H. G. Rammensee, R. M. Zinkernagel, and H. Hengartner. 1998b. A protective cytotoxic T cell response to a subdominant epitope is influenced by the stability of the MHC class I/peptide complex and the overall spectrum of viral peptides generated within infected cells. *Eur. J. Immunol.* **28:**3301–3311.

Ganta, R. R., L. Peddireddi, G. M. Seo, S. E. Dedonder, C. Cheng, and S. K. Chapes. 2009. Molecular characterization of *Ehrlichia* interactions with tick cells and macrophages. *Front. Biosci.* **14:**3259–3273.

Gat, O., H. Grosfeld, N. Ariel, I. Inbar, G. Zaide, Y. Broder, A. Zvi, T. Chitlaru, Z. Altboum, D. Stein, S. Cohen, and A. Shafferman. 2006. Search for *Bacillus anthracis* potential vaccine candidates by a functional genomic-serologic screen. *Infect. Immun.* **74:**3987–4001.

Ge, Y., and Y. Rikihisa. 2007a. Identification of novel surface proteins of *Anaplasma phagocytophilum* by affinity purification and proteomics. *J. Bacteriol.* **189:**7819–7828.

Ge, Y., and Y. Rikihisa. 2007b. Surface-exposed proteins of *Ehrlichia chaffeensis*. *Infect. Immun.* **75:** 3833–3841.

Gillespie, J. J., K. A. Brayton, K. P. Williams, M. A. Diaz, W. C. Brown, A. F. Azad, and B. W. Sobral. 2010. Phylogenomics reveals a diverse *Rickettsiales* type IV secretion system. *Infect. Immun.* **78:**1809–1823.

Gorham, J. R., W. J. Foreyt, and J. E. Sykes. 2012. *Neorickettsia helminthoeca* infection (salmon poisoning disease), p. 220–226. *In* C. E. Green (ed.), *Infectious Diseases of the Dog and Cat*, 4th ed. Saunders, St. Louis, MO.

Granato, D., G. E. Bergonzelli, R. D. Pridmore, L. Marvin, M. Rouvet, and I. E. Corthesy-Theulaz. 2004. Cell surface-associated elongation factor Tu mediates the attachment of *Lactobacillus johnsonii* NCC533 (La1) to human intestinal cells and mucins. *Infect. Immun.* **72:**2160–2169.

Granquist, E. G., S. Stuen, L. Crosby, A. M. Lundgren, A. R. Alleman, and A. F. Barbet. 2010. Variant-specific and diminishing immune responses towards the highly variable MSP2(P44) outer membrane protein of *Anaplasma phagocytophilum* during persistent infection in lambs. *Vet. Immunol. Immunopathol.* **133:**117–124.

Granquist, E. G., S. Stuen, A. M. Lundgren, M. Braten, and A. F. Barbet. 2008. Outer membrane protein sequence variation in lambs experimentally infected with *Anaplasma phagocytophilum*. *Infect. Immun.* **76:**120–126.

Greene, C. E., L. Kidd, and E. B. Breitschwerdt. 2012. Rocky mountain and mediterranean spotted fevers, cat-flea typhuslike illness, rickettsialpox and typhus, p. 259–270. *In* C. E. Green (ed.), *Infectious Diseases of the Dog and Cat*, 4th ed. Saunders, St. Louis, MO.

Greene, C. E., and M. Vandevelde. 2012. Canine distemper, p. 25–42. *In* C. E. Green (ed.), *Infectious Diseases of the Dog and Cat*, 4th ed. Saunders, St. Louis, MO.

Griffin, D. E. 2010. Measles virus-induced suppression of immune responses. *Immunol. Rev.* **236:**176–189.

Gueye, A., F. Jongejan, M. Mbengue, A. Diouf, and G. Uilenberg. 1994. Field trial of an attenuated vaccine against heartwater disease. *Rev. Elev. Med. Vet. Pays Trop.* **47:**401–404. (In French.)

Han, S., J. Norimine, K. A. Brayton, G. H. Palmer, G. A. Scoles, and W. C. Brown. 2010. *Anaplasma marginale* infection with persistent high-load bacteremia induces a dysfunctional memory $CD4^+$ T lymphocyte response but sustained high IgG titers. *Clin. Vaccine Immunol.* **17:**1881–1890.

Han, S., J. Norimine, G. H. Palmer, W. Mwangi, K. K. Lahmers, and W. C. Brown. 2008. Rapid deletion of antigen-specific $CD4^+$ T cells following infection represents a strategy of immune evasion and persistence for *Anaplasma marginale*. *J. Immunol.* **181:**7759–7769.

Han, T., and W. A. Marasco. 2011. Structural basis of influenza virus neutralization. *Ann. N. Y. Acad. Sci.* **1217:**178–190.

Harrus, S., T. Waner, H. Bark, F. Jongejan, and A. W. Cornelissen. 1999. Recent advances in determining the pathogenesis of canine monocytic ehrlichiosis. *J. Clin. Microbiol.* **37:**2745–2749.

Headley, S. A., D. G. Scorpio, O. Vidotto, and J. S. Dumler. 2011. *Neorickettsia helminthoeca* and salmon poisoning disease: a review. *Vet. J.* **187:**165–173.

Heeb, H. L., M. J. Wilkerson, R. Chun, and R. R. Ganta. 2003. Large granular lymphocytosis, lymphocyte subset inversion, thrombocytopenia, dysproteinemia, and positive *Ehrlichia* serology in a dog. *J. Am. Anim. Hosp. Assoc.* **39:**379–384.

Hilleman, M. R. 1992. The dilemma of AIDS vaccine and therapy. Possible clues from comparative pathogenesis with measles. *AIDS Res. Hum. Retroviruses* **8:**1743–1747.

Hilleman, M. R. 2001. Current overview of the pathogenesis and prophylaxis of measles with focus on practical implications. *Vaccine* **20:**651–665.

Hise, A. G., K. Daehnel, I. Gillette-Ferguson, E. Cho, H. F. McGarry, M. J. Taylor, D. T. Golenbock, K. A. Fitzgerald, J. W. Kazura, and E. Pearlman. 2007. Innate immune responses to endosymbiotic *Wolbachia* bacteria in *Brugia malayi* and *Onchocerca volvulus* are dependent on TLR2,

TLR6, MyD88, and Mal, but not TLR4, TRIF, or TRAM. *J. Immunol.* **178:**1068–1076.

IJdo, J. W., W. Sun, Y. Zhang, L. A. Magnarelli, and E. Fikrig. 1998. Cloning of the gene encoding the 44-kilodalton antigen of the agent of human granulocytic ehrlichiosis and characterization of the humoral response. *Infect. Immun.* **66:**3264–3269.

IJdo, W., C. Wu, S. R. Telford III, and E. Fikrig. 2002. Differential expression of the p44 gene family in the agent of human granulocytic ehrlichiosis. *Infect. Immun.* **70:**5295–5298.

IJdo, W., Y. Zhang, E. Hodzic, L. A. Magnarelli, M. L. Wilson, S. R. Telford III, S. W. Barthold, and E. Fikrig. 1997. The early humoral response in human granulocytic ehrlichiosis. *J. Infect. Dis.* **176:**687–692.

Ingolotti, M., O. Kawalekar, D. J. Shedlock, K. Muthumani, and D. B. Weiner. 2010. DNA vaccines for targeting bacterial infections. *Expert Rev. Vaccines* **9:**747–763.

Ismail, N., K. C. Bloch, and J. W. McBride. 2010. Human ehrlichiosis and anaplasmosis. *Clin. Lab. Med.* **30:**261–292.

Ismail, N., E. C. Crossley, H. L. Stevenson, and D. H. Walker. 2007. Relative importance of T-cell subsets in monocytotropic ehrlichiosis: a novel effector mechanism involved in *Ehrlichia*-induced immunopathology in murine ehrlichiosis. *Infect. Immun.* **75:**4608–4620.

Ismail, N., L. Soong, J. W. McBride, G. Valbuena, J. P. Olano, H. M. Feng, and D. H. Walker. 2004. Overproduction of TNF-α by $CD8^+$ type 1 cells and down-regulation of IFN-γ production by $CD4^+$ Th1 cells contribute to toxic shock-like syndrome in an animal model of fatal monocytotropic ehrlichiosis. *J. Immunol.* **172:**1786–1800.

Jacobson, G. R., and J. P. Rosenbusch. 1976. Abundance and membrane association of elongation factor Tu in *E. coli*. *Nature* **261:**23–26.

Jones, S. L. 2004. Inflammatory diseases of the gastrointestinal tract causing diarrhea, p. 884–913. *In* S. M. Reed, W. M. Bayly, and D. C. Sellon (ed.), *Equine Internal Medicine*. Sanders, St. Louis, MO.

Jongejan, F. 1991. Protective immunity to heartwater (*Cowdria ruminantium* infection) is acquired after vaccination with in vitro-attenuated rickettsiae. *Infect. Immun.* **59:**729–731.

Kester, K. E., J. F. Cummings, O. Ofori-Anyinam, C. F. Ockenhouse, U. Krzych, P. Moris, R. Schwenk, R. A. Nielsen, Z. Debebe, E. Pinelis, L. Juompan, J. Williams, M. Dowler, V. A. Stewart, R. A. Wirtz, M. C. Dubois, M. Lievens, J. Cohen, W. R. Ballou, and D. G. Heppner, Jr. 2009. Randomized, double-blind, phase 2a trial of falciparum malaria vaccines RTS,S/AS01B and RTS,S/AS02A in malaria-naive adults: safety, efficacy, and immunologic associates of protection. *J. Infect. Dis.* **200:**337–346.

Kim, H. Y., J. Mott, N. Zhi, T. Tajima, and Y. Rikihisa. 2002. Cytokine gene expression by peripheral blood leukocytes in horses experimentally infected with *Anaplasma phagocytophila*. *Clin. Diagn. Lab. Immunol.* **9:**1079–1084.

Kimberling, C. V. 1988. *Jensen and Swift's Diseases of Sheep*, p. 43–45. Lea and Febiger, Philadelphia, PA.

Kramer, L. H., F. Tamarozzi, R. Morchon, J. Lopez-Belmonte, C. Marcos-Atxutegi, R. Martin-Pacho, and F. Simon. 2005. Immune response to and tissue localization of the *Wolbachia* surface protein (WSP) in dogs with natural heartworm (*Dirofilaria immitis*) infection. *Vet. Immunol. Immunopathol.* **106:**303–308.

Krigel, Y., E. Pipano, and V. Shkap. 1992. Duration of carrier state following vaccination with live *Anaplasma centrale*. *Trop. Anim. Health Prod.* **24:**209–210.

Kubota-Koketsu, R., H. Mizuta, M. Oshita, S. Ideno, M. Yunoki, M. Kuhara, N. Yamamoto, Y. Okuno, and K. Ikuta. 2009. Broad neutralizing human monoclonal antibodies against influenza virus from vaccinated healthy donors. *Biochem. Biophys. Res. Commun.* **387:**180–185.

Ladbury, G. A., S. Stuen, R. Thomas, K. J. Bown, Z. Woldehiwet, E. G. Granquist, K. Bergstrom, and R. J. Birtles. 2008. Dynamic transmission of numerous *Anaplasma phagocytophilum* genotypes among lambs in an infected sheep flock in an area of anaplasmosis endemicity. *J. Clin. Microbiol.* **46:**1686–1691.

Lepidi, H., J. E. Bunnell, M. E. Martin, J. E. Madigan, S. Stuen, and J. S. Dumler. 2000. Comparative pathology, and immunohistology associated with clinical illness after *Ehrlichia phagocytophila*-group infections. *Am. J. Trop. Med. Hyg.* **62:**29–37.

Leroux-Roels, G. 2010. Unmet needs in modern vaccinology: adjuvants to improve the immune response. *Vaccine* **28**(Suppl. 3)**:**C25–C36.

Li, J. S., F. Chu, A. Reilly, and G. M. Winslow. 2002. Antibodies highly effective in SCID mice during infection by the intracellular bacterium *Ehrlichia chaffeensis* are of picomolar affinity and exhibit preferential epitope and isotype utilization. *J. Immunol.* **169:**1419–1425.

Lin, M., and Y. Rikihisa. 2003. *Ehrlichia chaffeensis* and *Anaplasma phagocytophilum* lack genes for lipid A biosynthesis and incorporate cholesterol for their survival. *Infect. Immun.* **71:**5324–5331.

Lin, Q., Y. Rikihisa, N. Ohashi, and N. Zhi. 2003. Mechanisms of variable *p44* expression by *Anaplasma phagocytophilum*. *Infect. Immun.* **71:**5650–5661.

Little, S. E. 2010. Ehrlichiosis and anaplasmosis in dogs and cats. *Vet. Clin. North Am. Small Anim. Pract.* **40:**1121–1140.

Löhr, C. V., K. A. Brayton, V. Shkap, T. Molad, A. F. Barbet, W. C. Brown, and G. H. Palmer. 2002. Expression of *Anaplasma marginale* major surface protein 2 operon-associated proteins during mammalian and arthropod infection. *Infect. Immun.* **70:**6005–6012.

Lopez, J. E., P. A. Beare, R. A. Heinzen, J. Norimine, K. K. Lahmers, G. H. Palmer, and W. C. Brown. 2008. High-throughput identification of T-lymphocyte antigens from *Anaplasma marginale* expressed using *in vitro* transcription and translation. *J. Immunol. Methods* **332:**129–141.

Lopez, J. E., G. H. Palmer, K. A. Brayton, M. J. Dark, S. E. Leach, and W. C. Brown. 2007. Immunogenicity of *Anaplasma marginale* type IV secretion system proteins in a protective outer membrane vaccine. *Infect. Immun.* **75:**2333–2342.

Lopez, J. E., W. F. Siems, G. H. Palmer, K. A. Brayton, T. C. McGuire, J. Norimine, and W. C. Brown. 2005. Identification of novel antigenic proteins in a complex *Anaplasma marginale* outer membrane immunogen by mass spectrometry and genomic mapping. *Infect. Immun.* **73:**8109–8118.

Luo, T., X. Zhang, A. Wakeel, V. L. Popov, and J. W. McBride. 2008. A variable-length PCR target protein of *Ehrlichia chaffeensis* contains major species-specific antibody epitopes in acidic serine-rich tandem repeats. *Infect. Immun.* **76:**1572–1580.

Macmillan, H., K. A. Brayton, G. H. Palmer, T. C. McGuire, G. Munske, W. F. Siems, and W. C. Brown. 2006. Analysis of the *Anaplasma marginale* major surface protein 1 complex protein composition by tandem mass spectrometry. *J. Bacteriol.* **188:**4983–4991.

Macmillan, H., J. Norimine, K. A. Brayton, G. H. Palmer, and W. C. Brown. 2008. Physical linkage of naturally complexed bacterial outer membrane proteins enhances immunogenicity. *Infect. Immun.* **76:**1223–1229.

Madigan, J. E., S. Hietala, S. Chalmers, and E. DeRock. 1990. Seroepidemiologic survey of antibodies to *Ehrlichia equi* in horses of northern California. *J. Am. Vet. Med. Assoc.* **196:**1962–1964.

Mahan, S. M. 1995. Review of the molecular biology of *Cowdria ruminantium*. *Vet. Parasitol.* **57:** 51–56.

Mahan, S. M., H. R. Andrew, N. Tebele, M. J. Burridge, and A. F. Barbet. 1995. Immunisation of sheep against heartwater with inactivated *Cowdria ruminantium*. *Res. Vet. Sci.* **58:**46–49.

Mahan, S. M., T. C. McGuire, S. M. Semu, M. V. Bowie, F. Jongejan, F. R. Rurangirwa, and A. F. Barbet. 1994. Molecular cloning of a gene encoding the immunogenic 21 kDa protein of *Cowdria ruminantium*. *Microbiology* **140:**2135–2142.

Marques, M. A., S. Chitale, P. J. Brennan, and M. C. Pessolani. 1998. Mapping and identification of the major cell wall-associated components of *Mycobacterium leprae*. *Infect. Immun.* **66:**2625–2631.

Martin, M. E., J. E. Bunnell, and J. S. Dumler. 2000. Pathology, immunohistology, and cytokine responses in early phases of human granulocytic ehrlichiosis in a murine model. *J. Infect. Dis.* **181:**374–378.

Martin, M. E., K. Caspersen, and J. S. Dumler. 2001. Immunopathology and ehrlichial propagation are regulated by interferon-γ and interleukin-10 in a murine model of human granulocytic ehrlichiosis. *Am. J. Pathol.* **158:**1881–1888.

Martinez, D., J. C. Maillard, S. Coisne, C. Sheikboudou, and A. Bensaid. 1994. Protection of goats against heartwater acquired by immunisation with inactivated elementary bodies of *Cowdria ruminantium*. *Vet. Immunol. Immunopathol.* **41:**153–163.

Maxie, M. G., and W. F. Robinson. 2007. Heartwater (cowdriosis), p. 84–86. *In* M. G. Maxie (ed.), *Jubb, Kennedy, and Palmer's Pathology of Domestic Animals*, 5th ed., vol. 3. Saunders, Philadelphia, PA.

McBride, J. W., R. E. Corstvet, S. D. Gaunt, C. Boudreaux, T. Guedry, and D. H. Walker. 2003. Kinetics of antibody response to *Ehrlichia canis* immunoreactive proteins. *Infect. Immun.* **71:**2516–2524.

McBride, J. W., C. K. Doyle, X. Zhang, A. M. Cardenas, V. L. Popov, K. A. Nethery, and M. E. Woods. 2007. Identification of a glycosylated *Ehrlichia canis* 19-kilodalton major immunoreactive protein with a species-specific serine-rich glycopeptide epitope. *Infect. Immun.* **75:**74–82.

Murphy, C. I., J. R. Storey, J. Recchia, L. A. Doros-Richert, C. Gingrich-Baker, K. Munroe, J. S. Bakken, R. T. Coughlin, and G. A. Beltz. 1998. Major antigenic proteins of the agent of human granulocytic ehrlichiosis are encoded by members of a multigene family. *Infect. Immun.* **66:**3711–3718.

Murphy, K., P. Travers, and M. Walport. 2008. *Janeway's Immunobiology*, 7th ed., p. 379–495. Garland Science, New York, NY.

Mwangi, D. M., S. M. Mahan, J. K. Nyanjui, E. L. Taracha, and D. J. McKeever. 1998. Immunization of cattle by infection with *Cowdria ruminantium* elicits T lymphocytes that recognize autologous, infected endothelial cells and monocytes. *Infect. Immun.* **66:**1855–1860.

Mwangi, W., W. C. Brown, H. A. Lewin, C. J. Howard, J. C. Hope, T. V. Baszler, P. Caplazi, J. Abbott, and G. H. Palmer. 2002.

DNA-encoded fetal liver tyrosine kinase 3 ligand and granulocyte macrophage-colony-stimulating factor increase dendritic cell recruitment to the inoculation site and enhance antigen-specific CD4$^+$ T cell responses induced by DNA vaccination of outbred animals. *J. Immunol.* **169:**3837–3846.

Nandi, B., K. Hogle, N. Vitko, and G. M. Winslow. 2007. CD4 T-cell epitopes associated with protective immunity induced following vaccination of mice with an ehrlichial variable outer membrane protein. *Infect. Immun.* **75:**5453–5459.

Neer, T. M., and S. Harrus. 2006. Ehrlichiosis, neorickettsiosis, anaplasmosis, and *Wolbachia* infection, p. 203–232. *In* C. E. Greene (ed.), *Infectious Diseases of the Dog and Cat*, 3rd ed. Saunders, St. Louis, MO.

Neitz, W. O., and R. A. Alexander. 1941. The immunization of calves against heartwater. *J. S. Afr. Vet. Med. Assoc.* **12:**103–111.

Neitz, W. O., R. A. Alexander, and T. F. Adelaar. 1947. Studies on immunity in heartwater. *Onderstepoort J. Vet. Sci. Anim. Ind.* **21:**243–249.

Newton, P. N., J. M. Rolain, B. Rasachak, M. Mayxay, K. Vathanatham, P. Seng, R. Phetsouvanh, T. Thammavong, J. Zahidi, Y. Suputtamongkol, B. Syhavong, and D. Raoult. 2009. Sennetsu neorickettsiosis: a probable fish-borne cause of fever rediscovered in Laos. *Am. J. Trop. Med. Hyg.* **81:**190–194.

Niu, H., V. Kozjak-Pavlovic, T. Rudel, and Y. Rikihisa. 2010. *Anaplasma phagocytophilum* Ats-1 is imported into host cell mitochondria and interferes with apoptosis induction. *PLoS Pathog.* **6:** e1000774.

Noh, S. M., K. A. Brayton, W. C. Brown, J. Norimine, G. R. Munske, C. M. Davitt, and G. H. Palmer. 2008. Composition of the surface proteome of *Anaplasma marginale* and its role in protective immunity induced by outer membrane immunization. *Infect. Immun.* **76:**2219–2226.

Noh, S. M., K. A. Brayton, D. P. Knowles, J. T. Agnes, M. J. Dark, W. C. Brown, T. V. Baszler, and G. H. Palmer. 2006. Differential expression and sequence conservation of the *Anaplasma marginale msp2* gene superfamily outer membrane proteins. *Infect. Immun.* **74:**3471–3479.

Noh, S. M., Y. Zhuang, J. E. Futse, W. C. Brown, K. A. Brayton, and G. H. Palmer. 2010. The immunization-induced antibody response to the *Anaplasma marginale* major surface protein 2 and its association with protective immunity. *Vaccine* **28:**3741–3747.

Nyika, A., A. F. Barbet, M. J. Burridge, and S. M. Mahan. 2002. DNA vaccination with *map1* gene followed by protein boost augments protection against challenge with *Cowdria ruminantium*, the agent of heartwater. *Vaccine* **20:**1215–1225.

Nyindo, M. B., M. Ristic, G. E. Lewis, Jr., D. L. Huxsoll, and E. H. Stephenson. 1978. Immune response of ponies to experimental infection with *Ehrlichia equi. Am. J. Vet. Res.* **39:**15–18.

Ohashi, N., N. Zhi, Y. Zhang, and Y. Rikihisa. 1998. Immunodominant major outer membrane proteins of *Ehrlichia chaffeensis* are encoded by a polymorphic multigene family. *Infect. Immun.* **66:**132–139.

Olano, J. P., W. Hogrefe, B. Seaton, and D. H. Walker. 2003. Clinical manifestations, epidemiology, and laboratory diagnosis of human monocytotropic ehrlichiosis in a commercial laboratory setting. *Clin. Diagn. Lab. Immunol.* **10:**891–896.

Oukka, M., J. C. Manuguerra, N. Livaditis, S. Tourdot, N. Riche, I. Vergnon, P. Cordopatis, and K. Kosmatopoulos. 1996. Protection against lethal viral infection by vaccination with nonimmunodominant peptides. *J. Immunol.* **157:**3039–3045.

Paddock, C. D., and J. E. Childs. 2003. *Ehrlichia chaffeensis*: a prototypical emerging pathogen. *Clin. Microbiol. Rev.* **16:**37–64.

Palmer, G. H. 2009. Anaplasmosis, p. 1155–1157. *In* B. P. Smith (ed.), *Large Animal Internal Medicine*. Elsevier, St. Louis, MO.

Palmer, G. H., T. Bankhead, and S. A. Lukehart. 2009. 'Nothing is permanent but change'—antigenic variation in persistent bacterial pathogens. *Cell. Microbiol.* **11:**1697–1705.

Palmer, G. H., A. F. Barbet, W. C. Davis, and T. C. McGuire. 1986. Immunization with an isolate-common surface protein protects cattle against anaplasmosis. *Science* **231:**1299–1302.

Palmer, G. H., J. E. Futse, C. K. Leverich, D. P. Knowles, Jr., F. R. Rurangirwa, and K. A. Brayton. 2007. Selection for simple major surface protein 2 variants during *Anaplasma marginale* transmission to immunologically naive animals. *Infect. Immun.* **75:**1502–1506.

Palmer, G. H., D. P. Knowles, Jr., J. L. Rodriguez, D. P. Gnad, L. C. Hollis, T. Marston, and K. A. Brayton. 2004. Stochastic transmission of multiple genotypically distinct *Anaplasma marginale* strains in a herd with high prevalence of *Anaplasma* infection. *J. Clin. Microbiol.* **42:**5381–5384.

Palmer, G. H., and T. F. McElwain. 1995. Molecular basis for vaccine development against anaplasmosis and babesiosis. *Vet. Parasitol.* **57:**233–253.

Palmer, G. H., and T. C. McGuire. 1984. Immune serum against *Anaplasma marginale* initial bodies neutralizes infectivity for cattle. *J. Immunol.* **133:**1010–1015.

Palmer, G. H., S. M. Oberle, A. F. Barbet, W. L. Goff, W. C. Davis, and T. C. McGuire. 1988. Immunization of cattle with a 36-kilodalton surface protein induces protection against homologous and heterologous *Anaplasma marginale* challenge. *Infect. Immun.* **56:**1526–1531.

Palmer, G. H., F. R. Rurangirwa, K. M. Kocan, and W. C. Brown. 1999. Molecular basis for vaccine development against the ehrlichial pathogen *Anaplasma marginale*. *Parasitol. Today* **15:**281–286.

Peddireddi, L., C. Cheng, and R. R. Ganta. 2009. Promoter analysis of macrophage- and tick cell-specific differentially expressed *Ehrlichia chaffeensis p28-Omp* genes. *BMC Microbiol.* **9:**99.

Pipano, E. 1995. Live vaccines against hemoparasitic diseases in livestock. *Vet. Parasitol.* **57:**213–231.

Poitout, F. M., J. K. Shinozaki, P. J. Stockwell, C. J. Holland, and S. K. Shukla. 2005. Genetic variants of *Anaplasma phagocytophilum* infecting dogs in western Washington State. *J. Clin. Microbiol.* **43:**796–801.

Polack, F. P., P. G. Auwaerter, S. H. Lee, H. C. Nousari, A. Valsamakis, K. M. Leiferman, A. Diwan, R. J. Adams, and D. E. Griffin. 1999. Production of atypical measles in rhesus macaques: evidence for disease mediated by immune complex formation and eosinophils in the presence of fusion-inhibiting antibody. *Nat. Med.* **5:**629–634.

Porksakorn, C., S. Nuchprayoon, K. Park, and A. L. Scott. 2007. Proinflammatory cytokine gene expression by murine macrophages in response to *Brugia malayi Wolbachia* surface protein. *Mediators Inflamm.* **2007:**84318.

Potgieter, F. T., and L. Van Rensburg. 1983. Infectivity virulence and immunogenicity of *Anaplasma centrale* live blood vaccine. *Onderstepoort J. Vet. Res.* **50:**29–31.

Pretorius, A., N. E. Collins, H. C. Steyn, F. van Strijp, M. van Kleef, and B. A. Allsopp. 2007. Protection against heartwater by DNA immunisation with four *Ehrlichia ruminantium* open reading frames. *Vaccine* **25:**2316–2324.

Pretorius, A., J. Liebenberg, E. Louw, N. E. Collins, and B. A. Allsopp. 2010. Studies of a polymorphic *Ehrlichia ruminantium* gene for use as a component of a recombinant vaccine against heartwater. *Vaccine* **28:**3531–3539.

Pretorius, A., M. van Kleef, N. E. Collins, N. Tshikudo, E. Louw, F. E. Faber, M. F. van Strijp, and B. A. Allsopp. 2008. A heterologous prime/boost immunisation strategy protects against virulent *E. ruminantium* Welgevonden needle challenge but not against tick challenge. *Vaccine* **26:**4363–4371.

Punkosdy, G. A., D. G. Addiss, and P. J. Lammie. 2003. Characterization of antibody responses to *Wolbachia* surface protein in humans with lymphatic filariasis. *Infect. Immun.* **71:**5104–5114.

Pusterla, N., and J. E. Madigan. 2007a. *Anaplasma phagocytophila*, p. 354–357. *In* D. C. Sellon and M. T. Long (ed.), *Equine Infectious Diseases*. Saunders, St. Louis, MO.

Pusterla, N., and J. E. Madigan. 2007b. *Neorickettsia risticii*, p. 357–362. *In* D. C. Sellon and M. T. Long (ed.), *Equine Infectious Diseases*. Saunders, St. Louis, MO.

Qazi, O., A. Brailsford, A. Wright, J. Faraar, J. Campbell, and N. Fairweather. 2007. Identification and characterization of the surface-layer protein of *Clostridium tetani*. *FEMS Microbiol. Lett.* **274:**126–131.

Reardon, M. J., and K. R. Pierce. 1981. Acute experimental canine ehrlichiosis. II. Sequential reaction of the hemic and lymphoreticular system of selectively immunosuppressed dogs. *Vet. Pathol.* **18:**384–395.

Reddy, G. R., and C. P. Streck. 1999. Variability in the 28-kDa surface antigen protein multigene locus of isolates of the emerging disease agent *Ehrlichia chaffeensis* suggests that it plays a role in immune evasion. *Mol. Cell Biol. Res. Commun.* **1:**167–175.

Reddy, G. R., C. R. Sulsona, A. F. Barbet, S. M. Mahan, M. J. Burridge, and A. R. Alleman. 1998. Molecular characterization of a 28 kDa surface antigen gene family of the tribe *Ehrlichiae*. *Biochem. Biophys. Res. Commun.* **247:**636–643.

Reddy, G. R., C. R. Sulsona, R. H. Harrison, S. M. Mahan, M. J. Burridge, and A. F. Barbet. 1996. Sequence heterogeneity of the major antigenic protein 1 genes from *Cowdria ruminantium* isolates from different geographical areas. *Clin. Diagn. Lab. Immunol.* **3:**417–422.

Rikihisa, Y. 1999. Clinical and biological aspects of infection caused by *Ehrlichia chaffeensis*. *Microbes Infect.* **1:**367–376.

Rikihisa, Y. 2010. *Anaplasma phagocytophilum* and *Ehrlichia chaffeensis*: subversive manipulators of host cells. *Nat. Rev. Microbiol.* **8:**328–339.

Rikihisa, Y., S. A. Ewing, and J. C. Fox. 1994. Western immunoblot analysis of *Ehrlichia chaffeensis*, *E. canis*, or *E. ewingii* infections in dogs and humans. *J. Clin. Microbiol.* **32:**2107–2112.

Rikihisa, Y., M. Lin, and H. Niu. 2010. Type IV secretion in the obligatory intracellular bacterium *Anaplasma phagocytophilum*. *Cell. Microbiol.* **12:**1213–1221.

Rikihisa, Y., C. I. Pretzman, G. C. Johnson, S. M. Reed, S. Yamamoto, and F. Andrews. 1988. Clinical, histopathological, and immunological responses of ponies to *Ehrlichia sennetsu* and subsequent *Ehrlichia risticii* challenge. *Infect. Immun.* **56:**2960–2966.

Rima, B. K., J. A. Earle, K. Baczko, P. A. Rota, and W. J. Bellini. 1995a. Measles virus strain variations. *Curr. Top. Microbiol. Immunol.* **191:**65–83.

Rima, B. K., J. A. Earle, R. P. Yeo, L. Herlihy, K. Baczko, V. ter Meulen, J. Carabana, M. Caballero, M. L. Celma, and R. Fernandez-Munoz. 1995b. Temporal and geographical distribution of measles virus genotypes. *J. Gen. Virol.* **76:**1173–1180.

Ristic, M. R., and C. J. Holland. 1993. Canine ehrlichiosis, p. 169–186. *In* Z. Woldehiwet and M. R. Ristic (ed.), *Rickettsial and Chlamydial Diseases of Domestic Animals*. Pergamon Press, Oxford, United Kingdom.

Rodriguez, J. L., G. H. Palmer, D. P. Knowles, Jr., and K. A. Brayton. 2005. Distinctly different *msp2* pseudogene repertoires in *Anaplasma marginale* strains that are capable of superinfection. *Gene* **361:**127–132.

Saint André, A., N. M. Blackwell, L. R. Hall, A. Hoerauf, N. W. Brattig, L. Volkmann, M. J. Taylor, L. Ford, A. G. Hise, J. H. Lass, E. Diaconu, and E. Pearlman. 2002. The role of endosymbiotic *Wolbachia* bacteria in the pathogenesis of river blindness. *Science* **295:** 1892–1895.

Sallusto, F., A. Lanzavecchia, K. Araki, and R. Ahmed. 2010. From vaccines to memory and back. *Immunity* **33:**451–463.

Saridaki, A., and K. Bourtzis. 2010. *Wolbachia*: more than just a bug in insects genitals. *Curr. Opin. Microbiol.* **13:**67–72.

Sebatjane, S. I., A. Pretorius, J. Liebenberg, H. Steyn, and M. Van Kleef. 2010. *In vitro* and *in vivo* evaluation of five low molecular weight proteins of *Ehrlichia ruminantium* as potential vaccine components. *Vet. Immunol. Immunopathol.* **137:**217–225.

Seo, G. M., C. Cheng, J. Tomich, and R. R. Ganta. 2008. Total, membrane, and immunogenic proteomes of macrophage- and tick cell-derived *Ehrlichia chaffeensis* evaluated by liquid chromatography-tandem mass spectrometry and MALDI-TOF methods. *Infect. Immun.* **76:**4823–4832.

Sher, A. 1988. Strategies for vaccination against parasites, p. 169–172. *In* P. T. Englund and A. Sher (ed.), *The Biology of Parasitism*. Alan R. Liss, Inc., New York, NY.

Shkap, V., B. Leibovitz, Y. Krigel, T. Molad, L. Fish, M. Mazuz, L. Fleiderovitz, and I. Savitsky. 2008. Concomitant infection of cattle with the vaccine strain *Anaplasma marginale* ss *centrale* and field strains of *A. marginale*. *Vet. Microbiol.* **130:**277–284.

Shkap, V., T. Molad, L. Fish, and G. H. Palmer. 2002. Detection of the *Anaplasma centrale* vaccine strain and specific differentiation from *Anaplasma marginale* in vaccinated and infected cattle. *Parasitol. Res.* **88:**546–552.

Sotomayor, E. A., V. L. Popov, H. M. Feng, D. H. Walker, and J. P. Olano. 2001. Animal model of fatal human monocytotropic ehrlichiosis. *Am. J. Pathol.* **158:**757–769.

Stich, R. W., G. A. Olah, K. A. Brayton, W. C. Brown, M. Fechheimer, K. Green-Church, S. Jittapalapong, K. M. Kocan, T. C. McGuire, F. R. Rurangirwa, and G. H. Palmer. 2004. Identification of a novel *Anaplasma marginale* appendage-associated protein that localizes with actin filaments during intraerythrocytic infection. *Infect. Immun.* **72:**7257–7264.

Storey, J. R., L. A. Doros-Richert, C. Gingrich-Baker, K. Munroe, T. N. Mather, R. T. Coughlin, G. A. Beltz, and C. I. Murphy. 1998. Molecular cloning and sequencing of three granulocytic *Ehrlichia* genes encoding high-molecular-weight immunoreactive proteins. *Infect. Immun.* **66:**1356–1363.

Stuen, S., and D. Longbottom. 2011. Treatment and control of chlamydial and rickettsial infections in sheep and goats. *Vet. Clin. North Am. Food Anim. Pract.* **27:**213–233.

Sulsona, C. R., S. M. Mahan, and A. F. Barbet. 1999. The *map1* gene of *Cowdria ruminantium* is a member of a multigene family containing both conserved and variable genes. *Biochem. Biophys. Res. Commun.* **257:**300–305.

Sutten, E. L., J. Norimine, P. A. Beare, R. A. Heinzen, J. E. Lopez, K. Morse, K. A. Brayton, J. J. Gillespie, and W. C. Brown. 2010. *Anaplasma marginale* type IV secretion system proteins VirB2, VirB7, VirB11, and VirD4 are immunogenic components of a protective bacterial membrane vaccine. *Infect. Immun.* **78:**1314–1325.

Taylor, M. J., H. F. Cross, L. Ford, W. H. Makunde, G. B. Prasad, and K. Bilo. 2001. *Wolbachia* bacteria in filarial immunity and disease. *Parasite Immunol.* **23:**401–409.

Tebele, N., T. C. McGuire, and G. H. Palmer. 1991. Induction of protective immunity by using *Anaplasma marginale* initial body membranes. *Infect. Immun.* **59:**3199–3204.

Telford, J. L. 2008. Bacterial genome variability and its impact on vaccine design. *Cell Host Microbe* **3:**408–416.

Theiler, A. 1910. *Report of the Government Veterinary Bacteriologist, 1908–1909*, p. 7–64. Department of Agriculture, Union of South Africa, Transvaal, South Africa.

Theiler, A. 1911. *First Report of the Director of Veterinary Research, Union of South Africa*, p. 7–46. Department of Agriculture, Union of South Africa, Johannesburg, South Africa.

Theiler, A. 1912. Gallsickness of imported cattle and the protective inoculation against this disease. *Agric. J. Union S. Afr.* **3:**7–46.

Totté, P., A. Bensaid, S. M. Mahan, D. Martinez, and D. J. McKeever. 1999. Immune responses to *Cowdria ruminantium* infections. *Parasitol. Today* **15:**286–290.

Totté, P., F. Jongejan, A. L. de Gee, and J. Wérenne. 1994. Production of alpha interferon in *Cowdria ruminantium*-infected cattle and its effect on infected endothelial cell cultures. *Infect. Immun.* **62:**2600–2604.

Totté, P., N. Vachiery, D. Martinez, I. Trap, K. T. Ballingall, N. D. MacHugh, A. Bensaid, and J. Wérenne. 1996. Recombinant bovine interferon gamma inhibits the growth of *Cowdria ruminantium* but fails to induce major histocompatibility complex class II following infection of endothelial cells. *Vet. Immunol. Immunopathol.* **53:**61–71.

Turton, J. A., T. C. Katsande, M. B. Matingo, W. K. Jorgensen, U. Ushewokunze-Obatolu, and R. J. Dalgliesh. 1998. Observations on the use of *Anaplasma centrale* for immunization of cattle against anaplasmosis in Zimbabwe. *Onderstepoort J. Vet. Res.* **65:**81–86.

van der Most, R. G., A. Sette, C. Oseroff, J. Alexander, K. Murali-Krishna, L. L. Lau, S. Southwood, J. Sidney, R. W. Chesnut, M. Matloubian, and R. Ahmed. 1996. Analysis of cytotoxic T cell responses to dominant and subdominant epitopes during acute and chronic lymphocytic choriomeningitis virus infection. *J. Immunol.* **157:**5543–5554.

van Heerden, H., N. E. Collins, K. A. Brayton, C. Rademeyer, and B. A. Allsopp. 2004. Characterization of a major outer membrane protein multigene family in *Ehrlichia ruminantium*. *Gene* **330:**159–168.

Van Kleef, M., N. J. Gunter, H. Macmillan, B. A. Allsopp, V. Shkap, and W. C. Brown. 2000. Identification of *Cowdria ruminantium* antigens that stimulate proliferation of lymphocytes from cattle immunized by infection and treatment or with inactivated organisms. *Infect. Immun.* **68:**603–614.

Van Kleef, M., H. Macmillan, N. J. Gunter, E. Zweygarth, B. A. Allsopp, V. Shkap, D. H. Du Plessis, and W. C. Brown. 2002. Low molecular weight proteins of *Cowdria ruminantium* (Welgevonden isolate) induce bovine $CD4^+$-enriched T-cells to proliferate and produce interferon-γ. *Vet. Microbiol.* **85:**259–273.

Vidotto, M. C., T. C. McGuire, T. F. McElwain, G. H. Palmer, and D. P. Knowles, Jr. 1994. Intermolecular relationships of major surface proteins of *Anaplasma marginale*. *Infect. Immun.* **62:**2940–2946.

Vidotto, M. C., E. J. Venancio, and O. Vidotto. 2008. Cloning, sequencing and antigenic characterization of rVirB9 of *Anaplasma marginale* isolated from Parana State, Brazil. *Genet. Mol. Res.* **7:**460–466.

Wakeel, A., X. Zhang, and J. W. McBride. 2010. Mass spectrometric analysis of *Ehrlichia chaffeensis* tandem repeat proteins reveals evidence of phosphorylation and absence of glycosylation. *PLoS One* **5:**e9552.

Werren, J. H., L. Baldo, and M. E. Clark. 2008. *Wolbachia*: master manipulators of invertebrate biology. *Nat. Rev. Microbiol.* **6:**741–751.

Whist, S. K., A. K. Storset, G. M. Johansen, and H. J. Larsen. 2003. Modulation of leukocyte populations and immune responses in sheep experimentally infected with *Anaplasma* (formerly *Ehrlichia*) *phagocytophilum*. *Vet. Immunol. Immunopathol.* **94:**163–175.

Whist, S. K., A. K. Storset, and H. J. Larsen. 2002. Functions of neutrophils in sheep experimentally infected with *Ehrlichia phagocytophila*. *Vet. Immunol. Immunopathol.* **86:**183–193.

Winslow, G. M., E. Yager, K. Shilo, D. N. Collins, and F. K. Chu. 1998. Infection of the laboratory mouse with the intracellular pathogen *Ehrlichia chaffeensis*. *Infect. Immun.* **66:**3892–3899.

Winslow, G. M., E. Yager, K. Shilo, E. Volk, A. Reilly, and F. K. Chu. 2000. Antibody-mediated elimination of the obligate intracellular bacterial pathogen *Ehrlichia chaffeensis* during active infection. *Infect. Immun.* **68:**2187–2195.

Woldehiwet, Z. 2006. *Anaplasma phagocytophilum* in ruminants in Europe. *Ann. N. Y. Acad. Sci.* **1078:**446–460.

Woldehiwet, Z., and G. R. Scott. 1982. Immunological studies on tick-borne fever in sheep. *J. Comp. Pathol.* **92:**457–467.

Yager, E., C. Bitsaktsis, B. Nandi, J. W. McBride, and G. Winslow. 2005. Essential role for humoral immunity during *Ehrlichia* infection in immunocompetent mice. *Infect. Immun.* **73:**8009–8016.

Zepp, F. 2010. Principles of vaccine design—lessons from nature. *Vaccine* **28**(Suppl. 3)**:**C14–C24.

Zhi, N., N. Ohashi, Y. Rikihisa, H. W. Horowitz, G. P. Wormser, and K. Hechemy. 1998. Cloning and expression of the 44-kilodalton major outer membrane protein gene of the human granulocytic ehrlichiosis agent and application of the recombinant protein to serodiagnosis. *J. Clin. Microbiol.* **36:**1666–1673.

Zhi, N., Y. Rikihisa, H. Y. Kim, G. P. Wormser, and H. W. Horowitz. 1997. Comparison of major antigenic proteins of six strains of the human granulocytic ehrlichiosis agent by Western immunoblot analysis. *J. Clin. Microbiol.* **35:**2606–2611.

Zhuang, Y., J. E. Futse, W. C. Brown, K. A. Brayton, and G. H. Palmer. 2007. Maintenance of antibody to pathogen epitopes generated by segmental gene conversion is highly dynamic during long-term persistent infection. *Infect. Immun.* **75:**5185–5190.

Zweygarth, E., A. I. Josemans, M. F. Van Strijp, L. Lopez-Rebollar, M. Van Kleef, and B. A. Allsopp. 2005. An attenuated *Ehrlichia ruminantium* (Welgevonden stock) vaccine protects small ruminants against virulent heartwater challenge. *Vaccine* **23:**1695–1702.

PERSISTENCE AND ANTIGENIC VARIATION

Kelly A. Brayton

12

PERSISTENCE

Self-propagation is a driving force in biological systems, and for obligate intracellular pathogens this means establishing mechanisms of persistence so that there are increased opportunities for transmission and continued propagation beyond the life span of a single host. Consider the infection dynamics of a pathogen in the blood of the infected host—three scenarios are possible: (i) mutually assured destruction, i.e., an increase in pathogen load, leading to severe disease and resulting in death of the host; (ii) clearance of the pathogen by the immune system after an acute period of infection, typically manifested as disease; or (iii) entry into a persistent phase of infection (Fig. 1). While pathogens may be acquired (and subsequently transmitted) during the acute period of infection prior to clearance or entry into the persistent phase of infection, it is the third scenario, that of persistence, that provides for the greatest opportunities for vector-borne transmission and thus self-propagation; organisms that persist for short periods will have fewer opportunities to be acquired by the vector and thus perpetuated by transmission to a new host (Borst, 1991; Palmer and Brayton, 2007). Therefore, persistence is an adaptive survival mechanism and must circumvent the development of immune responses against the pathogen. Among rickettsial pathogens, persistence is a defining feature of multiple genera; this chapter focuses on the genera *Anaplasma* and *Ehrlichia*. While numerous immune evasion strategies have evolved for bacterial persistence (Young et al., 2002), the most striking of these among rickettsial pathogens is antigenic variation in *Anaplasma* spp., an example that will be used to highlight the molecular mechanisms and host-pathogen interactions responsible for efficient transmission and thus self-propagation.

ANTIGENIC VARIATION

In its simplest terms, antigenic variation is the process by which an organism changes epitopes. Typically, antigenically variable proteins are immunodominant and thus allow evasion of the predominant immune responses. Antigenic variation sensu stricto involves a rapid change in the pathogen population that allows evasion of the immune response; in contrast, antigenic variation sensu lato (antigenic variability or "antigenic drift" in influenza terms) is the

Kelly A. Brayton, Department of Veterinary Microbiology and Pathology, Paul G. Allen School for Global Animal Health, Washington State University, Pullman, WA 99164-7040.

Intracellular Pathogens II: Rickettsiales
Edited by G. H. Palmer and A. F. Azad © 2012 ASM Press, Washington, DC
10.1128/9781555817336.ch12

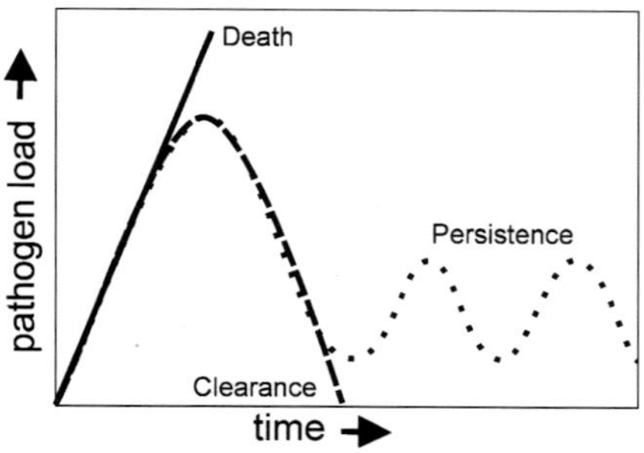

FIGURE 1 Infection scenarios. Upon infection, pathogen load can increase in an uncontrolled fashion, ultimately causing death (solid line). Alternatively, a controlling immune response can be established, resulting in clearance of the pathogen and resolution of disease (dashed line). Finally, in the face of an adaptive immune response, the pathogen is able to evade recognition and enter a persistent phase of infection (dotted line). In persistence mediated by antigenic variation, the infection scenario is typically characterized by waves of parasitemia, reflecting control of variants, followed by emergence of new variants that are not immediately recognized by the existing immune response. doi:10.1128/9781555817336.ch12.f1

process of random mutations that slowly accumulate over time (Fig. 2). The latter occurs commonly in a broad set of parasites, while the former occurs in a more limited set of parasites and is effected to allow persistence. Several related mechanisms have evolved for true sensu stricto antigenic variation among bacteria and protozoa; however, all require the presence of a multigene family for the variant protein and most require a recombinatorial strategy for creating new variants (Borst, 1991; Barbour and Restrepo, 2000; Palmer and Brayton, 2007). A nonrecombinatorial strategy is expression-site switching, which is employed by eukaryotic pathogens, such as *Trypanosoma brucei*, and takes advantage of chromosome remodeling to turn specific expression sites on or off (Rudenko, 2010). This mechanism is not well-known in bacteria.

Additional factors to consider for antigenic variation to be effective are that the repertoire of variants created must be large enough for immune evasion over a significant time period to allow onward transmission, and that the switch mechanism must be employed before a clearing immune response is fully developed. If the organism made the switch only in the face of immune effectors capable of clearance, the population would die out before a new variant could be generated and replicate. However, it is important to emphasize that while the immune response exerts a powerful selection on the variant population, it does not itself direct switching. This mechanism provides powerful costs and benefits: most of the organisms in a population are killed as an immune response develops to the major pathogen variants present in the population (in *T. brucei*, it has been estimated that >99.9% of the organisms are killed), but the trade-off is that the small fraction that survive can rapidly expand to become the dominant variant and allow the population to persist long enough to enable transmission (Borst, 1991).

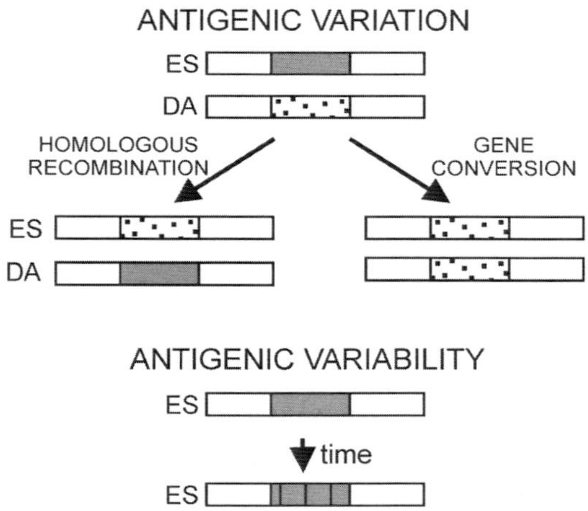

FIGURE 2 Paradigms of antigenic variation. Antigenic shift is the result of a recombination event; in bacteria it can result either from a classical homologous recombination event, wherein the donor allele (DA) is exchanged with the expressed copy of the gene (ES), resulting in a change in the donor allele repertoire, or from a specialized type of homologous recombination called gene conversion where the donor allele is "copied and pasted" into the expression site, thus maintaining the donor allele repertoire. Antigenic variation sensu stricto is a rapid process that requires one or more expression sites and a pool of donor alleles. Antigenic variability is the accumulation of mutations in the expression-site copy of a gene over time, typically a slower process. doi:10.1128/9781555817336.ch12.f2

Much of what we know about antigenic variation sensu stricto in pathogens other than viruses has been elucidated in the African trypanosome, *T. brucei*, and as such, it serves as a comparative model (Taylor and Rudenko, 2006). *T. brucei* is a flagellated unicellular protozoan organism transmitted by the tsetse fly that causes sleeping sickness when it infects humans or nagana in animals. In this system the variable surface glycoprotein (VSG) forms a dense surface coat that protects other conserved membrane proteins from being recognized by the host immune response (Morrison et al., 2009). Immune evasion and persistence are achieved by antigenic variation of VSG employing a large archive of >1,500 alleles and ~20 telomerically located bloodstream-form expression sites, with only a single expression site being active at a given time (Berriman et al., 2005; Taylor and Rudenko, 2006). Trypanosomes utilize expression-site switching (turning off the active genes and turning on a silent gene), telomere exchange (retaining the same active promoter, but switching the VSG coding region with another telomerically located expression site), and gene conversion, which has been shown to be the most commonly used method for variant switching (Robinson et al., 1999; Taylor and Rudenko, 2006).

Gene conversion is a type of homologous recombination and is a common method for antigenic variation sensu stricto, involving a duplicative transposition of a donor allele into the expression site (Fig. 2). The donor allele repertoire remains unchanged during antigenic variation, while the expression-site variant is lost. This method is employed by some rickettsiae in the family *Anaplasmataceae* and allows for lifelong persistence in the host with donor allele repertoires 10- to 100-fold smaller than those found in African trypanosomes. Among the *Anaplasmataceae*, this process has been best studied in *Anaplasma marginale*, the mechanism of which is detailed below. Following the detailed discussion of *A. marginale*, available information on related genes that appear to undergo this process in other *Anaplasmataceae* is presented.

A. MARGINALE

Antigenic Variation of *A. marginale*

Early work demonstrated that major surface proteins (MSP) 2 and 3 are immunodominant, antigenically variable proteins that are instrumental in evading the host immune response (Palmer et al., 1994; Eid et al., 1996; Alleman et al., 1997; French et al., 1998, 1999; Meeus et al., 2003). Each of these proteins contains conserved amino and carboxy termini flanking a central hypervariable region (HVR). New variants arise throughout acute and persistent infection, and antibody present at a given time point will not recognize variants found at a later time point (French et al., 1999). While direct evidence of immune evasion is lacking for MSP3, it has been shown to be antigenically variable and it is presumed to function with MSP2 in immune evasion (Alleman and Barbet, 1996; Alleman et al., 1997; Meeus et al., 2003). Antigenic switching of these two proteins occurs simultaneously, and they are presumed to work as one to create a large set of epitopes to attract and evade the host immune response in much the same way as the VSG in the monomolecular coat of trypanosomes (Brayton et al., 2003). Interestingly, although quite different in size (MSP2 is 40 to 42 kDa, while MSP3 varies from 80 to 90 kDa) and sequence composition (~30 to 40% sequence identity), these two proteins appear to have arisen from a common ancestor, as the carboxy-terminal 138 amino acids are identical and there are short stretches of amino acid similarity between the amino termini (Fig. 3) (Meeus et al., 2003).

The central HVR of MSP2 varies from 60 to 70 amino acids in length and contains only seven universally conserved residues, including two cysteines that appear to be involved in disulfide bond formation. The HVR begins with what has been described as the "block 1 region," which displays one of five potential amino acid triplets in this position: KAV, NAV, NAI, TTV, or TTI. A given strain may contain only a few of these variants; for example, sequencing of the St. Maries strain

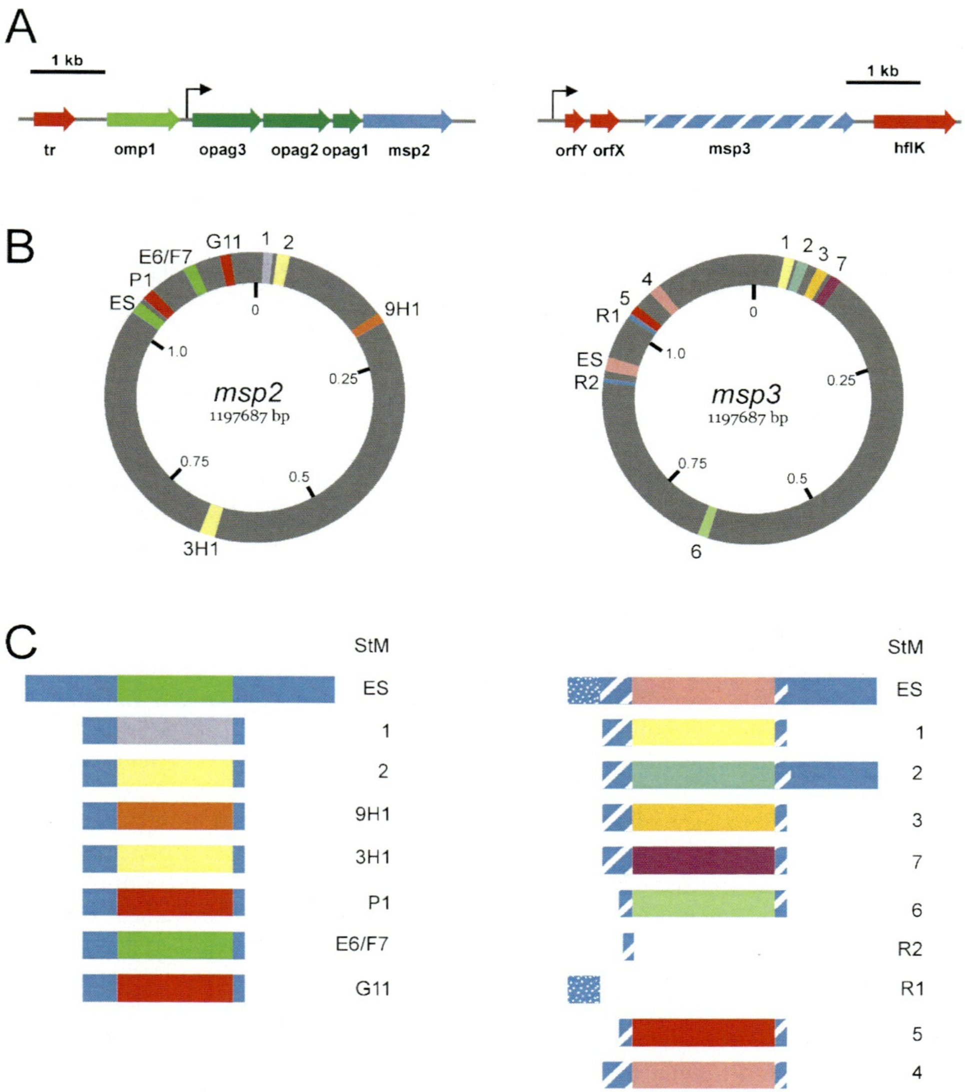

FIGURE 3 *A. marginale msp2* and *msp3* expression sites (ES), genomic arrangement, and donor allele repertoires. (A) The *msp2* (left) and *msp3* (right) operons that are the sole expression sites for these genes. The promoter for each operon is indicated by the bent arrowhead. *msp2* is transcribed from an operon of four genes with *msp2* at the 3′ end. The other genes in the operon (*opag1* to *-3*) are members of PF01617, as is the gene immediately upstream (*omp1*), and these are depicted in green. A putative transcriptional regulator (*tr*) that has been used as an indicator of synteny resides upstream and is shown in red. The *msp3* operon contains three genes with *msp3* at the 3′ end. The other two genes in the operon were annotated as *orfX* and *-Y* (red), and recently have been implicated as being homologs of *virB2*, a component of the type IV secretion system. *hflK*, 3′ to the operon, has been used to identify the syntenic locus in related organisms, and is shown in red. (B) The genomic arrangement of the *msp2* and *msp3* expression sites and donor allele repertoires. Alleles depicted in identical colors (e.g., 3H1 and 2) have identical HVR sequences. The genome backbone is shown as gray. (C) The donor allele and pseudogene repertoires for *msp2* and *msp3*, showing the relative portion of the expression-site molecule that each encodes. The *msp3* remnant sequences (R1 and R2) do not contain any portion of the HVR and could not serve as donor alleles. *msp3* has a 3′ end identical to *msp2* (solid blue), regions flanking the HVR that have similarity to *msp2* (diagonal stripes), and a unique 5′ end (stippled). doi:10.1128/9781555817336.ch12.f3

shows that only NAV, TTV, and TTI are represented in the donor alleles for this strain (Brayton et al., 2005). Immunologic analysis of this region shows that $CD4^{+}$-T-cell epitopes are contained in this block 1 region, and switching from one triplet to another provides immune escape (Brown et al., 2003).

In addition to MSP2 and MSP3, there are several other closely related outer membrane proteins (OMPs) encoded by *A. marginale*; together these constitute PF01617. Pfam is a collection of protein families represented by multiple sequence alignments and hidden Markov models (Finn et al., 2010), and establishes relationships at the family or superfamily level. Other PF01617 members include MSP4, OpAG1 to -3, and OMP1 to -15. Interestingly, aside from MSP2 and MSP3, these PF01617 proteins do not seem to vary rapidly (Noh et al., 2006). The current hypothesis is that the highly immunodominant MSP2 and MSP3 allow evasion from the effective immune response with only nonprotective responses developing against the other proteins during natural infection.

Gene Conversion Strategy of *A. marginale*

Genome sequencing of the St. Maries strain of *A. marginale* revealed the presence of multigene families for both MSP2 and MSP3 (Brayton et al., 2005). Both MSP2 and MSP3 are expressed from operons (Fig. 3): *msp2* is located at the 3′ end of an operon of four genes, with the upstream genes being called operon-associated genes (*opag1* to *3*); these single-copy genes are members of PF01617 and related to *msp2* at the superfamily level (Barbet et al., 2000). MSP3 is expressed from an operon of three genes, with the upstream genes named *orfX* and *orfY*, which recently have been implicated as being members of the type IV secretion system (Meeus et al., 2003). *orfX* and *orfY* are multicopy genes (with 12 and 7 copies, respectively) that, in addition to appearing in the *msp3* expression site, are associated with the *msp2* and *msp3* donor alleles (Brayton et al., 2005). Aside from the single expression site, each gene family has seven donor alleles (the *msp3* family also contains two short remnant sequences that cannot act as donor alleles) (Fig. 3). The donor alleles were initially identified as "functional pseudogenes," in that these are truncated copies of the full-length gene that cannot in their present location be transcribed but can create new variants when recombined into the expression site (Fig. 4) (Brayton et al., 2005).

Both the whole donor allele and short oligonucleotide segments of donor alleles have been shown to recombine into the expression site by gene conversion, the latter producing a combinatorial number of variants sufficient for lifelong immune evasion (Brayton et al., 2001, 2002; Meeus et al., 2003; Futse et al., 2009). In an in-depth study examining over 1,300 variants during the course of infection (Futse et al., 2005), it was determined that there was some promiscuity during recombination. In this model, defined as the "anchoring model" of segmental gene conversion, the recombination complex is anchored in position by sequence identity in the conserved flanking regions 5′ and 3′ of the HVR (Fig. 4). One recombination event occurs within one of the conserved flanking ends, identical between the donor and the expression site, while the other recombination event occurs within the HVR, independent of sequence identity (Futse et al., 2005). This novel mechanism is in stark contrast to studies in *Escherichia coli* that show that ~20 nucleotides of sequence identity are required for recombination events (Watt et al., 1985). This comprehensive *A. marginale* study also revealed that in a small percentage (1.2%) of variants, a deletion occurred at the recombination site or, alternatively, a small insertion of 3 to 18 nucleotides occurred to generate a novel sequence at the recombination site. Even more rarely (0.2% of variants), substitutions can be incorporated throughout the expression-site variant that are not templated by a donor allele, suggesting that additional antigenic changes are generated de novo during infection.

The new MSP2 variants that are generated in the expression site by gene conversion

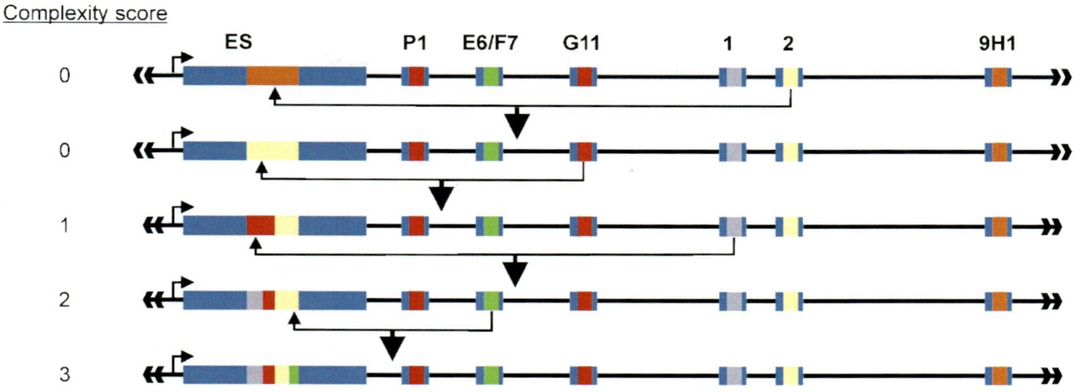

FIGURE 4 The anchoring model for gene conversion. A segment of the genome containing the *msp2* expression site (ES) and several donor alleles is shown (not to scale). The first recombination event illustrates the complete HVR from donor allele 2 being recombined into the expression site, with both recombination sites occurring in the conserved flanking regions. The donor allele repertoire remains unchanged. The second recombination event incorporates a segment of donor allele G11 into the expression site. Importantly, one recombination event has occurred in the 5′ conserved flanking region, while the other has occurred in the HVR without the requirement for sequence identity at the recombination site. In the anchoring model, one end of the newly recombined segment will always be juxtaposed to the conserved flanking regions, as the sequence identity in these regions anchors the recombination complex. doi:10.1128/9781555817336.ch12.f4

are selected by the immune response of the host. The donor alleles themselves have been selected to encode structurally and antigenically unique variants: the *A. marginale* population at a given time point is not recognized by antibodies from earlier stages of infection (Eid et al., 1996; French et al., 1999). Later studies also demonstrated the presence of $CD4^+$-T-cell epitopes in the MSP2 HVR (Brown et al., 2003; Abbott et al., 2004) and thus the capacity to vary both B- and T-cell epitopes. However, it is tremendous combinatorial diversity generated by segmental gene conversion that allows continual evasion of clearance and long-term persistence. Unlike the donor alleles, which are permanently maintained in the genome, expression-site mosaics, having been assembled from more than one donor, are transient. Immunologic studies have confirmed that even a small segmental change is sufficient to evade recognition (French et al., 1999) and allow continued *A. marginale* persistence. The expression-site mosaics serve as the template for additional segmental gene conversion events, resulting in increasingly complex MSP2 variants as infection persists (Futse et al., 2005).

Dual Selection Pressures

As noted above, the MSP2 donor alleles have been evolutionarily selected to encode structurally and antigenically diverse HVRs, while expression-site mosaics derived from multiple alleles are transient and under only the immediate selection of the existing immune response. The early acute bacteremia, in which pathogen levels reach 10^9 bacteria per ml, is composed of primarily "simple" variants derived from recombination of a single donor allele into the expression site (Futse et al., 2005). In contrast, the mosaic or "complex" variants derived from multiple donor alleles emerge during persistence, in which bacteremia levels fluctuate between 10^2 and 10^6 bacteria/ml. The marked difference in bacteremia levels suggested that the mosaic variants were at a growth fitness disadvantage. This hypothesis was tested by inoculating a mixture of organisms bearing complex variant mosaic MSP2 molecules on their surface, obtained from long-term persistently infected animals, and demonstrated that the first variants that grew in the immunologically naïve recipient animal, prior to the onset of immunity, expressed simple MSP2 variants derived from the archived donor allele repertoire (Palmer et al., 2006). This observation was reinforced by analysis of the variant population in ticks allowed to acquire *A. marginale* by feeding on persistently infected animals: the complex mosaic variant population in the blood during tick feeding was rapidly replaced by replication of simple variants within the tick midgut. Thus, the mosaic variants, transient in the expression site, retained their selective advantage only in the face of adaptive immunity in the mammalian host but were at a severe disadvantage in the absence of immune selection. The ramification of this observation is that there are competing selection pressures on the variants—pressure to structurally diversify and thus evade the immune response and pressure to maintain growth fitness. While the basis for the growth fitness is not well understood, the MSP2 homolog in *Anaplasma phagocytophilum* has been reported to have porin activity (Huang et al., 2007), suggesting that structural changes that allow immune escape may well compromise porin function and thus overall growth fitness.

A. marginale msp2 and *msp3* Donor Allele Repertoire in Fully Sequenced Strains

The gene conversion mechanism implies that the donor allele repertoire will remain relatively constant over time. On a rare occasion, instead of the unidirectional gene conversion event, a bidirectional, classical homologous recombination event could take place, exchanging the expression-site copy with the donor allele copy, which would result in an altered donor allele repertoire. Accumulation of this type of event would eventually result in generation of strains with different *msp2* donor allele repertoires. The donor alleles that are maintained in the genome reflect balanced selection for both diversity in the allelic repertoire and for continued immune evasion and growth fitness. Accordingly, analysis of the donor allele repertoire is informative for strain structure and epidemiology.

The St. Maries strain of *A. marginale* was the first to be fully sequenced and contains seven donor alleles for *msp2* (Fig. 3); however, two of the donor alleles have exact copies (P1 and G11, and 2 and 3H1), such that there are only five unique *msp2* donor allele sequences to contribute to gene conversion events (Brayton et al., 2005). This observation demonstrates that the small repertoire of five donor alleles coupled with the combinatorial gene conversion mechanism is sufficient for continual immune evasion and lifelong persistence. The second fully sequenced strain of *A. marginale* was the Florida strain, which was shown to contain eight *msp2* donor alleles (Dark et al., 2009). The contextual positions (loci) of the donor alleles are conserved between the two strains, with the exception of *msp2* donor allele 2, which is shifted ~30 kb as a result of a genomic inversion mediated by *msp3* donor alleles 1 and 2. The additional *msp2* donor allele in the Florida strain is syntenic to the R2 *msp3* remnant sequence in the St. Maries strain. The syntenic conservation of the donor alleles is an indication that the repertoire is not evolving rapidly through production of new insertion sites for these sequences. The Florida strain contains two sets of identical donor alleles. The first set of identical donor alleles (KAV4F15 and KAV1F20) is syntenic to one of the St. Maries sets of identical donor alleles (2 and 3H1). The second set of identical alleles (TTV1O6 and TTV4F15) includes the allele that is positionally unique (TTV1O6) in the Florida strain (Fig. 3).

The *msp3* gene family in the St. Maries strain consists of the single expression site, seven donor alleles, and two remnant sequences (Brayton et al., 2005). The remnant sequences are short sequences with identity to the conserved flanking regions of *msp3* and are unlikely to contribute to gene conversion events. R1 corresponds to the 705-bp 5′ end of full-length *msp3*. R2 is just 210 bp and corresponds to the 5′ conserved flanking region upstream from the HVR. Each of the seven donor alleles contains a unique HVR sequence. Genome sequencing of the Florida strain of *A. marginale* revealed that each of the donor alleles is positionally conserved, except that as stated above, a genomic inversion event has been mediated by *msp3* donor alleles 1 and 2. The Florida strain contains only one of the short remnant sequences, R1, as the second (R2 in the St. Maries strain) has been replaced by an *msp2* donor allele. None of the Florida strain *msp3* donor alleles are identical; however, comparing between strains reveals that two sets of identical donor alleles are present, St. Maries 1 and Florida C 4F15, and St. Maries 6 and Florida 1F20 (Dark et al., 2009).

msp2 Donor Allele Repertoire from Additional Strains

In addition to the St. Maries and Florida strains, the *msp2* donor allele repertoire has been examined in another six strains. Nine donor alleles have been deposited in GenBank for the South Idaho strain (Brayton et al., 2001, 2002), the most detected for any *A. marginale* strain to date; however, two of these sequences, A-2 and A10, only differ by two nucleotides, and thus these may represent the same donor allele. The remaining five strains were analyzed in a study that determined the *msp1α* genotypes (*msp1α* contains repeats at the 5′ end that differ in both sequence and number and has been used to differentiate strains) for *A. marginale* strains isolated from a cattle herd in Kansas that found the presence of three families of strains based on *msp1α* genotypes: the B family, which contained from two to six B repeats; the DE family, which contained two to nine D repeats and usually ended with a single E repeat; and the EMΦ family, which was monotypic with three repeats of sequence EMΦ (Palmer et al., 2001, 2004). To test whether strains with identical or similar *msp1α* genotypes (within a family) had varying compositions of *msp2* donor alleles, two strains with identical genotypes (5B) and one strain with a closely related genotype (6B) were chosen. Conversely, whether strains with disparate *msp1α* genotypes had varying *msp2* donor allele repertoires was tested by analyzing two strains with very different *msp1α* genotypes (EMΦ and 6DE). The results revealed that strains with closely related or identical *msp1α* genotypes (i.e., belonging to the same *msp1α* genotype family) had identical *msp2* donor allele repertoires and, likewise, strains with dissimilar genotypes had donor allele repertoires that differed to varying degrees (Rodriguez et al., 2005). All B family genotypes had identical *msp2* donor allele repertoires. The 6DE strain had three donor alleles in common with the B family strains; however, EMΦ had no donor alleles that were identical to any of the other Kansas herd strains or to either of the other strains for which *msp2* donor alleles have been extensively analyzed (Table 1).

The evolution of the *msp1α* genotype appears to be on a shorter time scale than that of the *msp2* donor allele repertoire, as evidenced by two herd studies that showed that there were families of closely related *msp1α* genotypes circulating within the herds (Palmer et al., 2001, 2004). One explanation would be that an *A. marginale msp1α* genotype becomes established into an animal or tick population, and over time the *msp1α* gene would diverge to closely related forms of repeats, but that in this same time frame the *msp2* donor allele repertoire would not undergo any changes. Therefore, we expect that the relationship between *msp1α* genotype and *msp2* donor allele repertoire holds only within a spatially and/or temporally defined ecological setting. Strains may undergo convergent evolution to have similar *msp1α* genotypes, but not have similar *msp2* donor allele repertoires. For example, the South Idaho strain has an *msp1α* genotype (5DE) similar to that of the Kansas 6DE strain, but has only a single donor allele with 100% sequence identity (Table 1). These two strains were isolated 20 years apart from different geographic regions of the United States (McGuire et al., 1984; Palmer et al., 2004).

Donor Allele Usage

Analysis of the usage of the *msp2* and *msp3* donor alleles in the St. Maries strain showed preferential usage of certain alleles (Futse et al., 2009). *msp2* donor allele 1 and *msp3* donor alleles 3 and 7 were overrepresented during acute infection in an analysis of infection in four animals, while *msp2* donor allele 9H1 and *msp3* donor alleles 2, 4, and 6 were underrepresented. These results prompted further analysis of the role of locus position on donor allele usage. Distance from the origin of replication and distance from the expression site were tested for an effect on the usage of donor alleles, and were found not to significantly affect usage of both *msp2* and *msp3* donor alleles. It was noticed early on that some of the donor alleles occur in a "donor allele complex," with a pairing of one *msp3* donor allele with one *msp2* donor allele in a tail-to-tail arrangement (Brayton et al., 2001). Whole-genome sequencing subsequently revealed that four such complexes exist in the Florida and St. Maries strain genomes (Brayton et al., 2005; Dark et al., 2009). The complex is flanked by repeat sequences, and it was hypothesized that this complex could facilitate recombination events. However, analysis of preferential usage of donor alleles found in the donor allele complex revealed that these alleles were not used preferentially. These data indicate that the donor allele loci have been selected as efficient mediators of gene conversion events and do not contribute to preferential allele usage and an ordered appearance of variants, as seen in *T. brucei* (Morrison et al., 2005).

Epidemiological Consequences of the Allelic Repertoire

The consequences of the composition of the *msp2* donor allele repertoire may be significant in shaping the *A. marginale* strain structure in the host population. Animals carrying more than a single strain were initially detected by identification of unique *msp1*α genotypes (Allred et al., 1990; Palmer et al., 2001, 2004). In these coinfected animals, the EMΦ *msp1*α genotype was present with a second strain corresponding to a B or DE family genotype. As the EMΦ strain was shown to have a unique MSP2 repertoire, this led to the hypothesis that a unique *msp2* donor allele repertoire is required for "strain superinfection" to occur (Rodriguez et al., 2005). Strain superinfection specifically describes infection with a second strain following infection and development of an immune response to a primary strain. The driving concept is that immunity against the MSP2 variants of the primary strain (defined by the donor allele repertoire) prevents superinfection with a second strain with a similar allelic repertoire—remembering that the tick transmits "simple" variants that would be recognized by the immune response developed during acute infection with the primary strain. However, if the second strain has a unique *msp2* donor allele repertoire, then superinfection is established, as the tick-transmitted simple variants will not be recognized by the existing immune response. In this manner, there would be selection pressure for divergence of donor allele repertoires among strains so that a given strain would have the greatest opportunity to infect a host population despite preexisting immunity to other strains.

This hypothesis was tested with reciprocal superinfection experiments (Futse et al., 2008). Either the St. Maries, 6DE, or EMΦ strain was used to establish long-term persistent infection in the bovine host, which was subsequently challenged by tick transmission of one of the other strains. These strains were selected for the study as they contained entirely novel *msp2* donor allele repertoires (St. Maries and EMΦ) or overlapping donor allele repertoires (St. Maries has a single novel *msp2* donor allele compared with 6DE, while 6DE has two unique alleles compared with St. Maries). The data showed that a single unique *msp2* donor allele was sufficient for superinfection to be established in the presence of a robust immune response to the initial infecting strain, provided that the unique allele was used

TABLE 1 *msp2* donor allele repertoires

Pseudogene name	Strain (*msp1α* genotype)[a]								Reference(s)
	StM (JBB)	FL (A7B)	SI (5DE)	Ks 3201 (EMΦ)	Ks 6192 (5B)	Ks 7072 (5B)	Ks 9038 (6B)	Ks 8416 (6DE)	
Ψ1-NAI4F15	X	X			X	X	X	X	Brayton et al., 2001; Dark et al., 2009
Ψ2-3H1	X							X	Brayton et al., 2001, 2005
G11-P1 -NAV2G15	X	X						X	Brayton et al., 2005; Dark et al., 2009
E6/F7-TTV-TTV4F15-1O6	X	X	X		X	X	X	X	Brayton et al., 2002, 2005; Dark et al., 2009
9H1-SI 13	X		X						Brayton et al., 2005; unpublished: GenBank accession no. AF402257
A2			X						Brayton et al., 2001
A3			X						Brayton et al., 2001
A10			X						Brayton et al., 2001
B10			X						Brayton et al., 2001
C1-1			X						Brayton et al., 2001
D2			X						Brayton et al., 2001
SI 4			X						Unpublished: GenBank accession no. AF402258
Ks 4				X					Rodriguez et al., 2005
Ks 8				X					Rodriguez et al., 2005
Ks 20				X					Rodriguez et al., 2005
Ks 22					X	X	X		Rodriguez et al., 2005
Ks 35					X	X	X	X	Rodriguez et al., 2005
Ks 37				X					Rodriguez et al., 2005
Ks 39					X	X	X		Rodriguez et al., 2005
Ks 41					X	X	X		Rodriguez et al., 2005
Ks 42								X	Rodriguez et al., 2005
Ks 46				X					Rodriguez et al., 2005
FL KAV4F15-1F20		X							Dark et al., 2009
FL NAV4F15		X							Dark et al., 2009
FL NAI1O11		X							Dark et al., 2009

[a]Strain designations are given along with *msp1α* genotype in parentheses. StM, St. Maries; FL, Florida; SI, South Idaho; Ks, Kansas.

in the expression site of the second infecting strain at the time that superinfection was established. Once established, the ability to generate unique mosaic variants allows continuation of the superinfection.

A. MARGINALE SUBSP. *CENTRALE*

Antigenic Variation in *A. marginale* subsp. *centrale*

A. marginale subsp. *centrale* was initially classified as a subtype of *A. marginale* based on its morphology and reduced virulence (Theiler, 1910, 1911). Work in the early 1900s by Sir Arnold Theiler demonstrated the ability of this organism to be used as a live vaccine to mitigate the severity of infection with *A. marginale* sensu stricto strains (Theiler, 1912), and it has been used as a vaccine in Israel, southern Africa, South America, and Australia (Potgieter, 1979; Losos, 1986; Pipano et al., 1986; Jorgensen et al., 1989) (see chapter 11 for detailed discussion of live vaccine mechanisms). Immunized animals maintain long-term persistent infection with the vaccine strain (Krigel et al., 1992; Shkap et al., 2002, 2008; Galletti et al., 2009), consistent with sequential generation of MSP2 and MSP3 antigenic variants. Genome sequencing of the Israel vaccine strain of *A. marginale* subsp. *centrale* (which is derived from the original Theiler isolate) revealed the complete structure of both *msp2* and *msp3* gene families (Herndon et al., 2010; Molad et al., 2010). As seen from the genome map (compare Fig. 3 and 5), the synteny of these gene families is not conserved between *A. marginale* subsp. *centrale* and *A. marginale* sensu stricto strains. Furthermore, comparative genome analysis revealed that these genes have been sites of genome recombination, leading to inversions around the origin of replication. The *A. marginale* subsp. *centrale msp2* family consists of a single expression site and eight donor alleles, with one set of identical alleles (G1 and G2). The structure of the molecules is similar to the sensu stricto strains, with conserved flanking ends and a central HVR that maintains the two conserved cysteine residues. Interestingly, all but one of the donor alleles encode either TAV or QAV in the block 1 region of the HVR, sequences not seen in the sensu stricto strains. The remaining donor allele encodes a KAV triplet sequence in this position.

The *msp3* gene family consists of a putative expression site, eight donor alleles, and a short remnant sequence that corresponds to the 5′ conserved flanking region upstream from the HVR (Fig. 5). Definitive identification of the *msp3* expression site has not been established; however, ACIS00617 was nominated as the putative expression site as it encodes an amino-terminal extension of 58 amino acids relative to the donor alleles (shorter than the amino terminus seen in *A. marginale* sensu stricto strains) as well as the complete carboxy terminus. This locus is not syntenic to the sensu stricto strain *msp3* expression site, with the *A. marginale* subsp. *centrale* gene being flanked by *nuoI* and *ankA*, rather than *orfX* and *hflK*. *A. marginale* subsp. *centrale msp3* donor allele 3 sits in the locus flanked by *orfX* and *hflK*, and although it also encodes the complete carboxy terminus, it is truncated at the 5′ end relative to other donor alleles and is thought not to encode the full-length protein. None of the *msp3* HVRs are identical within the *A. marginale* subsp. *centrale* genome, nor between the *A. marginale* subsp. *centrale* and sensu stricto genomes. The tail-to-tail *msp2*-*msp3* donor allele complex seen in sensu stricto strains is not found in *A. marginale* subsp. *centrale*; instead, two of the donor alleles are found in an overlapping head-to-head arrangement. Whether this is a recurring arrangement in other sensu lato strains is unknown.

The *msp2* Superfamily in *A. marginale* subsp. *centrale*

Analysis of the PF01617/MSP2 superfamily (Löhr et al., 2002; Brayton et al., 2005; Noh et al., 2006) reveals that *opag1* to *3*, associated with the *msp2* expression site, and *msp4* are present. However, *omp1* to *15*, found in sensu stricto strains (Brayton et al., 2005; Noh et al., 2006; Dark et al., 2009), represent a family reduction in *A. marginale* subsp. *centrale*: the closely related *omp7* to *9* are collapsed to a single open reading frame and homologs for *omp2*, *3*, *6*, and *15* are missing. Within the nine maintained OMPs, 57 to 75% amino acid identity exists between the homologs.

ANAPLASMA OVIS

A. ovis, a small ruminant pathogen, has been shown to establish persistent infections in goats (Palmer et al., 1998). Investigation demonstrated the presence of both the *msp2* and *msp3* gene families, with Southern anal-

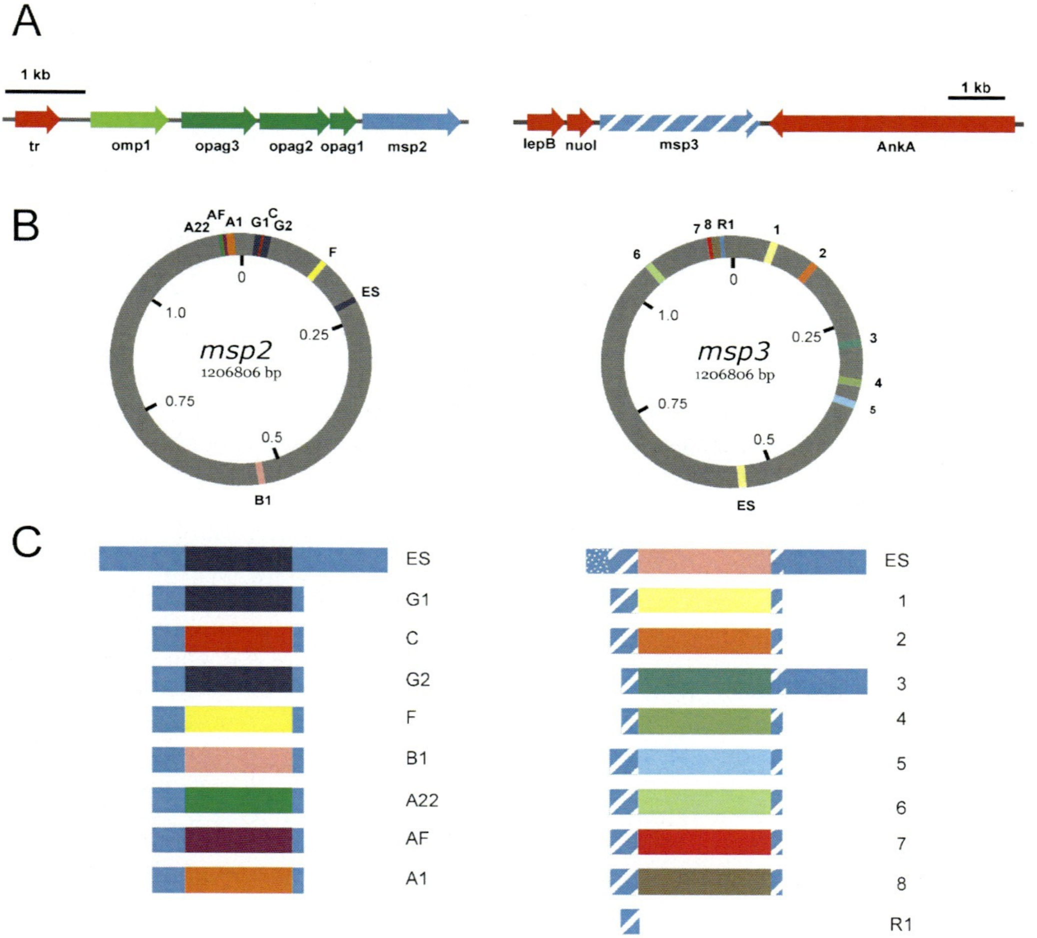

FIGURE 5 Schematic representation of *A. marginale* subsp. *centrale msp2* and *msp3* expression sites (ES), genomic arrangement, and donor allele repertoires. (A) The *msp2* (left) and *msp3* (right) operons that are the sole expression sites for these genes. *msp2* is the 3′ gene in an operon with *opag1* to *3*. The putative *tr* and *omp1* genes reside upstream and are depicted in red and green, respectively. The candidate *msp3* expression site is flanked by *nuoI* and *ankA* (red). (B) The genomic arrangement of the *msp2* and *msp3* expression site and donor allele repertoires. Alleles depicted in identical colors (e.g., G1 and G2) have identical HVR sequences. The genome backbone is shown as gray. (C) The donor allele and remnant sequence repertoires for *msp2* and *msp3*, showing the relative portion of the expression-site molecule that each encodes. The *msp3* remnant sequence (R1) does not contain any portion of the HVR. *msp3* has a 3′ end identical to *msp2* (solid blue), regions flanking the HVR that have similarity to *msp2* (diagonal stripes), and a unique 5′ end (stippled) that is shorter than the 5′ end sequence found in *A. marginale msp3*. Comparison with Fig. 3 shows that the synteny of these sequences has not been maintained. doi:10.1128/9781555817336.ch12.f5

ysis indicating each gene family has several copies similar to *A. marginale*. *A. ovis* and *A. marginale* MSP2 share epitopes recognized by $CD4^+$ T lymphocytes from *A. marginale*-immunized cattle (Brown et al., 2001). The *msp2* gene was shown to reside in an operon identical in structure to that of *A. marginale* (Löhr et al., 2002). Just eight MSP2 sequences have been reported to GenBank, representing seven different variants, revealing that *A. ovis* MSP2, like *A. marginale* MSP2, is composed of a central HVR flanked by conserved regions (Brown et al., 2001; Löhr et al., 2002). These sequences appear to represent segmental variants of each other, and show block 1 sequences encoding NAV, TAV, and TVV triplets. To date there has been no analysis of how these genes vary, but it is presumed that they undergo gene conversion. *A. ovis* has not been subjected to whole-genome sequencing, so the extent of PF01617 genes is not known; however, the genome is known to contain *msp4*, which has been the target of several studies examining strain variation (de la Fuente et al., 2002, 2007; Psaroulaki et al., 2009).

A. PHAGOCYTOPHILUM

Antigenic Variation in *A. phagocytophilum*

A. phagocytophilum is a zoonotic agent that persists in a broad variety of animal species, including but not limited to sheep, several species of mice, white-tailed deer, horses, cattle, dogs, and likely humans when untreated (Telford et al., 1996; Egenvall et al., 2000; Massung et al., 2005; Franzen et al., 2009; Granquist et al., 2010a). Persistence of *A. phagocytophilum* in small-mammal reservoirs is thought to play a key role in the epidemiology of human disease, although the interplay of host predilection of strains, reservoir host, and duration of infection is not well understood (Massung et al., 2005; Dumler et al., 2007; Woldehiwet, 2010). Antigenic variation in *A. phagocytophilum* is effected through the homolog of *A. marginale msp2* (35). Notably, an ortholog to *A. marginale msp3* is absent from the fully sequenced HZ strain (Dunning Hotopp et al., 2006). *A. phagocytophilum* MSP2 (also known as p44) is an immunodominant multimeric molecule containing a central HVR flanked by conserved ends (Asanovich et al., 1997; IJdo et al., 1997, 1998; Zhi et al., 1997; Park et al., 2003b; Granquist et al., 2010b). The antibody response to the MSP2 HVR is variant specific and diminishes rapidly, consistent with the idea that antigenic variation of MSP2 is responsible for persistence (Granquist et al., 2010b). Although other novel methods of antigenic switching have been described for *msp2* (Zhi et al., 2002), the predominant method appears to be gene conversion (Barbet et al., 2003; Lin and Rikihisa, 2005) utilizing a RecF recombination pathway (Lin et al., 2006). Early data suggested that the changes in *msp2* corresponded to segmental gene conversion; however, these studies were done without the benefit of the complete genome sequence (Barbet et al., 2003). In a study that fully utilized the genome sequence (using the same strain for the study as was sequenced), all expression-site variants mapped back to functional donor alleles or complete genes (Lin and Rikihisa, 2005). One explanation for the lack of segmental gene conversion events may be the short time frame of the experiment—data from *A. marginale* indicate that whole donor alleles tend to be utilized early in infection (resulting in "simple" variants) and segmental variants occur later (resulting in "complex" mosaic variants). An alternative explanation is that because *A. phagocytophilum* has such a large repertoire of *msp2* donor alleles, persistence can be achieved without the need for additional segmental variants. Antigenic variation must be a proactive, dynamic process that occurs prior to immune clearance (Borst, 1991). Studies in SCID mice demonstrate that gene conversion of *msp2* occurs in the absence of selection pressure (Lin and Rikihisa, 2005), as would be necessary for a true antigenic variation event.

Strain Definition of the MSP2 Repertoire

Genome sequencing revealed that the *msp2* gene family is greatly expanded in *A. phagocytophilum* (Fig. 6). This gene family has also been referred to as *p44* (Caspersen et al., 2002; Barbet et al., 2003; Park et al., 2003a), and this terminology was used when annotating the *A. phagocytophilum* HZ strain genome, resulting in a total of 111 sequences being annotated as *p44* (Dunning Hotopp et al., 2006). In addition, a gene (APH_1361) was annotated as *msp2* based on work of Rikihisa and coworkers (Lin et al., 2004a), and two additional genes were annotated as *msp2* family members (APH_1017 and APH_1325). While it was correctly noted that this "*msp2*" gene is distinct from the *p44* genes (Dunning Hotopp et al., 2006), the *p44* genes (similarity, ~40%) rather than "*msp2*" (similarity, ~20%) appear to be the true orthologs of *A. marginale msp2*. *p44* as a true *A. marginale msp2* ortholog is further corroborated by the description of synteny between the *A. marginale msp2* and *A. phagocytophilum msp2* (*p44*) expression sites (Barbet et al., 2003; Lin et al., 2003) (compare Fig. 4 and 6). The gene originally reported as *msp2* (Lin et al., 2004a) would more accurately be described as an *msp2* superfamily member; this gene and the two genes initially annotated as *msp2* family genes are referred to here as *msp2* SF (for superfamily). The complement of "true" *msp2* genes includes 10 genes that are full length or close to full length, 13 short 3′ fragments, 10 fragments that start in the HVR and extend into the 3′ conserved end, three short 5′ fragments, and the remaining 75 genes that contain an HVR but are truncated at the 5′ or 3′ end or truncated at both ends and could serve as donor alleles (Fig. 6). Additional PF01617 members include the three *msp2* SF genes, *msp4*, and three genes annotated as *omp1X*, *omp1N*, and *omp1A* (APH_1219, APH_1220, and APH_1359, respectively). Figure 7 shows an alignment of the full-length or nearly full-length *A. phagocytophilum* MSP2 sequences with *A. phagocytophilum* MSP2SF and *A. marginale* MSP2. Several of

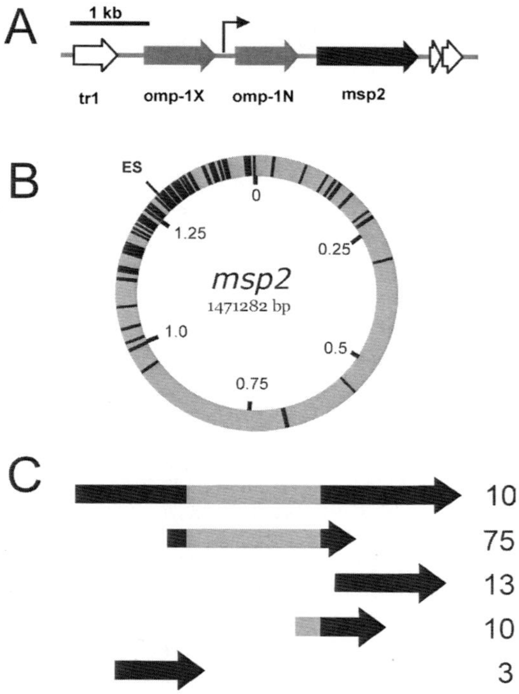

FIGURE 6 Schematic representation of *A. phagocytophilum msp2* expression site (ES), genomic arrangement, and donor allele/pseudogene repertoire. (A) Genomic arrangement of the *msp2* operon. *msp2* is shown in black at the 3′ end of the two-gene operon. A bent arrow indicates the position of the operon promoter. *omp1N* is also known as *p44ESup1* (Barbet et al., 2003), and this gene and *omp1X* (shown in gray) are both members of PF01617 and therefore related to *msp2*. The putative transcriptional regulator (*tr1*, in white) is found upstream from the *msp2* operon. (B) Genomic arrangement of the *msp2* family. Sequences with identity to *msp2* are shown as black bars, with the *msp2* expression site marked with a longer bar. The genome backbone is gray. (C) Diagram of the *msp2* gene family, with conserved regions shown as black and the HVR shown as gray. There are 10 full-length genes, 75 sequences that contain the HVR and could act as donor alleles, and 26 pseudogenes that are unlikely to contribute to variation as they are lacking the necessary components for gene conversion events, i.e., conserved flanking regions and HVR. doi:10.1128/9781555817336.ch12.f6

the full-length genes (APH_0662, APH_0663, APH_0664, and APH_1355) are less similar to the "typical" *msp2* and may actually represent distinct orthologs (i.e., related at the superfamily level) rather than true paralogs in much the same way that *A. marginale* has *msp2*-related genes (i.e., OMP1 to 15) that are distinct from *msp2* but are members of PF01617. In fact, three of these genes (APH_0662 to APH_0664) are closely positioned and may be transcribed in a polycistron, as are some of the *A. marginale* superfamily genes. In addition, three genes (APH_1275, APH_1215, and APH_1269) have start sites that would yield a protein of a similar size to the protein synthesized from the known expression site, but each have different carboxy-terminal ends. Thus, if there is a requirement for "MRKG..." at the amino terminus and "...FAF" at the carboxy terminus, there is only one expression site for *msp2* in the *A. phagocytophilum* genome. While an early report prior to the availability of the genome sequence indicated that *msp2* transcription from the *p44-1/p44-18* locus (APH_1194/ APH_1195) involved mRNA splicing such that a full-length *msp2-18* gene was generated from the truncated donor allele (Zhi et al., 2002), it is now thought that the polycistronic expression site encoding APH_1221 is the sole expression site for *msp2* (Barbet et al., 2003; Lin et al., 2003; Lin and Rikihisa, 2005).

Donor Allele Usage

A study designed to determine the usage of the donor alleles mapped 199 expression-site variants to the HZ strain genome (Foley et al., 2009). Importantly, the expression-site variants that were employed in the study were from a range of strains infecting a variety of host species, which may have different donor allele repertoires. However, most of the expression-site variants (109/199) had ≥90% sequence identity to one of the HZ strain donor alleles. Mapping of these sequences with high similarity values to the donor allele repertoire shows that donor alleles that cluster near the expression site have a higher likelihood of being recombined into the expression site.

A. phagocytophilum msp2 Repertoire Variation

Relatively little is known about how the functional donor allele repertoire varies in *A. phagocytophilum*, as to date only a single strain of *A. phagocytophilum* (HZ) has been completely sequenced, and the highly repetitive nature of the genome, due in large part to the *msp2* gene family, makes genome assembly of next-generation sequencing data extremely difficult. Newer, "third-generation" sequencing technology, such the Pacific Biosciences sequencer, providing longer individual read lengths (currently ~3 kb) may facilitate elucidation of the *msp2* repertoire from additional strains (McCarthy, 2010). Many studies have now examined *msp2* variation, but in the absence of a genomic context for appropriate analysis of the variants, questions remain regarding the mechanism of variation. There has been some debate as to whether the *A. phagocytophilum* gene conversion mechanism uses only whole donor alleles or if segmental gene conversion is employed, as in *A. marginale* (Lin et al., 2003). If expression-site variants used in the study (Foley et al., 2009) described above differed by 3%, and were obtained from a strain other than HZ, these may represent either segmental variants or strain differences. This question is impossible to answer in the absence of known donor allele repertoires.

Because *msp2* was thought to play a role in pathogenesis and host tropism, several studies targeted *msp2* for strain typing or simply to understand the extent of variation of this gene; however, many of these studies were undertaken prior to a full knowledge of the extent of the functional donor allele repertoire (Murphy et al., 1998; Caspersen et al., 2002; Lin et al., 2003, 2004b; Casey et al., 2004; Felek et al., 2004; Teglas and Foley, 2006). Only one of these early studies compared a genomic locus containing *msp2* genes between strains (Lin et al., 2004b). This study found that the *p44-1/p44-18* genomic locus

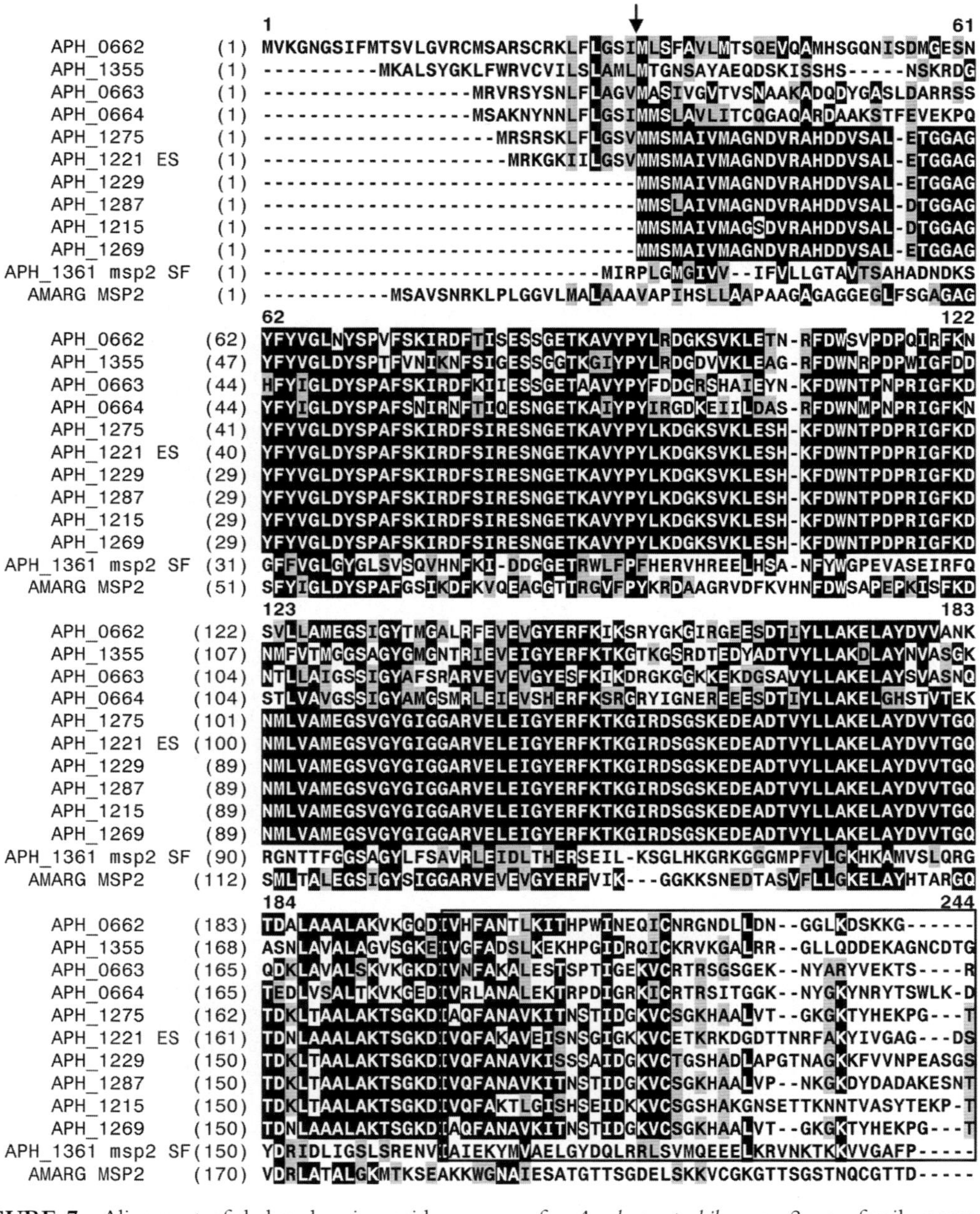

FIGURE 7 Alignment of deduced amino acid sequences for *A. phagocytophilum msp2* superfamily genes and *A. marginale msp2*. Each *A. phagocytophilum* sequence is labeled with its genome identifier number (taken from GenBank accession no. CP000235) (Dunning Hotopp et al., 2006) and represents an *msp2* sequence that is full length or nearly full length. The *msp2* expression site sequence is denoted with "ES." All genes are annotated as *p44* paralogs except APH_1361, which is annotated as *msp2*. The central HVR is boxed. Conserved residues are highlighted in black, while blocks of similar residues are shown on a gray background. A positionally conserved methionine residue, indicated with an arrow, may be the start codon for these genes. Numbers above the alignment indicate alignment position number, while numbers in parentheses indicate position for each sequence. The *A. marginale msp2* sequence used for alignment is from GenBank accession no. U07862 (Palmer et al., 1994).
doi:10.1128/9781555817336.ch12.f7

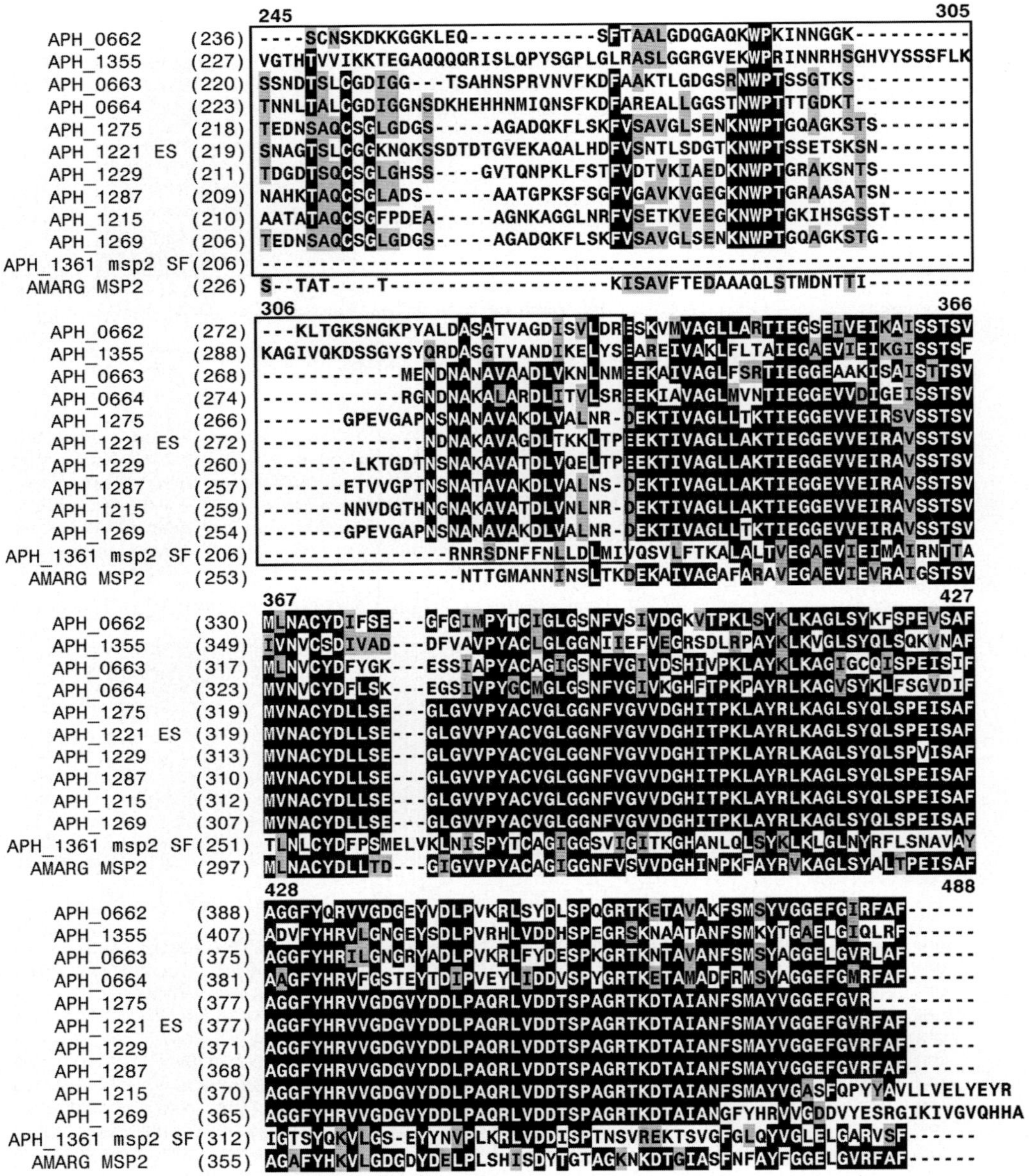

FIGURE 7 *Continued.*

varied between strains from different geographic regions, including different HVRs and even deletion of one *msp2* paralog. A comparative genomic microarray analysis employing 70-mer oligonucleotides suggested that all but four unique segments representing *msp2* HVRs from the HZ strain were detected in two additional strains (Dunning Hotopp et al., 2006). Whether these HVR oligonucleotide segments are maintained in the same context in the other strains or are present, but in an alternative context, representing a novel HVR is unknown. Analysis of *msp2* sequences obtained from RNA from three human patient-derived strains shows that some of the sequences matched to *msp2* donor alleles present in the sequenced HZ strain, while most were unique (Lin et al., 2002). It is unknown if these unique sequences represent strain differences in the *msp2* donor allele repertoire or if they are segmental variants generated by gene conversion events in the expression site of the patient strains.

Implications of the Donor Allele Repertoire

A. phagocytophilum displays a host species predilection that varies by geographic region, and may actually be a genogroup of closely related species (Dumler et al., 2001; Teglas and Foley, 2006). The suggestion that *msp2* gene content could play a role in species-specific infection is intriguing, yet difficult to determine with the available data. Given the large repertoire of *msp2* genes, differences in gene content will be very difficult to determine by high-throughput methodologies, such as microarray or pyrosequencing, and may require complete genome sequencing.

ANAPLASMA PLATYS

A. platys is a pathogen of dogs, phylogenetically most closely positioned to *A. phagocytophilum* (Dumler et al., 2001). Recently, the *msp2* expression-site locus was cloned and shown to be most similar to that of *A. phagocytophilum*, with the locus containing the putative transcriptional regulator *tr1* at the 5′ end, followed by *omp1X* and the *msp2* gene (Lai et al., 2011). The *omp1N* gene is missing from the locus (Fig. 6). The size of the donor allele repertoire has not been elucidated.

THE MSP2 SUPERFAMILY IN *EHRLICHIA*

Anaplasma MSP2 has an immunodominant homolog that is a member of Pfam01617 within *Ehrlichia* species. This homolog is called OMP1/p28 (*E. chaffeensis*, *E. ewingii*) (Zhang et al., 2008), p30/p28 (*E. canis*) (Reddy et al., 1998; McBride et al., 1999; Yu et al., 2000), or MAP1 (*E. ruminantium*) (van Vliet et al., 1994). The genes encoding these proteins are arranged in a tandem array in each genome, with 16 to 22 genes, depending on the ehrlichial species (Ohashi et al., 2001; van Heerden et al., 2004; Collins et al., 2005; Dunning Hotopp et al., 2006; Mavromatis et al., 2006). The array is bordered by *ndk* and a putative transcriptional regulator at the 5′ end and *secA* at the 3′ end (Fig. 8), and is syntenic to the *msp2* expression site found in *Anaplasma* species (Löhr et al., 2004). In *E. canis*, there is a duplication of three of the genes from the array into a separate locus in the genome. Multiple genes within the array appear to be transcriptionally active at the same time, suggesting that differential transcription is not a mechanism for expressing different alleles of this variable gene family (Ohashi et al., 2001; Unver et al., 2001, 2002; van Heerden et al., 2004; Bekker et al., 2005). Multistrain analysis of *p30-10* reveals a high degree of conservation (zero to three nucleotide changes). In addition, a study examining variation of the *map1* gene sequence in 30 isolates of *E. ruminantium* showed no evidence for positive selection, as would be expected if the molecule was involved in immune evasion (Allsopp et al., 2001). Similarly to PF01617 genes contained within *Anaplasma* species that do not undergo antigenic shift, such as *msp4* or *omp1* to *15*, the *Ehrlichia* spp. genes appear more likely to undergo relatively slow divergence rather than true antigenic variation associated with immune evasion. *E. canis*, *E. chaffeensis*, and *E. ruminantium* are well documented to persist in their respective hosts, and it is assumed that closely related ehrlichial species also establish persistent infections (Codner and Farris-Smith, 1986; Andrew and Norval, 1989; Dumler et al., 1993; Wen et al., 1997; Harrus et al., 1998; Davidson et al., 2001). However, whether antigenic variation has a role in persistence by pathogens in the genus

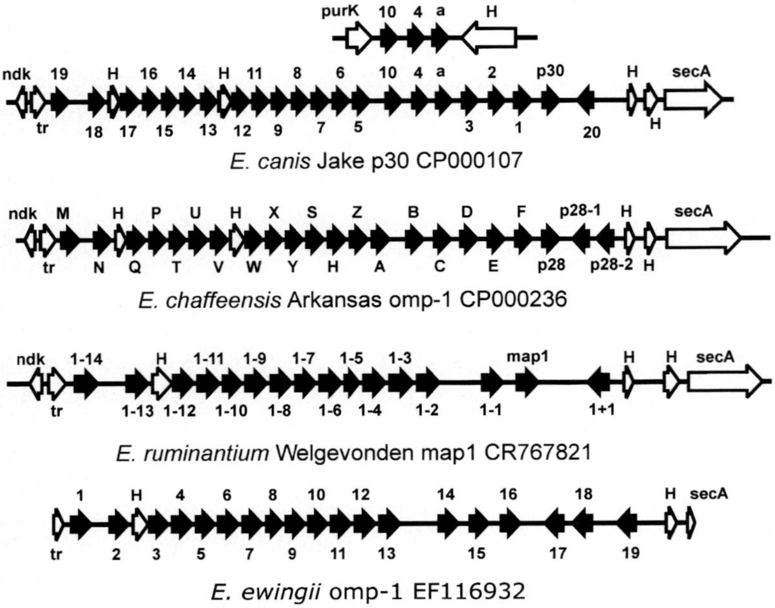

FIGURE 8 Schematic representation of *Ehrlichia* species *msp2* homolog tandem gene arrays. Genome sequences were used to identify the genomic context of sequences for *E. canis* (strain Jake; GenBank accession no. CP000107), *E. chaffeensis* (strain Arkansas; GenBank accession no. CP000236), and *E. ruminantium* (strain Welgevonden; GenBank accession no. CR767821); however, nomenclature was taken from published analyses of these gene loci, as was information on *E. ewingii* (Ohashi et al., 2001; van Heerden et al., 2004; Collins et al., 2005; Dunning Hotopp et al., 2006; Mavromatis et al., 2006; Zhang et al., 2008). The genome backbone is indicated by a gray line; PF01617 genes are shown in black, and other unrelated genes are shown in white. Gene names are indicated or shown as "H" for hypothetical. *E. canis* contains a second smaller locus wherein *p30* genes *10*, *4*, and *a* are duplicated. doi:10.1128/9781555817336.ch12.f8

Ehrlichia has not been definitively proven nor excluded.

SUMMARY

The immunodominant surface proteins of *Anaplasma* and *Ehrlichia* species are members of PF01617 and thus share a degree of sequence similarity, and perhaps functionality. Although the role of these proteins in persistence is unknown for *Ehrlichia* species, the *Anaplasma* spp. proteins are responsible for immune evasion and persistence of these organisms in their respective hosts. The research highlighted here shows that the implications of donor allele repertoires go beyond antigenic variation in the individual hosts, but that they also play critical roles in the epidemiology of pathogen strain structure and possibly host tropism.

REFERENCES

Abbott, J. R., G. H. Palmer, C. J. Howard, J. C. Hope, and W. C. Brown. 2004. *Anaplasma marginale* major surface protein 2 $CD4^+$-T-cell epitopes are evenly distributed in conserved and hypervariable regions (HVR), whereas linear B-cell epitopes are predominantly located in the HVR. *Infect. Immun.* **72:**7360–7366.

Alleman, A. R., and A. F. Barbet. 1996. Evaluation of *Anaplasma marginale* major surface protein 3 (MSP3) as a diagnostic test antigen. *J. Clin. Microbiol.* **34:**270–276.

Alleman, A. R., G. H. Palmer, T. C. McGuire, T. F. McElwain, L. E. Perryman, and A. F. Barbet. 1997. *Anaplasma marginale* major surface protein 3 is encoded by a polymorphic, multigene family. *Infect. Immun.* **65:**156–163.

Allred, D. R., T. C. McGuire, G. H. Palmer, S. R. Leib, T. M. Harkins, T. F. McElwain, and A. F. Barbet. 1990. Molecular basis for surface antigen size polymorphisms and conservation of a neutralization-sensitive epitope in *Anaplasma marginale*. *Proc. Natl. Acad. Sci. USA* **87:**3220–3224.

Allsopp, M. T., C. M. Dorfling, J. C. Maillard, A. Bensaid, D. T. Haydon, H. van Heerden, and B. A. Allsopp. 2001. *Ehrlichia ruminantium* major antigenic protein gene (*map1*) variants are not geographically constrained and show no evidence of having evolved under positive selection pressure. *J. Clin. Microbiol.* **39:**4200–4203.

Andrew, H. R., and R. A. Norval. 1989. The carrier status of sheep, cattle and African buffalo recovered from heartwater. *Vet. Parasitol.* **34:**261–266.

Asanovich, K. M., J. S. Bakken, J. E. Madigan, M. Aguero-Rosenfeld, G. P. Wormser, and J. S. Dumler. 1997. Antigenic diversity of granulocytic *Ehrlichia* isolates from humans in Wisconsin and New York and a horse in California. *J. Infect. Dis.* **176:**1029–1034.

Barbet, A. F., A. Lundgren, J. Yi, F. R. Rurangirwa, and G. H. Palmer. 2000. Antigenic variation of *Anaplasma marginale* by expression of MSP2 mosaics. *Infect. Immun.* **68:**6133–6138.

Barbet, A. F., P. F. Meeus, M. Belanger, M. V. Bowie, J. Yi, A. M. Lundgren, A. R. Alleman, S. J. Wong, F. K. Chu, U. G. Munderloh, and S. D. Jauron. 2003. Expression of multiple outer membrane protein sequence variants from a single genomic locus of *Anaplasma phagocytophilum*. *Infect. Immun.* **71:**1706–1718.

Barbour, A. G., and B. I. Restrepo. 2000. Antigenic variation in vector-borne pathogens. *Emerg. Infect. Dis.* **6:**449–457.

Bekker, C. P., M. Postigo, A. Taoufik, L. Bell-Sakyi, C. Ferraz, D. Martinez, and F. Jongejan. 2005. Transcription analysis of the major antigenic protein 1 multigene family of three in vitro-cultured *Ehrlichia ruminantium* isolates. *J. Bacteriol.* **187:**4782–4791.

Berriman, M., E. Ghedin, C. Hertz-Fowler, G. Blandin, H. Renauld, D. C. Bartholomeu, N. J. Lennard, E. Caler, N. E. Hamlin, B. Haas, U. Bohme, L. Hannick, M. A. Aslett, J. Shallom, L. Marcello, L. Hou, B. Wickstead, U. C. Alsmark, C. Arrowsmith, R. J. Atkin, A. J. Barron, F. Bringaud, K. Brooks, M. Carrington, I. Cherevach, T. J. Chillingworth, C. Churcher, L. N. Clark, C. H. Corton, A. Cronin, R. M. Davies, J. Doggett, A. Djikeng, T. Feldblyum, M. C. Field, A. Fraser, I. Goodhead, Z. Hance, D. Harper, B. R. Harris, H. Hauser, J. Hostetler, A. Ivens, K. Jagels, D. Johnson, J. Johnson, K. Jones, A. X. Kerhornou, H. Koo, N. Larke, S. Landfear, C. Larkin, V. Leech, A. Line, A. Lord, A. Macleod, P. J. Mooney, S. Moule, D. M. Martin, G. W. Morgan, K. Mungall, H. Norbertczak, D. Ormond, G. Pai, C. S. Peacock, J. Peterson, M. A. Quail, E. Rabbinowitsch, M. A. Rajandream, C. Reitter, S. L. Salzberg, M. Sanders, S. Schobel, S. Sharp, M. Simmonds, A. J. Simpson, L. Tallon, C. M. Turner, A. Tait, A. R. Tivey, S. Van Aken, D. Walker, D. Wanless, S. Wang, B. White, O. White, S. Whitehead, J. Woodward, J. Wortman, M. D. Adams, T. M. Embley, K. Gull, E. Ullu, J. D. Barry, A. H. Fairlamb, F. Opperdoes, B. G. Barrell, J. E. Donelson, N. Hall, C. M. Fraser, S. E. Melville, and N. M. El-Sayed. 2005. The genome of the African trypanosome *Trypanosoma brucei*. *Science* **309:**416–422.

Borst, P. 1991. Molecular genetics of antigenic variation. *Immunol. Today* **12:**A29–A33.

Brayton, K. A., L. S. Kappmeyer, D. R. Herndon, M. J. Dark, D. L. Tibbals, G. H. Palmer, T. C. McGuire, and D. P. Knowles, Jr. 2005. Complete genome sequencing of *Anaplasma marginale* reveals that the surface is skewed to two superfamilies of outer membrane proteins. *Proc. Natl. Acad. Sci. USA* **102:**844–849.

Brayton, K. A., D. P. Knowles, T. C. McGuire, and G. H. Palmer. 2001. Efficient use of a small genome to generate antigenic diversity in tick-borne ehrlichial pathogens. *Proc. Natl. Acad. Sci. USA* **98:**4130–4135.

Brayton, K. A., P. F. Meeus, A. F. Barbet, and G. H. Palmer. 2003. Simultaneous variation of the immunodominant outer membrane proteins, MSP2 and MSP3, during *Anaplasma marginale* persistence in vivo. *Infect. Immun.* **71:**6627–6632.

Brayton, K. A., G. H. Palmer, A. Lundgren, J. Yi, and A. F. Barbet. 2002. Antigenic variation of *Anaplasma marginale msp2* occurs by combinatorial gene conversion. *Mol. Microbiol.* **43:**1151–1159.

Brown, W. C., K. A. Brayton, C. M. Styer, and G. H. Palmer. 2003. The hypervariable region of *Anaplasma marginale* major surface protein 2 (MSP2) contains multiple immunodominant $CD4^+$ T lymphocyte epitopes that elicit variant-specific proliferative and IFN-γ responses in MSP2 vaccinates. *J. Immunol.* **170:**3790–3798.

Brown, W. C., T. C. McGuire, D. Zhu, H. A. Lewin, J. Sosnow, and G. H. Palmer. 2001. Highly conserved regions of the immunodominant major surface protein 2 of the genogroup II ehrlichial pathogen *Anaplasma marginale* are rich in naturally derived $CD4^+$ T lymphocyte epitopes that elicit strong recall responses. *J. Immunol.* **166:** 1114–1124.

Casey, A. N., R. J. Birtles, A. D. Radford, K. J. Bown, N. P. French, Z. Woldehiwet, and N. H. Ogden. 2004. Groupings of highly similar major surface protein (p44)-encoding paralogues:

a potential index of genetic diversity amongst isolates of *Anaplasma phagocytophilum*. *Microbiology* **150:**727–734.

Caspersen, K., J. H. Park, S. Patil, and J. S. Dumler. 2002. Genetic variability and stability of *Anaplasma phagocytophila msp2* (p44). *Infect. Immun.* **70:**1230–1234.

Codner, E. C., and L. L. Farris-Smith. 1986. Characterization of the subclinical phase of ehrlichiosis in dogs. *J. Am. Vet. Med. Assoc.* **189:**47–50.

Collins, N. E., J. Liebenberg, E. P. de Villiers, K. A. Brayton, E. Louw, A. Pretorius, F. E. Faber, H. van Heerden, A. Josemans, M. van Kleef, H. C. Steyn, M. F. van Strijp, E. Zweygarth, F. Jongejan, J. C. Maillard, D. Berthier, M. Botha, F. Joubert, C. H. Corton, N. R. Thomson, M. T. Allsopp, and B. A. Allsopp. 2005. The genome of the heartwater agent *Ehrlichia ruminantium* contains multiple tandem repeats of actively variable copy number. *Proc. Natl. Acad. Sci. USA* **102:**838–843.

Dark, M. J., D. R. Herndon, L. S. Kappmeyer, M. P. Gonzales, E. Nordeen, G. H. Palmer, D. P. Knowles, Jr., and K. A. Brayton. 2009. Conservation in the face of diversity: multistrain analysis of an intracellular bacterium. *BMC Genomics* **10:**16.

Davidson, W. R., J. M. Lockhart, D. E. Stallknecht, E. W. Howerth, J. E. Dawson, and Y. Rechav. 2001. Persistent *Ehrlichia chaffeensis* infection in white-tailed deer. *J. Wildl. Dis.* **37:**538–546.

de la Fuente, J., M. W. Atkinson, V. Naranjo, I. G. Fernandez de Mera, A. J. Mangold, K. A. Keating, and K. M. Kocan. 2007. Sequence analysis of the *msp4* gene of *Anaplasma ovis* strains. *Vet. Microbiol.* **119:**375–381.

de la Fuente, J., J. C. Garcia-Garcia, E. F. Blouin, J. T. Saliki, and K. M. Kocan. 2002. Infection of tick cells and bovine erythrocytes with one genotype of the intracellular ehrlichia *Anaplasma marginale* excludes infection with other genotypes. *Clin. Diagn. Lab. Immunol.* **9:**658–668.

Dumler, J. S., A. F. Barbet, C. P. Bekker, G. A. Dasch, G. H. Palmer, S. C. Ray, Y. Rikihisa, and F. R. Rurangirwa. 2001. Reorganization of genera in the families *Rickettsiaceae* and *Anaplasmataceae* in the order *Rickettsiales*: unification of some species of *Ehrlichia* with *Anaplasma*, *Cowdria* with *Ehrlichia* and *Ehrlichia* with *Neorickettsia*, descriptions of six new species combinations and designation of *Ehrlichia equi* and 'HGE agent' as subjective synonyms of *Ehrlichia phagocytophila*. *Int. J. Syst. Evol. Microbiol.* **51:**2145–2165.

Dumler, J. S., J. E. Madigan, N. Pusterla, and J. S. Bakken. 2007. Ehrlichioses in humans: epidemiology, clinical presentation, diagnosis, and treatment. *Clin. Infect. Dis.* **45**(Suppl. 1)**:**S45–S51.

Dumler, J. S., W. L. Sutker, and D. H. Walker. 1993. Persistent infection with *Ehrlichia chaffeensis*. *Clin. Infect. Dis.* **17:**903–905.

Dunning Hotopp, J. C., M. Lin, R. Madupu, J. Crabtree, S. V. Angiuoli, J. Eisen, R. Seshadri, Q. Ren, M. Wu, T. R. Utterback, S. Smith, M. Lewis, H. Khouri, C. Zhang, H. Niu, Q. Lin, N. Ohashi, N. Zhi, W. Nelson, L. M. Brinkac, R. J. Dodson, M. J. Rosovitz, J. Sundaram, S. C. Daugherty, T. Davidsen, A. S. Durkin, M. Gwinn, D. H. Haft, J. D. Selengut, S. A. Sullivan, N. Zafar, L. Zhou, F. Benahmed, H. Forberger, R. Halpin, S. Mulligan, J. Robinson, O. White, Y. Rikihisa, and H. Tettelin. 2006. Comparative genomics of emerging human ehrlichiosis agents. *PLoS Genet.* **2:**e21.

Egenvall, A., I. Lilliehook, A. Bjoersdorff, E. O. Engvall, E. Karlstam, K. Artursson, M. Heldtander, and A. Gunnarsson. 2000. Detection of granulocytic *Ehrlichia* species DNA by PCR in persistently infected dogs. *Vet. Rec.* **146:**186–190.

Eid, G., D. M. French, A. M. Lundgren, A. F. Barbet, T. F. McElwain, and G. H. Palmer. 1996. Expression of major surface protein 2 antigenic variants during acute *Anaplasma marginale* rickettsemia. *Infect. Immun.* **64:**836–841.

Felek, S., S. Telford III, R. C. Falco, and Y. Rikihisa. 2004. Sequence analysis of p44 homologs expressed by *Anaplasma phagocytophilum* in infected ticks feeding on naive hosts and in mice infected by tick attachment. *Infect. Immun.* **72:**659–666.

Finn, R. D., J. Mistry, J. Tate, P. Coggill, A. Heger, J. E. Pollington, O. L. Gavin, P. Gunasekaran, G. Ceric, K. Forslund, L. Holm, E. L. Sonnhammer, S. R. Eddy, and A. Bateman. 2010. The Pfam protein families database. *Nucleic Acids Res.* **38**(Database issue)**:**D211–D222.

Foley, J. E., N. C. Nieto, A. Barbet, and P. Foley. 2009. Antigen diversity in the parasitic bacterium *Anaplasma phagocytophilum* arises from selectively-represented, spatially clustered functional pseudogenes. *PLoS One* **4:**e8265.

Franzen, P., A. Aspan, A. Egenvall, A. Gunnarsson, E. Karlstam, and J. Pringle. 2009. Molecular evidence for persistence of *Anaplasma phagocytophilum* in the absence of clinical abnormalities in horses after recovery from acute experimental infection. *J. Vet. Intern. Med.* **23:**636–642.

French, D. M., W. C. Brown, and G. H. Palmer. 1999. Emergence of *Anaplasma marginale* antigenic variants during persistent rickettsemia. *Infect. Immun.* **67:**5834–5840.

French, D. M., T. F. McElwain, T. C. McGuire, and G. H. Palmer. 1998. Expression of *Anaplasma marginale* major surface protein 2 variants during persistent cyclic rickettsemia. *Infect. Immun.* **66:**1200–1207.

Futse, J. E., K. A. Brayton, M. J. Dark, D. P. Knowles, Jr., and G. H. Palmer. 2008. Superinfection as a driver of genomic diversification in antigenically variant pathogens. *Proc. Natl. Acad. Sci. USA* **105:**2123–2127.

Futse, J. E., K. A. Brayton, D. P. Knowles, Jr., and G. H. Palmer. 2005. Structural basis for segmental gene conversion in generation of *Anaplasma marginale* outer membrane protein variants. *Mol. Microbiol.* **57:**212–221.

Futse, J. E., K. A. Brayton, S. D. Nydam, and G. H. Palmer. 2009. Generation of antigenic variants via gene conversion: evidence for recombination fitness selection at the locus level in *Anaplasma marginale. Infect. Immun.* **77:**3181–3187.

Galletti, M. F., M. W. Ueti, D. P. Knowles, Jr., K. A. Brayton, and G. H. Palmer. 2009. Independence of *Anaplasma marginale* strains with high and low transmission efficiencies in the tick vector following simultaneous acquisition by feeding on a superinfected mammalian reservoir host. *Infect. Immun.* **77:**1459–1464.

Granquist, E. G., K. Bårdsen, K. Bergström, and S. Stuen. 2010a. Variant- and individual dependent nature of persistent *Anaplasma phagocytophilum* infection. *Acta Vet. Scand.* **52:**25.

Granquist, E. G., S. Stuen, L. Crosby, A. M. Lundgren, A. R. Alleman, and A. F. Barbet. 2010b. Variant-specific and diminishing immune responses towards the highly variable MSP2(P44) outer membrane protein of *Anaplasma phagocytophilum* during persistent infection in lambs. *Vet. Immunol. Immunopathol.* **133:**117–124.

Harrus, S., T. Waner, I. Aizenberg, J. E. Foley, A. M. Poland, and H. Bark. 1998. Amplification of ehrlichial DNA from dogs 34 months after infection with *Ehrlichia canis. J. Clin. Microbiol.* **36:**73–76.

Herndon, D. R., G. H. Palmer, V. Shkap, D. P. Knowles, Jr., and K. A. Brayton. 2010. Complete genome sequence of *Anaplasma marginale* subsp. *centrale. J. Bacteriol.* **192:**379–380.

Huang, H., X. Wang, T. Kikuchi, Y. Kumagai, and Y. Rikihisa. 2007. Porin activity of *Anaplasma phagocytophilum* outer membrane fraction and purified P44. *J. Bacteriol.* **189:**1998–2006.

IJdo, J. W., W. Sun, Y. Zhang, L. A. Magnarelli, and E. Fikrig. 1998. Cloning of the gene encoding the 44-kilodalton antigen of the agent of human granulocytic ehrlichiosis and characterization of the humoral response. *Infect. Immun.* **66:**3264–3269.

IJdo, W., Y. Zhang, E. Hodzic, L. A. Magnarelli, M. L. Wilson, S. R. Telford III, S. W. Barthold, and E. Fikrig. 1997. The early humoral response in human granulocytic ehrlichiosis. *J. Infect. Dis.* **176:**687–692.

Jorgensen, W. K., A. J. de Vos, and R. J. Dalgliesh. 1989. Infectivity of cryopreserved *Babesia bovis, Babesia bigemina* and *Anaplasma centrale* for cattle after thawing, dilution and incubation at 30 degrees C. *Vet. Parasitol.* **31:**243–251.

Krigel, Y., E. Pipano, and V. Shkap. 1992. Duration of carrier state following vaccination with live *Anaplasma centrale. Trop. Anim. Health Prod.* **24:**209–210.

Lai, T. H., N. G. Orellana, Y. Yuasa, and Y. Rikihisa. 2011. Cloning of the major outer membrane protein expression locus in *Anaplasma platys* and seroreactivity of a species-specific antigen. *J. Bacteriol.* **193:**2924–2930.

Lin, Q., and Y. Rikihisa. 2005. Establishment of cloned *Anaplasma phagocytophilum* and analysis of *p44* gene conversion within an infected horse and infected SCID mice. *Infect. Immun.* **73:**5106–5114.

Lin, Q., Y. Rikihisa, S. Felek, X. Wang, R. F. Massung, and Z. Woldehiwet. 2004a. *Anaplasma phagocytophilum* has a functional *msp2* gene that is distinct from *p44. Infect. Immun.* **72:**3883–3889.

Lin, Q., Y. Rikihisa, R. F. Massung, Z. Woldehiwet, and R. C. Falco. 2004b. Polymorphism and transcription at the *p44-1/p44-18* genomic locus in *Anaplasma phagocytophilum* strains from diverse geographic regions. *Infect. Immun.* **72:**5574–5581.

Lin, Q., Y. Rikihisa, N. Ohashi, and N. Zhi. 2003. Mechanisms of variable *p44* expression by *Anaplasma phagocytophilum. Infect. Immun.* **71:**5650–5661.

Lin, Q., C. Zhang, and Y. Rikihisa. 2006. Analysis of involvement of the RecF pathway in *p44* recombination in *Anaplasma phagocytophilum* and in *Escherichia coli* by using a plasmid carrying the *p44* expression and *p44* donor loci. *Infect. Immun.* **74:**2052–2062.

Lin, Q., N. Zhi, N. Ohashi, H. W. Horowitz, M. E. Aguero-Rosenfeld, J. Raffalli, G. P. Wormser, and Y. Rikihisa. 2002. Analysis of sequences and loci of *p44* homologs expressed by *Anaplasma phagocytophila* in acutely infected patients. *J. Clin. Microbiol.* **40:**2981–2988.

Löhr, C. V., K. A. Brayton, A. F. Barbet, and G. H. Palmer. 2004. Characterization of the *Anaplasma marginale msp2* locus and its synteny with the *omp1/p30* loci of *Ehrlichia chaffeensis* and *E. canis. Gene* **325:**115–121.

Löhr, C. V., K. A. Brayton, V. Shkap, T. Molad, A. F. Barbet, W. C. Brown, and G. H. Palmer. 2002. Expression of *Anaplasma marginale* major surface protein 2 operon-associated proteins during mammalian and arthropod infection. *Infect. Immun.* **70:**6005–6012.

Losos, G. J. 1986. Anaplasmosis, p. 743–795. *In* G. J. Losos (ed.), *Infectious Tropical Diseases of Domestic Animals*. Longman Press, Essex, United Kingdom.

Massung, R. F., J. W. Courtney, S. L. Hiratzka, V. E. Pitzer, G. Smith, and R. L. Dryden. 2005. *Anaplasma phagocytophilum* in white-tailed deer. *Emerg. Infect. Dis.* **11:**1604–1606.

Mavromatis, K., C. K. Doyle, A. Lykidis, N. Ivanova, M. P. Francino, P. Chain, M. Shin, S. Malfatti, F. Larimer, A. Copeland, J. C. Detter, M. Land, P. M. Richardson, X. J. Yu, D. H. Walker, J. W. McBride, and N. C. Kyrpides. 2006. The genome of the obligately intracellular bacterium *Ehrlichia canis* reveals themes of complex membrane structure and immune evasion strategies. *J. Bacteriol.* **188:**4015–4023.

McBride, J. W., X. Yu, and D. H. Walker. 1999. Molecular cloning of the gene for a conserved major immunoreactive 28-kilodalton protein of *Ehrlichia canis*: a potential serodiagnostic antigen. *Clin. Diagn. Lab. Immunol.* **6:**392–399.

McCarthy, A. 2010. Third generation DNA sequencing: Pacific Biosciences' single molecule real time technology. *Chem. Biol.* **17:**675–676.

McGuire, T. C., G. H. Palmer, W. L. Goff, M. I. Johnson, and W. C. Davis. 1984. Common and isolate-restricted antigens of *Anaplasma marginale* detected with monoclonal antibodies. *Infect. Immun.* **45:**697–700.

Meeus, P. F., K. A. Brayton, G. H. Palmer, and A. F. Barbet. 2003. Conservation of a gene conversion mechanism in two distantly related paralogues of *Anaplasma marginale*. *Mol. Microbiol.* **47:**633–643.

Molad, T., B. Leibovich, M. Mazuz, L. Fleiderovich, L. Fish, and V. Shkap. 2010. Identification of *Anaplasma centrale* major surface protein-2 pseudogenes. *Vet. Microbiol.* **143:**277–283.

Morrison, L. J., P. Majiwa, A. F. Read, and J. D. Barry. 2005. Probabilistic order in antigenic variation of *Trypanosoma brucei*. *Int. J. Parasitol.* **35:** 961–972.

Morrison, L. J., L. Marcello, and R. McCulloch. 2009. Antigenic variation in the African trypanosome: molecular mechanisms and phenotypic complexity. *Cell. Microbiol.* **11:**1724–1734.

Murphy, C. I., J. R. Storey, J. Recchia, L. A. Doros-Richert, C. Gingrich-Baker, K. Munroe, J. S. Bakken, R. T. Coughlin, and G. A. Beltz. 1998. Major antigenic proteins of the agent of human granulocytic ehrlichiosis are encoded by members of a multigene family. *Infect. Immun.* **66:**3711–3718.

Noh, S. M., K. A. Brayton, D. P. Knowles, J. T. Agnes, M. J. Dark, W. C. Brown, T. V. Baszler, and G. H. Palmer. 2006. Differential expression and sequence conservation of the *Anaplasma marginale msp2* gene superfamily outer membrane proteins. *Infect. Immun.* **74:**3471–3479.

Ohashi, N., Y. Rikihisa, and A. Unver. 2001. Analysis of transcriptionally active gene clusters of major outer membrane protein multigene family in *Ehrlichia canis* and *E. chaffeensis*. *Infect. Immun.* **69:**2083–2091.

Palmer, G. H., J. R. Abbott, D. M. French, and T. F. McElwain. 1998. Persistence of *Anaplasma ovis* infection and conservation of the *msp-2* and *msp-3* multigene families within the genus *Anaplasma*. *Infect. Immun.* **66:**6035–6039.

Palmer, G. H., and K. A. Brayton. 2007. Gene conversion is a convergent strategy for pathogen antigenic variation. *Trends Parasitol.* **23:**408–413.

Palmer, G. H., G. Eid, A. F. Barbet, T. C. McGuire, and T. F. McElwain. 1994. The immunoprotective *Anaplasma marginale* major surface protein 2 is encoded by a polymorphic multigene family. *Infect. Immun.* **62:**3808–3816.

Palmer, G. H., J. E. Futse, C. K. Leverich, D. P. Knowles, Jr., F. R. Rurangirwa, and K. A. Brayton. 2007. Selection for simple major surface protein 2 variants during *Anaplasma marginale* transmission to immunologically naive animals. *Infect. Immun.* **75:**1502–1506.

Palmer, G. H., D. P. Knowles, Jr., J. L. Rodriguez, D. P. Gnad, L. C. Hollis, T. Marston, and K. A. Brayton. 2004. Stochastic transmission of multiple genotypically distinct *Anaplasma marginale* strains in a herd with high prevalence of *Anaplasma* infection. *J. Clin. Microbiol.* **42:**5381–5384.

Palmer, G. H., F. R. Rurangirwa, and T. F. McElwain. 2001. Strain composition of the ehrlichia *Anaplasma marginale* within persistently infected cattle, a mammalian reservoir for tick transmission. *J. Clin. Microbiol.* **39:**631–635.

Park, J., K. S. Choi, and J. S. Dumler. 2003a. Major surface protein 2 of *Anaplasma phagocytophilum* facilitates adherence to granulocytes. *Infect. Immun.* **71:**4018–4025.

Park, J., K. J. Kim, D. J. Grab, and J. S. Dumler. 2003b. *Anaplasma phagocytophilum* major surface protein-2 (Msp2) forms multimeric complexes in the bacterial membrane. *FEMS Microbiol. Lett.* **227:**243–247.

Pipano, E., Y. Krigel, M. Frank, A. Markovics, and E. Mayer. 1986. Frozen *Anaplasma centrale* vaccine against anaplasmosis in cattle. *Br. Vet. J.* **142:**553–556.

Potgieter, F. T. 1979. Epizootiology and control of anaplasmosis in South africa. *J. S. Afr. Vet. Assoc.* **50:**367–372.

Psaroulaki, A., D. Chochlakis, V. Sandalakis, I. Vranakis, I. Ioannou, and Y. Tselentis. 2009. Phylogentic analysis of *Anaplasma ovis* strains isolated from sheep and goats using *groEL* and *msp4* genes. *Vet. Microbiol.* **138:**394–400.

Reddy, G. R., C. R. Sulsona, A. F. Barbet, S. M. Mahan, M. J. Burridge, and A. R. Alleman. 1998. Molecular characterization of a 28 kDa surface antigen gene family of the tribe *Ehrlichiae*. *Biochem. Biophys. Res. Commun.* **247:**636–643.

Robinson, N. P., N. Burman, S. E. Melville, and J. D. Barry. 1999. Predominance of duplicative *VSG* gene conversion in antigenic variation in African trypanosomes. *Mol. Cell. Biol.* **19:**5839–5846.

Rodriguez, J. L., G. H. Palmer, D. P. Knowles, Jr., and K. A. Brayton. 2005. Distinctly different *msp2* pseudogene repertoires in *Anaplasma marginale* strains that are capable of superinfection. *Gene* **361:**127–132.

Rudenko, G. 2010. Epigenetics and transcriptional control in African trypanosomes. *Essays Biochem.* **48:**201–219.

Shkap, V., B. Leibovitz, Y. Krigel, T. Molad, L. Fish, M. Mazuz, L. Fleiderovitz, and I. Savitsky. 2008. Concomitant infection of cattle with the vaccine strain *Anaplasma marginale* ss *centrale* and field strains of *A. marginale*. *Vet. Microbiol.* **130:**277–284.

Shkap, V., T. Molad, K. A. Brayton, W. C. Brown, and G. H. Palmer. 2002. Expression of major surface protein 2 variants with conserved T-cell epitopes in *Anaplasma centrale* vaccinates. *Infect. Immun.* **70:**642–648.

Taylor, J. E., and G. Rudenko. 2006. Switching trypanosome coats: what's in the wardrobe? *Trends Genet.* **22:**614–620.

Teglas, M. B., and J. Foley. 2006. Differences in the transmissibility of two *Anaplasma phagocytophilum* strains by the North American tick vector species, *Ixodes pacificus* and *Ixodes scapularis* (Acari: *Ixodidae*). *Exp. Appl. Acarol.* **38:**47–58.

Telford, S. R., III, J. E. Dawson, P. Katavolos, C. K. Warner, C. P. Kolbert, and D. H. Persing. 1996. Perpetuation of the agent of human granulocytic ehrlichiosis in a deer tick-rodent cycle. *Proc. Natl. Acad. Sci. USA* **93:**6209–6214.

Theiler, A. 1910. *Report of the Government Veterinary Bacteriologist, 1908-1909*, p. 7–64. Department of Agriculture, Union of South Africa, Transvaal, South Africa.

Theiler, A. 1911. *First Report of the Director of Veterinary Research, Union of South Africa*, p. 7–46. Department of Agriculture, Union of South Africa, Johannesburg, South Africa.

Theiler, A. 1912. Gallsickness of imported cattle and the protective inoculation against this disease. *Agric. J. Union S. Afr.* **3:**7–46.

Unver, A., N. Ohashi, T. Tajima, R. W. Stich, D. Grover, and Y. Rikihisa. 2001. Transcriptional analysis of *p30* major outer membrane multigene family of *Ehrlichia canis* in dogs, ticks, and cell culture at different temperatures. *Infect. Immun.* **69:**6172–6178.

Unver, A., Y. Rikihisa, R. W. Stich, N. Ohashi, and S. Felek. 2002. The *omp-1* major outer membrane multigene family of *Ehrlichia chaffeensis* is differentially expressed in canine and tick hosts. *Infect. Immun.* **70:**4701–4704.

van Heerden, H., N. E. Collins, K. A. Brayton, C. Rademeyer, and B. A. Allsopp. 2004. Characterization of a major outer membrane protein multigene family in *Ehrlichia ruminantium*. *Gene* **330:**159–168.

van Vliet, A. H., F. Jongejan, M. van Kleef, and B. A. van der Zeijst. 1994. Molecular cloning, sequence analysis, and expression of the gene encoding the immunodominant 32-kilodalton protein of *Cowdria ruminantium*. *Infect. Immun.* **62:**1451–1456.

Watt, V. M., C. J. Ingles, M. S. Urdea, and W. J. Rutter. 1985. Homology requirements for recombination in *Escherichia coli*. *Proc. Natl. Acad. Sci. USA* **82:**4768–4772.

Wen, B., Y. Rikihisa, J. M. Mott, R. Greene, H. Y. Kim, N. Zhi, G. C. Couto, A. Unver, and R. Bartsch. 1997. Comparison of nested PCR with immunofluorescent-antibody assay for detection of *Ehrlichia canis* infection in dogs treated with doxycycline. *J. Clin. Microbiol.* **35:**1852–1855.

Woldehiwet, Z. 2010. The natural history of *Anaplasma phagocytophilum*. *Vet. Parasitol.* **167:**108–122.

Young, D., T. Hussell, and G. Dougan. 2002. Chronic bacterial infections: living with unwanted guests. *Nat. Immunol.* **3:**1026–1032.

Yu, X., J. W. McBride, X. Zhang, and D. H. Walker. 2000. Characterization of the complete transcriptionally active *Ehrlichia chaffeensis* 28 kDa outer membrane protein multigene family. *Gene* **248:**59–68.

Zhang, C., Q. Xiong, T. Kikuchi, and Y. Rikihisa. 2008. Identification of 19 polymorphic major outer membrane protein genes and their immunogenic peptides in *Ehrlichia ewingii* for use in a serodiagnostic assay. *Clin. Vaccine Immunol.* **15:**402–411.

Zhi, N., N. Ohashi, and Y. Rikihisa. 2002. Activation of a *p44* pseudogene in *Anaplasma phagocytophila* by bacterial RNA splicing: a novel mechanism for post-transcriptional regulation of a multigene family encoding immunodominant major outer membrane proteins. *Mol. Microbiol.* **46:**135–145.

Zhi, N., Y. Rikihisa, H. Y. Kim, G. P. Wormser, and H. W. Horowitz. 1997. Comparison of major antigenic proteins of six strains of the human granulocytic ehrlichiosis agent by Western immunoblot analysis. *J. Clin. Microbiol.* **35:**2606–2611.

TRANSMISSION AND THE DETERMINANTS OF TRANSMISSION EFFICIENCY

Shane M. Ceraul

13

INTRODUCTION

Vector-borne diseases as a group are on the rise. An analysis by Jones et al. (2008) shows that vector-borne diseases comprise 22.8% of emerging infectious diseases. Interestingly, 54.3% of the pathogens involved in emerging infectious diseases are characterized as bacterial and rickettsial diseases (Jones et al., 2008). Bacteria within the order *Rickettsiales* would have little impact on human and veterinary medicine in the absence of the arthropod vector. Indeed, bacteria within the *Rickettsiales* are acquired, maintained, and transmitted to humans, nonhuman mammals, and birds by hematophagous arthropods. To understand these complex rickettsial zoonotic cycles, we must investigate what drives the vector-host and vector-microbe interactions.

Once the vector-host encounter initiates, a hematophagous arthropod attaches and prepares to imbibe a blood meal that takes place over a period of seconds for fleas and lice and 3 to 14 days for ticks and mites. The arthropod salivates into the wound, releasing immunomodulatory and anti-inflammatory chemicals that increase the blood flow into the bite site and thwart the host's attempts to reject the parasite (Fig. 1). The purpose of generating such trauma is to acquire a blood meal. However, this area of low immune activity has been coadapted by pathogens to serve as a highway for the rickettsiae harbored by either vector or host.

In general, microbes can be transmitted vertically, as from female to progeny, or horizontally, as when the vector takes another blood meal. Each partner in an established vector-rickettsiae relationship is in a struggle for survival that is characterized by a balance between proactive and reactive stance. As a microbe is imbibed with the blood meal, it encounters the midgut epithelium of the arthropod. The midgut epithelium is a physical and chemical barrier to microbes, some of which have developed the ability to pass through to the hemocoel of the vector (Fig. 1). This breach of the midgut barrier stimulates an innate immune response locally and possibly systemically in other tissues (Fig. 1). The antimicrobial peptides produced in response to microbial challenge work to control or prevent infection. However, microbes like those in the order *Rickettsiales* exist as endosymbionts and have evolved strategies that allow them to

Shane M. Ceraul, Department of Microbiology and Immunology, University of Maryland School of Medicine, Baltimore, MD 21201.

Intracellular Pathogens II: Rickettsiales
Edited by G. H. Palmer and A. F. Azad © 2012 ASM Press, Washington, DC
10.1128/9781555817336.ch13

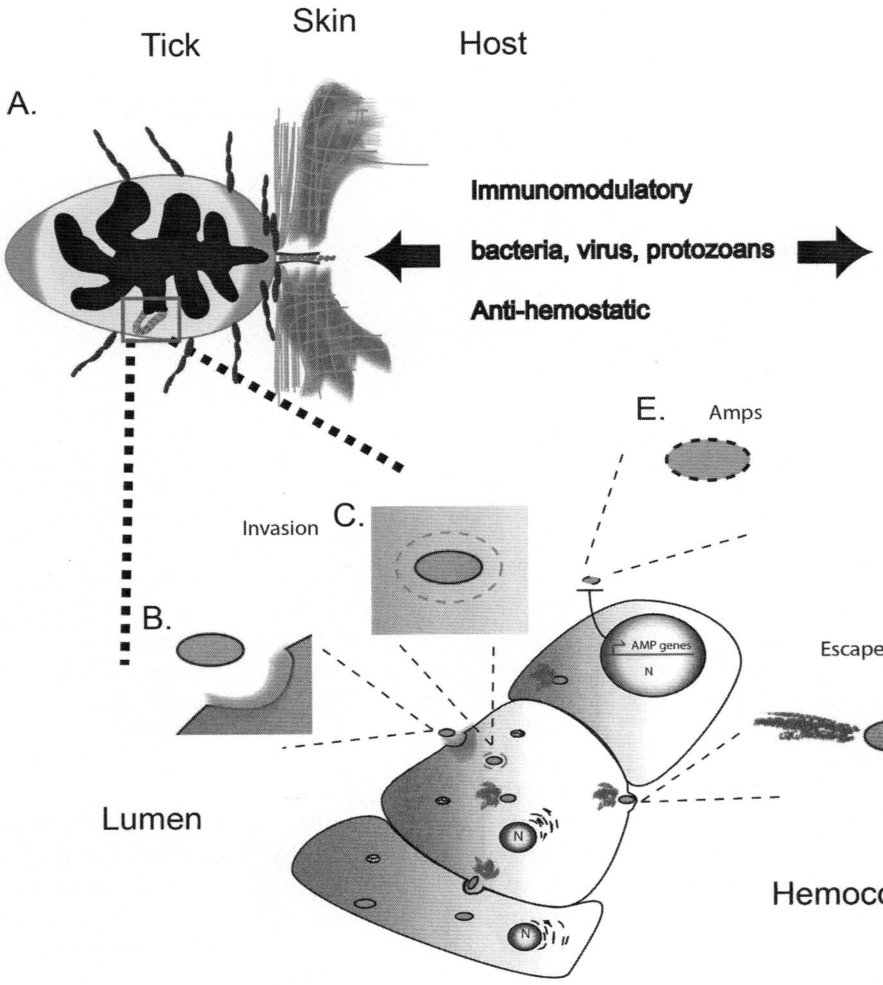

FIGURE 1 Tick-host-rickettsiae interaction. (A) Ticks secrete a number of antihemostatic and immunomodulatory substances into the bite site to increase blood flow and reduce immune activation to the tick during feeding. Rickettsiae are imbibed with the blood meal or transmitted to the host during this stage of the interaction. As rickettsiae enter the midgut lumen, they attach to the host cell (B) and induce phagocytosis so as to be enveloped in a vacuole that is rapidly degraded (C) to lie in direct contact with the cytoplasm. (D) Spotted fever group rickettsiae move between and within each host cell using actin tail polymerization. (E) Ticks respond to rickettsiae by mounting an innate immune response that involves antimicrobial peptide gene transcription. Recent work suggests that the rickettsial invasion process is limited by tick antimicrobial peptides. doi:10.1128/9781555817336.ch13.f1

evade these immune defenses. The order as a whole transmigrate the midgut and colonize the salivary gland and/or the ovary. Bacterial replication in the salivary gland is a critical event that determines perpetuation of the transmission cycle in nature. Little is known regarding the dynamics of bacterial infection of the ovary. However, it is clear that some microbes within the *Rickettsiales* fail to infect the ovary or infect below the threshold limit for maintenance of the pathogen within the vector. In some instances, egg production decreases or progeny numbers are reduced due to the lethality that some rickettsiae have for the tick vector. Interestingly, the influence of primary infections with one *Rickettsia* sp. can influence the success of transovarial transmission (female to progeny) of a second. This chapter details some fascinating trends observed regarding vertical and horizontal (between different life stages and between same life stages, respectively) transmission.

Vector-borne transmission cycles are complex interactions that involve multiple factors. For the natural cycle to exist, the vector, mammal, and microbe must be present in a niche that supports development of all three. The very concept of zoonotic transmission cycles depends heavily on the ecology where these transmission cycles exist. Indeed, vector and host reproduction, vector and host competency, and vector and host distribution are altered by the changing ecology. Biotic (e.g., vector competence) and abiotic (e.g., temperature) factors determine the stability of any sylvatic or zoonotic transmission cycle.

The chapter centers on a discussion of the attributes of successful pathogen transmission in the context of the vector's ability to modulate (i) the mammalian host's response during acquisition and transmission and (ii) microbial growth within the vector during the maintenance phase. The discussion in these two sections essentially defines the environment and competency of both the vector and mammalian host as determinants of transmission and transmission efficiency. This chapter ends with a survey of fluctuating ecological trends that can enhance or diminish the potency of vector-borne rickettsial zoonotic cycles.

VECTORS OF THE *RICKETTSIALES*

Ticks

Ticks belong to the class *Arachnida*, subclass *Acari*, and superorder *Parasitiformes*, which comprises three main families (Sonenshine, 1991). Organisms within the *Rickettsiales* are transmitted by the family of hard ticks known as the *Ixodidae* (Fig. 2A). The other families are the *Argasidae* (soft ticks) and the *Nuttalliellidae*, of which there is only one species, *Nuttalliella namaqua* (Sonenshine, 1991). Worldwide there are an estimated 18 genera and 850 species of ticks, which are found on every continent except at the poles (Sonenshine et al., 2002).

Ixodid ticks possess a one-, two-, or three-host life cycle. All life stages (except egg hatching) of *Rhipicephalus microplus*, a vector for *Anaplasma marginale*, take place on one host, and therefore it is considered a one-host tick. The *Ixodes* spp., *Dermacentor* spp., and *Amblyomma* spp. involved in transmission of rickettsial pathogens are three-host ticks. The larvae and nymphal life stages require a blood meal to molt to the next stage. The female adult requires a blood meal for egg production. Egg clutches are oviposited in the environment, where they hatch, and the emerging larvae once again search for a host (Sonenshine, 1991). Ticks quest (sit and wait) on vegetation and latch onto passing hosts, at which time they attach and feed for a period of minutes (*Argasidae*) to days (*Ixodidae*). In general, the immature life stages quest at lower levels and are found on small mammals, lizards, and birds. Adult life stages are frequently found to feed on medium-size to large mammals.

Rickettsiae are imbibed with a blood meal and most likely infect most tissues within the tick (Baldridge et al., 2007). Infection of salivary glands and ovaries is critical to the transmission process. Most rickettsiae are maintained in a tick population through transstadial transmission (life stage to life stage), vertically through transovarial transmission or horizontally between ticks feeding on the same host. The transmission process involves one or more of the aforementioned mechanisms.

Fleas

Fleas belong to the class *Insecta*, subclass *Pterygota*, and superorder *Endopterygota* (order: *Siphonaptera*) (Bitam et al., 2010). There are approximately 2,000 species and subspecies of fleas found throughout the world (Fig. 2B) (Azad et al., 1997). Most notable among these species are the Oriental rat flea, *Xenopsylla cheopis*, and the cat flea, *Ctenocephalides felis*. *X. cheopis* has gained notoriety as the vector for the etiologic agent of plague, *Yersinia pestis*.

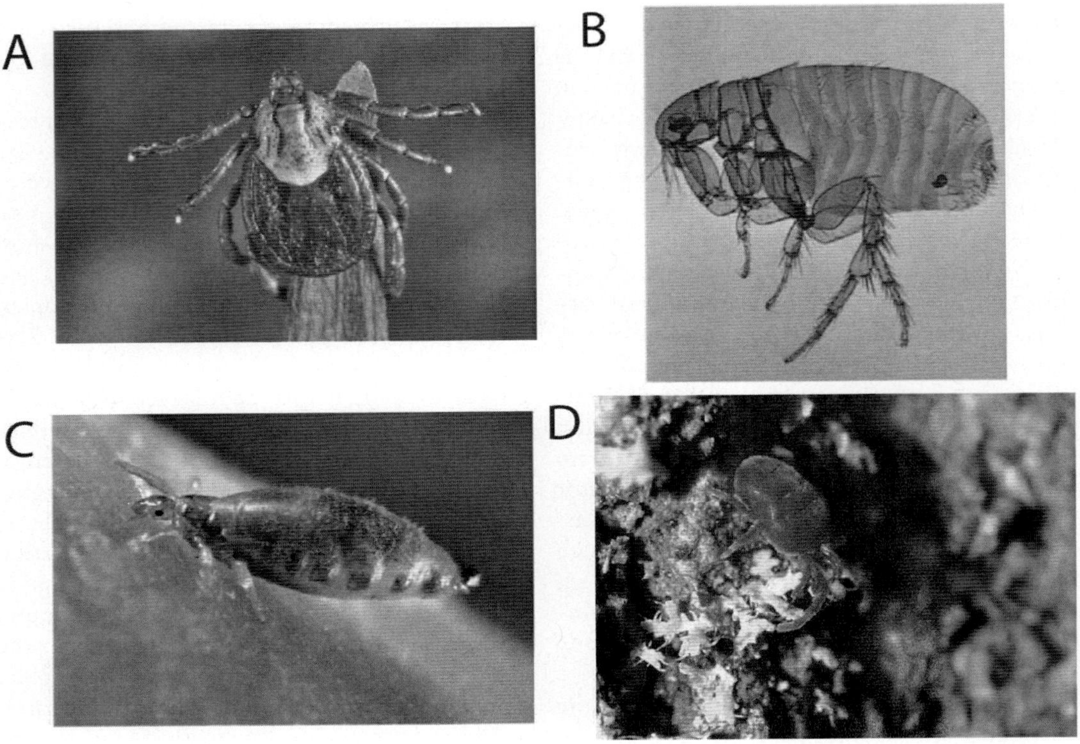

FIGURE 2 Vectors of the *Rickettsiales*. (A) *Ixodidae* tick. (B) Oriental rat flea, *X. cheopis*. (C) Body louse, *P. humanus humanus*. (D) Trombiculid larval mite (chigger). (Images A, B, and C from the Centers for Disease Control and Prevention Public Health Image Library. Image A modified from CDC image 10865; image B modified from CDC image 4633; image C modified from CDC image 9208. Image D from Wikipedia Commons. Photo credit: Luc Viatour; www.lucnix.be.) doi:10.1128/9781555817336.ch13.f2

Both *X. cheopis* and *C. felis* are vectors of endemic or murine typhus caused by *Rickettsia typhi* (Azad et al., 1997). *C. felis* is also a vector of *Rickettsia felis* (McElroy et al., 2010).

The life cycle of the flea includes the egg, larva, pupa, and imago (adult) (Bitam et al., 2010). After feeding, a female lays her eggs on the host. The eggs hatch into larvae, which do not take a blood meal but feed on fecal matter (digested hemoglobin) or other organic material (dead skin) nearby (Bitam et al., 2010). Pupae coat themselves in a cocoon and develop into adults over a period of 5 to 14 days (Bitam et al., 2010). Once the cocoon is disturbed by movement or vibration, the adult emerges. Even though the common names suggest host specificity, most fleas feed on mammals of all sizes, including humans.

Rickettsiae are excreted with and remain viable in flea feces. Fecal material contaminates skin abrasions or wounds as it is rubbed into the skin once humans or mammals scratch. Humans are dead-end hosts for both *R. typhi* and *R. felis*.

Lice

Lice belong to the class *Insecta*, subclass *Pterygota*, and superorder *Psocodea*, which branches into two orders: *Pscocoptera* and *Phthiraptera* (Raoult and Roux, 1999). The lice responsible for transmission of the etiologic agent of epidemic typhus, *Rickettsia prowazekii*, belong to the

Anoplura group (sucking lice) (Fig. 2C) (Raoult and Roux, 1999). Of the 3,000 documented lice species, 540 belong to the *Anoplura* group, which parasitize mammals to feed. Lice are found worldwide and disproportionately afflict people of lower socioeconomic status in both underdeveloped and developed countries (Raoult and Roux, 1999). Indeed, outbreaks of epidemic typhus are always associated with louse infestations that are precipitated by cold weather, poor sanitation, and overcrowding (Raoult and Roux, 1999). Most notable was the outbreak that affected Napoleon's campaign of 1812 to Russia. The cold, harsh conditions of the march exacerbated louse infestations among Napoleon's Grand Army. Along with the louse infestations came *R. prowazekii* and *Bartonella quintana* (Raoult et al., 2006).

Humans are susceptible to infestation with the body louse, *Pediculus humanus humanus*; the head louse, *Pediculus humanus capitis*; and the pubic louse, *Phthirus pubis* (Raoult and Roux, 1999). The body louse is most commonly associated with transmission of *R. prowazekii*. The females lay eggs in the clothing of an individual, which require 6 to 9 days to hatch. Upon hatching, the nymph feeds on humans and molts. The nymphal stage includes three instars before reaching the adult stage. The adult body louse feeds up to five times a day, at which time it acquires *R. prowazekii*. *R. prowazekii* replicates to high levels in the midgut, eventually killing the louse (Raoult and Roux, 1999). Similarly to flea-borne rickettsiae, *R. prowazekii* is shed in the feces of the louse, which contaminate the abrasions or wounds made as the host scratches.

If humans survive infection with *R. prowazekii* (30% fatality without treatment), they are considered lifelong carriers (Raoult and Roux, 1999). Disease spreads rapidly when these individuals intersect deteriorating conditions, poor sanitation, and louse infestations (Raoult et al., 1997, 1998; Tarasevich et al., 1998). In the United States, *R. prowazekii* may also be maintained in a sylvatic cycle that includes flying squirrels and their fleas (Bozeman et al., 1981; Duma et al., 1981).

Mites

Trombiculid larval mites (chiggers) (Fig. 2D) are responsible for the transmission of the scrub typhus agent, *Orienta tsutsugamushi*. Trombiculid mites belong to the class *Arachnida*, subclass *Acari*, and family *Trombiculidae*. Collectively, mite species are found worldwide (Sonenshine, 1991). The life cycle of trombiculid mites includes the egg, larva, nymph, and adult. Once the larvae emerge, they quest on vegetation awaiting a host (Traub and Wisseman, 1974). Once the larvae attack a host, they attach and inject digestive enzymes in a process that forms a tube known as a stylostome. Larvae chew on the skin cells that line the inside of the stylostome, causing swelling and the leakage of tissue exudates (Frances et al., 2000; Traub and Wisseman, 1974). Larvae are not hematophagous but do feed on the tissue exudates that result during the formation and maintenance of the stylostome. The larval stage is the only stage to parasitize mammalian hosts. The nymphal and adult stages of the trombiculid mites obtain nutrition by feeding on arthropod eggs or nascent soft-bodied insects (Traub and Wisseman, 1974).

O. tsutsugamushi is maintained through transovarial transmission (Frances et al., 2000; Traub and Wisseman, 1974). Emerging uninfected larvae can acquire *O. tsutsugamushi* by cofeeding on a mammal with infected larvae (Frances et al., 2000; Traub and Wisseman, 1974). Mites are considered both the vector and the reservoir for *O. tsutsugamushi* (Demma et al., 2006). Although a clear rodent reservoir is not documented, serological titers of *O. tsutsugamushi* of >1:64 were found in the serum of both Norwegian rats (*Rattus norvegicus*) and black rats (*Rattus rattus*) in the Republic of Palau (Demma et al., 2006).

Scrub typhus is prevalent in rural southern and southeastern Asia and the western Pacific (Jensenius et al., 2004). As with most rickettsial

TABLE 1 List of rickettsial diseases[a]

Group	Disease	Agent	Location	Vector	Reservoir host[b]	Other animals exposed[c]
Spotted fever group rickettsiae	Rocky Mountain spotted fever	*Rickettsia rickettsii*	North, Central, and South America	*Dermacentor andersoni, Dermacentor variabilis, Dermacentor occidentalis, Amblyomma cajennense, Amblyomma aureolatum, Rhipicephalus sanguineus*	Small mammals, dogs, rabbits, birds	Domestic goats, red foxes, coyotes, lemurs, raccoons, white-footed mice, Virginia opossums
	Rickettsialpox	*Rickettsia akari*	United States and former Soviet Union	House mouse mite (*Liponyssoides sanguineus*)	House mouse (*Mus musculus*)	
	Mediterranean spotted fever (boutonneuse fever)	*Rickettsia conorii*	Mediterranean countries, Africa, Southwest Asia, and India	*Dermacentor reticulatus, Ixodes ricinus, R. sanguineus*	Rodents, dogs	Red deer, roe deer, fallow deer, horses, canines, cattle, sheep, bank voles, wood mice, yellow-necked mice, common shrews
	Siberian tick typhus	*Rickettsia sibirica*	Siberia, Mongolia, and northern China	*Haemaphysalis concinna, Dermacentor nuttalli, Dermacentor marginatus, Dermacentor silvarum*	Rodents	Canines
	Australian tick typhus	*Rickettsia australis*	Australia	*Ixodes holocyclus, Ixodes tasmani*	Rodents	
	Japanese spotted fever	*Rickettsia japonica*	Japan	*Haemaphysalis flava, Haemaphysalis longicornis, Ixodes ovatus, Dermacentor taiwanensis*	Rodents, dogs	Cattle
	Maculatum disease	*Rickettsia parkeri*	Coastal regions of southern and southeastern United States	*Amblyomma maculatum*		Domestic goats, domestic dogs, red foxes, coyotes, lemurs, raccoons, white-footed mice, Virginia opossums
Transitional group rickettsiae	Flea-borne spotted fever	*Rickettsia felis*	North and South America, southern Europe, and Australia	*Ctenocephalides felis*	Opossums?	Felines
Typhus group rickettsiae	Epidemic typhus, Brill-Zinsser disease	*Rickettsia prowazekii*	Worldwide	*Pediculus humanus humanus*	Humans, eastern flying squirrel (*Glaucomys volans volans*), southern flying squirrel (*Glaucomys volans*)	

	Murine typhus (epidemic typhus)	*Rickettsia typhi*	Worldwide	*Xenopsylla cheopis*	Rats, opossums	Felines
Scrub typhus group	Scrub typhus	*Orientia tsustsugamushi*	Asia and Australia	Trombiculid mites (*Leptotrombidium deliense*)	Trombiculid mites (*L. deliense*)	Rodents, birds
Anaplasmataceae	Human granulocytic anaplasmosis	*Anaplasma phagocytophilum*	Europe, North America, and Asia	*Ixodes scapularis*, *Ixodes pacificus*	Deer	Cattle, donkey, birds, sheep, European wild boar, rodents, rabbits
	Bovine anaplasmosis	*Anaplasma marginale*	North, Central, and South America; southern Europe; Africa; Asia; and Australia	*D. variabilis* and *D. andersoni* in the United States (*Boophilus*, *Dermacentor*, *Rhipicephalus*, *Ixodes*, *Hyalomma*, and *Ornithodoros* species have also been implicated); biting flies and dirty farm practices; mechanical transmission	Cattle	Bighorn sheep, bison, goats, domestic sheep, deer, elk
	Human monocytic ehrlichiosis	*Ehrlichia chaffeensis*	North, Central, and South America	*A. americanum*	White-tailed deer	Domestic goats, domestic dogs, red foxes, coyotes, lemurs (PCR[d]), raccoons, white-footed mice, Virginia opossums (IFA[d])
	Ehrlichia ewingii ehrlichiosis	*Ehrlichia ewingii*	Midwestern United States	*A. americanum*	Canines?	
	Heartwater	*Ehrlichia ruminantium*		*Amblyomma* spp.	Ruminants	
	Tick-borne fever	*Anaplasma ovis*		*Dermacentor* spp.	Cattle, sheep	
	Canine monocytic ehrlichiosis	*Ehrlichia canis*	North, Central, and South America	*R. sanguineus*, *D. variabilis*	Wild canids	Foxes
	Potomac horse fever	*Neorickettsia risticii*	North America	Fluke by way of caddis fly	Caddis flies; horses are dead end	Dogs and cats (experimental only)
	Salmon poisoning	*Neorickettsia helminthoeca*		Fluke by way of raw fish	Salmonid fish	Infects dogs; makes them very sick
	Sennetsu neorickettsiosis	*Neorickettsia sennetsu*	Asia	Fluke by way of raw fish		

[a]The following sources were used to compile this table: Anderson and Magnarelli, 2008; Azad and Beard, 1998; Jongejan and Uilenberg, 2004; Krusell et al., 2002; McElroy et al., 2010; Paddock et al., 2004; Parola et al., 2005; Raoult and Roux, 1999; Traub and Wisseman, 1974; Walker, et al., 2008.

[b]Reservoirs are identified through isolation or detection in field-collected samples.

[c]Exposed animals are those that serve as host for ticks but may not contract the rickettsiae. When no information was available the blocks were left blank.

[d]Detection method.

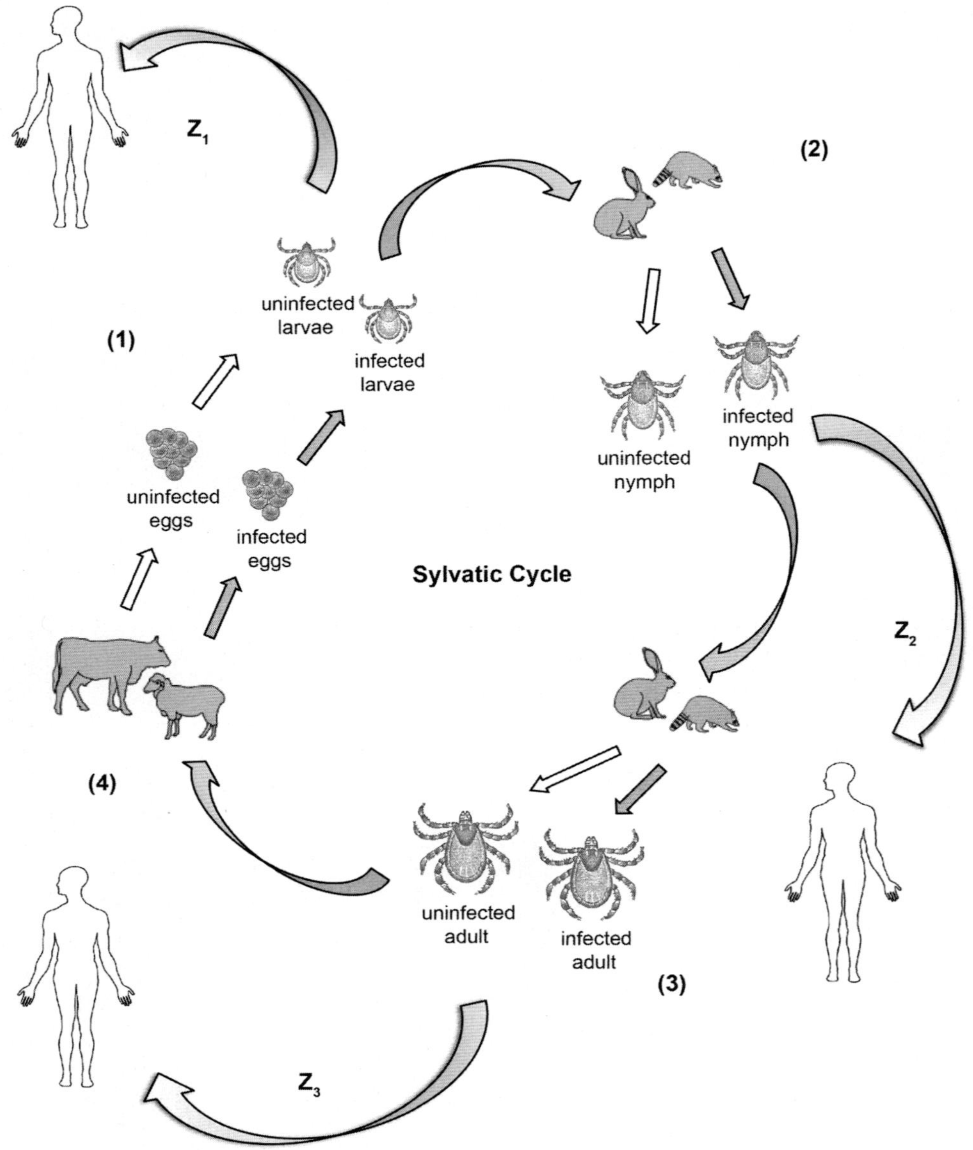

FIGURE 3 Sylvatic and zoonotic transmission. Rickettsiae are transmitted in sylvatic cycles that involve a vector, in this case a tick, and their mammalian, reptilian, or avian hosts. Humans are accidentally infected when they encroach into the habitat where the sylvatic cycle exists. The cycle begins (1) and ends (4) when infected and uninfected ticks feed on large mammals. Horizontal transmission between infected and uninfected ticks can occur at this stage through cofeeding. Uninfected adults can also contract the pathogen by feeding on infected large mammals. If the bacterium is transmitted vertically (transovarial transmission), the egg clutch will be infected. Otherwise, uninfected egg masses will be oviposited. If transovarial transmission occurs, infected larvae will perpetuate pathogen transmission by feeding on small to medium-size hosts (2). Uninfected larvae can become infected at step 2 through cofeeding with infected larvae or by feeding on infected hosts. Infected larvae can also feed on humans (Z_1), representing the first point where humans can be infected. Transmission continues to uninfected hosts and ticks in the same manner at stage 3, perpetuating the pathogen in nature. Human infection can occur at Z_2 and Z_3. Solid gray curved arrows follow the sylvatic cycle. The gray-to-white gradient curved arrows indicate accidental human infection. Solid gray straight arrows denote the infected path, while the open straight arrow denotes the uninfected path. (Portions of this figure adapted from images from the Centers for Disease Control and Prevention Public Health Image Library. Ticks modified from CDC image 6005; human modified from CDC image 3425; small mammals modified from CDC image 3381; large mammals modified from CDC image 3392.) doi: 10.1128/9781555817336.ch13.f3

diseases, there is an occupational hazard for individuals who work in rice fields or military personnel deployed to an endemic region. An estimated 1 billion individuals are at risk, and approximately 1 million cases occur each year (Jensenius et al., 2004; Nakayama et al., 2008). General flulike symptoms develop approximately 6 to 21 days after a chigger bite (Demma et al., 2006). More advanced symptoms such as pneumonitis, meningoencephalitis, jaundice, renal failure, and myocarditis occur in untreated individuals, in whom case fatality rates can reach 30% (Demma et al., 2006).

Table 1 lists selected rickettsiae and their geographical distribution, vectors, reservoir hosts, and animals that may be exposed to rickettsiae in nature.

SYLVATIC AND ZOONOTIC CYCLES OF THE ORDER *RICKETTSIALES*

Transmission cycles are subject to the influence of climate and human encroachment. The effectiveness of a cycle can be described by both its epidemiological and zoonotic potency, which are defined by the frequency of human and animal infections, respectively (Mather and Ginsberg, 1994). Transmission cycles are categorized as sylvatic or zoonotic (Fig. 3). Sylvatic cycles involve a reservoir host, which can be the vector itself or a separate mammal, possibly a human (Mather and Ginsberg, 1994; Randolph and Rogers, 2010). Zoonotic cycles are established once recreational activities or employment requirements (e.g., logging, park ranger duties) bring humans into close contact with an ecological niche where sylvatic cycles exist. Arthropod vectors that are generalist in their feeding behavior with an affinity for any animal will bite and may infect humans with whatever they harbor. In most instances, humans are dead-end or accidental hosts, so most rickettsial infections will not be transmitted from human to human or human to nonhuman mammals, with few exceptions. Research suggests that humans infected with the etiologic agent of epidemic typhus, *R. prowazekii*, maintain a subclinical infection for the remainder of their lives (Raoult et al., 1998; Raoult and Roux, 1999). The infection may recrudesce in later life as Brill-Zinsser disease as a result of stress or a weakened immune system (Raoult et al., 1998; Raoult and Roux, 1999). These human reservoirs of *R. prowazekii* can initiate an epidemic in the presence of the louse vector and deteriorating conditions (e.g., war, famine, disaster, or overcrowded jailhouses) (Raoult et al., 1997, 1998; Tarasevich et al., 1998).

Two types of hosts are observed. Reproductive hosts are those that "amplify" the vector population by supporting and promoting the life cycle of the vector. Reproductive hosts are not infected by the pathogen and therefore do not perpetuate the organism in nature. Reservoir hosts not only support the vector's life cycle but also maintain the pathogen (Mather and Ginsberg, 1994). The ehrlichiosis transmission cycle involves both reproductive and reservoir hosts. The etiologic agent for human monocytic ehrlichiosis, *Ehrlichia chaffeensis*, is transmitted by the lone star tick, *Amblyomma americanum*, to the white-tailed deer, *Odocoileus virginianus*. The white-tailed deer supports all life stages of the tick and maintains an *E. chaffeensis* infection, rendering it the officially recognized reservoir host. *A. americanum* displays voracious generalist biting behavior, attacking and feeding on birds, both large and small nonhuman mammals, and humans (Paddock and Childs, 2003). Small to medium-size mammals such as the white-footed mouse, red fox, raccoon, and opossum host immature life stages of *A. americanum*. The reservoir competency of these mammals for *E. chaffeensis* is not clear or has been largely discounted, classifying these hosts as reproductive hosts (Lockhart et al., 1998; Paddock and Childs, 2003).

COFEEDING, MODULATION OF HOST IMMUNITY, AND ACQUISITION

Cofeeding

Transmission between vectors and hosts is described as being horizontally or vertically maintained. Horizontal transmission is the transmission of a pathogen between a vector and a host or between two vectors (of the same or different genera) that feed on a single host (Mather and Ginsberg, 1994). Vertical transmission involves the transmission of a microbe from the female to the progeny in a process called transovarial transmission. Interesting trends for transovarial transmission are discussed below (see "Interference Phenomenon and Exclusion").

One interesting variation of horizontal transfer occurs between an infected and an uninfected vector that are simultaneously feeding on the same uninfected host. Transmission in this instance is said to occur through cofeeding and can be illustrated nicely using the mite-scrub typhus zoonosis.

The etiologic agent of scrub typhus, *O. tsutsugamushi*, is transmitted from adult to larvae (vertical transmission) by trombiculid mites (Frances et al., 2000). There is little evidence that mites acquire *O. tsutsugamushi* horizontally. However, perpetuation of a pathogen within a natural cycle requires some horizontal transmission even if the primary mechanism of transfer is thought to occur vertically (Fine, 1981; Mather and Ginsberg, 1994). Frances et al. (2000) demonstrated that uninfected mite larvae can acquire *O. tsutsugamushi* from infected mite larvae while cofeeding on a nonrickettsemic host.

The multiexperimental study began by demonstrating that a single infected *Leptotrombidium deliense* mite could transfer *O. tsutsugamushi* to a group of uninfected mites of the same *Leptotrombidium* sp. while both were feeding on an uninfected rat, with an overall rate of acquisition of 1.6% (Frances et al., 2000). The researchers also demonstrated that intergenera transmission of *O. tsutsugamushi* appears to have lower transmission efficiency but that the organism is transferred between mite genera nonetheless. Transmission of *O. tsutsugamushi* from a single infected *L. deliense* to a group of *Blankaartia acuscutellaris* mites feeding on the same rat was unsuccessful; however, *B. acuscutellaris* successfully acquired *O. tsutsugamushi* while cofeeding with multiple infected *L. deliense* mites, with an overall rate of acquisition of 2.0% (Frances et al., 2000). Even though acquisition rates were similar for each transmission experiment, intergenera transmission required cofeeding of multiple infected mites with uninfected mites. Thus, the transmission efficiency appears to be less for the intergenera transmission experiment. The difference in transmission efficiency in this study may be linked to discrepancies in the number of *O. tsutsugamushi* organisms present in the bite site or the infectious dose required for acquisition by each mite, which is related to the vector competence of each mite (Mather and Ginsberg, 1994).

Cofeeding transmission is not universal for all rickettsial agents. Uninfected male *Dermacentor andersoni* ticks were unable to acquire *A. marginale* when placed on the same ear of nonrickettsemic rabbits with *A. marginale*-infected male ticks of the same species (Kocan and de la Fuente, 2003). The salivary glands of the infected ticks were infected with *A. marginale* and were transmission competent (Kocan and de la Fuente, 2003). The studies demonstrate nicely that cofeeding transmission is not universally critical to maintenance of pathogens in nature, further highlighting the unique complexities that can exist for each rickettsial transmission cycle.

Saliva-Activated Transmission and Modulation of Host Immunity

Transmission through cofeeding is supported by a phenomenon termed saliva-activated transmission (Kovár, 2004; Nuttall and Labuda, 2004). Saliva-activated transmission and feeding in general are mediated by factors found in the saliva of the vector (flea, tick, or mite). Tick saliva has commanded a considerable amount of attention regarding its promotion of feeding and/or pathogen transmission. The main function of the salivary glands is to concentrate the blood meal by secreting the water from the blood that is imbibed back into the host (Bowman et al., 1997). Many transcriptomes have been published that aim to elucidate the vector-host interaction that is mediated by the saliva (Dreher-Lesnick et al., 2009; Ribeiro et al., 2006; Ribeiro and Francischetti, 2003; Valenzuela et al., 2002, 2003). The relevance of saliva-activated transmission to rickettsial transmission can be appreciated through a survey of studies on saliva from various tick vectors (Bowman et al., 1997; Kovár, 2004; Nuttall and Labuda, 2004; Ribeiro and Francischetti, 2003).

Once a tick embeds its mouthparts into a host, glue is secreted that cements ixodid (hard) ticks into the skin for the duration of feeding (Bowman et al., 1997; Sonenshine, 1991). Soon after, a number of factors are released from the salivary glands to prepare the site for feeding. Prostaglandins function as platelet inhibitors and vasodilators that initiate and maintain the flow of blood into the bite site, anticoagulants like ixolaris prevent blood coagulation so the tick can imbibe a blood meal unimpeded, and antihistamine functions to reduce or prevent edema at the attachment site (Bowman et al., 1997; Brossard and Wikel, 2004; Nuttall and Labuda, 2004). There is also evidence that saliva from *Ixodes ricinus* and *Ixodes scapularis* promotes the growth of *Borrelia garinii* and possesses chemotactic properties for *Borrelia burgdorferi* in vitro, respectively (Nuttall and Labuda, 2004). This latter point is quite interesting because an increase in pathogen numbers at the bite site will theoretically increase transmission potential, especially for those zoonoses where transmission requires a higher infectious dose.

Tick-borne rickettsiae are secreted with saliva into the bite site, where immune activity is suppressed. Pathogens of all types, including those in the *Rickettsiales*, probably exploit this opportunity to infect the host in a low-stress environment. For instance, lymphocytes from lymph nodes that drain the tick attachment site produced low levels of gamma interferon (IFN-γ) when stimulated with salivary gland extract (Brossard and Wikel, 2004; Mejri et al., 2001). This finding was corroborated with a finding that low IFN-γ transcript abundance was observed in the lymph nodes that drain the *I. ricinus* attachment site in BALB/c mice (Brossard and Wikel, 2004). This area of low IFN-γ activity would benefit rickettsiae injected into the bite site, as IFN-γ is known to be important for the production of rickettsicidal nitric oxide radicals (Walker, 2007).

NK-cell activity is known to be important for an early response to rickettsial infection (Walker, 2007). Interestingly, tick saliva from 6-day-fed *Dermacentor reticulatus*, a potential vector for the emerging pathogen *Rickettsia helvetica*, reduces NK-cell activity (Kubes et al., 2002).

Rickettsial pathogens will also be vulnerable to an increase in immune-functioning cells and factors that have been drawn to the bite site. It is the immunomodulatory factors secreted with tick saliva that function to promote rickettsial transmission by increasing the number of infectious organisms at the site of tick attachment. Ultimately, this attribute of tick saliva will increase the probability that transmission to the host or acquisition by the tick will occur.

Bacterial Load in the Tick and Transmission—*A. marginale*-Tick Model

REPLICATION AND ACQUISITION BY TICKS

Transmission of rickettsiae is dependent on the vector-bacterium interaction as much as it is the vector-host interaction. In fact, vector competency may be defined as a function of the vector's ability to modulate the host's immune response and the suitability of the vector as a replicative niche for the bacterium. Transmission of a bacterium from host to tick is dependent on the level of host bacteremia and the overlap between duration of infectivity and the host-vector interaction (Mather and Ginsberg, 1994). Research suggests that *A. marginale* levels in the bovine host correlate to tick infection rate. An average host rickettsemia of <4.0 and 5.84 log infected erythrocytes/ml of blood results in an intrastadial male tick infection rate that reaches 27 and 81%, respectively (Eriks et al., 1993). Regardless of the differences in the tick infection rate, the log mean *A. marginale* organisms per infected tick were similar; i.e., *A. marginale* replication in the tick abolished the differences in tick infection rate that host rickettsemia dictates during the acquisition feed (Eriks et al., 1993).

REPLICATION AND TRANSMISSION TO THE HOST

A. marginale replication during transmission from the tick to the host upon subsequent feeding events is as important to maintenance of the pathogen in nature as is replication in the tick following acquisition. A replication advantage may be attributable to *A. marginale* strain-tick compatibility. For instance, the tropical tick *R. (Boophilus) microplus* supports and transmits both the tropical Puerto Rico strain and the temperate St. Maries strain of *A. marginale* even though the Puerto Rico strain replicates with a 1.8-fold advantage over the St. Maries strain (Futse et al., 2003). *R. microplus* has had little contact with the temperate St. Maries strain because of its eradication from the United States for more than 60 years (Futse et al., 2003). The sympatric existence of the Puerto Rico strain and *R. microplus* is hypothesized to contribute to the replicative advantage that the Puerto Rico strain has over the St. Maries strain. These data suggest that tick-strain compatibility is a determinant of the epidemiological and zoonotic potency of *A. marginale* in nature.

REPLICATION AND MAINTENANCE IN THE TICK

A. marginale strain-tick compatibility may also be attributable to the ability of the bacterium to establish itself within the tick. Research suggests that strain differences determine *A. marginale* replication in the tick (Futse et al., 2003). During acquisition feeding, the mean number of *A. marginale* Mississippi strain organisms per *D. andersoni* midgut is 100- or 1,000-fold less than the number of Israel vaccine strain or St. Maries strain organisms, respectively (Ueti et al., 2007). Upon transmission feeding, only 2% of midguts were infected with the Mississippi strain and the number of organisms in the salivary glands was below the quantifiable threshold (Ueti et al., 2007). The inability of the Mississippi strain to replicate in the midgut or infect and replicate in the salivary glands partially demonstrates the incompatibility between tick and bacterium.

THE TICK AS A BARRIER TO TRANSMISSION

At the risk of anthropomorphizing the bacterium, the tick is a filter that *Anaplasma* spp. must negotiate to insure their own survival in nature. Replication competence may not suffice to overcome the tick as an obstruction to infection. Recent research establishing the midgut and salivary glands as true barriers to transmission supports this "tick as a filter" observation. One hundred percent of *D. andersoni* ticks that acquisition fed on calves infected with either the St. Maries or Israel vaccine strain of *A. marginale* possessed *A. marginale* in their midguts (lumen and epithelium) (Ueti et al., 2007). The mean number of St. Maries strain ($10^{(6.36 \pm 0.85)}$) and Israel vaccine strain ($10^{(5.4 \pm 0.69)}$) organisms per midgut varied during acquisition; however, during transmission feeding the mean number of organisms in the salivary glands was comparable for each strain [St. Maries, $10^{(7.7 \pm 0.45)}$; Israel vaccine, $10^{(7.4 \pm 0.08)}$] (Ueti et al., 2007). The difference came during measurement of transmission of each strain to naïve calves. Regardless of the equal burden of each strain in the salivary glands, only the St. Maries strain was transmitted (Ueti et al., 2007).

In a recent study, the same researchers addressed this discrepancy between load and transmission for *A. marginale* strains. Three major differences were observed between the strains during infection of the tick: (i) the Israel vaccine strain colonized and replicated both the midgut and salivary glands at a reduced rate compared with the St. Maries strain; (ii) immunohistochemistry revealed that the St. Maries strain formed multiple colonies during infection of the salivary gland, while the vaccine strain formed large single colonies; and (iii) >90% of the saliva was positive for St. Maries, while only 30% of saliva from Israel vaccine strain-infected ticks was positive, albeit with a 10-fold-lower abundance of organisms (Ueti et al., 2009). These studies illustrate how replication and development within the tick and delivery of organisms by the tick to the host must complement one another for transmission to occur. Given the obstacles presented, we must marvel at the complexities of the *Anaplasma* spp.-tick interaction and that select strains (St. Maries) can transmit with greater efficiency.

The aforementioned studies were and are critical to establishment of the tick as a filter for *A. marginale* transmission in nature. Collectively, the data indicate that adaptation of a particular *A. marginale* strain to its tick vector is a function of sympatry; however, many factors besides proximity of a tick to an individual *Anaplasma* spp. contribute to the potency of the sylvatic cycle. This raises several other questions for these and other bacteria in the *Rickettsiales*. How well do the organisms replicate within their hosts and vectors, and what is role of the tick as a barrier to transmission? The *A. marginale-D. andersoni* tick model epitomizes the complexities associated with host infectivity and vector competency while highlighting the importance of strain differences and their collective effect on perpetuation of the sylvatic cycle.

MAINTENANCE OF RICKETTSIAL AGENTS BY ARTHROPOD VECTORS

The success of any species is marked by its ability to reproduce. The longevity of any species through evolutionary time is also a testament to its success. In general, survival while under the constant threat of microbial invasion constitutes success in vector biology. The driving question may be, how are the majority of vector arthropods able to acquire, maintain, and transmit bacterial, viral, or protozoan pathogens without suffering any short-term physiological maladies? This isn't true in all instances, especially if one considers the lethal effects of *R. prowazekii* on the body louse. There are also reports on the lethal effects of *Rickettsia rickettsii* on its tick vector. Infection of *D. andersoni* females with *R. rickettsii* resulted in high mortality (96.8%), reduced transovarial transmission (39.0%), and reduced number of hatched larvae derived from infected females (Niebylski et al., 1999). Regardless of these observations, mortality, although close to 100%, is not 100%. Is the percentage of ticks that survive to perpetuate *R. rickettsii* in nature resistant to those lethal effects? This question can be addressed by investigating the arthropods' response to rickettsial infection. It is important to note that the arthropods' defense response to rickettsial infection will be opposed by rickettsial evasion mechanisms. This balance between action and reaction is a hallmark of a successful vector-rickettsiae relationship and its perpetuation in nature. Thus, the mechanisms behind an antirickettsial defense response will be bacteriostatic at best.

Early observations on the arthropod response to infection were made using *Drosophila melanogaster* as a model (Dushay and Eldon, 1998). The innate immune response in *D. melanogaster* does not appear to be driven by somatic hypermutation and acquired immunity. Instead, the immune response in invertebrates is considered to be innate and possesses a cellular as well as a humoral component (Dushay and Eldon, 1998).

Cell-Mediated Response

Experimentation in ticks revealed that immune hemocytes monitor the hemocoel for microbes that may have breached the midgut barrier. Foreign bodies introduced into the hemocoel are rapidly surrounded and walled off to prevent further infiltration in a process termed encapsulation. The original studies describing this phenomenon employed the *Rickettsiales* vector *Dermacentor variabilis* to demonstrate that the cellular response is stimulated to encapsulate large inanimate foreign bodies (Epon). Encapsulation of the Epon sliver initiated quite rapidly. Within 1 hour of implantation, hemocytes begin to attach and degranulate (Eggenberger et al., 1990). The process snowballs quickly and is complete by 72 hours postimplantation. The capsule is composed of many hemocytes that have degranulated and formed a multilayer cellular matrix (Eggenberger et al., 1990). In a similar study, *Escherichia coli* inoculated intrahemocoelically into *D. variabilis* was observed to form aggregates and was encapsulated by hemocytes as early as 1 hour postinoculation (Ceraul et al., 2002). As in the studies by Eggenberger et al. (1990), the aggregated bacteria were increasingly associated with flattened hemocytes. Interestingly, viable *E. coli* was not detected using the CFU assay as early as 6 hours postinoculation (Ceraul et al., 2002). These observations suggest that encapsulation is one mechanism employed to control growth of infiltrating bacteria.

There are five main types of hemocytes defined in ticks: (i) prohemocytes, (ii) plasmatocytes, (iii) granulocytes, (iv) spherulocytes, and (v) oenocytoids (Sonenshine, 1991). Immune tick hemocytes are classified in broad terms as plasmatocytes (phagocytes) and granulocytes (phagocytic and lipid- and fibrillar material-containing) (Sonenshine, 1991). Studies suggest that the phagocytic activity of granulocytes increases in response to fetal bovine serum. Individual granulocytes from the argasid tick *Ornithodoros moubata* engulfed a greater number of fluorescent polystyrene beads compared with controls when treated with fetal bovine serum (Inoue et al., 2001). Phagocytosis and reactive oxygen species (ROS) production also appear to be important to immunity in the *A. marginale* vector *R.* (*Boophilus*) *microplus*. ROS production increases in phagocytic hemocytes when stimulated with bacteria and decreases in the presence of superoxide dismutase and catalase (Pereira et al., 2001). Interestingly, cytokine- or RANTES-stimulated human umbilical vein endothelial cell, THP-1, or freshly isolated human monocyte-mediated cytotoxicity for *Rickettsia conorii* was abolished by catalase treatment, indicating that hydrogen peroxide is the rickettsicidal factor responsible (Feng and Walker, 2000). Taken together, bacteria-induced ROS production in tick hemocytes and the cytotoxic effects of ROS species for *R. conorii* argue strongly for the role of ROS in limiting rickettsial infection in ticks as well as humans. To date, no direct correlation has been drawn between ROS and rickettsial infection in ticks.

For both vertical and horizontal transmission to occur, organisms within the *Rickettsiales* must transmigrate the midgut to colonize the salivary glands and/or ovaries, which places them in direct contact with hemocytes. To date, we have no data on the importance of cell-mediated immunity and rickettsial infection of vectors. However, it seems probable that a mechanism like encapsulation may be important to limiting rickettsial infiltration.

Humoral (Soluble) Response

The foundation of the humoral response is the numerous soluble factors that are liberated from hemocytes or the epithelial tissue layers found in the arthropod. The humoral arm of the innate immune system in *D. melanogaster* has been the most intensely studied model. Bacterial insult results in a rapid production of antimicrobial factors (≤30 minutes to hours) in the fat body, which are released into the hemolymph (bloodlike fluid that bathes hemocoel and the organs) to combat the foreign organisms that elicited the response (Hultmark, 2003). It has been determined that the arthropod innate immune system is able to differentiate between gram-negative and gram-positive bacteria and fungi, regulating antimicrobial peptide production accordingly (Hultmark, 2003; Lemaitre et al., 1997). These studies have been extended to other arthropod vectors and invertebrates such as mosquitoes, the tsetse fly, and *Caenorhabditis elegans*, although studies in ticks are lacking.

Early studies on the humoral arm of tick immune function focused on the antibacterial properties of hemolymph from part-fed or fully engorged ticks. Investigative studies showed that tick hemolymph was a potent antimicrobial substance. Initially, studies were directed to identify the factor(s) responsible for the ability of *D. variabilis*, a vector for the *Rickettsiales*, to survive intrahemocoelic inoculation with

Bacillus subtilis, *E. coli*, or *Staphylococcus aureus*. Priming the tick with intrahemocoelic inoculation of bacteria enhanced the constitutive antimicrobial activity of the hemolymph (Johns et al., 1998). The antimicrobial activity of the hemolymph led these same researchers to ask questions about this activity as an attribute of vector competency. The hemolymph from *D. variabilis* (noncompetent for *B. burgdorferi*) was cytotoxic to spirochetes, whereas the hemolymph from *I. scapularis* (competent vector for *B. burgdorferi*) was not (Johns et al., 2001). It was later discovered that defensin stored in hemocytes was released into the hemolymph and probably contributed to the anti-*B. burgdorferi* activity of *D. variabilis* hemolymph (Ceraul et al., 2003; Johns et al., 2001).

There has been an explosion of reports on the identification and characterization of antimicrobial peptides from numerous tick species, including those involved in transmission of rickettsial agents (Fogaca et al., 2004; Nakajima et al., 2001, 2002, 2003a, 2003c; Simser et al., 2004a, 2004b). Most of these studies characterize antimicrobial peptide activity using laboratory strains of bacteria such as *E. coli* or *Micrococcus luteus*. This is important for basic characterization and provides data on the specificity of an antimicrobial peptide. The relevance of the defense response to rickettsial colonization of a vector is beginning to be addressed.

Response of the Vector to Rickettsial Infection and Transmission

Early studies assessed whether challenge with rickettsiae causes a shift in transcriptional patterns in the salivary glands, midgut, or ovaries of vector ticks and fleas. The identified transcripts are normally classified on the basis of how they mediate the vector-rickettsiae interaction. By comparing transcriptional profiles of *Rickettsia montanensis*-infected and uninfected ovaries from *D. variabilis*, transcripts were classified according to their role as part of the tick's immune response, as adhesin receptors possibly utilized by spotted fever group rickettsiae or the tick's stress response (Mulenga et al., 2003; Macaluso et al., 2003). A number of transcripts related to the innate immune response, vesicular trafficking, and blood meal digestion were differentially regulated in the midgut of *C. felis* in response to feeding alone or *R. typhi* infection. *R. typhi* infection of the midgut increased expression of Rab5, a GTPase localized to the early endosome (Dreher-Lesnick et al., 2009). Interestingly, a number of Rabs that regulate endocytic recycling and endoplasmic reticulum-to-Golgi trafficking are recruited to the *Anaplasma phagocytophilum*-occupied vacuole of infected host cells (Huang et al., 2010). Modification of the *A. phagocytophilum*-containing vacuole by Rab may be important to maintenance of infection of the host cell. Alternatively, *R. typhi* escapes the early endosome and replicates in the cytoplasm of the host cell. An increase in transcript abundance of Rab5 in the midgut of *R. typhi*-infected fleas may signify its involvement in early endosomal escape.

Functional Characterization in Terms of Rickettsial Infection

Functional characterization of tick immune response genes in the context of *R. montanensis* infection quite nicely supports the hypothesis that ticks control rickettsial infection. This hypothesis relates back to the idea that ticks are a successful species because they have adapted to deal with microbial challenge during feeding. In general, competent vectors will not survive if bacterial infection and growth proceed without boundaries set by the defense response. The midgut is the first organ encountered by rickettsiae as they are imbibed with the blood meal. It is feasible to hypothesize that control of bacterial infection will be most effective if it occurs early, before rickettsiae invade the midgut epithelium. It is likely that rickettsiae encounter rickettsiostatic and rickettsicidal factors before invasion, as feeding alone induces a defense response in the midgut (Nakajima et al., 2001, 2002; Sonenshine et al., 2005). Interestingly, a by-product of hemoglobin digestion, β-hemoglobin, is observed to be antibacterial (Fogaca et al., 1999; Nakajima et al., 2003b; Sonenshine et al., 2005). Some research suggests that immune activity in the hemolymph and midgut is constitutively maintained in the absence of stimulation but can be enhanced when ticks are stimulated with bacteria (Ceraul et al., 2007; Johns et al., 1998; Nakajima et al., 2001, 2002, 2003a).

D. variabilis defensin and lysozyme transcript abundance is significantly upregulated above that of controls in the midgut as early as 24 hours postinfection in response to *R. montanensis* infection (Ceraul et al., 2007). This was some of the first evidence that immune-specific antimicrobial peptides of ticks were differentially regulated in response to rickettsial infection. However, the data, while informative, provide no indication of the functional capacity of defensin or lysozyme as rickettsicidal proteins. The data simply indicate that ticks respond immunologically to rickettsial infection.

Movement toward identification of immune-functional rickettsicidal or rickettsiostatic molecules comes from an unexpected source. Kunitz-type protease inhibitors (KPIs) are primarily expressed in the salivary glands of hematophagous arthropods and function as anticoagulants, keeping the blood fluid, so a blood meal can be imbibed (Francischetti et al., 2004). *D. variabilis* KPI (DvKPI) was identified through a serine protease inhibitor homology cloning project. DvKPI is predominantly transcribed in the midgut and secreted into the midgut lumen (Ceraul et al., 2011). An in vitro antibacterial assay shows that *R. montanensis* burden is reduced ~60-fold in the presence of DvKPI (Ceraul et al., 2008). *R. montanensis* burden increases in ticks that are treated with small interfering RNA specific to DvKPI, corroborating the in vitro antibacterial assay results (Ceraul et al., 2011). Further studies show that DvKPI function is targeted to the invasion process of *R. montanensis* (Ceraul et al., 2011).

In symbiotic relationships between nodulating bacteria and legumes, it is hypothesized that KPIs contribute to the symbiotic homeostasis by limiting bacterial invasion (Lievens et al., 2004; Manen et al., 1991; Vasse et al., 1993). While the mechanism is still unclear, the data suggest that the KPI associates with nodulating bacteria (Lievens et al., 2004; Manen et al., 1991). Extending these observations to their own system, researchers hypothesized that DvKPI associates with *R. montanensis* organisms as they enter the midgut lumen, thereby limiting rickettsial invasion. Immunofluorescence assays and electron microscopy studies demonstrate that DvKPI does indeed associate with *R. montanensis*, offering one possibility for how DvKPI exerts its rickettsiostatic effect (Ceraul et al., 2011).

Experimentation has informed us that arthropod vectors are not passive to rickettsial infection. Overgrowth of nodulating bacteria on a legume root system physiologically stresses the plant (Marx, 2004) and could lead to death. Similarly to the nodulating bacteria-legume symbiosis, if ticks were void of a defense response, they may well have not survived for the millions of years that they are thought to have existed (Oliver, 1989). Similarly, the defense response is one factor that mediates vector competency for rickettsial transmission. Unfortunately, humans and domestic and companion animals suffer at the expense of rickettsial zoonotic success. By understanding what stabilizes the vector-rickettsiae interaction, we can develop strategies to interrupt acquisition and transmission.

Interference Phenomenon and Exclusion

Competition for survival is observed on macrobiological and microbiological scales. The chapter has focused thus far on how tick physiology affects acquisition and maintenance of rickettsial infections. Interspecific competition between rickettsiae colonizing the same host can also play a role in determining rickettsial transmission in nature. Researchers who submitted a report on "spotted fever" made the interesting observation that individuals who contracted spotted fever did so on the western side of the Bitterroot River in the Bitterroot Valley (Wilson and Chowning, 1904). In 1981, the interference phenomenon, which adheres to Gause's law of competitive exclusion, was put forth (Burgdorfer et al., 1981; Telford, 2009). Burgdorfer et al. (1981) reported that 80% of *D. andersoni* ticks from the east side of the Bitterroot River possessed ovary infections with a nonpathogenic rickettsia, now known as *Rickettsia peacockii* (East Side agent). Experimentation revealed that some females laid eggs that were infected with both *R. peacockii* and *R. rickettsii* (Burgdorfer et al., 1981). Consequently, questions have been raised regarding the importance of interference as a basis for low *R. rickettsii* infection rates in nature (Telford, 2009). A more inclusive perspective would certainly include geographical barriers and climate. Yet there are data in the literature on rickettsiae to support the interference phenomenon in the maintenance and exclusion of rickettsial infections. When *D. variabilis* ticks constitutively infected with *R. montanensis* or *Rickettsia rhipicephali* were reciprocally challenged with the opposing rickettsia, only the original rickettsia was detected in their progeny (Macaluso et al., 2002). Moreover, ticks that lost the primary *R. rhipicephali* infection were proven to be susceptible to infection by *R. montanensis* (Macaluso et al., 2002).

The exclusion phenomenon is also observed in ticks, tick cells, and bovine erythrocytes between *A. marginale* genotypes. Inoculation of host cells (bovine or tick) with equal doses of the Virginia and Oklahoma isolates resulted in the establishment of the Virginia isolate only. Varying the ratios used for inoculation resulted in the establishment of the strain delivered at the higher concentration. A primary infection of bovine erythrocytes with *Anaplasma ovis* excluded both the Virginia and Oklahoma *A. marginale* isolates (de la Fuente et al., 2002). Finally, ticks with a primary Virginia isolate infection failed to pick up the Oklahoma isolate through acquisition feeding on an Oklahoma isolate-infected calf (de la Fuente et al., 2003).

Elucidating the molecular correlates that drive rickettsial transmission is constantly progressing. From transcriptome studies to characterization of individual genes and their products, we have begun to define the molecular foundation upon which rickettsial symbiotic homeostasis stands. The real task is to correlate the microscale gene (vector and microbe) function to the macroscale interaction between the organisms involved. We now turn our attention to the environment and global trends that may affect the potency of rickettsial zoonoses.

EFFECTS OF CLIMATE CHANGE ON RICKETTSIAL CHALLENGE

It is safe to state that without vectors, rickettsial diseases would not affect humans. Sylvatic (natural) transmission cycles that involve a microbe, arthropod, and host are complex and are dependent on the environment and sympatric existence of competent vectors and hosts/reservoirs for maintenance of the pathogen. Zoonotic cycles develop once humans encroach on the habitat that supports the sylvatic cycle. Conversely, human disturbances of a particular niche or environmental changes can cause vectors or their hosts and the diseases that they carry to migrate closer to humans and their domestic animals. Expansion of the opossum-cat flea (*C. felis*) cycle (suburban cycle) toward humans and their domestic pets may lead to human murine typhus infections in the absence of the well-documented rat-rat flea (*X. cheopis*) cycle (urban cycle) (Azad et al., 1997). The discovery that the suburban murine typhus cycle may serve to increase human *R. typhi* infections is a good example of how shifts in cycle distributions can lead to an increase in human or animal infection.

Diseases caused by bacteria in the order *Rickettsiales* are frequently listed under the umbrella term "emerging and reemerging diseases." The basic question of what causes a disease to emerge and reemerge often leads to the discovery of a shift in conditions that disturbs ecological homeostasis. Vector reproduction, distribution, biting intensity, pathogen development within the vector, and acquisition and transmission competence of a vector all affect the degree of endemicity of a zoonotic cycle (Kovats et al., 2001; Lafferty, 2009; Randolph, 2009). Of course, the success of transmission cycles rests equally on the mammals, which are subject to the same criteria as vectors, to provide a source of food for the vector and in some instances act as a reservoir for the pathogen. These biological shifts are best understood in terms of the effects that abiotic and biotic factors (see "Cofeeding, Modulation of Host Immunity, and Acquisition" and "Maintenance of Rickettsial Agents by Arthropod Vectors," above) have on a transmission cycle.

Abiotic factors include those that involve the climate, such as temperature, humidity, and rainfall (Ellis and Wilcox, 2009; Randolph, 2009; Randolph and Rogers, 2007). Arthropod vectors are sensitive to environmental changes brought on by climatic events that alter the soil moisture, humidity, and vegetation growth (Randolph, 2009). Temperature is a hot topic in climate change research and has been proposed to influence development rates, increase distribution, and affect seasonal activity of ticks (Gray et al., 2009). Indeed, with a rise in temperature, ticks may modify their behavior to thermoregulate (Lafferty, 2009). The geographical range of *D. reticulatus*, a vector for *Rickettsia* sp. RpA4 and possibly *A. marginale*, is documented to have expanded over the last 3 decades (Gray et al., 2009). From 1976 to 2003, *D. reticulatus* spread from 4 known sites to 26 previously unidentified sites in Germany (Gray et al., 2009). Although studies are required to qualify this finding, the authors speculate that tick movement occurred because the soil and air temperature did not meet the requirements to support oviposition as well as egg and larval development (Dautel et al., 2006; Gray et al., 2009).

The seasonal activity of vectors is another variable that is affected by temperature. *I. ricinus*, a tick vector for *A. marginale* and *A. phagocytophilum* in Europe, quests on vegetation waiting to ambush the next host to walk by. In central Europe, nymphs and adults quest from March to October, while larvae begin to quest in May (Gray et al., 2009). During the mild winter of 2006-2007, fall temperatures were only 3.4°C and winter temperatures 4.6°C higher than those measured from 1961 to 1990; however, this increase was enough to prolong questing activity (Gray et al., 2009). Prolonged questing activity increases the probability that humans and nonhuman mammals will be infested by infected ticks.

Vector-host contact is crucial to the spread or emergence of disease. Questing behavior mediates tick-host contact. Vegetation height can indirectly dictate the size of the animal that is infested by ticks. The higher a tick quests in the microclimate, the greater the probability that larger mammals (including humans) will be encountered (Randolph and Rogers, 2007; Sonenshine, 1993). This behavior is critical to the acquisition and transmission of *A. marginale*, *A. phagocyophilum*, *A. ovis*, *E. chaffeensis*, *Ehrlichia ruminantium*, and *Ehrlichia canis* to new hosts (de la Fuente et al., 2005; Paddock and Childs, 2003; Satta et al., 2011). Because the vegetation is greater in height, larvae and nymphs (immature stages) will encounter larger mammals. Larger animals act as a reservoir source of the pathogens. By feeding on larger animals, uninfected immature stages will become infected while infected immature stages will transmit pathogens to uninfected hosts, thereby perpetuating the sylvatic transmission cycle (Fig. 3). Additionally, mammals may also serve as amplification hosts, providing a food source for ticks and thereby increasing the tick population within a disease focus.

The degree of moisture present in the microclimate as a result in rainfall can affect both flea and tick populations. Ticks spend a major part of their life cycle in this microclimate because desiccation is detrimental to their development (Sonenshine, 1993). A relative humidity of less than 33% is lethal to the cat flea (*C. felis*) vector for *R. typhi*, the etiologic agent of murine typhus (Silverman et al., 1981). The disruption of microclimate conditions like humidity is hypothesized to reduce the cat flea population (Parmenter et al., 1999). In theory, a reduction of flea populations can reduce human exposure to infectious fecal material and the disease that results from such exposure.

One must also consider the anthropogenic disturbances in natural habitats as a separate factor in disease transmission or emergence. The degree of disturbance imparted by human activity can have varying effects on the potential for disease transmission. One intriguing study attempted to correlate the degree of human disturbance to flea infestation as a predictor of the risk for flea-borne disease. Interestingly, the number of fleas per mammal and the number of mammals infected by fleas considered to be generalist in their host preference increased with the level of disturbance (Friggens and Beier, 2010). This is an important note because an increase in generalist feeding behavior suggests that the geographical range of the pathogen will increase, especially if competent hosts and vectors exist in the new range (Friggens and Beier, 2010).

Most of the discussion regarding the ecology and transmission of bacteria within the order *Rickettsiales* is based on research that was performed with a clear goal of elucidating the biology of the vector. Clearly, the climate is not static, nor are the population dynamics, distribution, and activity (vector-host contact) of ticks, fleas, lice, mites, and the mammals that they infest. For transmission cycles, the behavior modifications of animals to climatic fluctuations correlate to the degree of human exposure and the prevalence of disease.

PERSPECTIVES ON THE ROLE OF INNATE IMMUNITY AND TRANSMISSION OF RICKETTSIAE

The data generated in our field are advancing toward a consensus on how ticks and fleas control rickettsial infections. Transcriptional profiling shows clearly that the immune response in arthropods is activated once a rickettsiae-infected blood meal is imbibed. However, the signaling mechanisms that drive immune activation are not well defined like they are for model systems such as *Drosophila*. The challenge in tick and flea research, given the lack of adequate genomes, will be to elucidate the molecules and pathways that mediate immune activation in response to rickettsial infection.

The fact that both bacteriostatic and bactericidal mechanisms are in place suggests that control may be a common theme with regard to vector-rickettsiae interactions. This may transfer to more longer-term maintenance of nonpathogenic endosymbionts. While immune-dependent control can lead to stable symbioses, immune activation directed toward invading microbes can have its costs. Pathogenic effects may arise from long-term heightened immune activation and/or immune senescence.

Recently, a long-form peptidoglycan recognition protein (PGRP-LB) was identified in a midgut transcript library from the flea vector for *R. typhi*, *C. felis* (Dreher-Lesnick et al., 2009). Likewise, PGRP-LB was annotated within the genome of *I. scapularis*, the tick vector for *B. burgdorferi* (the etiologic agent of Lyme disease) and *A. phagocytophilum* (the etiologic agent of anaplasmosis) (Gerardo et al., 2010). To date, we are unsure if PGRP-LB plays a role in detecting rickettsial infection and activating the immune response in the tick or flea.

Interestingly, PGRP-LB expression is positively correlated with the density of *Wigglesworthia glossinidiae*, a mutualistic symbiont of the tsetse fly, the vector of *Trypanosoma brucei* (the etiologic agent of African sleeping sickness) (Wang et al., 2009). When PGRP-LB expression is suppressed by RNA interference, *Wigglesworthia* density decreases. This is due to the apparent negative effect that PGRP-LB expression has on immune activation. Also, the susceptibility of tsetse flies to trypanosomes increases when PGRP-LB expression is suppressed (Wang et al., 2009). The authors hypothesize that PGRP-LB suppression allows for an increase in immune activation with a concomitant decrease in *Wigglesworthia* density and an increase in trypanosome susceptibility.

The authors hypothesize that the amidase activity of PGRP-LB may scavenge peptidoglycan from *Wigglesworthia*, thereby suppressing the immune response (Wang et al., 2009; Zaidman-Rémy et al., 2006). Furthermore, it appears from the data that the presence of *Wigglesworthia* has a negative effect on trypanosome infection. Collectively, PGRP-LB seems to be a major player in conferring tsetse fly vector competency for trypanosomes.

A similar biological theme may be driving vector competence of the tick in the presence of endosymbiotic primary rickettsial infections. As mentioned earlier, a primary infection of hard ticks limits or prevents infection of the ovaries by a second rickettsia (Burgdorfer et al., 1981; Macaluso et al., 2002). The primary infection protects the rickettsia from dual infection at the level of the ovary, which has implications for transovarial transmission

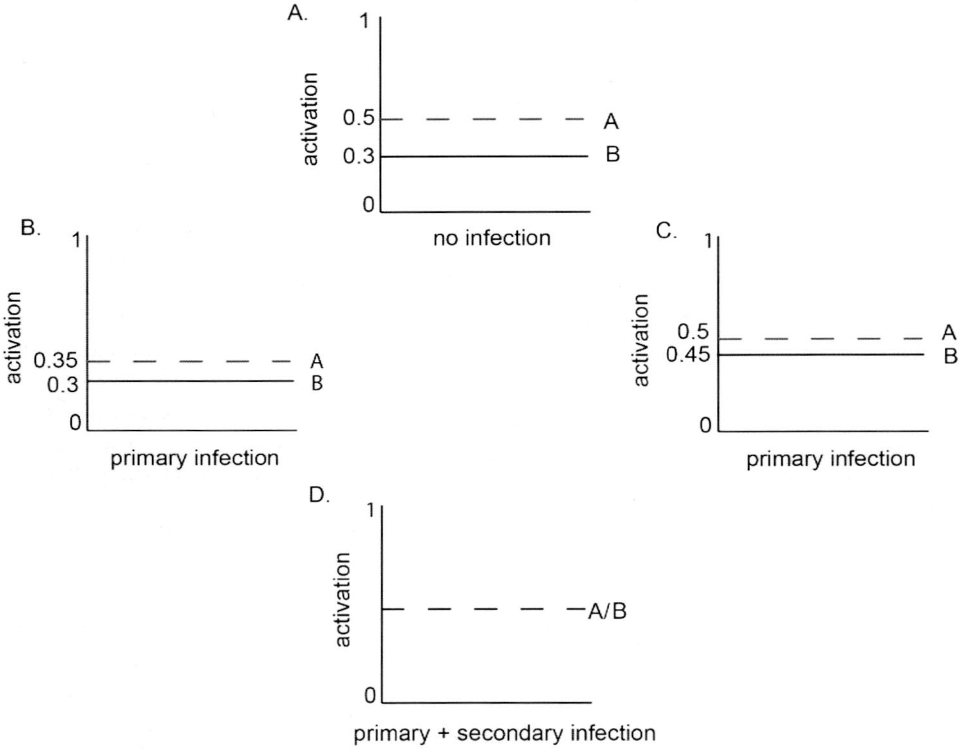

FIGURE 4 Endosymbiotic rickettsial modulation of tick immune activation. Endosymbiotic rickettsiae like *R. peacockii* may recalibrate immune homeostatic norms, thereby increasing the sensitivity of ticks to secondary rickettsial infection. Primary rickettsial infections may increase the immunological sensitivity of ticks to secondary rickettsial infections. Although this hypothesis has not been tested, it is possible that by-products of replication like free peptidoglycan functionally saturate recognition proteins such as PGRP-LB, which normally prevents activation by degrading immunostimulatory peptidoglycan. Alternatively, rickettsiae may modulate host gene transcription or posttranslational modification of immune effectors to make them more potent. We hypothesize that endosymbiotic rickettsiae effectively lower the activation threshold (B) or increase basal homeostatic immune activity, bringing it closer to the activation threshold (C). Scenario A represents immune activation in ticks that possess no infection. Scenario D represents the immune activation that occurs when ticks possess a primary endosymbiont and imbibe a secondary rickettsia. We hypothesize that rapid and possibly sustained immune activation in scenario D limits infection of the ovary by a second rickettsia. Dashed lines represent immune activation threshold; solid lines represent baseline (constitutive) immune activity. doi:10.1128/9781555817336.ch13.f4

and perpetuation of the second rickettsia in nature. The driving question is, how does a primary infection prevent a dual infection of the ovary? One scenario would involve modulation of the tick's immune response by the primary rickettsia. In this instance, the primary rickettsia suppresses immune activation by reduction of the immune activation threshold, thereby increasing sensitivity to the secondarily imbibed rickettsia (Fig. 4B). Alternatively, a heightened level of immunity could exist in the presence of the primary infection (Fig. 4C). That is, the primary rickettsia infection raises the baseline closer to the activation threshold. This too creates an environment of increased sensitivity to the incoming secondary rickettsia. Of course, both scenarios have to be empirically tested in the context of primary, secondary, and simultaneous dual rickettsial infection over time. Additionally, the discovery or identification of a number of immune-related signaling pathways and effectors in the *I. scapularis* genome will allow us to test immune activation in response to different rickettsial pathogens.

CONCLUSION

The complexity of a transmission cycle is determined by the multiple moving parts of which it is composed. Being obligate intracellular bacteria, rickettsiae are completely dependent on their host for survival. The micro- and macroclimate of an environment must support and "concentrate" the rickettsial vectors and mammalian and nonmammalian hosts. Thus, the incidence of rickettsial disease is dependent on the stability of the ecology, reservoirs, and vectors. As scientists, we must appreciate how the adaptations acquired by rickettsiae and their vectors contribute to the perpetuation of each transmission cycle.

In general, humans are not required for the success of the sylvatic cycle. We care about transmission of rickettsiae because we are invested in our own survival. Rickettsial diseases have the potential to change the outcomes of war and prey on the unfortunate circumstances that arise from disaster. Our ability to "live with" the threat of rickettsial infection requires that we look for clues in nature to help determine how each piece of the zoonotic puzzle fits together so that we may exploit the gaps to our benefit.

ACKNOWLEDGMENTS

I thank Jennifer McClure for providing critical comments on content of this chapter and compiling the data used in Table 1. Work for this chapter was supported by funds from the National Institutes of Health/National Institute of Allergy and Infectious Diseases (R01AI043006 and R01AI017828). The content is solely the responsibility of the author and does not represent the official views of the National Institute of Allergy and Infectious Diseases or the National Institutes of Health.

REFERENCES

Anderson, J. F., and L. A. Magnarelli. 2008. Biology of ticks. *Infect. Dis. Clin. North Am.* **22:** 195–215, v.

Azad, A. F., and C. B. Beard. 1998. Rickettsial pathogens and their arthropod vectors. *Emerg. Infect. Dis.* **4:**179–186.

Azad, A. F., S. Radulovic, J. A. Higgins, B. H. Noden, and J. M. Troyer. 1997. Flea-borne rickettsioses: ecologic considerations. *Emerg. Infect. Dis.* **3:**319–327.

Baldridge, G. D., T. J. Kurtti, N. Burkhardt, A. S. Baldridge, C. M. Nelson, A. S. Oliva, and U. G. Munderloh. 2007. Infection of *Ixodes scapularis* ticks with *Rickettsia monacensis* expressing green fluorescent protein: a model system. *J. Invertebr. Pathol.* **94:**163–174.

Bitam, I., K. Dittmar, P. Parola, M. F. Whiting, and D. Raoult. 2010. Fleas and flea-borne diseases. *Int. J. Infect. Dis.* **14:**e667–e676.

Bowman, A. S., L. B. Coons, G. R. Needham, and J. R. Sauer. 1997. Tick saliva: recent advances and implications for vector competence. *Med. Vet. Entomol.* **11:**277–285.

Bozeman, F. M., D. E. Sonenshine, M. S. Williams, D. P. Chadwick, D. M. Lauer, and B. L. Elisberg. 1981. Experimental infection of ectoparasitic arthropods with *Rickettsia prowazekii* (GvF-16 strain) and transmission to flying squirrels. *Am. J. Trop. Med. Hyg.* **30:**253–263.

Brossard, M., and S. K. Wikel. 2004. Tick immunobiology. *Parasitology* **129**(Suppl.)**:**S161–S176.

Burgdorfer, W., S. F. Hayes, and A. J. Mavros. 1981. Nonpathogenic rickettsiae in *Dermacentor andersoni*: a limiting factor for the distribution of *Rickettsia rickettsii*, p. 585–594. *In* W. Burgdorfer and R. L. Anacker (ed.), *Rickettsiae and Rickettsial Diseases*. Academic Press, New York, NY.

Ceraul, S. M., A. Chung, K. T. Sears, V. L. Popov, M. Beier-Sexton, M. S. Rahman, and A. F. Azad. 2011. A Kunitz protease inhibitor from *Dermacentor variabilis*, a vector for spotted fever group rickettsiae, limits *Rickettsia montanensis* invasion. *Infect. Immun.* **79:**321–329.

Ceraul, S. M., S. M. Dreher-Lesnick, J. J. Gillespie, M. S. Rahman, and A. F. Azad. 2007. New tick defensin isoform and antimicrobial gene expression in response to *Rickettsia montanensis* challenge. *Infect. Immun.* **75:**1973–1983.

Ceraul, S. M., S. M. Dreher-Lesnick, A. Mulenga, M. S. Rahman, and A. F. Azad. 2008. Functional characterization and novel rickettsiostatic effects of a Kunitz-type serine protease inhibitor from the tick *Dermacentor variabilis*. *Infect. Immun.* **76:**5429–5435.

Ceraul, S. M., D. E. Sonenshine, and W. L. Hynes. 2002. Resistance of the tick *Dermacentor variabilis* (Acari: Ixodidae) following challenge with the bacterium *Escherichia coli* (Enterobacteriales: Enterobacteriaceae). *J. Med. Entomol.* **39:**376–383.

Ceraul, S. M., D. E. Sonenshine, R. E. Ratzlaff, and W. L. Hynes. 2003. An arthropod defensin expressed by the hemocytes of the American dog tick, *Dermacentor variabilis* (Acari: Ixodidae). *Insect Biochem. Mol. Biol.* **33:**1099–1103.

Dautel, H., C. Dippel, R. Oehme, K. Hartelt, and E. Schettler. 2006. Evidence for an increased geographical distribution of *Dermacentor reticulatus* in Germany and detection of *Rickettsia* sp. RpA4. *Int. J. Med. Microbiol.* **296**(Suppl. 40)**:**149–156.

de la Fuente, J., E. F. Blouin, and K. M. Kocan. 2003. Infection exclusion of the rickettsial pathogen *Anaplasma marginale* in the tick vector *Dermacentor variabilis*. *Clin. Diagn. Lab. Immunol.* **10:**182–184.

de la Fuente, J., J. C. Garcia-Garcia, E. F. Blouin, J. T. Saliki, and K. M. Kocan. 2002. Infection of tick cells and bovine erythrocytes with one genotype of the intracellular ehrlichia *Anaplasma marginale* excludes infection with other genotypes. *Clin. Diagn. Lab. Immunol.* **9:**658–668.

de la Fuente, J., V. Naranjo, F. Ruiz-Fons, U. Höfle, I. G. Fernández De Mera, D. Villanúa, C. Almazán, A. Torina, S. Caracappa, K. M. Kocan, and C. Gortázar. 2005. Potential vertebrate reservoir hosts and invertebrate vectors of *Anaplasma marginale* and *A. phagocytophilum* in central Spain. *Vector Borne Zoonotic Dis.* **5:**390–400.

Demma, L. J., J. H. McQuiston, W. L. Nicholson, S. M. Murphy, P. Marumoto, M. Sengebau-Kingzio, S. Kuartei, A. M. Durand, and D. L. Swerdlow. 2006. Scrub typhus, Republic of Palau. *Emerg. Infect. Dis.* **12:**290–295.

Dreher-Lesnick, S. M., S. M. Ceraul, S. C. Lesnick, J. J. Gillespie, J. M. Anderson, R. C. Jochim, J. G. Valenzuela, and A. F. Azad. 2009. Analysis of *Rickettsia typhi*-infected and uninfected cat flea (*Ctenocephalides felis*) midgut cDNA libraries: deciphering molecular pathways involved in host response to *R. typhi* infection. *Insect Mol. Biol.* **19:**229–241.

Duma, R. J., D. E. Sonenshine, F. M. Bozeman, J. M. Veazey, Jr., B. L. Elisberg, D. P. Chadwick, N. I. Stocks, T. M. McGill, G. B. Miller, Jr., and J. N. MacCormack. 1981. Epidemic typhus in the United States associated with flying squirrels. *JAMA* **245:**2318–2323.

Dushay, M. S., and E. D. Eldon. 1998. *Drosophila* immune responses as models for human immunity. *Am. J. Hum. Genet.* **62:**10–14.

Eggenberger, L. R., W. J. Lamoreaux, and L. B. Coons. 1990. Hemocytic encapsulation of implants in the tick *Dermacentor variabilis*. *Exp. Appl. Acarol.* **9:**279–287.

Ellis, B. R., and B. A. Wilcox. 2009. The ecological dimensions of vector-borne disease research and control. *Cad. Saude Publica* **25**(Suppl. 1)**:** S155–S167.

Eriks, I. S., D. Stiller, and G. H. Palmer. 1993. Impact of persistent *Anaplasma marginale* rickettsemia on tick infection and transmission. *J. Clin. Microbiol.* **31:**2091–2096.

Feng, H. M., and D. H. Walker. 2000. Mechanisms of intracellular killing of *Rickettsia conorii* in infected human endothelial cells, hepatocytes, and macrophages. *Infect. Immun.* **68:**6729–6736.

Fine, P. E. M. 1981. Epidemiological principles of vector-mediated transmission, p. 77–91. *In* J. J. McKelvey, B. F. Eldridge, and K. Maramorosche (ed.), *Vectors of Disease Agents*. Praeger, New York, NY.

Fogaca, A. C., P. I. da Silva, Jr., M. T. Miranda, A. G. Bianchi, A. Miranda, P. E. Ribolla, and S. Daffre. 1999. Antimicrobial activity of a bovine hemoglobin fragment in the tick *Boophilus microplus*. *J. Biol. Chem.* **274:**25330–25334.

Fogaca, A. C., D. M. Lorenzini, L. M. Kaku, E. Esteves, P. Bulet, and S. Daffre. 2004. Cysteine-rich antimicrobial peptides of the cattle tick *Boophilus microplus*: isolation, structural characterization and tissue expression profile. *Dev. Comp. Immunol.* **28:**191–200.

Frances, S. P., P. Watcharapichat, D. Phulsuksombati, and P. Tanskul. 2000. Transmission of *Orientia tsutsugamushi*, the aetiological agent for scrub typhus, to co-feeding mites. *Parasitology* **120:**601–607.

Francischetti, I. M., T. N. Mather, and J. M. Ribeiro. 2004. Penthalaris, a novel recombinant five-Kunitz tissue factor pathway inhibitor (TFPI) from the salivary gland of the tick vector of Lyme disease, *Ixodes scapularis*. *Thromb. Haemost.* **91:**886–898.

Friggens, M. M., and P. Beier. 2010. Anthropogenic disturbance and the risk of flea-borne disease transmission. *Oecologia* **164:**809–820.

Futse, J. E., M. W. Ueti, D. P. Knowles, Jr., and G. H. Palmer. 2003. Transmission of *Anaplasma marginale* by *Boophilus microplus*: retention of vector competence in the absence of vector-pathogen interaction. *J. Clin. Microbiol.* **41:**3829–3834.

Gerardo, N. M., B. Altincicek, C. Anselme, H. Atamian, S. M. Barribeau, M. de Vos, E. J. Duncan, J. D. Evans, T. Gabaldón, M. Ghanim, A. Heddi, I. Kaloshian, A. Latorre, A. Moya, A. Nakabachi, B. J. Parker, V. Pérez-Brocal, M. Pignatelli, Y. Rahbé, J. S. Ramsey, C. J. Spragg, J. Tamames, D. Tamarit, C. Tamborindeguy, C. Vincent-Monegat, and A. Vilcinskas. 2010. Immunity and other defenses in pea aphids, *Acyrthosiphon pisum*. *Genome Biol.* **11:**R21.

Gray, J. S., H. Dautel, A. Estrada-Peña, O. Kahl, and E. Lindgren. 2009. Effects of climate change on ticks and tick-borne diseases in Europe. *Interdiscip. Perspect. Infect. Dis.* **2009:**593232.

Huang, B., A. Hubber, J. A. McDonough, C. R. Roy, M. A. Scidmore, and J. A. Carlyon. 2010. The *Anaplasma phagocytophilum*-occupied vacuole selectively recruits Rab-GTPases that are predominantly associated with recycling endosomes. *Cell. Microbiol.* **12:**1292–1307.

Hultmark, D. 2003. *Drosophila* immunity: paths and patterns. *Curr. Opin. Immunol.* **15:**12–19.

Inoue, N., K. Hanada, N. Tsuji, I. Igarashi, H. Nagasawa, T. Mikami, and K. Fujisaki. 2001. Characterization of phagocytic hemocytes in *Ornithodoros moubata* (Acari: Ixodidae). *J. Med. Entomol.* **38:**514–519.

Jensenius, M., P. E. Fournier, and D. Raoult. 2004. Rickettsioses and the international traveler. *Clin. Infect. Dis.* **39:**1493–1499.

Johns, R., D. E. Sonenshine, and W. L. Hynes. 1998. Control of bacterial infections in the hard tick *Dermacentor variabilis* (Acari: Ixodidae): evidence for the existence of antimicrobial proteins in tick hemolymph. *J. Med. Entomol.* **35:**458–464.

Johns, R., D. E. Sonenshine, and W. L. Hynes. 2001. Identification of a defensin from the hemolymph of the American dog tick, *Dermacentor variabilis*. *Insect Biochem. Mol. Biol.* **31:**857–865.

Jones, K. E., N. G. Patel, M. A. Levy, A. Storeygard, D. Balk, J. L. Gittleman, and P. Daszak. 2008. Global trends in emerging infectious diseases. *Nature* **451:**990–993.

Jongejan, F., and G. Uilenberg. 2004. The global importance of ticks. *Parasitology* **129**(Suppl.)**:**S3–S14.

Kocan, K. M., and J. de la Fuente. 2003. Co-feeding studies of ticks infected with *Anaplasma marginale*. *Vet. Parasitol.* **112:**295–305.

Kovár, L. 2004. Tick saliva in anti-tick immunity and pathogen transmission. *Folia Microbiol. (Praha)* **49:**327–336.

Kovats, R. S., D. H. Campbell-Lendrum, A. J. McMichael, A. Woodward, and J. S. Cox. 2001. Early effects of climate change: do they include changes in vector-borne disease? *Philos. Trans. R. Soc. Lond. B Biol. Sci.* **356:**1057–1068.

Krusell, A., J. A. Comer, and D. J. Sexton. 2002. Rickettsialpox in North Carolina: a case report. *Emerg. Infect. Dis.* **8:**727–728.

Kubes, M., P. Kocakova, M. Slovak, M. Slavikova, N. Fuchsberger, and P. A. Nuttall. 2002. Heterogeneity in the effect of different ixodid tick species on human natural killer cell activity. *Parasite Immunol.* **24:**23–28.

Lafferty, K. D. 2009. The ecology of climate change and infectious diseases. *Ecology* **90:**888–900.

Lemaitre, B., J. M. Reichhart, and J. A. Hoffmann. 1997. *Drosophila* host defense: differential induction of antimicrobial peptide genes after infection by various classes of microorganisms. *Proc. Natl. Acad. Sci USA* **94:**14614–14619.

Lievens, S., S. Goormachtig, and M. Holsters. 2004. Nodule-enhanced protease inhibitor gene: emerging patterns of gene expression in nodule development on *Sesbania rostrata*. *J. Exp. Bot.* **55:**89–97.

Lockhart, J. M., W. R. Davidson, D. E. Stallknecht, and J. E. Dawson. 1998. Lack of seroreactivity to *Ehrlichia chaffeensis* among rodent populations. *J. Wildl. Dis.* **34:**392–396.

Macaluso, K. R., A. Mulenga, J. A. Simser, and A. F. Azad. 2003. Differential expression of genes in uninfected and *Rickettsia*-infected *Dermacentor variabilis* ticks as assessed by differential-display PCR. *Infect. Immun.* **71:**6165–6170.

Macaluso, K. R., D. E. Sonenshine, S. M. Ceraul, and A. F. Azad. 2002. Rickettsial infection in *Dermacentor variabilis* (Acari: Ixodidae) inhibits transovarial transmission of a second rickettsia. *J. Med. Entomol.* **39:**809–813.

Manen, J. F., P. Simon, J. C. Van Slooten, M. Østerås, S. Frutiger, and G. J. Hughes. 1991. A nodulin specifically expressed in senescent nodules of winged bean is a protease inhibitor. *Plant Cell* **3:**259–270.

Marx, J. 2004. The roots of plant-microbe collaborations. *Science* **304:**234–236.

Mather, T. N., and H. S. Ginsberg. 1994. Vector-host-pathogen relationships: transmission dynamics of tick-borne infections, p. 68–90. *In* D. E. Sonenshine and T. N. Mather (ed.), *Ecological Dynamics of Tick-Borne Zoonoses*. Oxford University Press, New York, NY.

McElroy, K. M., B. L. Blagburn, E. B. Breitschwerdt, P. S. Mead, and J. H. McQuiston. 2010. Flea-associated zoonotic diseases of cats in the USA: bartonellosis, flea-borne rickettsioses, and plague. *Trends Parasitol.* **26:**197–204.

Mejri, N., N. Franscini, B. Rutti, and M. Brossard. 2001. Th2 polarization of the immune response of BALB/c mice to *Ixodes ricinus* instars, importance of several antigens in activation of specific Th2 subpopulations. *Parasite Immunol.* **23:**61–69.

Mulenga, A., K. R. Macaluso, J. A. Simser, and A. F. Azad. 2003. Dynamics of *Rickettsia*-tick interactions: identification and characterization of differentially expressed mRNAs in uninfected and infected *Dermacentor variabilis. Insect Mol. Biol.* **12:**185–193.

Nakajima, Y., J. Ishibashi, F. Yukuhiro, A. Asaoka, D. Taylor, and M. Yamakawa. 2003a. Antibacterial activity and mechanism of action of tick defensin against Gram-positive bacteria. *Biochim. Biophys. Acta* **1624:**125–130.

Nakajima, Y., K. Ogihara, D. Taylor, and M. Yamakawa. 2003b. Antibacterial hemoglobin fragments from the midgut of the soft tick, *Ornithodoros moubata* (Acari: Argasidae). *J. Med. Entomol.* **40:**78–81.

Nakajima, Y., H. Saido-Sakanaka, D. Taylor, and M. Yamakawa. 2003c. Up-regulated humoral immune response in the soft tick, *Ornithodoros moubata* (Acari: Argasidae). *Parasitol. Res.* **91:**476–481.

Nakajima, Y., A. Van der Goes van Naters-Yasui, D. Taylor, and M. Yamakawa. 2002. Antibacterial peptide defensin is involved in midgut immunity of the soft tick, *Ornithodoros moubata. Insect Mol. Biol.* **11:**611–618.

Nakajima, Y., A. Van der Goes van Naters-Yasui, D. Taylor, and M. Yamakawa. 2001. Two isoforms of a member of the arthropod defensin family from the soft tick, *Ornithodoros moubata* (Acari: Argasidae). *Insect Biochem. Mol. Biol.* **31:**747–751.

Nakayama, K., K. Kurokawa, M. Fukuhara, H. Urakami, S. Yamamoto, K. Yamazaki, Y. Ogura, T. Ooka, and T. Hayashi. 2008. Genome comparison and phylogenetic analysis of *Orientia tsutsugamushi* strains. *DNA Res.* **17:** 281–291.

Niebylski, M. L., M. G. Peacock, and T. G. Schwan. 1999. Lethal effect of *Rickettsia rickettsii* on its tick vector (*Dermacentor andersoni*). *Appl. Environ. Microbiol.* **65:**773–778.

Nuttall, P. A., and M. Labuda. 2004. Tick-host interactions: saliva-activated transmission. *Parasitology* **129**(Suppl.)**:**S177–S189.

Oliver, J. H., Jr. 1989. Biology and systematics of ticks (Acari:Ixodida). *Annu. Rev. Ecol. Syst.* **20:**397–430.

Paddock, C. D., and J. E. Childs. 2003. *Ehrlichia chaffeensis*: a prototypical emerging pathogen. *Clin. Microbiol. Rev.* **16:**37–64.

Paddock, C. D., J. W. Sumner, J. A. Comer, S. R. Zaki, C. S. Goldsmith, J. Goddard, S. L. McLellan, C. L. Tamminga, and C. A. Ohl. 2004. *Rickettsia parkeri*: a newly recognized cause of spotted fever rickettsiosis in the United States. *Clin. Infect. Dis.* **38:**805–811.

Parmenter, R. R., E. P. Yadav, C. A. Parmenter, P. Ettestad, and K. L. Gage. 1999. Incidence of plague associated with increased winter-spring precipitation in New Mexico. *Am. J. Trop. Med. Hyg.* **61:**814–821.

Parola, P., C. D. Paddock, and D. Raoult. 2005. Tick-borne rickettsioses around the world: emerging diseases challenging old concepts. *Clin. Microbiol. Rev.* **18:**719–756.

Pereira, L. S., P. L. Oliveira, C. Barja-Fidalgo, and S. Daffre. 2001. Production of reactive oxygen species by hemocytes from the cattle tick *Boophilus microplus. Exp. Parasitol.* **99:**66–72.

Randolph, S. E. 2009. Perspectives on climate change impacts on infectious diseases. *Ecology* **90:** 927–931.

Randolph, S. E., and D. J. Rogers. 2007. Ecology of tick-borne disease and the role of climate, p. 167–186. *In* O. Ergonul and C. A. Whitehouse (ed.), *Crimean-Congo Hemorrhagic Fever.* Springer, Dordrecht, The Netherlands.

Randolph, S. E., and D. J. Rogers. 2010. The arrival, establishment and spread of exotic diseases: patterns and predictions. *Nat. Rev. Microbiol.* **8:**361–371.

Raoult, D., O. Dutour, L. Houhamdi, R. Jankauskas, P. E. Fournier, Y. Ardagna, M. Drancourt, M. Signoli, V. D. La, Y. Macia, and G. Aboudharam. 2006. Evidence for louse-transmitted diseases in soldiers of Napoleon's Grand Army in Vilnius. *J. Infect. Dis.* **193:**112–120.

Raoult, D., J. B. Ndihokubwayo, H. Tissot-Dupont, V. Roux, B. Faugere, R. Abegbinni, and R. J. Birtles. 1998. Outbreak of epidemic typhus associated with trench fever in Burundi. *Lancet* **352:**353–358.

Raoult, D., and V. Roux. 1999. The body louse as a vector of reemerging human diseases. *Clin. Infect. Dis.* **29:**888–911.

Raoult, D., V. Roux, J. B. Ndihokubwayo, G. Bise, D. Baudon, G. Marte, and R. Birtles. 1997. Jail fever (epidemic typhus) outbreak in Burundi. *Emerg. Infect. Dis.* **3:**357–360.

Ribeiro, J. M., F. Alarcon-Chaidez, I. M. Francischetti, B. J. Mans, T. N. Mather, J. G.

Valenzuela, and S. K. Wikel. 2006. An annotated catalog of salivary gland transcripts from *Ixodes scapularis* ticks. *Insect Biochem. Mol. Biol.* **36:**111–129.

Ribeiro, J. M., and I. M. Francischetti. 2003. Role of arthropod saliva in blood feeding: sialome and post-sialome perspectives. *Annu. Rev. Entomol.* **48:**73–88.

Satta, G., V. Chisu, P. Cabras, F. Fois, and G. Masala. 2011. Pathogens and symbionts in ticks: a survey on tick species distribution and presence of tick-transmitted micro-organisms in Sardinia, Italy. *J. Med. Microbiol.* **60:**63–68.

Silverman, J., M. K. Rust, and D. A. Reierson. 1981. Influence of temperature and humidity on survival and development of the cat flea, *Ctenocephalides felis* (*Siphonaptera: Pulicidae*). *J. Med. Entomol.* **18:**78–83.

Simser, J. A., K. R. Macaluso, A. Mulenga, and A. F. Azad. 2004a. Immune-responsive lysozymes from hemocytes of the American dog tick, *Dermacentor variabilis* and an embryonic cell line of the Rocky Mountain wood tick, *D. andersoni*. *Insect Biochem. Mol. Biol.* **34:**1235–1246.

Simser, J. A., A. Mulenga, K. R. Macaluso, and A. F. Azad. 2004b. An immune responsive factor D-like serine proteinase homologue identified from the American dog tick, *Dermacentor variabilis*. *Insect Mol. Biol.* **13:**25–35.

Sonenshine, D. E. 1991. *Biology of Ticks*, vol. 1. Oxford University Press, New York, NY.

Sonenshine, D. E. 1993. *Biology of Ticks*, vol. 2. Oxford University Press, New York, NY.

Sonenshine, D. E., W. L. Hynes, S. M. Ceraul, R. Mitchell, and T. Benzine. 2005. Host blood proteins and peptides in the midgut of the tick *Dermacentor variabilis* contribute to bacterial control. *Exp. Appl. Acarol.* **36:**207–223.

Sonenshine, D. E., R. S. Lane, and W. L. Nicholson. 2002. Ticks (*Ixodida*), p. 597. *In* G. Mullen and L. Durden (ed.), *Medical and Veterinary Entomology*. Academic Press, San Diego, CA.

Tarasevich, I., E. Rydkina, and D. Raoult. 1998. Outbreak of epidemic typhus in Russia. *Lancet* **352:**1151.

Telford, S. R., III. 2009. Status of the "East Side Hypothesis" (transovarial interference) 25 years later. *Ann. N. Y. Acad. Sci.* **1166:**144–150.

Traub, R., and C. L. Wisseman, Jr. 1974. The ecology of chigger-borne rickettsiosis (scrub typhus). *J. Med. Entomol.* **11:**237–303.

Ueti, M. W., D. P. Knowles, C. M. Davitt, G. A. Scoles, T. V. Baszler, and G. H. Palmer. 2009. Quantitative differences in salivary pathogen load during tick transmission underlie strain-specific variation in transmission efficiency of *Anaplasma marginale*. *Infect. Immun.* **77:**70–75.

Ueti, M. W., J. O. Reagan, Jr., D. P. Knowles, Jr., G. A. Scoles, V. Shkap, and G. H. Palmer. 2007. Identification of midgut and salivary glands as specific and distinct barriers to efficient tick-borne transmission of *Anaplasma marginale*. *Infect. Immun.* **75:**2959–2964.

Valenzuela, J. G., I. M. Francischetti, V. M. Pham, M. K. Garfield, T. N. Mather, and J. M. Ribeiro. 2002. Exploring the sialome of the tick *Ixodes scapularis*. *J. Exp. Biol.* **205:**2843–2864.

Valenzuela, J. G., I. M. Francischetti, V. M. Pham, M. K. Garfield, and J. M. Ribeiro. 2003. Exploring the salivary gland transcriptome and proteome of the *Anopheles stephensi* mosquito. *Insect Biochem. Mol. Biol.* **33:**717–732.

Vasse, J., F. d. Billy, and G. Truchet. 1993. Abortion of infection during the Rhizobium meliloti-alfalfa symbiotic interaction is accompanied by a hypersensitive reaction. *The Plant Journal* **4:**555–566.

Walker, D. H. 2007. Rickettsiae and rickettsial infections: the current state of knowledge. *Clin. Infect. Dis.* **45**(Suppl. 1)**:**S39–S44.

Walker, D. H., C. D. Paddock, and J. S. Dumler. 2008. Emerging and re-emerging tick-transmitted rickettsial and ehrlichial infections. *Med. Clin. North Am.* **92:**1345–1361, x.

Wang, J., Y. Wu, G. Yang, and S. Aksoy. 2009. Interactions between mutualist *Wigglesworthia* and tsetse peptidoglycan recognition protein (PGRP-LB) influence trypanosome transmission. *Proc. Natl. Acad. Sci. USA* **106:**12133–12138.

Wilson, L. B., and W. M. Chowning. 1904. Studies in pyroplasmosis hominis ('spotted fever' or 'tick fever' of the Rocky Mountains). *J. Infect. Dis.* **I:**31–33.

Zaidman-Rémy, A., M. Hervé, M. Poidevin, S. Pili-Floury, M. S. Kim, D. Blanot, B. H. Oh, R. Ueda, D. Mengin-Lecreulx, and B. Lemaitre. 2006. The *Drosophila* amidase PGRP-LB modulates the immune response to bacterial infection. *Immunity* **24:**463–473.

THE WAY FORWARD: IMPROVING GENETIC SYSTEMS

Ulrike G. Munderloh, Roderick F. Felsheim, Nicole Y. Burkhardt, Michael J. Herron, Adela S. Oliva Chávez, Curtis M. Nelson, and Timothy J. Kurtti

14

INTRODUCTION

In this chapter, we review the research that has, cumulatively, led to the first successes in genetic transformation of arthropod-borne bacteria belonging to the order *Rickettsiales* in the *Alphaproteobacteria* subdivision. With this relatively newly gained knowledge, rickettsiologists are taking steps toward a future of more predictable rickettsial genetic manipulation that allows us to address the many questions about rickettsial physiology and pathogenicity that have not conclusively been answered. Thus far, only species in the genera *Anaplasma* and *Rickettsia* have been transformed successfully, and we will focus on them for that reason.

Members of the *Rickettsiales* include extremely virulent microbes capable of killing their hosts, e.g., spotted fever rickettsiae such as *Rickettsia rickettsii*; as well as benign symbionts restricted to their arthropod hosts and unable to infect vertebrates, e.g., *Rickettsia peacockii*. The tremendous diversity in lifestyles of the *Alphaproteobacteria*, from free-living to obligately intracellular, is reflected in their extraordinarily diverse genomes, which vary in size (from ~1 to ~9 Mbp), GC content (Ettema and Andersson, 2009), presence or absence of extrachromosomal elements (plasmids) (Ogata et al., 2005; Baldridge et al., 2010a), and ability to cause diseases (Gillespie et al., 2007). Clearly, dramatic changes in gene content, loss as well as gain, are associated with significant shifts in the environment occupied by *Alphaproteobacteria* (Boussau et al., 2004). Species in the genus *Rickettsia* have small (1.1 to ~2 Mb), AT-rich genomes, ranging from 29% GC content in *Rickettsia prowazekii* to 32.5% in *R. rickettsii* (Ellison et al., 2008), which is thought to have arisen, in part, due to a strong mutational bias for transition of GC to AT. They also harbor a surprisingly large amount of noncoding DNA, amounting to 24% in *R. prowazekii* (Andersson and Andersson, 1999). In *Anaplasmataceae*, the divergence of these values is even greater, with a 27% GC content and 66% coding density in *Ehrlichia ruminantium*, a 41% GC content and 68% coding density in *Anaplasma phagocytophilum*, and a 49% GC content and 85% coding density in *Anaplasma marginale* (Brayton et al., 2005; Collins et al., 2005; Dunning Hotopp et al., 2006). These genomic characteristics theoretically affect the success rate of transformations, as certain transposon/transposase pairs may have a bias toward

Ulrike G. Munderloh, Roderick F. Felsheim, Nicole Y. Burkhardt, Michael J. Herron, Adela S. Oliva Chávez, Curtis M. Nelson, and Timothy J. Kurtti, Department of Entomology, University of Minnesota, St. Paul, MN 55108.

Intracellular Pathogens II: Rickettsiales
Edited by G. H. Palmer and A. F. Azad © 2012 ASM Press, Washington, DC
10.1128/9781555817336.ch14

GC-rich genomic regions and thus reduce the rate of transformations in rickettsiae (e.g., the Tn*5* transposon) (Herron et al., 2004; Wu et al., 2006). Also, in our experience, about two-thirds of transposase-facilitated insertions recovered fall into noncoding regions of *A. phagocytophilum* (Felsheim et al., 2006). If we assume that *Himar1*-mediated transposition occurs randomly across the genome, this bias would indicate that the majority of generated hits are into essential genes, the functionality of which is indispensible to the bacteria. This suggests that noncoding sequences could be successfully targeted for site-directed gene insertions with minimal injury to *A. phagocytophilum*, and could explain why transposition has not been achieved in *A. marginale*, a species with a much tighter genome than *A. phagocytophilum* (Brayton et al., 2005; Felsheim et al., 2010).

In general, there is phylogenetic support for the idea that with adaptation to the intracellular niche, loss and degradation of genes necessary for a free-living existence began, and *Rickettsiales* came to rely increasingly on metabolites imported from the host cell. This trend is evident from the small size of their genomes, which nevertheless may contain substantial amounts of seemingly degraded, noncoding DNA and numerous split genes (Fuxelius et al., 2007, 2008; Gillespie et al., 2008). Compaction of genomes is proceeding with generation of overlapping genes to maximize coding capacity, a process that evolves from degeneration of open reading frame (ORF) ends, resulting in use of initiation and/or stop codons of upstream or downstream chromosomal regions for transcription (Sakharkar et al., 2005). However, evolution toward optimal adaptation to intracellular residence (Boussau et al., 2004) does not follow a straight path of gene elimination, as the *Rickttsiales* continually tweak the balance of their coexistence with the host. For example, both nonpathogenic symbionts as well as pathogens have retained pathways for production of cofactors, including B vitamins, lipoate, ubiquinone, and others. This is especially true for the *Anaplasmataceae* that encode a nearly complete set, whereas *R. prowazekii* appears limited to production of protoheme, lipoate, and ubiquinone (Dunning Hotopp et al., 2006). The authors suggested that this capability might provide a benefit to the vector, in which pathogens could fulfill a symbiont role. It is also possible that these pathways must be retained to enable *Anaplasma* and *Ehrlichia* spp. to survive in mammalian blood cells, an environment poor in these nutrients. Accelerated genetic degradation continues during forced adaptation to unnatural environments, such as repeated passage in the yolk sac of hen's eggs, the standard laboratory system for rickettisae until well into the 20th century, or duck embryonic cell cultures (Kenyon et al., 1972). Even so, restoration of function in several *R. prowazekii* genes, notably following passage in mouse L929 cells and restoration of *recO*, which is involved in gene repair, is a reminder of the surprising genetic plasticity of rickettsiae (Bechah et al., 2010), and should be kept in mind when analyzing potential mutants. Genomic reduction typical of the obligate intracellular rickettsiae (Fournier et al., 2009) can apparently sometimes go too far, but function can be rescued by acquisition of genes/gene clusters from divergent taxa. This capability has become particularly obvious with the characterization of rickettsial plasmids that appear to be preferred targets of lateral gene transfer (Baldridge et al., 2010a; Felsheim et al., 2009) without running the risk of critical chromosomal genes being inactivated by such events. A striking example supporting this notion is *R. peacockii*, closely related to *R. rickettsii* but unable to infect vertebrates, probably following acquisition of an insertion element that resulted in extensive genome reshuffling with inactivation of numerous genes, at least some of which are involved in host invasion and motility (Niebylski et al., 1997; Simser et al., 2005). *R. peacockii* possesses a plasmid that carries a group of five genes related to ORFs in the "glycosylation island" of *Pseudomonas aeruginosa* (Arora et al., 2001). These are involved in fatty acid biosynthesis and possibly make up for lost function of the mutated glycerol-3-phosphate dehydrogenase gene in *R. peacockii* (Felsheim et al., 2009). Although these genes are not essential for in vitro survival of *R. peacockii*, as demonstrated by the eventual loss of the entire plasmid following more than 70 serial passages in tick cell culture (Baldridge et al., 2008), they may provide an important advantage to the rickettsiae during their existence in tick hosts.

Among bacterial pathogens, members of the *Rickettsiales* have been very difficult to manipulate genetically. This characteristic is in part directly attributable to their genomic structure and composition, and also in part to their obligate intracellular lifestyle, which requires several distinct steps for successful colonization of host cells, likely involving expression of multiple genes, that must be correctly executed in the right sequence. It follows, then, that this process is highly sensitive to disruption: if even one of the steps in the sequence is blocked, subsequent processes may be severely impaired to the point that infection, and thus pathogenesis, is abolished. This should make the infection process an attractive target for disruption; however, we have only a vague picture of the events, and few details are known. It is assumed that ticks inoculate extracellular rickettsiae into the bite site with saliva during the blood meal (Munderloh and Kurtti, 1995), and this is followed by lymphatic or hematogenous dissemination to their targets—endothelial cells or blood cells—although specific processes are not well understood (Brouqui et al., 1997; Fournier et al., 2005; de Sousa et al., 2008). In *Rickettsia* species, several dominant surface proteins have been implicated in mediating adhesion and subsequent invasion of host cells in vitro using specific antibodies or *Escherichia coli* transformed to express cloned rickettsial genes. These include rickettsial outer membrane proteins (OMPs) A (Sca0) and B (Sca5), RickA, Sca2 (surface cell antigen 2), and rickettsial adhesins Adr1 and -2 (Li and Walker, 1998; Uchiyama, 2003; Martinez et al., 2005; Renesto et al., 2006; Uchiyama et al., 2006; Cardwell and Martinez, 2009; Chan et al., 2009; Riley et al., 2010; Vellaiswamy et al., 2011). Much less information is available for the genera *Anaplasma* and *Ehrlichia*; the bovine pathogen *A. marginale* has been reported to differentially use subunits of the heterodimeric, immunodominant major surface protein 1 (MSP1) for adhesion to bovine (both MSP1a and b) or tick cells (MSP1b only) (McGarey et al., 1994; de la Fuente et al., 2001; Garcia-Garcia et al., 2004). *A. phagocytophilum*, the human granulocytic anaplasmosis agent, binds P-selectin glycoprotein ligand-1 (PSGL-1) to gain entry into its target, human neutrophil granulocytes (Herron et al., 2000). This is the ligand that these cells otherwise use to engage P-selectin displayed on inflamed microvascular endothelium during rolling (Doré et al., 1993). Except for Ku70 and components of lipid raft microdomains (Chan et al., 2009), no other host cell receptors for *Rickettsiales* have been identified (Walker et al., 2003).

Despite their small genomes, a large proportion of *Rickettsiales* ORFs encode hypothetical proteins, e.g., about 40% or more of the *A. phagocytophilum* and *Ehrlichia chaffeensis* genomes (Dunning Hotopp et al., 2006), many of which may be involved in colonization of the tick vector (Nelson et al., 2008; Ramabu et al., 2010; Lin et al., 2011). In such cases comparative genomic analyses may not always provide clues for identification of the specific roles of encoded proteins, especially because steps taken during adhesion and invasion may each be enabled by expression of multiple genes. Therefore, single-gene knockouts, even if they cause a phenotypic change, will be difficult to exactly define by function. Likewise, screening for genes required for successful infection of vectors and hosts in vivo (Mahan et al., 1993) is fraught with difficulties; e.g., at least some *A. phagocytophilum* mutants are soon lost upon continued in vitro passage from pools of transformants initially recovered after electroporation (M. J. Herron, unpublished data), making it problematic to distinguish genes required in vivo from those whose deletion resulted in inherently unstable mutants.

GENETIC TRANSFORMATION SYSTEMS FOR *RICKETTSIALES*

Table 1 provides a summary of the methods and techniques used that have resulted in successful transformation of *Rickettsia* and *Anaplasma* species. This table may serve as a quick reference guide to the history of genetic manipulations that are referred to in the following sections.

Homologous Recombination

Over a decade has elapsed since the first reports of successful genetic modification of *Rickettsia* species. These seminal results provided the very first evidence that direct genetic manipulation of rickettsiae was achievable, and that these bacteria were able to maintain and express foreign genetic sequences inserted via allelic exchange/homologous recombination under control of rickettsial or *E. coli* promoters. This work set the stage for all subsequent research efforts aspiring to manipulate and analyze rickettsiae in a manner that is nearly commonplace in extracellular bacteria (Rachek et al., 1998, 2000; Troyer et al., 1999; Renesto et al., 2002). While highly significant, these early transformants were difficult to obtain, unstable during long-term maintenance, and not useful for functional gene analysis since insertion was not designed to result in gene inactivation and detectable defective phenotype. Nevertheless, these early success stories stimulated and revived the field and identified some of the markers, selection systems, and transformation technologies still in use today, such as GFPuv (Crameri et al., 1996), a variant of green fluorescent protein (GFP) that fluoresces more brightly in bacteria; resistance to rifampin; and electroporation conditions.

Site-Directed Mutagenesis

Methodically knocking out particular target genes one by one is an important step toward identification or confirmation of gene function. This would ideally be followed by complementation in a manner that does not introduce further genomic "hits" to avoid generating (and inadvertently analyzing) the effects of new and unknown potential knockouts. Disrupting the function of specific genes by site-directed mutagenesis has many advantages, e.g., simplification of the processes for detection of the insert, since it can be assumed that recombination occurred into the target site as intended. Therefore, assays such as PCR-mediated recovery and sequencing, cloning of the insert along with diagnostic flanking regions, or Southern analysis of restricted DNA using enzymes that cut in nearby flanking sites can quickly confirm anticipated results, especially when genomic sequences are available for the bacteria. Nevertheless, it is a slow process for arriving at a comprehensive assessment of gene function, even in small-genome microbes such as the *Rickettsiales* that encode "only" about 1,000 or so genes. Judicious selection of targets is probably essential, but may leave hypothetical (and possibly important) genes that make up around 40% of rickettsial genomes unexplored.

Although the first successful transformation experiments with *R. prowazekii* and *Rickettsia typhi* utilized homologous recombination (Rachek et al., 1998, 2000; Troyer et al., 1999; Renesto et al., 2002), only a single report using site-directed mutagenesis for analysis of a putative rickettsial virulence gene has been published (Driskell et al., 2009). Using *R. prowazekii*, these authors targeted *pld*, encoding phospholipase D, which is thought to enable rickettsial escape from the phagosome early after uptake into a host cell. Essential features of the linear construct included the *pld* gene in which an internal 93-bp fragment had been replaced by an antibiotic (rifampin) resistance cassette driven by the rickettsial *rpsL* promoter. Upstream and downstream portions of flanking regions provided the sequences necessary for the double-crossover event to occur at the target site. The resulting mutant was indeed generated as predicted, confirming that this important strategy of rickettsial mutagenesis is feasible. Interestingly, though, the mutant remained unimpaired and fully capable of exiting the phagosome in cultured macrophagelike cells. But while the mutant rickettsiae were able

TABLE 1 *Rickettsiales* genetic transformation systems

Bacterial strain	Vector	Construct reporter gene	Promoter	Delivery	Selection	Result	Reference
Rickettsia							
R. prowazekii Madrid E Rif^S (rifampin-sensitive strain)	pMW1027	*rpoB* (Rif^r)[a]	Native	Electroporation	Rifampin	Homologous recombination	Rachek et al., 1998
R. prowazekii Madrid E Em^S (erythromycin sensitive)	pMW1047	*ereB*[b]	glt^P	Electroporation	Erythromycin	Homologous recombination	Rachek et al., 2000
R. typhi Wilmington	*rpoB*-GFPuv amplicon	GFPuv	$rpoB^P$	Electroporation	FACS	Homologous recombination	Troyer et al., 1999
R. conorii Malish 7	*rpoB*-GFP amplicon	GFP	sr^{P}[c]	Electroporation	FACS	Homologous recombination	Renesto et al., 2002
R. prowazekii Madrid Evir	Δ*pld;* PCR amplicon of pMW1628	*Rparr-2*[d]	$rpsL^P$	Electroporation	Rifampin	Site-directed mutagenesis	Driskell et al., 2009
R. prowazekii Madrid E	EZ-TN transposome	*Rparr-2*[d]	$rpsL^P$	Electroporation	Rifampin	Transposon mutagenesis	Qin et al., 2004
R. monacensis IrR/Munich	EZ-TN transposome pMOD658 pMOD61	CAT[e] GFPuv	$rompA^P$	Electroporation	Chloram-phenicol	Transposon mutagenesis	Baldridge et al., 2005, 2007a
R. prowazekii Madrid E	pMW1650 *Himar1*	GFPuv *Rparr-2*	$rompA^P$ $rpsL^P$	Electroporation	Rifampin	Transposon mutagenesis	Liu et al., 2007
R. rickettsii R	pMW1650 *Himar1*	GFPuv *Rparr-2*	$rompA^P$ $rpsL^P$	Electroporation	Rifampin	Transposon mutagenesis	Kleba et al., 2010

R. montanensis M5/6	EZ-TN transposome pMOD700	CAT[e] GFPuv	*rompA*^{P} *rompB*^{P}	Electroporation	Chloramphenicol	Transposon mutagenesis	Baldridge et al., 2010b
R. montanensis M5/6	pRAM18dRGA	GFPuv *Rparr-2*	*rompA*^{P} *rpsL*^{P}	Electroporation	Rifampin	Shuttle plasmid	Burkhardt et al., 2011
Anaplasma							
A. phagocytophilum	pHimar cis- A7 C-SS pET28Am tr A7 Himar	GFPuv *aadA*	Amtr^{P}	Electroporation	Spectinomycin, streptomycin	Transposon mutagenesis	Felsheim et al., 2006
	pHimar cis- A7 C-SS	mCherry *aadA*	Amtr^{P}	Electroporation	Spectinomycin	Transposon mutagenesis	R. F. Felsheim, unpublished data
	pHimar C-Bar and pET28Am tr A7 Himar	mCherry	Amtr^{P} *bar* (synthetic)	Electroporation	Phosphinothricin	Transposon mutagenesis	Felsheim et al., 2007
A. marginale St. Maries	pHimar-Amtr Turbo-GPS SS	TurboGFP Spec-Strep	Amtr^{P}	Electroporation	Spectinomycin, streptomycin	Homologous recombination	Felsheim et al., 2010

[a]Rifampin-resistant mutant gene of *R. prowazekii* Madrid E.
[b]*E. coli ereB* gene.
[c]*E. coli srp* promoter.
[d]*R. prowazekii*-adapted rifampin resistance *arr-2* gene.
[e]CAT, chloramphenicol acetyltransferase.

to infect guinea pigs, they did not cause disease, and protected the animals from challenge with virulent *R. prowazekii*. Although the result did not unambiguously identify the function of the *R. prowazekii pld* gene product, it does suggest a role in pathogenesis that needs to be explored further. In light of the genomic flexibility of *R. prowazekii* referred to above, it is possible that partial *pld* gene activity was retained, but due to host cell background, this scenario could not be tested. Perhaps more likely, redundant function provided by *pla*, encoding phospholipase A in *R. prowazekii*, made up for the *pld* knockout (Sahni and Rydkina, 2009), and serves to emphasize that interpretation of even single-gene knockouts may be difficult.

Transposition

When Barbara McClintock first identified "jumping genes" as the cause of the color mosaic of Indian corn ear kernels (McClintock, 1950), there was little anticipation of the role transposons would hold in future biotechnology. By now, it is clear that transposons are very likely present in all genomes, including those of *Rickettsiales* (Shapiro, 1999; Simser et al., 2005). Unlike site-directed mutagenesis methods, transposon mutagenesis does not require knowledge of the genome sequence of an organism, although some transposases exhibit preference for base composition of genomic DNA. For example, the Tn*5* transposase used in the Epicentre EZ-TN5 transposome system (Epicentre Technologies, Madison, WI) preferentially inserts into GC-rich DNA (Herron et al., 2004), which could explain the low transposition frequencies obtained with this system in initial attempts at transposon mutagenesis of rickettsiae (Qin et al., 2004; Baldridge et al., 2005, 2010b). More recently, a *mariner*-type transposase from horn flies, *Haematobia irritans* (*Himar1*), was reconstructed as a consensus sequence from several clones to display hyperactivity in *E. coli* (Lampe et al., 1999). The *mariner*-element transposons, of which *Himar1* and *Mos1* are the only two active ones known, are especially suited for a wide range of organisms as they do not require host-provided elements in order to function, and occur across kingdoms. The transposase recognizes short, inverted, 31-bp repeats flanking the transposon (5′-ACA GGT TGG CTG ATA AGT CCC CGG TCT GAC A-3′) and mobilizes especially successfully from plasmids, but only poorly from chromosomes (Lampe et al., 1996, 1998). Unlike Tn*5*, *Himar1* does not require specific target sequences for insertion, outside of the target TA dinucleotide, and has a predilection for bent or flexible DNA regions (Lampe et al., 1998). *Himar1* requires the presence of $MgCl_2$, which is assured in an intracellular environment. Additionally, temperature and transposon size are the parameters that affect transposition efficiency most prominently, with a temperature optimum between 15 and 28°C and a sharp decline in transposition of transposons larger than 2 kb. This restricted "payload" size is one of the drawbacks of the *Himar1* transposon system and makes it unsuitable for delivery of large constructs. Nonetheless, *Himar1* has sufficient activity in *R. rickettsii* to readily allow complementation of single-gene mutants that fall within the size range of transposons that can be mobilized by *Himar1*. Clark et al. (2011) took advantage of this capability to restore a functional *relA/spoT* gene copy to clear-plaque mutants of *R. rickettsii* Iowa and *R. rickettsii* R that had been shown by sequence analysis to each have a mutation within the ORF of *relA/spoT*. Both complemented rickettsial strains displayed the opaque plaque type typical of rickettsiae that naturally have an intact *relA/spoT* copy. This elegant study clearly demonstrated for the first time the utility of *Himar1* for functional genomics of spotted fever rickettsiae, and it is conceivable that similar studies can be performed with *Rickettsiales* species that are amenable to *Himar1* mutagenesis.

Use and Optimization of the *Himar1* Transposase System

We have successfully used the *Himar1* transposase allele A7 (but not C9) for mutagenesis of *A. phagocytophilum* (Felsheim et al., 2006) and spotted fever group rickettsial tick symbionts (our unpublished results) using fluorescent markers and antibiotic (spectinomycin) resistance. Although we have yet to achieve saturating mutagenesis, as has been reported for Lyme disease spirochetes (*Borrelia burgdorferi*) (Stewart and Rosa, 2008), this technology has enabled reproducible transformation of *A. phagocytophilum* at a much greater efficiency than with Tn*5* in rickettsiae. The *Himar1* transposon system that yielded the first successful *A. phagocytophilum* transformants employed *trans* constructs in which the transposase and the tranposon were encoded on separate plasmids. An important feature of the transposase-encoding plasmid (pET28Am tr A7 Himar) (Felsheim et al., 2006) was in-

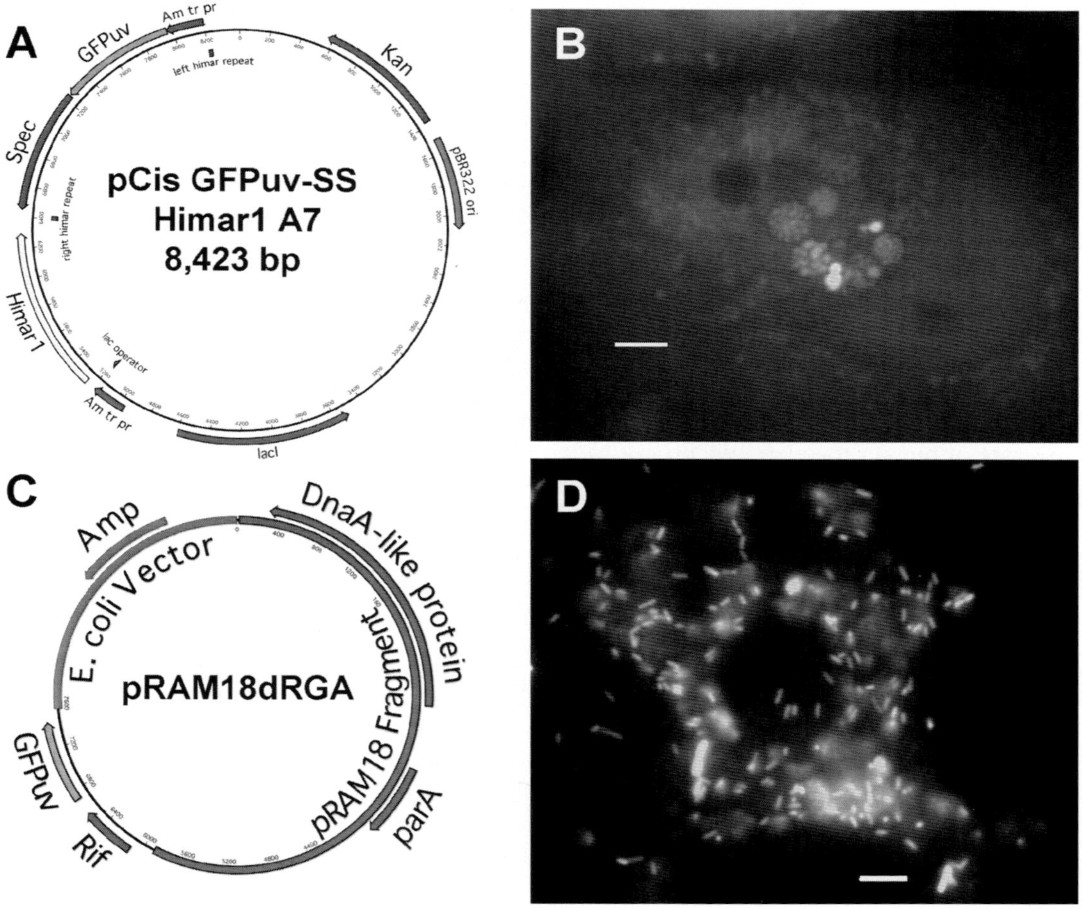

FIGURE 1 General features of plasmid constructs being developed for use in *Rickettsiales* genetics. (A) Construct for transposon-mediated mutagenesis of *Anaplasma*. Depicted is a *cis*-plasmid construct: both transposase (*Himar1*) and a transposon containing a *bar* (spectinomycin resistance) gene and GFPuv are encoded on a single plasmid (pCis GFPuv Himar1 A7) in order to improve efficiency. In this construct two *A. marginale* Am tr promoters are used to drive expression of *Himar1* and the reporter and resistance genes. (B) Confocal fluorescent microscopic image of GFPuv-expressing *A. phagocytophilum* infecting a rhesus cell (RF/6A) expressing DsRed2, a red fluorescent protein. (C) Shuttle vector for transformation of rickettsiae. Depicted is a generalized example of the essential elements of shuttle vectors based on pRAM18 from *R. amblyommii*. A fragment of pRAM18 containing *parA* and the DnaA-like protein gene was ligated into an *E. coli* cloning vector containing genes for rifampin resistance and GFPuv. (D) Fluorescent microscopic image of pRAM18dRGA-transformed *R. bellii* strain 369 in an *I. scapularis* cell (cell line ISE6). Bars (B and D), 5 μm. doi:10.1128/9781555817336.ch14.f1

corporation of the *lac* operator between the Am tr promoter (the *A. marginale* transcription regulator promoter) (Barbet et al., 2005) and the *Himar1* gene, as this reduced unwanted transposition during replication in *E. coli*. This design allowed for maximum flexibility in the choice of transposons encoding a variety of selectable and fluorescent markers, and independent assessment of improvements to either transposase or transposon constructs. However, as protocols became more standardized, we moved toward use of a *cis*-plasmid construct in which both transposase and transposon were encoded on a single plasmid, in order to improve efficiency (Fig. 1A). In this example, *A. phagocytophilum* expressing GFPuv is shown infecting the rhesus cell line RF/6A (Munderloh et al., 2004) which expresses mCherry (Shaner et al., 2005), a red fluorescent protein (Fig. 1B). Since only the transposon is flanked by the *Himar1* inverted repeats, self-mobilization of the transposase does not occur. Because of the relatively low rate of mutant recovery, we do not have hard data to support the idea that the single-plasmid *Himar1* construct is more efficient. However, the likelihood that both plasmids from the *trans* system will land inside the same rickettsial cell is probably much smaller than for a single-plasmid *cis* construct.

Purification of *Rickettsiales* and electroporation conditions must be optimized for each bacterial species, and it is good practice to reevaluate the procedures occasionally in light of new findings from one's own and other laboratories. An essential consideration is maintaining the viability and infectiousness of obligate intracellular bacteria during their manipulations in the extracellular milieu. Therefore, reducing the time required, starting with release from the host cell through washing steps through electroporation in the presence of the transformation construct, is important, as is using solutions and conditions that minimally compromise the bacteria. As an example, bacterial purification using filtration is faster than gradient centrifugation and yields *Anaplasma* and *Rickettsia* of sufficient purity (Felsheim et al., 2006).

Conditions following electroporation must likewise be carefully optimized to facilitate adhesion to and successful colonization of host cells. Different host cell types may be more or less supportive of rickettsiae, and choosing a highly susceptible host can improve recovery of mutants. Steps aimed at bringing bacteria and host cells into close contact, such as coincubation in a reduced volume or centrifuging bacteria onto host cell layers, may improve the success rate of infection by mutants (Kurtti et al., 2005). Other factors known to affect invasiveness and subsequent proliferation include pH, osmotic pressure, and temperature, and these differ with the rickettsial species as well as the host cell type (Munderloh et al., 1999).

Selection, Markers, Controlled Expression

To improve recovery rates, it is helpful that mutants express selectable markers. The first transformed rickettsiae carried genes conferring resistance to rifampin or erythromycin and later also the gene for GFP. These selection markers were problematic due to generation of nonspecifically resistant rickettsiae that outgrew transformants or due to low expression of GFP that could only be detected by fluorescence-activated cell sorting (FACS) (Rachek et al., 1998, 2000; Troyer et al., 1999; Renesto et al., 2002). How well *Rickettsiales* express foreign markers varies for reasons that are not always clear. However, the AT-rich genomes of many rickettsiae impose constraints on expression of foreign DNA if genes are not compatible with rickettsial methylation or codon usage or are too GC rich. For example, the DsRed gene from a coral is not expressed, whereas GFPuv, selected for brightness in *E. coli*, is expressed strongly (Baldridge et al., 2005). Choice of promoter also affects expression levels of foreign genes (Baldridge et al., 2010b), and choice of a less potent promoter may be desirable when the gene product is toxic to the transfomant or has other undesirable effects. Overall, control of gene expression in the *Rickettsiales* is poorly understood, and much remains to be learned from the rich literature on the subject in other prokaryotes (Bassler, 1999; Grundy and Henkin, 2006). Rickettsial chromosomes and plasmids encode genes that presumably are regulated by environmental conditions—e.g., *hsp* genes, which encode chaperonins protecting sensitive cellular components from damage; and *spoT*, which is part of the stringent response system (Baldridge et al., 2007a, 2008)—but their promoters have not been tested for controlled transgene expression in rickettsiae. Likewise, *A. phagocytophilum* encodes many genes that are conditionally expressed, i.e., in either human or tick cells, but not both. These features are clearly discernible when examining gene expression profiles of *A. phagocytophilum* in different host environments (Nelson et al., 2008). Inducible gene expression, e.g., using *tet* regulation (Bertram and Hillen, 2008; Stary et al., 2010), is a strategy successfully employed in other bacteria. Whether this system would be useful in *Rickettsiales* needs to be evaluated.

Antibiotics for Selection

Selection of mutants using growth inhibitors is an indispensible strategy to recover mutants from the background of nontransformed bacteria. By far the most common strategy is incorporation of an antibiotic resistance gene in the transformation cassette. When working with human or animal pathogens, this is a potentially sensitive issue, as introduction of resistance to antibiotics used to treat disease induced by the pathogen to be transformed is not encouraged. This poses a dilemma, as the most effective selection is likely achieved by using the clinically most effective antibiotics. This is the case with tetracyclines, which are considered the treatment of choice for rickettsial diseases (Purvis and Edwards, 2000; Holman et al., 2001; Donovan et al., 2002). There has been renewed debate about how laboratory research could potentially be misused for acts of bioterrorism, referred to as "dual-use" research. Security agencies and policy makers have conducted much of this debate, and the scientific community has unfortunately been relatively quiet (Suk et al., 2011). The obstacles to mass application of research techniques developed through the use of sophisticated, expensive, and sensitive technology by highly trained and experienced scientists are not trivial. Certainly, this is not to say that scenarios of abuse should not be given serious consideration; they definitely should, but researchers need to be more actively engaged in the discussion. Fortunately, the debate has fueled the search for alternate selection methods, e.g., those based on fluorescent markers being expressed sufficiently to enable collection of mutants using FACS (Troyer et al., 1999), although this has not been a robust selection method for *Rickettsiales*, possibly due to their small size, which may be obscured by host cell autofluorescence. Metabolic inhibitors that result in the accumulation of toxic products offer an attractive solution. An example is phosphinothricin, the active ingredient of a widely used herbicide, a glutamate analog and potent inhibitor of glutamine synthetase (Gill and Eisenberg, 2001) that causes intracellular accumulation of toxic ammonia. This effect can be counteracted by expression of the *bar* gene, encoding phosphinothricin acetyltransferase (Lutz et al., 2001; Felsheim et al., 2007). However, this approach is currently not widely used, in part because metabolic pathways are not well understood in the *Rickettsiales*, in which a large percentage of genes encode proteins of unknown function. Another source of selectable markers is antibiotics that have little clinical use due to poor efficacy or toxicity in vivo. One such antibiotic is spectinomycin, which has limited application as a nonchoice alternative among 21 drugs used to treat uncomplicated gonorrhea (Moran and Levine, 1995), and is not effective against any diseases caused by *Rickettsiales*. For the recovery of *A. phagocytophilum* mutants from HL-60 cells, spectinomycin is sufficient by itself as a selecting agent (Felsheim et al., 2006). However, in tick cell culture, where *A. phagocytophilum* replicates more slowly, double selection with streptomycin (using 100 μM each of spectinomycin and streptomycin) is helpful to eliminate nonspecifically resistant bacteria (U. G. Munderloh, unpublished results). Unfortunately, antibiotics that do not have clinical applications are usually not manufactured and are unavailable. Spectinomycin is commonly used in plant molecular genetics, and thus should remain readily available.

Limitations of Transposon Mutagenesis

The *Himar1* element is limited with respect to the size of the transposon that can be efficiently mobilized (Lampe et al., 1998). Transposons 1 to 2 kb in size showed variably reduced mobilization rates of 5 to about 30%, while for every 1 kb in size of the construct over 2 kb there was a nearly 40% decrease in relative transposition efficiency, down to 20% with use of a 4-kb transposon (Lampe et al., 1999). Although this capacity is sufficient for the average *Rickettsiales* gene of about 0.8 to 1 kb, it would not accommodate the larger genes or multigene operons that are common in bacteria. Also, the ability to incorporate genes along with selectable markers into a single transposon would be limited by the maximum practical size of *Himar1*.

With the currently used designs that involve constitutively expressed transposons driven by strong promoters, it is difficult to identify essential genes, as insertions that disable gene function cannot be identified if they result in lethal phenotypes. The same is likely true for disruption of important regulatory regions. These drawbacks can be partially overcome when more than one host cell system is available to recover potential mutants. As we have shown, genomic expression patterns in *A. phagocytophilum* differ depending on the host cell type and species origin they infect. The differences are especially pronounced when comparing *A. phagocytophilum* organisms residing in the human promyelocyte cell line HL-60 to those from the *Ixodes scapularis* tick cell line ISE6 (Nelson et al., 2008). We predict that *A. phagocytophilum* mutants with insertions into ORFs expressed exclusively in one host type can be recovered in a host cell type in which the ORF is not utilized. This way it may be possible to identify vector-specific versus vertebrate-specific rickettsial genes that could be targeted to disrupt the *A. phagocytophilum* life cycle at specific, strategic points. In support of this hypothesis, we have identified *A. phagocytophilum* mutants with single-transposon insertions that abolish the ability of the bacteria to replicate in HL-60 or ISE6 cells (Oliva Chávez et al., 2010). To achieve this, we exploited our ability to recover transformants of *A. phagocytophilum* in both of these cell types. A similar approach could potentially be applied to any arthropod-borne intracellular pathogen for which suitable in vitro culture systems utilizing vector cell lines are available.

Rickettsial Plasmids

Mobile genetic elements that are naturally occurring in the *Rickettsiales* could be an attractive source and target for engineering shuttle vectors capable of replicating in *E. coli* strains as well as in a wide range of rickettsiae. The discovery that many *Rickettsia* spp. harbor plasmids is still relatively new, but information is rapidly accumulating to indicate that rickettsial plasmids are not uncommon. The plasmids of several species have now been sequenced, either separately or as by-products of whole-genome sequencing of the *Rickettsia* species hosting them (Ogata et al., 2005; Blanc et al., 2007; Baldridge et al., 2007a, 2008, 2010a; Fournier et al., 2009). For an in-depth comparative discussion of the characteristics of plasmids of the genus *Rickettsia* the reader is referred to Baldridge et al. (2010a).

Interestingly, some rickettsiae are able to support several plasmid species, a finding that has implications for the use of plasmids as transforming elements. Known rickettsial plasmids appear to exist as low-copy-number replicons and vary in size considerably, from the 12.4-kb pRAF (Fournier et al., 2009) to the 63-kb pRF (Ogata et al., 2005), and plasmids in other species at intermediate sizes (Baldridge et al., 2010a). Except for genes necessary for plasmid maintenance (*parA*- and *dnaA*-like genes), they share little sequence homology, with the notable exception of *hsp2*, suggesting a role in adaptation to environmental changes that are an inherent feature of arthropod-borne pathogen life cycles. Of significance to their use as shuttle vectors is the fact that rickettsiae can support multiple plasmids, and this provides a basis for the design of shuttle vectors that could be compatible with the majority of rickettsial species. Likewise, the presence on plasmids of genes or even operons seemingly acquired by lateral transfer is an indicator of their potential to accept and maintain introduced, foreign genetic sequences. The 26-kb *R. peacockii* plasmid, pRPR, harbors five ORFs of an operon most closely related to the *P. aeruginosa* glycosylation island (Arora et al., 2001), and specifically those genes that encode proteins involved in phospholipid synthesis. *R. peacockii* has a mutated glycerol-3-phosphate dehydrogenase gene and probably relies entirely on import of glycerol-3-phosphate from the host, likely using GlpT, a transmembrane glycerol-3-phosphate transporter. The presence of the five genes on ~10 copies of pRPR, which is higher than the copy number of other rickettsial plasmids, suggests an important need to alleviate this shortcoming in *R. peacockii* (Felsheim et al., 2009). The presence of these genes, which were apparently acquired from an unrelated bacterium on the *R. peacockii* plasmid, clearly supports the notion that plasmids may be important acceptors of laterally acquired genes and operons that replace or otherwise complement gene loss suffered through transposase activity.

Our initial attempts to transform rickettsiae using plasmid-derived constructs focused on the *Rickettsia monacensis* pRM plasmid, which we cloned and sequenced as overlapping fragments. Even though we were able to rescue-clone fragments in *E. coli*, attempts to transform rickettsiae with those clones were unsuccessful. It is likely that our failure is related to rickettsial plasmid incompatibility groups that prevent comaintenance of plasmids sharing the same plasmid-partitioning genes (Funnell, 2005; Bouet et al., 2007; Cevallos et al., 2008). We reasoned that plasmids from *Rickettsia amblyommii* that appear to exist in single rickettsiae as a mixture of three or possibly four different plasmids could be successfully maintained in recipient rickettsiae. We cloned plasmid pRAM18 (GenBank accession no. GU322808) in its entirety, and following ligation with a 1- to 6-kb cassette from pMW1650 (Liu et al., 2007) containing the *R. prowazekii arr-2* rifampin resistance (*RrpsLp-arr-2*$_{Rp}$; RIF) and GFPuv genes, used it to successfully transform *Rickettsia bellii*, a symbiotic rickettsia in the ancestral group (Gillespie et al., 2008; Burkhardt et al., 2010). This construct has formed the basis for subsequently derived versions to develop a smaller transformation vector, and work to optimize a rickettsial shuttle vector is currently ongoing in our lab. Figure 1C gives a generalized example of the essential elements of pRAM18-derived constructs, and Fig. 1D shows the resultant transformant *R. bellii* under UV-light illumination. The rickettsial *parA* and *dnaA* genes are necessary and required for maintenance of this plasmid in recipients. Since some rickettsiae carry plasmids of considerable size, we anticipate that such transformation vectors would be able to accommodate a much larger payload than *Himar1*. We envision that shuttle vectors would be useful for testing the function of genes that are naturally defective in certain *Rickettsia* spp., in complementation assays, overexpression of native or foreign genes, testing gene regulation, and other applications that are now carried out in *E. coli*. However, because rickettsial plasmids can be unstable, it may be necessary to maintain plasmid transformants under antibiotic selection, making them perhaps less useful for long-term in vivo applications. In this regard, *Himar1* transformants with insertions into the chromosome appear to remain stable even in the absence of selection (Felsheim et al., 2006).

CONCLUSIONS

During the past decade, rickttsiologists have made great strides toward reproducible genetic manipulation of *Rickettsiales*. In many cases, significant advances have been achieved through the use of nonpathogenic rickettsial species that provide useful models for virulent rickettsiae in which complex manipulations may be difficult to execute. Transposon mutagenesis using the broad-host-range *mariner* element *Himar1* has proven useful for insertion of fluorescent markers as well as for complementation of defective genes with restoration of wild-type phenotype (Felsheim et al., 2006; Clark et al., 2011). Random gene knockouts generated by *Himar1* activity may provide a method for genome-wide mutational analysis if the system can be optimized for greater efficiency in *Rickettsiales*. With more researchers using this tool, it is likely that progress will accelerate. Rickettsial plasmids have significant potential for the development of shuttle vectors able to deliver a greater diversity of larger constructs than *Himar1* (Baldridge et al., 2008), but are potentially less stable. This potential instability could, however, also prove to be an advantage as removal of markers carried on plasmids may sometimes be desirable. By contrast, removal of a transposon would be more challenging. Homologous recombination for analysis of gene function has recently been reported in *R. prowazekii* (Driskell et al., 2009), suggesting that this technique, a staple with other bacteria, has the potential for wider application in rickettsial genomics.

REFERENCES

Andersson, J. O., and S. E. G. Andersson. 1999. Genome degradation is an ongoing process in *Rickettsia*. *Mol. Biol. Evol.* **16:**1178–1191.

Arora, S. K., M. Bangera, S. Lory, and R. Ramphal. 2001. A genomic island in *Pseudomonas aeruginosa* carries the determinants of flagellin glycosylation. *Proc. Natl. Acad. Sci. USA* **98:** 9342–9347.

Baldridge, G. D., N. Y. Burkhardt, R. E. Felsheim, T. J. Kurtti, and U. G. Munderloh. 2007a. Transposon insertion reveals pRM, a plasmid of *Rickettsia monacensis*. *Appl. Environ. Microbiol.* **73:**4984–4995.

Baldridge, G. D., N. Y. Burkhardt, R. F. Felsheim, T. J. Kurtti, and U. G. Munderloh. 2008. Plasmids of the pRM/pRF family occur in diverse *Rickettsia* species. *Appl. Environ. Microbiol.* **74:**645–652.

Baldridge, G. D., N. Burkhardt, M. J. Herron, T. J. Kurtti, and U. G. Munderloh. 2005. Analysis of fluorescent protein expression in transformants of *Rickettsia monacensis*, an obligate intracellular tick symbiont. *Appl. Environ. Microbiol.* **71:**2095–2105.

Baldridge, G. D., N. Y. Burkhardt, M. B. Labruna, R. C. Pacheco, C. D. Paddock, P. C. Williamson, P. M. Billingsley, R. F. Felsheim, T. J. Kurtti, and U. G. Munderloh. 2010a. Wide dispersal and possible multiple origins of low-copy-number plasmids in *Rickettsia* species associated with blood-feeding arthropods. *Appl. Environ. Microbiol.* **76:**1718–1731.

Baldridge, G. D., N. Y. Burkhardt, A. S. Oliva, T. J. Kurtti, and U. G. Munderloh. 2010b. Rickettsial *ompB* promoter regulated expression of GFP_{uv} in transformed *Rickettsia montanensis*. *PLoS One* **5:**e8965.

Baldridge, G. D., T. J. Kurtti, N. Burkhardt, A. S. Baldridge, C. M. Nelson, A. S. Oliva, and U. G. Munderloh. 2007b. Infection of *Ixodes scapularis* ticks with *Rickettsia monacensis* expressing green fluorescent protein: a model system. *J. Invertebr. Pathol.* **94:**163–174.

Barbet, A. F., J. T. Agnes, A. L. Moreland, A. M. Lundgren, A. R. Alleman, S. M. Noh, K. A. Brayton, U. G. Munderloh, and G. H. Palmer. 2005. Identification of functional promoters in the *msp2* expression loci of *Anaplasma marginale* and *Anaplasma phagocytophilum*. *Gene* **353:**89–97.

Bassler, L. 1999. How bacteria talk to each other: regulation of gene expression by quorum sensing. *Curr. Opin. Microbiol.* **2:**582–587.

Bechah, Y., K. El Karkouri, O. Mediannikov, Q. Leroy, N. Pelletier, C. Robert, C. Médigue, J. L. Mege, and D. Raoult. 2010. Genomic, proteomic, and transcriptomic analysis of virulent and avirulent *Rickettsia prowazekii* reveals its adaptive mutation capabilities. *Genome Res.* **20:**655–663.

Bertram, R., and W. Hillen. 2008. The application of Tet repressor in prokaryotic gene regulation and expression. *Microb. Biotechnol.* **1:**2–16.

Blanc, G., H. Ogata, C. Robert, S. Audic, J. M. Claverie, and D. Raoult. 2007. Lateral gene transfer between obligate intracellular bacteria: evidence from the *Rickettsia massiliae* genome. *Genome Res.* **17:**1657–1664.

Bouet, J.-Y., K. Nordstrom, and D. Lane. 2007. Plasmid partition and incompatibility—the focus shifts. *Mol. Microbiol.* **65:**1405–1414.

Boussau, B., E. O. Karlberg, A. C. Frank, B. A. Legault, and S. G. E. Andersson. 2004. Computational inference of scenarios for α-proteobacterial genome evolution. *Proc. Natl. Acad. Sci. USA* **101:**9722–9727.

Brayton, K. A., L. S. Kappmeyer, D. R. Herndon, M. J. Dark, D. L. Tibbals, G. H. Palmer, T. C. McGuire, and D. P. Knowles, Jr. 2005. Complete genome sequencing of *Anaplasma marginale* reveals that the surface is skewed to two superfamilies of outer membrane proteins. *Proc. Natl. Acad. Sci. USA* **102:**844–849.

Brouqui, P., J. R. Harle, J. Delmont, C. Frances, P. J. Weiller, and D. Raoult. 1997. African tick-bite fever. An imported spotless rickettsiosis. *Arch. Intern. Med.* **157:**119–124.

Burkhardt, N. Y., G. D. Baldridge, P. C. Williamson, P. M. Billingsley, R. F. Felsheim, T. J. Kurtti, and U. G. Munderloh. 2010. Development of shuttle vectors for transformation of diverse *Rickettsia* species, *PLoS One* **6:**e29511.

Cardwell, M. M., and J. J. Martinez. 2009. The Sca2 autotransporter protein from *Rickettsia conorii* is sufficient to mediate adherence to and invasion of cultured mammalian cells. *Infect. Immun.* **77:**5272–5280.

Cevallos, M. A., R. Cervantes-Rivera, and R. M. Gutiérrez-Ríos. 2008. The *repABC* plasmid family. *Plasmid* **60:**19–37.

Chan, Y. G., M. M. Cardwell, T. M. Hermanas, T. Uchiyama, and J. J. Martinez. 2009. Rickettsial outer-membrane protein B (rOmpB) mediates bacterial invasion through Ku70 in an actin, c-Cbl, clathrin and caveolin 2-dependent manner. *Cell. Microbiol.* **11:**629–644.

Clark, T. R., D. W. Ellison, B. Kleba, and T. Hackstadt. 2011. Complementation of *Rickettsia rickettsii* RelA/SpoT restores a non-lytic plaque phenotype. *Infect. Immun.* **79:**1631–1637.

Collins, N. E., J. Liebenberg, E. P. de Villiers, K. A. Brayton, E. Louw, A. Pretorius, F. E. Faber, H. van Heerden, A. Josemans, M. van Kleef, H. C. Steyn, M. F. van Strijp, E. Zweygarth, F. Jongejan, J. C. Maillard, D. Berthier, M. Botha, F. Joubert, C. H. Corton, N. R. Thomson, M. T. Allsopp, and B. A. Allsopp. 2005. The genome of the heartwater agent *Ehrlichia ruminantium* contains multiple tandem repeats of actively variable copy number. *Proc. Natl. Acad. Sci. USA* **102:**838–843.

Crameri, A., E. A. Whitehorn, E. Tate, and W. P. Stemmer. 1996. Improved green fluorescent protein by molecular evolution using DNA shuffling. *Nat. Biotechnol.* **14:**315–319.

de la Fuente, J., J. C. Garcia-Garcia, E. F. Blouin, and K. M. Kocan. 2001. Differential adhesion of major surface proteins 1a and 1b of the ehrlichial cattle pathogen *Anaplasma marginale* to bovine erythrocytes and tick cells. *Int. J. Parasitol.* **31:**145–153.

de Sousa, R., L. Duque, J. Poças, J. Torgal, F. Bacellar, J. P. Olano, and D. H. Walker. 2008. Lymphangitis in a Portuguese patient infected with *Rickettsia sibirica*. *Emerg. Infect. Dis.* **14:**529–531.

Donovan, B. J., D. J. Weber, J. C. Rublein, and R. H. Raasch. 2002. Treatment of tick-borne diseases. *Ann. Pharmacother.* **36:**1590–1597.

Doré, M., R. J. Korthuis, D. N. Granger, M. L. Entman, and C. W. Smith. 1993. P-selectin mediates spontaneous leukocyte rolling in vivo. *Blood* **82:**1308–1316.

Driskell, L. O., X.-J. Yu, L. Zhang, Y. Liu, V. L. Popov, D. H. Walker, A. M. Tucker, and D. O. Wood. 2009. Directed mutagenesis of the *Rickettsia prowazekii pld* gene encoding phospholipase D. *Infect. Immun.* **77:**3244–3248.

Dunning Hotopp, J. C., M. Lin, R. Madupu, J. Crabtree, S. V. Angiuoli, J. Eisen, R. Seshadri, Q. Ren, M. Wu, T. R. Utterback, S. Smith, M. Lewis, H. Khouri, C. Zhang, H. Niu, Q. Lin, N. Ohashi, N. Zhi, W. Nelson, L. M. Brinkac, R. J. Dodson, M. J. Rosovitz, J. Sundaram, S. C. Daugherty, T. Davidsen, A. S. Durkin, M. Gwinn, D. H. Haft, J. D. Selengut, S. A. Sullivan, N. Zafar, L. Zhou, F. Benahmed, H. Forberger, R. Halpin, S. Mulligan, J. Robinson, O. White, Y. Rikihisa, and H. Tettelin. 2006. Comparative genomics of emerging human ehrlichiosis agents. *PLoS Genet.* **2:**e21.

Ellison, D. W., T. R. Clark, D. E. Sturdevant, K. Virtaneva, S. F. Porcella, and T. Hackstadt. 2008. Genomic comparison of virulent *Rickettsia rickettsii* Sheila Smith and avirulent *Rickettsia rickettsii* Iowa. *Infect. Immun.* **76:**542–550.

Ettema, T. J. G., and S. G. E. Andersson. 2009. The α-proteobacteria: the Darwin finches of the bacterial world. *Biol. Lett.* **5:**429–432.

Felsheim, R. F., M. J. Herron, C. and U. G. Munderloh. 2007. Selection of *Anaplasma phagocytophilum* transformants with an herbicide resistance system, abstr. 3. *21st Meet. Am. Soc. Rickettsiol.*, Colorado Springs, CO, 8 to 11 September 2007.

Felsheim, R. F., M. J. Herron, C. M. Nelson, N. Y. Burkhardt, A. F. Barbet, T. J. Kurtti, and U. G. Munderloh. 2006. Transformation of *Anaplasma phagocytophilum*. *BMC Biotechnol.* **6:**42.

Felsheim, R. F., T. J. Kurtti, and U. G. Munderloh. 2009. Genome sequence of the endosymbiont *Rickettsia peacockii* and comparison with virulent *Rickettsia rickettsii*: identification of virulence factors. *PLoS One* **4:**e8361.

Felsheim, R. F., A. S. Oliva Chávez, G. H. Palmer, L. Crosby, A. F. Barbet, T. J. Kurtti, and U. G. Munderloh. 2010. Transformation of *Anaplasma marginale*. *Vet. Parasitol.* **167:**167–174.

Fournier, P. E., K. El Karkouri, Q. Leroy, C. Robert, B. Giumelli, P. Renesto, C. Socolovschi, P. Parola, S. Audic, and D. Raoult. 2009. Analysis of the *Rickettsia africae* genome reveals that virulence acquisition in *Rickettsia* species may be explained by genome reduction. *BMC Genomics* **10:**166.

Fournier, P. E., F. Gouriet, P. Brouqui, F. Lucht, and D. Raoult. 2005. Lymphangitis-associated rickettsiosis, a new rickettsiosis caused by *Rickettsia sibirica mongolotimonae*: seven new cases and review of the literature. *Clin. Infect. Dis.* **40:**1435–1444.

Funnell, B. 2005. Partition-mediated plasmid pairing. *Plasmid* **53:**119–125.

Fuxelius, H. H., A. C. Darby, N. H. Cho, and S. G. E. Andersson. 2008. Visualization of pseudogenes in intracellular bacteria reveals the different tracks to gene destruction. *Genome Biol.* **9:**R42.

Fuxelius, H. H., A. Darby, C. K. Min, N. M. Cho, and S. G. E. Andersson. 2007. The genomic and metabolic diversity of *Rickettsia*. *Res. Microbiol.* **158:**745–753.

Garcia-Garcia, J. C., J. de la Fuente, G. Bell-Eunice, E. F. Blouin, and K. M. Kocan. 2004. Glycosylation of *Anaplasma marginale* major surface protein 1a and its putative role in adhesion to tick cells. *Infect. Immun.* **72:**3022–3030.

Gill, H. S., and D. Eisenberg. 2001. The crystal structure of phosphinothricin in the active site of glutamine synthetase illuminates the mechanism of enzymatic inhibition. *Biochemistry* **40:** 1903–1912.

Gillespie, J. J., M. S. Beier, M. S. Rahman, N. C. Ammerman, J. M. Shallom, A. Purkayastha, B. S. Sobral, and A. F. Azad. 2007. Plasmids and rickettsial evolution: insight from *Rickettsia felis*. *PLoS One* **2:**e266.

Gillespie, J. J., K. Williams, M. Shukla, E. E. Snyder, E. K. Nordberg, S. M. Ceraul, C. Dharmanolla, D. Rainey, J. Soneja, J. M. Shallom, N. D. Vishnubhat, R. Wattam, A. Purkayastha, M. Czar, O. Crasta, J. C. Setubal, A. F. Azad, and B. S. Sobral. 2008. *Rickettsia* phylogenomics: unwinding the intricacies of obligate intracellular life. *PLoS One* **3:** e2018.

Grundy, F. J. and T. M. Henkin. 2006. From ribosome to riboswitch: control of gene expression in bacteria by RNA structural rearrangements. *Crit. Rev. Biochem. Mol. Biol.* **41:**329–338.

Herron, M. J., C. M. Nelson, J. Larson, K. R. Snapp, G. S. Kansas, and J. L. Goodman. 2000. Intracellular parasitism by the human granulocytic ehrlichiosis bacterium through the P-selectin ligand, PSGL-1. *Science* **288:**1653–1656.

Herron, P. R., G. Hughes, G. Chandra, S. Fielding, and P. J. Dyson. 2004. Transposon Express, a software application to report the identity of insertions obtained by comprehensive transposon mutagenesis of sequenced genomes: analysis of the preference for *in vitro* Tn*5* transposition into GC-rich DNA. *Nucleic Acids Res.* **32:**e113.

Holman, R. C., C. D. Paddock, A. T. Curns, J. W. Krebs, J. H. McQuiston, and J. E. Childs. 2001. Analysis of risk factors for fatal Rocky Mountain spotted fever: evidence for superiority of tetracyclines for treatment. *J. Infect. Dis.* **184:**1437–1444.

Kenyon, R. H., W. M. Acree, G. G. Wright, and F. W. Melchior, Jr. 1972. Preparation of vaccines for Rocky Mountain spotted fever from rickettsiae propagated in cell culture. *J. Infect. Dis.* **125:**146–152.

Kurtti, T. J., J. A. Simser, G. D. Baldridge, A. T. Palmer, and U. G. Munderloh. 2005. Factors influencing in vitro infectivity and growth of *Rickettsia peacockii* (Rickettsiales: Rickettsiaceae), an endosymbiont of the Rocky Mountain wood tick, *Dermacentor andersoni* (Acari, Ixodidae). *J. Invertebr. Pathol.* **90:**177–186.

Lampe, D. J., B. J. Akerley, E. J. Rubin, J. J. Mekalanos, and H. M. Robertson. 1999. Hyperactive transposase mutants of the *Himar1 mariner* transposon. *Proc. Natl. Acad. Sci. USA* **96:** 11428–11433.

Lampe, D. J., M. E. Churchill, and H. M. Robertson. 1996. A purified *mariner* transposase is sufficient to mediate transposition *in vitro*. *EMBO J.* **15:**5470–5479.

Lampe, D. J., T. E. Grant, and H. M. Robertson. 1998. Factors affecting transposition of the *Himar1 mariner* transposon *in vitro*. *Genetics* **149:**179–187.

Li, H., and D. H. Walker. 1998. rOmpA is a critical protein for the adhesion of *Rickettsia rickettsii* to host cells. *Microb. Pathog.* **24:**289–298.

Lin, M., T. Kikuchi, H. M. Brewer, A. D. Norbeck, and Y. Rikihisa. 2011. Global proteomic analysis of two tick-borne emerging zoonotic agents: *Anaplasma phagocytophilum* and *Ehrlichia chaffeensis*. *Front. Microbiol.* **2:**24.

Liu, Z.-M., A. M. Tucker, L. O. Driskell, and D. O. Wood. 2007. *Mariner*-based transposon mutagenesis of *Rickettsia prowazekii*. *Appl. Environ. Microbiol.* **73**:6644–6649.

Lutz, K. A., J. E. Knapp, and P. Maliga. 2001. Expression of *bar* in the plastid genome confers herbicide resistance. *Plant Physiol.* **125:**1585–1590.

Mahan, M. J., J. M. Slauch, and J. J. Mekalanos. 1993. Selection of bacterial virulence genes that are specifically induced in host tissues. *Science* **259:**686–688.

Martinez, J. J., S. Seveau, E. Veiga, S. Matsuyama, and P. Cossart. 2005. Ku70, a component of DNA-dependent protein kinase, is a mammalian receptor for *Rickettsia conorii*. *Cell* **123:** 1013–1023.

McClintock, B. 1950. The origin and behavior of mutable loci in maize. *Proc. Natl. Acad. Sci. USA* **36:**344–355.

McGarey, D. J., A. F. Barbet, G. H. Palmer, T. C. McGuire, and D. R. Allred. 1994. Putative adhesins of *Anaplasma marginale*: major surface polypeptides 1a and 1b. *Infect. Immun.* **62:**4594–4601.

Moran, J. S., and W. C. Levine. 1995. Drugs of choice for the treatment of uncomplicated gonococcal infections. *Clin. Infect. Dis.* **1:**S47–S65.

Munderloh, U. G., S. D. Jauron, V. Fingerle, L. Leitritz, S. F. Hayes, J. M. Hautman, C. M. Nelson, B. W. Huberty, T. J. Kurtti, G. G. Ahlstrand, B. Greig, M. A. Mellencamp, and J. L. Goodman. 1999. Invasion and intracellular development of the human granulocytic ehrlichiosis agent in tick cell culture. *J. Clin. Microbiol.* **37:**2518–2524.

Munderloh, U. G., and T. J. Kurtti. 1995. Cellular and molecular interrelationships between ticks and prokaryotic tick-borne pathogens. *Annu. Rev. Entomol.* **40:**221–243.

Munderloh, U. G., M. J. Lynch, M. J. Herron, A. T. Palmer, T. J. Kurtti, R. D. Nelson, and J. L. Goodman. 2004. Infection of endothelial cells with *Anaplasma marginale* and *A. phagocytophilum*. *Vet. Microbiol.* **101:**53–64.

Nelson, C. M., M. J. Herron, R. F. Felsheim, B. R. Schloeder, S. M. Grindle, A. O. Chavez, T. J. Kurtti, and U. G. Munderloh. 2008. Whole genome transcription profiling of *Anaplasma phagocytophilum* in human and tick host cells by tiling array analysis. *BMC Genomics* **9:**364.

Niebylski, M. L., M. E. Schrumpf, W. Burgdorfer, E. R. Fischer, K. L. Gage, and T. G. Schwan. 1997. *Rickettsia peacockii* sp. nov., a new species infecting wood ticks, *Dermacentor andersoni*, in western Montana. *Int. J. Syst. Bacteriol.* **47:**446–452.

Ogata, H., P. Renesto, S. Audic, C. Robert, G. Blanc, P. E. Fournier, H. Parinello, J. M. Claverie, and D. Raoult. 2005. The genome sequence of *Rickettsia felis* identifies the first putative conjugative plasmid in an obligate intracellular parasite. *PLoS Biol.* **3:**e248.

Oliva Chávez, A. S., R. F. Felsheim, N. Y. Burkhardt, T. J. Kurtti, and U. G. Munderloh. 2010. Identification of genes involved in the infection process of tick (ISE6) and human (HL-60) cells by the tick-borne pathogen *Anaplasma phagocytophilum*. *42nd Annu. Conf. Soc. Vector Ecol.*, Raleigh, NC, 26 to 30 September 2010.

Purvis, J. J., and M. S. Edwards. 2000. Doxycycline use for rickettsial disease in pediatric patients. *Pediatr. Infect. Dis. J.* **19:**871–874.

Qin, A., A. M. Tucker, A. Hines, and D. O. Wood. 2004. Transposon mutagenesis of the obligate intracellular pathogen *Rickettsia prowazekii*. *Appl. Environ. Microbiol.* **70:**2816–2822.

Rachek, L. I., A. Hines, A. M. Tucker, H. H. Winkler, and D. O. Wood. 2000. Transformation of *Rickettsia prowazekii* to erythromycin resistance encoded by the *Escherichia coli ereB* gene. *J. Bacteriol.* **182:**3289–3291.

Rachek, L. I., A. M. Tucker, H. H. Winkler, and D. O. Wood. 1998. Transformation of *Rickettsia prowazekii* to rifampin resistance. *J. Bacteriol.* **180:**2118–2124.

Ramabu, S. S., M. W. Ueti, K. A. Brayton, T. V. Baszler, and G. H. Palmer. 2010. Identification of *Anaplasma marginale* proteins specifically upregulated during colonization of the tick vector. *Infect. Immun.* **78:**3047–3052.

Renesto, P., E. Gouin, and D. Raoult. 2002. Expression of green fluorescent protein in *Rickettsia conorii*. *Microb. Pathog.* **33:**17–21.

Renesto, P., L. Samson. H. Ogata, S. Azza, P. Fourquet, J. P. Gorvel, R. A. Heinzen, and D. Raoult. 2006. Identification of two putative rickettsial adhesins by proteomic analysis. *Res. Microbiol.* **157:**605–612.

Riley, S. P., K. C. Goh, T. M. Hermanas, M. M. Cardwell, Y. G. Chan, and J. J. Martinez. 2010. The *Rickettsia conorii* autotransporter protein Sca1 promotes adherence to nonphagocytic mammalian cells. *Infect. Immun.* **78:**1895–1904.

Sahni, S. K. and E. Rydkina. 2009. Progress in the functional analysis of rickettsial genes through directed mutagenesis of *Rickettsia prowazekii* phospholipase D. *Future Microbiol.* **4:**1249–1253.

Sakharkar, K. R., M. K. Sakharkar, C. Verma, and V. T. K. Chow. 2005. Comparative study of overlapping genes in bacteria, with special reference to *Rickettsia prowazekii* and *Rickettsia conorii*. *Int. J. Syst. Evol. Microbiol.* **55:**1205–1209.

Shaner, N. C., P. A. Steinbach, and R. Y. Tsien. 2005. A guide to choosing fluorescent proteins. *Nat. Methods* **2:**905–909.

Shapiro, J. A. 1999. Transposable elements as the key to a 21st century view of evolution. *Genetica* **107:**171–179.

Simser, J. A., M. S. Rahman, S. M. Dreher-Lesnick, and A. F. Azad. 2005. A novel and naturally occurring transposon, ISRpe1 in the *Rickettsia peacockii* genome disrupting the *rickA* gene involved in actin-based motility. *Mol. Microbiol.* **58:**71–79.

Stary, E., R. Gaupp, S. Lechner, M. Leibig, E. Tichy, M. Kolb, and R. Bertram. 2010. New architectures for Tet-on and Tet-off regulation in *Staphylococcus aureus*. *Appl. Environ. Microbiol.* **76:**680–687.

Stewart, P. E., and P. A. Rosa. 2008. Transposon mutagenesis of the Lyme disease agent *Borrelia burgdorferi*. *Methods Mol. Biol.* **431:**85–95.

Suk, J. E., A. Zmorzynska, A. Hunger, W. Biederbick, J. Sasse, H. Maidhof, and J. C. Semenza. 2011. Dual-use research and technological diffusion: reconsidering the bioterrorism threat spectrum. *PLoS Pathog.* **7:**e1001253.

Troyer, J. M., S. Radulovic, and A. F. Azad. 1999. Green fluorescent protein as a marker in *Rickettsia typhi* transformation. *Infect. Immun.* **67:** 3308–3311.

Uchiyama, T. 2003. Adherence to and invasion of Vero cells by recombinant *Escherichia coli* expressing the outer membrane protein rOmpB of *Rickettsia japonica. Ann. N. Y. Acad. Sci.* **990:**585–590.

Uchiyama, T., H. Kawano, and Y. Kusuhara. 2006. The major outer membrane protein rOmpB of spotted fever group rickettsiae functions in the rickettsial adherence to and invasion of Vero cells. *Microbes Infect.* **8:**801–809.

Vellaiswamy, M., M. Kowalczewska, V. Merhej, C. Nappez, R. Vincentelli, P. Renesto, and D. Raoult. 2011. Characterization of rickettsial adhesin Adr2 belonging to a new group of adhesins in α-proteobacteria. *Microb. Pathog.* **50:**233–242.

Walker, D. H., G. A. Valbuena, and J. P. Olano. 2003. Pathogenic mechanisms of diseases caused by *Rickettsia. Ann. N. Y. Acad. Sci.* **990:**1–11.

Wu, Q., J. Pei, C. Turse, and T. A. Ficht. 2006. Mariner mutagenesis of *Brucella melitensis* reveals genes with previously uncharacterized roles in virulence and survival. *BMC Microbiol.* **6:**102.

INDEX

A

B

F

G

S